The Semiconductor Memory Book

The Semiconductor Memory Book

INTEL MARKETING COMMUNICATIONS

A WILEY-INTERSCIENCE PUBLICATION

JOHN WILEY & SONS, New York • Chichester • Brisbane • Toronto

Copyright © 1975, 1976, 1977 by Intel Corporation.
Copyright © 1978 by John Wiley & Sons, Inc.

All rights reserved. Published simultaneously in Canada.

Reproduction or translation of any part of this work
beyond that permitted by Sections 107 or 108 of the
1976 United States Copyright Act without the permission
of the copyright owner is unlawful. Requests for
permission or further information should be addressed to
the Permissions Department, John Wiley & Sons, Inc.

Intel Corporation assumes no responsibility for the use of any circuitry
other than embodied in an Intel product. No other circuit patent licenses are
implied.

Library of Congress Cataloging in Publication Data:

Intel Marketing Communications.
   The semiconductor memory book.

   "A Wiley-Interscience publication."
   Includes index.
   1. Semiconductor storage devices. I. Title.
TK7895.M4 I57   1978   621.3819′58′33   78-3741
ISBN 0-471-03567-X

Printed in the United States of America

10 9 8 7 6 5 4 3 2 1

The Semiconductor Memory Book represents the first in a series of Wiley/Intel publications covering large-scale integrated semiconductor products. Intel was organized to develop the technology of integrated electronics, from which the corporation derives its name. During Intel's brief history, it has become the world's largest supplier of memories and microprocessors.

This book represents the most extensive collection of Intel memory application and specification information available in a single volume. Subsequent Wiley/Intel publications will cover microprocessors, microcomputers, and related subjects.

ROB WALKER
Manager, Marketing Communications
Intel Corporation

Preface

Developments in the semiconductor industry during the last six years have resulted in a major shift in the type of storage technology used in digital systems. Semiconductor memories used today are lower in cost, higher in density, faster in access and cycle time, higher in reliability, and more modular in incremental size than comparable core memories. The curves shown in Figure 1 show the change in the memory and the increasing importance of semiconductor memory.

Semiconductor memories are divided into three broad categories, as shown in Figure 2. With two exceptions, each of these generic categories can be implemented with either of the two major semiconductor technologies: MOS or bipolar. These exceptions are the CCDs (charge-coupled devices) and EPROMs (erasable-programmable read-only memories), which are uniquely implemented with MOS technology.

RANDOM-ACCESS MEMORIES

No other area of semiconductor memory has grown as rapidly and as large as that of random-access memories. Leading the way in the explosive growth of RAMs are the MOS devices. One of the reasons for their wide acceptance has been the increasing bit density of MOS devices. The density has been quadrupling on the average of every two years, as shown by the graph in Figure 3.

In 1969 Intel introduced the 1101, a 256 × 1-bit static MOS random-access memory (RAM). This device was designed primarily for small buffer-storage applications where 256-word modularity, low overhead-support cost, and ease of use were important design objectives.

In 1971 Intel introduced the 1103, a 1024 × 1-bit dynamic MOS RAM. The 1103 offered a 4 : 1 density improvement along with a 4 : 1 speed improvement over the 1101. The 1103 was the first semiconductor memory element to be speed- and cost-competitive with core-memory systems.

In 1973 Intel introduced the first 4K dynamic NMOS memory, the 2107. This product was subsequently improved in the 2107B and 2107C versions. It became the

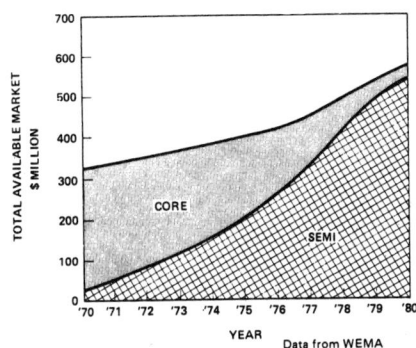

Figure 1. Total Available Memory Market.

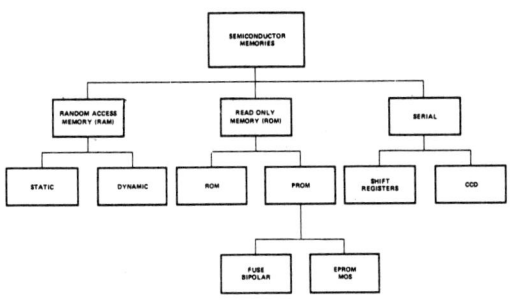

Figure 2. Semiconductor Memory Family Tree.

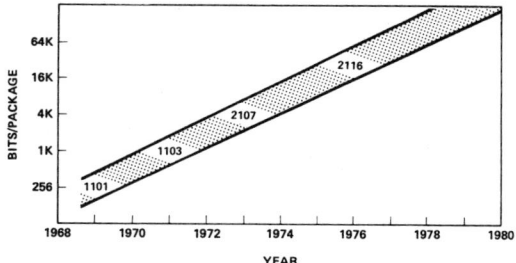

Figure 3. Dynamic RAM Evolution.

industry standard for 4K RAMs in 22-pin packages. In addition, Intel now offers the 2104A, which has reduced the package requirement for a 4K RAM to a standard 16-pin package. The 2104A, in integrating many of the support circuits internal to the device, has produced an improvement in ease of use. P-channel 1K RAMs, with their MOS level inputs, required high-voltage TTL-MOS drivers on all input pins. Their low-level signal output required the use of external sense amplifiers. These overhead devices have been integrated onto the 4K 2104A chip such that all inputs and outputs are fully TTL-compatible. The trend has been and continues to be toward denser, faster, and easier-to-use semiconductor memory devices.

In 1977 Intel introduced the 2116, a 16-pin 16,384 dynamic RAM, thus continuing the evolution shown in Figure 3. This RAM is also TTL-compatible on all inputs and outputs and can be plugged directly into 2104A sockets, providing a 4 : 1 increase in density.

While dramatic improvements have been made in MOS dynamic memories (such as the 1103, 2107, 2104A, and 2116), equal improvements have been made in high-density static, TTL-compatible memories. This family of devices, such as the 2102A, 2101A, 2111A, 2112A, 2114, 2115A, 2147, and the CMOS 5101, has greatly increased the ease of use of memories in systems that do not require a large amount of memory. These static RAMs (like the dynamic RAMs) are continuing to expand to include faster devices.

READ-ONLY MEMORIES

The read-only memory, like the random-access memory, has gone through evolutionary changes in a short period of time. Innovations in bipolar and MOS technology have resulted in programmable and erasable-programmable ROMs, called PROMs and EPROMs respectively. These two types of devices have greatly increased the usefulness and acceptability of read-only memories in system applications.

One of the most unique technologies in the ROM family is the erasable PROM (EPROM) such as the Intel® 1702A, 2708, and 2716. These devices, which have bit densities of 2K, 8K, and 16K respectively, offer system designers maximum flexibility in changing program instructions and so forth in the development of their systems.

Other user-programmable device types (not erasable) are the Intel® 3601, 3602, and 3604 family of bipolar PROMs. These devices offer the system designer very fast access times along with the ability to change programs "in-house" by merely replacing an old PROM with a newly programmed PROM.

Since their introduction in 1971 MOS EPROMs have undergone evolution similar to dynamic RAMs, only at a somewhat slower rate. Figure 4 indicates that their density doubles approximately every two years.

To maintain compatibility with the new generation of microprocessors that have a 5V technology, in 1977 Intel introduced the 2716, a 2K by 8-bit UV erasable PROM, which requires only a single power supply for normal operation. In addition, programming was simplified and now resembles a bipolar PROM type of programming; after raising programming supply to +26 volts, addresses can be programmed in random order, with all signals being TTL-compatible, including the address data and program pulse-inputs. Erasure requirements remain the same as the 2708, 15w sec/cm².

In addition to the 2716, mask ROM replacement is available in the form of the Intel 2316E. When the programmable Chip-Select inputs are selected in accordance with the suggested pinout, the 2316E can plug directly into the 2716 socket with no need to relayout the board. In addition, a system designed for use with the 2316E can be "customized" for OEM special systems by programming them with the custom data pattern and inserting them in the 2316E sockets, either at time of manufacture or in the field.

The entire Intel PROM family of devices has a counterpart in ROM (nonalterable or mask-programmable) form. These devices are generally used in systems that are in mass production.

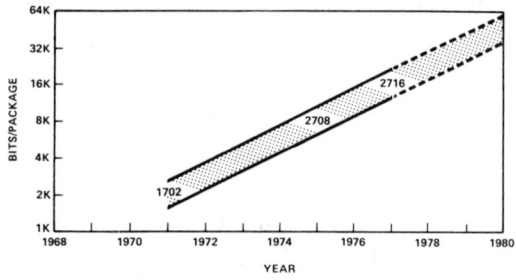

Figure 4. MOS EPROM Evolution

Preface

SERIAL MEMORIES

One of the most exciting new memory products to be recently introduced is the 2416, a 16K charge-coupled device (CCD). The high density and low cost of this device make it very attractive for use in "drum" replacement-type systems as well as terminal and minicomputer applications.

To facilitate the use of the 2416, Intel offers the 5244, a clock driver that minimizes the problems of manipulating the four clock inputs by providing (1) TTL inputs, (2) output rise-time control, and (3) "cross-coupling" control to minimize intercoupling between phases. One 5244 will drive four 2416s or provide storage and transfer clock-control for a total of 64K bits.

This handbook contains a detailed explanation of the use of the 2416 in a system environment. The significance of the unique organization of the 2416 is also fully explored so that the designer may take maximum advantage of its characteristics.

BOB GREENE

Santa Clara, California
February 1978

Contents

CHAPTER 1 STATIC RAMs	1
Designing with Static MOS Rams	2
Designing Non-Volatile Semiconductor Memory Systems	20
Bipolar/MOS Static RAM Compatibility	38
2101A 256 × 4 Bit Static RAM	44
2102A, 2102AL 1K × 1 Bit Static RAM	48
2111A 256 × 4 Static RAM	52
2112A 256 × 4 Bit Static RAM	56
2114 1024 × 4 Bit Static RAM	61
2115A, 2125A High Speed 1K × 1 Bit Static RAM	65
2141 4096 × 1 Bit Static RAM	70
2142 1024 × 4 Bit Static RAM	74
2147 4096 × 1 Bit Static RAM	78
5101 256 × 4 Bit Static CMOS RAM	84
CHAPTER 2 DYNAMIC RAMs	88
Designing with 16 Pin, 4096 Dynamic RAMs	89
Designing with 22 Pin, 4096 Dynamic RAMs	103
Designing with 16K Dynamic RAMs	130
Dynamic RAMs Used with Microprocessors	139
1103 1024 × 1 Bit Dynamic RAM	182
1103-1 1024 × 1 Bit Dynamic RAM	187
1103A 1024 × 1 Bit Dynamic RAM	190
1103A-1 1024 × 1 Bit Dynamic RAM	195
1103A-2 1024 × 1 Bit Dynamic RAM	200
2104A 4096 × 1 Bit Dynamic RAM	204
2107A 4096 × 1 Bit Dynamic RAM	212
2107B 4096 × 1 Bit Dynamic RAM	218
2107C 4096 × 1 Bit Dynamic RAM	224
2108 8192 × 1 Bit Dynamic RAM	229
2116 16,384 × 1 Bit Dynamic RAM	237
2117 16,384 × 1 Bit Dynamic RAM	245
CHAPTER 3 PROMs and ROMs	257
Designing with PROMs and ROMs	258
Designing with 2708 EPROM	294
Designing with 2716 EPROM	308
1702A 2K (256 × 8) UV EPROM	322
1702AL 2K (256 × 8) Low Power UV EPROM	326
2308 8192 Bit Static MOS ROM	329
2316E 16,384 Bit Static ROM	333
2708 8K and 4K UV EPROM	336
2716 16K (2K × 8) UV EPROM	341
2758 8K (1K × 8) Low Power +5V UV EPROM	345
3602A, 3622A, 3602, 3622, 2048 Bit (512 × 4) High Speed PROM	351
3604A, 3624A and 3604, 3624, 4096 Bit (512 × 8) High Speed PROM	354
3605, 3625 4K Bipolar PROM	357
3608, 3628 8K (1K × 8) Bipolar PROM	360
ROM/PROM Programming Instructions	363

CHAPTER 4 CHARGED COUPLED DEVICES	373
Design and Application of 2416 CCD Memory	374
2416 16,384 Bit CCD Serial Memory	403
CHAPTER 5 MEMORY SUPPORT CIRCUITS FOR DYNAMIC RAMs	411
Using Support Circuits for Dynamic RAMs	412
3205 High Speed 1 out of 8 Binary Decoder, 3404 High Speed 6 Bit Latch	435
3207A Quad Bipolar-to-MOS Level Shifter and Driver	439
3207A-1 Quad Bipolar-to-MOS Level Shifter and Driver	443
3208A, 3408A Hex Bipolar Sense Amplifiers for MOS Circuits	445
3222 Refresh Controller for 4K Dynamic RAMs	451
3232 Address Multiplexer and Refresh Counter for 4K Dynamic RAMs	457
3242 Address Multiplexer and Refresh Counter for 16K Dynamic RAMs	461
3245 Quad TTL-to-MOS Driver	465
5235, 5235-1 Quad TTL-to-MOS Driver	469
5244 Quad CCD Clock Driver	473
CHAPTER 6 RELIABILITY REPORTS	478
2107 N-Channel Silicon Gate MOS 4K Static RAMs	479
2115/2125 N-Channel Silicon Gate MOS 1K Static RAMs	498
2708 8K UV EPROM	507
2416 16K CCD Memory	514
Index	521

The Semiconductor Memory Book

CHAPTER ONE
Static RAMs

Designing with Static MOS RAMs	2
Designing Non-Volatile Semiconductor Memory Systems	20
Bipolar/MOS Static Ram Compatibility	38
2101A 256 × 4 Bit Static RAM	44
2102A, 2102AL 1K × 1 Bit Static Ram	48
2111A 256 × 4 Static RAM	52
2112A 256 × 4 Bit Static RAM	56
2114 1024 × 4 Bit Static RAM	61
2115A, 2125A High Speed 1K × 1 Bit Static RAM	65
2141 4096 × 1 Bit Static RAM	70
2142 1024 × 4 Bit Static RAM	74
2147 4096 × 1 Bit Static RAM	78
5101 256 × 4 Bit Static CMOS RAM	84

DESIGNING WITH STATIC MOS RAMS

JIM OLIPHANT

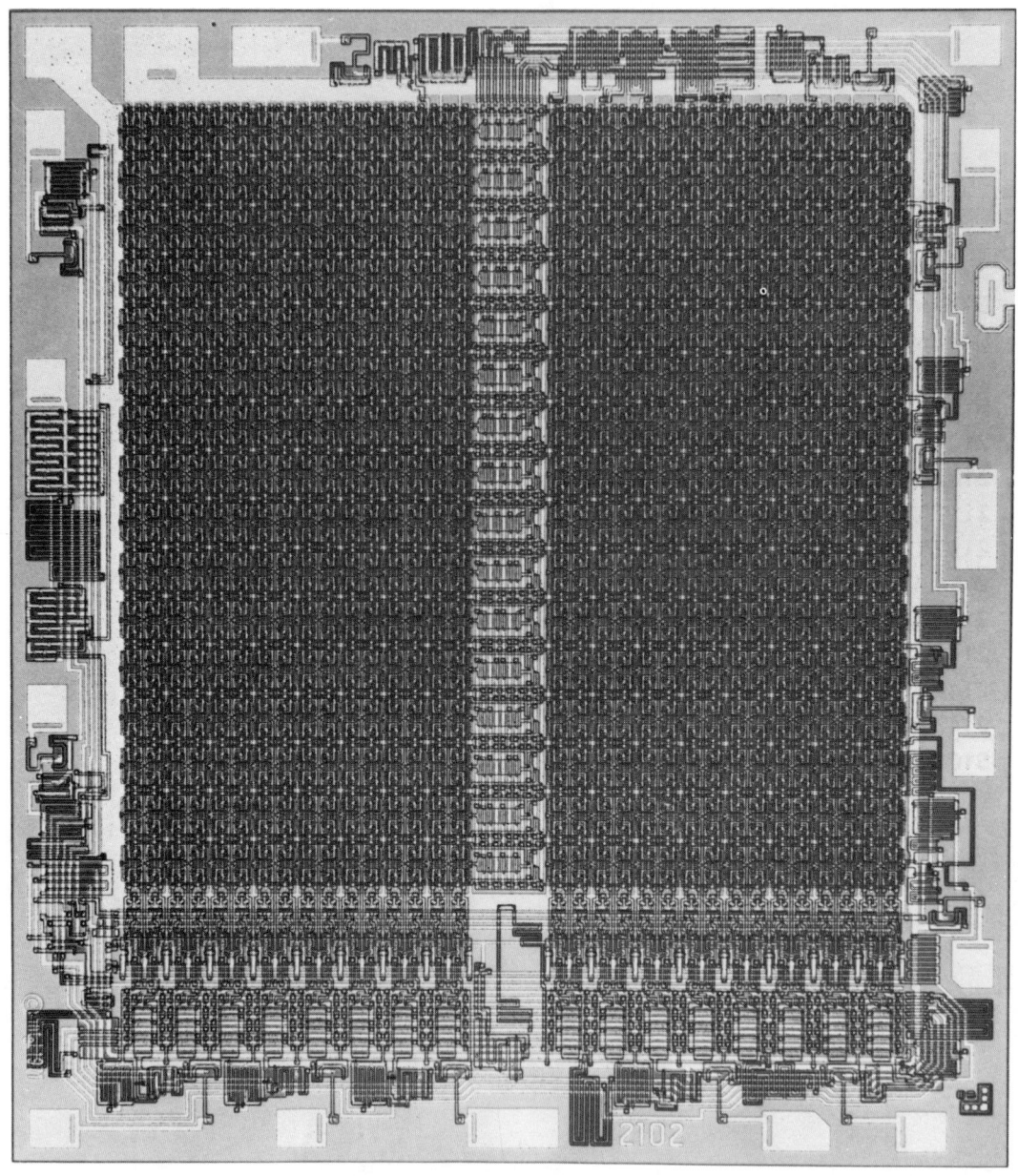

Photomicrograph of 1024 Word x 1 Bit 2102 Static MOS RAM

DESIGNING WITH STATIC MOS RAMS

INTRODUCTION

Intel's introduction of reliable, low cost, static, high-density MOS RAMs has done much to stimulate the use of semiconductor memory in new and unique applications. These RAMs, which do not require special refresh or timing circuitry, are used with as much ease as the standard TTL gates.

The Intel family of static high-density MOS RAMs, shown in Table I, offers the system designer flexibility in memory configuration, speed and ease of use. The devices in this family are directly TTL compatible in all respects: inputs, output(s) and power supply. Internal circuits are designed for full DC stability requiring no clocks or refreshing to operate. These static RAMs are manufactured with Intel's reliability proven N-channel silicon gate and CMOS silicon gate process.

The purpose of this application note is to outline the internal operation of these static RAMs, how they are used, and to present system design considerations in their use. In addition, suggested layout configurations for larger memory systems and techniques for reducing power dissipation during standby will be discussed.

DEVICE DESCRIPTIONS

As shown in Table I, there are two data organizations in the Intel static RAM family—1024 words x 1 bit and 256 words x 4 bits. The memory devices organized as 256 words x 4 bits are available with separate data input and output pins with an output disable pin (22 pin DIP), combined input/output pins with an output disable pin (18 pin DIP), and combined input/output with no output disable pin (16 pin DIP).

The sections on Device Operation detailed below describe the internal circuits which are common to both the 1024 word x 1 bit devices and the 256 word x 4 bit devices. The operational differences between the devices in the static RAM family are limited to the logic state and timing of chip enable(s) and data I/O lines and are discussed separately under the heading for each device type.

General Device Operation

Each of the Intel N-channel static RAMs utilize a DC stable six transistor cell configuration for the storage medium. The storage cells are arranged in a 32 x 32 matrix as shown in Figure 1. Data selection on the 1024 x 1 devices is accomplished by the coincidence of a row select (A_0-A_4, 1 of 32) and column select (A_5-A_9, 1 of 32). For the four bit wide configured RAMs, the selection is made by a row select (A_0-A_4, 1 of 32) and four column selects (A_5-A_7, 4 of 32). The data contained in the selected cell(s) is sensed, buffered, and presented to the data out pin D_O. In all devices the polarity of data read from memory is the same polarity as the data written into memory.

Storage Cell Operation

The two types of storage cells used in the Intel static RAM family are shown in Figure 2A and 2B. Static RAMs suffixed by "A" (e.g., 2102A) utilize

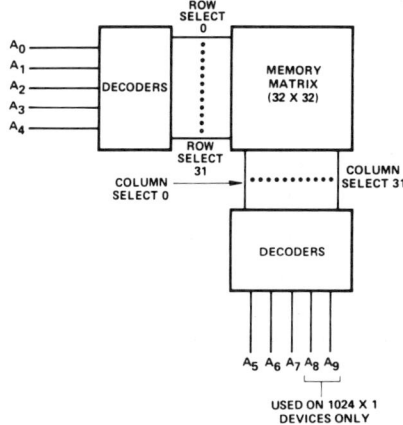

Figure 1. Simplified Memory Block Diagram

Table I. Intel Static MOS RAM Family

INTEL PART NUMBER	CONFIGURATION (WORDS X BITS)	DATA INPUT/OUTPUT	OUTPUT DISABLE	NUMBER CHIP ENABLES	POWER DOWN CAPABILITY	NUMBER PACKAGE PINS
2101A	1024 X 1	SEPARATE	N/A	1	NO	16
2102AL	1024 X 1	SEPARATE	N/A	1	YES	16
2101A	256 X 4	SEPARATE	YES	2	NO	22
2111A	256 X 4	COMMON	YES	2	NO	18
2112A	256 X 4	COMMON	NO	1	NO	16
5101[1]	256 X 4	SEPARATE	YES	2	YES*	22

*Extremely low standby current at 15.4 ua total.
(1) (CMOS)

depletion mode load devices, which are normally "on" (Fig. 2B). (Earlier product designated without the "A" suffix (e.g., 2102, 2101, etc.) utilize enhancement mode load devices which are normally "off" [Fig. 2A].) The basic operation of these two types of cells is similar in the manner in which data is written, stored, and retrieved. The differences between these cells will be discussed later.

Consider the storage cell shown in Figure 2A. Data is stored as a charge on the gate of either Q_3 or Q_4 (which determines the logic state of the cell). The voltage on the charged node is approximately $V_{CC} - V_{TH}$ (where V_{TH} is the effective threshold of the load devices) and turns Q_3 or Q_4 on. By definition a logic "0" is stored in the cell if Q_3 is on and a logic "1" is stored if Q_4 is on. If it is assumed that Q_3 is on (logic "0" stored) then current will flow from the load on Q_3 (device Q_2) through Q_3 to ground (V_{SS}). This current will cause the voltage at node (1) to assume a value near V_{SS} (the voltage is proportional to the effective on resistance of Q_2 and Q_3). The resultant low voltage on node (1) turns device Q_4 off. Device Q_5 maintains the charge on the gate of Q_3 by replacing charge leaked off through the high impedance parasitic leakage resistor $R_{LEAKAGE}$. (This leakage is typically in the picoampere range.) The storage cell will remain in this logic state until an external forcing function is applied (write cycle).

Operation of the storage cell shown in Figure 2B is similar to that described above except for the implementation of the load device. A brief discussion of differences between enhancement and depletion type devices will aid the understanding of the storage cell operation.

A depletion type MOS device has a channel implanted between the source and drain (see Fig. 3). The effect of this conducting channel is to shift the threshold of a standard enhancement device such that it is on at lower gate voltages. The basic operation of these two types of devices can be summarized as follows for N-channel technology: An enhancement mode device requires a positive gate voltage (relative to the source) to turn the device on. A depletion mode device requires a negative gate voltage (relative to source) to turn it off. These two conditions are shown in Figure 3.

(The actual threshold of the depletion mode device can be controlled by the degree of channel doping used in the fabrication process.)

Operation of the storage cell is as follows: assume the gate of Q_3 is high turning Q_3 on causing current to flow in Q_3 and Q_2. Since devices Q_2 and Q_3 are ratioed (that is, Q_2 has a higher impedance than Q_3) the voltage at node 1 will drop close to V_{SS}. Note that the gate of Q_2 is tied to node 1; therefore as node 1 decreases in voltage, the voltage drive on Q_2 is reduced, making the effective impedance of Q_2 higher. This allows the voltage at node 1 to move even closer to V_{SS}.

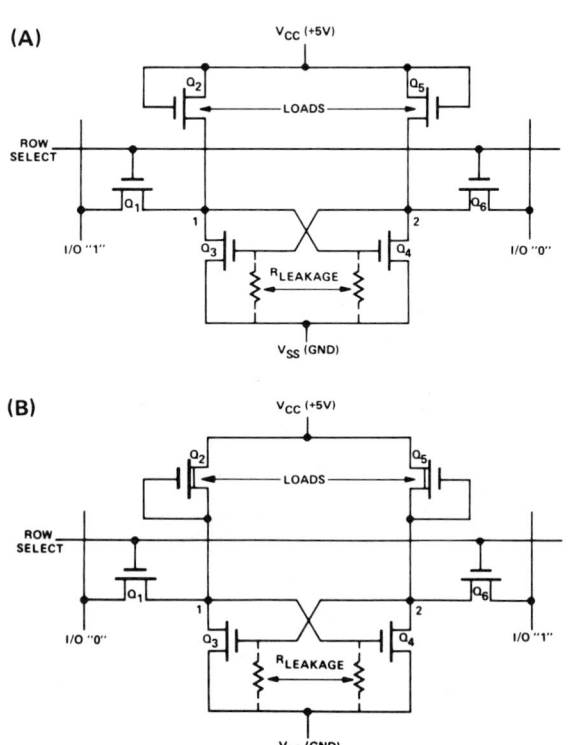

Figure 2. Storage Cell

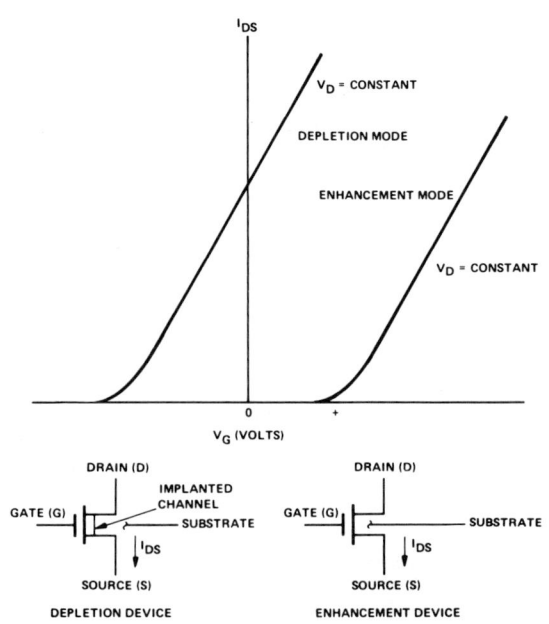

Figure 3. Enhancement/Depletion Characteristics

DESIGNING WITH STATIC MOS RAMS

Since node 1 is low and is tied to the gate of Q_4, device Q_4 is off. The charge on Q_3 is maintained by the load device Q_5. Note that only leakage currents flow through device Q_5 which has a minimal effect on the voltage at node 2. Since increased positive voltage at node 2 increases the voltage drive on Q_5, device Q_5 turns on hard. The voltage at node 2 is therefore equal to V_{CC} (note that there is no threshold drop across device Q_5 since it is a depletion mode device).

Accessing the Storage Cell

The storage cell is interrogated for a read or write operation by activating the proper row select line which turns devices Q_1 and Q_6 on (Fig. 2A, 2B). For a read operation, a sense amplifier (see Fig. 4) connected to both the I/O "0" and I/O "1" outputs of each column detects the state of the selected storage cell in that column. If Q_3 is on (logic "0") then current will flow in the I/O "0" line. If Q_4 is on (logic "1"), current will flow in the I/O "1" line. A write buffer (Fig. 4) places a high level ($\sim V_{CC}$) on the I/O "0" line to write a logic "0", and a high level on the I/O "1" to write a logic "1". For both write conditions, the opposite line is held low (V_{SS}).

As is shown in Figure 4, there are internal data-in/data-out buses. Data is gated to/from the appropriate columns by column select. Note that chip enable(s) gate the output data to a three state buffer and then to the output pin. Therefore, if a chip is not selected, the output pin goes to a high impedance state (allowing the output pins to be OR tied).

Address Buffers/Decoders

Typical address buffers and decoders, for the static RAM family are shown in Figures 5 and 6 respectively. As is shown in these figures, the address buffers and decoders are static requiring no precharging for operation. The buffers/decoders respond to changes on the address lines and do not latch the input addresses. Therefore, in those systems where the address lines are not stable throughout the cycle, it may be necessary to buffer them with external latches. An example of such a system is discussed later.

It should be noted however that in many systems requiring memory the addition of external address latches is not required. This is particularly true of microprocessor systems.

2102A OPERATION

As discussed before, the 2102A device is organized as 1024 words x 1 bit having separate data-in/data-out pins. The memory is organized internally in a 32 row by 32 column matrix as shown in Figure 7. The pin configuration and logic symbol are shown in Figure 8.

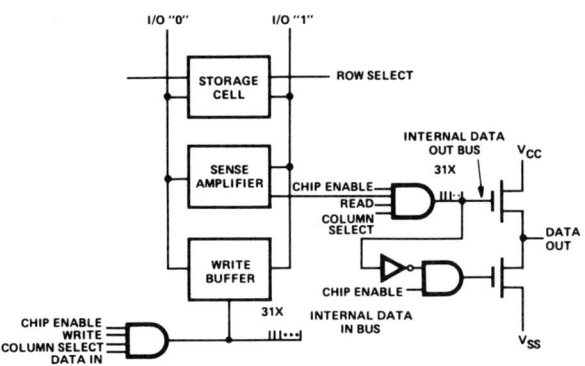

Figure 4. Internal Data Path

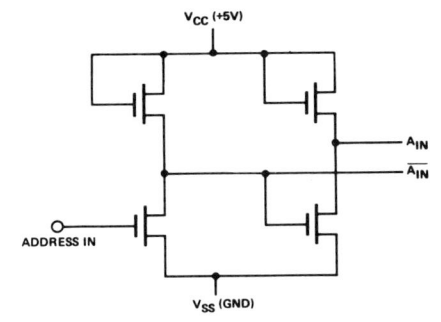

Figure 5. Address Input Buffer

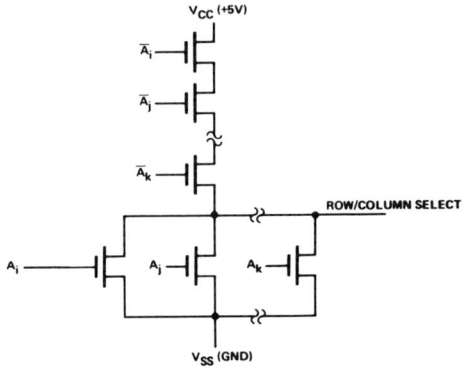

Figure 6. Address Decoder

There are only two control inputs to the 2102A: Read/write and $\overline{\text{chip enable}}$. For unselected devices ($\overline{\text{chip enable}}$ high), the data-in input is electrically disconnected from the input data bus internal to the 2102A and the data-out buffer goes to a high impedance state. The addresses, however, are buffered and decoded (generating an internal row/column select) independent of chip enable.

BLOCK DIAGRAM

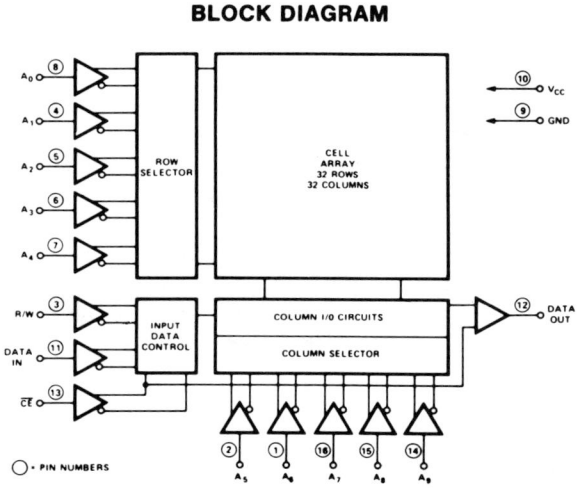

Figure 7. 2102A Block Diagram

PIN CONFIGURATION

2102A

LOGIC SYMBOL

2102A

PIN NAMES

D_{IN}	DATA INPUT	$\overline{CE}$	CHIP ENABLE
$A_0 - A_9$	ADDRESS INPUTS	D_{OUT}	DATA OUTPUT
R/W	READ/WRITE INPUT	V_{CC}	POWER (+5V)

Figure 8. 2102A Pin Configuration, Logic Symbol

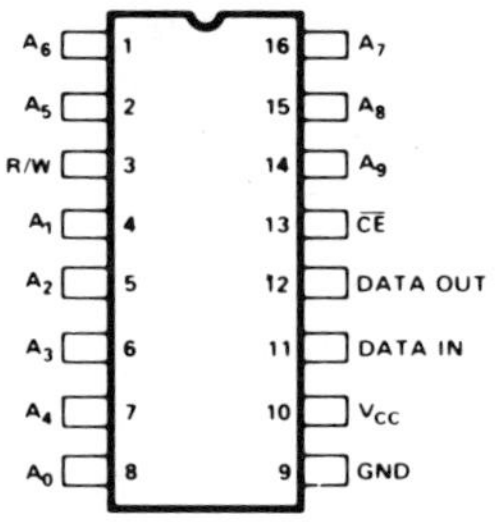

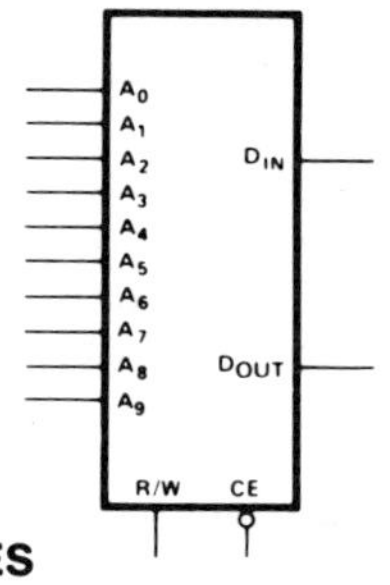

Read Cycle

Basic read cycle timing is shown in Figure 9. Note that although chip enable is shown as a pulse occurring after the address changes, there is no specified time at which it must occur (either before or after address change). It is therefore permissible to tie the chip enable input low if the data-out pin is not OR tied with other outputs and operate the memory device with only the read/write line and address inputs.

For example, if a series of read cycles are to be performed (such as for CRT displays), and the data-out pin is not OR-tied with another output, chip enable may be held low and the addresses may be cycled in any order to access data. During this time, however, the read/write input must always be in the high state. For this case, output data will be valid at T_A as shown in Figure 9 and specified in Table II.

A second method may be used to read data from the memory. If the addresses are set up before a read decision can be made, then chip enable may be brought low at the read decision time. Output data will be valid at t_{CO} (Table II) for this condition.

Write Cycle

Basic timing for a write cycle is shown in Figure 10. In the write cycle it is *not* permissible to perform a series of write cycles by holding chip enable and read/write low and cycling through the desired addresses. However, chip enable can be held low for continuous writes if the read/write input is timed per Figure 10. For the 2102A, a minimum write to address set up time, t_{AW}, must be observed per Table III. The minimum data hold time, t_{DH}, beyond read/write is 0 ns.

READ CYCLE

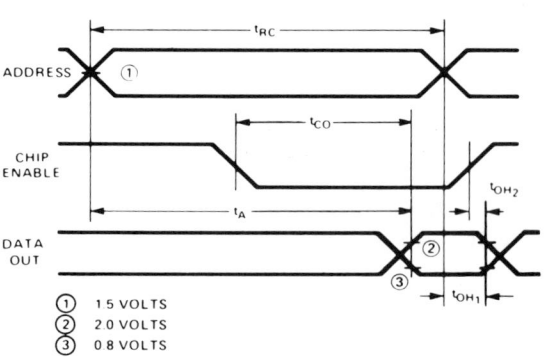

Figure 9. 2102A Read Cycle

WRITE CYCLE

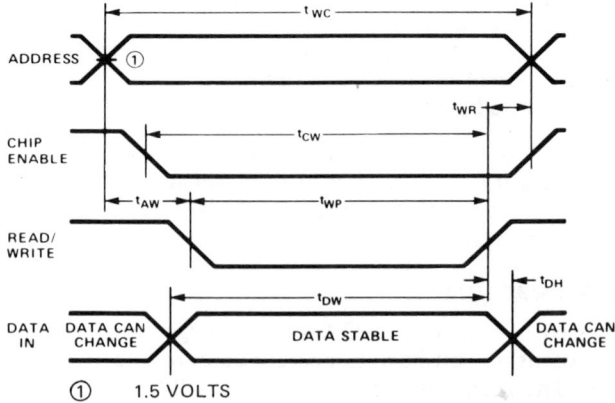

Figure 10. 2102A Write Cycle

Table II. 2102A Read Timing
READ CYCLE

Symbol	Parameter	Min.	Typ.[1]	Max.	Unit
t_{RC}	Read Cycle	350			ns
t_A	Access Time			350	ns
t_{CO}	Chip Enable to Output Time			180	ns
t_{OH1}	Previous Read Data Valid with Respect to Address	40			ns
t_{OH2}	Previous Read Data Valid with Respect to Chip Enable	0			ns

Table III. 2102A Write Timing
WRITE CYCLE

Symbol	Parameter	Min.	Typ.[1]	Max.	Unit
t_{WC}	Write Cycle	350			ns
t_{AW}	Address to Write Setup Time	20			ns
t_{WP}	Write Pulse Width	250			ns
t_{WR}	Write Recovery Time	0			ns
t_{DW}	Data Setup Time	250			ns
t_{DH}	Data Hold Time	0			ns
t_{CW}	Chip Enable to Write Setup Time	250			ns

NOTE: 1. Typical values are for $T_A = 25°C$ and nominal supply voltage.

Note that the minimum write cycle may be obtained by using the minimum times associated with t_{AW}, t_{WP} and t_{WR} (Fig. 10), that is:

$$t_{WC} (MIN) = t_{AW} + t_{WP} + t_{WR}$$

Read-Modify-Write

A read-modify-write cycle is merely a combination of a read cycle and a read/write pulse, t_{WP}. The minimum read-modify-write cycle time is therefore $t_{RC} + t_{WP}$. The timing associated with the 2102A read-modify-write cycle is shown in Figure 11.

D.C. and Operating Characteristics

The D.C. and operating characteristics for the 2102A is given in Table IV. Power supply current versus V_{CC} supply voltage is shown in Figure 12 for the 2102A. Power supply current as a function of temperature is shown in Figure 13 for the 2102A.

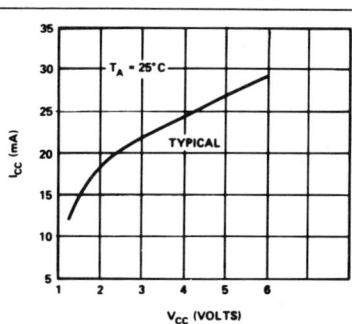

Figure 12. 2102A Power Supply vs. Supply Voltage

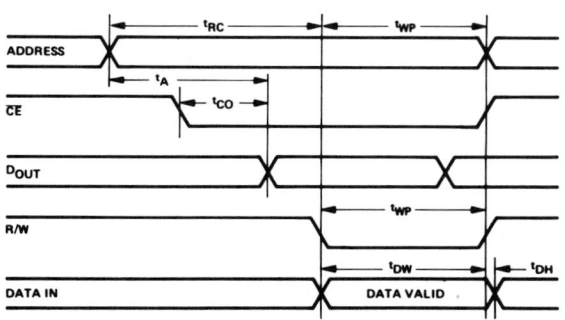

Figure 11. 2102A Read-Modify-Write Cycle

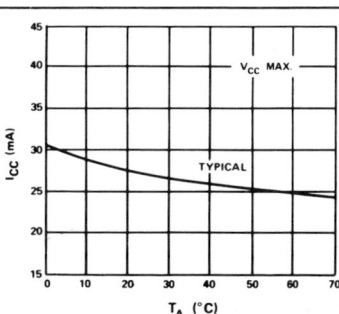

Figure 13. Power Supply Current vs. Ambient Temperature.

For reference, typical A.C. and D.C characteristics for the 2102A are shown in the graphs of Figure 14. In particular, note the relative insensitivity of access time as a function of load capacitance.

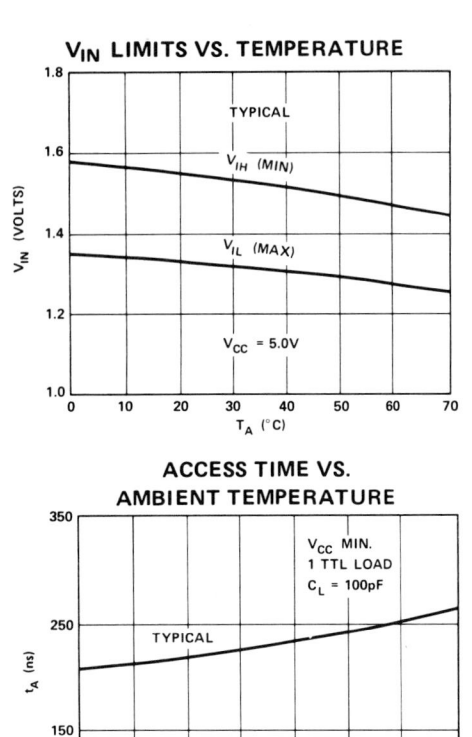
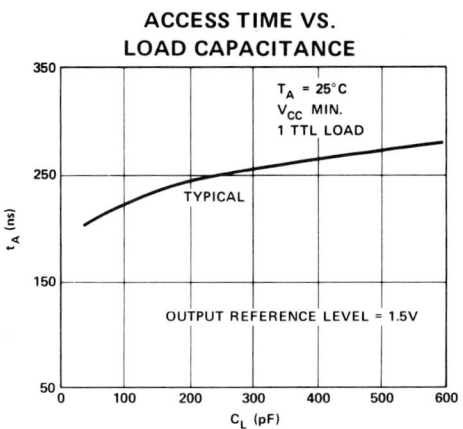

Figure 14. 2102A Typical D.C. and A.C. Characteristics

Table IV. 2102A D.C. and Operating Characteristics.
$T_A = 0°C$ to $70°C$, $V_{CC} = 5V \pm 5\%$ unless otherwise specified.

Symbol	Parameter	2102A, 2102A-4 2102AL, 2102AL-4 Limits			2102A-2, 2102AL-2 Limits			2102A-6 Limits			Unit	Test Conditions
		Min.	Typ.[1]	Max.	Min.	Typ.[1]	Max.	Min.	Typ.[1]	Max.		
I_{LI}	Input Load Current		1	10		1	10		1	10	μA	$V_{IN} = 0$ to 5.25V
I_{LOH}	Output Leakage Current		1	5		1	5		1	5	μA	$\overline{CE} = 2.0V$, $V_{OUT} = V_{OH}$
I_{LOL}	Output Leakage Current		−1	−10		−1	−10		−1	−10	μA	$\overline{CE} = 2.0V$, $V_{OUT} = 0.4V$
I_{CC}	Power Supply Current		33	Note 2		45	65		33	55	mA	All Inputs = 5.25V, Data Out Open, $T_A = 0°C$
V_{IL}	Input Low Voltage	−0.5		0.8	−0.5		0.8	−0.5		0.65	V	
V_{IH}	Input High Voltage	2.0		V_{CC}	2.0		V_{CC}	2.2		V_{CC}	V	
V_{OL}	Output Low Voltage			0.4			0.4			0.45	V	$I_{OL} = 2.1mA$
V_{OH}	Output High Voltage	2.4			2.4			2.2			V	$I_{OH} = -100\mu A$

Notes: 1. Typical values are for $T_A = 25°C$ and nominal supply voltage.
2. The maximum I_{CC} value is 55mA for the 2102A and 2102A-4, and 33mA for the 2102AL and 2102AL-4.

DESIGNING WITH STATIC MOS RAMS

Power Down Standby Operation

The 2102AL may be placed in a power down mode where data is maintained with greatly reduced power dissipation (maximum of 42 mW vs. 368 mW). Data is maintained at a reduced power setting because the load devices in the storage cell (Q_2 and Q_5) shown in Figure 2B are implemented with depletion load devices. As mentioned previously, a depletion mode device is normally "on" and requires a negative voltage (below ground) to reach cut-off in operation (see Fig. 3). The only requirement, therefore, to assure data retention is to guarantee that the minimum V_{CC} voltage allowed in standby operation is sufficient to bias the gate of the appropriate storage node (Q_3 or Q_4 Fig. 2B) on. (Recall that there is no threshold drop across the depletion load device supplying the on drive to the storage device.)

A summary of the power down requirements and characteristics for the 2102AL family are shown in Figure 15 and Table V. As is shown in this figure, there is a requirement that chip enable be brought to a level of 2.0V, or higher, a T_{CP} time (minimum 0 nsec) before V_{CC} drops below its minimum value (4.75V).

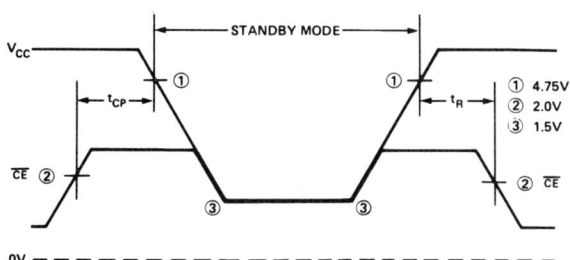

Figure 15. 2102AL Family Standby Characteristics.

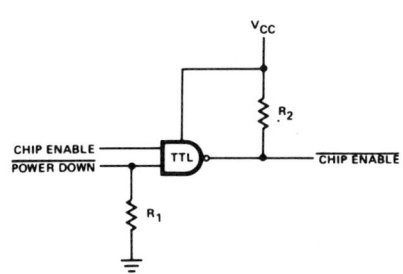

Figure 16. Chip Enable Generator for Power Down Mode

To assure that stored data is not over-written, the chip enable input must either be held at a level of 1.5V, or higher, or allowed to track with V_{CC} at the same or higher voltage level and same or slower discharge rate as V_{CC}.

A circuit that implements the tracking of chip enable with V_{CC} is shown in Figure 16. In this figure, the chip enable NAND circuit is powered by the same supply, V_{CC}, that is discharging. A power down signal ($\overline{\text{power down}}$) is generated to set the chip enable signal high at the appropriate time (discussed later). As V_{CC} begins to discharge beyond the limit of TTL operation, resistors R_1 and R_2 are used to assure that the chip enable output stays high and tracks the discharging V_{CC} independent of the chip enable input.

The power down signal can occur at any time if the memory is in a read cycle or is inactive. However, if a write cycle is being executed and power loss is detected, then the power down signal must be delayed until the write cycle is complete. In most

Table V. 2101AL Family D.C. Standby Characteristics.
$T_A = 0°C$ to $70°C$

Symbol	Parameter	2102AL, 2102AL-4 Limits			2102AL-2 Limits			Unit	Test Conditions
		Min.	Typ.[1]	Max.	Min.	Typ.[1]	Max.		
V_{PD}	V_{CC} in Standby	1.5			1.5			V	
V_{CES}[2]	$\overline{CE}$ Bias in Standby	2.0			2.0			V	$2.0V \leq V_{PD} \leq V_{CC}$ Max.
		V_{PD}			V_{PD}			V	$1.5V \leq V_{PD} < 2.0V$
I_{PD1}	Standby Current		15	23		20	28	mA	All Inputs = V_{PD1} = 1.5V
I_{PD2}	Standby Current		20	30		25	38	mA	All Inputs = V_{PD2} = 2.0V
t_{CP}	Chip Deselect to Standby Time	0			0			ns	
t_R[3]	Standby Recovery Time	t_{RC}			t_{RC}			ns	

NOTES:
1. Typical values are for $T_A = 25°C$ and nominal supply voltage.
2. Consider the test conditions as shown: If the standby voltage (V_{PD}) is between 5.25V (V_{CC} Max.) and 2.0V, then CE must be held at 2.0V Min. (V_{IH}). If the standby voltage is less than 2.0V but greater than 1.5V (V_{PD} Min.), then CE and standby voltage must be at least the same value or, if they are different, CE must be the more positive of the two.
3. $t_R = t_{RC}$ (READ CYCLE TIME).

systems this is entirely feasible since the decoupling capacitors will hold V_{CC} power long enough to finish a complete write cycle at full power.

A schematic representation of a sudden loss of normal V_{CC} power is shown in Figure 17. If at t = 0 the V_{CC} supply is removed, power will be supplied to the load (L) from the capacitor (C) until the load voltage (V_L) is a diode drop below the battery voltage (V_B), after which power to the load is supplied by the battery. There is no requirement to control the rate at which V_{CC} discharges.

Similarly, when the supply voltage is restored, the voltage at the load (V_L) will begin to rise when the supply voltage becomes greater than the battery voltage.

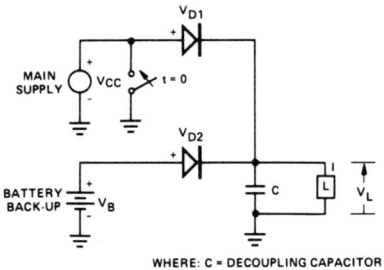

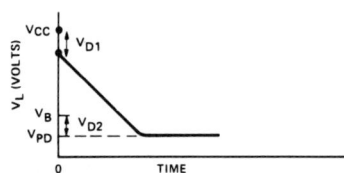

Figure 17. Battery Backup Characteristics

Note that the 2102A is capable of retaining data in a power down mode over a V_{CC} voltage range of 1.5 to 4.75 *if the chip enable input is always a high level equal to or higher than V_{PD}* during standby. This allows maximum flexibility in the selection of batteries for standby. Remember that chip enable tracking requirements to V_{CC} are required *only* if the state of the chip enable input can not be guaranteed to be a high level during the entire standby period.

256 WORD x 4-BIT STATIC RAMs

The introduction of static, high density MOS RAMs organized as 256 words x 4 bits has significantly reduced the complexity, size and component count of systems not requiring large storage capacity.

With the Intel family of 256 word x 4 bit RAMs it is now possible to realize more benefits of "distributed" memory using MOS devices with their attendant low power and simple interface.

designated as 2101A, 2111A, and 2112A. In summary the 2101A is packaged in a 22 pin DIP, has four data-in and four data-out lines, two chip enables and an output disable. The 2111A is packaged in an 18 pin DIP, has four common data-in/data-out lines, two chip enables, and an output disable. The 2112A is packaged in a 16 pin DIP, has four common data-in/data-out lines, one chip enable and does *not* have an output disable. These selections allow the system designer almost any configuration he might desire.

2101A Operation

Internal operation of the 2101A is similar to that outlined for the 2102A. The storage cell is shown in Figure 2A; the address input buffers and internal decoders are shown in Figures 5 and 6 respectively.

Maximum system design flexibility is achieved with the 2101A for those applications requiring 256 word x 4 bit memory devices. Since the input/output lines are separated, it is not necessary to multiplex these lines unless required by the system. The two chip enables of opposite logic polarity simplify system interface design (especially with the 8080 microprocessor as discussed in the *Systems* section).

The pin configuration and logic symbol for the 2101A are shown in Figure 18. The block diagram is shown in Figure 19.

The 2101A may be operated in the same operating modes as the 2102A. For example, a series of reads may be performed on a given device with the chip enables at the proper selected state and the output disable line held low. The write and read-modify-write cycles may be performed per the 2102A description. For reference, the read and write waveforms are shown in Figure 20 with A.C. characteristics given in Table VI. D.C. and operating characteristics are shown in Table VII.

As discussed previously, the 2101A has separate input and output pins for data. When operating the device with the output OR-tied, it is permissible to tie the output disable pin low for all operations. If the data output pins are OR-tied with other devices, the chip enable inputs are used to electrically disconnect the unselected devices from the output data buses. In this unselected state ($\overline{CE1}$ high or CE2 low) the output devices are placed in the high impedance state.

The 2101A may also be operated with the corresponding input and output lines tied together. In this mode of operation output disable must be used to place the output devices in the high impedance state during a write cycle or the write portion of a read-modify-write cycle.

DESIGNING WITH STATIC MOS RAMS

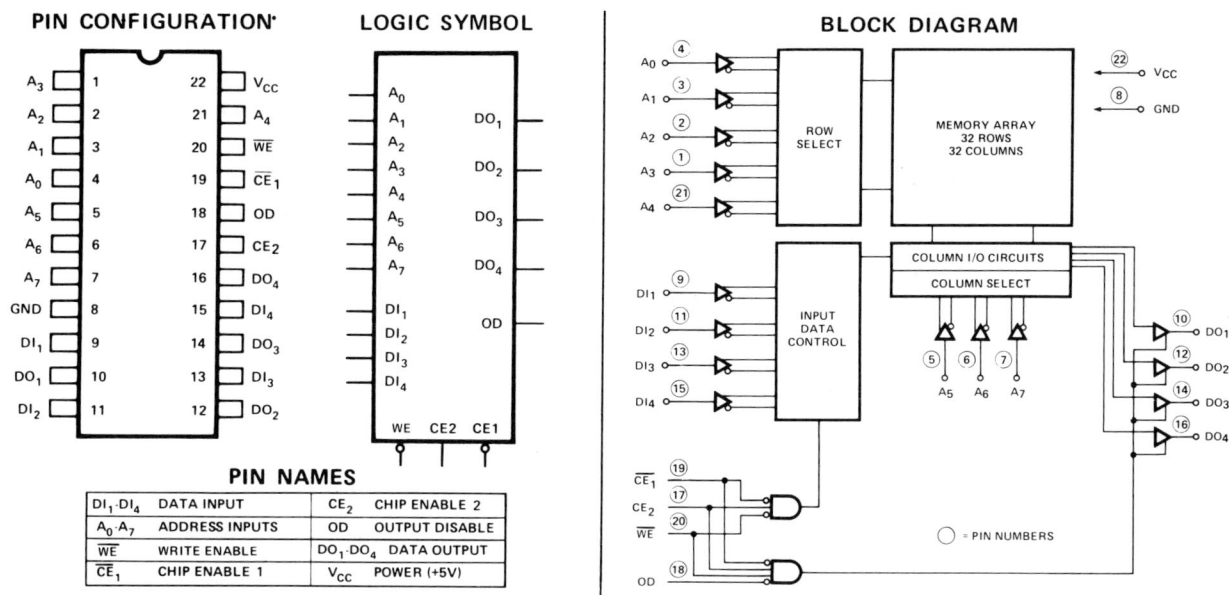

Figure 18. 2101A Pin Configuration/Logic Symbol.

Figure 19. 2101A Block Diagram.

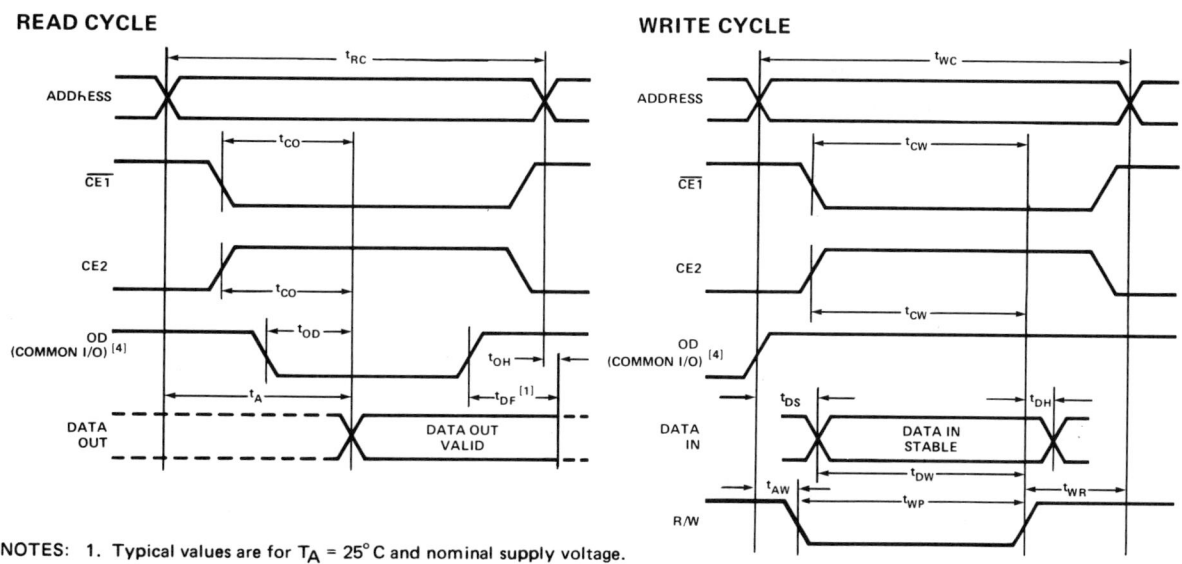

NOTES:
1. Typical values are for $T_A = 25°C$ and nominal supply voltage.
2. This parameter is periodically sampled and is not 100% tested.
3. t_{DF} is with respect to the trailing edge of $\overline{CE}_1$, CE_2, or OD, whichever occurs first.
4. OD should be tied low for separate I/O operation.

Figure 20. 2101A Read/Write Waveforms.

Table VI. 2101A A.C. Characteristics.

READ CYCLE $T_A = 0°C$ to $70°C$, $V_{CC} = 5V \pm 5\%$, unless otherwise specified.

Symbol	Parameter	Min.	Typ.[1]	Max.	Unit	Test Conditions
t_{RC}	Read Cycle	250			ns	
t_A	Access Time			250	ns	$t_r, t_f = 20ns$
t_{CO}	Chip Enable To Output			180	ns	Input Levels = 0.8V or 2.0V
t_{OD}	Output Disable To Output			130	ns	Timing Reference = 1.5V
t_{DF}[3]	Data Output to High Z State	0		180	ns	Load = 1 TTL Gate
t_{OH}	Previous Read Data Valid after change of Address	40			ns	and $C_L = 100pF$.

WRITE CYCLE

Symbol	Parameter	Min.	Typ.[1]	Max.	Unit	Test Conditions
t_{WC}	Write Cycle	170			ns	
t_{AW}	Write Delay	20			ns	$t_r, t_f = 20ns$
t_{CW}	Chip Enable To Write	150			ns	Input Levels = 0.8V or 2.0V
t_{DW}	Data Setup	150			ns	Timing Reference = 1.5V
t_{DH}	Data Hold	0			ns	Load = 1 TTL Gate
t_{WP}	Write Pulse	150			ns	and $C_L = 100pF$.
t_{WR}	Write Recovery	0			ns	
t_{DS}	Output Disable Setup	20			ns	

Table VII. 2101A D.C. and Operating Characteristics

$T_A = 0°C$ to $70°C$, $V_{CC} = 5V \pm 5\%$ unless otherwise specified.

Symbol	Parameter		Min.	Typ.[1]	Max.	Unit	Test Conditions
I_{LI}	Input Current			1	10	μA	$V_{IN} = 0$ to 5.25V
I_{LOH}	Data Output Leakage Current			1	10	μA	Output Disabled, $V_{OUT}=4.0V$
I_{LOL}	Data Output Leakage Current			-1	-10	μA	Output Disabled, $V_{OUT}=0.45V$
I_{CC1}	Power Supply Current	2101A, 2101A-4		35	55	mA	$V_{IN} = 5.25V$, $I_O = 0mA$
		2101A-2		45	65		$T_A = 25°C$
I_{CC2}	Power Supply Current	2101A, 2101A-4			60	mA	$V_{IN} = 5.25V$, $I_O = 0mA$
		2101A-2			70		$T_A = 0°C$
V_{IL}	Input "Low" Voltage		-0.5		+0.8	V	
V_{IH}	Input "High" Voltage		2.0		V_{CC}	V	
V_{OL}	Output "Low" Voltage				+0.45	V	$I_{OL} = 2.0mA$
V_{OH}	Output "High" Voltage	2101A, 2101A-2	2.4			V	$I_{OH} = -200\mu A$
		2101A-4	2.4			V	$I_{OH} = -150\mu A$

DESIGNING WITH STATIC MOS RAMS

2111A Operation

The 2111A has common input/output data buses and operates in a manner similar to that described for the 2101A with the data bus made common. The only logical difference between the two devices is the logic level of the two chip enables. For the 2111A both chip enables are true in the low state.

If either or both of the chip enables are high the internal input output data buffers are electrically disconnected from the external data bus.

The pin configuration and logic symbol for the 2111A are shown in Figure 21. The block diagram is shown in Figure 22.

As indicated previously, the read/write, address, and data in timing requirements for the 2111A are the same as for the 2101A operating in the

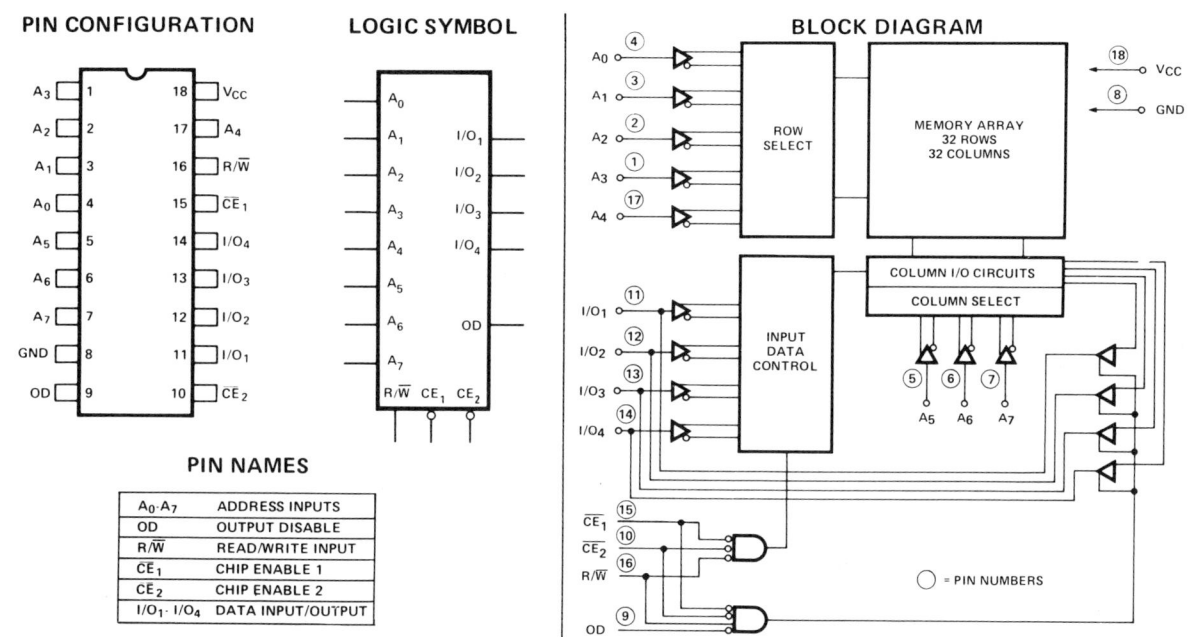

Figure 21. 2111A Pin Configuration/Logic Symbol

Figure 22. 2111A Block Diagram

Intel Memory Systems Division 4K x 8 Memory Card

common data bus mode. For reference, however, the read/write waveforms for the 2111A are shown in Figure 23 with the A.C. characteristics shown in Table VIII. D.C. characteristics are given in Table IX.

2112A Operation

The 2112A operates in a manner very similar to the 2111A. The major difference is that no output disable pin is available and one (instead of two) chip enables is used. Pin configuration and logic symbol for the 2112A are shown in Figure 24. The block diagram is shown in Figure 25.

Since no output disable pin is available for the 2112A, care should be exercised to assure that the data in bus is not activated any time the read/write line is high (read cycle). When operating the memory in a write, read-modify-write, or write-verify-read cycle, the read/write input is used to perform the function of an output disable. The output is disabled on the chip according to the following logical equation:

$$OD = \overline{R/W} + \overline{CE}$$

where:

OD = OUTPUT DISABLE

The basic read/write timing waveforms are shown in Figure 26 with the A.C. characteristics given in Table X.

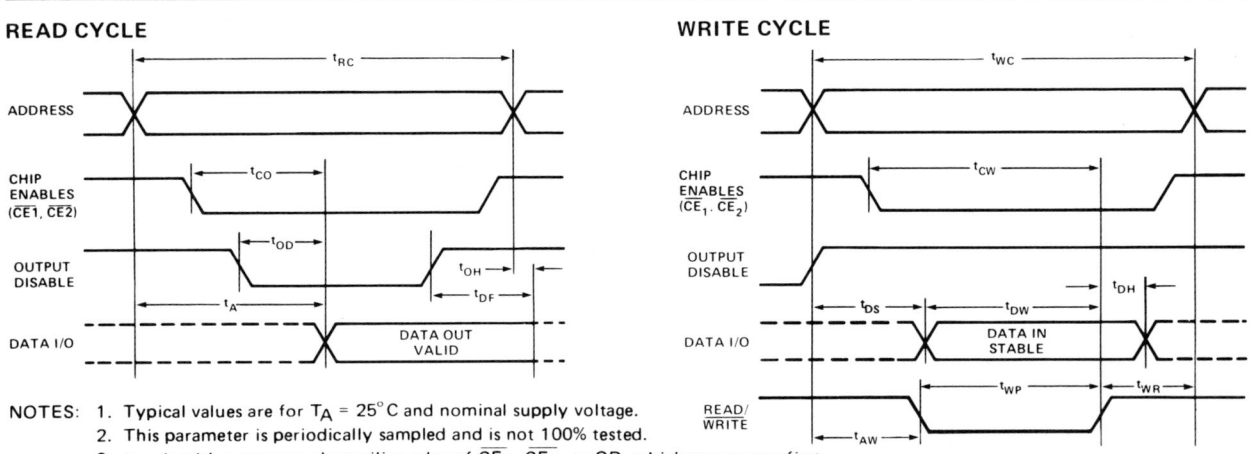

NOTES: 1. Typical values are for $T_A = 25°C$ and nominal supply voltage.
2. This parameter is periodically sampled and is not 100% tested.
3. t_{DF} is with respect to the trailing edge of $\overline{CE_1}$, $\overline{CE_2}$, or OD, whichever occurs first.

Figure 23. 2111A Read/Write Waveforms.

Table VIII. 2111A A.C. Characteristics

READ CYCLE $T_A = 0°C$ to $70°C$, $V_{CC} = 5V \pm 5\%$, unless otherwise specified.

Symbol	Parameter	Min.	Typ.[1]	Max.	Unit	Test Conditions
t_{RC}	Read Cycle	250			ns	t_r, t_f = 20ns Input Levels = 0.8V or 2.0V Timing Reference = 1.5V Load = 1 TTL Gate and C_L = 100pF.
t_A	Access Time			250	ns	
t_{CO}	Chip Enable To Output			180	ns	
t_{OD}	Output Disable To Output			130	ns	
t_{DF} [3]	Data Output to High Z State	0		180	ns	
t_{OH}	Previous Read Data Valid after change of Address	40			ns	

WRITE CYCLE

Symbol	Parameter	Min.	Typ.[1]	Max.	Unit	Test Conditions
t_{WC}	Write Cycle	170			ns	t_r, t_f = 20ns Input Levels = 0.8V or 2.0V Timing Reference = 1.5V Load = 1 TTL Gate and C_L = 100pF.
t_{AW}	Write Delay	20			ns	
t_{CW}	Chip Enable To Write	150			ns	
t_{DW}	Data Setup	150			ns	
t_{DH}	Data Hold	0			ns	
t_{WP}	Write Pulse	150			ns	
t_{WR}	Write Recovery	0			ns	
t_{DS}	Output Disable Setup	20			ns	

DESIGNING WITH STATIC MOS RAMS

Table IX. 2111A D.C. Characteristics.
$T_A = 0°C$ to $70°C$, $V_{CC} = 5V \pm 5\%$, unless otherwise specified.

Symbol	Parameter		Min.	Typ.[1]	Max.	Unit	Test Conditions
I_{LI}	Input Load Current			1	10	μA	V_{IN} = 0 to 5.25V
I_{LOH}	I/O Leakage Current			1	10	μA	Output Disabled, $V_{I/O}$ = 4.0V
I_{LOL}	I/O Leakage Current			−1	−10	μA	Output Disabled, $V_{I/O}$ = 0.45V
I_{CC1}	Power Supply Current	2111A, 2111A-4		35	55	mA	V_{IN} = 5.25V
		2111A-2		45	65		$I_{I/O}$ = 0mA, T_A = 25°C
I_{CC2}	Power Supply Current	2111A, 2111A-4			60	mA	V_{IN} = 5.25V
		2111A-2			70		$I_{I/O}$ = 0mA, T_A = 0°C
V_{IL}	Input Low Voltage		−0.5		0.8	V	
V_{IH}	Input High Voltage		2.0		V_{CC}	V	
V_{OL}	Output Low Voltage				0.45	V	I_{OL} = 2.0mA
V_{OH}	Output High Voltage	2111A, 2111A-2	2.4			V	I_{OH} = −200μA
		2111A-4	2.4			V	I_{OH} = −150μA

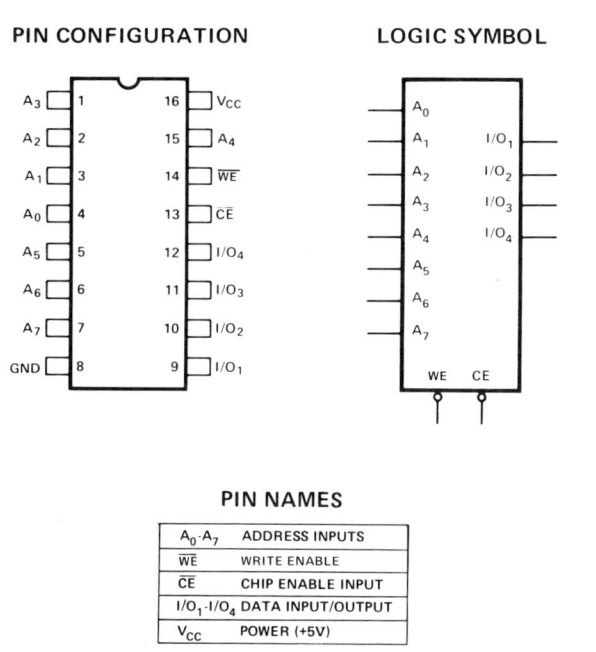

PIN NAMES

A_0-A_7	ADDRESS INPUTS
$\overline{WE}$	WRITE ENABLE
$\overline{CE}$	CHIP ENABLE INPUT
I/O_1-I/O_4	DATA INPUT/OUTPUT
V_{CC}	POWER (+5V)

Figure 24. 2112A Pin Configuration/Logic Symbol.

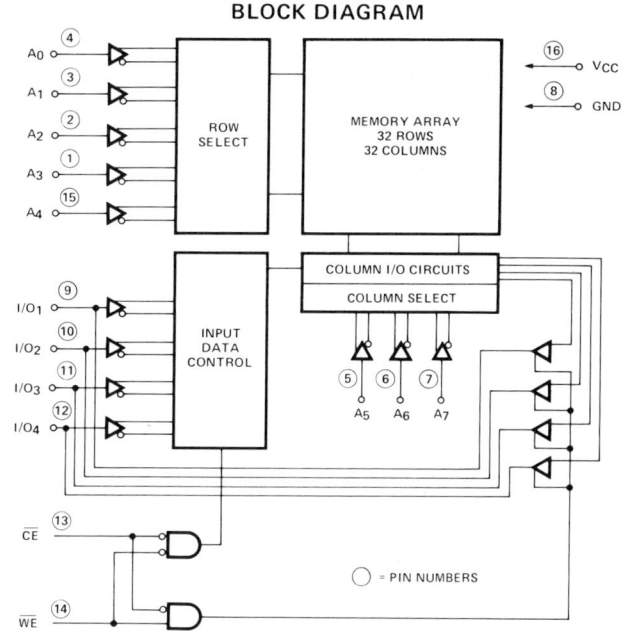

Figure 25. 2112A Block Diagram.

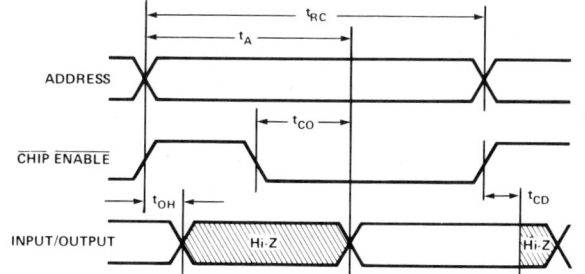

NOTE 1: Data Hold Time (T_{OH}) is referenced to the trailing edge of CHIP ENABLE (CE) or READ/WRITE (R/W) whichever comes first.

Figure 26. 2112A Read/Write Waveforms.

WRITE CYCLE #1

WRITE CYCLE #2

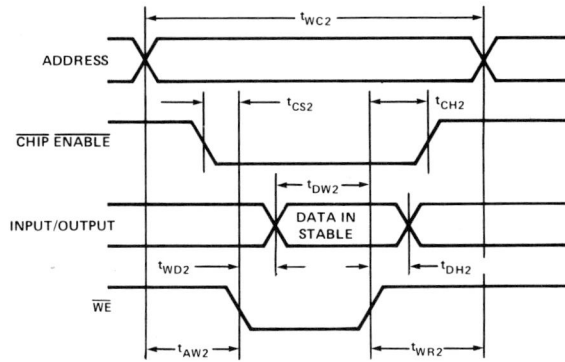

NOTE: 1. Typical values are for $T_A = 25°C$ and nominal supply voltage.

Figure 26. (cont'd)

Table X. 2112A A.C. Characteristics.

READ CYCLE $T_A = 0°C$ to $70°C$, $V_{CC} = 5V \pm 5\%$ unless otherwise specified.

Symbol	Parameter	Min.	Typ.[1]	Max.	Unit	Test Conditions
t_{RC}	Read Cycle	350			ns	$t_r, t_f = 20ns$
t_A	Access Time			350	ns	Input Levels = 0.8V or 2.0V
t_{CO}	Chip Enable To Output Time			240	ns	Timing Reference = 1.5V
t_{CD}	Chip Enable To Output Disable Time	0		200	ns	Load = 1 TTL Gate
t_{OH}	Previous Read Data Valid After Change of Address	40			ns	and $C_L = 100pF$.

WRITE CYCLE #1 $T_A = 0°C$ to $70°C$, $V_{CC} = 5V \pm 5\%$

Symbol	Parameter	Min.	Typ.[1]	Max.	Unit	Test Conditions
t_{WC1}	Write Cycle	270			ns	$t_r, t_f = 20ns$
t_{AW1}	Address To Write Setup Time	20			ns	Input Levels = 0.8V or 2.0V
t_{DW1}	Write Setup Time	250			ns	Timing Reference = 1.5V
t_{WP1}	Write Pulse Width	250			ns	Load = 1 TTL Gate
t_{CS1}	Chip Enable Setup Time	0			ns	and $C_L = 100pF$.
t_{CH1}	Chip Enable Hold Time	0			ns	
t_{WR1}	Write Recovery Time	0			ns	
t_{DH1}	Data Hold Time	0			ns	
t_{CW1}	Chip Enable to Write Setup Time	250			ns	

WRITE CYCLE #2 $T_A = 0°C$ to $70°C$, $V_{CC} = 5V \pm 5\%$

Symbol	Parameter	Min.	Typ.[1]	Max.	Unit	Test Conditions
t_{WC2}	Write Cycle	470			ns	$t_r, t_f = 20ns$
t_{AW2}	Address To Write Setup Time	20			ns	Input Levels = 0.8V or 2.0V
t_{DW2}	Write Setup Time	250			ns	Timing Reference = 1.5V
t_{WD2}	Write To Output Disable Time	200			ns	Load = 1 TTL Gate
t_{CS2}	Chip Enable Setup Time	0			ns	and $C_L = 100pF$.
t_{CH2}	Chip Enable Hold Time	0			ns	
t_{WR2}	Write Recovery Time	0			ns	
t_{DH2}	Data Hold Time	0			ns	

NOTE: 1. Typical values are for $T_A = 25°C$ and nominal supply voltage.

DESIGNING WITH STATIC MOS RAMS

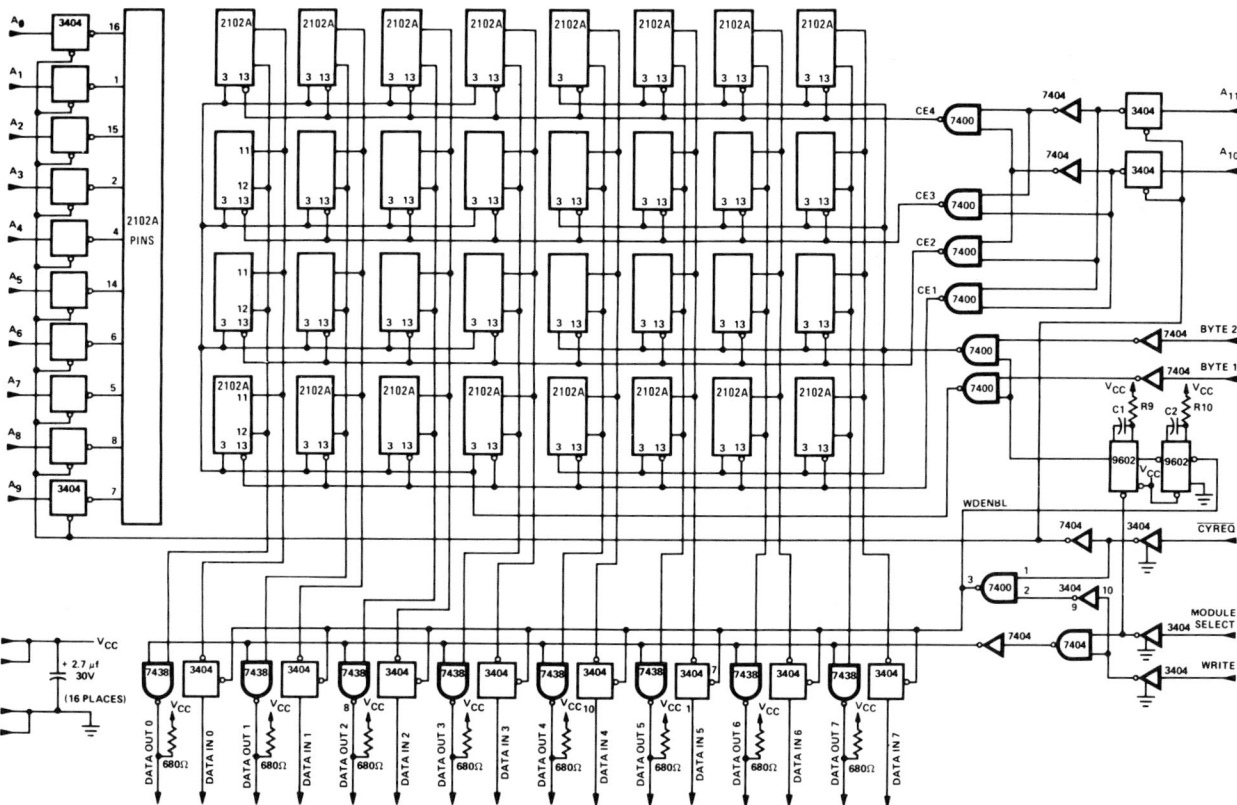

Figure 27. 4K x 8 Memory System.

SYSTEM DESIGN/OPERATION

The design of timing and interface circuits for memory systems utilizing Intel static RAMs is simple and straightforward. In this section, details of system designs using these static RAMs will be discussed.

Consider first the 4K x 8 system shown in Figure 27. This system, Intel's in-26 self-contained memory card, is expandable in both the number of words and number of bits/word directions. (Expanding the number of words per system is accomplished with the module select input.) Note that there are only three input control lines: write, module select and cycle request (CYREQ) (with byte control provided). Operation of the 4K x 8 memory system is explained with the aid of the timing diagram is shown in Figure 28.

At time T_O the 100 nsec $\overline{\text{CYREQ}}$ pulse is applied, the addresses made valid, the write input, and data is made valid (for write mode only). The module select input is set low no later than 80 nsec after T_O. (If only one board is used, the module select line may be permanently tied to ground.) Output data is available at time defined by timing diagram (650 nsec) with cycle completed for both read and write at 650 nsec after start of cycle. (Note

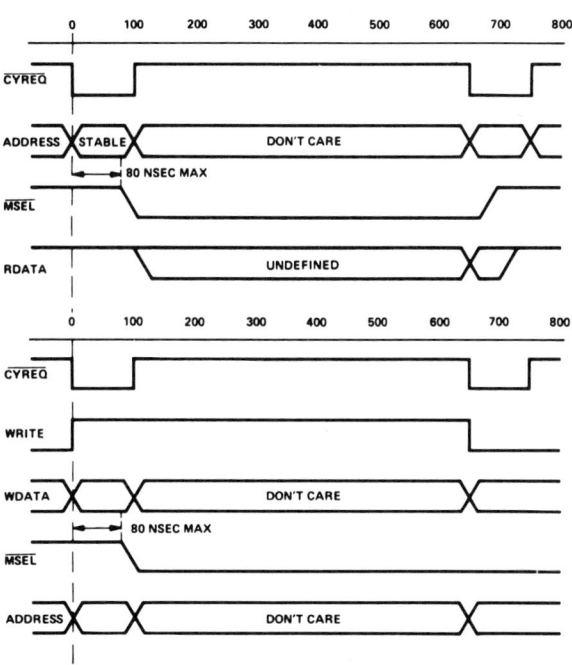

Figure 28. Timing Diagram, 4K x 8/9 System.

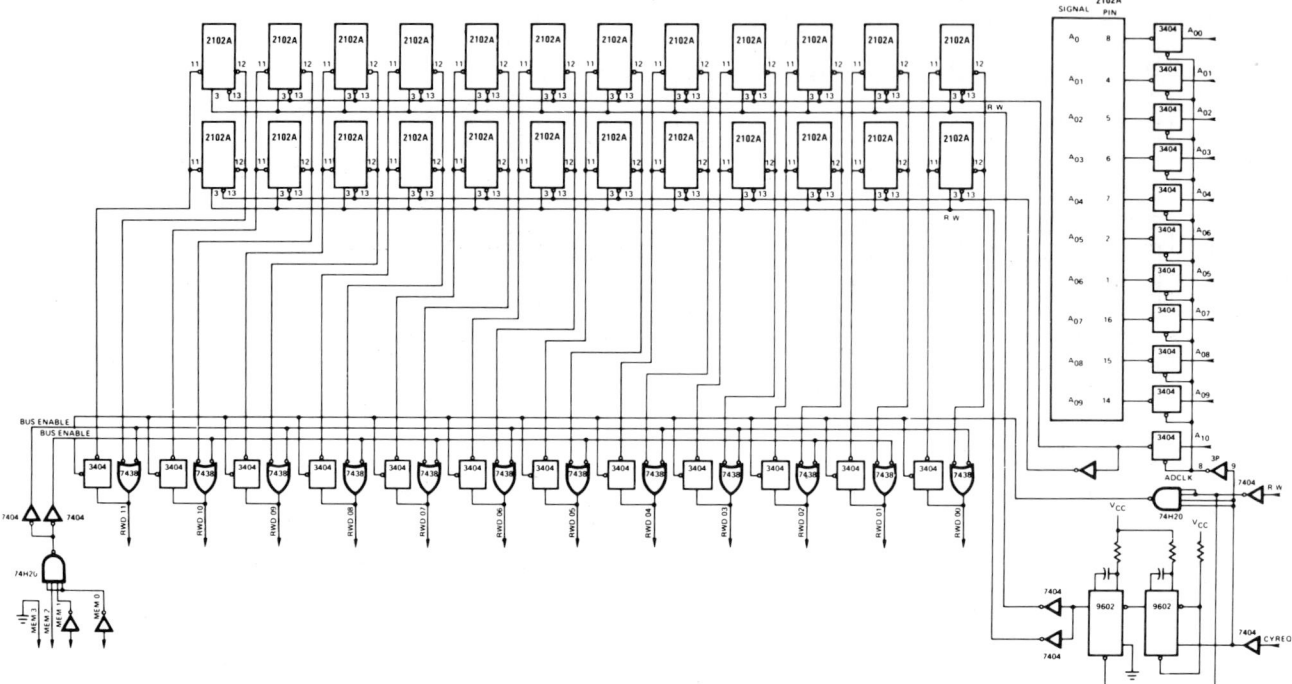

Figure 29. 2K x 12 Memory System.

that the in-26 is designed for use with standard 2102-1 devices. Faster system access/cycle times can be obtained by using 2102A devices and 74H or 74S series instead of 74 series gates in the $\overline{\text{CYREQ}}$ and module select data paths.)

Note that the input data is latched at the end of cycle request ($\overline{\text{CYREQ}}$). It is possible to remove the two monostable multivibrators from the system and control the memory device read/write line externally further simplifying the memory system.

An example of a 2K x 12 single card memory system (IN-24) with full control and interfacing is shown in Figure 29. In this system, the input/output data is on a single bus. Note that data output enable is provided by an address selection for this system. Operation of the in-24 is similar to the in-26.

STATIC RAM MEMORY ARRAY

A layout of the memory array for the 2102A static RAM is shown in Figure 30. Note that there are no layout constraints as a result of noise considerations caused by high level clocks. Power busing is greatly simplified over other MOS RAMs because only one supply plus ground is required for the memory.

Memory array layout for the 2101A, 2111A, and 2112A is entirely similar to the layout shown above. The exception, of course, is the number of data input/output lines in the array. Decoupling is handled in a manner identical to that shown in Figure 31.

INTERFACING WITH MICROPROCESSORS

The Intel static RAM family is ideal for use in microprocessor applications. Control and timing functions are all performed by the microprocessor itself so that additional timing is not required by the memory.

An example of a microprocessor system utilizing both read only (ROM) and random access memory is shown in Figure 32. Although it is not the purpose of this application note to explain microprocessor systems, several comments on the operation of the system are in order.

The buffered 16 bit address bus is tapped (as shown) to provide both chip select and memory address to the static RAMs. (In this case the 2101A equivalent for microprocessors, the 8101 is used.) A control circuit used with the 8080 generates a memory read signal which enables the output on the 8101. Since both chip enables and output disable are used to gate data out of the 8101, the data out bus from these sources is in a high impedance state whenever the ROM is being addressed. This allows OR tying of the data out lines to the data bus. Note that in this case the data in/data out pins of the 8101 are tied together (see discussion of 2101A operation).

Care should be taken when connecting P-channel

DESIGNING WITH STATIC MOS RAMS

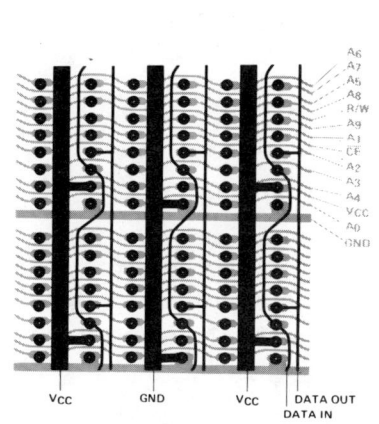

Figure 30. 2102A Memory Array Layout.

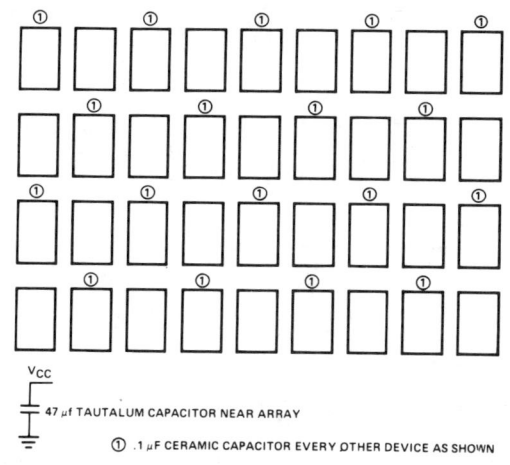

Figure 31. Memory System Decoupling.

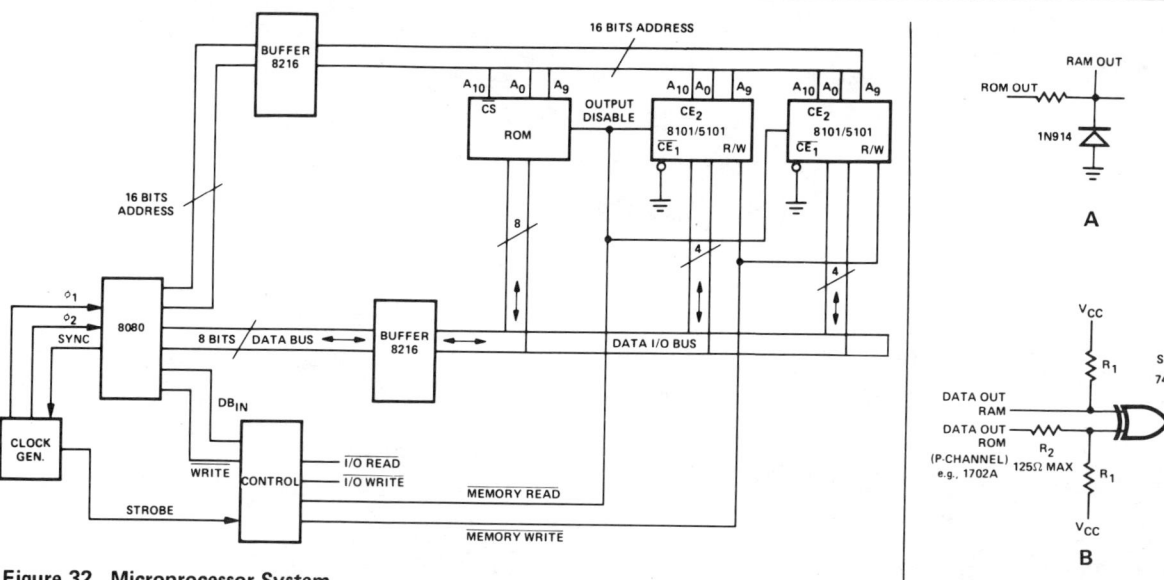

Figure 32. Microprocessor System.

Figure 33. Output Data Bus.

ROMs such as the 1602A and 1702A to the data bus as shown in Figure 32 to assure that the minimum output low level is compatible with the N-channel RAMs being used. It is necessary to protect the data line of the static RAMs if the output level of the ROMs attached to the line can drop below V_{SS} −0.8V. This protection can be done by using a diode to ground and current limiting resistor on those data lines effected (see Fig. 33). It is also permissible to use an exclusive OR which has an internal clamping diode on its input configured per Figure 34B. Note that series 74L86 cannot be used in this application because it does not have a terminating diode.

In the figure shown in Figure 33B, resistors R_1 are pull up resistors to the unselected data out line. R_2 is a current limiting resistor connected to the output of the P-channel ROM. The maximum value permissible for this resistor is determined by the maximum sink current drawn by the ROM device and the maximum acceptable (most positive) down level required for the input of the exclusive OR.

SUMMARY

The Intel static RAM family is a broad and expanding line of simple to use high density MOS RAMs. This application note has detailed those portions of the internal MOS circuits of these RAMs which are of primary concern to the system designer. Through a better understanding of the internal workings of the device, the designer is able to take full advantage of the capability of these RAMs.

A summary of some of the more important technical specifications for each device and device spec type is given in the Product Selection Guide at the end of this section.

DESIGNING NON-VOLATILE SEMICONDUCTOR MEMORY SYSTEMS

JIM OLIPHANT

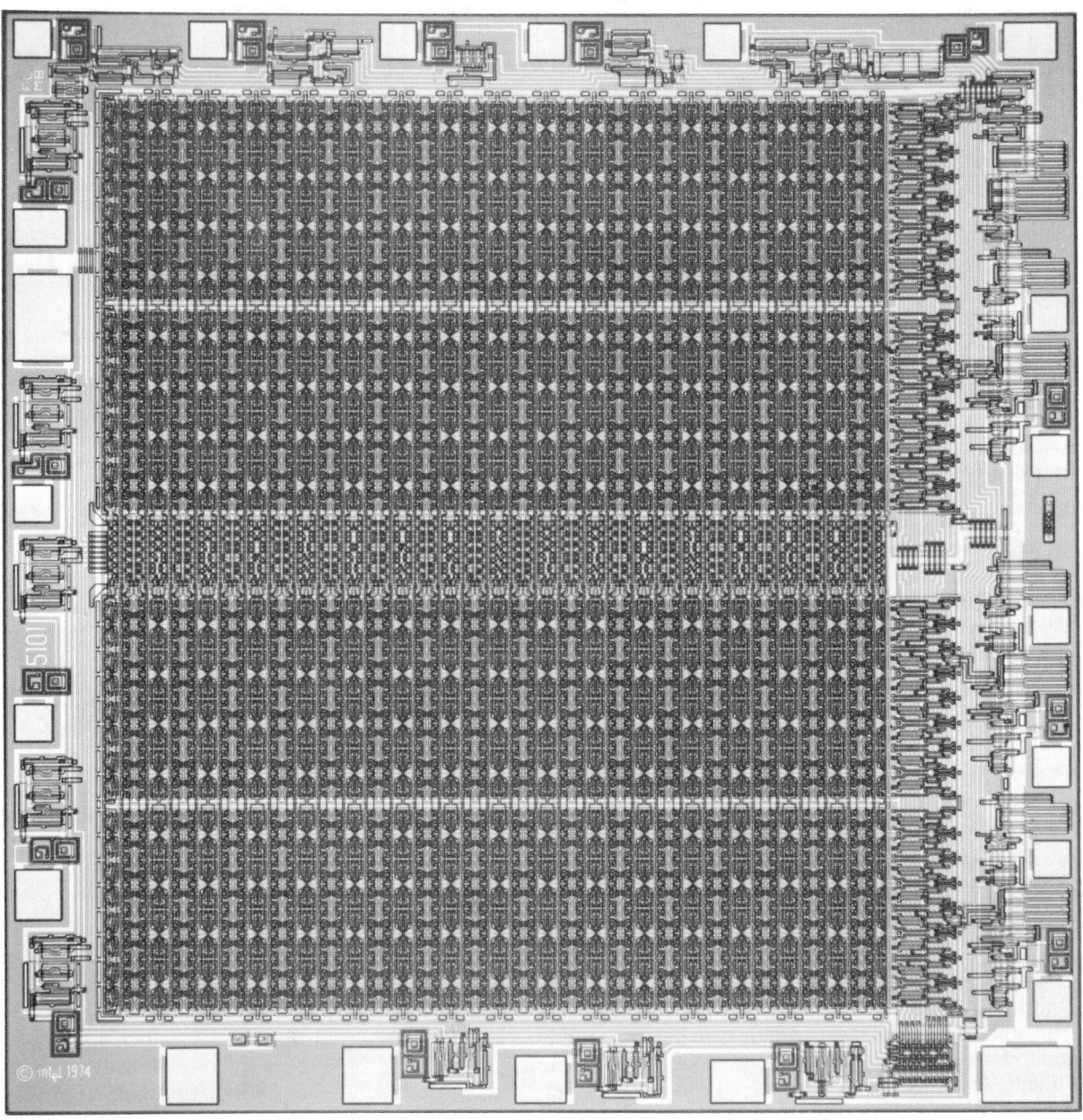

Photomicrograph of 5101 CMOS RAM

DESIGNING NON-VOLATILE SEMICONDUCTOR MEMORY SYSTEMS

INTRODUCTION

The Intel® 5101 is an ultra-low power 1024 bit static RAM organized as 256 words X 4 bits. It is fabricated with an advanced ion-implanted silicon gate CMOS technology. The 5101 is fully TTL compatible, uses only a single supply voltage V_{CC} (+5V) and does not require a clocking operation on the chip enable input. This device is ideally suited for low power and high speed applications where battery support for non-volatility is required.

The purpose of this application note is to describe the internal circuitry and operation of the 5101 and to outline various circuit techniques for battery supported non-volatile operation. In addition, designs using the 5101 will be described and the interface discussed.

DEVICE DESCRIPTION

The 5101 is pin compatible with the Intel® 2101 n-channel silicon gate static MOS RAM. The internal circuitry, however, differs from the 2101 in that the 5101 is implemented with CMOS technology and the 2101 is implemented with n-channel technology. (However, both the 5101 and 2101 are TTL compatible.) The pin configuration and logic symbol for the 5101 are shown in Figure 1. Memory expansion is simplified by the use of two chip enables $\overline{CE}_1$ and CE_2. CE_2 may be used to place the memory in the ultra low power standby mode completely independent of the state of *all* other inputs. In addition, an output disable pin is provided to place the internal data output buffers in a high impedance state. This is particularly useful in those systems which have a common data bus. Both the output disable and chip enable features will be discussed in more detail in the Systems Considerations section.

A block diagram for the 5101 is shown in Figure 2. The memory array is arranged in a 32 X 32 matrix. The five low order addresses A_0-A_4 select 1 of 32 rows; the three high order addresses A_5-A_7 select 1 of 8 column select lines. Each of the column select lines enable 4 of the 32 columns. Figure 3 shows a selection matrix for the selection of a given address to the 5101.

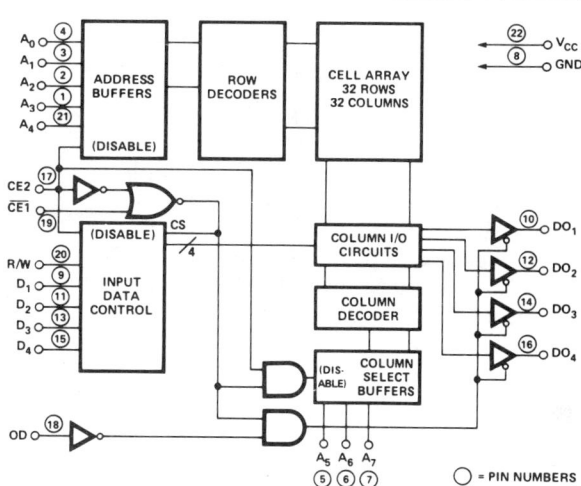

Figure 2. 5101 Block Diagram

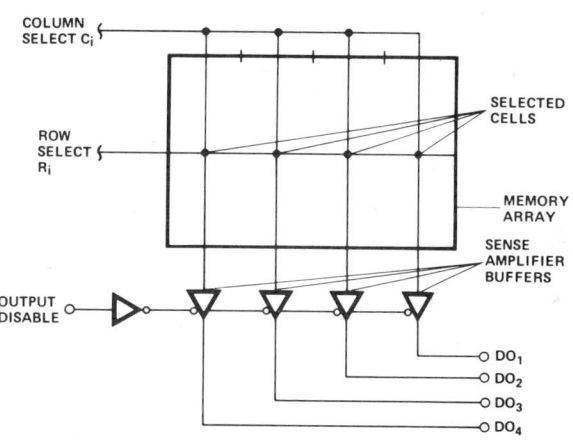

Figure 3. 5101 Selection Matrix

As shown in the block diagram, $\overline{CE_1}$ and CE_2 control the input data buffers and output data buffers. If either $\overline{CE_1}$ is high or CE_2 is low, the data-in and read/write buffers are disabled and the memory is isolated from the data in inputs. Likewise when either or both of the chip selects are in the non-select state (see Table I) the output buffers are placed in a high impedance state. When the chip is selected (i.e., $\overline{CE_1}$ is low and CE_2 is high), the output disable pin (OD) can be used to place the output buffers in a high impedance state.

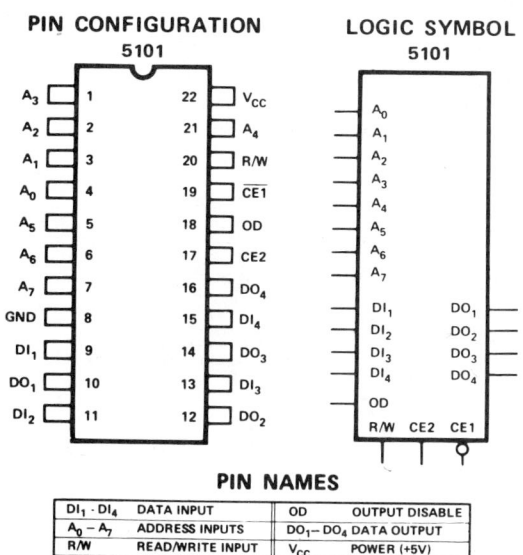

Figure 1. 5101 Pin Configuration and Logic Symbol

STATIC RAMS

Table I. 5101 Output State & Selection Matrix

OD	$\overline{CE}_1$	CE_2	Selection	Output
H	H	H	Deselected	High Imp.
H	H	L	Deselected	High Imp.
H	L	H	Selected	High Imp.
H	L	L	Deselected	High Imp.
L	H	H	Deselected	High Imp.
L	H	L	Deselected	High Imp.
L	L	H	Selected	Enabled
L	L	L	Deselected	High Imp.

Device Operation

STORAGE CELL

The storage cell used in the 5101 is implemented with 6 MOS transistors as shown in Figure 4. The six transistors are connected to form a cross-coupled latch which acts as the memory element. Note that the logic and gating transistors Q_1, Q_3, Q_5, and Q_6 are n-channel enhancement mode (normally off) MOS devices. The load transistors Q_2 and Q_4 are p-channel enhancement mode devices.

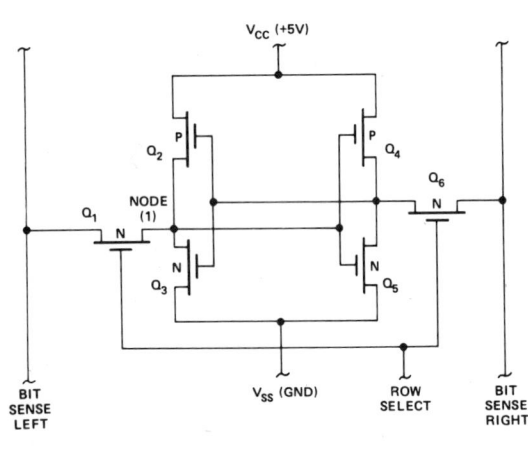

P = P-CHANNEL DEVICE
N = N-CHANNEL DEVICE

Figure 4. 5101 Storage Cell

In the following discussion of storage cell operation, remember that an n-channel device will be "on" if the gate is at a high level ($\sim V_{CC}$). A p-channel device will be "on" if its gate is at a low level ($\sim$GND). Operation of the storage cell is as follows:

Assume that the gate of Q_3 is at a high level (V_{CC}), device Q_3 is therefore turned on (it is an n-channel device) while device Q_2 is turned off (it is a p-channel device). Node (1) is therefore pulled to V_{SS} (ground) and cross-coupled back to the gates of devices Q_4 and Q_5. This low level on node (1) will turn device Q_4 on and Q_5 off. Since the output of Q_4-Q_5 is fed back to the gates of Q_2 and Q_3, an initial charge of V_{CC} on the gate of Q_3 will hold the latch in the above state. This logic state (node 1 at GND) is defined as a "1". The cell contains a logic "0" if the gate of Q_4 and Q_5 is high (V_{CC}) which puts node (1) at V_{CC}. Table II summarizes the state of the memory cell for a logic "1" and logic "0".

Table II. 5101 Memory Cell State

Cell State	Q_2	Q_3	Q_4	Q_5
Logic "0"	On	Off	Off	On
Logic "1"	Off	On	On	Off

Note that in the above discussion no mention was made of any d.c. currents flowing to set the proper voltage levels in the latch. This is because there aren't any. For the example given, the gate of Q_3 is held high (V_{CC}) by device Q_4 (the p-channel load). Since Q_5 is off there is no d.c. path for the current to take in the quiescent state. The only current flowing is the junction leakage currents associated with the source/drain of the MOS devices. This current is typically in the nano-ampere range.

The memory cell is accessed for a read or write operation by activating the appropriate row select line (i.e. row select is brought to V_{CC}). This turns on devices Q_1 and Q_6 and allows data on the bit sense lines to be written into the cell or the state of the cell to be interrogated (read) by a sense amplifier placed on the bit sense lines. For a write operation Bit Sense right is set high (V_{CC}) to write a "1" or Bit Sense left is set high (V_{CC}) to write a "0". The opposite Bit Sense line is held low (V_{SS}).

ADDRESS BUFFER

The address buffers translate the low level TTL address inputs (V_{IL} max. = 0.65V, V_{IH} min. = 2.2V) to a CMOS level (high = V_{CC}, low = V_{SS}) for internal use. The buffer configuration used is shown in Figure 5.

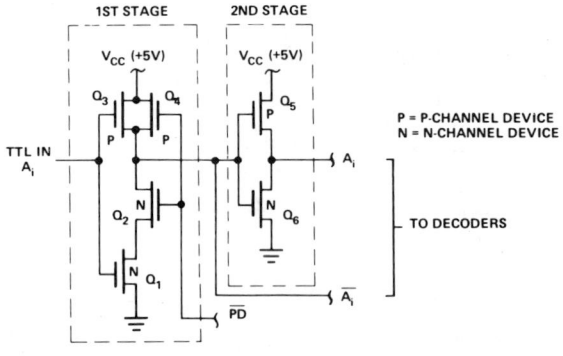

Figure 5. 5101 TTL Address Buffer

The first stage of the address buffer consists of a NAND gate (Q_1, Q_3) with control gates (Q_2, Q_4) added to *disconnect* the TTL address from the decoders when the device is not selected (that is, CE_2 is low). This places the address buffers in a standby mode (only leakage currents flowing) and eliminates the need to control the state of the addresses during standby. The internally generated signal $\overline{PD}$ which blocks the input addresses from the internal decoders is generated from CE_2. The second stage is an inverting buffer to provide increased drive for the A_i addresses.

Note that when the device is in a quiescent state no d.c. current is being drawn by the buffer. Therefore, the power dissipated during operation is very small and amounts to only the leakage current associated with the source/drain p-n junctions.

DECODERS

The row decoders (selecting 1 of 32 rows) on the 5101 use an AND gate of the type shown in Figure 6. To activate the selected row decode line, the five addresses going to that particular decoder must be at a high level and the internal chip select (CS) must be high. (Chip select is formed by the logical AND of CE_2 and $\overline{CE_1}$). If all address inputs to the decoder are high, a low is placed on the gates of the inverter buffer (devices Q_1 and Q_2) which will turn Q_1 off and Q_2 on. The selected row decode line is thereby brought high, turning on the appropriate gates in the selected memory cells. If the device is not selected, then CS forces all row decoder lines low which disables any access to all memory cells.

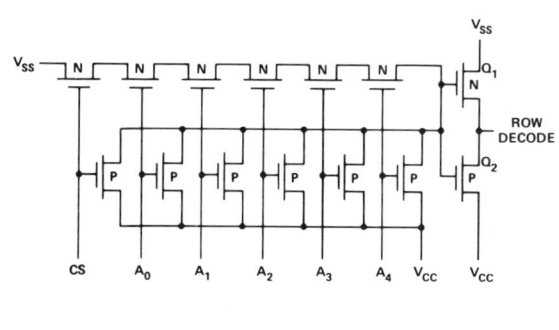

P = P-CHANNEL DEVICE
N = N-CHANNEL DEVICE
CS = $\overline{CE1}$ · CE2

Figure 6. 5101 Row Decode

The column decoders use a NOR type gate shown in Figure 7. The selected column decode line goes high if the 3 addresses (A_5-A_7) being decoded are all at a low level. Note that the internal column decoder uses only three address inputs. These three inputs select 1 of 8 separate decode lines. Each of the 8 decode lines select 4 columns for each word addressed.

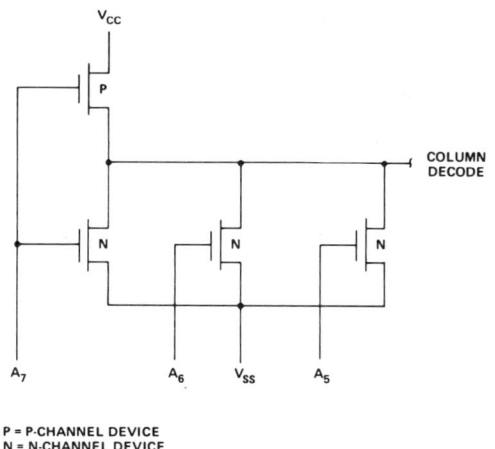

P = P-CHANNEL DEVICE
N = N-CHANNEL DEVICE

Figure 7. 5101 Column Decode

INTERNAL DATA SENSING

A simplified schematic of the 5101 column sense amplifier is shown in Figure 8. The sense amplifier is constructed in an AND configuration with the I/O left line of each column of memory cells AND'ed with a particular column decode (devices Q_1 and Q_2). The line to the output buffer, O_B, is held at V_{CC} by device Q_3 unless both the I/O left line *and* the column decode line (for that particular column) are both high (logic "0" in the memory cell). In this case, the output of the sense amplifier O_B will be driven to a low level (slightly above GND). For example, if memory cell "M" shown in Figure 8 contained a "0" (I/O left high) then O_B would be low. However, if "M" contained a "1" (I/O left low) then O_B would remain high.

Devices Q_4 and Q_6 shown in Figure 8 are used as a load on the particular I/O line. The n-channel devices (Q_5 and Q_7) are used to limit the logic swing on the I/O lines.

Data is written into the memory cell by the circuitry shown in Figure 8. Note that the I/O right line goes high only when a logic "1" (high level) is applied to the data-in input on a selected device during a write cycle. The I/O left line, however, goes high when either a low is on the data-in input or the chip is non-selected. (Recall that for a non-selected device, all row selects are at the non-selected state, i.e. low level.)

OUTPUT BUFFER

A simplified schematic for the 5101 output buffer is shown in Figure 9. As shown in this figure the output buffer is implemented with complementary n-channel and p-channel drivers. For this type of driver, the gates of devices Q_1 and Q_2 must be at the same logic level (high or low) so that one of these devices is on while the other is off for a normal read operation. However, when the chip is deselected (the internal CS is low) or output disable is high, both Q_1 and Q_2 are turned off and the Data Out output goes to a high impedance state.

Device Specifications

READ CYCLE

Minimum timing for a 5101 read cycle is shown in Figure 10. This timing diagram shows the relationship of all necessary control signals required for a read cycle and is for a general application. However, if the user has certain flexibilities in his system, other modes of operation are possible.

For those systems which have separate data inputs and outputs in the memory array, the output disable input (pin 18) may be tied low. Also, if the input and output pins of the 5101 are not OR tied to any other device both chip enable inputs may be held true (i.e. $\overline{CE_1}$ is held low and CE_2 is held high) while the addresses are being cycled in any order for a series of read cycles. For this case, the read/write input must be held high throughout the read operations. However, when operating the 5101 with CE_2 held high, it is necessary to control the voltage level of all inputs if ultra low power dissipation is desired. The ultra low standby power can be achieved with $\overline{CE_1}$ *only* deselected (i.e. at a high level) by holding all address, chip enable, data-in and read/write inputs to *one* of the following levels:

1. $V_{in} \leq 0.2V$
2. $V_{in} \geq V_{CC} - 0.2V$

Note that $\overline{CE_1}$ may be tied low, if so desired, and the ultra low standby power controlled only with CE_2 (i.e. $CE_2 \leq 0.2V$, all other inputs in a "don't care" state). The definition of terms outlined in Figure 10 is contained in Table III.

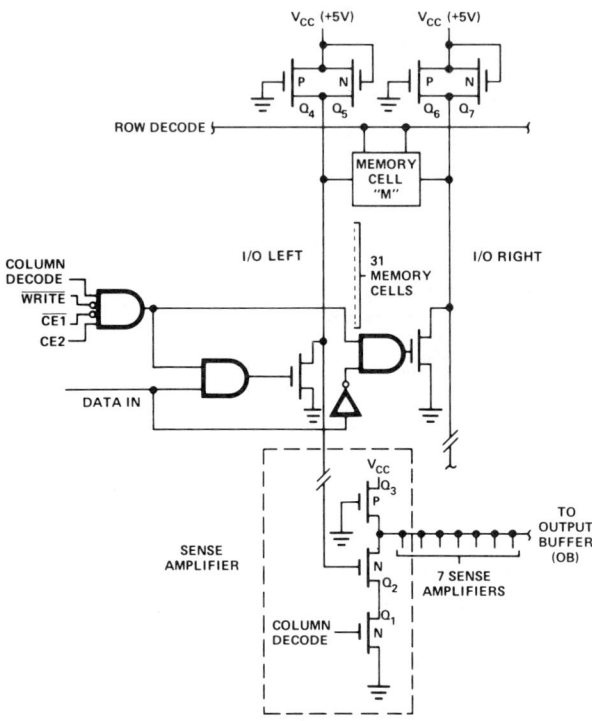

Figure 8. 5101 Column Sense Amplifier

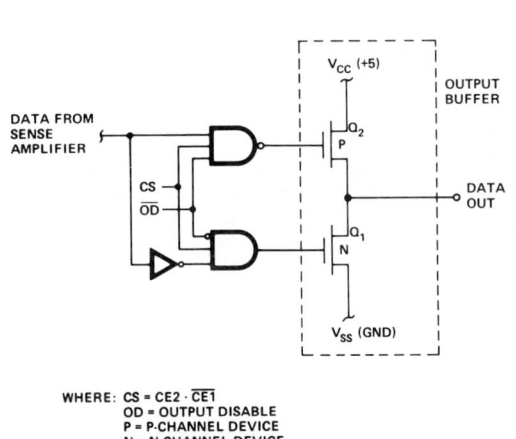

Figure 9. 5101 Output Buffer

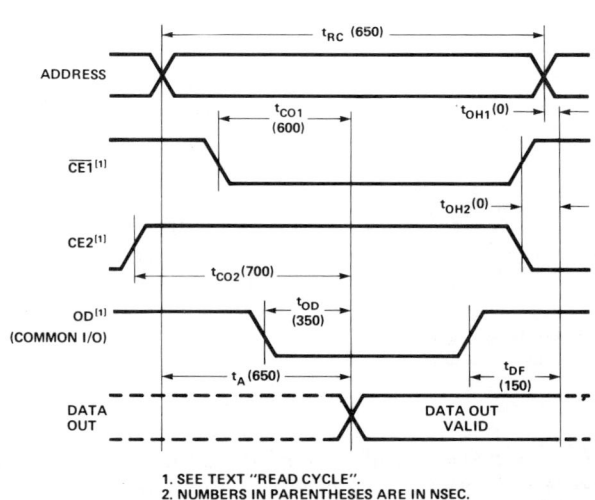

Figure 10. 5101 Read Cycle

Table III. 5101 Read Cycle A.C. Characteristics. $T_A = 0°C$ to $70°C$, $V_{CC} = 5V \pm 5\%$ unless otherwise specified.

Symbol	Parameter	Min.	Typ.	Max.	Unit	Test Conditions
t_{RC}	Read Cycle	650			ns	Input Pulse Levels +0.65V to 2.2V. Input Pulse Rise and Fall Times 20 nsec. Timing Measurement Reference Level 1.5V. Output Load 1 TTL Gate and $C_L = 100pF$.
t_A	Access Time			650	ns	
t_{CO1}	Chip Enable ($\overline{CE1}$) to Output			600	ns	
t_{CO2}	Chip Enable (CE2) to Output			700	ns	
t_{OD}	Output Disable To Output			350	ns	
t_{DF}	Data Output to High Z State	0		150	ns	
t_{OH1}	Previous Read Data Valid with Respect to Address Change	0			ns	
t_{OH2}	Previous Read Data Valid with Respect to Chip Enable	0			ns	

Table IV. 5101 Write Cycle A.C. Characteristics. $T_A = 0°C$ to $70°C$, $V_{CC} = 5V \pm 5\%$ unless otherwise specified.

Symbol	Parameter	Min.	Typ.	Max.	Unit	Test Conditions
t_{WC}	Write Cycle	650			ns	Input Pulse Levels +0.65V to 2.2V. Input Pulse Rise and Fall Times 20nsec. Timing Measurement Reference Level 1.5V. Output Load 1 TTL Gate and $C_L = 100pF$.
t_{AW}	Write Delay	150			ns	
t_{CW1}	Chip Enable ($\overline{CE1}$) To Write	550			ns	
t_{CW2}	Chip Enable (CE2) To Write	550			ns	
t_{DW}	Data Setup	400			ns	
t_{DH}	Data Hold	100			ns	
t_{WP}	Write Pulse	400			ns	
t_{WR}	Write Recovery	50			ns	
t_{DS}	Output Disable Setup	150			ns	

WRITE CYCLE

Minimum timing for a 5101 write cycle is shown in Figure 11. The waveforms shown in Figure 11 are for a general application of the 5101 during a write cycle and may be modified to some degree depending on the users requirements. For example, if no other data inputs or outputs are OR tied to the 5101, $\overline{CE_1}$ may be held low, CE_2 held high and output disable held low.

However, it is *not* permissible to hold the read/write line low while cycling through addresses for a series of write cycles. Attempting to perform a series of write cycles in this manner will result in writing into multiple address locations during address transitions.

Although it is not necessary to conform exactly to the waveforms shown in Figure 11 for a write cycle, care should be taken to assure all minimum timing constraints, listed in Table IV, are adhered to. Particular attention should be paid to T_{aw} (address to write set-up time), T_{cw1} and T_{cw2}.

Since the 5101 is a completely static random access memory, it does not require an edge on any input line (e.g. chip enable or address) to initiate a cycle. Therefore, when a device is enabled (i.e. $\overline{CE_1}$ is low and CE_2 is high) and addresses are changed, time must be provided for the row and column decoders to settle (T_{aw}) before commencing a write to make sure undesired address locations are not partially rewritten by the data on the data input line.

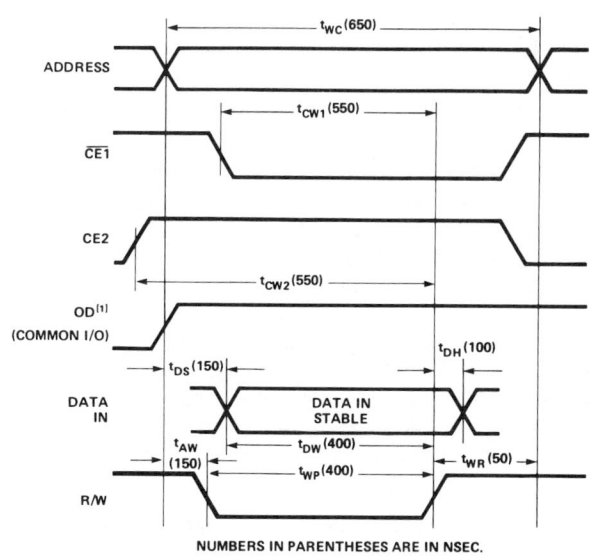

NUMBERS IN PARENTHESES ARE IN NSEC.
1. FOR SEPARATE I/O OPERATION OD MAY BE TIED LOW.

Figure 11. 5101 Write Cycle

Table V. D.C. Operating Characteristics. T_A = 0°C to 70°C, V_{CC} = 5V ±5% unless otherwise specified.

Symbol	Parameter	Min.	Typ.[1]	Max.	Unit	Test Conditions
I_{LI}	Input Current		5		nA	V_{IN} = 0 to 5.25V
I_{LOH}	Output High Leakage			1	μA	$\overline{CE1}$ = 2.2V, V_{OUT} = V_{CC}
I_{LOL}	Output Low Leakage			-1	μA	$\overline{CE1}$ = 2.2V, V_{OUT} = 0.0V
I_{CC1}	Operating Current		9	22	mA	V_{IN} = V_{CC} Except $\overline{CE1}$ ≤ 0.01V Outputs Open
I_{CC2}	Operating Current		13	27	mA	V_{IN} = 2.2V Except $\overline{CE1}$ ≤ 0.65V Outputs Open
I_{CCL} [2]	Standby Current			15	μA	V_{IN} = 0 to V_{CC}, Except CE2 ≤ 0.2V
V_{IL}	Input "Low" Voltage	-0.3		0.65	V	
V_{IH}	Input "High" Voltage	2.2		V_{CC}	V	
V_{OL}	Output "Low" Voltage			0.4	V	I_{OL} = 2.0mA
V_{OH}	Output "High" Voltage	2.4			V	I_{OH} = -1.0mA

NOTES: 1. Typical values are T_A = 25°C and nominal supply voltage.
2. Current through all inputs and outputs included in I_{CCL}.

Table VI. 5101L Low V_{CC} Data Retention Characteristics. T_A = 0°C to 70°C, V_{CC} = 5V ±5% unless otherwise specified.

Symbol	Parameter	Min.	Typ.[1]	Max.	Unit	Test Conditions	
V_{DR}	V_{CC} for Data Retention	2.0			V	CE2 ≤ 0.2V	V_{DR} = 2.0V
I_{CCDR}	Data Retention Current			15	μA		
t_{CDR}	Chip Deselect to Data Retention Time	0			ns		
t_R	Operation Recovery Time	t_{RC}[2]			ns		

NOTES: 1. Typical values are T_A = 25°C and nominal supply voltage.
2. t_{RC} = Read Cycle Time.

D.C. OPERATING CHARACTERISTICS

The D.C. operating characteristics of the 5101 are given in Table V. I_{CCL} (standby current) in Table V is emphasized because of its importance in standby battery back-up operation. Note that the maximum value of the standby power supply current is an extremely low 15μA (and is typically only 0.2μA). If CE_2 is used to control the low power state (i.e. CE_2 ≤ 0.2V), then the state of all other inputs is a "don't care." If $\overline{CE_1}$ is used to control the low power state, all inputs must be either high or low (as defined in Read Cycle section). As is shown later, (in the Systems Considerations section) this allows the designer maximum flexibility in the design of simple battery interfaces to implement a battery back-up system.

As shown in Table V, the 5101 is capable of driving a maximum TTL load of 2mA at an output voltage V_{OL} of 0.4V. Attempting to sink more than 2mA will result in an increased V_{OL}.

LOW V_{CC} DATA RETENTION

The 5101L family of RAMs has ultra low standby current and requires only that V_{CC} be between 2.0V ≤ V_{CC} ≤ 5.25V to maintain data. As shown, these devices are guaranteed to operate in a standby mode with V_{CC} a minimum of 2.0V. Table VI gives the low V_{CC} data retention characteristics of the 5101L. The waveforms for low V_{CC} data retention operation are shown in Figure 12.

As shown in Figure 12, CE_2 must be brought low (≤0.2V) at or before the V_{CC} supply drops to 4.75V. In addition, CE_2 must remain in the low

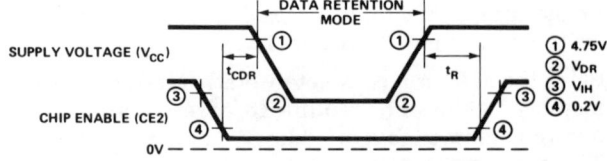

Figure 12. Low V_{CC} Data Retention Waveforms

DESIGNING NON-VOLATILE SEMICONDUCTOR MEMORY SYSTEMS

state for a period equal to a read cycle time after V_{CC} has reached a minimum of 4.75V after power-up. It is important to note that the supply voltage V_{CC} does *not* have to be reduced below 4.75V as shown in Figure 12. Remember that the standby current I_{CCL} is a maximum of 15µA up to V_{CC} = 5.25V. The typical data retention current as a function of V_{DR} (V_{CC} in data retention mode) is shown in Figure 13.

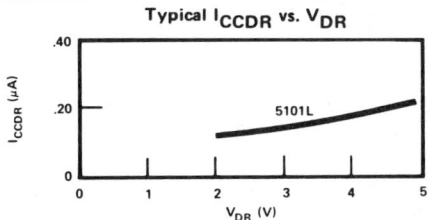

Figure 13. 5101L Data Retention Current Vs. V_{DR}

SYSTEMS CONSIDERATIONS

Since the 5101 is a completely static TTL compatible random access memory device requiring no clocks, refresh or special drivers/sense circuitry, the designer can treat the 5101 as any other TTL compatible device. Because of the ease with which the 5101 can be used, this section on Systems Considerations will concentrate on circuitry associated with battery-supported standby operation. Discussions of any interface buffers (if required) to a 5101 system will be relative to the effect these buffers have on the standby power source (e.g. battery) and what can be done to minimize the adverse effects of the buffers. Additional information regarding buffers for static TTL compatible RAMs can be found in the next section, "Designing with Intel's Static MOS RAMs".

Low Power Standby Operation

When designing a non-volatile semiconductor memory system, the basic requirements can be outlined as follows:

1. Maximum data retention time-battery back-up.
2. Maximum load current during standby-data retention mode.
3. Physical size requirement of battery.
4. Access/cycle time (operating mode).

Access time is important as it effects the selection of address and data buffers required by the system. If high speed operation is desired, it may be necessary to use series 74S type gates for the buffers. If speed is not of primary interest then CMOS type buffers may be used. Clearly, TTL type buffers will draw considerably more power than CMOS buffers if left connected to the battery supply during the data retention mode. The battery interface to both TTL and CMOS buffers will be discussed in the battery section.

The required data retention time for battery supported standby operation is of primary importance in the selection of a battery. The usual trade-offs associated with data retention time are:

1. Memory size (number of words that must be non-volatile).
2. Physical battery size desired (determines if the battery is to be placed on a printed circuit card or is external to the card).

Within reasonable constraints of memory size and data retention time, there are many types and configurations of batteries that can be used.

Power Switching

Two basic types of power switching circuits (switching between the main supply and the battery) are described which are simple and inexpensive.

These two types are:

1. Diode Coupled
2. Switch Coupled

These two types of switching circuits are shown in Figure 14. The diode coupled circuit requires the main d.c. supply to be above the required V_{CC} voltage by the amount of drop through the diode. The diode used should have a low forward drop (such as found in Germanium diodes) and low series resistance.

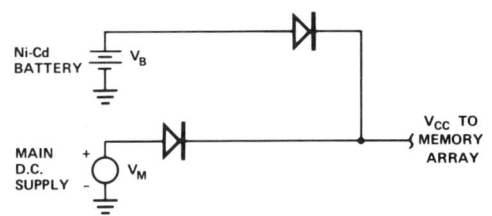

A. Diode Coupled

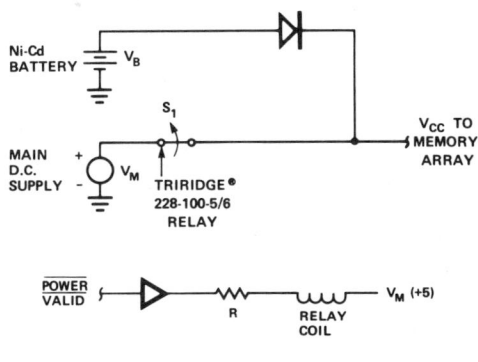

B. Switch Coupled

Figure 14. Power Switching Circuits

If it is not desirable to have a power supply voltage above V_{CC} (e.g. existing +5.0V supplies are to be used), then a normally open switch can be used in place of the diode. The switch is held closed by a simple TTL buffer gate as shown in Figure 14 as long as $\overline{\text{POWER VALID}}$ is held low. When power loss is detected (see Power Loss Detect Section) the switch is opened and the battery automatically supplies power to the memory array. (Note that if the memory is to be used for a short period to load memory, etc., after $\overline{\text{POWER VALID}}$ goes high a delay must be included in the switch line to take power from the supply before it drops below 4.75V.)

Power Loss Detect

In memory systems which have TTL interface and other control circuitry, it is usually necessary to have advanced warning that A.C. power has been lost. This allows the orderly shut down on the system and can provide time to store data/records in the non-volatile portion of memory. Such a circuit is shown in Figure 15.

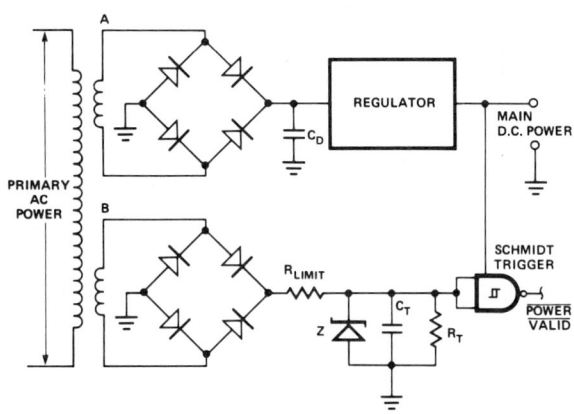

Figure 15. Power Loss Detect Circuit

The detect circuit uses a separate transformer winding (available on many power supplies) to provide a positive ($\approx$ +5V) voltage reference to a schmidt trigger. A separate winding is used so as not to interfere with the regulation of the main d.c. power source.

Operation of the detect circuit is as follows:
A high level ($\approx$ 5V) is established at the input of a schmidt trigger (e.g. 7414) by the diode bridge network and zener Z. Resistor R_{LIMIT} is a current limiting resistor between the bridge and schmidt trigger input network. The $R_T C_T$ combination controls the discharge rate of the input voltage to the schmidt trigger when reference power is lost. The time constant is used to prevent short (a few cycles) a.c. power loss from shutting down the system. The only restriction on the maximum value of the time constant is the $\overline{\text{POWER VALID}}$ signal must go high before the main d.c. power source drops below the minimum allowable operating voltage of the main d.c. source.

In general it is not desirable to combine the power loss detect circuitry with the main d.c. power source for two reasons:

1. Adverse effect on d.c. output regulation by R_{LIMIT} resistor, and
2. The large decoupling capacitor, C_D, on most d.c. supplies.

The large decoupling capacitor C_D will cause a time constant which is too large and may not allow sufficient time between $\overline{\text{POWER VALID}}$ going high and the main d.c. power dropping below acceptable minimums

Batteries For Non-Volatile Semiconductor Memories

The first place to begin in the selection of a battery for a particular application is to analyze those factors dictated by system requirements and fit the battery to the requirement. Some of these important criteria are:

1. Load current imposed on battery.
2. Battery voltage-full charge.
3. Battery voltage-end of life.
4. Life of battery under maximum load conditions.
5. Environment-temperature range (operating, non-operating).
6. Physical factors (size, weight).
7. Battery operation.

Of the seven criteria listed above, the one most likely to be overlooked is the effect of temperature on the capacity and life of the battery. For many batteries commerical grade temperature requirements (0°C to 70°C) may adversely effect both the capability and life of the battery.

Criteria seven, battery operation, refers to the operating schedule the battery is expected to meet. For example, if the battery is expected to maintain data *only* on a.c. power outages which are assumed to be rare, then a rechargeable battery (secondary cell) with a slow recharge rate may be selected. For this case, it may even be desirable to use a non-rechargeable battery (primary cell) with battery replacement scheduled at appropriate intervals (six months to 1 year).

However, if the system is operated in a mode where power is turned on in the morning and off in the evening then fast rechargeable batteries (with appropriate recharging circuitry) may be required.

DESIGNING NON-VOLATILE SEMICONDUCTOR MEMORY SYSTEMS

In the evaluation of the seven criteria listed previously, one of the first things to be determined is what type of battery is to be used in the system. The chart shown in Table VII outlines the characteristics of various storage cell types. Consider first the primary type.

PRIMARY BATTERIES (NON-RECHARGEABLE)

The use of primary batteries in a memory system is usually limited to those systems which require standby data retention infrequently or where very high battery capacities (mA-hr) and very small battery physical size are required. For these cases, both mercury and silver-oxide batteries offer large capacity combined with very small phsyical size. The small size of these batteries is shown in Figure 16 for a silver-oxide battery (110mA-hr).

Figure 16. Silver-Oxide Button Cell

Typical voltage discharge curves for silver-oxide and mercury batteries are shown in Figures 17 and 18 respectively. Note that in both cells, the cell voltage remains nearly constant during discharge — a highly desirable characteristic. In addition, the mercury cell generally has greater capacity for a given size as compared with a silver-oxide battery.

Carbon-zinc batteries offer the lowest cost of any primary battery described, but suffer from a severe degradation of output voltage as a function of use. This characteristic makes carbon-zinc batteries undesirable for most standby power applications.

Alkaline batteries have a much better discharge characteristic than carbon-zinc, but are not quite as

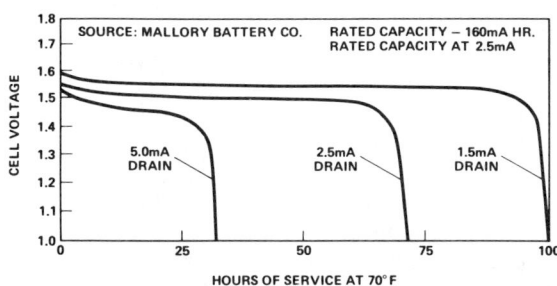

Figure 17. Silver-Oxide Cell Typical Voltage Discharge Characteristics

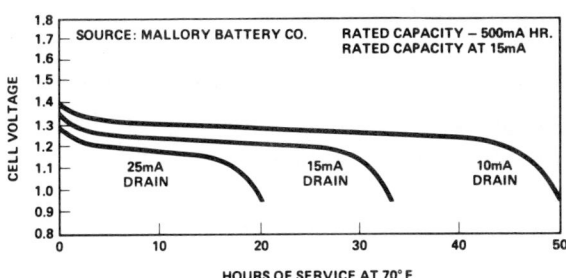

Figure 18. Mercury Cell Typical Voltage Discharge Characteristics

good as mercury or silver-oxide. The relative low cost of these batteries can make them attractive for use in some systems applications.

Both carbon-zinc and alkaline batteries are discussed in detail in the Eveready Battery Applications Engineering Data handbook (see Bibliography 3). It is emphasized that adequate attention should be directed to the output voltage characteristics of these two batteries before using them in a standby power application.

Because of printed circuit board area limitations small battery size is usually the reason for selecting a primary cell for battery support. In this case, it may be desirable to limit the number of cells in a particular system to one. However, this requires that a voltage boost circuit be used in the system to achieve a minimum sustaining voltage of 2.0V at end of battery life to operate the 5101. Such a boost circuit is shown in Figure 19.

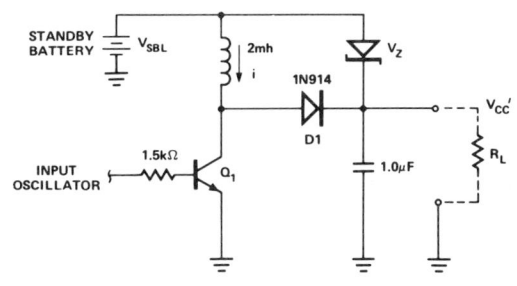

Figure 19. Basic Voltage Boost Circuit

Voltage Boost

Operation of the voltage boost circuit is as follows:

The input to Q_1 is a low duty-cycle signal. This signal turns Q_1 on forcing current through inductor L. When Q_1 is turned off, current i cannot change instantaneously and is diverted through diode D,

Table VII. Battery Characteristics

I. PRIMARY TYPE (NON-RECHARGEABLE)[1]

Cell Construction	Cell Voltage (Typ.)[2]	Comments
Carbon-Zinc (Leclanche)	1.5	Lowest cost; discharge characteristics may be inadequate for some systems
Silver-Oxide	~ 1.6	Good for low temperature operation, discharge characteristic excellent for most system requirements
Mercury	1.4	Good for high temperature operation, discharge characteristic excellent for most systems, long shelf life
Alkaline	1.5	Good efficiency for use with systems requiring total battery operation

II. SECONDARY TYPE (RECHARGEABLE)

Cell Construction	Cell Voltage (Typ.)	Comments
Nickel-Cadmium	1.2	Excellent all around characteristics for battery back-up, widely used
Lead-Calcium	2.0	Excellent all around characteristics for battery back-up

[1] Some information in this table condensed from "EVEREADY" Battery Applications and Engineering Data Handbook copyrighted 1971 by Union Carbide Corporation.

[2] Cells can be put in series to obtain multiples of basic cell voltage.

---

charging capacitor C. The voltage to which C charges is a function of the capacitance C, load R_L and zener V_Z.

In order to minimize the load on the battery and to maximize the efficiency of the boost circuit, it is necessary to turn on Q_1 only for the minimum amount of time which will still maintain the desired output standby voltage. Such a circuit is useful only if there is a way to power the input oscillator required for Q_1 off the same battery V_{SB}. Such a circuit is shown in Figure 20. The 5801, shown in Figure 20, is a low voltage CMOS oscillator made by Intel (used extensively by Microma, an Intel subsidiary). The output of this oscillator triggers a CMOS single-shot which in turn drives the voltage boost circuit shown in Figure 19. Note that the 5801 is powered by the battery (Cell voltage = 1.5V) and the 4047 (CMOS single-shot) is powered by the boosted V_{CC}'. It is, therefore, necessary to

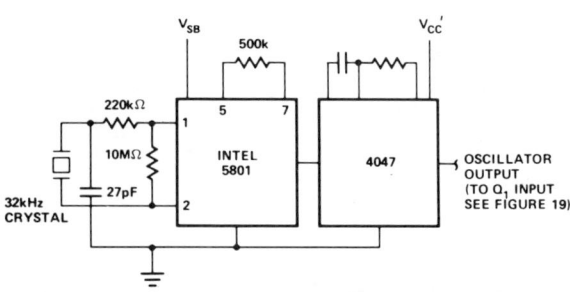

Figure 20. Voltage Boost Oscillator Circuit

assure that the standby voltage V_{CC} does not fall below 3V (minimum 4047 operating voltage) before starting the oscillator circuit. A power loss detect circuit can be used to warn of an impending power down condition and allow the boost circuit to be turned on in time to hold the V_{CC} voltage to $\cong 3V$.

As stated previously, the power conversion efficiency of the oscillator and voltage boost circuits should be maximized to minimize the current drain on the battery. The efficiency of these circuits is largely a function of the duty cycle of the oscillator. Figure 21 shows the waveforms of the input signal to the voltage boost circuit and the current i through inductor L. A summary of the data in Figure 21 is shown in the graphs of Figure 22. Note that the curve of V_{CC}' levels out at 4.0V, this is the result of the clamp zener V_Z (Figure 19). Also note that the efficiency is markedly decreased as the input pulse to Q_1 is lengthened. For a given duty cycle on Q_1, the efficiency of the voltage boost circuit will increase if the load current is reduced.

SECONDARY BATTERIES (RECHARGEABLE)

As outlined in Table VII there are two basic types of rechargeable batteries ideally suited for memory system standby power-down operation. Nickel-Cadmium (Ni-Cd) and Lead-Calcium (such as Gel/Cell®). This section will outline some of the salient features of each type. No attempt will be made to compare the two for general operation. It is recommended that the system designer interface directly with the battery manufacturer to obtain guaranteed specification data, operating limitations and safety precautions (if any).

DESIGNING NON-VOLATILE SEMICONDUCTOR MEMORY SYSTEMS

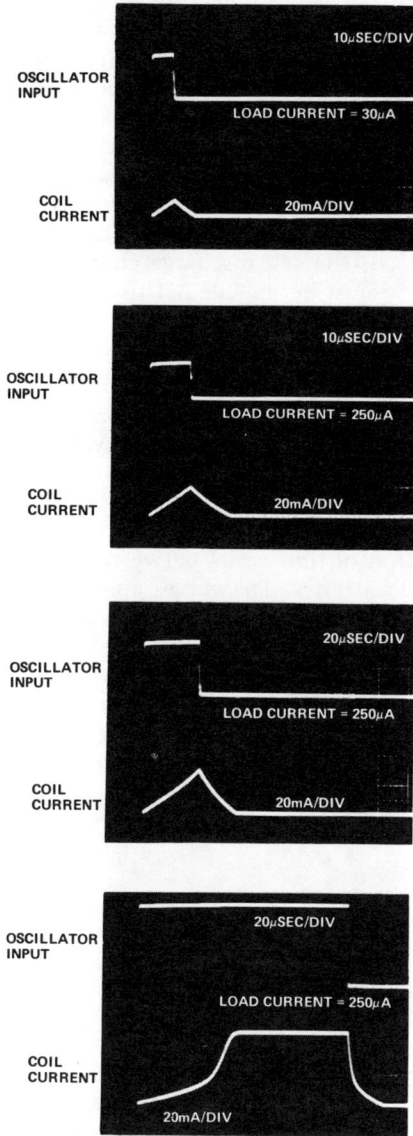

Figure 21. Voltage Boost Waveforms

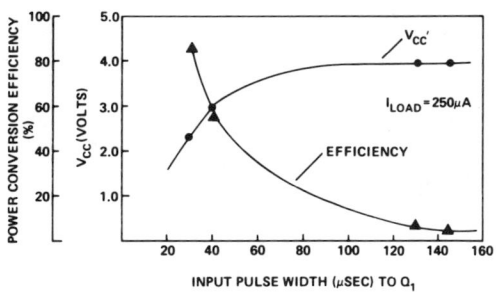

Figure 22. Boost Circuit Output Voltage and Efficiency Vs. Input Pulse Width

Figure 23. Sizes of Selected Cylindrical Ni-Cd Batteries

Nickel-Cadmium

Nickel-Cadmium batteries are available in a wide variety of capacities (mA-hr) sizes and styles (see Figure 23). The styles include button, cylindrical and rectangular cells and may be placed on the memory printed circuit card or in the same enclosure as the main d.c. power supply. (Enclosing the batteries in with the main d.c. supply is usually done only in large back-up capacity systems.)

There are many manufacturers of Ni-Cd batteries who can supply the desired battery configuration. A useful place to begin looking for a battery supplier is <u>Electronic Buyers Guide</u> (McGraw-Hill publications.) (Also see bibliography.)

Electrical Characteristics

Electrical characteristics such as capacity (mA-hr) and cell voltage as a function of discharge rate for Ni-Cd batteries are temperature dependent. It is important for the designer to realize that high system operating temperatures may have an adverse effect on battery life and capacitance even though the battery is not expected to be called on to provide standby power at those temperatures.

An example of the effect of temperature on capacity (based on GE battery specifications) is shown in Figure 24 for two types of General Electric Ni-Cd batteries. Note that two types of usage are given: one for infrequent discharge with extended periods

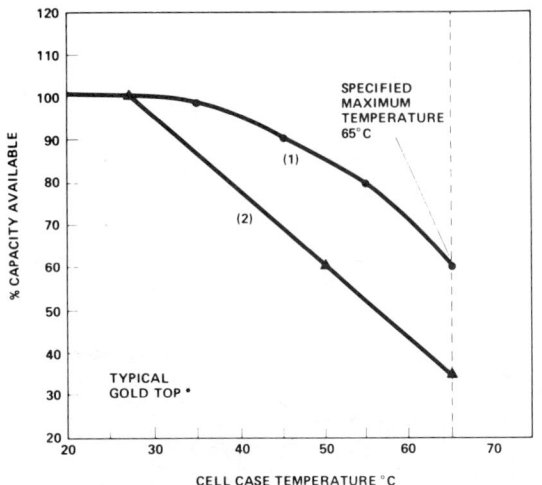

* GOLD TOP IS A REGISTERED TRADEMARK OF GENERAL ELECTRIC.

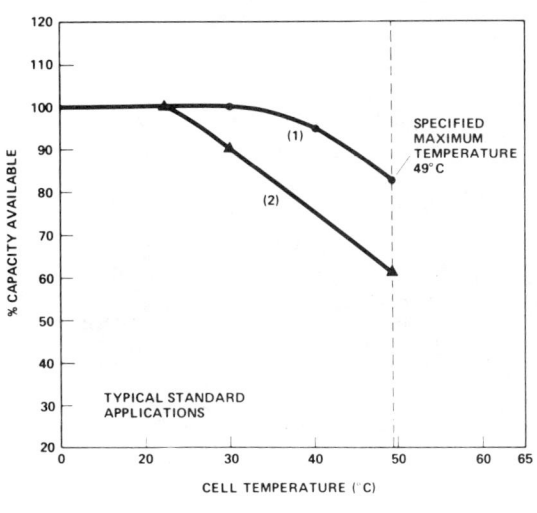

NOTES:
(1) INFREQUENT DISCHARGE, EXTENDED PERIOD OF OVERCHARGE.
(2) FREQUENT DISCHARGE.

Figure 24. Ni-Cd Battery Capacity as a Function of Temperature

of overcharging, condition (1), the other for frequent discharge, condition (2). In most memory applications condition (1) will apply, which is the condition for maximum capacity as a function of temperature.

Ni-Cd batteries also have a self discharge characteristic which is a function of temperature. One result of this characteristic is that these types of batteries should not be stored in a charged condition for an appreciable length of time. Therefore, before inserting these batteries into a system or after the system has been powered down for an extended time (i.e. no trickle charge available and batteries disconnected from load) care must be exercised to assure that the batteries have sufficient charge to perform in a power down standby mode. In addition, when calculating the capacity of the battery for a particular load, the self discharge characteristic of the battery must be included. This self discharge rate can be as high as approximately 7%/day loss of capacity at 50°C to an average of 1%/day at room temperature (25°C) for Ni-Cd batteries.

The discharge characteristics of Ni-Cd batteries are flat, making them ideal for use in memory systems requiring standby power. The general shape of such characteristics is shown in Figure 25. Note that no scales are given in this figure because the output voltage as a function of time varies between manufacturers of Ni-Cd batteries. The curves are shown to demonstrate the flat voltage characteristics at end of battery capacity for Ni-Cd batteries.

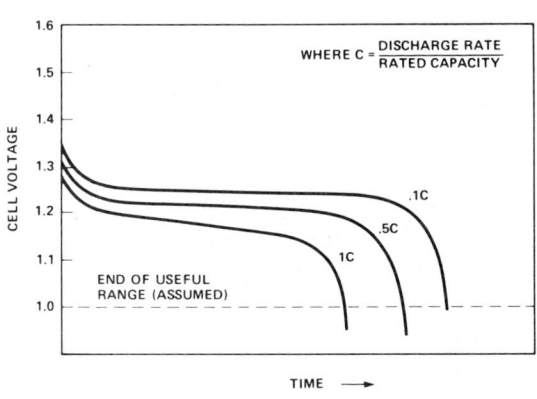

Figure 25. Ni-Cd Voltage Discharge Characteristic

Since the single cell voltage of a Ni-Cd battery is approximately 1.2 volts (as shown in Figure 25), it is necessary to boost the voltage with external circuitry (discussed previously) or stack the cells in series to obtain the proper operating voltage for the 5101L. If it is desired to stack the cells in series to obtain a higher voltage, care should be exercised to assure that the cells are reasonably matched. Cells which are not matched can cause problems during charging when placed more than three in series. Most manufacturers will provide Ni-Cd stacks of the desired size which should be adequately matched to avoid any charging problems. It is important to discuss this phenomonon with the battery manufacturer if several Ni-Cd batteries are to be used in series.

Trickle Charge Nickel-Cadmium

Ni-Cd batteries used as standby support power for memory systems should be provided with a continuous charging current from the main power source. This assures that the self discharge characteristic of the battery does not deplete battery capacity. The trickle charge current should be a constant current at a rate of one-tenth the total battery capacity (e.g. a 400mA-hr Ni-Cd should be trickle charged with 40mA current). A simple trickle charger is shown in Figure 26. If the system is to be operated at high temperature, care should be taken to assure that the maximum battery cell temperature is not exceeded during charging.

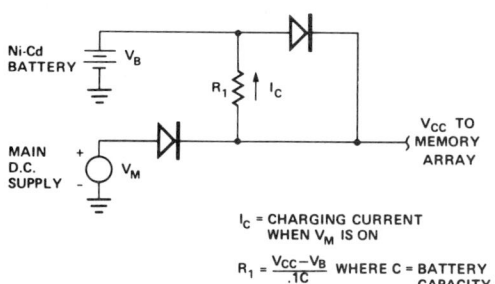

A. Diode Coupled

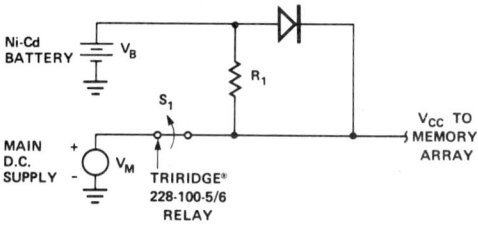

B. Switch Coupled

Figure 26. Ni-Cd Trickle Charger

Fast Charge Nickel-Cadmium

Some Ni-Cds can be charged at a much faster rate than that described above. However, the charging current must be monitored and reduced to trickle charge when the battery is fully charged. Failure to properly handle the charging of Ni-Cd batteries can present safety problems. The manufacturer should be consulted for recommended fast charge techniques.

Lead-Calcium

Lead-calcium batteries are also ideally suited for use as a standby power source for semiconductor memories. A popular brand of lead-calcium cell is the Gel/Cell® made by Globe Battery (Gel/Cell® is a registered trademark of Globe-Union). These types of batteries have several highly desirable characteristics such as:

1. Small size-to-capacity ratio.
2. Low standby self-discharge characteristics.
3. Flat operating discharge characteristics.
4. No permanent cell reversal.
5. Good operating temperature range.

Several manufacturers supply lead-calcium type batteries. However, for the purpose of this application note only the characteristics of the Gel/Cell® will be discussed.

Gel/Cell® Characteristics

A small Gel/Cell® battery is shown in Figure 27. The nominal cell voltage of this type of battery is 2.0 volts (the capacity of the battery shown is 1 amp-hr). This battery is ideal for use in those systems having a relatively high discharge load (~1mA). The output discharge characteristics are shown in Figure 28.

Figure 27. Gel/Cell® Lead-Calcium Battery

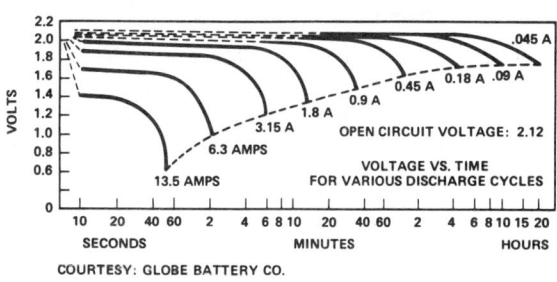

Figure 28. Gel/Cell® Voltage Discharge Characteristics

Note that the minimum discharge rate shown in Figure 28 is a hefty 45mA. At this rate the battery can supply power for 20 hours. At the rate of 1mA, this battery will last 1000 hours or approximately 6 weeks. Lower discharge rates will of course increase the battery life time proportionally.

These types of batteries are optimally used in systems having a large current drain. Although the batteries are indeed very small for a given capacity, the cell voltage is a nominal 2.0 volts which is the minimum acceptable for maintaining data in the 5101L. Therefore, either two batteries are required (series connection) or the voltage boost circuit described earlier must be used. In those systems having very small current drains in standby, the addition of a second battery will most likely take up too much room on the p.c. board. For these systems other types of batteries are recommended (such as Ni-Cd, Mercury, etc.).

Capacity of a Gel/Cell® as a function of temperature is shown in Figure 29. As is shown in this figure, at low temperatures the battery loses a great deal of capacity. Therefore, when designing systems using this type of battery, proper attention will have to be paid to the environmental temperature extremes expected in the system and proper battery selection made.

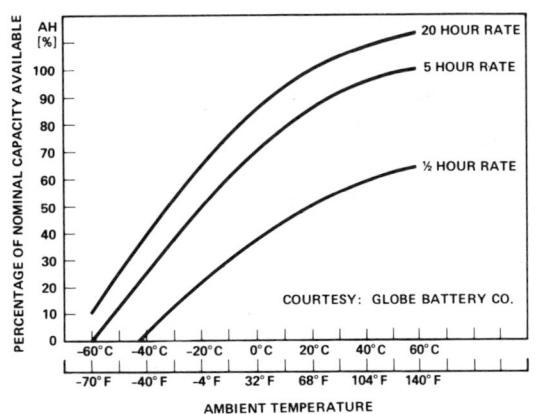

Figure 29. Gel/Cell® Capacity Vs. Temperature

Trickle Charge Lead-Calcium

Unlike Ni-Cd batteries, which accept a constant current trickle charge, the lead-calcium battery is trickle charged by a constant voltage source. The voltage required is 2.25 to 2.30 volts per cell. At this voltage, referred to as the float voltage, a Gel/Cell® will accept only the amount of charge necessary to maintain capacity.

The implementation of a trickle charger for lead-calcium batteries in a system is not as straight forward as for Ni-Cd batteries. A simple charger is shown in Figure 30. In this figure, the "float" voltage is maintained by zener Z, and potentiometer P. The potentiometer is used to adjust the voltage at node (1) to the proper level (2.25 to 2.30 volts per cell). Most zeners are accurate to no more than ±5% which is not adequate for the desired "float" voltage. Diodes D_1 and D_2 isolate the battery and power supply from each other.

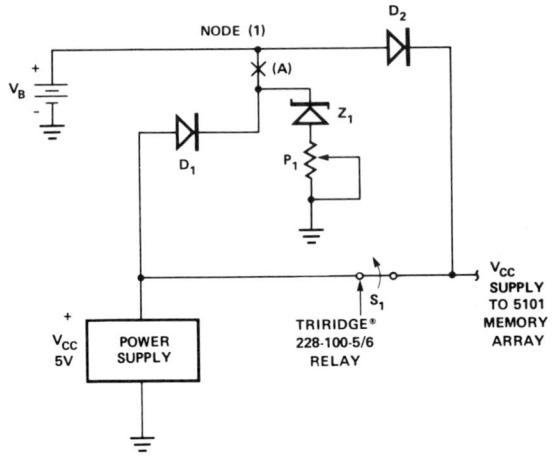

Figure 30. Lead-Calcium Trickle Charger

It is important to select a very low leakage zener (Z_1) to minimize the parasitic load on the battery during power down operations. Indeed, it may be desirable to insert a normally open switch at location (A), Figure 30, to disconnect the zener from the battery during standby operation. The switch would be controlled identically to S_1, as shown in Figure 30.

It is clear from the example given that providing a trickle charge to a lead-calcium battery and having the battery ready for instantaneous operation is more complicated than for Ni-Cd batteries. Other charging methods are available (see bibliography 4) but they all require that the battery and system voltage be identical and in 2 volt increments. Since operation of the 5101 is $4.75 \leq V_{CC} \leq 5.25$ and standby operation is $2.0 \leq V_{CC} \leq 5.25$, the charger/supply combinations described in the reference have limited value for the present applications.

Summary: Lead-Calcium

Lead-calcium batteries are particularly useful with those systems which have a relatively high standby discharge rate (greater than 1mA). The high energy density of these batteries also lend themselves to providing power in normal operation (taking into account V_{CC} requirements of the 5101L) of some systems.

DESIGNING NON-VOLATILE SEMICONDUCTOR MEMORY SYSTEMS

The primary disadvantage of lead-calcium cells is the relative complexity of supplying a trickle charge to the batteries in those systems where the standby voltage is lower than the operating supply voltage V_{CC}.

System Implementation

The 5101 is an extremely easy to use static RAM. No refresh timing is required, only one power supply (+5V) is needed, and the device is fully TTL compatible. In addition, current transients on the V_{CC} (+5V) pin are minimal and require no special decoupling techniques. Therefore, this section will concentrate on interface techniques to the 5101 in order to minimize the power in power down/standby applications.

1K X 16 MEMORY SYSTEM

The discussion on interface techniques to the 5101 is illustrated with a 1024 word X 16 bit system shown in Figure 33. The memory array is configured as shown in Figure 33. Note that for a read/write access one of four columns is enabled by one CE_2 (CE_2A, CE_2B, CE_2C, CE_2D). The other chip enable ($\overline{CE}_1$) and the output disable pin are tied to ground to simplify the layout. All corresponding addresses are bused together and driven by one buffer as are the read/write inputs. Data in and data out pins are OR tied along a given row. Access is then simply a matter of providing the correct address (A_0-A_7), selecting a read or write function and enabling the proper row. Two simple methods for providing the proper CE_2 signals are shown in Figure 31.

TTL Interface

Interface circuits shown in Figure 33 can be implemented with either CMOS or TTL devices. If access/cycle time of the memory system is to be minimized, then series 74 or 74S type TTL can be used. However, for power down operations where a battery is used for back-up power the V_{CC} (+5V) supply to these TTL devices must be independent of the V_{CC} supply to the memory array. This is most easily accomplished by a slight modification to the power supply diagram shown in Figures 26 and 30 as modified in Figure 32. As shown, when the main supply V_m goes off, switch S_1 is opened (isolating V_m from the memory devices).

The state of the addresses, read/write, $\overline{CE}_1$, output disable and data-in to the 5101 memory array are in a "don't care" condition for standby/power down operation. Only CE_2 is required to be low ($\leq 0.2V$) for the low power state. For CE_2 TTL interface drivers, a resistor to ground is required to maintain CE_2 at the proper level when power is removed from the series 74/74S gates. The resistor value required is calculated by considering two requirements:

1. CE_2 high ($V_{IH} \geq 2.2V$) during operation.
2. CE_2 low ($V_{IL} \leq 0.2V$) during standby/power down.

The first requirement above is determined by the maximum source current capability of the TTL drivers (I_{OH}) allowed which guarantees the proper high level output. Requirement 2 above is a function of the maximum leakage on the CE_2 line from the four 5101 devices driven by the CE_2 line. The range of values for the pull down resistor is $6.2k\Omega \leq R \leq 50k\Omega$ for Series 74 drivers.

The POWER VALID input signal (shown in Figure 33) is derived from a power loss detect circuit. The power loss detect circuit should be able to detect a power loss before the output V_{CC} falls below 4.75V (lowest guaranteed operating power level for TTL circuits). A power loss detect circuit to implement the POWER VALID signal is discussed in the POWER LOSS DETECT section.

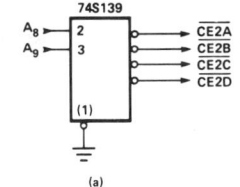

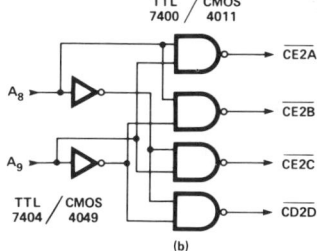

Figure 31. Chip Enable Generators 1K X 16 Memory

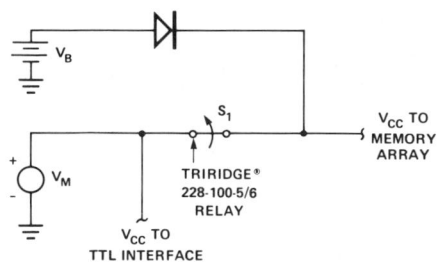

Figure 32. Power Distribution for TTL and CMOS Interface

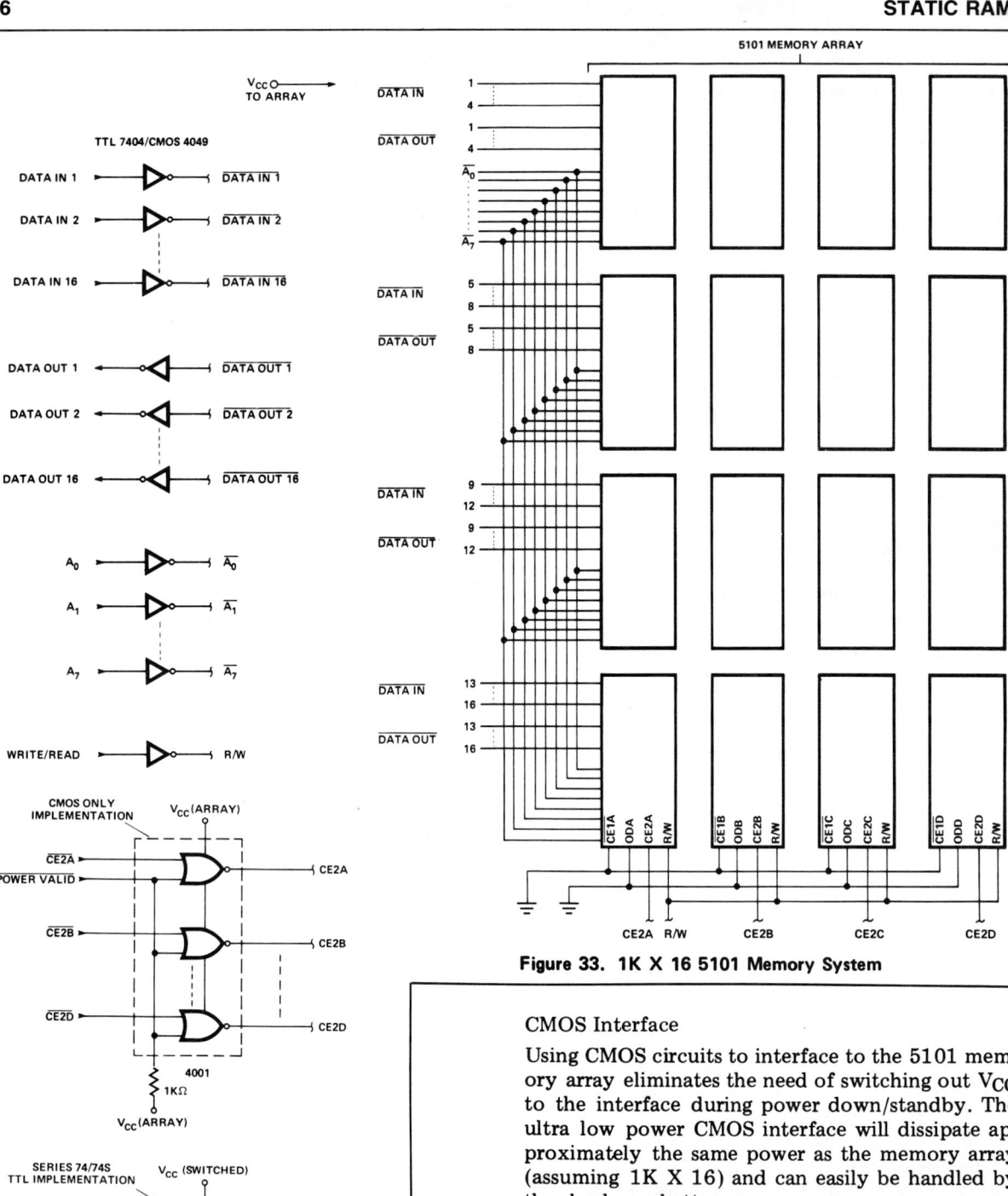

Figure 33. 1K X 16 5101 Memory System

CMOS Interface

Using CMOS circuits to interface to the 5101 memory array eliminates the need of switching out V_{CC} to the interface during power down/standby. The ultra low power CMOS interface will dissipate approximately the same power as the memory array (assuming 1K X 16) and can easily be handled by the back-up battery.

Photos of CMOS waveforms driving the 1K X 16 5101 memory array are shown in Figure 34. Also included in the photo is the noise generated on the V_{CC} supply during operation. As is shown, noise on the power line is virtually non-existent.

1K X 16 MEMORY ARRAY LAYOUT AND CARD ASSEMBLY

The layout used on the 5101 1K X 16 system described previously is shown in Figure 35. Note that V_{CC} and ground are distributed in a grided matrix and decoupled as shown. More decoupling was used

DESIGNING NON-VOLATILE SEMICONDUCTOR MEMORY SYSTEMS

CMOS Interface

Using CMOS circuits to interface to the 5101 memory array eliminates the need of switching out V_{CC} to the interface during power down/standby. The ultra low power CMOS interface will dissipate approximately the same power as the memory array (assuming 1K X 16) and can easily be handled by the back-up battery.

Photos of CMOS waveforms driving the 1K X 16 5101 memory array are shown in Figure 34. Also included in the photo is the noise generated on the V_{CC} supply during operation. As is shown, noise on the power line is virtually non-existent.

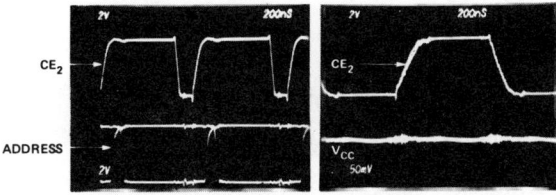

Figure 34. CMOS Interface Driver Waveforms

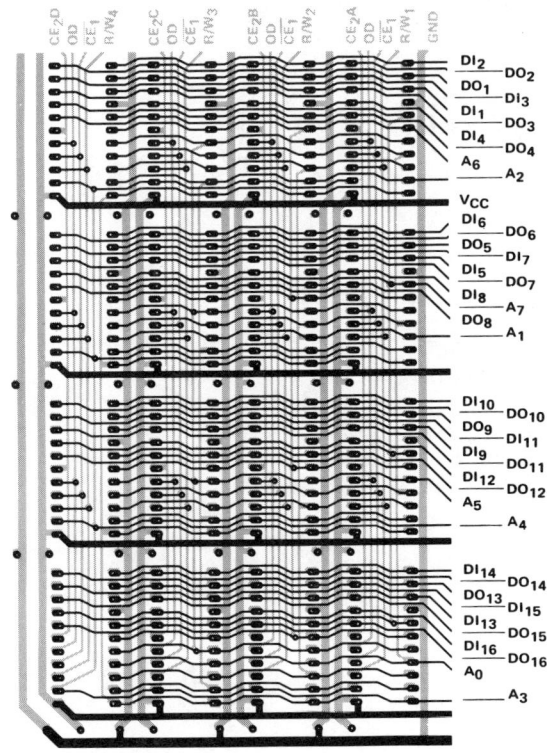

Figure 35. 5101 1K X 16 Array Layout

1K X 16 MEMORY ARRAY LAYOUT AND CARD ASSEMBLY

The layout used on the 5101 1K X 16 system described previously is shown in Figure 35. Note that V_{CC} and ground are distributed in a grided matrix and decoupled as shown. More decoupling was used in this system than is ordinarily required so the designer can use his own judgement in this regard.

The 1K X 16 memory card used was configured per the diagram in Figure 36. Notice that the card is completely self contained for standby/power down operation with the battery included on the card. With this configuration the card can be unplugged, transported to another location (with data being maintained by batteries) and operation resumed.

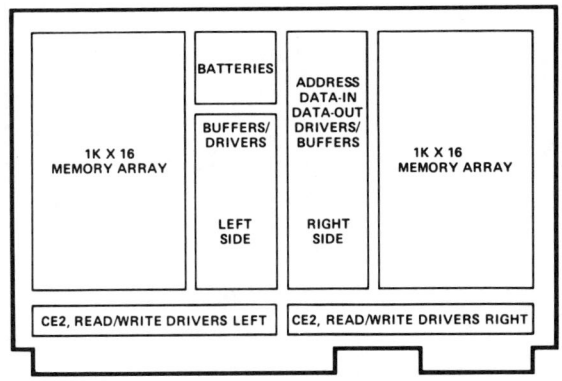

Figure 36. Dual 1K X 16 5101 Memory Card

5101 ORGANIZATION ADVANTAGES

The organization of the 5101 as 256 words X 4 bits has distinct advantages over memory devices organized as 1024 words X 1 bit in many systems applications. These applications include terminals, CRT displays, microprocessors and others which have most (or at least a portion) of their memory expandable in 256 or 512 word increments. For these cases, the number of devices required for a 256 X 4 memory device is much smaller than for a 1024 X 1 memory device.

SUMMARY

There are many selections of 5101 256 word X 4 bit devices available. Table VIII is the product selection guide for this family of devices. As shown, the designer has a wide range of choices in selecting the device most suited to his particular requirements.

BIPOLAR/MOS STATIC RAM COMPATIBILITY

DENNIS GALLOWAY

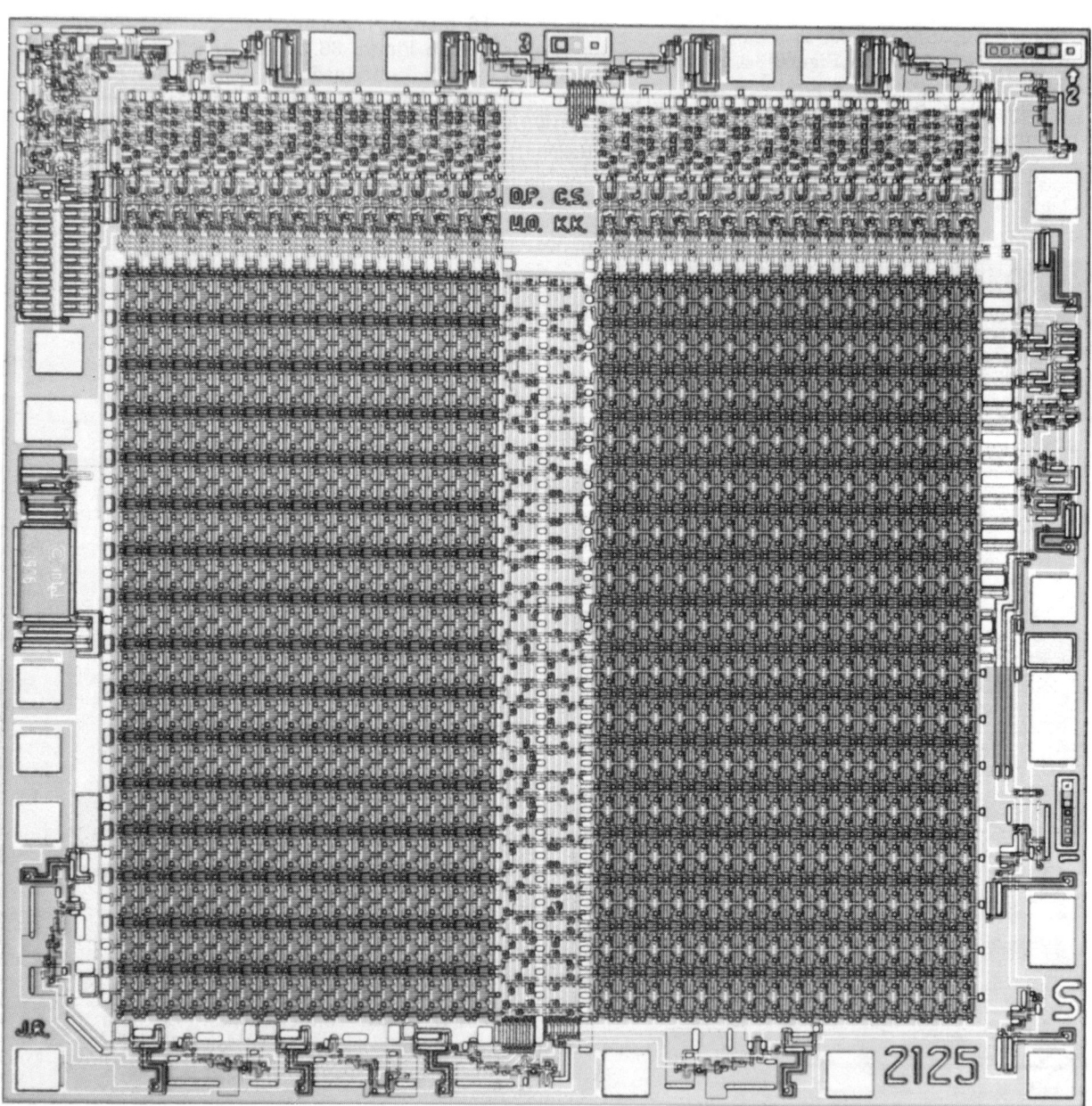

Photomicrograph of 1024 × 1 2125 Static RAM.

BIPOLAR/MOS STATIC RAM COMPATIBILITY

INTRODUCTION

High speed memory systems (system access in the 50 - 100nsec range) have in the past only been implemented with bipolar devices. Bipolar RAMs are generally characterized as very high speed, relatively high powered, and high bit cost devices. The speed characteristic of the bipolar device, necessary for system operation, usually outweighed the power penalty that was paid to obtain the system speed.

MOS technology is characterized as having very high bit densities, very low operating and standby power, with relatively slow access time and low bit cost, However, recent advances have now allowed bipolar speeds to be achieved and at the same time retaining MOS power dissipation with low cost.

Now, the best of both worlds, in the design of the memory system with Bipolar speeds and MOS power dissipation combined, is possible.

In new designs of memory systems, the user is able to take advantage of low power characteristic of MOS with minimum power supplies, thereby reducing overall system cost. In addition, the lower bit cost of MOS can help to reduce system costs further.

The advantage of replacing Bipolar devices with memory devices in existing designs is to reduce device costs in addition to savings on lower power requirements.

When upgrading existing systems to benefit from MOS characteristics, the user begins to realize possible potential problems in interfacing Bipolar and MOS devices simultaneously. For MOS and Bipolar to be truly compatible, not only must access and cycle timings be equal, but all intermediate timing considerations for MOS must be at least equal or better to those of the Bipolar.

In this article we will describe the operation of a static RAM memory board using both Bipolar and MOS memory devices together.

Intel's 2115A/2125A is designed to be pin for pin and timing compatible with Fairchild's 93415A/25A. We begin with a summary of the device characteristics (similarities and differences), compatibilities operating together at the board level, and summarize the various present and future possibilities of high speed MOS memory systems.

BIPOLAR/MOS SPECIFICATION SUMMARY

The specifications for Intel's 2125A and Fairchild's 93425A are similar and shown below. (Fairchild's 93425A data sheet dated 8/74.)

The Similarities:

t_{AA}	(Address Access Time)	45ns
t_{ACS}	(Chip Select Time)	30ns
t_{RCS}	(Chip Select Recovery Time)	30ns
t_{WSCS}	(Chip Select Set-Up Time)	5ns
t_{WHCS}	(Chip Select Hold Time)	5ns
t_{WSA}	(Address Set-Up Time)	5ns
t_{WHA}	(Address Hold Time)	5ns

	Differences	93425A	2125A
t_{WS}	(Write Enable Time)	30ns	25ns
t_{WSD}	(Data Set-Up Prior to Write)	0ns	-10ns
t_W	(Write Pulse Width)	35ns	30ns
I_{IL}	(Input Low Current)	-400µA	-40µA
I_{CC}	(Power Supply Current)	130mA	80mA
Power Dissipation		0.5 mW/bit	0.3mW/bit

MOS/BIPOLAR SYSTEM COMPATIBILITY

In evaluating the compatibility between MOS and Bipolar devices, it is necessary to examine the characteristics of a memory system which uses both devices on the same board, in the evaluation primary emphasis is placed on analyzing those characteristics found in the data sheet specifications which may have a negative impact on system level compatibility.

The memory system used to perform this MOS/Bipolar system compatibility is shown in Figure 1.

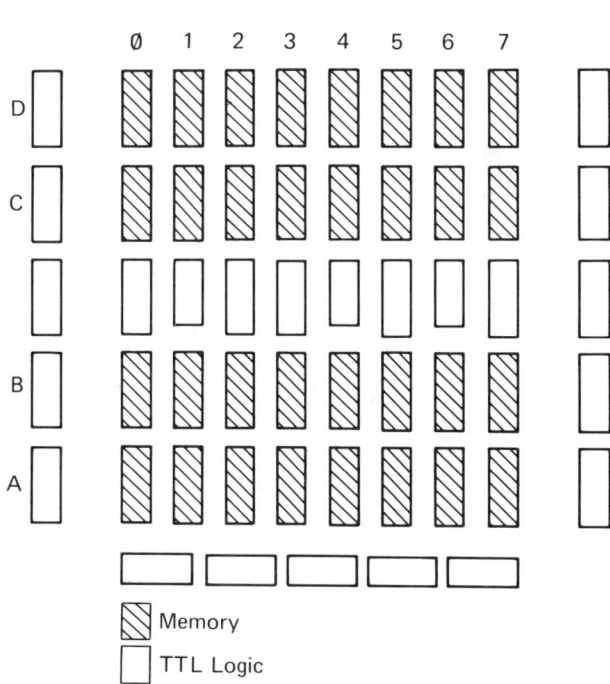

Figure 1. MOS/Bipolar Evaluation Board.

The system analysis performed considers the following system related characteristics:
1) Device placement
2) Timing requirements
3) Power characteristics

Each of the above characteristics is explained as to their effect on a system in the following sections.

DEVICE PLACEMENT

The ultimate test of system level compatibility between devices is whether or not these devices can be treated as "black boxes". For MOS high speed devices to be universally accepted in new and existing systems, they must be able to replace existing bipolar at any location on the memory card. The configurations of interest are shown in Figure 2 and Figure 3.

The configurations in Figure 2 are as follows (label numbers refer to configuration shown in Figure 2):

label
Number   Comments

(1) Determines compatibility along the Dout OR tie between devices, e.g., if bipolar device turns off (or vice versa) excessive transient current will be drawn in power supply and data out line.

(2) (3) Established to provide a "standard" for a row of MOS and bipolar during testing. Determine loading change (if any) on inputs — Particularly $\overline{CS}$.

(4) Examines capacitive and leakage loading effects on data out line. Effects, if any, on access time are evaluated.

(5) Determines address line capacitive and leakage loading effect. Provides information on possible system level sensitivities to either "too" fast or "too" slow address transition time effecting address valid time (for t_{AA} measurements).

Figure 3 is a configuration likely to be encountered when a MOS device is used to replace a defective bipolar device in an existing storage card. (Of course it is always possible that a MOS device will fail and have to be replaced by a bipolar device ... even that possibility is considered.)

The condition shown by label (1) Figure 2 is best analyzed by considering the circuit configuration shown in Figure 4.

Opposite data is loaded into devices (1) and (2) in Figure 4. A read operation is performed by toggling chip select as shown in Figure 4b. If the device being deselected goes to the high impedance state too slowly while the device being selected accesses data too rapidly, a large transient current spike in the data-out output will occur during the time both devices are ON. Time t_0 represents the case where a slow bipolar device may cause a transient spike while time t_1 represents the case where a slow MOS device may cause a similar spike. Photos showing these conditions as a function of $\overline{CS}$, $\overline{CS}_2$ overlap are shown in Figure 5.

The width and height of the current spike are a function of the amount of overlap of the respective $\overline{CS}$. Normal system timing requirements should be such that neither device is turned ON (or OFF) while the neighbor device, on the data-out line is still active.

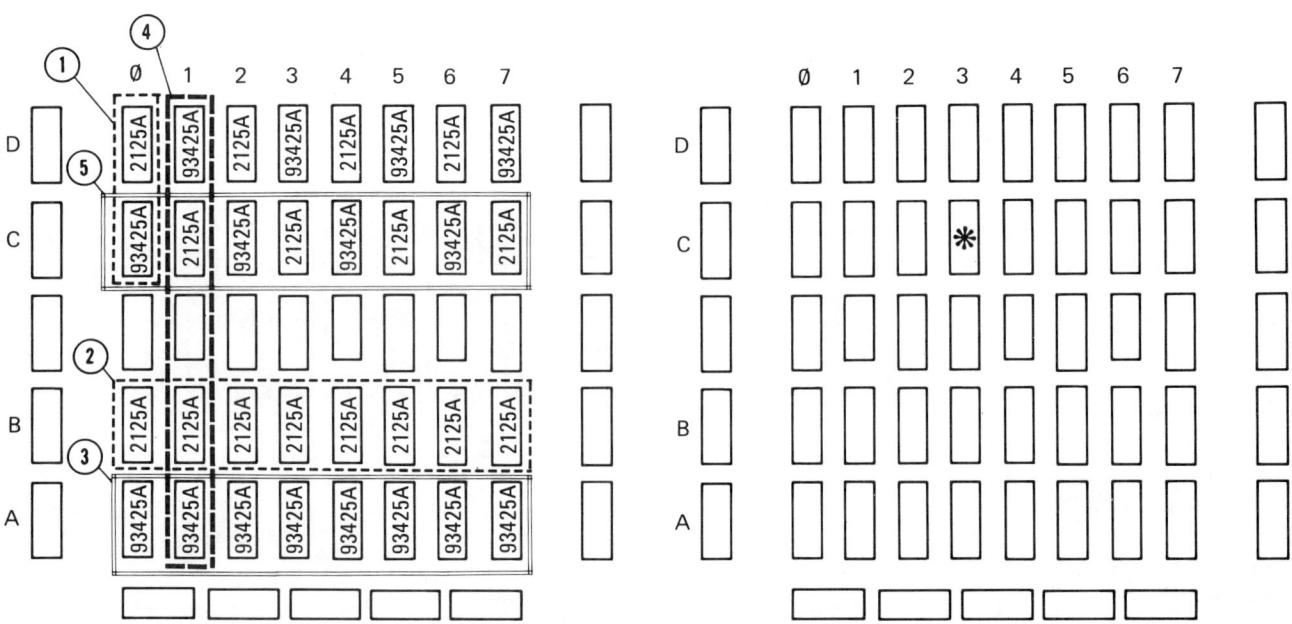

Figure 2. MOS/Bipolar Evaluation Board Population.

Figure 3. Isolated MOS Device On Bipolar Storage Card.

BIPOLAR/MOS STATIC RAM COMPATIBILITY

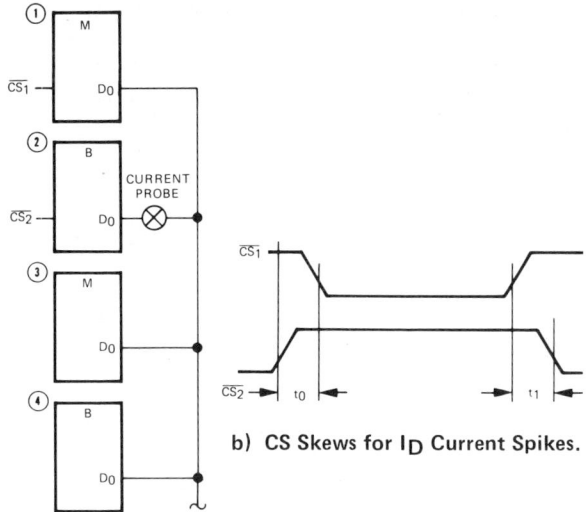

a) Current Probe Location for I_D Spike.

b) CS Skews for I_D Current Spikes.

Figure 4. I_D Circuit Configuration for Measurement.

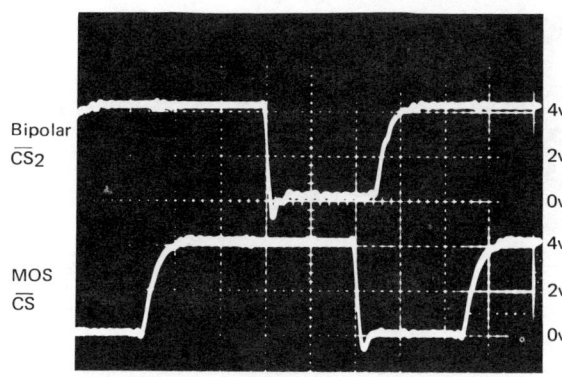

a) Bipolar/MOS $\overline{CS}$ With Bipolar Turning Off Late.

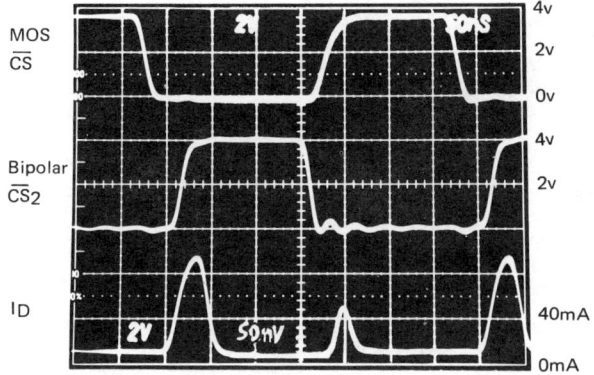

b) Bipolar/MOS Data Out Current Spike.

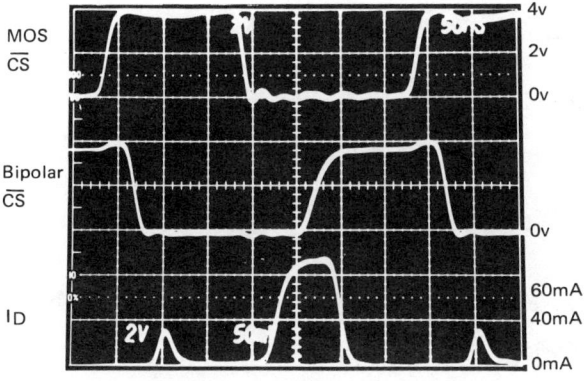

c) Bipolar/MOS Data Out Current Spike.

Figure 5. Bipolar OFF/MOS ON Current Transients.

However, when the system timing is designed such that a row is deselected at the same time another row is selected, the designer must assure himself that the current spike generated is not excessive. The photos shown in Figures 5 and 6 illustrate these conditions. Figure 5a, b, c, show the condition for a bipolar device turning OFF while a MOS device is turning ON. The transient current through the data out buffers is shown in Figures 5b and c for different deselection times. It is important to note that for deselection times shown it makes no difference whether or not the devices used are all bipolar, all MOS, or a mixture of the two. The transient spike shown is similar for all three conditions.

Figures 6a, b, and c show the condition for a MOS device turning OFF while a bipolar device is turning ON. The transient spikes shown are again independent of the use of bipolar, MOS, or a combination of devices.

The analysis and photos showing the effect on the data-out line between MOS and bipolar devices has demonstrated their compatibility in a data out OR tie condition. Subsequent analysis of timing parameters will show that, as far as system timing is concerned, there is no difference between MOS and bipolar (both are equally fast).

LOADING EFFECTS

There are two basic loading effects to be considered: "horizontal" and "vertical". A horizontal loading effect refers to a typical printed circuit board layout where the address CS and write enable lines are distributed horizontally. The "vertical loading effect is along the data in and data out directions; usually laid out vertically.

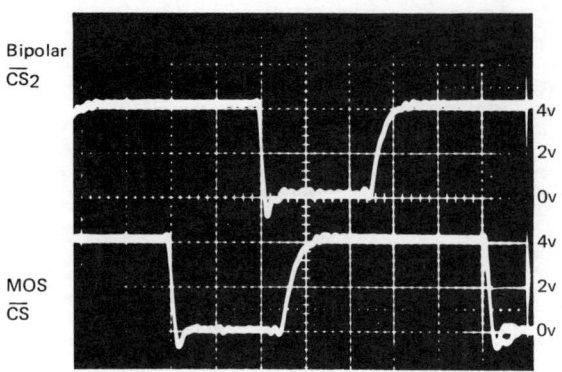

a) Bipolar /MOS $\overline{CS}$ With MOS Turning Off Late.

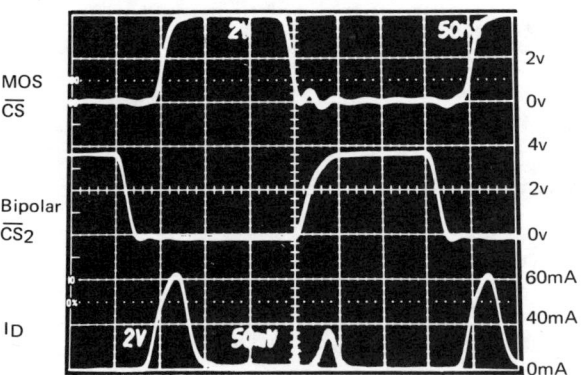

b) Bipolar /MOS Data Out Current Spike.

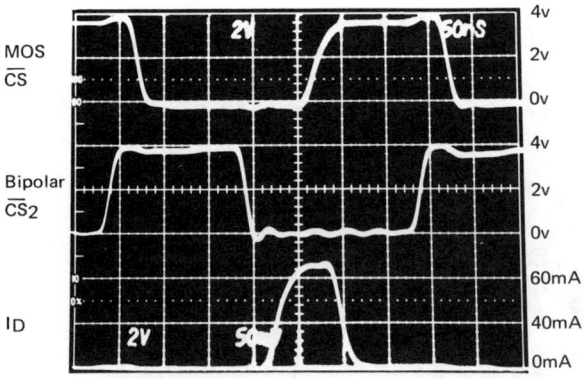

c) Bipolar /MOS Data Out Current Spike.

Figure 6. MOS OFF/Bipolar ON Current Transients.

The horizontal loading effects and changes (if any) between MOS and bipolar are demonstrated by configurations 2 and 3 in Figure 2. $\overline{CS}$ is chosen to show this effect. If the capacitance loading of $\overline{CS}$ is significantly different between MOS and bipolar, than a difference in the $\overline{CS}$ waveform would be expected to occur. Figure 7 shows the $\overline{CS}$ waveform for a row of all MOS and a row of all bipolar devices. From this figure, it is evident that the capacitive loading effect of MOS and bipolar are comparable. However, due to the higher loading currents, the low level input can be expected to be typically higher for bipolar than for MOS on the inputs. The noise margin for a bipolar system is therefore, somewhat reduced from that of a MOS system. Configuration 4 in Figure 2 considers the differences (if any) in the loading effects of OR tieing data out lines. The primary difference due to capacitance loading would be a change in access time due to this loading. Specifications for both bipolar and MOS devices indicate comparable capacitive loads. The effect of these two devices on each other in a system environment, does not impair the system.

TIMING REQUIREMENTS

When considering device specifications for a memory board, one of the most critical aspects is timing. If changes to the timing were required, when changing to MOS from bipolar (even with the same access time) device compatibility could not be claimed.

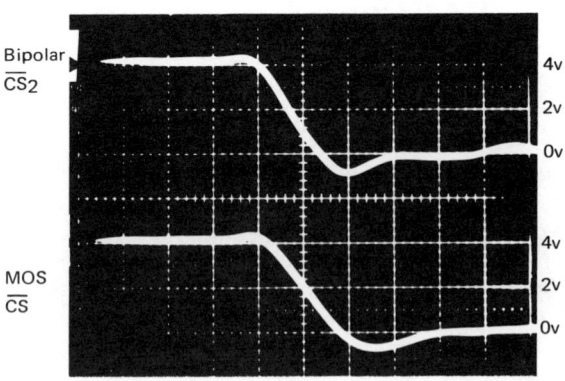

Figure 7. Bipolar /MOS $\overline{CS}$ Accessing Two Rows Simultaneously.

BIPOLAR/MOS STATIC RAM COMPATIBILITY

Selected plots showing critical timing parameter tested for different device configurations on a board is shown in Figure 8. As is shown in these figures, both MOS and bipolar are well within worst case device specifications over temperature and voltage.

POWER CHARACTERISTICS

The use of the 2115A/2125A in a high speed memory system replacing bipolar devices causes no change in system characteristics .... except one. That of course is power dissipated in the system. The 2115A/2125A devices are much lower in power than their bipolar counterparts. Table 1 summarizes the system level power when using all bipolar and all MOS devices. Reducing power, of course, reduces system operating temperature. This reduction in operating temperature improves semiconductor reliability. Equally important to lowering temperature is that lower power results in lower power supply costs.

Table I. MOS/Bipolar Power Dissipation.

Device	Pwr/bit	Total bits	Total/Pwr
93425A	0.5mW	32,768	16.384watts
2125A	0.3mW	32,758	9.830watts

SUMMARY

The compatibility between the 2115A/2125A and the 93425A family has been demonstrated. The designer now has the opportunity to achieve very high speed operation and simultaneously achieve low memory system power dissipation.

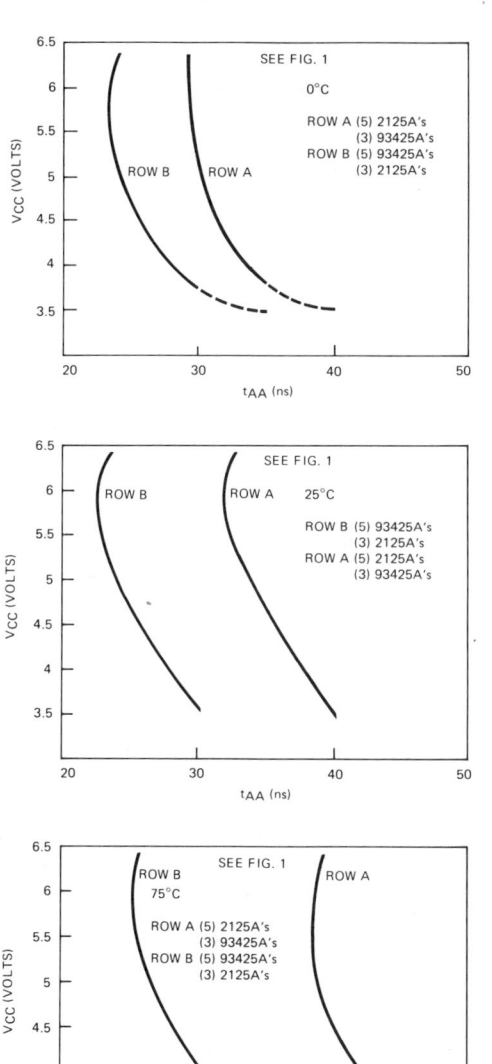

Figure 8. V_{CC} vs t_{AA} for Various Temperatures.

2101A
256 X 4 BIT STATIC RAM

2101A-2	250 ns Max.
2101A	350 ns Max.
2101A-4	450 ns Max.

- **256 x 4 Organization to Meet Needs for Small System Memories**
- **Single +5V Supply Voltage**
- **Directly TTL Compatible: All Inputs and Output**
- **Statis MOS: No Clocks or Refreshing Required**
- **Simple Memory Expansion: Chip Enable Input**
- **Inputs Protected: All Inputs Have Protection Against Static Charge**
- **Low Cost Packaging: 22 Pin Plastic Dual In-Line Configuration**
- **Low Power: Typically 150 mW**
- **Three-State Output: OR-Tie Capability**
- **Output Disable Provided for Ease of Use in Common Data Bus Systems**

The Intel® 2101A is a 256 word by 4-bit static random access memory element using N-channel MOS devices integrated on a monolithic array. It uses fully DC stable (static) circuitry and therefore requires no clocks or refreshing to operate. The data is read out nondestructively and has the same polarity as the input data.

The 2101A is designed for memory applications where high performance, low cost, large bit storage, and simple interfacing are important design objectives.

It is directly TTL compatible in all respects: inputs, outputs, and a single +5V supply. Two chip-enables allow easy selection of an individual package when outputs are OR-tied. An output disable is provided so that data inputs and outputs can be tied for common I/O systems. The output disable function eliminates the need for bi-directional logic in a common I/O system.

The Intel® 2101A is fabricated with N-channel silicon gate technology. This technology allows the design and production of high performance, easy-to-use MOS circuits and provides a higher functional density on a monolithic chip than either conventional MOS technology or P-channel silicon gate technology.

Intel's silicon gate technology also provides excellent protection against contamination. This permits the use of low cost plastic packaging.

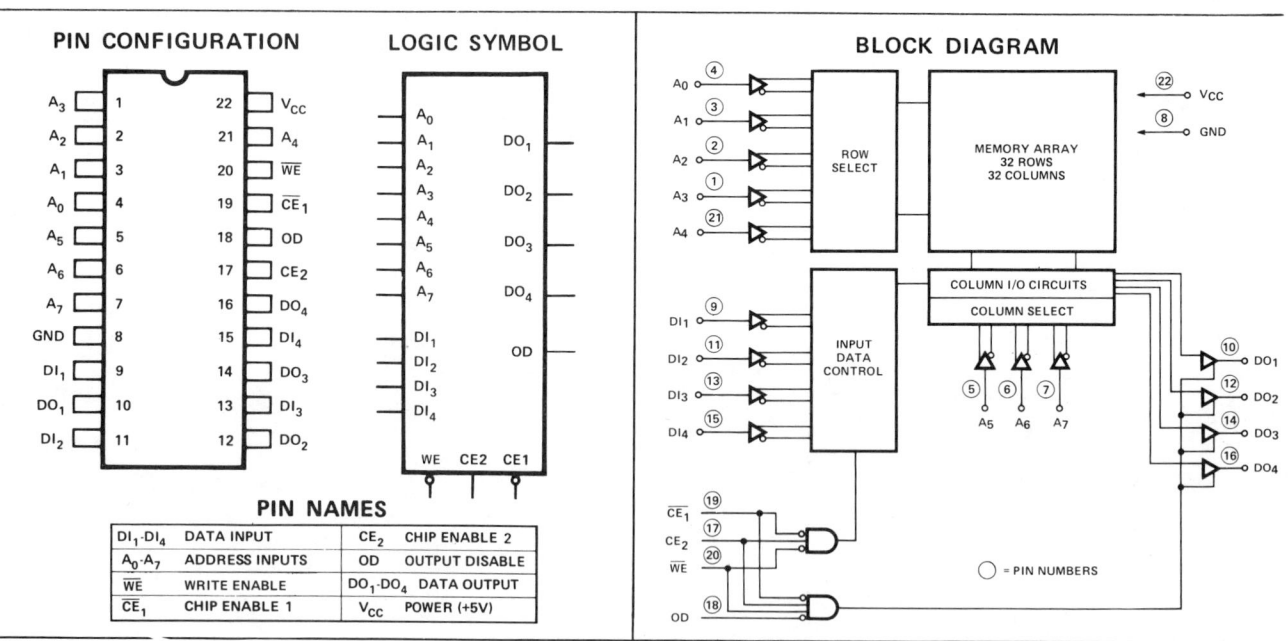

2101A

ABSOLUTE MAXIMUM RATINGS*

Ambient Temperature Under Bias . . . . . -10°C to 80°C
Storage Temperature . . . . -65°C to +150°C
Voltage On Any Pin
  With Respect to Ground . . . . -0.5V to +7V
Power Dissipation . . . . 1 Watt

*COMMENT:

Stresses above those listed under "Absolute Maximum Rating" may cause permanent damage to the device. This is a stress rating only and functional operation of the device at these or at any other condition above those indicated in the operational sections of this specification is not implied. Exposure to absolute maximum rating conditions for extended periods may affect device reliability.

D.C. AND OPERATING CHARACTERISTICS

T_A = 0°C to 70°C, V_{CC} = 5V ±5% unless otherwise specified.

Symbol	Parameter		Min.	Typ.[1]	Max.	Unit	Test Conditions
I_{LI}	Input Current			1	10	μA	V_{IN} = 0 to 5.25V
I_{LOH}	Data Output Leakage Current			1	10	μA	Output Disabled, V_{OUT}=4.0V
I_{LOL}	Data Output Leakage Current			-1	-10	μA	Output Disabled, V_{OUT}=0.45V
I_{CC1}	Power Supply Current	2101A, 2101A-4		35	55	mA	V_{IN} = 5.25V, I_O = 0mA
		2101A-2		45	65		T_A = 25°C
I_{CC2}	Power Supply Current	2101A, 2101A-4			60	mA	V_{IN} = 5.25V, I_O = 0mA
		2101A-2			70		T_A = 0°C
V_{IL}	Input "Low" Voltage		-0.5		+0.8	V	
V_{IH}	Input "High" Voltage		2.0		V_{CC}	V	
V_{OL}	Output "Low" Voltage				+0.45	V	I_{OL} = 2.0mA
V_{OH}	Output "High" Voltage	2101A, 2101A-2	2.4			V	I_{OH} = -200μA
		2101A-4	2.4			V	I_{OH} = -150μA

TYPICAL D.C. CHARACTERISTICS

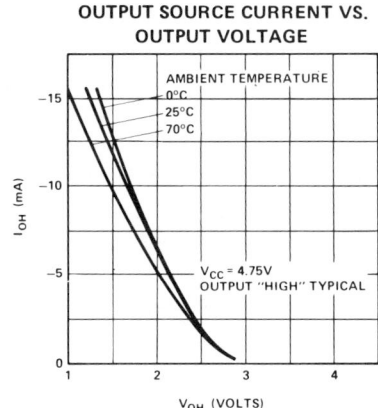

OUTPUT SOURCE CURRENT VS. OUTPUT VOLTAGE

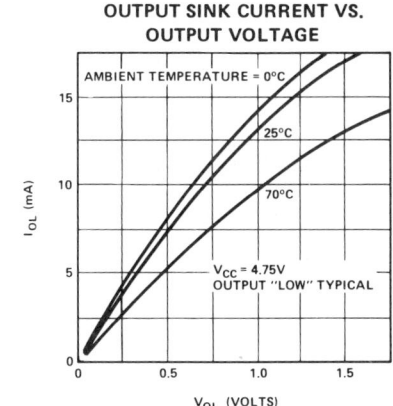

OUTPUT SINK CURRENT VS. OUTPUT VOLTAGE

NOTES: 1. Typical values are for T_A = 25°C and nominal supply voltage.

A.C. CHARACTERISTICS FOR 2101A-2 (250 ns ACCESS TIME)

READ CYCLE $T_A = 0°C$ to $70°C$, $V_{CC} = 5V \pm 5\%$, unless otherwise specified.

Symbol	Parameter	Min.	Typ.[1]	Max.	Unit	Test Conditions
t_{RC}	Read Cycle	250			ns	
t_A	Access Time			250	ns	t_r, t_f = 20ns
t_{CO}	Chip Enable To Output			180	ns	Input Levels = 0.8V or 2.0V
t_{OD}	Output Disable To Output			130	ns	Timing Reference = 1.5V
t_{DF} [3]	Data Output to High Z State	0		180	ns	Load = 1 TTL Gate
t_{OH}	Previous Read Data Valid after change of Address	40			ns	and C_L = 100pF.

WRITE CYCLE

Symbol	Parameter	Min.	Typ.[1]	Max.	Unit	Test Conditions
t_{WC}	Write Cycle	170			ns	
t_{AW}	Write Delay	20			ns	t_r, t_f = 20ns
t_{CW}	Chip Enable To Write	150			ns	Input Levels = 0.8V or 2.0V
t_{DW}	Data Setup	150			ns	Timing Reference = 1.5V
t_{DH}	Data Hold	0			ns	Load = 1 TTL Gate
t_{WP}	Write Pulse	150			ns	and C_L = 100pF.
t_{WR}	Write Recovery	0			ns	
t_{DS}	Output Disable Setup	20			ns	

CAPACITANCE [2] $T_A = 25°C$, f = 1MHz

Symbol	Test	Typ.[1] (pF)	Max. (pF)
C_{IN}	Input Capacitance (All Input Pins) V_{IN} = 0V	4	8
C_{OUT}	Output Capacitance V_{OUT} = 0V	8	12

WAVEFORMS

READ CYCLE

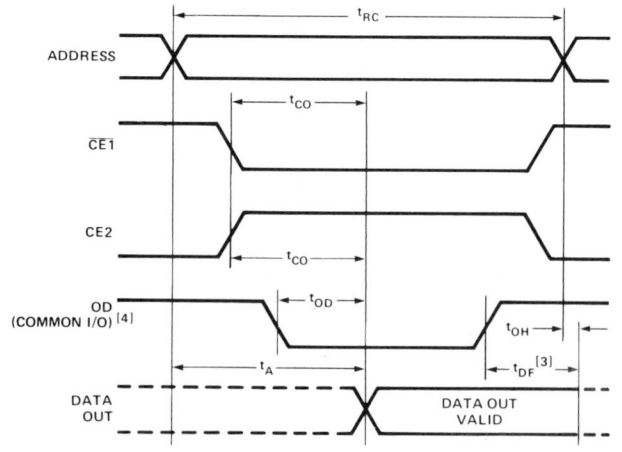

WRITE CYCLE

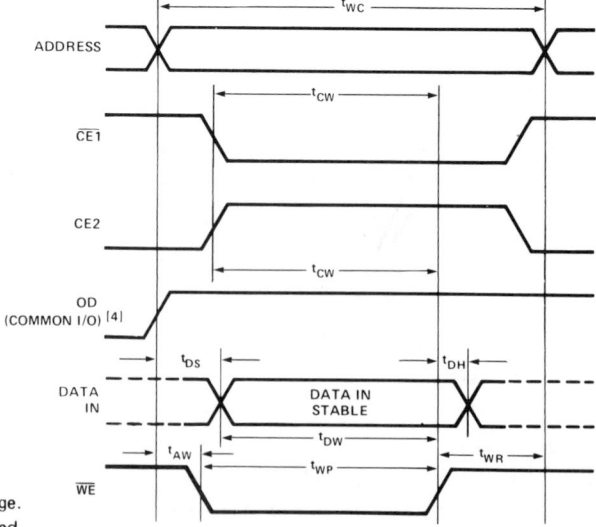

NOTES:
1. Typical values are for $T_A = 25°C$ and nominal supply voltage.
2. This parameter is periodically sampled and is not 100% tested.
3. t_{DF} is with respect to the trailing edge of $\overline{CE_1}$, CE_2, or OD, whichever occurs first.
4. OD should be tied low for separate I/O operation.

2101A (350 ns ACCESS TIME)
A.C. CHARACTERISTICS

READ CYCLE $T_A = 0°C$ to $70°C$, $V_{CC} = 5V \pm 5\%$, unless otherwise specified.

Symbol	Parameter	Min.	Typ.[1]	Max.	Unit	Test Conditions
t_{RC}	Read Cycle	350			ns	
t_A	Access Time			350	ns	t_r, t_f = 20ns
t_{CO}	Chip Enable To Output			240	ns	Input Levels = 0.8V or 2.0V
t_{OD}	Output Disable To Output			180	ns	Timing Reference = 1.5V
t_{DF} [2]	Data Output to High Z State	0		150	ns	Load = 1 TTL Gate
t_{OH}	Previous Read Data Valid after change of Address	40			ns	and C_L = 100pF.

WRITE CYCLE

Symbol	Parameter	Min.	Typ.[1]	Max.	Unit	Test Conditions
t_{WC}	Write Cycle	220			ns	
t_{AW}	Write Delay	20			ns	t_r, t_f = 20ns
t_{CW}	Chip Enable To Write	200			ns	Input Levels = 0.8V or 2.0V
t_{DW}	Data Setup	200			ns	Timing Reference = 1.5V
t_{DH}	Data Hold	0			ns	Load = 1 TTL Gate
t_{WP}	Write Pulse	200			ns	and C_L = 100pF.
t_{WR}	Write Recovery	0			ns	
t_{DS}	Output Disable Setup	20			ns	

2101A-4 (450 ns ACCESS TIME)
A.C. CHARACTERISTICS

READ CYCLE $T_A = 0°C$ to $70°C$, $V_{CC} = 5V \pm 5\%$, unless otherwise specified.

Symbol	Parameter	Min.	Typ.[1]	Max.	Unit	Test Conditions
t_{RC}	Read Cycle	450			ns	
t_A	Access Time			450	ns	t_r, t_f = 20ns
t_{CO}	Chip Enable To Output			310	ns	Input Levels = 0.8V or 2.0V
t_{OD}	Output Disable To Output			250	ns	Timing Reference = 1.5V
t_{DF} [2]	Data Output to High Z State	0		200	ns	Load = 1 TTL Gate
t_{OH}	Previous Read Data Valid after change of Address	40			ns	and C_L = 100pF.

WRITE CYCLE

Symbol	Parameter	Min.	Typ.[1]	Max.	Unit	Test Conditions
t_{WC}	Write Cycle	270			ns	
t_{AW}	Write Delay	20			ns	t_r, t_f = 20ns
t_{CW}	Chip Enable To Write	250			ns	Input Levels = 0.8V or 2.0V
t_{DW}	Data Setup	250			ns	Timing Reference = 1.5V
t_{DH}	Data Hold	0			ns	Load = 1 TTL Gate
t_{WP}	Write Pulse	250			ns	and C_L = 100pF.
t_{WR}	Write Recovery	0			ns	
t_{DS}	Output Disable Setup	20			ns	

NOTES: 1. Typical values are for $T_A = 25°C$ and nominal supply voltage.
         2. t_{DF} is with respect to the trailing edge of $\overline{CE_1}$, CE_2, or OD, whichever occurs first.

2102A, 2102AL
1K x 1 BIT STATIC RAM

P/N	Standby Pwr. (mW)	Operating Pwr. (mW)	Access (ns)
2102AL-4	35	174	450
2102AL	35	174	350
2102AL-2	42	342	250
2102A-2	——	342	250
2102A	——	289	350
2102A-4	——	289	450
2102A-6	——	289	650

- **Single +5 Volts Supply Voltage**
- **Directly TTL Compatible: All Inputs and Output**
- **Standby Power Mode (2102AL)**
- **Three-State Output: OR-Tie Capability**
- **Inputs Protected: All Inputs Have Protection Against Static Charge**
- **Low Cost Packaging: 16 Pin Dual-In-Line Configuration**

The Intel® 2102A is a high speed 1024 word by one bit static random access memory element using N-channel MOS devices integrated on a monolithic array. It uses fully DC stable (static) circuitry and therefore requires no clocks or refreshing to operate. The data is read out nondestructively and has the same polarity as the input data.

The 2102A is designed for memory applications where high performance, low cost, large bit storage, and simple interfacing are important design objectives. *A low standby power version (2102AL) is also available. It has all the same operating characteristics of the 2102A with the added feature of 35mW maximum power dissipation in standby and 174mW in operations.*

It is directly TTL compatible in all respects: inputs, output, and a single +5 volt supply. A separate chip enable ($\overline{CE}$) lead allows easy selection of an individual package when outputs are OR-tied.

The Intel® 2102A is fabricated with N-channel silicon gate technology. This technology allows the design and production of high performance easy to use MOS circuits and provides a higher functional density on a monolithic chip than either conventional MOS technology or P-channel silicon gate technology.

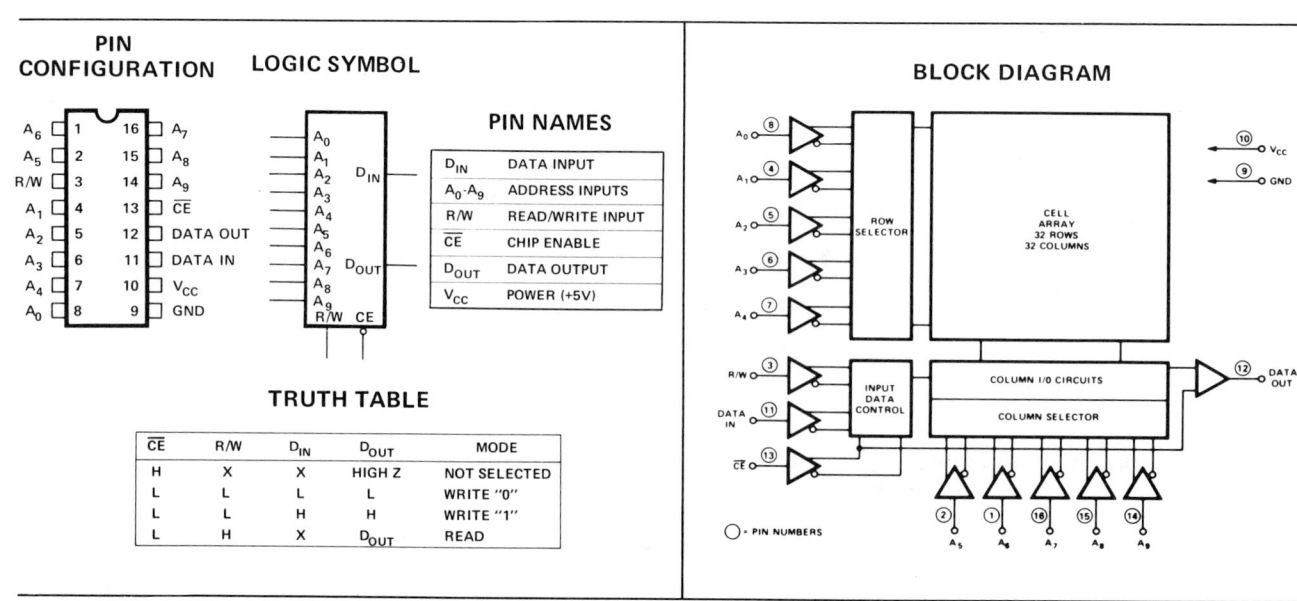

2102A

Absolute Maximum Ratings*

Ambient Temperature Under Bias	−10°C to 80°C
Storage Temperature	−65°C to +150°C
Voltage On Any Pin With Respect To Ground	−0.5V to +7V
Power Dissipation	1 Watt

*COMMENT:
Stresses above those listed under "Absolute Maximum Rating" may cause permanent damage to the device. This is a stress rating only and functional operation of the device at these or at any other condition above those indicated in the operational sections of this specification is not implied. Exposure to absolute maximum rating conditions for extended periods may affect device reliability.

D. C. and Operating Characteristics

$T_A = 0°C$ to $70°C$, $V_{CC} = 5V \pm 5\%$ unless otherwise specified.

Symbol	Parameter	2102A, 2102A-4 2102AL, 2102AL-4 Limits			2102A-2, 2102AL-2 Limits			2102A-6 Limits			Unit	Test Conditions
		Min.	Typ.[1]	Max.	Min.	Typ.[1]	Max.	Min.	Typ.[1]	Max.		
I_{LI}	Input Load Current		1	10		1	10		1	10	μA	V_{IN} = 0 to 5.25V
I_{LOH}	Output Leakage Current		1	5		1	5		1	5	μA	$\overline{CE}$ = 2.0V, $V_{OUT} = V_{OH}$
I_{LOL}	Output Leakage Current		−1	−10		−1	−10		−1	−10	μA	$\overline{CE}$ = 2.0V, V_{OUT} = 0.4V
I_{CC}	Power Supply Current		33	Note 2		45	65		33	55	mA	All Inputs = 5.25V, Data Out Open, $T_A = 0°C$
V_{IL}	Input Low Voltage	−0.5		0.8	−0.5		0.8	−0.5		0.65	V	
V_{IH}	Input High Voltage	2.0		V_{CC}	2.0		V_{CC}	2.2		V_{CC}	V	
V_{OL}	Output Low Voltage			0.4			0.4			0.45	V	I_{OL} = 2.1mA
V_{OH}	Output High Voltage	2.4			2.4			2.2			V	I_{OH} = −100μA

Notes: 1. Typical values are for $T_A = 25°C$ and nominal supply voltage.
2. The maximum I_{CC} value is 55mA for the 2102A and 2102A-4, and 33mA for the 2102AL and 2102AL-4.

Standby Characteristics 2102AL, 2102AL-2, and 2102AL-4 (Available only in the Plastic Package)

$T_A = 0°C$ to $70°C$

Symbol	Parameter	2102AL, 2102AL-4 Limits			2102AL-2 Limits			Unit	Test Conditions
		Min.	Typ.[1]	Max.	Min.	Typ.[1]	Max.		
V_{PD}	V_{CC} in Standby	1.5			1.5			V	
V_{CES}[2]	$\overline{CE}$ Bias in Standby	2.0			2.0			V	$2.0V \leq V_{PD} \leq V_{CC}$ Max.
		V_{PD}			V_{PD}			V	$1.5V \leq V_{PD} < 2.0V$
I_{PD1}	Standby Current		15	23		20	28	mA	All Inputs = V_{PD1} = 1.5V
I_{PD2}	Standby Current		20	30		25	38	mA	All Inputs = V_{PD2} = 2.0V
t_{CP}	Chip Deselect to Standby Time	0			0			ns	
t_R[3]	Standby Recovery Time	t_{RC}			t_{RC}			ns	

STANDBY WAVEFORMS

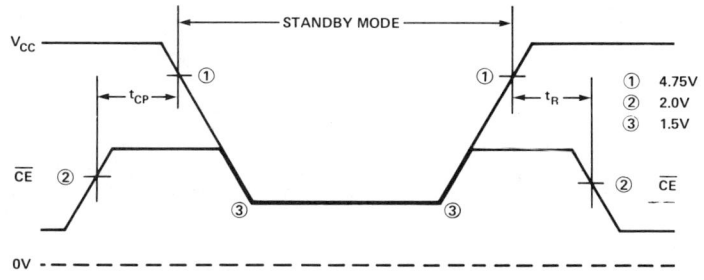

NOTES:
1. Typical values are for $T_A = 25°C$.
2. Consider the test conditions as shown: If the standby voltage (V_{PD}) is between 5.25V (V_{CC} Max.) and 2.0V, then $\overline{CE}$ must be held at 2.0V Min. (V_{IH}). If the standby voltage is less than 2.0V but greater than 1.5V (V_{PD} Min.), then $\overline{CE}$ and standby voltage must be at least the same value or, if they are different, $\overline{CE}$ must be the more positive of the two.
3. $t_R = t_{RC}$ (READ CYCLE TIME).

A. C. Characteristics
$T_A = 0°C$ to $70°C$, $V_{CC} = 5V \pm 5\%$ unless otherwise specified

READ CYCLE

Symbol	Parameter	2102A-2, 2102AL-2 Limits (ns) Min.	Max.	2102A, 2102AL Limits (ns) Min.	Max.	2102A-4, 2102AL-4 Limits (ns) Min.	Max.	2102A-6 Limits (ns) Min.	Max.
t_{RC}	Read Cycle	250		350		450		650	
t_A	Access Time		250		350		450		650
t_{CO}	Chip Enable to Output Time		130		180		230		400
t_{OH1}	Previous Read Data Valid with Respect to Address	40		40		40		50	
t_{OH2}	Previous Read Data Valid with Respect to Chip Enable	0		0		0		0	

WRITE CYCLE

Symbol	Parameter	2102A-2, 2102AL-2	2102A, 2102AL	2102A-4, 2102AL-4	2102A-6
t_{WC}	Write Cycle	250	350	450	650
t_{AW}	Address to Write Setup Time	20	20	20	200
t_{WP}	Write Pulse Width	180	250	300	400
t_{WR}	Write Recovery Time	0	0	0	50
t_{DW}	Data Setup Time	180	250	300	450
t_{DH}	Data Hold Time	0	0	0	20
t_{CW}	Chip Enable to Write Setup Time	180	250	300	550

A.C. CONDITIONS OF TEST

Input Pulse Levels:	0.8 Volt to 2.0 Volt
Input Rise and Fall Times:	10nsec
Timing Measurement Inputs:	1.5 Volts
Reference Levels Output:	0.8 and 2.0 Volts
Output Load:	1 TTL Gate and C_L = 100 pF

Capacitance[2] $T_A = 25°C$, $f = 1MHz$

SYMBOL	TEST	LIMITS (pF) TYP.[1]	MAX.
C_{IN}	INPUT CAPACITANCE (ALL INPUT PINS) V_{IN} = 0V	3	5
C_{OUT}	OUTPUT CAPACITANCE V_{OUT} = 0V	7	10

Waveforms

READ CYCLE

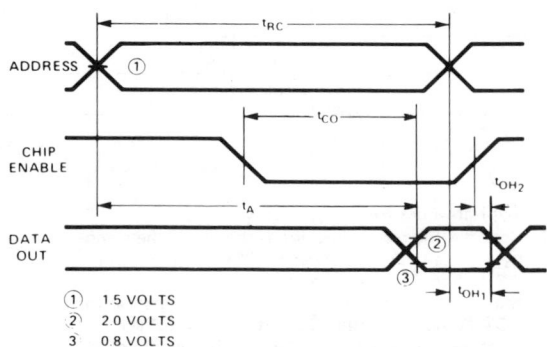

① 1.5 VOLTS
② 2.0 VOLTS
③ 0.8 VOLTS

WRITE CYCLE

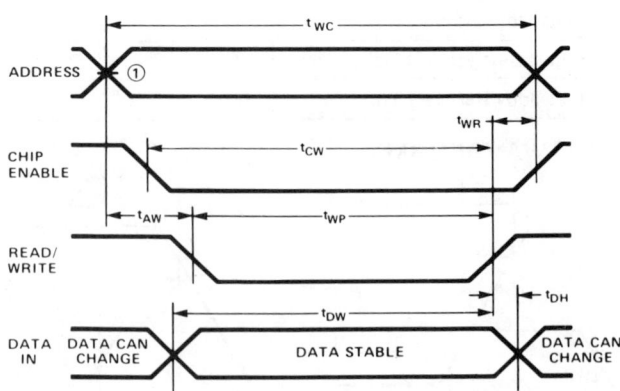

NOTES: 1. Typical values are for $T_A = 25°C$ and nominal supply voltage.
2. This parameter is periodically sampled and is not 100% tested.

Typical D. C. and A. C. Characteristics

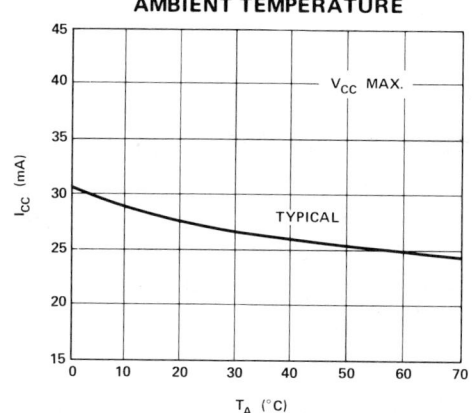

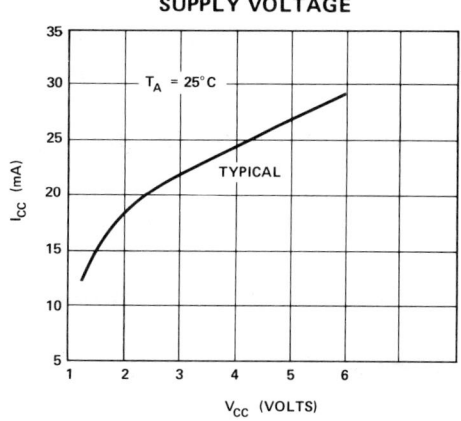

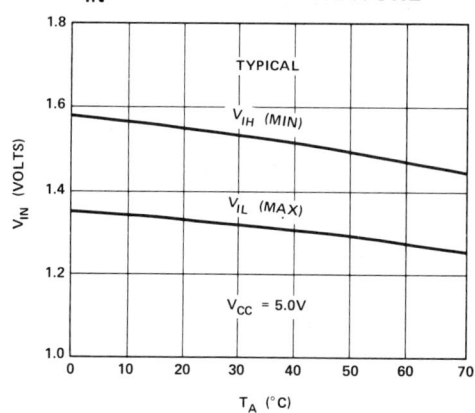

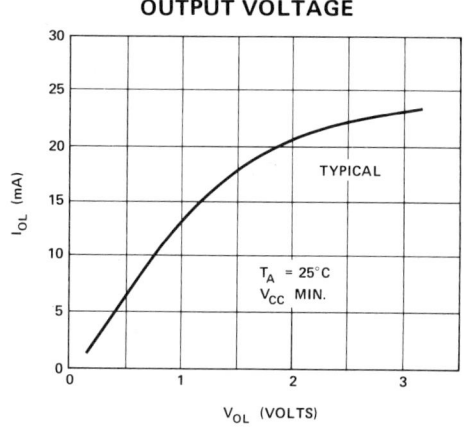

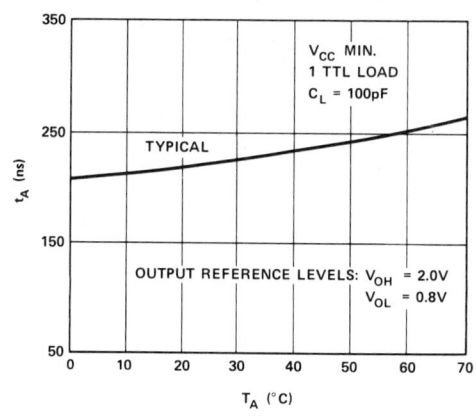

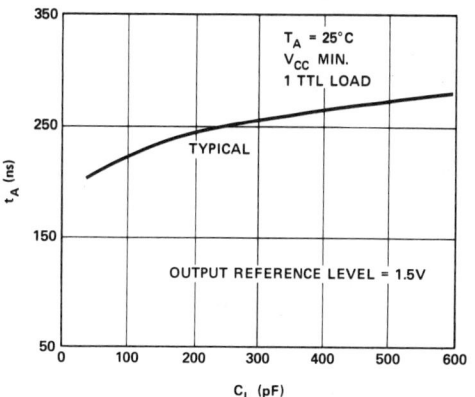

2111A
256 x 4 STATIC RAM

2111A-2	250 ns Max.
2111A	350 ns Max.
2111A-4	450 ns Max.

- **Common Data Input and Output**
- **Single +5V Supply Voltage**
- **Directly TTL Compatible: All Inputs and Output**
- **Static MOS: No Clocks or Refreshing Required**
- **Simple Memory Expansion: Chip Enable Input**
- **Fully Decoded: On Chip Address Decode**
- **Inputs Protected: All Inputs Have Protection Against Static Charge**
- **Low Cost Packaging: 18 Pin Plastic Dual In-Line Configuration**
- **Low Power: Typically 150 mW**
- **Three-State Output: OR-Tie Capability**

The Intel® 2111A is a 256 word by 4-bit static random access memory element using N-channel MOS devices integrated on a monolithic array. It uses fully DC stable (static) circuitry and therefore requires no clocks or refreshing to operate. The data is read out nondestructively and has the same polarity as the input data. Common input/output pins are provided.

The 2111A is designed for memory applications in small systems where high performance, low cost, large bit storage, and simple interfacing are important design objectives.

It is directly TTL compatible in all respects: inputs, outputs, and a single +5V supply. Separate chip enable ($\overline{CE}$) leads allow easy selection of an individual package when outputs are OR-tied.

The Intel® 2111A is fabricated with N-channel silicon gate technology. This technology allows the design and production of high performance, easy-to-use MOS circuits and provides a higher functional density on a monolithic chip than either conventional MOS technology or P-channel silicon gate technology.

Intel's silicon gate technology also provides excellent protection against contamination. This permits the use of low cost plastic packaging.

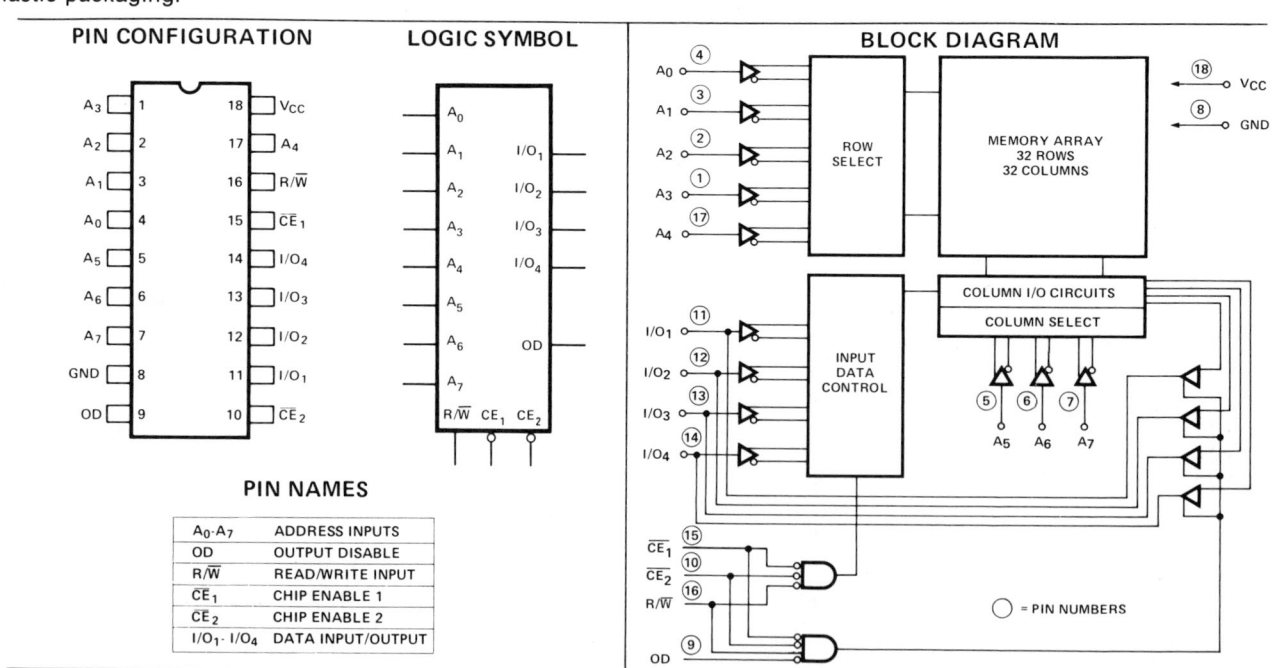

2111A

ABSOLUTE MAXIMUM RATINGS*

Ambient Temperature Under Bias . . . . . –10°C to 80°C
Storage Temperature . . . . –65°C to +150°C
Voltage On Any Pin
    With Respect to Ground . . . . –0.5V to +7V
Power Dissipation . . . . 1 Watt

*COMMENT:

Stresses above those listed under "Absolute Maximum Rating" may cause permanent damage to the device. This is a stress rating only and functional operation of the device at these or at any other condition above those indicated in the operational sections of this specification is not implied. Exposure to absolute maximum rating conditions for extended periods may affect device reliability.

D.C. AND OPERATING CHARACTERISTICS

T_A = 0°C to 70°C, V_{CC} = 5V ±5%, unless otherwise specified.

Symbol	Parameter		Min.	Typ.[1]	Max.	Unit	Test Conditions
I_{LI}	Input Load Current			1	10	µA	V_{IN} = 0 to 5.25V
I_{LOH}	I/O Leakage Current			1	10	µA	Output Disabled, $V_{I/O}$ = 4.0V
I_{LOL}	I/O Leakage Current			–1	–10	µA	Output Disabled, $V_{I/O}$ = 0.45V
I_{CC1}	Power Supply Current	2111A, 2111A-4		35	55	mA	V_{IN} = 5.25V
		2111A-2		45	65		$I_{I/O}$ = 0mA, T_A = 25°C
I_{CC2}	Power Supply Current	2111A, 2111A-4			60	mA	V_{IN} = 5.25V
		2111A-2			70		$I_{I/O}$ = 0mA, T_A = 0°C
V_{IL}	Input Low Voltage		–0.5		0.8	V	
V_{IH}	Input High Voltage		2.0		V_{CC}	V	
V_{OL}	Output Low Voltage				0.45	V	I_{OL} = 2.0mA
V_{OH}	Output High Voltage	2111A, 2111A-2	2.4			V	I_{OH} = –200µA
		2111A-4	2.4			V	I_{OH} = –150µA

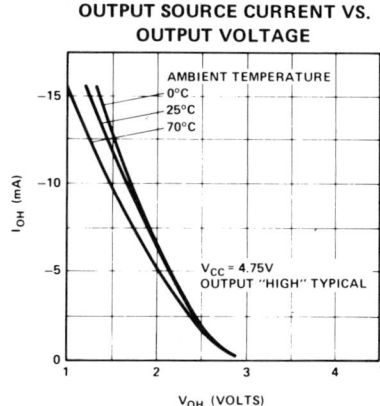

OUTPUT SOURCE CURRENT VS. OUTPUT VOLTAGE

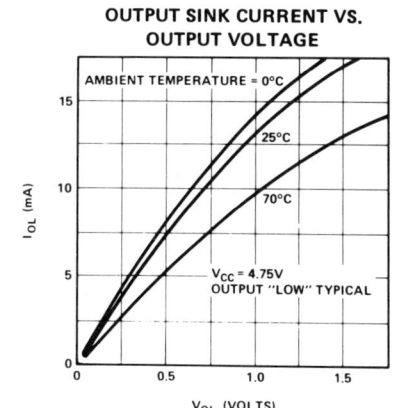

OUTPUT SINK CURRENT VS. OUTPUT VOLTAGE

NOTE: 1. Typical values are for T_A = 25°C and nominal supply voltage.

A.C. CHARACTERISTICS FOR 2111A-2 (250 ns ACCESS TIME)

READ CYCLE $T_A = 0°C$ to $70°C$, $V_{CC} = 5V \pm 5\%$, unless otherwise specified.

Symbol	Parameter	Min.	Typ.[1]	Max.	Unit	Test Conditions
t_{RC}	Read Cycle	250			ns	t_r, t_f = 20ns
t_A	Access Time			250	ns	Input Levels = 0.8V or 2.0V
t_{CO}	Chip Enable To Output			180	ns	Timing Reference = 1.5V
t_{OD}	Output Disable To Output			130	ns	Load = 1 TTL Gate
t_{DF} [3]	Data Output to High Z State	0		180	ns	and C_L = 100pF.
t_{OH}	Previous Read Data Valid after change of Address	40			ns	

WRITE CYCLE

Symbol	Parameter	Min.	Typ.[1]	Max.	Unit	Test Conditions
t_{WC}	Write Cycle	170			ns	
t_{AW}	Write Delay	20			ns	t_r, t_f = 20ns
t_{CW}	Chip Enable To Write	150			ns	Input Levels = 0.8V or 2.0V
t_{DW}	Data Setup	150			ns	Timing Reference = 1.5V
t_{DH}	Data Hold	0			ns	Load = 1 TTL Gate
t_{WP}	Write Pulse	150			ns	and C_L = 100pF.
t_{WR}	Write Recovery	0			ns	
t_{DS}	Output Disable Setup	20			ns	

CAPACITANCE[2] $T_A = 25°C$, $f = 1MHz$

Symbol	Test	Typ.[1] (pF)	Max. (pF)
C_{IN}	Input Capacitance (All Input Pins) $V_{IN} = 0V$	4	8
$C_{I/O}$	I/O Capacitance $V_{I/O} = 0V$	10	15

WAVEFORMS

READ CYCLE

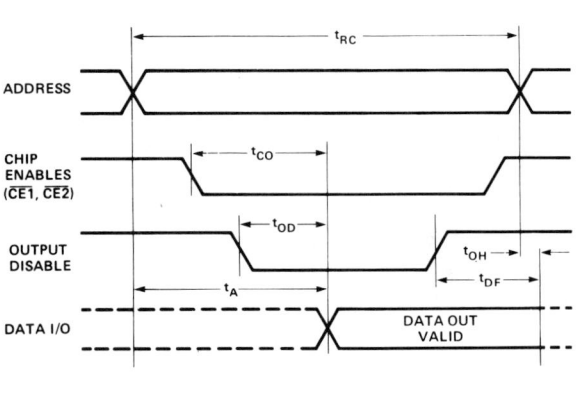

WRITE CYCLE

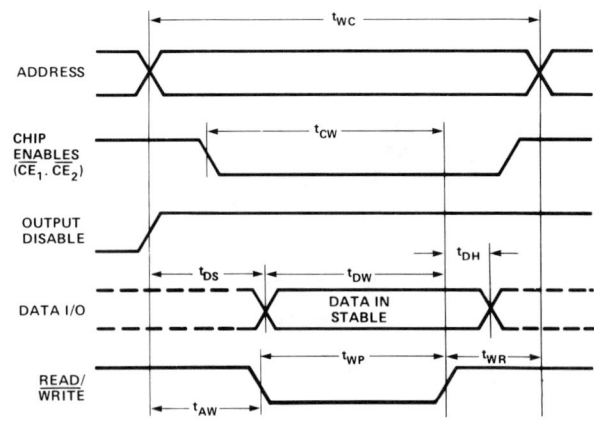

NOTES:
1. Typical values are for $T_A = 25°C$ and nominal supply voltage.
2. This parameter is periodically sampled and is not 100% tested.
3. t_{DF} is with respect to the trailing edge of $\overline{CE_1}$, $\overline{CE_2}$, or OD, whichever occurs first.

2111A (350 ns ACCESS TIME)
A.C. CHARACTERISTICS

READ CYCLE $T_A = 0°C$ to $70°C$, $V_{CC} = 5V \pm 5\%$, unless otherwise specified.

Symbol	Parameter	Min.	Typ.[1]	Max.	Unit	Test Conditions
t_{RC}	Read Cycle	350			ns	$t_r, t_f = 20ns$ Input Levels = 0.8V or 2.0V Timing Reference = 1.5V Load = 1 TTL Gate and $C_L = 100pF$.
t_A	Access Time			350	ns	
t_{CO}	Chip Enable To Output			240	ns	
t_{OD}	Output Disable To Output			180	ns	
t_{DF} [2]	Data Output to High Z State	0		150	ns	
t_{OH}	Previous Read Data Valid after change of Address	40			ns	

WRITE CYCLE

Symbol	Parameter	Min.	Typ.[1]	Max.	Unit	Test Conditions
t_{WC}	Write Cycle	220			ns	$t_r, t_f = 20ns$ Input Levels = 0.8V or 2.0V Timing Reference = 1.5V Load = 1 TTL Gate and $C_L = 100pF$.
t_{AW}	Write Delay	20			ns	
t_{CW}	Chip Enable To Write	200			ns	
t_{DW}	Data Setup	200			ns	
t_{DH}	Data Hold	0			ns	
t_{WP}	Write Pulse	200			ns	
t_{WR}	Write Recovery	0			ns	
t_{DS}	Output Disable Setup	20			ns	

2111A-4 (450 ns ACCESS TIME)
A.C. CHARACTERISTICS

READ CYCLE $T_A = 0°C$ to $70°C$, $V_{CC} = 5V \pm 5\%$, unless otherwise specified.

Symbol	Parameter	Min.	Typ.[1]	Max.	Unit	Test Conditions
t_{RC}	Read Cycle	450			ns	$t_r, t_f = 20ns$ Input Levels = 0.8V or 2.0V Timing Reference = 1.5V Load = 1 TTL Gate and $C_L = 100pF$.
t_A	Access Time			450	ns	
t_{CO}	Chip Enable To Output			310	ns	
t_{OD}	Output Disable To Output			250	ns	
t_{DF} [2]	Data Output to High Z State	0		200	ns	
t_{OH}	Previous Read Data Valid after change of Address	40			ns	

WRITE CYCLE

Symbol	Parameter	Min.	Typ.[1]	Max.	Unit	Test Conditions
t_{WC}	Write Cycle	270			ns	$t_r, t_f = 20ns$ Input Levels = 0.8V or 2.0V Timing Reference = 1.5V Load = 1 TTL Gate and $C_L = 100pF$.
t_{AW}	Write Delay	20			ns	
t_{CW}	Chip Enable To Write	250			ns	
t_{DW}	Data Setup	250			ns	
t_{DH}	Data Hold	0			ns	
t_{WP}	Write Pulse	250			ns	
t_{WR}	Write Recovery	0			ns	
t_{DS}	Output Disable Setup	20			ns	

NOTES: 1. Typical values are for $T_A = 25°C$ and nominal supply voltage.
2. t_{DF} is with respect to the trailing edge of $\overline{CE}_1$, $\overline{CE}_2$, or OD, whichever occurs first.

2112A
256 X 4 BIT STATIC RAM

2112A-2	250 ns Max.
2112A	350 ns Max.
2112A-4	450 ns Max.

- **Single +5V Supply Voltage**
- **Directly TTL Compatible: All Inputs and Outputs**
- **Static MOS: No Clocks or Refreshing Required**
- **Simple Memory Expansion: Chip Enable Input**
- **Fully Decoded: On Chip Address Decode**
- **Inputs Protected: All Inputs Have Protection Against Static Charge**
- **Low Cost Packaging: 16 Pin Plastic Dual In-Line Configuration**
- **Low Power: Typically 150 mW**
- **Three-State Output: OR-Tie Capability**

The Intel® 2112A is a 256 word by 4-bit static random access memory element using N-channel MOS devices integrated on a monolithic array. It uses fully DC stable (static) circuitry and therefore requires no clocks or refreshing to operate. The data is read out nondestructively and has the same polarity as the input data. Common input/output pins are provided.

The 2112A is designed for memory applications in small systems where high performance, low cost, large bit storage, and simple interfacing are important design objectives.

It is directly TTL compatible in all respects: inputs, outputs, and a single +5V supply. A separate chip enable ($\overline{CE}$) lead allows easy selection of an individual package when outputs are OR-tied.

The Intel® 2112A is fabricated with N-channel silicon gate technology. This technology allows the design and production of high performance, easy-to-use MOS circuits and provides a higher functional density on a monolithic chip than either conventional MOS technology or P-channel silicon gate technology.

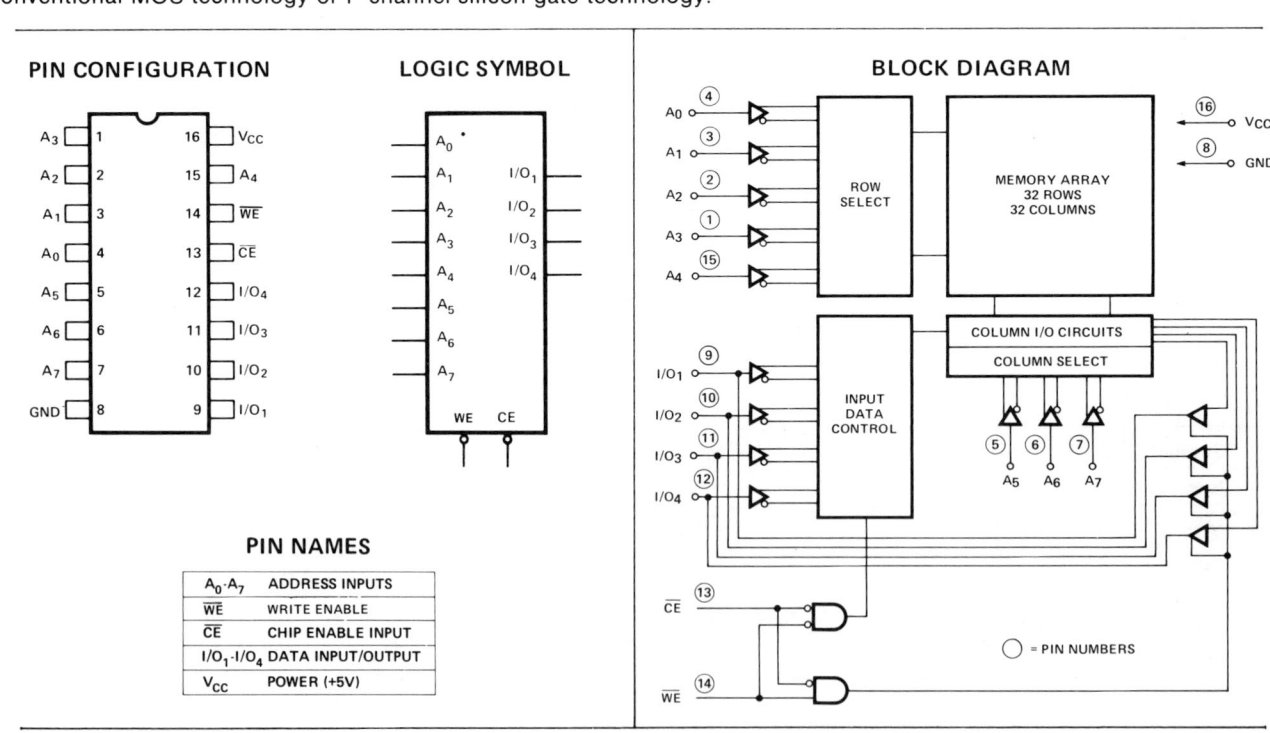

2112A

ABSOLUTE MAXIMUM RATINGS*

Ambient Temperature Under Bias . . . . . -10°C to 80°C
Storage Temperature . . . . -65°C to +150°C
Voltage On Any Pin
  With Respect to Ground . . . . -0.5V to +7V
Power Dissipation . . . . 1 Watt

*COMMENT:

Stresses above those listed under "Absolute Maximum Rating" may cause permanent damage to the device. This is a stress rating only and functional operation of the device at these or at any other condition above those indicated in the operational sections of this specification is not implied. Exposure to absolute maximum rating conditions for extended periods may affect device reliability.

D.C. AND OPERATING CHARACTERISTICS

T_A = 0°C to 70°C, V_{CC} = 5V ±5% unless otherwise specified.

Symbol	Parameter		Min.	Typ.[1]	Max.	Unit	Test Conditions
I_{LI}	Input Current			1	10	µA	V_{IN} = 0 to 5.25V
I_{LOH}	I/O Leakage Current			1	10	µA	Output Disabled, $V_{I/O}$ = 4.0V
I_{LOL}	I/O Leakage Current			-1	-10	µA	Output Disabled, $V_{I/O}$ = 0.45V
I_{CC1}	Power Supply Current	2112A, 2112A-4		35	55	mA	V_{IN} = 5.25V, $I_{I/O}$ = 0mA, T_A = 25°C
		2112A-2		45	65		
I_{CC2}	Power Supply Current	2112A, 2112A-4			60	mA	V_{IN} = 5.25V, $I_{I/O}$ = 0mA, T_A = 0°C
		2112A-2			70		
V_{IL}	Input "Low" Voltage		-0.5		0.8	V	
V_{IH}	Input "High" Voltage		2.0		V_{CC}	V	
V_{OL}	Output "Low" Voltage				+0.45	V	I_{OL} = 2.0 mA
V_{OH}	Output "High" Voltage	2112A, 2112A-2	2.4			V	I_{OH} = -200µA
		2112A-4	2.4			V	I_{OH} = -150µA

A.C. CHARACTERISTICS FOR 2112A-2

READ CYCLE T_A = 0°C to 70°C, V_{CC} = 5V ±5% unless otherwise specified.

Symbol	Parameter	Min.	Typ.[1]	Max.	Unit	Test Conditions
t_{RC}	Read Cycle	250			ns	t_r, t_f = 20ns
t_A	Access Time			250	ns	
t_{CO}	Chip Enable To Output Time			180	ns	Timing Reference = 1.5V
t_{CD}	Chip Enable To Output Disable Time	0		120	ns	Load = 1 TTL Gate
t_{OH}	Previous Read Data Valid After Change of Address	40			ns	and C_L = 100pF.

READ CYCLE WAVEFORMS

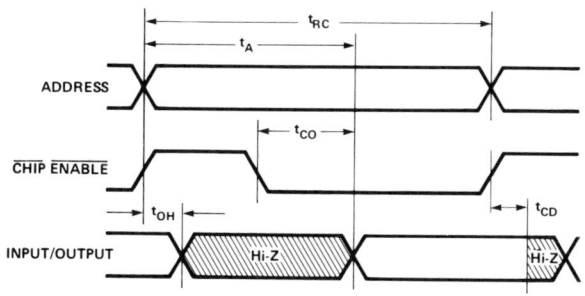

CAPACITANCE [2] T_A = 25°C, f = 1 MHz

Symbol	Test	Limits (pF)	
		Typ.[1]	Max.
C_{IN}	Input Capacitance (All Input Pins) V_{IN} = 0V	4	8
$C_{I/O}$	I/O Capacitance $V_{I/O}$ = 0V	10	15

NOTES:
1. Typical values are for T_A = 25°C and nominal supply voltage.
2. This parameter is periodically sampled and is not 100% tested.

A.C. CHARACTERISTICS FOR 2112A-2 (Continued)

WRITE CYCLE #1 $T_A = 0°C$ to $70°C$, $V_{CC} = 5V \pm 5\%$

Symbol	Parameter	Min.	Typ.[1]	Max.	Unit	Test Conditions
t_{WC1}	Write Cycle	200			ns	t_r, t_f = 20ns
t_{AW1}	Address To Write Setup Time	20			ns	Input Levels = 0.8V or 2.0V
t_{DW1}	Write Setup Time	180			ns	Timing Reference = 1.5V
t_{WP1}	Write Pulse Width	180			ns	Load = 1 TTL Gate
t_{CS1}	Chip Enable Setup Time	0			ns	and C_L = 100pF.
t_{CH1}	Chip Enable Hold Time	0			ns	
t_{WR1}	Write Recovery Time	0			ns	
t_{DH1}	Data Hold Time	0			ns	
t_{CW1}	Chip Enable To Write Setup Time	180			ns	

WRITE CYCLE #2 $T_A = 0°C$ to $70°C$, $V_{CC} = 5V \pm 5\%$

Symbol	Parameter	Min.	Typ.[1]	Max.	Unit	Test Conditions
t_{WC2}	Write Cycle	320			ns	t_r, t_f = 20ns
t_{AW2}	Address To Write Setup Time	20			ns	Input Levels = 0.8V or 2.0V
t_{DW2}	Write Setup Time	180			ns	Timing Reference = 1.5V
t_{WD2}	Write To Output Disable Time			120	ns	Load = 1 TTL Gate
t_{CS2}	Chip Enable Setup Time	0			ns	and C_L = 100pF.
t_{CH2}	Chip Enable Hold Time	0			ns	
t_{WR2}	Write Recovery Time	0			ns	
t_{DH2}	Data Hold Time	0			ns	

WRITE CYCLE WAVEFORMS

WRITE CYCLE #1

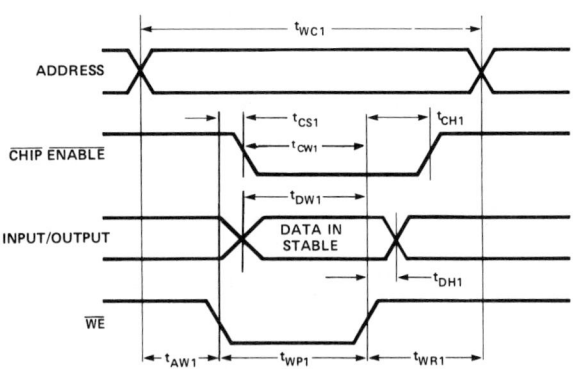

WRITE CYCLE #2

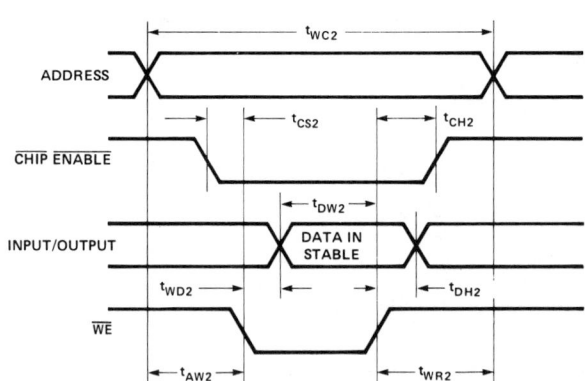

NOTE: 1. Typical values are for $T_A = 25°C$ and nominal supply voltage.

A.C. CHARACTERISTICS FOR 2112A

READ CYCLE $T_A = 0°C$ to $70°C$, $V_{CC} = 5V \pm 5\%$ unless otherwise specified.

Symbol	Parameter	Min.	Typ.[1]	Max.	Unit	Test Conditions
t_{RC}	Read Cycle	350			ns	t_r, t_f = 20ns
t_A	Access Time			350	ns	Input Levels = 0.8V or 2.0V
t_{CO}	Chip Enable To Output Time			240	ns	Timing Reference = 1.5V
t_{CD}	Chip Enable To Output Disable Time	0		200	ns	Load = 1 TTL Gate
t_{OH}	Previous Read Data Valid After Change of Address	40			ns	and C_L = 100pF.

WRITE CYCLE #1 $T_A = 0°C$ to $70°C$, $V_{CC} = 5V \pm 5\%$

Symbol	Parameter	Min.	Typ.[1]	Max.	Unit	Test Conditions
t_{WC1}	Write Cycle	270			ns	t_r, t_f = 20ns
t_{AW1}	Address To Write Setup Time	20			ns	Input Levels = 0.8V or 2.0V
t_{DW1}	Write Setup Time	250			ns	Timing Reference = 1.5V
t_{WP1}	Write Pulse Width	250			ns	Load = 1 TTL Gate
t_{CS1}	Chip Enable Setup Time	0			ns	and C_L = 100pF.
t_{CH1}	Chip Enable Hold Time	0			ns	
t_{WR1}	Write Recovery Time	0			ns	
t_{DH1}	Data Hold Time	0			ns	
t_{CW1}	Chip Enable to Write Setup Time	250			ns	

WRITE CYCLE #2 $T_A = 0°C$ to $70°C$, $V_{CC} = 5V \pm 5\%$

Symbol	Parameter	Min.	Typ.[1]	Max.	Unit	Test Conditions
t_{WC2}	Write Cycle	470			ns	t_r, t_f = 20ns
t_{AW2}	Address To Write Setup Time	20			ns	Input Levels = 0.8V or 2.0V
t_{DW2}	Write Setup Time	250			ns	Timing Reference = 1.5V
t_{WD2}	Write To Output Disable Time	200			ns	Load = 1 TTL Gate
t_{CS2}	Chip Enable Setup Time	0			ns	and C_L = 100pF.
t_{CH2}	Chip Enable Hold Time	0			ns	
t_{WR2}	Write Recovery Time	0			ns	
t_{DH2}	Data Hold Time	0			ns	

NOTE: 1. Typical values are for $T_A = 25°C$ and nominal supply voltage.

A.C. CHARACTERISTICS FOR 2112A-4

READ CYCLE $T_A = 0°C$ to $70°C$, $V_{CC} = 5V \pm 5\%$ unless otherwise specified.

Symbol	Parameter	Min.	Typ.[1]	Max.	Unit	Test Conditions
t_{RC}	Read Cycle	450			ns	$t_r, t_f = 20$ns
t_A	Access Time			450	ns	Input Levels = 0.8V or 2.0V
t_{CO}	Chip Enable To Output Time			310	ns	Timing Reference = 1.5V
t_{CD}	Chip Enable To Output Disable Time	0		260	ns	Load = 1 TTL Gate
t_{OH}	Previous Read Data Valid After Change of Address	40			ns	and $C_L = 100$pF.

WRITE CYCLE #1 $T_A = 0°C$ to $70°C$, $V_{CC} = 5V \pm 5\%$

Symbol	Parameter	Min.	Typ.[1]	Max.	Unit	Test Conditions
t_{WC1}	Write Cycle	320			ns	$t_r, t_f = 20$ns
t_{AW1}	Address To Write Setup Time	20			ns	Input Levels = 0.8V or 2.0V
t_{DW1}	Write Setup Time	300			ns	Timing Reference = 1.5V
t_{WP1}	Write Pulse Width	300			ns	Load = 1 TTL Gate
t_{CS1}	Chip Enable Setup Time	0			ns	and $C_L = 100$pF.
t_{CH1}	Chip Enable Hold Time	0			ns	
t_{WR1}	Write Recovery Time	0			ns	
t_{DH1}	Data Hold Time	0			ns	
t_{CW1}	Chip Enable to Write Setup Time	300			ns	

WRITE CYCLE #2 $T_A = 0°C$ to $70°C$, $V_{CC} = 5V \pm 5\%$

Symbol	Parameter	Min.	Typ.[1]	Max.	Unit	Test Conditions
t_{WC2}	Write Cycle	580			ns	$t_r, t_f = 20$ns
t_{AW2}	Address To Write Setup Time	20			ns	Input Levels = 0.8V or 2.0V
t_{DW2}	Write Setup Time	300			ns	Timing Reference = 1.5V
t_{WD2}	Write To Output Disable Time	260			ns	Load = 1 TTL Gate
t_{CS2}	Chip Enable Setup Time	0			ns	and $C_L = 100$pF.
t_{CH2}	Chip Enable Hold Time	0			ns	
t_{WR2}	Write Recovery Time	0			ns	
t_{DH2}	Data Hold Time	0			ns	

NOTE: 1. Typical values are for $T_A = 25°C$ and nominal supply voltage.

2114
1024 X 4 BIT STATIC RAM

	2114-2	2114-3	2114	2114L2	2114L3	2114L
Max. Access Time (ns)	200	300	450	200	300	450
Max. Power Dissipation (mw)	525	525	525	370	370	370

- **High Density 18 Pin Package**
- **Identical Cycle and Access Times**
- **Single +5V Supply**
- **No Clock or Timing Strobe Required**
- **Completely Static Memory**
- **Directly TTL Compatible: All Inputs and Outputs**
- **Common Data Input and Output Using Three-State Outputs**
- **Pin-Out Compatible with 3605 and 3625 Bipolar PROMs**

The Intel® 2114 is a 4096-bit static Random Access Memory organized as 1024 words by 4-bits using N-channel Silicon-Gate MOS technology. It uses fully DC stable (static) circuitry throughout — in both the array and the decoding — and therefore requires no clocks or refreshing to operate. Data access is particularly simple since address setup times are not required. The data is read out nondestructively and has the same polarity as the input data. Common input/output pins are provided.

The 2114 is designed for memory applications where high performance, low cost, large bit storage, and simple interfacing are important design objectives. The 2114 is placed in an 18-pin package for the highest possible density.

It is directly TTL compatible in all respects: inputs, outputs, and a single +5V supply. A separate Chip Select ($\overline{CS}$) lead allows easy selection of an individual package when outputs are or-tied.

The 2114 is fabricated with Intel's N-channel Silicon-Gate technology — a technology providing excellent protection against contamination permitting the use of low cost plastic packaging.

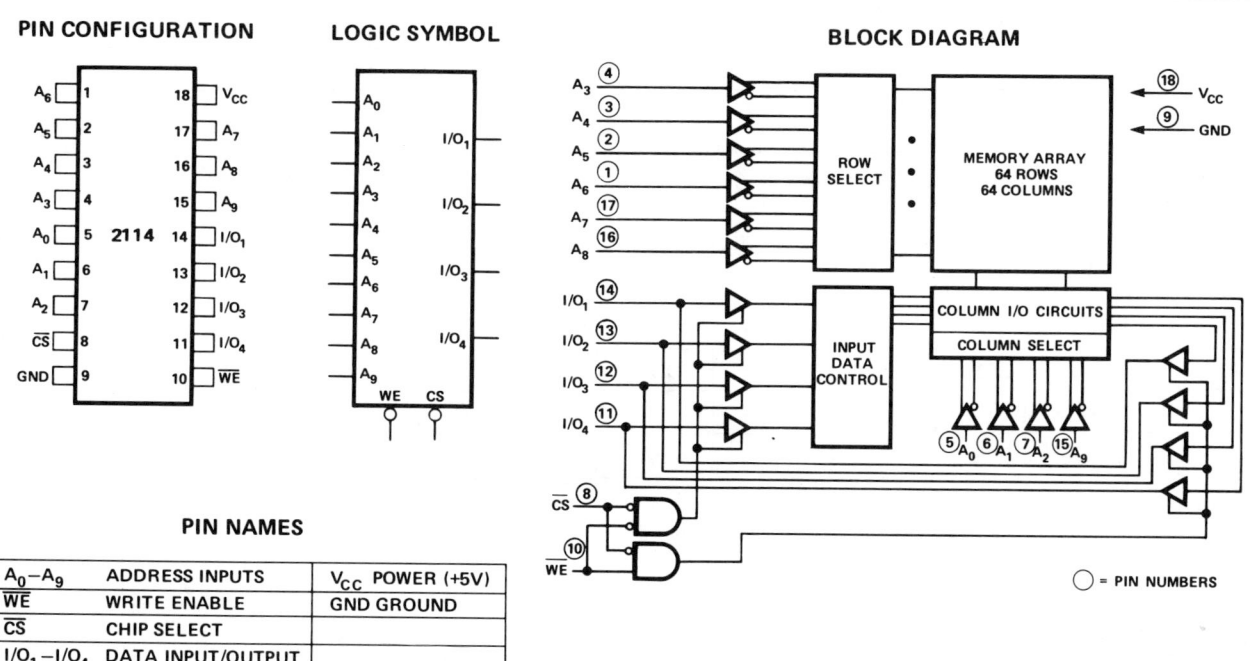

PIN CONFIGURATION

LOGIC SYMBOL

BLOCK DIAGRAM

PIN NAMES

A_0–A_9	ADDRESS INPUTS	V_{CC} POWER (+5V)
$\overline{WE}$	WRITE ENABLE	GND GROUND
$\overline{CS}$	CHIP SELECT	
I/O_1–I/O_4	DATA INPUT/OUTPUT	

ABSOLUTE MAXIMUM RATINGS*

Temperature Under Bias	–10°C to 80°C
Storage Temperature	–65°C to +150°C
Voltage on Any Pin With Respect to Ground	–0.5V to +7V
Power Dissipation	1.0W
D.C. Output Current	5mA

COMMENT: Stresses above those listed under "Absolute Maximum Ratings" may cause permanent damage to the device. This is a stress rating only and functional operation of the device at these or any other conditions above those indicated in the operational sections of this specification is not implied. Exposure to absolute maximum rating conditions for extended periods may affect device reliability.

D.C. AND OPERATING CHARACTERISTICS

$T_A = 0°C$ to $70°C$, $V_{CC} = 5V \pm 5\%$, unless otherwise noted.

SYMBOL	PARAMETER	2114-2, 2114-3, 2114			2114L2, 2114L3, 2114L			UNIT	CONDITIONS
		Min.	Typ.[1]	Max.	Min.	Typ.[1]	Max.		
I_{LI}	Input Load Current (All Input Pins)			10			10	µA	V_{IN} = 0 to 5.25V
$|I_{LO}|$	I/O Leakage Current			10			10	µA	$\overline{CS}$ = 2.4V, $V_{I/O}$ = 0.4V to V_{CC}
I_{CC1}	Power Supply Current		80	95			65	mA	V_{IN} = 5.25V, $I_{I/O}$ = 0 mA, $T_A = 25°C$
I_{CC2}	Power Supply Current			100			70	mA	V_{IN} = 5.25V, $I_{I/O}$ = 0 mA, $T_A = 0°C$
V_{IL}	Input Low Voltage	–0.5		0.8	–0.5		0.8	V	
V_{IH}	Input High Voltage	2.0		6.0	2.0		6.0	V	
I_{OL}	Output Low Current	2.1	6.0		2.1	6.0		mA	V_{OL} = 0.4V
I_{OH}	Output High Current		–1.4	–1.0		–1.4	–1.0	mA	V_{OH} = 2.4V
I_{OS}[2]	Output Short Circuit Current			40			40	mA	

NOTE: 1. Typical values are for $T_A = 25°C$ and $V_{CC} = 5.0V$.
2. Duration not to exceed 30 seconds.

CAPACITANCE

$T_A = 25°C$, $f = 1.0$ MHz

SYMBOL	TEST	MAX	UNIT	CONDITIONS
$C_{I/O}$	Input/Output Capacitance	5	pF	$V_{I/O}$ = 0V
C_{IN}	Input Capacitance	5	pF	V_{IN} = 0V

NOTE: This parameter is periodically sampled and not 100% tested.

A.C. CONDITIONS OF TEST

Input Pulse Levels	0.8 Volt to 2.4 Volt
Input Rise and Fall Times	10 nsec
Input and Output Timing Levels	1.5 Volts
Output Load	1 TTL Gate and C_L = 100 pF

A.C. CHARACTERISTICS
T_A = 0°C to 70°C, V_{CC} = 5V ± 5%, unless otherwise noted.

READ CYCLE [1]

SYMBOL	PARAMETER	2114-2, 2114L2 Min.	Max.	2114-3, 2114L3 Min.	Max.	2114, 2114L Min.	Max.	UNIT
t_{RC}	Read Cycle Time	200		300		450		ns
t_A	Access Time		200		300		450	ns
t_{CO}	Chip Selection to Output Valid		70		100		120	ns
t_{CX}	Chip Selection to Output Active	20		20		20		ns
t_{OTD}	Output 3-state from Deselection		60		80		100	ns
t_{OHA}	Output Hold from Address Change	50		50		50		ns

WRITE CYCLE [2]

SYMBOL	PARAMETER	2114-2, 2114L2 Min.	Max.	2114-3, 2114L3 Min.	Max.	2114, 2114L Min.	Max.	UNIT
t_{WC}	Write Cycle Time	200		300		450		ns
t_W	Write Time	120		150		200		ns
t_{WR}	Write Release Time	0		0		0		ns
t_{OTW}	Output 3-state from Write		60		80		100	ns
t_{DW}	Data to Write Time Overlap	120		150		200		ns
t_{DH}	Data Hold From Write Time	0		0		0		ns

NOTES:
1. A Read occurs during the overlap of a low $\overline{CS}$ and a high $\overline{WE}$.
2. A Write occurs during the overlap of a low $\overline{CS}$ and a low $\overline{WE}$.

WAVEFORMS

READ CYCLE [3]

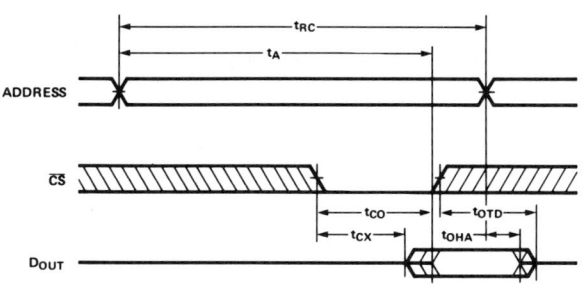

NOTES:
③ $\overline{WE}$ is high for a Read Cycle.
④ If the $\overline{CS}$ low transition occurs simultaneously with the $\overline{WE}$ low transition, the output buffers remain in a high impedance state.
⑤ $\overline{WE}$ must be high during all address transitions.

WRITE CYCLE

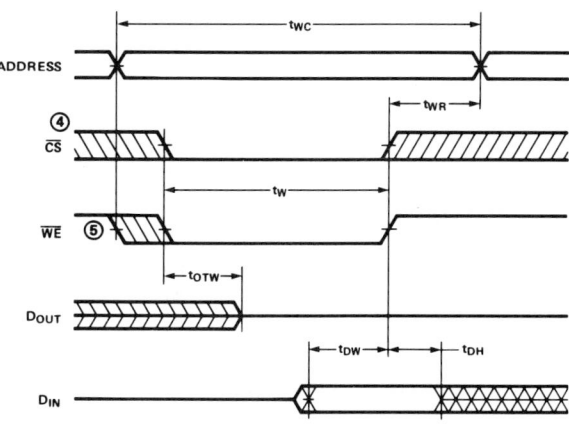

STATIC RAMS

TYPICAL D.C. AND A.C. CHARACTERISTICS

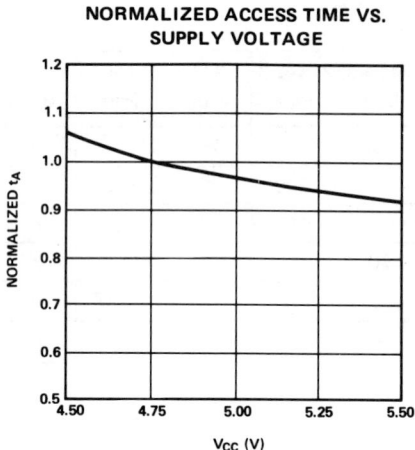

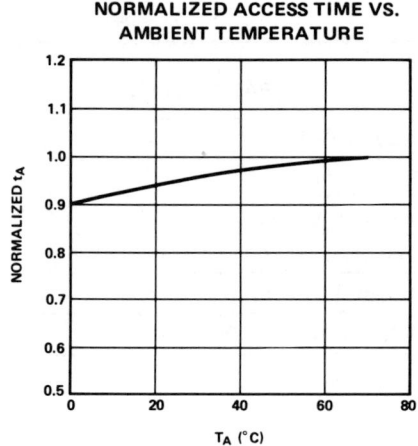

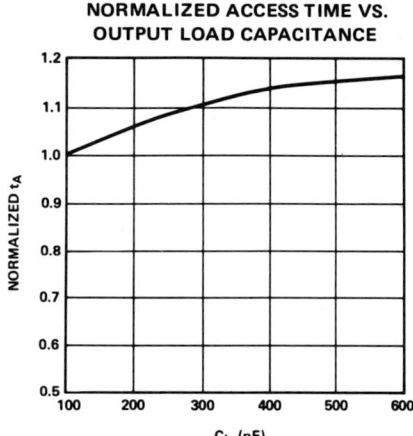

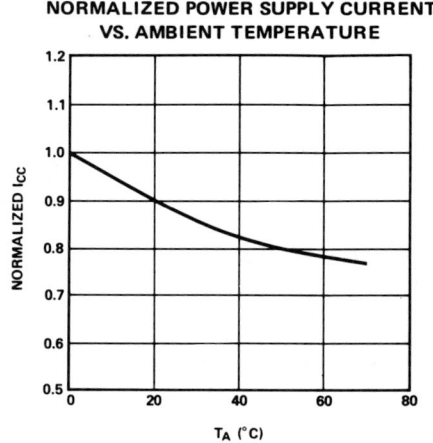

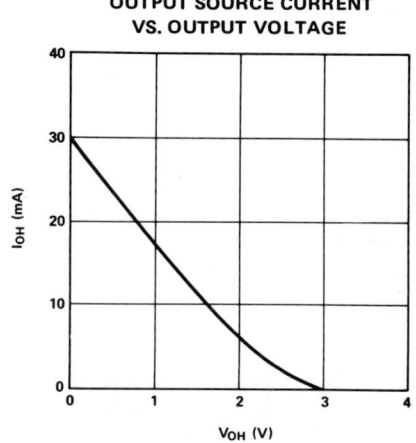

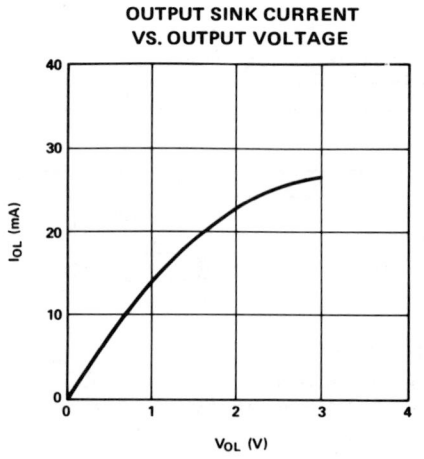

2115A, 2125A
HIGH SPEED 1K X 1 BIT STATIC RAM

	2115AL 2125AL	2115A 2125A	2115AL-2 2125AL-2	2115A-2 2125A-2
Max. T_{AA}(ns)	45	45	70	70
Max. I_{CC}(mA)	75	125	75	125

- Pin Compatible To 93415A (2115A) And 93425A (2125A)
- Fan-Out Of 10 TTL (2115A Family) -- 16mA Output Sink Current
- Low Operating Power Dissipation --Max. 0.39mW/Bit (2115AL, 2125AL)
- TTL Inputs And Outputs
- Single +5V Supply
- Uncommitted Collector (2115A) And Three-State (2125A) Output
- Standard 16-Pin Dual In-Line Package

The Intel® 2115A and 2125A families are high-speed, 1024 words by 1 bit random access memories. Both open collector (2115A) and three-state output (2125A) are available. The 2115A and 2125A use fully DC stable (static) circuitry throughout — in both the array and the decoding and, therefore, require no clocks or refreshing to operate. The data is read out non-destructively and has the same polarity as the input data.

The 2115AL/2125AL at 45 ns maximum access time and the 2115AL-2/2125AL-2 at 70 ns maximum access time are fully compatible with the industry-produced 1K bipolar RAMs, yet offer a 50% reduction in power of their bipolar equivalents. The power dissipation of the 2115AL/2125AL and 2115AL-2/2125AL-2 is 394 mW maximum as compared to 814 mW maximum of their bipolar equivalents. For systems already designed for 1K bipolar RAMs, the 2115A/2125A and the 2115A-2/2125A-2 at 45 ns and 70 ns maximum access times, respectively, offer complete compatibility with a 20% reduction in maximum power dissipation.

The devices are directly TTL compatible in all respects: inputs, outputs, and a single +5V supply. A separate select ($\overline{CS}$) lead allows easy selection of an individual package when outputs are OR-tied.

The 2115A and 2125A families are fabricated with Intel's N-channel MOS Silicon Gate Technology.

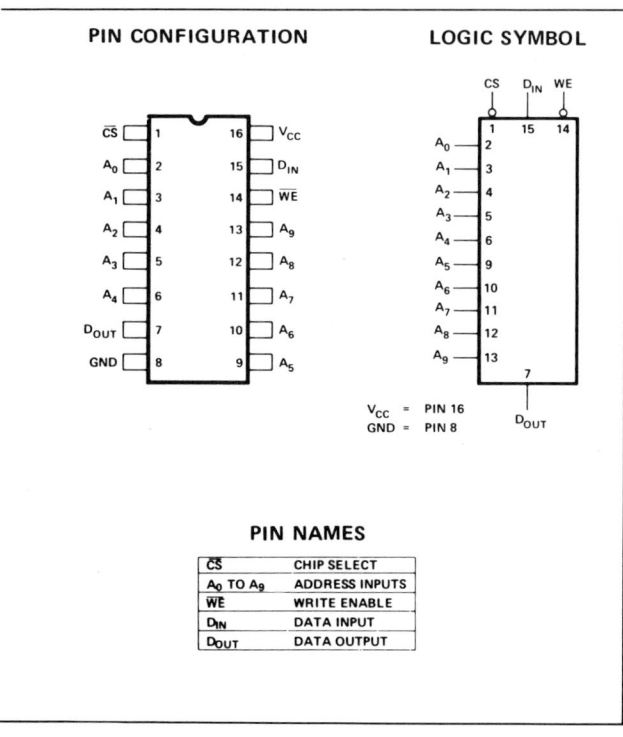

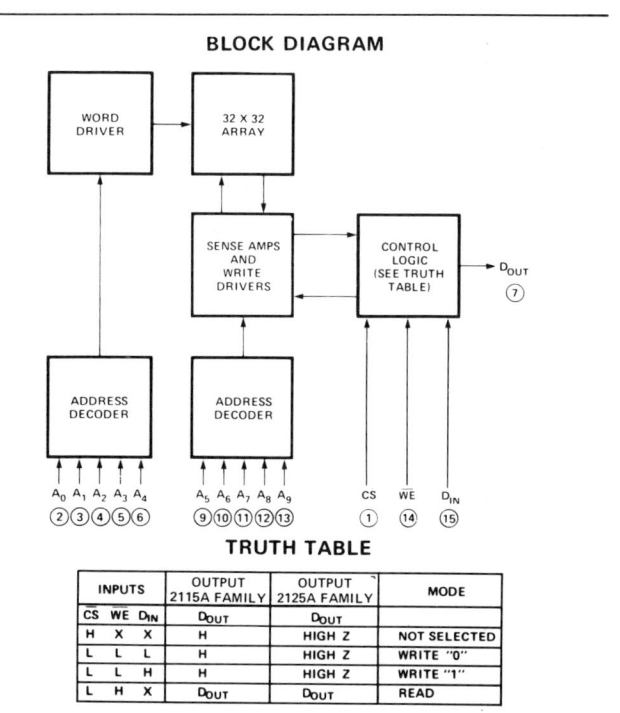

ABSOLUTE MAXIMUM RATINGS*

Temperature Under Bias	–10°C to +85°C
Storage Temperature	–65°C to +150°C
All Output or Supply Voltages	–0.5V to +7V
All Input Voltages	–0.5V to +5.5V
D.C. Output Current	20 mA

*COMMENT: Stresses above those listed under "Absolute Maximum Ratings" may cause permanent damage to the device. This is a stress rating only and functional operation of the device at these or at any other conditions above those indicated in the operational sections of this specification is not implied. Exposure to absolute maximum rating conditions for extended periods may affect device reliability.

D.C. CHARACTERISTICS[1,2]

V_{CC} = 5V ±5%, T_A = 0°C to 75°C

Symbol	Test	Min.	Typ.	Max.	Unit	Conditions
V_{OL1}	2115A Family Output Low Voltage			0.45	V	I_{OL} = 16 mA
V_{OL2}	2125A Family Output Lot Voltage			0.45	V	I_{OL} = 7 mA
V_{IH}	Input High Voltage	2.1			V	
V_{IL}	Input Low Voltage			0.8	V	
I_{IL}	Input Low Current		–0.1	–40	μA	V_{CC} = Max., V_{IN} = 0.4V
I_{IH}	Input High Current		0.1	40	μA	V_{CC} = Max., V_{IN} = 4.5V
I_{CEX}	2115A Family Output Leakage Current		0.1	100	μA	V_{CC} = Max., V_{OUT} = 4.5V
$|I_{OFF}|$	2125A Family Output Current (High Z)		0.1	50	μA	V_{CC} = Max., V_{OUT} = 0.5V/2.4V
I_{OS}[3]	2125A Family Current Short Circuit to Ground			–100	mA	V_{CC} = Max.
V_{OH}	Family Output High Voltage	2.4			V	I_{OH} = –3.2 mA
I_{CC}	Power Supply Current: I_{CC1}: 2115AL, 2115AL-2, 2125AL, 2125AL-2		60	75	mA	All Inputs Grounded, Output Open
	I_{CC2}: 2115A, 2115A-2, 2125A, 2125A-2		100	125	mA	

NOTES:

1. The operating ambient temperature ranges are guaranteed with transverse air flow exceeding 400 linear feet per minute and a two minute warm-up. Typical thermal resistance values of the package at maximum temperature are:

    θ_{JA} (@ 400 f$_{PM}$ air flow) = 45°C/W
    θ_{JA} (still air) = 60°C/W
    θ_{JC} = 25°C/W

2. Typical limits are at V_{CC} = 5V, T_A = +25°C, and maximum loading.
3. Duration of short circuit current should not exceed 1 second.

2115A FAMILY A.C. CHARACTERISTICS[1,2] V_{CC} = 5V ±5%, T_A = 0°C to 75°C

READ CYCLE

Symbol	Test	2115AL Limits Min. Typ. Max.	2115A Limits Min. Typ. Max.	2115AL-2 Limits Min. Typ. Max.	2115A-2 Limits Min. Typ. Max.	Units
t_{ACS}	Chip Select Time	5 15 30	5 15 30	5 15 30	5 15 40	ns
t_{RCS}	Chip Select Recovery Time	10 30	10 30	10 30	10 40	ns
t_{AA}	**Address Access Time**	30 45	30 45	40 70	40 70	**ns**
t_{OH}	Previous Read Data Valid After Change of Address	10	10	10	10	ns

WRITE CYCLE

Symbol	Test	Min. Typ. Max.	Min. Typ. Max.	Min. Typ. Max.	Min. Typ. Max.	Units
t_{WS}	Write Enable Time	10 25	10 30	10 25	10 40	
t_{WR}	Write Recovery Time	0 25	0 30	0 25	0 45	ns
t_W	Write Pulse Width	30 20	30 10	30 15	50 15	ns
t_{WSD}	Data Set-Up Time Prior to Write	0 −5	5 −5	0 −5	5 −5	ns
t_{WHD}	Data Hold Time After Write	5 0	5 0	5 0	5 0	ns
t_{WSA}	Address Set-Up Time	5 0	5 0	5 0	15 0	ns
t_{WHA}	Address Hold Time	5 0	5 0	5 0	5 0	ns
t_{WSCS}	Chip Select Set-Up Time	5 0	5 0	5 0	5 0	ns
t_{WHCS}	Chip Select Hold Time	5 0	5 0	5 0	5 0	ns

A.C. TEST CONDITIONS

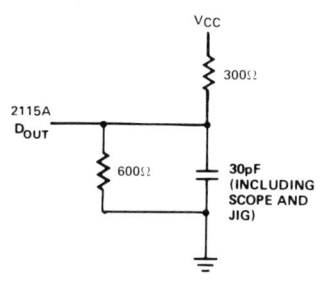
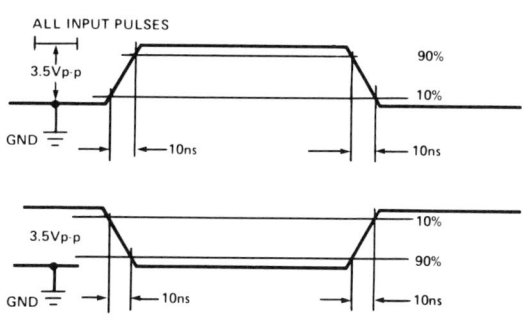

READ CYCLE

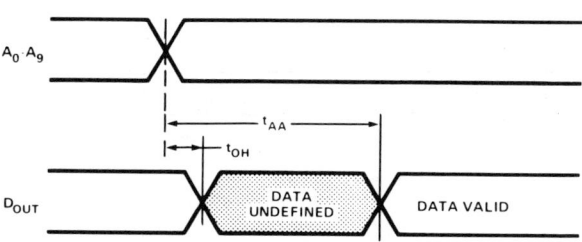

PROPAGATION DELAY FROM CHIP SELECT

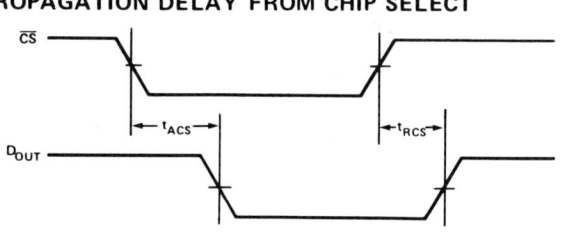

WRITE CYCLE

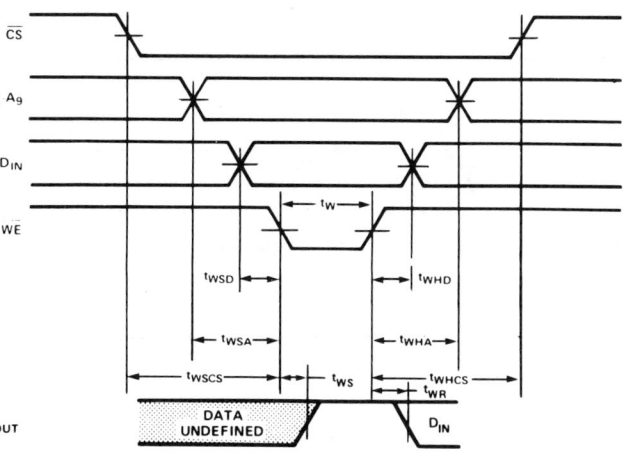

(ALL ABOVE MEASUREMENTS REFERENCED TO 1.5V)

2125 FAMILY A.C. CHARACTERISTICS[1,2] V_{CC} = 5V ±5%, T_A = 0°C to 75°C

READ CYCLE

Symbol	Test	2125AL Limits Min.	Typ.	Max.	2125A Limits Min.	Typ.	Max.	2125AL-2 Limits Min.	Typ.	Max.	2125A-2 Limits Min.	Typ.	Max.	Units
t_{ACS}	Chip Select Time	5	15	30	5	15	30	5	15	30	5	15	40	ns
t_{ZRCS}	Chip Select to HIGH Z		10	30		10	30		10	30		10	40	ns
t_{AA}	**Address Access Time**		30	45		30	45		40	70		40	70	ns
t_{OH}	Previous Read Data Valid After Change of Address	10			10			10			10			ns

WRITE CYCLE

Symbol	Test	Min.	Typ.	Max.	Min.	Typ.	Max.	Min.	Typ.	Max.	Min.	Typ.	Max.	Units
t_{ZWS}	Write Enable to HIGH Z		10	25		10	30		10	25		10	40	ns
t_{WR}	Write Recovery Time	0		25	0		30	0		25	0		45	ns
t_W	Write Pulse Width	30	20		30	10		30	10		50	15		ns
t_{WSD}	Data Set-Up Time Prior to Write	0	−5		5	−5		0	−5		5	−5		ns
t_{WHD}	Data Hold Time After Write	5	0		5	0		5	0		5	0		ns
t_{WSA}	Address Set-Up Time	5	0		5	0		5	0		15	0		ns
t_{WHA}	Address Hold Time	5	0		5	0		5	0		5	0		ns
t_{WSCS}	Chip Select Set-Up Time	5	0		5	0		5	0		5	0		ns
t_{WHCS}	Chip Select Hold Time	5	0		5	0		5	0		5	0		ns

A.C. TEST CONDITIONS

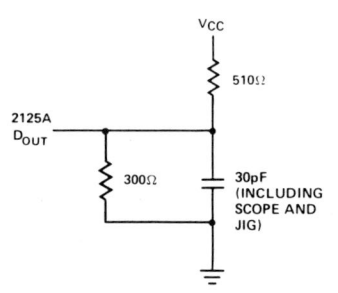
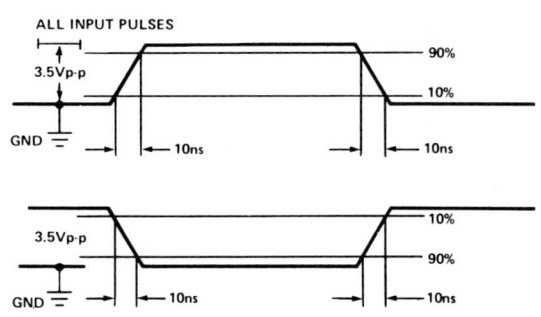

READ CYCLE

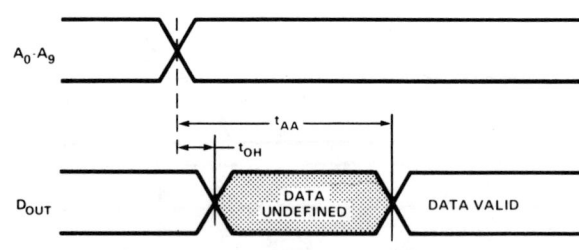

WRITE CYCLE

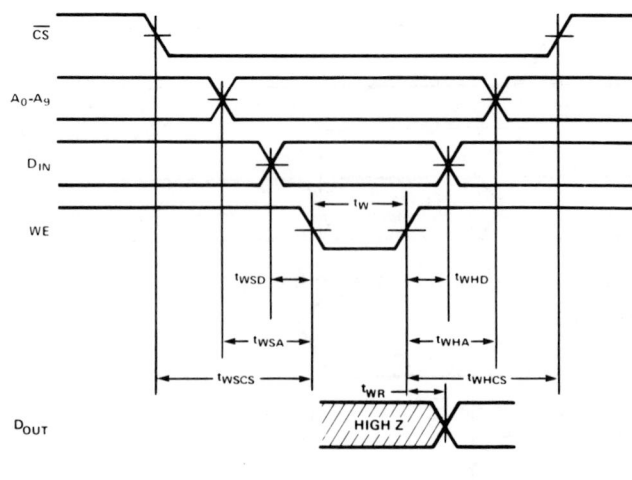

PROPAGATION DELAY FROM CHIP SELECT

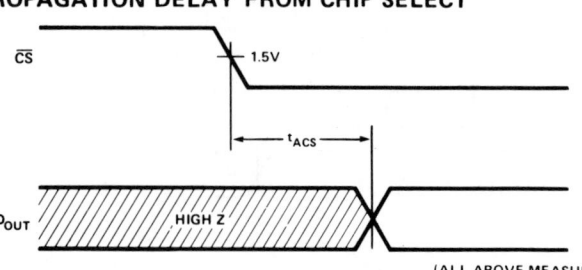

(ALL ABOVE MEASUREMENTS REFERENCED TO 1.5V)

2125A FAMILY WRITE ENABLE TO HIGH Z DELAY

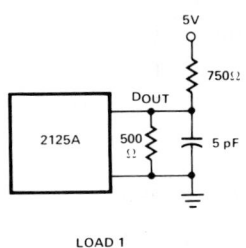

LOAD 1

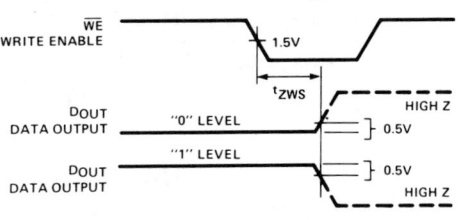

2125A FAMILY PROPAGATION DELAY FROM CHIP SELECT TO HIGH Z

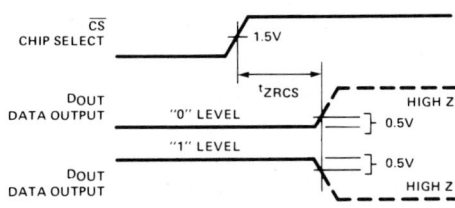

(ALL t_{ZXXX} PARAMETERS ARE MEASURED AT A DELTA OF 0.5V FROM THE LOGIC LEVEL AND USING LOAD 1.)

2115A/2125A FAMILY CAPACITANCE* V_{CC}= 5V, f = 1 MHz, T_A = 25°C

SYMBOL	TEST	2115A Family LIMITS		2125A Family LIMITS		UNITS	TEST CONDITIONS
		TYP.	MAX.	TYP.	MAX.		
C_I	Input Capacitance	3	5	3	5	pF	All Inputs = 0V, Output Open
C_O	Output Capacitance	5	8	5	8	pF	$\overline{CS}$ = 5V, All Other Inputs = 0V, Output Open

*This parameter is periodically sampled and is not 100% tested.

TYPICAL CHARACTERISTICS

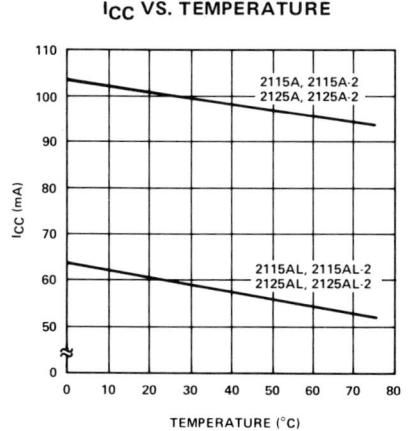

I_{CC} VS. TEMPERATURE

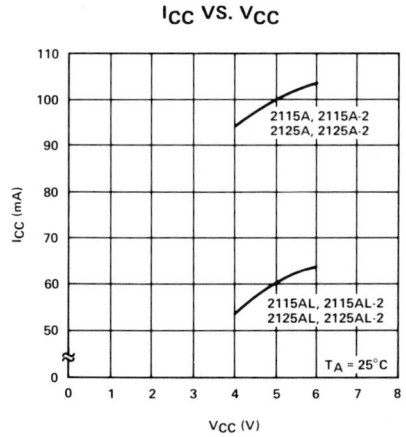

I_{CC} VS. V_{CC}

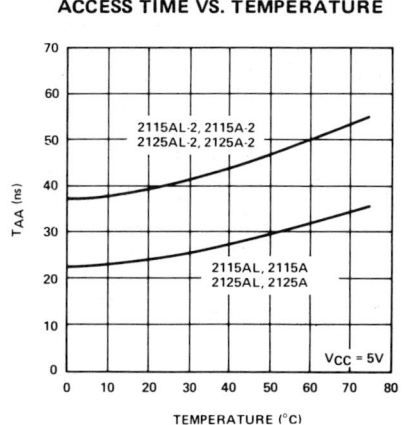

ACCESS TIME VS. TEMPERATURE

2141
4096 X 1 BIT STATIC RAM

	2141	2141-2
Max. Access Time (ns)	150	200
Max. Active Current (mA)	100	60
Max. Standby Current (mA)	12	8

- **Completely Static Memory — No Clock or Timing Strobe Required**
- **Equal Access and Cycle Times**
- **Single +5V Supply**
- **High Density 18-Pin Package**
- **Automatic Power-Down**
- **Directly TTL Compatible — All Inputs and Outputs**
- **Separate Data Input and Output**
- **Three-State Output**

The Intel® 2141 is a 4096-bit static Random Access Memory organized as 4096 words by 1-bit using N-channel Silicon-Gate MOS technology. It uses a uniquely innovative design approach which provides the ease-of-use features associated with non-clocked static memories and the reduced standby power dissipation associated with clocked static memories. To the user this means low standby power dissipation without the need for clocks, address setup and hold times, nor reduced data rates due to cycle times that are longer than access times.

$\overline{CS}$ controls the power-down feature. In less than a cycle time after $\overline{CS}$ goes high — deselecting the 2141 — the part automatically reduces its power requirements and remains in this low power standby mode as long as $\overline{CS}$ remains high. This device feature results in system power savings as great as 85% in larger systems, where the majority of devices are deselected. The automatic power-down feature causes no performance degradation as chip select access and address access are equal.

The 2141 is placed in an 18-pin package configured with the industry standard pinout. It is directly TTL compatible in all respects: inputs, outputs, and a single +5V supply. The data is read out nondestructively and has the same polarity as the input data. A data input and a separate three-state output are used.

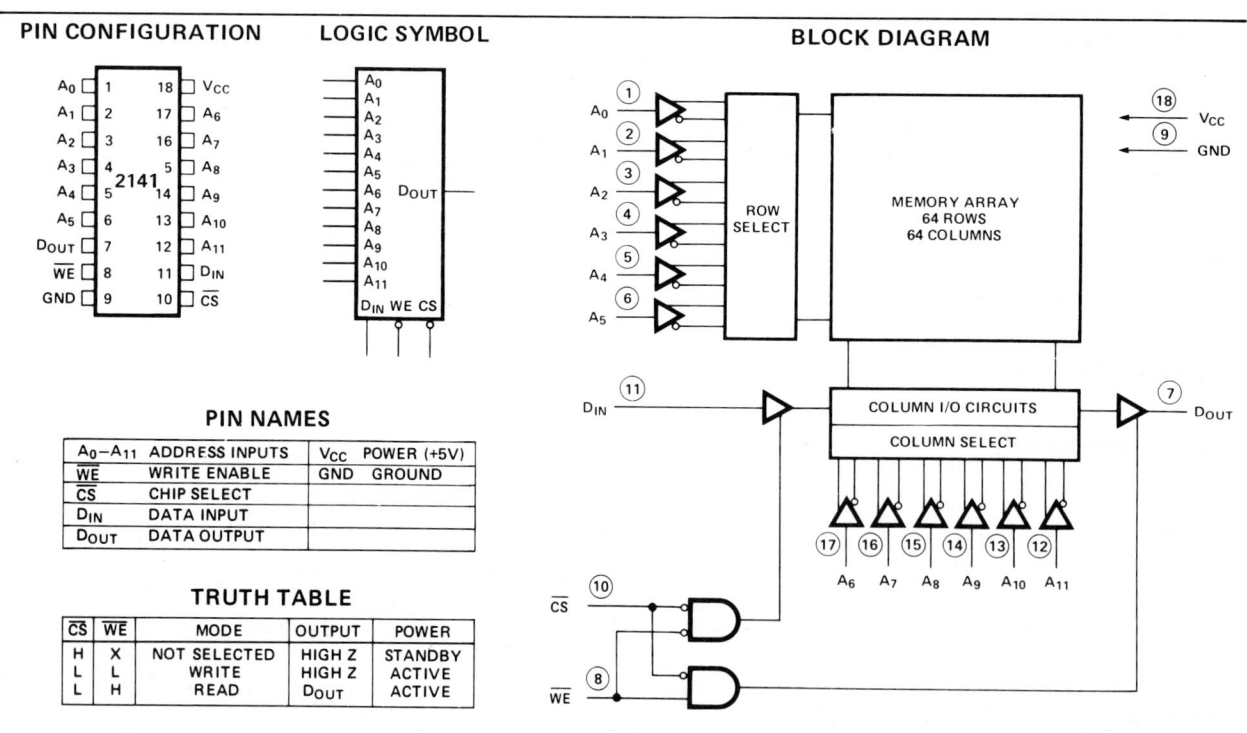

70

ABSOLUTE MAXIMUM RATINGS*

Temperature Under Bias .... -10°C to 85°C
Storage Temperature .... -65°C to +150°C
Voltage on Any Pin With
  Respect to Ground .... -0.5V to +7V
D.C. Output Current .... 20 mA

*COMMENT: Stresses above those listed under "Absolute Maximum Ratings" may cause permanent damage to the device. This is a stress rating only and functional operation of the device at these or any other conditions above those indicated in the operational sections of this specification is not implied. Exposure to absolute maximum rating conditions for extended periods may affect device reliability.

D.C. AND OPERATING CHARACTERISTICS

T_A = 0°C to 70°C, V_{CC} = +5V ±5%, unless otherwise noted.[2]

Symbol	Parameter	2141 Min.	2141 Typ.[1]	2141 Max.	2141-2 Min.	2141-2 Typ.[1]	2141-2 Max.	Unit	Conditions
I_{LI}	Input Load Current (All Input Pins)		0.01	10		0.01	10	μA	V_{CC}=Max, V_{IN}=GND to V_{CC}
$|I_{LO}|$	Output Leakage Current		0.1	50		0.1	50	μA	$\overline{CS}$=2.0V, V_{CC}=Max, V_{OUT}=GND to 4.5V
I_{CC}	Operating Current		65	90		40	55	mA	T_A=25°C — V_{CC}=Max., $\overline{CS}$=V_{IL}, Outputs Open
				100			60	mA	T_A=0°C
I_{SB}	Standby Current			12			8	mA	V_{CC}=Max, $\overline{CS}$=V_{IH} T_A=0°C to 70°C
V_{IL}	Input Low Voltage	-0.3		0.8	-0.3		0.8	V	
V_{IH}	Input High Voltage	2.0		6.0	2.0		6.0	V	
V_{OL}	Output Low Voltage			0.40			0.40	V	I_{OL} = 4.4 mA
V_{OH}	Output High Voltage	2.4			2.4			V	I_{OH} = 1.0 mA

Notes: 1. Typical limits are at V_{CC} = 5V, T_A = +25°C, and specified loading.
2. The operating ambient temperature ranges are guaranteed with transverse airflow exceeding 400 linear feet per minute and a two minute warm-up.

A.C. TEST CONDITIONS

Input Pulse Levels	GND to 3.5 Volts
Input Rise and Fall Times	10 nsec
Input and Output Timing Reference Levels	1.5 Volts
Output Load	1 TTL Load plus 100pF

CAPACITANCE[3]

T_A = 25°C, f = 1.0MHz

Symbol	Parameter	Max.	Unit	Conditions
C_{IN}	Input Capacitance	5	pF	V_{IN} = 0V
C_{OUT}	Output Capacitance	10	pF	V_{OUT} = 0V

Note 3. This parameter is sampled and not 100% tested.

A.C. CHARACTERISTICS

$T_A = 0°C$ to $70°C$, $V_{CC} = +5V \pm 5\%$, unless otherwise noted.

READ CYCLE

Symbol	Parameter	2141 Min.	2141 Max.	2141-2 Min.	2141-2 Max.	Unit
t_{RC}	Read Cycle Time	150		200		ns
t_{AA}	Address Access Time		150		200	ns
t_{ACS}	Chip Select Access Time		150		200	ns
t_{OH}	Output Hold from Address Change	10		10		ns
t_{LZ}	Chip Selection to Output in Low Z	10		10		ns
t_{HZ}	Chip Deselection to Output in High Z	0	60	0	60	ns
t_{PU}	Chip Selection to Power Up Time	0		0		ns
t_{PD}	Chip Deselection to Power Down Time		80		100	ns

WRITE CYCLE

Symbol	Parameter	2141 Min.	2141 Max.	2141-2 Min.	2141-2 Max.	Unit
t_{WC}	Write Cycle Time	150		180		ns
t_{CW}	Chip Selection to End of Write	90		120		ns
t_{AW}	Address Valid to End of Write	90		120		ns
t_{AS}	Address Setup Time	10		10		ns
t_{WP}	Write Pulse Width	70		100		ns
t_{WR}	Write Recovery Time	15		20		ns
t_{DW}	Data Valid to End of Write	60		90		ns
t_{DH}	Data Hold Time	15		15		ns
t_{WZ}	Write Enabled to Output in High Z	0	50	0	60	ns
t_{OW}	Output Active from End of Write	0		0		ns

WAVEFORMS

READ CYCLE NO. 1[1]

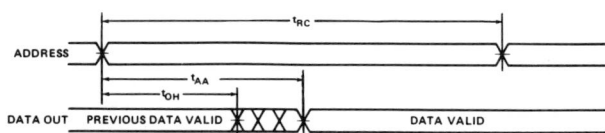

READ CYCLE NO. 2[1,2]

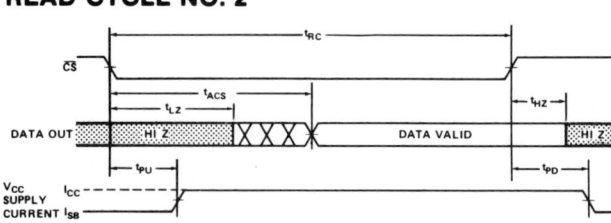

WRITE CYCLE

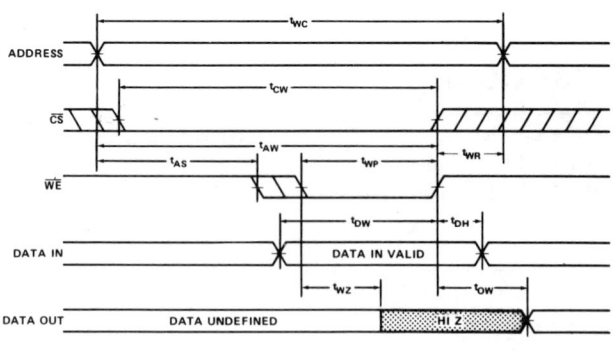

Notes: 1. $\overline{WE}$ is high for Read Cycles.
2. Addresses valid prior to or coincident with $\overline{CS}$ transition low.

DEVICE DESCRIPTION

The 2141 is produced with a new, high-performance MOS technology which incorporates on-chip substrate bias generation combined with device scaling to achieve a low speed power product. The speed-power product of this process has been measured at 1_{pj}, approximately four times better than previous MOS processes.

This process, combined with new design ideas, gives the 2141 its unique features. High speed, low power and ease-of-use have been obtained in a single part. The low-power standby feature is controlled with the Chip Select input. $\overline{CS}$ is not a clock and does not have to be cycled. This allows the user to tie $\overline{CS}$ directly to system addresses and use the line as part of the normal decoding logic. Whenever the 2141 is deselected, it automatically reduces its power requirements to a fraction of the active power, as shown in Figure 1. This is done without any need to lower the power supply voltage to the device.

The automatic power-down feature adds up to significant system power savings. Unselected devices draw low standby power. Only the active devices draw high active power and then only during the portion of the system cycle time that the devices are selected. On an average basis, the power consumed by a device actually declines as the system size increases, asymptotically approaching the standby power level as shown in Figure 2. Very large memories can be designed with the 2141 and still use economical power supplies.

The automatic power-down feature is obtained without any performance degradation, since chip select access is as fast or faster than address access. Performance is further enhanced by the fully static design of the 2141. Access time equals cycle time and multiple read or write operations are possible during a single select period.

The power switching characteristic of the 2141 requires more careful decoupling than would be required of a constant power device. It is recommended that a $0.1\mu F$ to $0.3\mu F$ ceramic capacitor be used on every other device, with a $11\mu F$ to $24\mu F$ bulk electrolytic decoupler every 16 devices. The actual values to be used will depend on board layout, trace widths and duty cycle. Power supply gridding is recommended for PC board layout. A very satisfactory grid can be developed on a two-layer board with vertical traces on one side and horizontal traces on the other, as shown in Figure 3.

Terminations are recommended on input signal lines to the 2141 devices. In high speed systems, fast drivers can cause significant reflections when driving the high impedance inputs of the 2141. Terminations may be required to match the impedance of the line to the driver. The type of termination used depends on designer preference and may be parallel resistive or resistive-capacitive. The latter reduces terminator power dissipation.

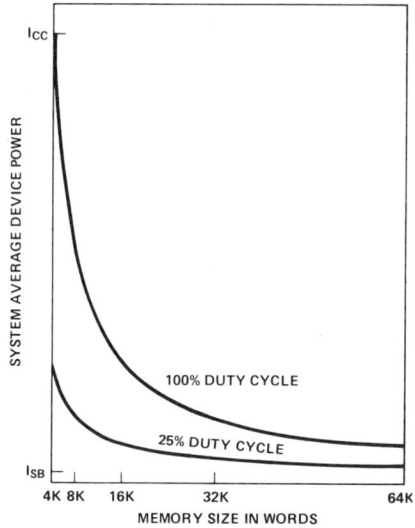

Figure 2. Average Device Dissipation vs. Memory Size.

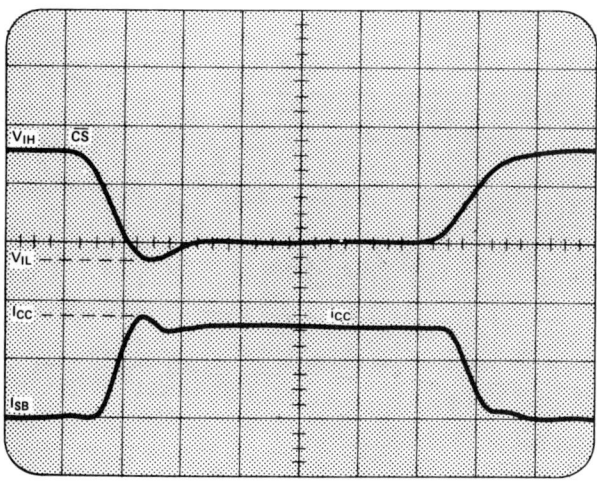

Figure 1. I_{CC} Waveform.

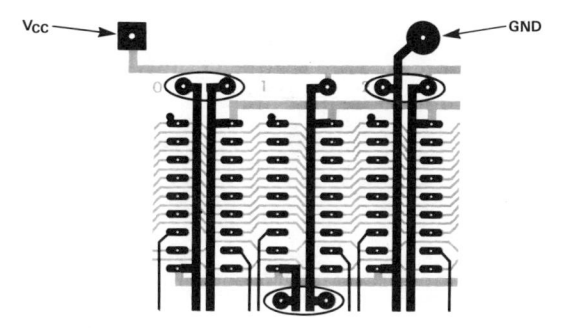

Figure 3. PC Layout.

2142
1024 X 4 BIT STATIC RAM

	2142-2	2142-3	2142	2142L2	2142L3	2142L
Max. Access Time (ns)	200	300	450	200	300	450
Max. Power Dissipation (mw)	525	525	525	370	370	370

- **High Density 20 Pin Package**
- **Access Time Selections From 200-450ns**
- **Identical Cycle and Access Times**
- **Low Operating Power Dissipation .1mW/Bit Typical**
- **Single +5V Supply**
- **No Clock or Timing Strobe Required**
- **Completely Static Memory**
- **Directly TTL Compatible: All Inputs and Outputs**
- **Common Data Input and Output Using Three-State Outputs**

The Intel® 2142 is a 4096-bit static Random Access Memory organized as 1024 words by 4-bits using N-channel Silicon-Gate MOS technology. It uses fully DC stable (static) circuitry throughout — in both the array and the decoding — and therefore requires no clocks or refreshing to operate. Data access is particularly simple since address setup times are not required. The data is read out nondestructively and has the same polarity as the input data. Common input/output pins are provided.

The 2142 is designed for memory applications where high performance, low cost, large bit storage, and simple interfacing are important design objectives. It is directly TTL compatible in all respects: inputs, outputs, and a single +5V supply.

The 2142 is placed in a 20-pin package. Two Chip Selects ($\overline{CS}_1$ and CS_2) are provided for easy and flexible selection of individual packages when outputs are OR-tied. An Output Disable is included for direct control of the output buffers.

The 2142 is fabricated with Intel's N-channel Silicon-Gate technology — a technology providing excellent protection against contamination permitting the use of low cost plastic packaging.

PIN CONFIGURATION   LOGIC SYMBOL   BLOCK DIAGRAM

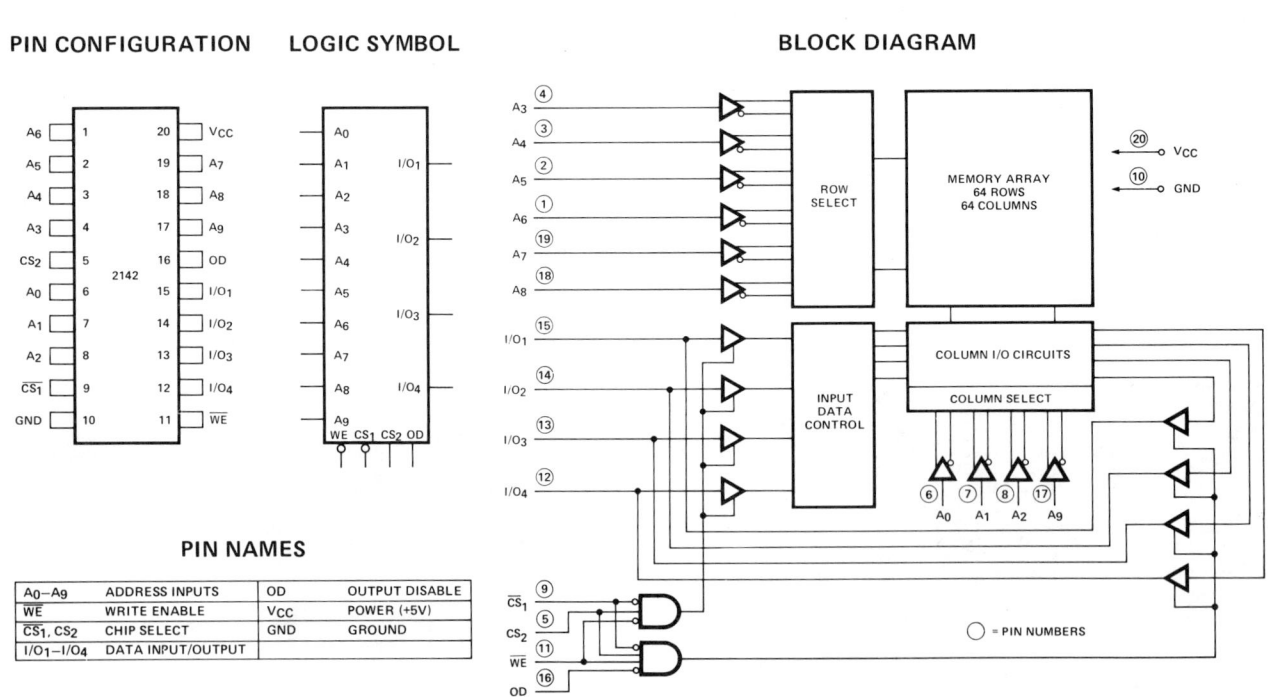

PIN NAMES

A_0–A_9	ADDRESS INPUTS	OD	OUTPUT DISABLE
$\overline{WE}$	WRITE ENABLE	V_{CC}	POWER (+5V)
$\overline{CS}_1$, CS_2	CHIP SELECT	GND	GROUND
I/O_1–I/O_4	DATA INPUT/OUTPUT		

ABSOLUTE MAXIMUM RATINGS*

Temperature Under Bias .... −10°C to 80°C
Storage Temperature .... −65°C to +150°C
Voltage on Any Pin
   With Respect to Ground .... −0.5V to +7V
Power Dissipation 1.0W
D.C. Output Current 10mA

*COMMENT: Stresses above those listed under "Absolute Maximum Ratings" may cause permanent damage to the device. This is a stress rating only and functional operation of the device at these or any other conditions above those indicated in the operational sections of this specification is not implied. Exposure to absolute maximum rating conditions for extended periods may affect device reliability.

D.C. AND OPERATING CHARACTERISTICS

$T_A = 0°C$ to $70°C$, $V_{CC} = 5V \pm 5\%$, unless otherwise noted.

SYMBOL	PARAMETER	2142-2, 2142-3, 2142 Min.	Typ.[1]	Max.	2142L2, 2142L3, 2142L Min.	Typ.[1]	Max.	UNIT	CONDITIONS
I_{LI}	Input Load Current (All Input Pins)			10			10	µA	V_{IN} = 0 to 5.25V
$|I_{LO}|$	I/O Leakage Current			10			10	µA	$\overline{CS}$ = 2.4V, $V_{I/O}$ = 0.4V to V_{CC}
I_{CC1}	Power Supply Current		80	95			65	mA	V_{IN} = 5.25V, $I_{I/O}$ = 0 mA, $T_A = 25°C$
I_{CC2}	Power Supply Current			100			70	mA	V_{IN} = 5.25V, $I_{I/O}$ = 0 mA, $T_A = 0°C$
V_{IL}	Input Low Voltage	−0.5		0.8	−0.5		0.8	V	
V_{IH}	Input High Voltage	2.0		6.0	2.0		6.0	V	
I_{OL}	Output Low Current	2.1		6.0	2.1		6.0	mA	V_{OL} = 0.4V
I_{OH}	Output High Current		−1.4	−1.0		−1.4	−1.0	mA	V_{OH} = 2.4V
I_{OS}[2]	Output Short Circuit Current			40			40	mA	$V_{I/O}$ = GND to V_{CC}

NOTE: 1. Typical values are for $T_A = 25°C$ and $V_{CC} = 5.0V$.
       2. Duration not to exceed 30 seconds.

CAPACITANCE

$T_A = 25°C$, f = 1.0 MHz

SYMBOL	TEST	MAX	UNIT	CONDITIONS
$C_{I/O}$	Input/Output Capacitance	5	pF	$V_{I/O}$ = 0V
C_{IN}	Input Capacitance	5	pF	V_{IN} = 0V

NOTE: This parameter is periodically sampled and not 100% tested.

A.C. CONDITIONS OF TEST

Input Pulse Levels .... 0.8 Volt to 2.4 Volt
Input Rise and Fall Times .... 10 nsec
Input and Output Timing Levels .... 1.5 Volts
Output Load .... 1 TTL Gate and C_L = 100 pF

STATIC RAMS

A.C. CHARACTERISTICS
$T_A = 0°C$ to $70°C$, $V_{CC} = 5V \pm 5\%$, unless otherwise noted.

READ CYCLE [1]

SYMBOL	PARAMETER	2142-2, 2142L2 Min.	2142-2, 2142L2 Max.	2142-3, 2142L3 Min.	2142-3, 2142L3 Max.	2142, 2142L Min.	2142, 2142L Max.	UNIT
t_{RC}	Read Cycle Time	200		300		450		ns
t_A	Access Time		200		300		450	ns
t_{OD}	Output Enable to Output Valid		70		100		120	ns
t_{ODX}	Output Enable to Output Active	20		20		20		ns
t_{CO}	Chip Selection to Output Valid		70		100		120	ns
t_{CX}	Chip Selection to Output Active	20		20		20		ns
t_{OTD}	Output 3-state from Disable		60		80		100	ns
t_{OHA}	Output Hold from Address Change	50		50		50		ns

WRITE CYCLE [2]

SYMBOL	PARAMETER	2142-2, 2142L2 Min.	2142-2, 2142L2 Max	2142-3, 2142L3 Min.	2142-3, 2142L3 Max.	2142, 2142L Min.	2142, 2142L Max.	UNIT
t_{WC}	Write Cycle Time	200		300		450		ns
t_W	Write Time	120		150		200		ns
t_{WR}	Write Release Time	0		0		0		ns
t_{OTD}	Output 3-state from Disable		60		80		100	ns
t_{DW}	Data to Write Time Overlap	120		150		200		ns
t_{DH}	Data Hold From Write Time	0		0		0		ns

NOTES:
1. A Read occurs during the overlap of a low $\overline{CS}$ and a high $\overline{WE}$.
2. A Write occurs during the overlap of a low $\overline{CS}$ and a low $\overline{WE}$.

WAVEFORMS

READ CYCLE [3]

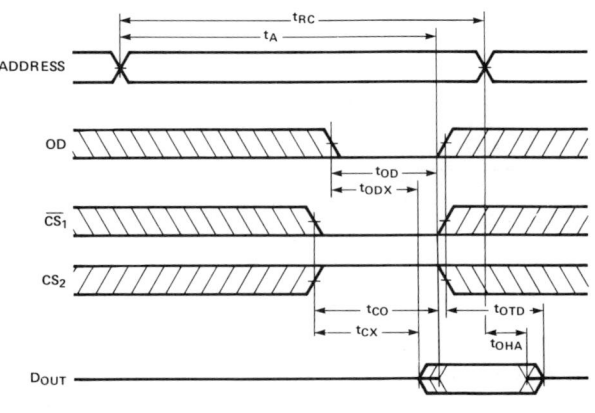

NOTES:
[3] $\overline{WE}$ is high for a Read Cycle.
[4] $\overline{WE}$ must be high during all address transitions.

WRITE CYCLE

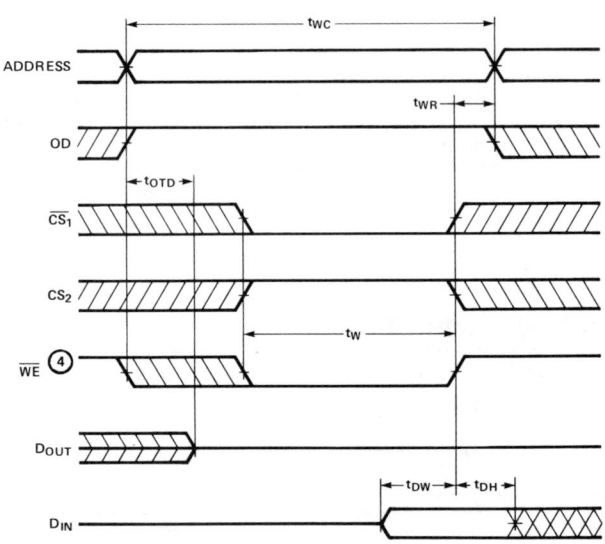

TYPICAL D.C. AND A.C. CHARACTERISTICS

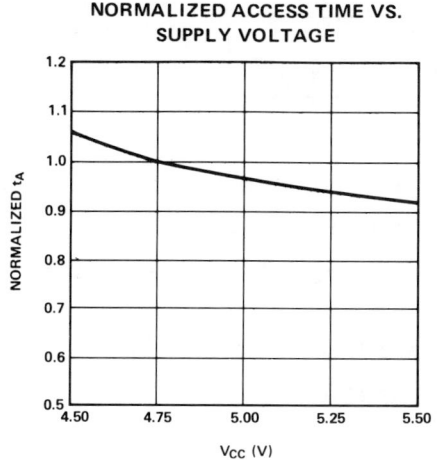

NORMALIZED ACCESS TIME VS. SUPPLY VOLTAGE

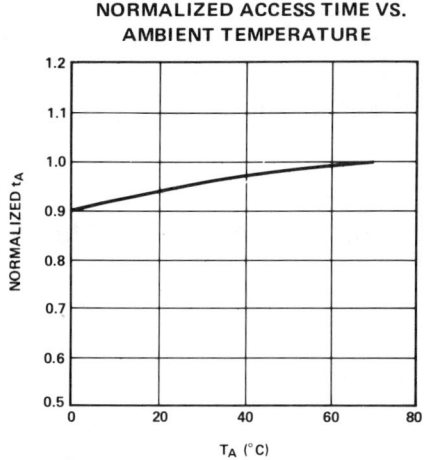

NORMALIZED ACCESS TIME VS. AMBIENT TEMPERATURE

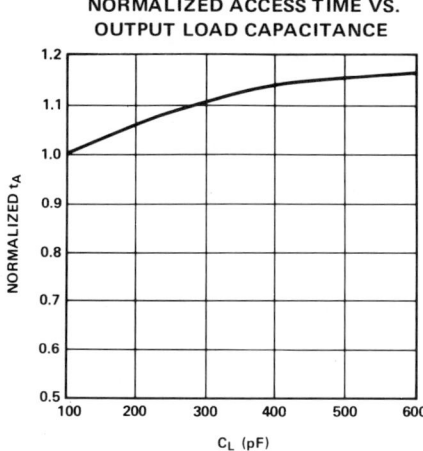

NORMALIZED ACCESS TIME VS. OUTPUT LOAD CAPACITANCE

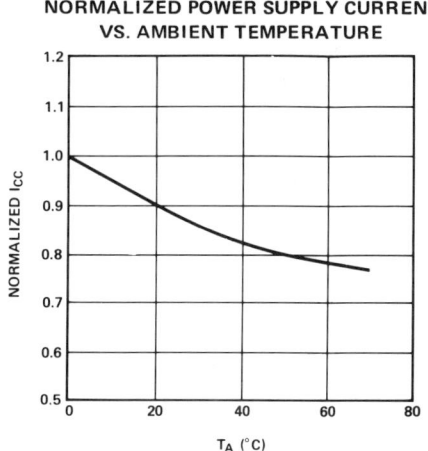

NORMALIZED POWER SUPPLY CURRENT VS. AMBIENT TEMPERATURE

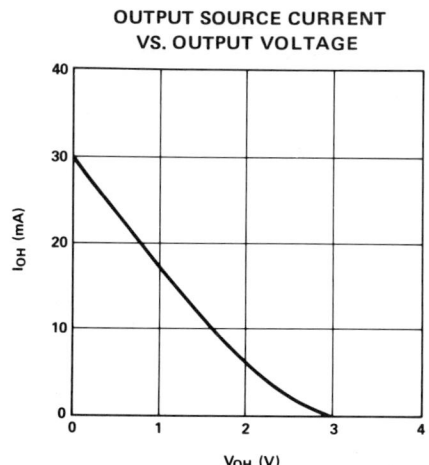

OUTPUT SOURCE CURRENT VS. OUTPUT VOLTAGE

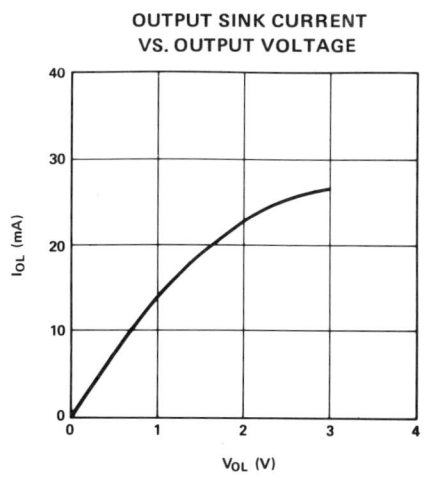

OUTPUT SINK CURRENT VS. OUTPUT VOLTAGE

2147
4096 X 1 BIT STATIC RAM

	2147-3	2147	2147L
Max. Access Time (ns)	55	70	70
Max. Active Current (mA)	180	160	140
Max. Standby Current (mA)	30	20	10

- **HMOS Technology**
- **Completely Static Memory — No Clock or Timing Strobe Required**
- **Equal Access and Cycle Times**
- **Single +5V Supply**
- **Automatic Power-Down**
- **High Density 18-Pin Package**
- **Directly TTL Compatible — All Inputs and Outputs**
- **Separate Data Input and Output**
- **Three-State Output**

The Intel® 2147 is a 4096-bit static Random Access Memory organized as 4096 words by 1-bit using HMOS, a high-performance MOS technology. It uses a uniquely innovative design approach which provides the ease-of-use features associated with non-clocked static memories and the reduced standby power dissipation associated with clocked static memories. To the user this means low standby power dissipation without the need for clocks, address setup and hold times, nor reduced data rates due to cycle times that are longer than access times.

$\overline{CS}$ controls the power-down feature. In less than a cycle time after $\overline{CS}$ goes high — deselecting the 2147 — the part automatically reduces its power requirements and remains in this low power standby mode as long as $\overline{CS}$ remains high. This device feature results in system power savings as great as 85% in larger systems, where the majority of devices are deselected.

The 2147 is placed in an 18-pin package configured with the industry standard pinout. It is directly TTL compatible in all respects: inputs, outputs, and a single +5V supply. The data is read out nondestructively and has the same polarity as the input data. A data input and a separate three-state output are used.

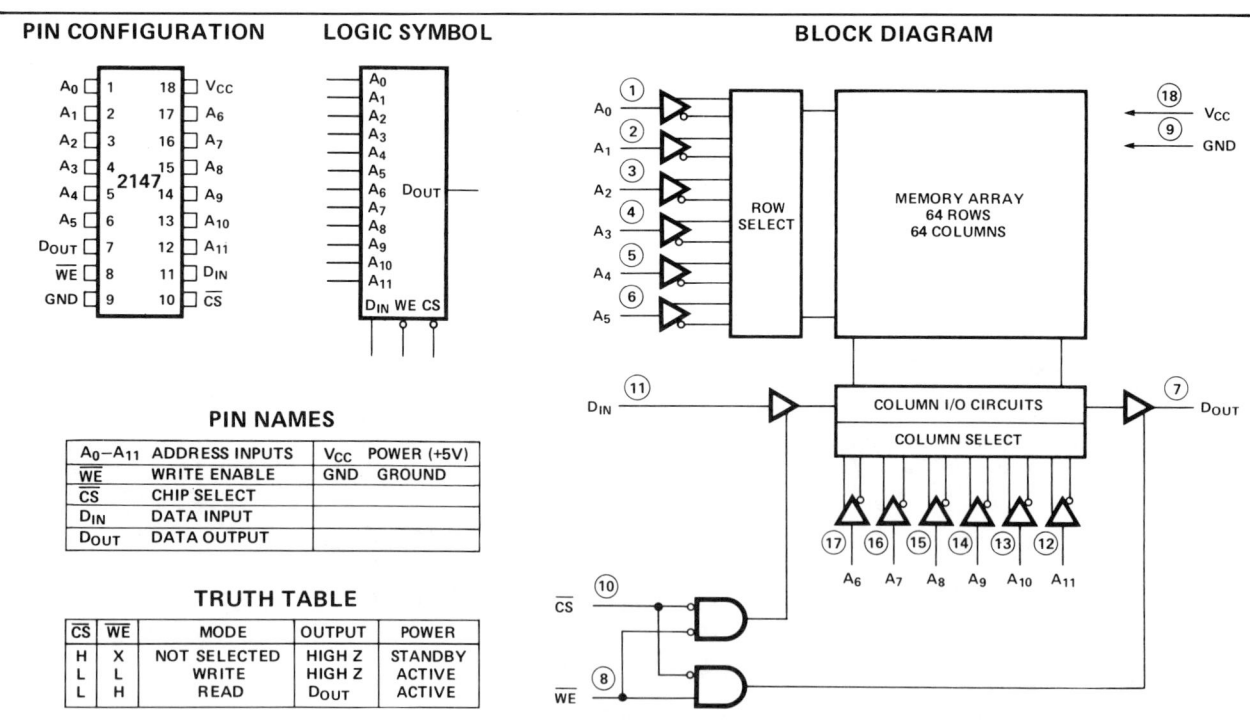

PIN NAMES

A_0–A_{11}	ADDRESS INPUTS	V_{CC}	POWER (+5V)
$\overline{WE}$	WRITE ENABLE	GND	GROUND
$\overline{CS}$	CHIP SELECT		
D_{IN}	DATA INPUT		
D_{OUT}	DATA OUTPUT		

TRUTH TABLE

$\overline{CS}$	$\overline{WE}$	MODE	OUTPUT	POWER
H	X	NOT SELECTED	HIGH Z	STANDBY
L	L	WRITE	HIGH Z	ACTIVE
L	H	READ	D_{OUT}	ACTIVE

ABSOLUTE MAXIMUM RATINGS*

Temperature Under Bias .... -10°C to 85°C
Storage Temperature .... -65°C to +150°C
Voltage on Any Pin With
 Respect to Ground .... -0.5V to +7V
D.C. Output Current .... 20 mA

*COMMENT: Stresses above those listed under "Absolute Maximum Ratings" may cause permanent damage to the device. This is a stress rating only and functional operation of the device at these or any other conditions above those indicated in the operational sections of this specification is not implied. Exposure to absolute maximum rating conditions for extended periods may affect device reliability.

D.C. AND OPERATING CHARACTERISTICS [1]

$T_A = 0°C$ to $70°C$, $V_{CC} = +5V \pm 5\%$, unless otherwise noted.

Symbol	Parameter	2147-3 Min.	2147-3 Typ.[2]	2147-3 Max.	2147 Min.	2147 Typ.[2]	2147 Max.	2147L Min.	2147L Typ.[2]	2147L Max.	Unit	Test Conditions	
I_{LI}	Input Load Current (All Input Pins)		0.01	10		0.01	10		0.01	10	μA	V_{CC}=MAX, V_{IN}=GND to V_{CC}	
$|I_{LO}|$	Output Leakage Current		0.1	50		0.1	50		0.1	50	μA	$\overline{CS}$=2.0V, V_{CC}=Max, V_{OUT}=GND to 4.5V	
I_{CC}	Operating Current		120	170		100	150		100	135	mA	T_A=25°C	V_{CC}=Max., $\overline{CS}$=V_{IL}, Outputs Open
				180			160			140	mA	T_A=0°C	
I_{SB}	Standby Current		15	30		10	20		7	10	mA	V_{CC}=Max, $\overline{CS}$=V_{IH} T_A=0°C to 70°C	
V_{IL}	Input Low Voltage	-0.3		0.8	-0.3		0.8	-0.3		0.8	V		
V_{IH}	Input High Voltage	2.0		6.0	2.0		6.0	2.0		6.0	V		
V_{OL}	Output Low Voltage			0.45			0.45			0.45	V	I_{OL} = 12mA	
V_{OH}	Output High Voltage	2.4			2.4			2.4			V	I_{OH} = -4.0mA	

Notes:
1. The operating ambient temperature range is guaranteed with transverse air flow exceeding 400 linear feet per minute.
2. Typical limits are at V_{CC} = 5V, T_A = +25°C, and specified loading.

A.C. TEST CONDITIONS

Input Pulse Levels	GND to 3.5 Volts
Input Rise and Fall Times	10 nsec
Input and Output Timing Reference Levels	1.5 Volts
Output Load	See Figure 1

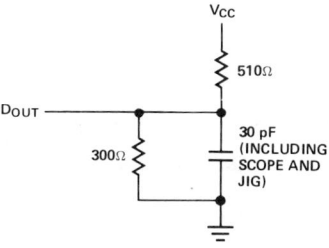

Figure 1. Output Load

CAPACITANCE [3]

$T_A = 25°C$, $f = 1.0$MHz

Symbol	Parameter	Max.	Unit	Conditions
C_{IN}	Input Capacitance	5	pF	V_{IN} = 0V
C_{OUT}	Output Capacitance	7	pF	V_{OUT} = 0V

Note 3. This parameter is periodically sampled and not 100% tested.

A.C. CHARACTERISTICS

$T_A = 0°C$ to $70°C$, $V_{CC} = +5V \pm 5\%$, unless otherwise noted.

READ CYCLE

Symbol	Parameter	2147-3 Min.	2147-3 Max.	2147, 2147L Min.	2147, 2147L Max.	Unit	Test Conditions
t_{RC}	Read Cycle Time	55		70		ns	
t_{AA}	Address Access Time		55		70	ns	
t_{ACS1}	Chip Select Access Time		55		70	ns	Note 1
t_{ACS2}	Chip Select Access Time		65		80	ns	Note 2
t_{OH}	Output Hold from Address Change	5		5		ns	
t_{LZ}	Chip Selection to Output in Low Z	10		10		ns	
t_{HZ}	Chip Deselection to Output in High Z	0	40	0	40	ns	
t_{PU}	Chip Selection to Power Up Time	0		0		ns	
t_{PD}	Chip Deselection to Power Down Time		30		30	ns	

WAVEFORMS

READ CYCLE NO. 1 [3]

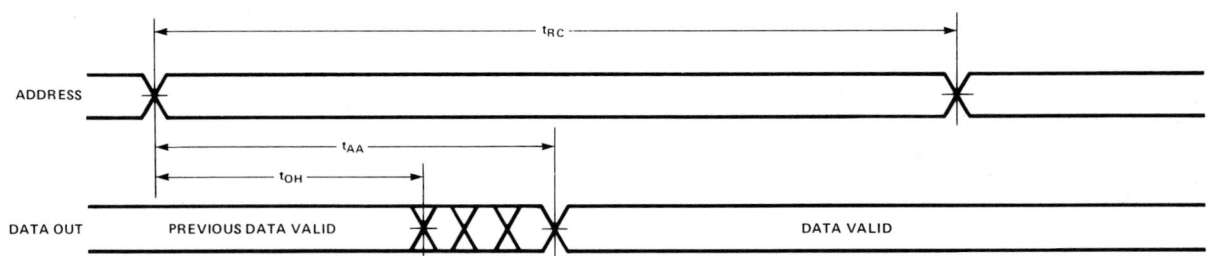

READ CYCLE NO. 2 [3,4]

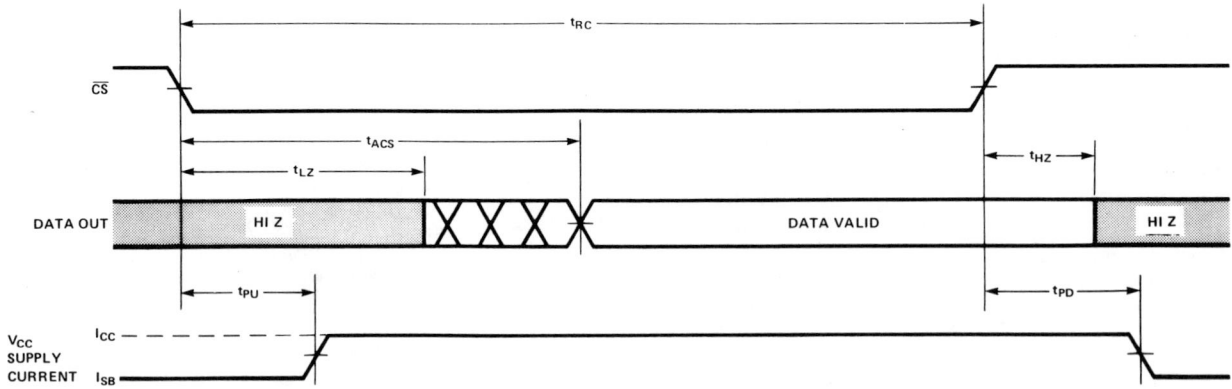

Notes:
1. Chip deselected for greater than 55ns prior to selection.
2. Chip deselected for a finite time that is less than 55ns prior to selection. (If the deselect time is 0ns, the chip is by definition selected and access occurs according to Read Cycle No. 1.)
3. $\overline{WE}$ is high for Read Cycles.
4. Addresses valid prior to or coincident with $\overline{CS}$ transition low.

A.C. CHARACTERISTICS (Continued)

WRITE CYCLE

Symbol	Parameter	2147-3 Min.	2147-3 Max.	2147, 2147L Min.	2147, 2147L Max.	Unit	Test Conditions
t_{WC}	Write Cycle Time	55		70		ns	
t_{CW}	Chip Selection to End of Write	45		55		ns	
t_{AW}	Address Valid to End of Write	45		55		ns	
t_{AS}	Address Setup Time	0		0		ns	
t_{WP}	Write Pulse Width	35		40		ns	
t_{WR}	Write Recovery Time	10		15		ns	
t_{DW}	Data Valid to End of Write	25		30		ns	
t_{DH}	Data Hold Time	10		10		ns	
t_{WZ}	Write Enabled to Output in High Z	0	30	0	35	ns	
t_{OW}	Output Active from End of Write	0		0		ns	

WRITE CYCLE

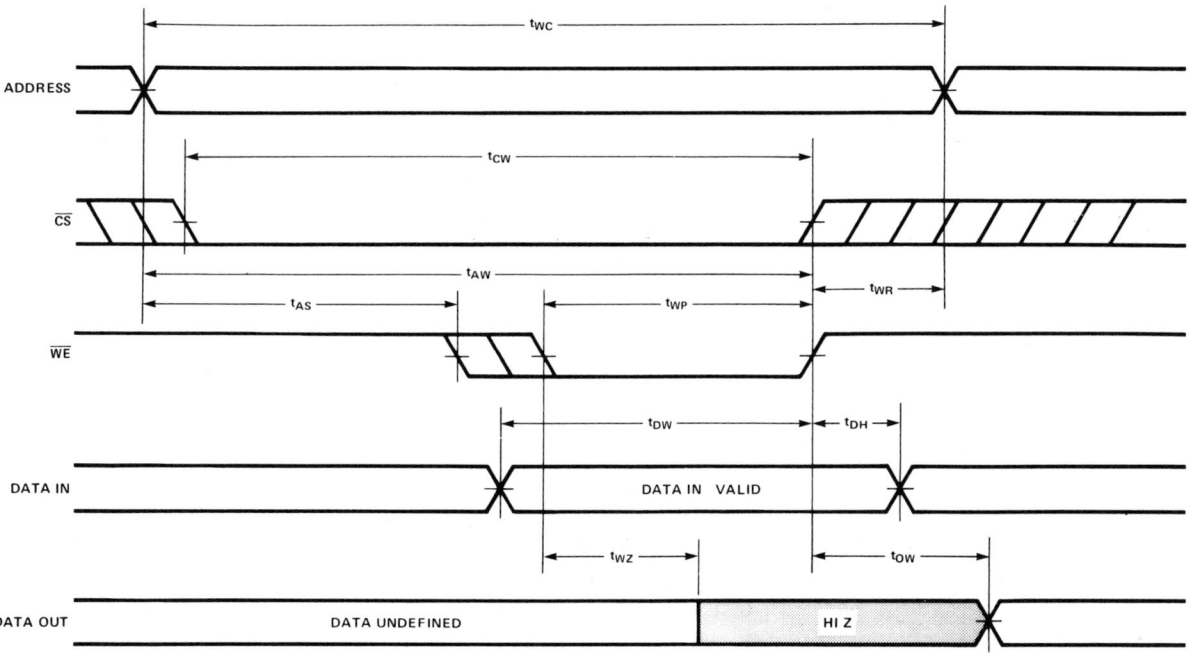

TYPICAL D.C. AND A.C. CHARACTERISTICS

SUPPLY CURRENT[1] VS. SUPPLY VOLTAGE

SUPPLY CURRENT[1] VS. AMBIENT TEMPERATURE

OUTPUT SOURCE CURRENT VS. OUTPUT VOLTAGE

NORMALIZED ACCESS TIME VS. SUPPLY VOLTAGE

NORMALIZED ACCESS TIME VS. AMBIENT TEMPERATURE

OUTPUT SINK CURRENT VS. OUTPUT VOLTAGE

ACCESS TIME CHANGE VS. INPUT VOLTAGE

ACCESS TIME CHANGE VS. OUTPUT LOADING

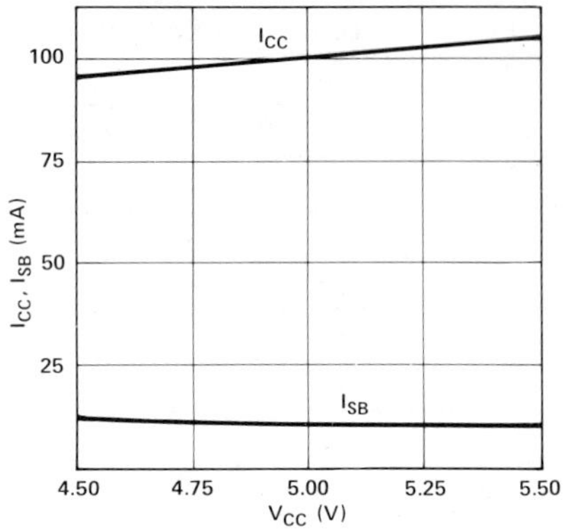

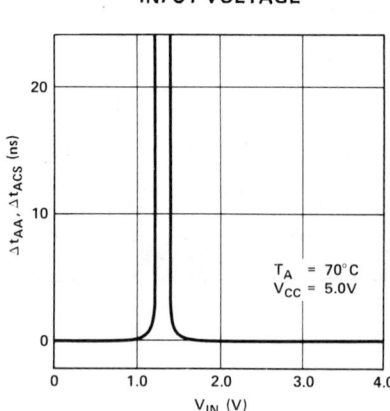

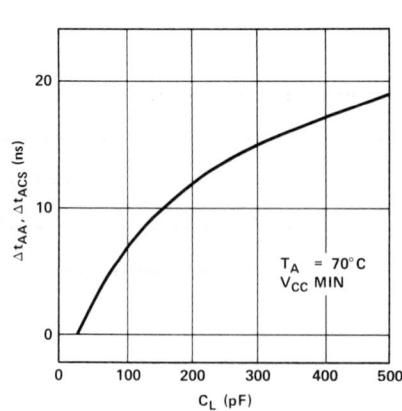

Note 1. The supply current curves shown in Figures 1 and 2 are for the 2147.
The supply current curves for the 2147L and 2147-3 can be calculated by scaling proportionately.

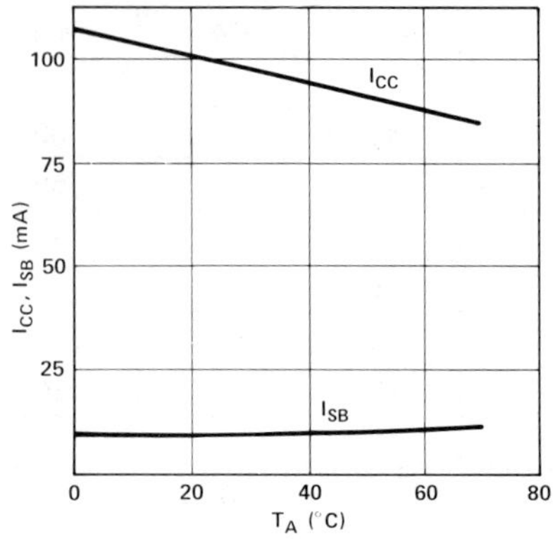

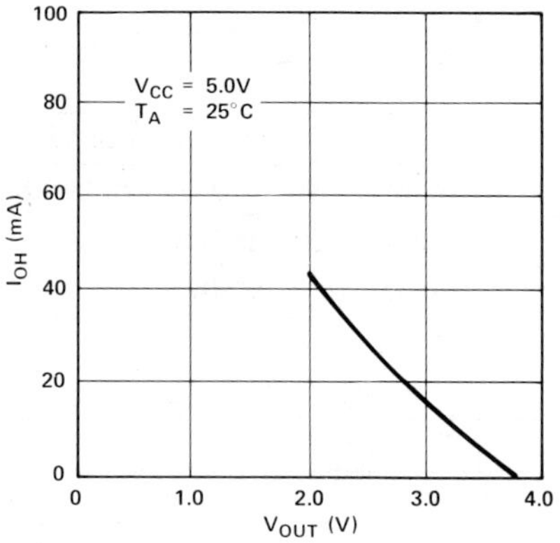

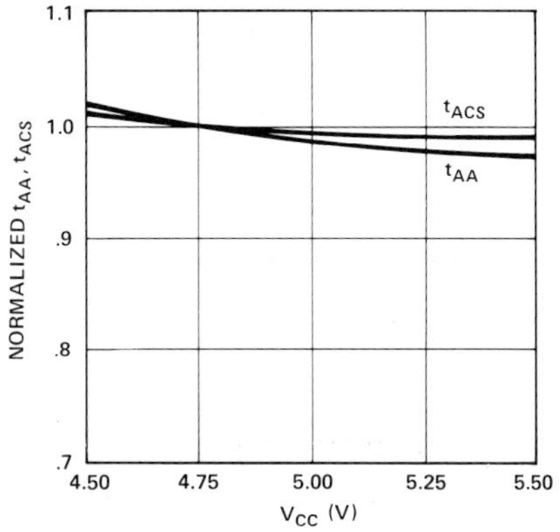

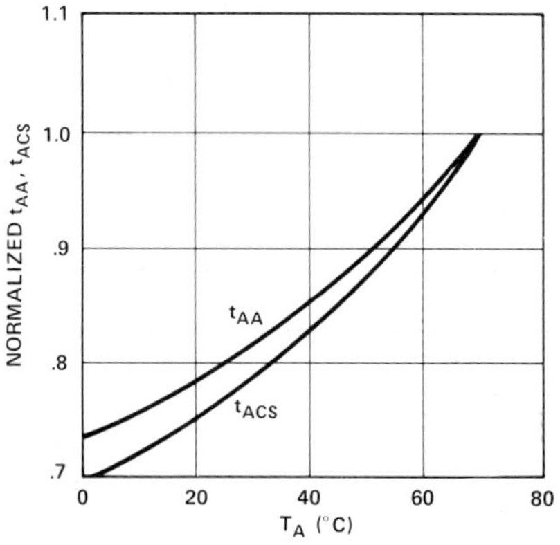

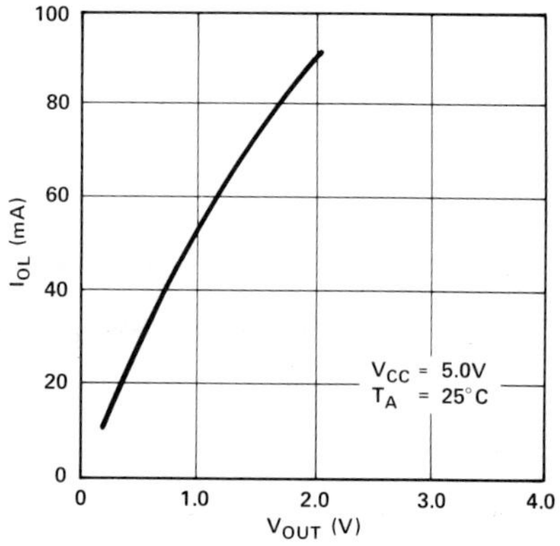

DEVICE DESCRIPTION

The 2147 is produced with HMOS, a new high-performance MOS technology which incorporates on-chip substrate bias generation combined with device scaling to achieve high-performance. The speed-power product of this process has been measured at 1pj, approximately four times better than previous MOS processes.

This process, combined with new design ideas, gives the 2147 its unique features. High speed, low power and ease-of-use have been obtained in a single part. The low-power feature is controlled with the Chip Select input, which is not a clock and does not have to be cycled. Multiple read or write operations are possible during a single select period. Access times are equal to cycle times, resulting in data rates of 14.3 MHz and 18 MHz for the 2147 and 2147-3, respectively. This is considerably higher performance than for clocked static designs.

Whenever the 2147 is deselected, it automatically reduces its power requirements to a fraction of the active power, as shown in Figure 1. This is achieved by switching off the power to unnecessary portions of the internal peripheral circuitry. This feature adds up to significant system power savings. The average power per device declines as system size grows because a continually higher portion of the memory is deselected. Device power dissipation asymptotically approaches the standby power level, as shown in Figure 2.

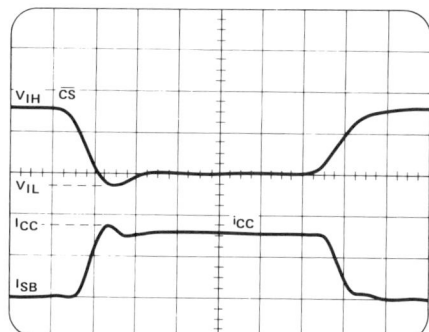

FIGURE 1. i_{CC} WAVEFORM.

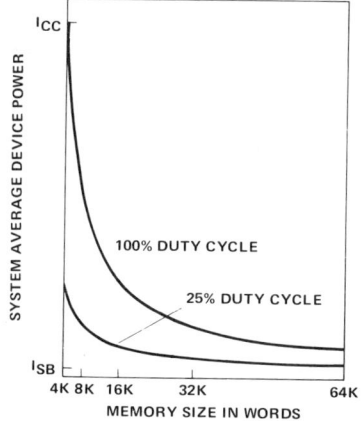

FIGURE 2. AVERAGE DEVICE DISSIPATION VS. MEMORY SIZE.

There is no functional constraint on the amount of time the 2147 is deselected. However, there is a relationship between deselect time and Chip Select access time. With no compensation, the automatic power switch would cause an increase in Chip Select access time, since some time is lost in repowering the device upon selection. A feature of the 2147 design is its ability to compensate for this loss. The amount of compensation is a function of deselect time, as shown in Figure 3. For short deselect times, Chip Select access time becomes slower than address access time, since full compensation typically requires 40ns. For longer deselect times, Chip Select access time actually becomes faster than address access time because the compensation more than offsets the time lost in powering up. The spec accounts for this characteristic by specifying two Chip Select access times, t_{ACS1} and t_{ACS2}.

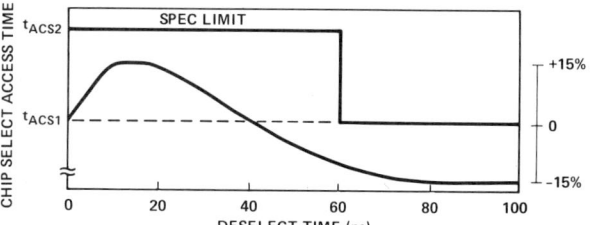

FIGURE 3. t_{ACS} VS. DESELECT TIME.

The power switching characteristic of the 2147 requires more careful decoupling than would be required of a constant power device. It is recommended that a 0.1μF to 0.3μF ceramic capacitor be used on every other device, with a 22μF to 47μF bulk electrolytic decoupler every 16 devices. The actual values to be used will depend on board layout, trace widths and duty cycle. Power supply gridding is recommended for PC board layout. A very satisfactory grid can be developed on a two-layer board with vertical traces on one side and horizontal traces on the other, as shown in Figure 4.

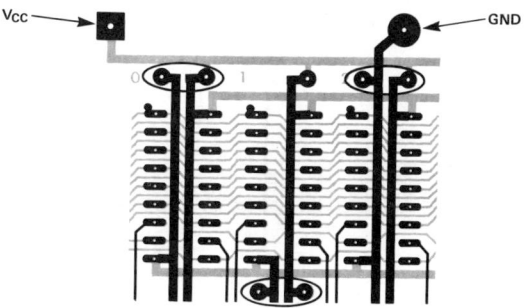

FIGURE 4. PC LAYOUT.

Terminations are recommended on input signal lines to the 2147 devices. In high speed systems, fast drivers can cause significant reflections when driving the high impedance inputs of the 2147. Terminations may be required to match the impedance of the line to the driver. The type of termination used depends on designer preference and may be parallel resistive or resistive-capacitive. The latter reduces terminator power dissipation.

5101
256 X 4 BIT STATIC CMOS RAM

P/N	Typ. Current @ 2V (µA)	Typ. Current @ 5V (µA)	Max Access (ns)
5101L	0.14	0.2	650
5101L-1	0.14	0.2	450
5101L-3	0.70	1.0	650
5101-8	— —	10.0	800

- **Single +5V Power Supply**
- **Ideal for Battery Operation (5101L)**
- **Directly TTL Compatible: All Inputs and Outputs**
- **Three-State Output**

The Intel® 5101 is an ultra-low power 1024-bit (256 words X 4 bits) static RAM fabricated with an advanced ion-implanted silicon gate CMOS technology. The device has two chip enable inputs. Minimum standby current is drawn by this device when CE2 is at a low level. When deselected the 5101 draws from the single 5-volt supply only 10 microamps. This device is ideally suited for low power applications where battery operation or battery backup for non-volatility are required.

The 5101 uses fully DC stable (static) circuitry; it is not necessary to pulse chip select for each address transition. The data is read out non-destructively and has the same polarity as the input data. All inputs and outputs are directly TTL compatible. The 5101 has separate data input and data output terminals. An output disable function is provided so that the data inputs and outputs may be wire OR-ed for use in common data I/O systems.

The 5101L has the additional feature of guaranteed data retention at a power supply voltage as low as 2.0 volts.

A pin compatible N-channel static RAM, the Intel® 2101A, is also available for low cost applications where a 256 X 4 organization is needed.

The Intel ion-implanted, silicon gate, Complementary MOS (CMOS) process allows the design and production of ultra-low power, high performance memories.

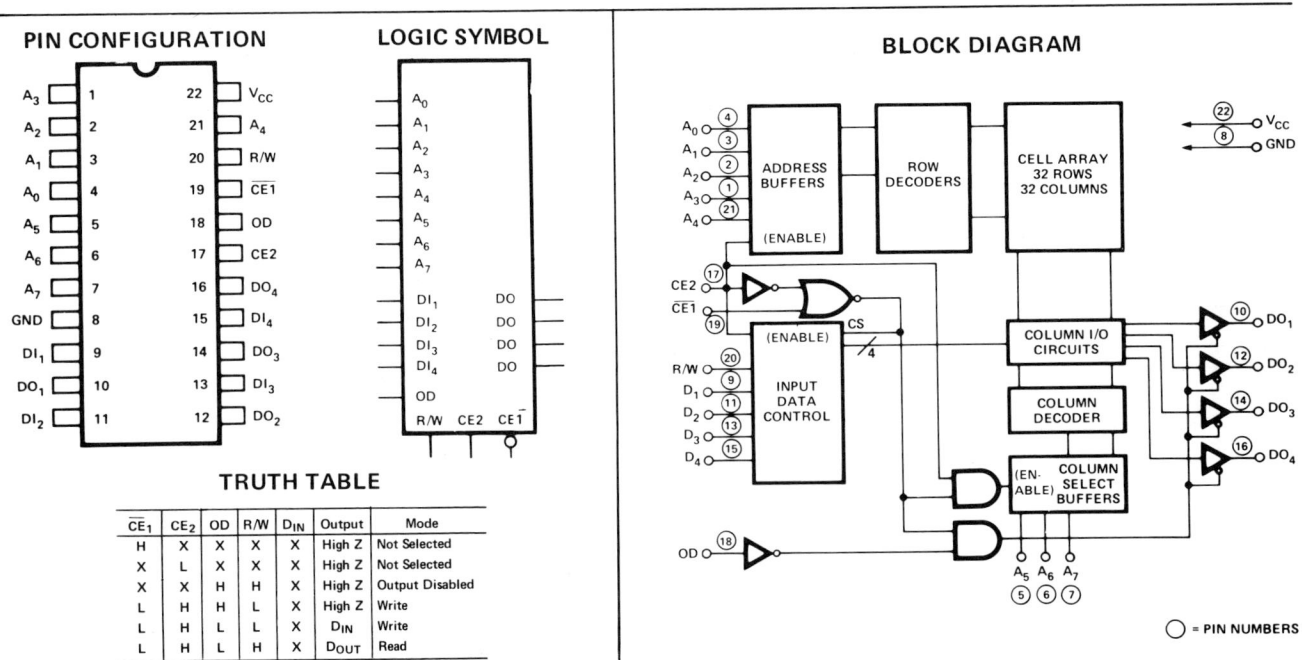

5101

Absolute Maximum Ratings *

Ambient Temperature Under Bias . . . . . $-10°C$ to $80°C$
Storage Temperature . . . . $-65°C$ to $+150°C$
Voltage On Any Pin
  With Respect to Ground . . . . $-0.3V$ to V_{CC} $+0.3V$
Maximum Power Supply Voltage . . . . $+7.0V$
Power Dissipation . . . . 1 Watt

*COMMENT:

Stresses above those listed under "Absolute Maximum Rating" may cause permanent damage to the device. This is a stress rating only and functional operation of the device at these or at any other condition above those indicated in the operational sections of this specification is not implied. Exposure to absolute maximum rating conditions for extended periods may affect device reliability.

D. C. and Operating Characteristics

$T_A = 0°C$ to $70°C$, $V_{CC} = 5V$ ±5% unless otherwise specified.

Symbol	Parameter	5101L and 5101L-1 Limits			5101L-3 Limits			5101-8 Limits			Units	Test Conditions
		Min.	Typ.[1]	Max.	Min.	Typ.[1]	Max.	Min.	Typ.[1]	Max.		
I_{L2}[2]	Input Current			5			5			5	nA	
I_{LO}[2]	Output Leakage Current			1			1			2	μA	$\overline{CE1}$=2.2V, V_{OUT}= 0 to V_{CC}
I_{CC1}	Operating Current		9	22		9	22		11	25	mA	$V_{IN}=V_{CC}$, Except $\overline{CE1} \leq 0.65V$, Outputs Open
I_{CC2}	Operating Current		13	27		13	27		15	30	mA	$V_{IN}=2.2V$, Except $\overline{CE1} \leq 0.65V$, Outputs Open
I_{CCL}[2]	Standby Current		10			200			500		μA	CE2 $\leq$ 0.2V, T_A= 70°C
V_{IL}	Input Low Voltage	-0.3		0.65	-0.3		0.65	-0.3		0.65	V	
V_{IH}	Input High Voltage	2.2		V_{CC}	2.2		V_{CC}	2.2		V_{CC}	V	
V_{OL}	Output Low Voltage			0.4			0.4			0.4	V	I_{OL}=2.0 mA
V_{OH}	Output High Voltage	2.4			2.4			2.4			V	I_{OH}= -1.0 mA

Low V_{CC} Data Retention Characteristics (For 5101L, 5101L-1 and 5101L-3) $T_A = 0°$ to $70°C$

Symbol	Parameter	Min.	Typ.[1]	Max.	Units	Test Conditions	
V_{DR}	V_{CC} for Data Retention	2.0			V		
I_{CCDR1}	5101L or 5101L-1 Data Retention Current		0.14	10	μA	CE2 $\leq$ 0.2V	V_{DR}=2.0V, T_A=70°C
I_{CCDR2}	5101L-3 Data Retention Current		0.70	200	μA		V_{DR}=2.0V, T_A=70°C
t_{CDR}	Chip Deselect to Data Retention Time	0			ns		
t_R	Operation Recovery Time	t_{RC}[3]			ns		

NOTES:
1. Typical values are T_A = 25°C and nominal supply voltage.
2. Current through all inputs and outputs included in I_{CCL} measurement.
3. t_{RC} = Read Cycle Time.

STATIC RAMS

Low V$_{CC}$ Data Retention Waveform

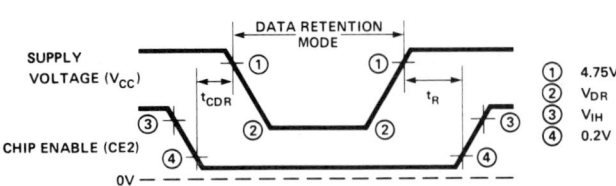

① 4.75V
② V$_{DR}$
③ V$_{IH}$
④ 0.2V

Typical I$_{CCDR}$ Vs. Temperature

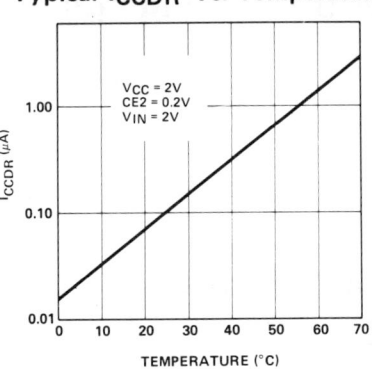

A.C. Characteristics
T$_A$ = 0°C to 70°C, V$_{CC}$ = 5V ±5%, unless otherwise specified.

READ CYCLE

Symbol	Parameter	5101L-1 Limits (ns) Min.	5101L-1 Limits (ns) Max.	5101L and 5101L-3 Limits (ns) Min.	5101L and 5101L-3 Limits (ns) Max.	5101-8 Limits (ns) Min.	5101-8 Limits (ns) Max.
t$_{RC}$	Read Cycle	450		650		800	
t$_A$	Access Time		450		650		800
t$_{CO1}$	Chip Enable ($\overline{CE\ 1}$) to Output		400		600		800
t$_{CO2}$	Chip Enable (CE 2) to Output		500		700		850
t$_{OD}$	Output Disable to Output		250		350		450
t$_{DF}$	Data Output to High Z State	0	130	0	150	0	200
t$_{OH1}$	Previous Read Data Valid with Respect to Address Change	0		0		0	
t$_{OH2}$	Previous Read Data Valid with Respect to Chip Enable	0		0		0	

WRITE CYCLE

t$_{WC}$	Write Cycle	450		650		800	
t$_{AW}$	Write Delay	130		150		200	
t$_{CW1}$	Chip Enable ($\overline{CE\ 1}$) to Write	350		550		650	
t$_{CW2}$	Chip Enable (CE 2) to Write	350		550		650	
t$_{DW}$	Data Setup	250		400		450	
t$_{DH}$	Data Hold	50		100		100	
t$_{WP}$	Write Pulse	250		400		450	
t$_{WR}$	Write Recovery	50		50		100	
t$_{DS}$	Output Disable Setup	130		150		200	

A. C. CONDITIONS OF TEST

Input Pulse Levels: +0.65 Volt to 2.2 Volt
Input Pulse Rise and Fall Times: 20 nsec
Timing Measurement Reference Level: 1.5 Volt
Output Load: 1 TTL Gate and C$_L$ = 100 pF

Capacitance [2] T$_A$ = 25°C, f = 1 MHz

Symbol	Test	Limits (pF) Typ.	Limits (pF) Max.
C$_{IN}$	Input Capacitance (All Input Pins) V$_{IN}$ = 0V	4	8
C$_{OUT}$	Output Capacitance V$_{OUT}$ = 0V	8	12

NOTES: 1. Typical values are for T$_A$ = 25°C and nominal supply voltage.
2. This parameter is periodically sampled and is not 100% tested.

Waveforms

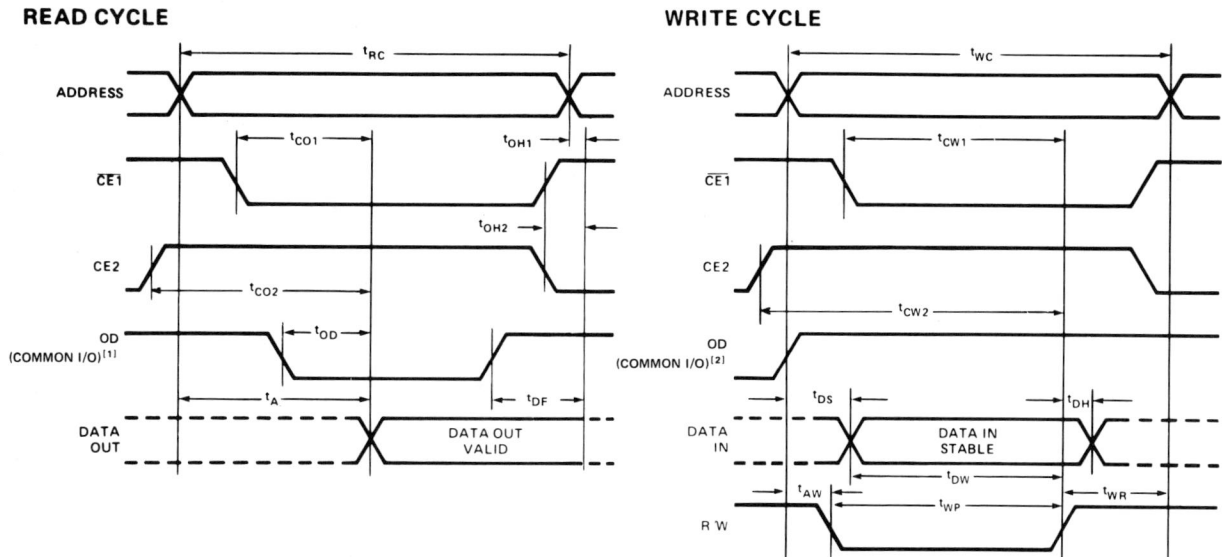

NOTES:
1. OD may be tied low for separate I/O operation.
2. During the write cycle, OD is "high" for common I/O and "don't care" for separate I/O operation.

CHAPTER TWO
Dynamic RAMs

Designing with 16 Pin, 4096 Dynamic RAMs	89
Designing with 22 Pin, 4096 Dynamic RAMs	103
Designing with 16K Dynamic RAMs	130
Dynamic RAMs Used with Microprocessors	139
1103 1024 × 1 Bit Dynamic RAM	182
1103–1 1024 × 1 Bit Dynamic RAM	187
1103A 1024 × 1 Bit Dynamic RAM	190
1103A–1 1024 × 1 Bit Dynamic RAM	195
1103A–2 1024 × 1 Bit Dynamic RAM	200
2104A 4096 × 1 Bit Dynamic RAM	204
2107A 4096 × 1 Bit Dynamic RAM	212
2107B 4096 × 1 Bit Dynamic RAM	218
2107C 4096 × 1 Bit Dynamic RAM	224
2108 8192 × 1 Bit Dynamic RAM	229
2116 16,384 × 1 Bit Dynamic RAM	237
2117 16,384 × 1 Bit Dynamic RAM	245

DESIGNING WITH 16 PIN, 4096 DYNAMIC RAMS

JIM COE

Photomicrograph of the Intel 2104A 4096 Bit Dynamic RAM.

DYNAMIC RAMS

INTRODUCTION

The Intel® 2104A is a 4096 word by 1 bit dynamic random access memory. The 2104A is fabricated using Intel's proven n-channel silicon gate MOS technology. The device is packaged in a standard 16-pin DIP. The pin configuration and logic symbol are shown in Figure 1.

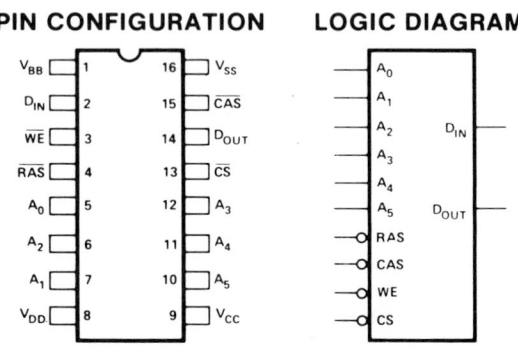

Figure 1. 2104A Pin Assignments

The combination of Intel's n-channel silicon gate process and circuit design has resulted in a part that is fast, easy to use, and economically produced in large volume. In addition, the combination of process and device design has resulted in a small device using conservative layout rules. The small size offers advantages in both large volume production and increased reliability.

The 2104A operates with three power supplies relative to ground: V_{DD} (+12V), V_{BB} (−5V), and V_{CC} (+5V). The V_{CC} (+5V) supply is connected only to the output buffer of the 2104A and may be turned off during power down operations.

The unique design of the 2104A allows it to be packaged in the industry standard 16-pin dual-in-line package. The 16-pin package provides the highest system bit densities and is compatible with widely available automated handling equipment.

The use of the 16-pin package is made possible by multiplexing the 12 address bits (required to address 1 of 4096 bits) into the 2104A on 6 address input pins. The two 6-bit address words are latched into the 2104A by the two TTL clocks, Row Address Strobe ($\overline{RAS}$) and Column Address Strobe ($\overline{CAS}$). Non-critical clock timing requirements allow use of the multiplexing technique while maintaining high performance.

The dynamic storage cell provides high speed along with low power dissipation. The memory cell requires refreshing for data retention. Refreshing is most easily accomplished by performing a read cycle at each of the 64 row addresses every 2 milliseconds.

The purpose of this application note is to describe the internal operation of the 2104A and outline those areas in system implementation to which the designer should pay particular attention.

DEVICE CIRCUIT OPERATION

Operation of the 2104A is most easily understood with the aid of the block diagram shown in Figure 2. As is shown in this figure, the memory array is arranged in a 64 row X 64 column matrix of storage cells. The storage cells are implemented with select transistors and "storage" capacitors. The operation of the storage cell will be discussed later. The cell is accessed by the coincidence of a row select (defined by addresses A_0-A_5) and a column select (defined by addresses A_6-A_{11}) signal at the desired address. On chip timing and control generators provide the internal timing signals for decoding, read/write strobing, data gating and output gating. All of the timing circuits in the 2104A are activated by the negative going edges of the two TTL clocks, $\overline{RAS}$ and $\overline{CAS}$.

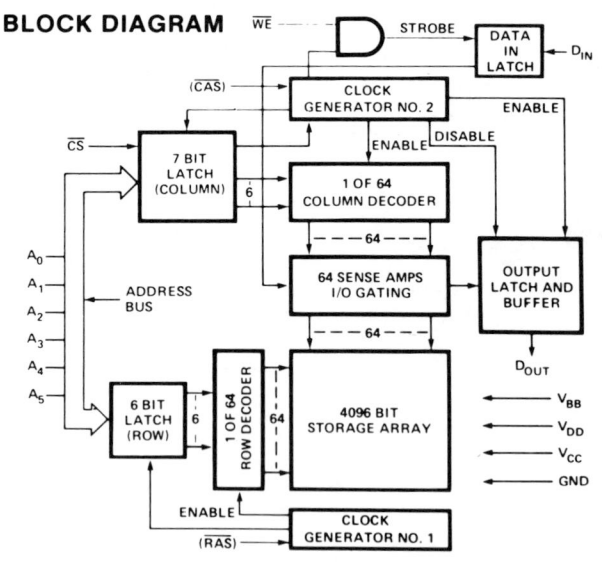

Figure 2. 2104A Block Diagram

Data Accessing

Prior to discussing the $\overline{RAS}/\overline{CAS}$ timing relationships, a discussion of the basic operation of dynamic 4K RAM devices is in order. Access of stored data from a dynamic memory device consists of two discrete retrieval operations. The first of these operations is the selection of the desired row of storage cells (1 of 64 rows of 64 cells in the 2104A), sensing the data stored in each of the cells with sense amplifiers (64 sense amplifiers in the 2104A), and restoring the sensed data back into the cells since the readout is destructive. When this operation is complete, the sensed data (64 bits) is available at the output of the sense amplifiers. This operation may be completed with only the row address and a clock ($\overline{RAS}$ with the 2104A) having been supplied to the memory device. This first operation fulfills the refresh requirement on the selected row since data has been restored in the cells on the row.

The second operation consists of connecting the output of one of the sense amplifiers to the device data output via a multiplexer (64 to 1 in the 2104A) and latching the data into the output data latch. In essence, this is accessing data from the sense amplifier outputs rather than from the data cells. This operation requires a column address and a clock ($\overline{CAS}$ with the 2104A). This second operation is the characteristic which makes page-mode operation possible. Page-mode will be discussed in the Applications Information Section.

The two access operations may occur in parallel as in the 18-pin and 22-pin 4K RAMs or in a time sequential manner in a 16-pin 4K RAM such as the 2104A. With proper design techniques such as used in the 2104A, the sequential mode of operation may be used, saving package pins and with no performance loss as compared to the parallel mode RAMs.

In the parallel mode RAMs (such as the Intel® 2107B) all address information is applied to the RAM at the same time and both access operations occur simultaneously. The cell data access is the slower of the two operations and is the limiting factor in device speed. The selection of the proper sense amplifier output for connection to the device output is completed prior to the time it is necessary.

In the sequential mode 2104A RAM, cell data access is begun first by the latching in of the row address information (6-bits for 1 of 64 row select) by the $\overline{RAS}$. The access of data from the sense amplifier outputs is faster and thus may be started later without impacting overall access time [up to 70 nanoseconds ($t_{RCL(max)}$) later in the 2104A-2]. The 6-bit, 1 of 64 sense amplifier data address (column address) is latched into the 2104A-2 by the CAS. As long as the sense amplifier output data access is started prior to 70 nanoseconds into the memory cycle, the limiting access time is t_{RAC}, the data cell access time plus the propagation time through the sense amplifier data select multiplexer. This access time is the same as the parallel mode access time would be.

If the column address latching is delayed until later than 70 nanoseconds into the memory cycle, the limiting access time will become the sense amplifier data access time. In this instance, the access time will be t_{CAC} (access time from $\overline{CAS}$ which includes the sense amplifier output data multiplexer propagation time) plus t_{RCL} actual (the actual $\overline{RAS}$ to $\overline{CAS}$ delay time). It is obvious that it is desirable to latch the column address into the 2104A-2 at or prior to the 70 nanosecond point in the memory cycle to preclude lengthening of data access time.

In the 2104A-2, a 45 nanosecond window is provided during which $\overline{CAS}$ may be switched while maintaining device access time. In other words, the $\overline{CAS}$ leading edge may occur at any time between 25 and 70 nanoseconds following $\overline{RAS}$ and the access time will be t_{RAC}. Timing accuracy required between $\overline{RAS}$ and $\overline{CAS}$ is thus reduced. The advantages of this timing "window" will be discussed in the Applications Information Section.

Clock Input Buffers

The two device clocks, $\overline{RAS}$ and $\overline{CAS}$, are TTL compatible, active-low signals. The clock input buffers are inverters which convert the TTL levels to the MOS (12 volt) levels required within the 2104A. The major design consideration for these buffers is speed since it is desirable to respond to the clock inputs as quickly as possible to obtain minimum data access. The speed is obtained by implementing the inverters with high gain (large geometry) devices operating at relatively high current levels. The inverter circuit is shown in Figure 3.

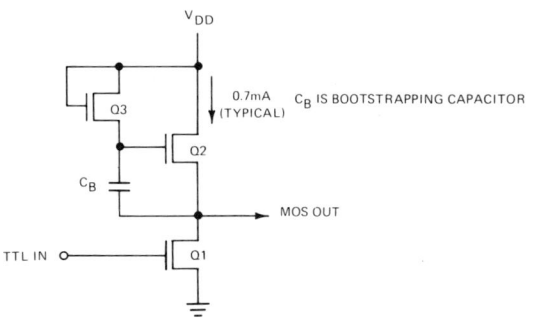

Figure 3. Simplified CLOCK Input Buffer

The inverter uses a bootstrapped, 0.7 milliampere (typical) load device. The bootstrapping is used to assure that the load device (Q2) is fully turned-on so that the drain voltage of Q1 reaches V_{DD}. Without the bootstrapping, the drain of Q1 would only reach $V_{DD}-V_T$ where V_T is the load device threshold voltage. This would slow down the inverter operation and affect the response time to the clock(s).

The current requirement of the input buffer accounts for the difference in standby power levels between the 16-pin TTL clock devices and the 18 or 22-pin MOS clock devices. When the $\overline{RAS}$ clock is inactive (high), the 2104A $\overline{RAS}$ buffer is on and the inverter load current (2.0mA maximum) is drawn from The V_{DD} supply yielding the 26.4mW maximum standby power specification. MOS clock devices (such as the Intel® 2107B) have inactive low clocks and no buffer is on during standby, yielding standby power specifications under 3mW maximum (leakage currents only). This standby power reduction at the memory device level is offset at the system level by the larger power dissipation levels of MOS level clock driver devices versus TTL level drivers. The 2104A TTL clock inputs are lower in capacitance than the MOS clock inputs (7 picofarads versus 25 picofarads). At a given speed, this means a typical TTL driver can drive 32 2104A clock inputs while a typical MOS clock driver can drive only 10 2107B clock inputs.

Address Buffer/Latch

The TTL-level compatible address buffer/latch circuit is shown in Figure 4. This circuit senses the input TTL level, translates it to MOS signal levels, and latches the address information. There are two groups of six input buffer/latches in the 2104A; one for the six row addresses and one for the six column addresses. The operation of each group of latches is the same except for the clock signals which control their function.

The operation of the address buffer/latch is as follows: During the clock ($\overline{RAS}$ or $\overline{CAS}$) off time (t_{RP} or t_{CP}) both sides of the latch (Nodes A and B) are precharged to V_X (≈ 10 volts) by devices Q7 and Q9. Device Q8 is turned on during the precharge period to assure that the two nodes charge to the same potential. Internal signal ϕ_P controls the precharge devices and is on while the $\overline{RAS}$ and $\overline{CAS}$ clocks are off (at V_{IH}). When the appropriate system clock ($\overline{RAS}$ or $\overline{CAS}$) goes low (active), ϕ_P turns off isolating the two precharged nodes and internal clock phase ϕ_A turns on connecting the TTL address (A_{IN}) to Node C (the gate of the input buffer device Q2 and capacitor C_1). Clock phase ϕ_A stays on for the address hold time (t_{AH}) and then turns off isolating Node C from the shared address input pin. The TTL level which was on the address pin during t_{AH} is still stored on capacitor C_1 allowing the address latch additional time to capture the address. This "sample and hold" technique allows short address hold times to be achieved.

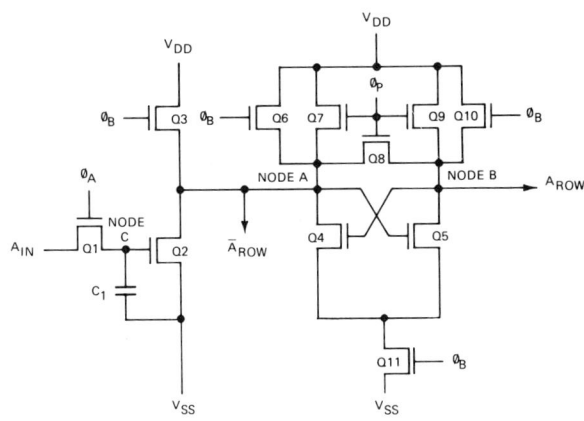

Figure 4. Simplified Address Buffer/Latch Schematic

Internal clock phase ϕ_B turns on after a slight delay from phase ϕ_A turning on. Phase ϕ_B enables the buffer/latch by turning on load devices Q3, Q6, and Q10. The buffer (Q2 and Q3) converts the TTL level address input to MOS levels (V_{SS} and V_{DD}) and drives Node A of the latch. The delay between ϕ_A and ϕ_B is to allow the voltage at Node C to stabilize prior to enabling the buffer/latch.

A TTL high level at the address input will force Node A to V_{SS}. The cross-coupled latch devices (Q4 and Q5) will then switch driving Node B to V_{DD}. Conversely, a TTL low level at the address input will force Node A to $\sim V_{DD}$ and Node B to V_{SS}. Since the buffer/latch is isolated from the address input after t_{AH}, the latched address will remain in the latch even though the TTL level at the address input may change due to the multiplexing of the addresses.

Data Sensing

A major contributor to the operating margins of the 2104A is the use of two single-transistor storage cells per bit of storage. The effect of using two cells per bit rather than one is best understood by comparison of the data sensing function when used with one and two cells per bit.

Figure 5 illustrates the commonly used sense amplifier and reference voltage scheme for single cell per bit 4K RAMs. The sense amplifier in Figure 5 senses data stored in a storage cell by comparing the voltage level in the storage cell capacitor to the voltage level in the cell capacitor of a "dummy" storage cell. The dummy cell capacitor contains a voltage which is less than the minimum high level and greater than the maximum low level which may be stored in the storage cell capacitor. The sense amplifier then senses the differential level between the storage and dummy cell capacitor voltages. The level stored in the dummy cell would ideally be equal to one-half the difference between a minimum written high and a maximum written low as this would yield a maximum differential across the sense amplifier during sensing. Unfortunately, leakage currents from the storage capacitors degrade the written high levels toward the written low levels. This normally requires that the designer set the dummy storage cell level lower (closer to a low level than a high level) to compensate for leakage degradation of a stored high level. Although this "lower" reference level tends to compensate for a leakage degraded high level, it also makes it more difficult to sense a ground (V_{SS}) noise degraded low level. Thus, designs with dummy reference cells must necessarily be a compromise in the maximum differential level between the storage and dummy cells and can never have a differential greater than one-half the difference between a high and low level.

The dummy cell technique is used (rather than simply developing a reference voltage level with a resistive divider) because it contributes to the capacitive balance of the sense amplifier.

The sense amplifer of Figure 5 will operate with maximum margins only when the capacitance seen by Node B is the same as the capacitance seen by Node A, i.e., the capacitances are equal or balanced. The capacitances of the left (BSLL) and right (BSLR) bit sense lines as well as the dummy cell and storage cell capacitances can be made approximately equal by layout constraints. The effect of the I/O line connection to the right bit sense line is to add capacitance on the right bit sense line which is not offset or balanced by capacitance on the left bit sense line. The placement of the dummy cell on the bit sense line also contributes to capacitance imbalance since its location is not a mirror image of the accessed storage cell. The resistive effects of the bit sense line magnify the effect of this placement disparity during the data sensing process. The ideal situation would be a dummy cell mirroring the placement of the accessed data cell and a balancing capacitance to the I/O connection capacitance.

This is essentially the technique used in the Intel® 2104A. Instead of a dummy cell containing a reference level of one-half a minimum high level, the 2104A stores the full opposite data level in a mirror image storage cell physically located near the accessed storage cell as shown in Figure 6. Not only does this mean that the storage cell and "image" cell capacitance and location with respect to the sense amplifier are equal, but the data level is now being sensed against the full opposite level rather than one-half of the minimum high level. Thus, the sense amplifier is seeing the maximum possible differential signal during the sensing operation.

Also, notice in Figure 6, that there is an I/O connection to each bit sense line rather than only to one. The I/O capacitance contribution to the bit sense line capacitance is therefore equal, contributing again to the overall balance of the sense amplifier.

The 2104A sense amplifier sees essentially equal capacitances at nodes A and B and this contributes greatly to the margins of the sensing operation. This balance of the sense amplifier and the sensing of data against a reference of a full opposite level allows the cell capacitors to be a smaller value than with the dummy cell approach (for equal margins) and allows 8192 cells to occupy only slightly more chip area than 4096 cells previously occupied in the Intel® 2104.

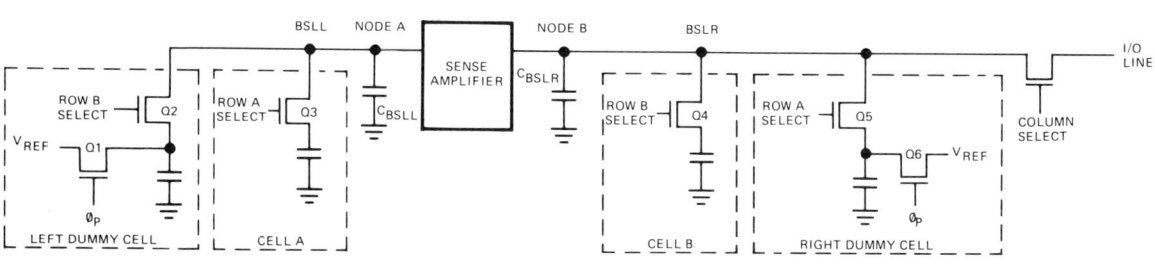

Figure 5. Old Dummy Cell Data Sensing Technique

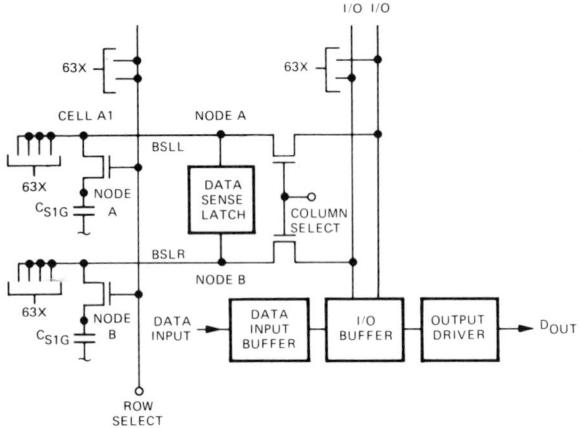

- Each storage location consists of two cells. Each cell is formed by single transistor and storage capacitor as shown.
- At row select time, two cells are turned on simultaneously to gate the data to the data sense latch.
- Data is stored in a given address location as both a high and a low on each of the two respective data cells.

Figure 6. 2104A Image Cell Data Sensing Technique

Data Sense Amplifier

The data sense amplifier of the 2104A is a cross-coupled static latch as shown in Figure 7. The state

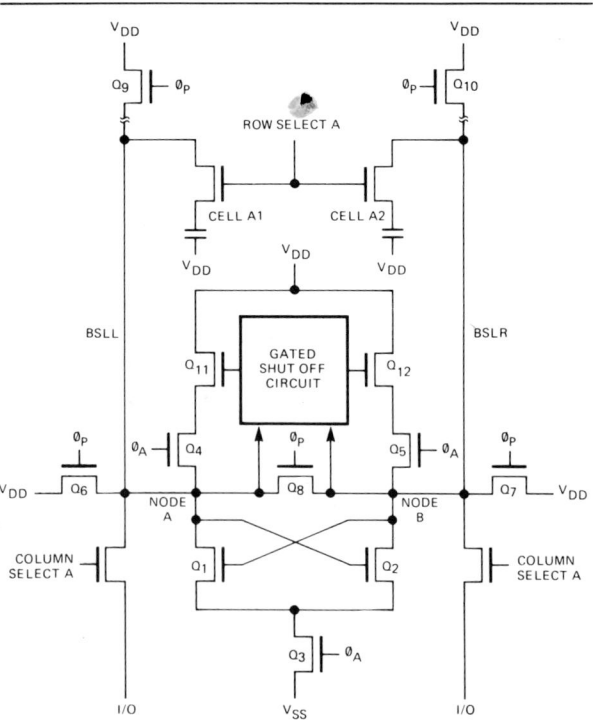

Figure 7. Simplified Data Sense Amplifier Schematic

the latch assumes during sensing of the stored data depends upon the differential voltage between nodes A and B. If the voltage on node A is higher than the voltage on node B, transistor Q_2 will turn on switching node B and the gate of transistor Q_1 to V_{SS} turning transistor Q_1 off. Conversely, if the voltage on node B is higher than the voltage on node A, transistor Q_1 will turn on while transistor Q_2 will be turned off. Devices Q_4 and Q_5 act as loads for the switching transistors Q_1 and Q_2, respectively. Additional transistors (Q_6 through Q_{10}) shown in the circuit diagram of Figure 7 serve to precharge nodes A and B to $\sim V_{DD}$ in preparation for the next memory cycle. This precharging assures that the sense amplifier and bit sense lines begin each memory cycle in the same known condition or state with no "history" or "memory" of data from previous cycles eliminating data pattern effects on the sensing function. Note that the 2104A has precharge transistors connected to both ends of the bit sense lines to speed up the precharging of the lines and sense amplifier. This increases the timing margin of the clock off time (t_{RP}) and enables the 2104A to run short memory cycles without degradation of the precharge function. The "folded", close proximity bit sense lines shown in Figure 6 and transistor Q_8 of Figure 7 assure that the precharge level of each pair of bit sense lines and the associated sense amplifier nodes reach the same precharge level contributing to the balance of the sensing function.

Transistor Q_3 in Figure 7 turns the sense amplifier on by completing the current path to V_{SS} when the row address bits have been decoded and the desired row of storage cells selected. The gated shutoff circuit controls transistors Q_{11} and Q_{12} to reduce the power dissipation of the sense amplifier following the sensing of the stored data. The shutoff circuitry senses the levels on Nodes A and B and turns off the load current to the switching transistor (Q_1 or Q_2) which is turned on to V_{SS}. This reduces the I_{DD} current drawn by the sense amplifier and contributes to the low power dissipation of the 2104A.

Output Data Latch/Driver

A simplified schematic of the 2104A output data latch/driver is shown in Figure 8. The three operational states for the output driver are:

1) "1" output (Q_1 on and Q_2 off)
2) "0" output (Q_1 off and Q_2 on)
3) Open output (Q_1 and Q_2 off)

Devices Q_1 and Q_2 are large geometry devices which allow the output of the 2104A to source and sink the relatively large current levels associated with TTL interfaces. Devices Q_3 through Q_6 control the output driver stage in conjunction with the data latch.

DESIGNING WITH 16 PIN, 4K DYNAMIC RAMS

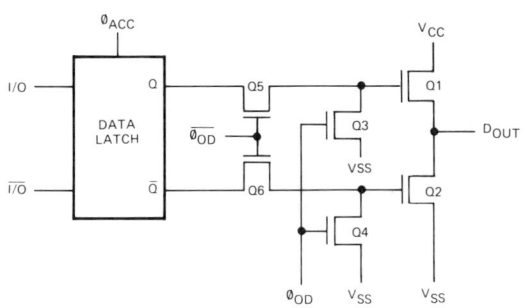

Figure 8. 2104A Output Data Latch and Buffer

Table I. Operational States of ∅OD

RAS	CAS	CS	∅OD	COMMENTS
LOW	LOW	HIGH	HIGH	Device Deselected by CS (D_{OUT} = HI-Z)
LOW	LOW	LOW	LOW	Device Selected by CS (D_{OUT} = Data)
HIGH	LOW	Don't Care	HIGH	Device Deselected by CAS (D_{OUT}=HI-Z)

The inputs to the latch are data from the selected cell and a clock phase (∅ACC) which is related to access time from the cell matrix. At the proper time after the memory cycle starts, ∅ACC will go high, clocking the data from the selected cell into the latch. The Q and $\overline{Q}$ outputs of the latch then drive the gates of devices Q_1 and Q_2 controlling the output level. The accessed data will remain in the latch until the next cycle when new data will be clocked into the latch. During a write cycle, the data input to the latch is the data on the I/O lines which is the data to be written into the selected cells. The data latch will, therefore, contain the input data following a write cycle. The first two of the three possible output data states are, therefore, related to the data level stored in the latch.

The third or open-circuited state occurs when devices Q_1 and Q_2 are both off (gates at V_{SS}). Internal signal ∅OD turns on devices Q_3 and Q_4 connecting the gates of devices Q_1 and Q_2 respectively to V_{SS} from shortly after the CAS input switches low until data access time. This signal forces the data output to the open-circuited condition following CAS in every memory cycle guarantying that no two OR-tied data outputs in a system will be on at the same time.

The control clock phase ∅OD is a logic function of CAS and CS. Table I lists the various combinations of CAS and CS and the corresponding states of ∅OD.

Devices Q_5 and Q_6 in Figure 8 are simply series switches which isolate the Q and $\overline{Q}$ outputs of the data latch from output devices Q_1 and Q_2 until the latch data has stablized. Q_5 and Q_6 are controlled by the inverse of ∅OD so that the latch is isolated from Q_1 and Q_2 when the gates of Q_1 and Q_2 are connected to V_{SS}.

APPLICATIONS INFORMATION

Addressing

The 2104A RAM combines the advantages of a very high speed RAM with the high packing density of the industry standard 16 pin dual-in-line package. The use of the 16 pin package is made possible by multiplexing the 12 address inputs (required to access 4096 words) on 6 external address pins. Two externally applied negative going clocks, Row Address Select (RAS), and Column Address Select (CAS), are used to strobe the two sets of 6 address bits into the internal address buffer registers. The first clock, RAS, strobes in the six low order address bits (A_0-A_5) which select one of 64 rows. The second clock, CAS, strobes in the six high order address bits (A_6-A_{11}) which select one of 64 columns and Chip Select (CS).

Note that CS and WE do not have to be valid until the second clock, CAS. It is, therefore, possible to start a memory cycle *before* it is known which device must be selected or what type of cycle is to be performed. This can result in a significant improvement in *system* access time since the decode time for chip selection does not enter into the calculation for access time.

Read Cycle

A memory cycle begins with addresses stable and a negative transition of RAS. The data-out pin of the selected device will unconditionally go to a high impedance state immediately following the leading edge of CAS and remain in this state until valid data appears at the output (refer to the Data Output Operation Section). The selected output data is internally latched and will remain valid until a subsequent CAS is given to the device by a Read, Write, Read-Modify-Write or Refresh cycle. Data-out goes to a high impedance state for all non-selected devices (CS high) that receive RAS and CAS.

Device access time, t_{ACC}, is the longer of two calculated intervals:

1) $t_{ACC} = t_{RAC}$

   OR

2) $t_{ACC} = t_{RCL} + t_T + t_{CAC}$

Access time from $\overline{RAS}$, t_{RAC}, and access time from $\overline{CAS}$, t_{CAC}, are device parameters. Row to column address strobe lead time, t_{RCL}, and transition time, t_T, are system dependent timing parameters.

Substituting the device parameters for the 2104A-2 and assuming a TTL level transition time of 5nS yields:

3) $t_{ACC} = t_{RAC} = 200$ns for $t_{RCL} + t_T \leq 70$ ns

OR

4) $t_{ACC} = t_{RCL} + t_T + t_{CAC} = t_{RCL} + t_T + 130$ns for $t_{RCL} + t_T > 70$ns

Note that if $t_{RCL} + t_T \leq t_{RCLmax}$, device access time is determined by equation 3 and is equal to t_{RAC}. If $t_{RCL} + t_T > t_{RCLmax}$, access time is determined by equation 4. A 45ns interval ($t_{RCLmax} - t_{RCLmin}$) in which the falling edge of $\overline{CAS}$ can occur without affecting the access time is provided to allow for system timing skew in the generation of $\overline{CAS}$. This "designed in" skew window at the device level allows minimum access times to be achieved in practical system designs.

Note that both the $\overline{RAS}$ and $\overline{CAS}$ clocks are TTL compatible and do not require external level shifting to high voltage MOS levels. Internal buffers in the 2104A convert the TTL level signals to MOS levels inside the device. Therefore, the delay associated with external TTL-MOS level converters is not added to the 2104A system access time.

Write Cycle

A Write Cycle is performed by bringing Write Enable ($\overline{WE}$) low before or during $\overline{CAS}$. If Write Enable goes low at or before $\overline{CAS}$ goes low, the input data must be valid at or before the $\overline{CAS}$ falling edge. If Write Enable goes low after $\overline{CAS}$, Data In must be valid at or before the falling edge of WE. If Write Enable is low before $\overline{CAS}$ goes low, the data-out buffer will contain the written data at access time. However, if Write Enable goes low *while* $\overline{CAS}$ is low, a read operation may also be performed and data-out will go either high or low depending on the state of the accessed cell before the write takes place (refer to the Data Output Operation Section).

Refresh

Each of the 64 rows internal to the 2104A must be refreshed every 2 msec to maintain data. Any data cycle (Read, Write, Read-Modify-Write) refreshes the entire selected row (defined by the 6-bit row address). The refresh operation is independent of the state of $\overline{CS}$. It is evident, of course, that if a Write or Read-Modify-Write cycle is used to refresh a row, the device should be deselected ($\overline{CS}$ high) if it is desired not to change the state of the selected cell. $\overline{RAS}$-only cycles may also be used to refresh the 2104A at a savings in power dissipation over data cycles.

Page Mode Operation

Page mode operation with the 2104A allows faster successive memory data operations at the 64 *column* locations in a single address *row*. Receipt of a $\overline{RAS}$ and a 6-bit row address byte causes the RAM to access the 64 data cells on the addressed row.

At access time all 64 data bits are available at the sense amplifier outputs as long as $\overline{RAS}$ is held active. By cycling the $\overline{CAS}$ clock and addressing the desired data bit with the 6-bit column address byte all 64 data bits may be brought to the data output of the device. Data access and cycle time in this mode, called page mode, is faster than normal data cycles. Page mode is an excellent way to transfer blocks of data to and from memory at high speed, but it is impacted by refreshing.

The refresh requirements of the device limits the number of consecutive page mode cycles that may be performed. The device may remain in the page mode for a period no longer than the time required between refresh cycles. As an example, recall that the distributed refresh mode requires a refresh cycle every 31 microseconds. $\overline{RAS}$ may then remain low (active) for 31 microseconds maximum before it must be cycled high to precharge and then perform a refresh cycle. System page mode cycle times of 485 nanoseconds or less will enable all 64 data bits in the selected row to be examined or written between refresh cycles, maximizing the usefulness of page mode.

Power Dissipation/Operating

The power dissipation of a continuously operating 2104A device is the sum of $V_{DD} \times I_{DD}$ and $V_{BB} \times I_{BB}$. For a cycle time of 320 ns (including a t_{RP} of 100ns) the typical power dissipation of the 2104A-1 is 289 mW.

Standby Power-Refresh Only

The standby power-refresh only is calculated by the following equation:

1) $P_{REF} = P_{OP} \left(64 \frac{t_{CYC}}{t_{REF}}\right) + P_{SB}\left[1 - \left(64 \frac{t_{CYC}}{t_{REF}}\right)\right]$

Where:

P_{REF} = Standby power-refresh only.
P_{OP} = Power dissipation-continuous operation.
t_{CYC} = Refresh cycle time.
t_{REF} = Refresh period.
P_{SB} = Standby power dissipation.

Table II. 2104A Family Current Specifications

Symbol	Parameter	Limits Typ	Limits Max	Units	Comments
I_{DD1}	V_{DD} Standby Current	0.7	2.0	mA	V_{DD} = 13.2 Volts
I_{DD2}	V_{DD} Data Cycle Operating Current	24	35	mA	2104A-1 t_{CYC} = 320 ns
		22	32	mA	2104A-2 t_{CYC} = 320 ns
		20	30	mA	2104A-3,-4 t_{CYC} = 375 ns
I_{DD3}	V_{DD} $\overline{RAS}$-Only Cycle Operating Current	12	25	mA	2104A-1,-2 t_{CYC} = 320 ns
		10	22	mA	2104A-3,-4 t_{CYC} = 375 ns
I_{BB1}	V_{BB} Standby Current	5	50	µA	
I_{BB2}	V_{BB} Operating Current	160	400	µA	Minimum Cycle Time

The standby power dissipation P_{SB} is given by:

2) $P_{SB} = V_{DD} \times I_{DD1} + V_{BB} \times I_{BB1}$

The operating power P_{OP} is given by:

3) $P_{OP} = V_{DD} \times I_{DD2} + V_{BB} \times I_{BB2}$

for read and write data cycles or by:

4) $P_{OP} = V_{DD} \times I_{DD3} + V_{BB} \times I_{BB2}$

for $\overline{RAS}$-only refresh cycles.

Table II lists the pertinent current values for the 2104A family of devices.

Calculating the standby-refresh only power dissipation for the 2104A-1 using equations 1 through 4 above and the data from Table II yields:

a) For $\overline{RAS}$-only Refresh:

P_{REF} = 330mW (0.01) + 26.7mW (0.99) = 29.7mW maximum

b) For Read or Write Cycle Refresh:

P_{REF} = 462mW (0.01) + 26.7mW (0.99) = 31.0mW maximum

at V_{DD} = 13.2 volts, V_{BB} = –5.5 volts and the specified maximum current levels.

Data Output Operation

The operation of the output data latch is controlled by the $\overline{CAS}$ clock. Figure 9 indicates the content of the data latch following access time during various types of 2104A memory cycles. Table III summarizes the information on data content shown in Figure 9.

POWER DISTRIBUTION/DECOUPLING

General

Typical I_{DD} and I_{BB} current waveforms for the 2104A are shown in Figure 10. Examination of these waveforms shows that transient current drawn from the memory circuit board power distri-

Table III. Data Latch Content at End of Cycle

Type of Cycle	Data Latch Content (D_{OUT})
Read Cycle	Data from Addressed Memory Cell
Write Cycle	Input Data (D_{IN})
$\overline{RAS}$-Only Cycle	Data from Previous Cycle or HI-Z if Device was Deselected in Previous Cycle
$\overline{CAS}$-Only Cycle	HI-Z ($\overline{CAS}$-Only Deselects Device and Turns Output Buffer Off)
R-M-W Cycle	Data Read from Addressed Memory Cell During Read Portion of Cycle
Page Mode Read and Entry Cycle	Data from Addresssed Memory Cell
Page Mode Read Cycle	Data from Addressed Memory Cell
Page Mode Write Cycle (or Page Mode Write and Exit Cycle)	Input Data (D_{IN})

bution system is a function of the two device clocks, $\overline{RAS}$ and $\overline{CAS}$. The peak amplitude of the V_{DD} current transients is approximately 60 milliamperes with rise and fall times in the 5 to 10 nanosecond range and widths of typically 20 nanoseconds. Rise and fall times of this magnitude generate significant harmonic noise components in the 10 MHz and above frequency region. The power distribution/decoupling techniques used to suppress these noise components must be effective at these higher frequencies. The series inductance of the circuit board traces and the decoupling capacitors must be minimized to reduce time constant response effects of the distribution/decoupling system.

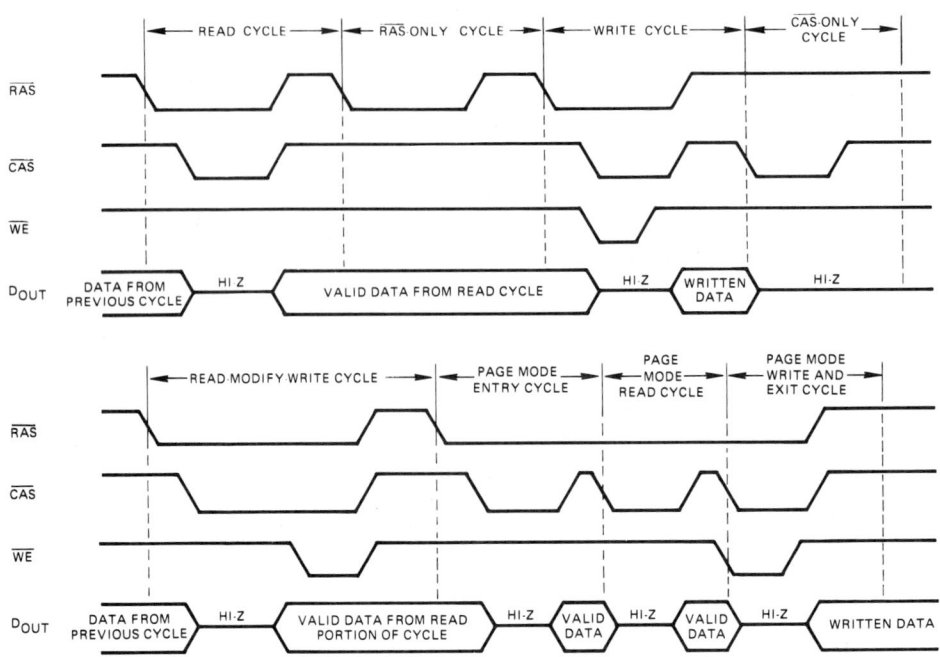

Figure 9. Operation of Data Output (D_{OUT})

Printed Circuit Board Trace Characteristics

Figure 11 shows the nominal lumped constant equivalent circuit of one-inch of 10 mil wide 2-ounce copper trace on a typical double sided printed circuit board with traces on both surfaces. The effect of the series resistance R_S is very small when compared to the series inductance L_S and can be ignored in practice. The series resistance is also non-reactive and its impedance is not frequency dependent. The following discussions will, therefore, not consider the minimal effects of R_S.

Decoupling Capacitor Characteristics

Capacitors used to decouple noise are not ideal devices and, therefore, exhibit inductive and resistive effects. Figure 12 shows the lumped constant equivalent circuit of a capacitor. The shunt resistance R_{SH} is a very high value (>10 MΩ) in capacitors of modern design and has minimal effects on the capacitor function. Therefore, the effect of R_{SH} will not be considered in the analysis of the decoupling capabilities of the capacitor.

The series inductance L_S in small disc ceramic and monolithic ceramic capacitors consists of lead inductance and is approximately 10nH/inch.

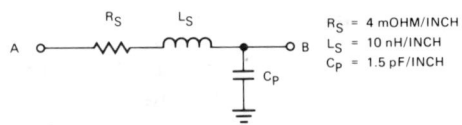

Figure 11. Lumped Constant Trace Equivalent Circuit

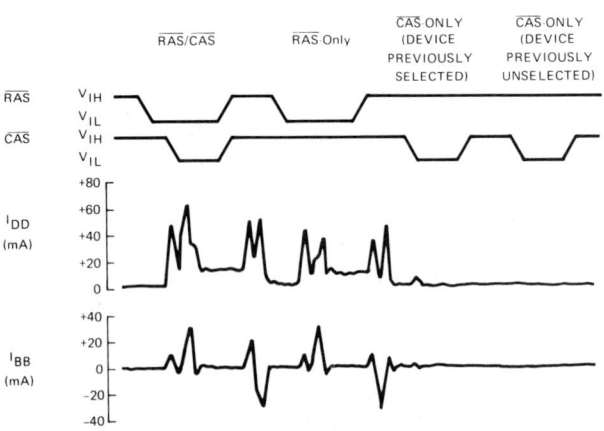

Figure 10. Typical Supply Current Waveforms

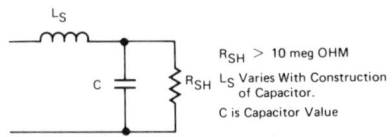

Figure 12. Lumped Constant Capacitor Equivalent Circuit

DESIGNING WITH 16 PIN, 4K DYNAMIC RAMS

The series inductance in bulk capacitors such as tantalum and aluminum electrolytics is much larger due to the construction of the capacitors. The internal series inductance of the electrolytic units varies widely with capacitance value, physical size, and construction type and is generally much greater than the lead inductance. For this reason, the effectiveness of electrolytic type capacitors as decoupling components for noise frequencies above 10 MHz is minimal. Their use in the power distribution/decoupling network is to provide a bulk power storage element located on the memory array board. This placement eliminates the inductive effects of the system backplane wiring on the power distribution to the memory array board.

Power Distribution System Characteristics

Now that models for the printed circuit board traces and decoupling capacitors have been generated, various power distribution/decoupling schemes can be compared for effectiveness in minimizing power supply noise levels.

Figure 13 shows a V_{DD} decoupling technique often used with dynamic RAMs. Total lead length external to the $0.1\mu F$ decoupling capacitor is approximately 1-inch. Add to that the 0.5 inch internal lead length of a typical disc ceramic capacitor and the lead length in series with the capacitor is 1.5 inches. This equates to a series inductance of (1.5 inches) (10nH/inch) = 15 nH. The impedance of L_S as a function of frequency is shown in Figure 14. When a current pulse occurs, current is drawn from the capacitor through the series inductance L_S.

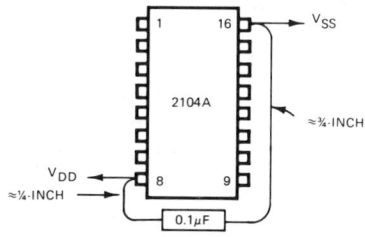

Figure 13. Commonly Used Capacitor Connection on V_{DD}

The impedance of the capacitor varies from 64 milliohms at 25 MHz to 16 milliohms at 100 MHz and is very small compared to the lead and trace impedance. Most of the impedance the current sees is in the inductance L_S and this is the impedance component of most concern to the system designer.

Fortunately, the energy spectrum of the current pulse is similar to that shown in Figure 15 which indicates that most of the energy is contained in the lower frequency components of the pulse.

Let's use Figures 14 and 15 as a hypothetical example and calculate the approximate noise generated by the leading edge of a 60 milliampere current pulse with a fast rise time. For simplicity, only the hypothetical components at 25, 50, 75, and 100 MHz will be included in the calculations. Also, the supply voltage will be assumed to be constant so that the vertical axis of Figure 15 represents current, I, as a percentage of total.

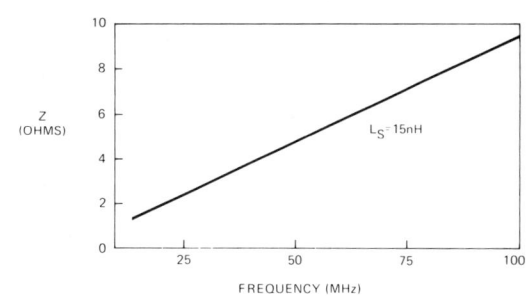

Figure 14. Impedance of L_S Versus Frequency

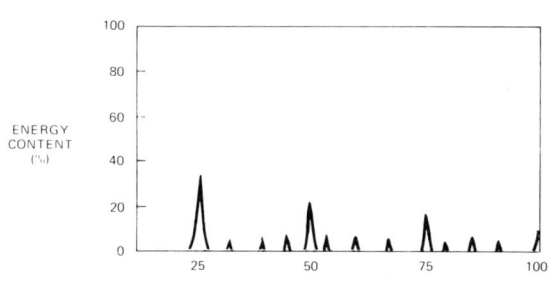

Figure 15. Hypothetical Energy Spectrum of Current Pulse

$$V_{noise} = (Z_{25})(I_{25}) + (Z_{50})(I_{50}) + (Z_{75})(I_{75}) + (Z_{100})(I_{100})$$
$$= (2.36)(35 \times 10^{-3}) + (4.71)(22 \times 10^{-3}) + (7.06)(15 \times 10^{-3}) + (9.4)(10 \times 10^{-3})$$
$$= 0.386 \text{ volts}$$

In other words, the voltage between the V_{DD} and V_{SS} pins on the memory device would drop by nearly 0.4 volt when the current pulse occurred. Considering all the current components would predictably increase this to 0.5 volt or more. Add to this the noise coupled into this LC circuit from the other similar circuits in the memory array and it becomes apparent that this memory device may see V_{DD} noise levels approaching 1.0 volt. While the device may operate with that noise level, operational margins of the device may well be reduced. Every practical effort should be made by the designer to reduce the overall noise level to 0.5 volts peak-to-peak or less.

One way this can be accomplished is by simply reducing the inductance in the circuit. Figure 16 shows a way to reduce the inductance. The equivalent inductance of the two traces from pin 16 to the capacitor is the parallel combination of the two paths since:

$$L_p = \frac{L_1 L_2}{L_1 + L_2} = \frac{(10nH)(10nH)}{20nH} = 5nH$$

Add that to the inductance of the capacitor lead length and the 0.25 inch trace from the capacitor to pin 8 of the device. The total inductance is then:

$$L_s = 5nH + (0.75 \text{ inch})(10nH/inch)$$
$$= 12.5nH$$

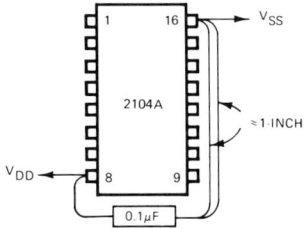

Figure 16. Reduction of Inductance by Paralleling Traces

The impedance of this inductance at 25 MHz and 100 MHz is 2.0-ohms and 7.8-ohms respectively. This compares to 2.4-ohms and 9.4-ohms for the original circuit at those frequencies and is a reduction in impedance of almost 17%.

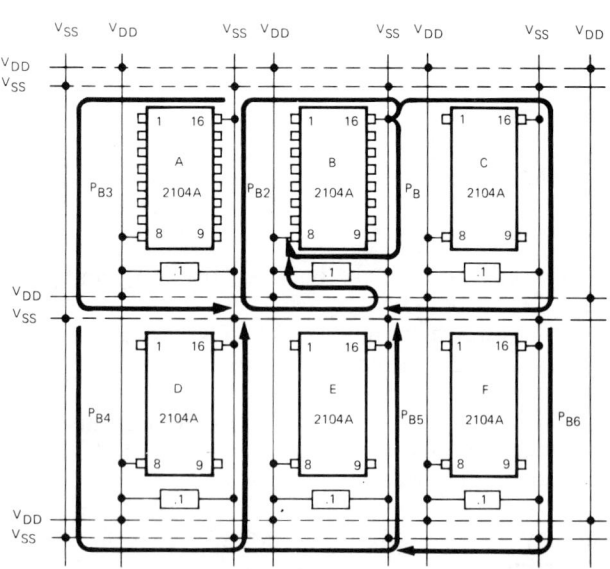

Figure 17. Gridded V_{DD} and V_{SS}

Additional reductions can be achieved by addition of more parallel traces or wider, lower inductance traces. Neither of these approaches is really practical, however, since board space is generally at a premium. A more practical, equally effective method of inductance reduction is the use of "gridded" power distribution. This involves bussing each power supply distribution network both horizontally and vertically on the circuit board. An example of this type of distribution for V_{DD} and V_{SS} is shown in Figure 17 for a matrix of six devices on a double sided printed circuit board. Consider the path of the current drawn by device B in Figure 17.

As indicated, in addition to the primary path P_B, there are no less than six other secondary, parallel paths (P_{B1} through P_{B6}) due to the gridding of the V_{SS} supply distribution system. Each of these secondary paths or traces is in parallel with the primary trace reducing the equivalent inductance of the current path between pin 16 and pin 8 of device B. Similar parallel paths exist for all the devices in the matrix.

Use of such a gridded power distribution network is recommended for all dynamic RAM systems due to the characteristics of the device current waveforms. Experience has shown the gridded distribution system to equal the performance of the more expensive multi-layer printed circuit board with internal power layers.

Additional power supply distribution traces may be added to the layout example of Figure 17 with only slightly greater side-to-side and end-to-end spacing between adjacent devices. It is recommended that V_{DD}, V_{BB}, and V_{SS} distribution systems be gridded. Figure 18 shows a recommended double-sided layout for 16-pin 4K and 16K RAMs. V_{CC} may be gridded but it is generally sufficient to distribute V_{CC} in one direction only since the 2104A itself does not draw power from the V_{CC} supply but only uses it to supply input levels to the peripheral TTL devices connected to the D_{OUT} pin (refer to the Device Circuit Operation Section).

Alternate power distribution layout techniques may be used with dynamic RAMs and some will show comparable results to the gridded system depending on memory array size, the number and placement of the decoupling capacitors, and the number of memory devices which are active in any given cycle. The gridded system has been proven in many production systems, however, and its use will result in predictable, workable power supply noise characteritistics. One commonly used distribution system is illustrated in Figure 19. This technique should definitely not be considered for use with dynamic RAMs simply due to the length of the current path between the V_{DD} and V_{SS} pins of any device in the matrix. As an example, device B

DESIGNING WITH 16 PIN, 4K DYNAMIC RAMS

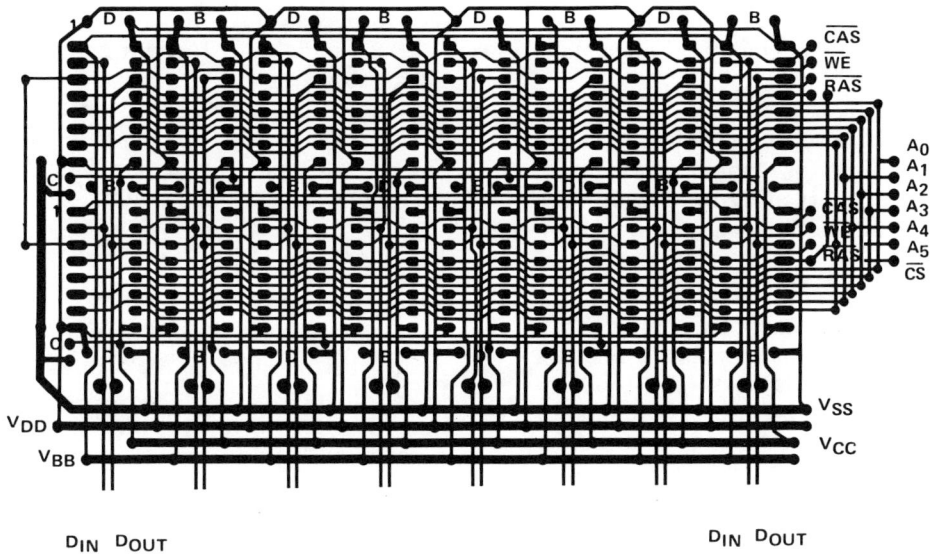

DECOUPLING CAPACITORS
D = 0.1 µF to V_{DD} TO V_{SS}
B = 0.1 µF V_{BB} TO V_{SS}
C = 0.01 µF V_{CC} TO V_{SS}

Figure 18. Recommended Two-Sided Board Layout for 2104A

in Figure 19 has an unnecessarily long and, therefore, high inductance current path between its V_{DD} and V_{SS} pins. Compounding the problem, it shares most of that current path with device C and all of it with device A. The magnitude of the noise between the V_{DD} and V_{SS} pins of device B is greatly dependent upon the noise generated by the other adjacent devices. This sytem is unacceptable at best and is to be avoided. Unfortunately, this is the power distribution scheme found on many of the "prototyping" printed circuit boards available on the market. Examine your prototyping boards carefully and avoid the use of this type for the memory array or add wiring to the board to grid the supplies.

Recommended Decoupling Values

The decoupling capacitors used in the memory array should be types which exhibit good high frequency characteristics as discussed earlier. It is recommended that a 0.1µF ceramic capacitor be connected between V_{DD} and V_{SS} at every other device in the memory array. It is also recommended that a 0.1µF ceramic capacitor be connected between V_{BB} and V_{SS} at every other device in the array, preferably the alternate devices to the V_{DD} decoupling. Smaller capacitor values such as 0.01µF may be substituted but noise levels will increase with any given distribution scheme due to the higher capacitive impedance. Empirical comparative data should be taken and decoupling efficiency considered with the distribution system being used before the smaller capacitors are used. The small

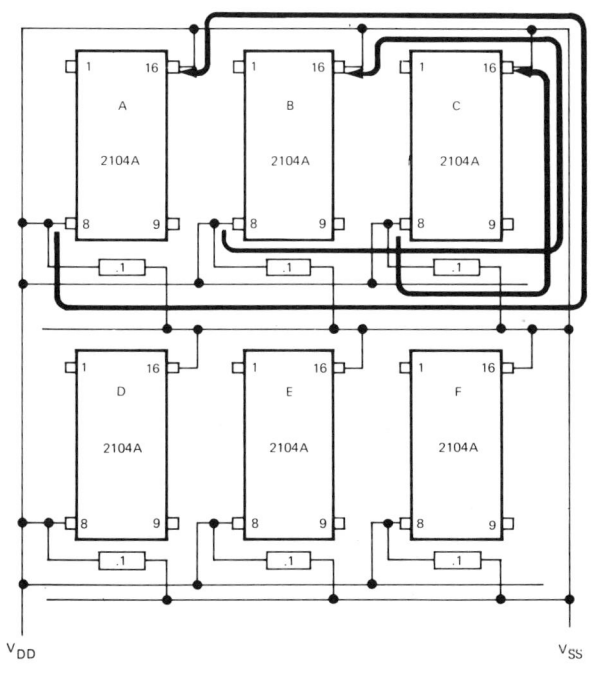

Figure 19. Unacceptable Power Distribution System

cost difference between 0.1μF and 0.01μF capacitors may be negated by degradation of system noise margins.

A 0.01μF ceramic capacitor is recommended between V_{CC} and V_{SS} for approximately each eight devices in the memory array to prevent noise coupled to the V_{CC} line in the memory array from affecting the peripheral TTL logic in the system.

In addition to the ceramic capacitors in the memory array, it is recommended that a 10μF tantalum or equivalent capacitor be connected between V_{DD} and V_{SS} adjacent to the array for each 16 memory devices in the array. An equal or slightly smaller value bulk capacitor is also recommended between V_{BB} and V_{SS} for each 32 memory devices on the array. These bulk capacitors eliminate the inductive and resistive effects of the memory system backplane wiring connecting the memory array boards to the system power supplies.

I_{BB} Characteristics

The high performance of the 2104A results from advanced design and processing techniques developed by Intel. These techniques yield slightly different characteristics in the I_{BB} parameter than with the previous Intel 4K Dynamic RAMs. These changes have little effect on the V_{BB} power supply requirements of a typical system but they do require that I_{BB} be specified in a different manner and for this reason, I_{BB} will be discussed here in detail for clarification.

In a typical MOSFET integrated circuit the current from the V_{BB} (substrate) supply when the device is turned off is essentially the leakage from the source and drain diffusions into the substrate. This leakage current is in the nanoampere range for each individual MOS transistor but when multiplied by the 6000 or so transistors in a typical 4K RAM, the total leakage is typically 50 to 60 microamperes. When the device turns on and current is conducted between the drain and source, charge carriers flow between the drain and source through the gate voltage induced channel between the two terminals. As the carriers move through the region of high electric field close to the drain, they generate additional carriers by impact ionization. Most of these carriers join the initial carriers and move between the source and drain terminals due to the influence of the electric field created by the potential difference between the source and gate terminals. Some of these carriers however, are accelerated through the boundaries of the channel into the substrate and add to the leakage currents from the drain and source diffusions. This increases the I_{BB} current slightly during the time the device is operational. The number of these additional carriers is relatively low in typical 4K RAM and results only in a 15 to 20 μA increase in I_{BB} during the time the device clocks ($\overline{RAS}$ and $\overline{CAS}$) are active. This is because the energy, i.e. speed, of the carriers in the channel is not high enough to generate many additional carriers via impact ionization. The 2104 I_{BB} specification was 100μA maximum.

In the 2104A, shallow diffusions are used for the source and drain and thin gate oxide is used for speed/performance reasons. This results in much higher energy, i.e., faster, carriers in the channel due to the high electric field in the channel. These higher energy carriers are capable of generating more carriers than in previous 4K RAMs. The increase in I_{BB} during this action is significant, typically 100μA or more. Importantly, however, this increase is only during the time the clocks are active.

As a result of this difference in I_{BB} during inactive (standby) and active conditions of the clock, the 2104A has two I_{BB} specifications. I_{BB1} is the I_{BB} current during standby and I_{BB2} is the I_{BB} current during a memory device cycle. Interestingly enough, the standby I_{BB} current for the 2104A is lower than for the earlier 2104 due mostly to processing and design improvements. Due to these same improvements, however, the operating I_{BB} specification, I_{BB2}, is typically 160μA (400μA maximum).

What does this higher I_{BB} during operation mean in a typical 16K by 8-bit system? Assuming a 50% duty cycle due to data and refresh cycles and a cycle time of 500nsec, the average I_{BB} per device during any 2 msec refresh period will be:

$$I_{BBavg} = \frac{(I_{BB2max})(1msec)}{2msec} + \frac{(I_{BB1max})(1msec)}{2msec} = 225\mu A \text{ max}$$

and typically:

$$I_{BBavg(typ)} = \frac{(160\times10^{-6})(1\times10^{-3})}{2\times10^{-3}} + \frac{(5\times10^{-6})(1\times10^{-3})}{2\times10^{-3}} = 83\mu A \text{ typ.}$$

Variations in duty cycle will decrease or increase these values but most systems will experience I_{BB} values not much different than with other 16-pin 4K RAMs. In the 16K X 8-bit system example, the total I_{BBavg} will be no greater than 7.2 milliamperes.

DESIGNING WITH 22 PIN, 4096 DYNAMIC RAMS

JIM OLIPHANT

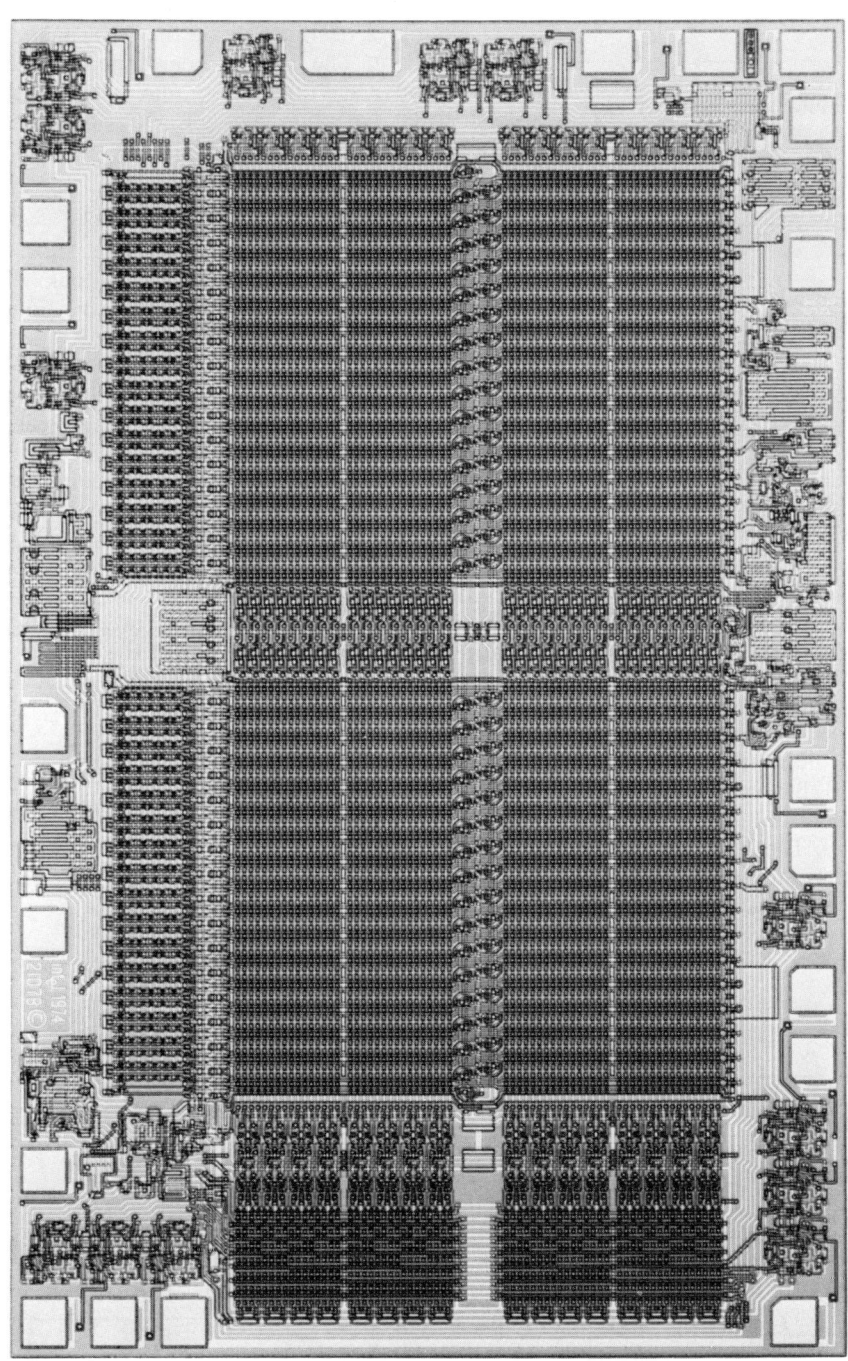

Photomicrograph of the Intel 4096 Word × 1 Bit 2107B Dynamic RAM.

INTRODUCTION

The Intel® 2107B is a 4096 word by 1 bit dynamic random access memory. The 2107B is fabricated using Intel's standard reliability proven n-channel silicon gate MOS technology. The device is packaged in a standard 22-pin DIP. The pin configuration and logic symbol are shown in Figure 1. Note that the 2107B can be used as a replacement for the 2107A.

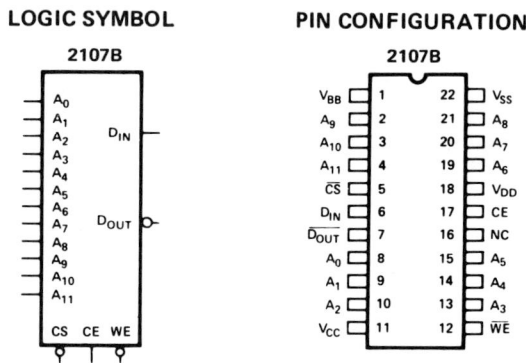

Figure 1. 2107B Logic Symbol and Pin Configuration

The combination of Intel's® n-channel silicon gate process and circuit design has resulted in a part that is very fast, easy to use, and economically produced in large volume. In addition, the combination of process and device design has resulted in a very small device (see Figure 2) using conservative layout rules (same as 2102A). The small size offers advantages in both large volume production and increased reliability. The 2107B operates with three power supplies relative to ground; V_{DD} (+12V), V_{BB} (−5V), and V_{CC} (+5V). The V_{CC} (+5V) supply is connected only to the output buffer of the 2107B and may be turned off during power down operations.

The 2107B has one MOS level clock (Chip Enable) with all other inputs being low level TTL compatible (+2.4V V_{IH} minimum). The output is capable of driving 1 TTL load.

The purpose of this chapter is to describe the internal operation of the 2107B, outline those areas in system implementation to which the designer should pay particular attention and to discuss typical examples of the uses of the 2107B in systems environment. The chapter is arranged so that each of the above sections can be read independently of each other without having to go through any unwanted detail in the other sections.

INTERNAL DEVICE OPERATION

Internal operation of the 2107B is most easily understood with the aid of the block diagram shown in Figure 3. As is shown in this figure, the memory array is arranged in a 64 row × 64 column matrix of storage cells. The storage cells are implemented with a single transistor and a "storage" capacitor and are called single transistor cells. The operation of the storage cell will be discussed later. The memory cell is accessed by the coincidence of a row select (defined by addresses $A_0 - A_5$) and a column select (defined by addresses $A_6 - A_{11}$) signal at the desired address. An on chip timing and control generator provides for the internal timing signals for decoding, read/write strobing, data gating and output gating. All of the timing circuits in the 2107B are activated by the positive-going edge of chip enable.

Chip select controls the data I/O gating circuits internal to the 2107B. When chip select is high the output data buffer is in a high impedance state and

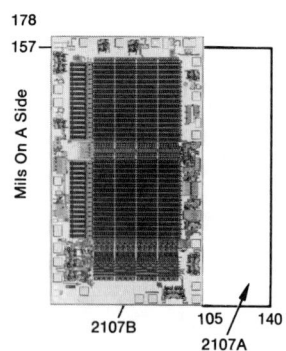

Figure 2. 2107B Comparative Die Size

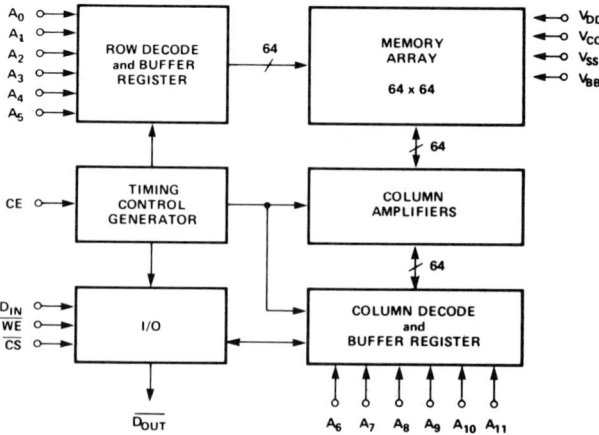

Figure 3. 2107B Block Diagram

the data-in buffer is electrically isolated from the data-in input pin. Since chip select controls only the internal data buffers and not the timing generators or address buffers internal to the 2107B, it is possible to refresh the 2107B with chip select high by initiating a read/refresh or write cycle.

The address buffer registers consist of latches activated at the leading edge of chip enable. Since the addresses are latched shortly after chip enable goes high, it is permissible to change the address long before the memory cycle is completed to set up for the next cycle.

The write enable input activates the data-in buffer gating data to the selected memory cell. Input data must be valid at the time write enable goes low to assure that the proper data is written into memory.

Circuit implementation and operation of each of the major input/output and storage portions are discussed below.

Storage Cell Operation

The storage cell used in the 2107B is implemented with a single transistor and storage capacitor as shown in Figure 4. From this figure it is shown that a charge on a storage cell is gated to the bit sense line by the MOS device connected to the column select line. (Note that for a given column select, 64 storage devices are gated to the respective 64 bit sense lines.)

Consider first a read operation and the case where the storage capacitor C_{STG} is discharged; i.e., node (1) is at V_{SS} (GND). Prior to chip enable going high, the bit sense lines have been precharged to V' by device Q_1. [V' is a voltage between V_{DD} (+12V) and V_{SS}.] After the address decoders have stabilized, the proper column select line is brought high, turning on device Q_2. The storage capacitor is then electrically connected to the bit sense line. At this time the charge on $C_{I/O}$ (proportional to the precharge voltage V') is redistributed between $C_{I/O}$ (parasitic capacitance of bit sense line) and C_{STG}. Since C_{STG} was initially discharged (node 1 at V_{SS}) the voltage will distribute between $C_{I/O}$ and C_{STG} according to the following relationship:

$$V_{BIT\ SENSE}(t_1) = V_{BIT\ SENSE}(t_0) \left(\frac{C_{I/O}}{C_{I/O} + C_{STG}}\right)$$

Since $C_{I/O}$ is very much larger than C_{STG} the change in the voltage on the bit sense line will be very small. The sense amplifier (S/A) is designed to detect very small changes in bit sense line voltage and to latch in a state near V_{SS} (GND) or V_{DD} (+12V), depending on the state of the storage cell.

Sensing an initial charge on C_{STG} (proportional to V_X where $V_X = V_{DD} - V_{TH}$, V_{TH} is the effective MOS threshold) is identical to the sequence described above. The only difference is that now the bit sense line is driven above the initial V' precharge voltage. Again the sense amplifier detects the small change in bit sense line voltage and latches in the appropriate state.

Note that during a read operation of the storage cell, the original charge (data) on the storage cell is changed (i.e., the read operation is effectively a destructive read). Data is rewritten back on the storage capacitor C_{STG} by the sense amplifier after it has latched in the proper state. For example, if C_{STG} was initially charged to V_X (~10V), the sense amplifier will latch the bit sense line to V_X and, since the column select line is on (high), the original data is automatically rewritten into C_{STG}. The entire operation is transparent to the user.

A plot of the voltage on the bit sense line for the two cases described above is shown in Figure 5.

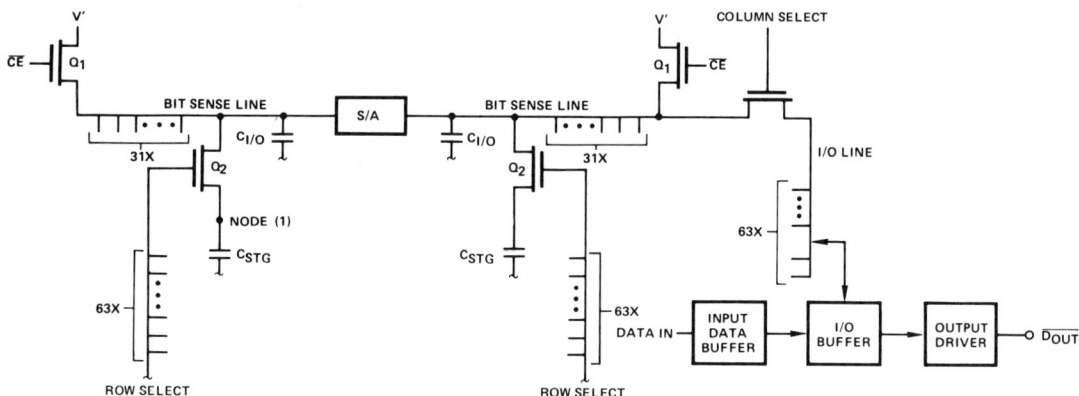

Figure 4. 2107B Memory Cell and Associated I/O Circuitry

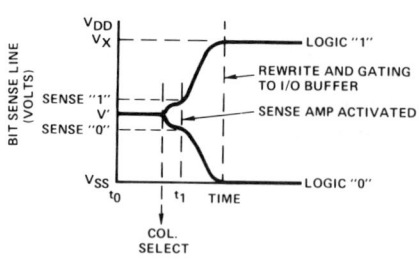

Figure 5. Bit Sense Line Voltage

A write operation is identical to the rewrite portion of a read cycle. In this case, however, the incoming data "overrides" the state of the sense amplifier (if different from the desired state) and writes into the selected cell. It is important to remember that the data-output at the output pin is the logical inverse of the data written into memory.

Data Sense/Latch

As discussed previously, a sense amplifier on the bit sense line is necessary to detect the low level data signals generated on the bit sense line during a read cycle. A simplified circuit schematic used for the sense amplifier is shown in Figure 6.

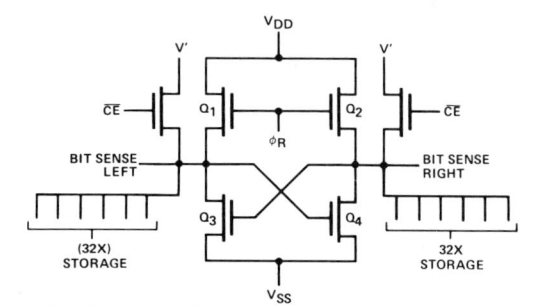

Figure 6. Data Sense/Latch

Before chip enable is brought high, both sides of the bit sense lines are precharged to V' (as discussed previously). At the proper time (after all data transients have subsided) devices Q_1 and Q_2 are turned on by ϕ_R going positive. At this time, the state of bit sense left is compared with bit sense right causing the latch to lock in the appropriate state. For example, if the right bit sense line is at a higher potential than the left bit sense line, device Q_3 will begin to conduct. The cross coupled latch will then fully switch with bit sense left going to V_{SS} and bit sense right to V_X.

Address Buffer/Latch

The address buffer/latch is shown in Figure 7. The input to the address buffer/latch is low voltage compatible which the circuit senses, translates to MOS level signals and latches.

Operation of the address buffers is as follows: During chip enable off time (CE low) both sides of the latch are precharged to V_X (~10V) by devices Q_1, Q_2, and Q_3. Device Q_3 is used to assure that the initial precharge on each side of the latch are equal.

When chip enable goes high, the input to the address buffer (A_{IN}) is gated to the cross coupled latch which latches the appropriate MOS level at A_I and $\overline{A_I}$. For example, if the TTL address input is high, then device Q_7 will turn on at ϕ_A time. The cross coupled latch then regenerates, turning Q_6 off. The quiescent state of the latch for this input is Q_6 off, Q_8 on, thereby setting A_I and $\overline{A_I}$ to MOS level high and low, respectively.

This type of latch is capable of triggering and latching at very high speeds which allows the addresses to be removed from the input as soon as possible. However, there are a few characteristics of this latch which have an effect when the device is placed in a system environment.

First note that node (1), Figure 7, has been precharged to a high MOS level of V_X (~10V). When

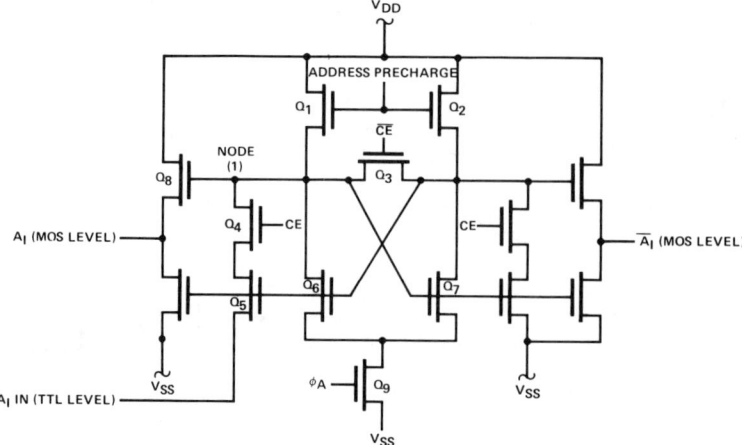

Figure 7. Address Buffer/Latch

chip enable goes high turning on Q_4, the charge on this node is connected to the address input node (A_{IN}) for a short period of time (until the latch switches). This results in a small positive voltage shift on the address input A_{IN}. It follows then that the more 2107B devices attached to a given address driver the larger the voltage excursion will be. This excursion has been found to cause no problem in any reasonable system environment (as described later) and amounts to no more than 9 mV positive shift for each 2107B. The amount of positive charge coupling depends upon the address driver and the address line impedance. As should be expected, the most sensitive address level is the low level (V_{IL}) since any positive coupling decreases the available noise margin.

Another characteristic to be aware of in this type address buffer is the input current drawn through the address driver when an address goes from a low state to a high state during chip enable high. This condition results from the latch being set in the state where Q_6 is on as well as Q_4 and Q_5. Current is then drawn through devices Q_4, Q_5, Q_6 and Q_9. This current is typically in the order of 0.5 mA. Note that although this may cause a load on the address driver and cause it to drop below 2.4V, there is no effect on the memory component since the desired address has been latched in. This current is drawn only as long as chip enable is high. When chip enable goes low, device Q_4 is turned off, opening the current path. This effect will be shown on various type drivers in a later section (Low Voltage Buffer/Drivers).

Output Driver

A schematic of the output buffer is shown in Figure 8. Note that the output is in a high impedance state if either chip select is high or chip enable is low. Further, the V_{CC} shown in Figure 8 is the only connection the V_{CC} makes on the 2107B. This allows V_{CC} to have a wide range of values (up to V_{DD}) if types of sensing other than TTL is desired.

2107B Bit Map

Figure 9 gives the location of each cell in the memory matrix for each address. As shown in this

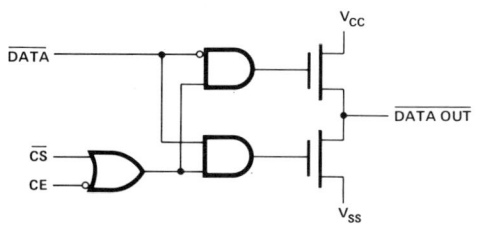

Figure 8. 2107B Output Driver

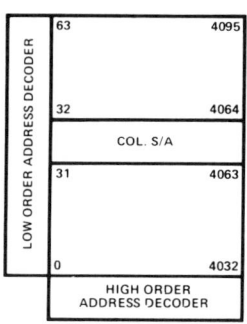

Figure 9. 2107B Bit Map

figure, the addresses run sequentially starting from the lower left corner (device oriented as shown).

2107B SPECIFICATION

Although the device specifications for the 2107B are concise and self explanatory, some sections are included here to emphasize those areas of most interest to the designer. Consider first the DC and operating characteristics shown in Table I marked with a [5].

The V_{DD} supply current during chip enable off is specified at 200 μa maximum with chip enable no higher than 0.6V. It is important to hold the low level of chip enable at or below this value (to a maximum of −1.0V) to assure that devices internal to the 2107B do not turn partially on. Note that considering only the AC operating environment, chip enable can go as high as 1.0V above V_{SS} and the device will still operate properly. This requirement on chip enable off is most important in those systems being placed in a low power refresh only standby mode.

The V_{BB} supply current load (I_{BB}) is maximum at 100 μa and includes all leakages. It is not necessary to add the other leakage currents (e.g., I_{LI}, I_{LC}) to I_{BB} to calculate supply drain on V_{BB}.

The input low voltage (for low level signals) V_{IL}, is specified as a function of the chip enable rise time and is referenced to a transition with $t_T = 20$ nsec. It is recognized that in some system applications, the load on the chip enable driver may result in transitions of 30 nsec or higher (to a maximum of 40 nsec). If the chip enable transition in the system is not 20 nsec or faster (to a minimum of 10 nsec), then the typical low level for the low level drivers is shown by the graph in Figure 10. It is important to include any noise which may be on the address line during t_{AH} (address hold) time. An example of the noise expected on an address line when chip enable goes high (during refresh) is shown in Figure 11. The noise shown here is the result of 36 devices attempting to raise the address driver (see Address Buffer/Latch) level during refresh time. Refresh

Table I. D.C. and Operating Characteristics

$T_A = 0°C$ to $70°C$, $V_{DD} = +12V \pm 5\%$, $V_{CC} = +5V \pm 10\%$, V_{BB}[1] $= -5V \pm 5\%$, $V_{SS} = 0V$, unless otherwise noted.

Symbol	Parameter	Limits Min.	Limits Typ.[2]	Limits Max.	Unit	Conditions
I_{LI}	Input Load Current (all inputs except CE)		.01	10	µA	$V_{IN} = V_{IL\,MIN}$ to $V_{IH\,MAX}$ $CE = V_{ILC}$ or V_{IHC}
I_{LC}	Input Load Current		.01	2	µA	$V_{IN} = V_{IL\,MIN}$ to $V_{IH\,MAX}$
$|I_{LO}|$	Output Leakage Current for high impedance state		.01	10	µA	$CE = V_{ILC}$ or $\overline{CS} = V_{IH}$ $V_O = 0V$ to $5.25V$
I_{DD1}[5]	V_{DD} Supply Current during CE off[3]		110	200	µA	$CE = -1V$ to $+.6V$
I_{DD2}	V_{DD} Supply Current during CE on			60	mA	$CE = V_{IHC}$, $\overline{CS} = V_{IL}$
$I_{DD\,AV}$[5]	Average V_{DD} Current		38	54	mA	Cycle time = 400ns, $t_{CE} = 230$ns $\overline{CS} = V_{IL}$; $T_A = 25°C$
I_{CC1}[4]	V_{CC} Supply Current during CE off		.01	10	µA	$CE = V_{ILC}$ or $\overline{CS} = V_{IH}$
I_{BB}[5]	V_{BB} Supply Current		5	100	µA	
V_{IL}[5]	Input Low Voltage	-1.0		0.6	V	$t_T = 20$ns — See Figure 10
V_{IH}	Input High Voltage	2.4		$V_{CC}+1$	V	$t_T = 20$ns
V_{ILC}	CE Input Low Voltage	-1.0		+1.0	V	
V_{IHC}[5]	CE Input High Voltage	$V_{DD}-1$		$V_{DD}+1$	V	
V_{OL}	Output Low Voltage	0.0		0.45	V	$I_{OL} = 2.0$mA
V_{OH}	Output High Voltage	2.4		V_{CC}	V	$I_{OH} = -2.0$mA

NOTES:

1. The only requirement for the sequence of applying voltage to the device is that V_{DD}, V_{CC}, and V_{SS} should never be 0.3V more negative than V_{BB}.
2. Typical values are for $T_A = 25°C$ and nominal power supply voltages.
3. The I_{DD} and I_{CC} currents flow to V_{SS}. The I_{BB} current is the sum of all leakage currents.
4. During CE on V_{CC} supply current is dependent on output loading, V_{CC} is connected to output buffer only.
5. See discussion — 2107B specifications.

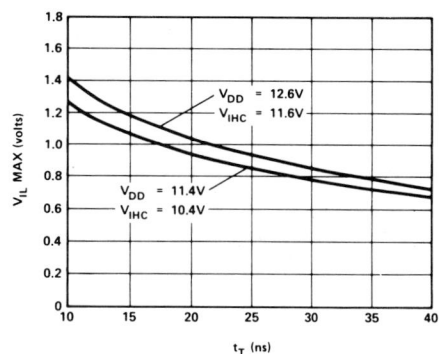

Figure 10. V_{IL} vs CE Rise Time

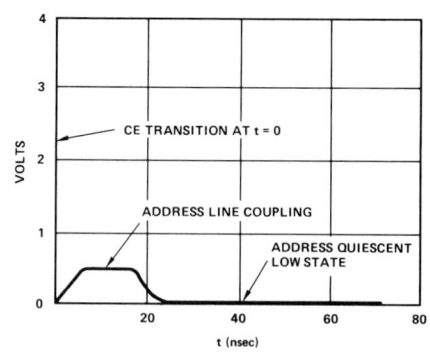

36 2107B DEVICES ON ADDRESS DRIVER
REFRESH CYCLE

Figure 11. Coupling to Address Line Caused by Chip Enable Transition

DESIGNING WITH 22 PIN, 4K DYNAMIC RAMS

is the worst case since all chip enable signals will be simultaneously decoded and driven at the same time.

The chip enable high voltage, V_{IHC}, can vary between $V_{DD}+1.0$ and $V_{DD}-1.0$ volts. This allows maximum flexibility in the driver design and provides for adequate noise margins.

The average V_{DD} current ($I_{DD\;AV}$) during a read/write cycle is specified to be a maximum of 54ma. This current is a function of both cycle time and temperature as shown in Figures 12 and 13, respectively. As shown by these curves, the maximum power occurs at low temperature and maximum duty cycle.

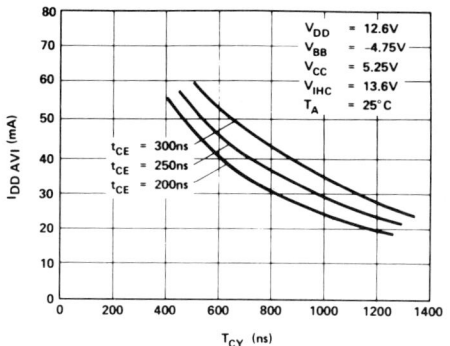

Figure 12. $I_{DD\;AV}$ vs Cycle Time

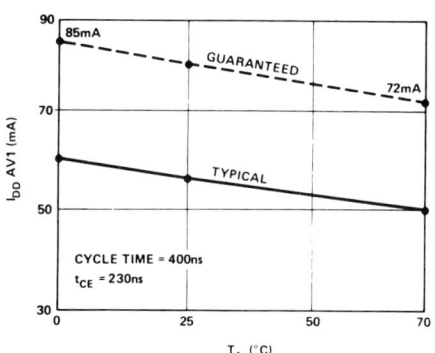

Figure 13. $I_{DD\;AV}$ vs Temperature

Timing

The timing relationship between the control, addresses, and data in/out is very straightforward as shown in the specification. For reference, the Read/Refresh, Write and Read-Modify-Write cycles are shown in Figures 14, 15, and 16, respectively, for minimum timing. Selected points are discussed which may cause the most problems if they are violated in a system environment.

For all cycles it is imperative to make certain that the address inputs are valid at or before chip enable reaches the $V_{SS}+2.0V$ level (t_{AC}). The high speed of the 2107B address buffer/latches means that if the address inputs are not valid until just after chip enable goes high, the wrong address is likely to be latched in the chip. Likewise, the input data must not change after write enable goes low while chip enable is high (t_{DW}). Again, violation of this requirement may result in incorrect data being written into memory.

Note that for all cycles, the data-out output goes to a low state shortly after chip enable goes high. This prohibits the output from being tied directly to a clear or preset input of a latch.

Problems can occur when one or more parts of these specifications are violated.

Transient Currents

Although the transient currents in the 2107B are easily handled, proper attention should be paid to the peak values and adequate decoupling provided to handle the expected transients. Figure 17 shows the transient currents present in the 2107B.

Consider first the transient current supplied by the chip enable driver I_{CE}. It is noted that this current does not have a resistive component but is strictly a charging current represented by the relationship:

$$I = C\left(\frac{dv}{dt}\right)$$

As expected, the largest transient current drawn by the 2107B is the V_{DD} supply and is represented by I_{DD}. The first portion of this curve shown as A is the result of internal nodes charging up for operation. The section shown as B is the result of the address buffers/decoders turning off. Portion C is the "steady state" current drawn by all internal circuits while chip enable is high.

Portion D of the transient current is the result of feedthrough capacitance (internal to the chip) coupling to V_{DD} when chip enable goes low. Portion E is the precharging of selected internal nodes by the chip enable generator (e.g., precharging bit sense line. See section on Internal Device Operation).

The transients associated with the V_{BB} supply I_{BB} should be reviewed closely. Note that the peak values are approximately 20 ma during a cycle with a time base as shown. Special attention is called to this because even though the *average* DC current is very small (maximum 100 μa) the peak currents can be two orders of magnitude higher.

(Numbers in parentheses are for minimum cycle timing in ns)

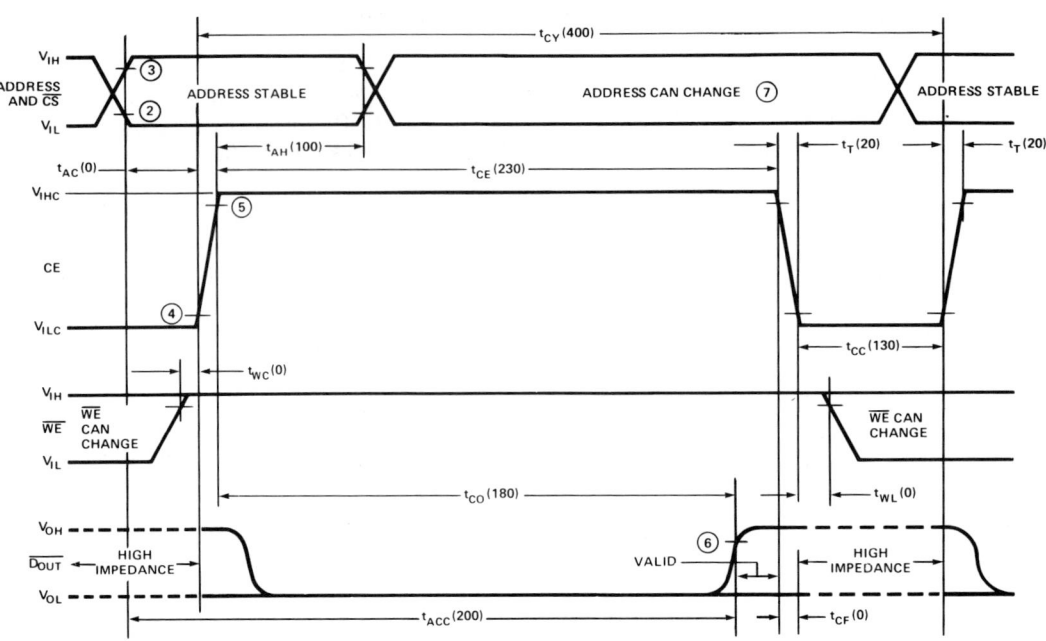

Figure 14. Read/Refresh Cycle[1]

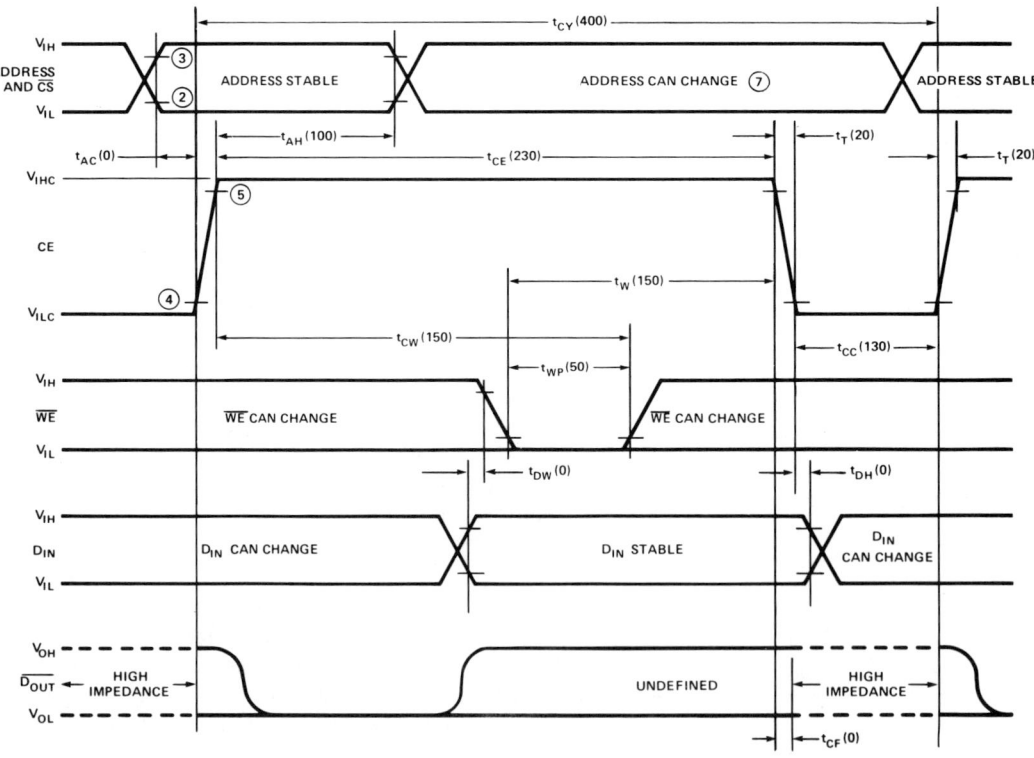

Figure 15. Write Cycle

NOTES:
1. For Refresh cycle row and column addresses must be stable before t_{AC} and remain stable for entire t_{AH} period.
2. V_{IL} MAX is the reference level for measuring timing of the addresses, $\overline{CS}$, $\overline{WE}$, and D_{IN}.
3. V_{IH} MIN is the reference level for measuring timing of the addresses, $\overline{CS}$, $\overline{WE}$, and D_{IN}.
4. V_{SS} +2.0V is the reference level for measuring timing of CE.
5. V_{DD} −2V is the reference level for measuring timing of CE.
6. V_{SS} +2.0V is the reference level for measuring the timing of $\overline{D_{OUT}}$.
7. During CE high typically 0.5mA will be drawn from any address pin which is switched from low to high.

DESIGNING WITH 22 PIN, 4K DYNAMIC RAMS

(Numbers in parentheses are for minimum cycle timing in ns.)

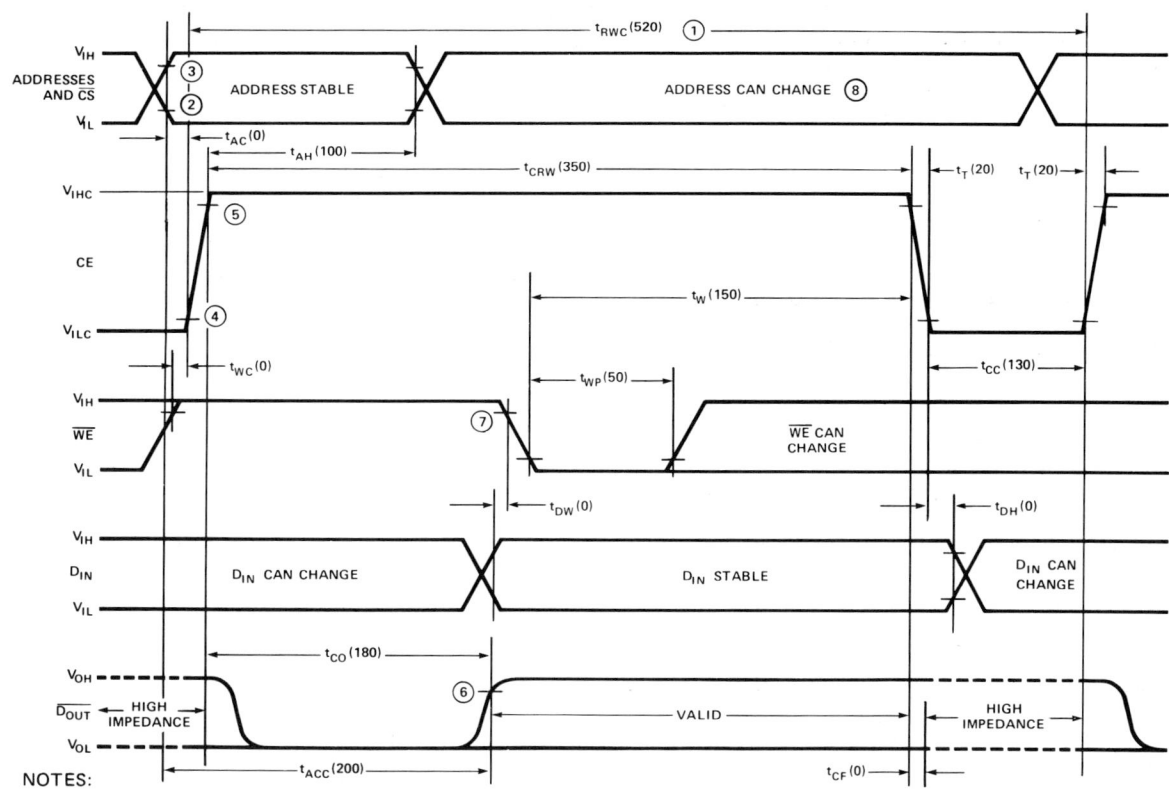

NOTES:
1. Minimum cycle timing is based on t_T of 20ns.
2. V_{IL} MAX is the reference level for measuring timing of the addresses, $\overline{CS}$, $\overline{WE}$, and D_{IN}.
3. V_{IH} MIN is the reference level for measuring timing of the addresses, $\overline{CS}$, $\overline{WE}$, and D_{IN}.
4. V_{SS} +2.0V is the reference level for measuring timing of CE.
5. V_{DD} −2V is the reference level for measuring timing of CE.
6. V_{SS} +2.0V is the reference level for measuring the timing of $\overline{D_{OUT}}$.
7. $\overline{WE}$ must be at V_{IH} until end of t_{CO}.
8. During CE high typically 0.5mA will be drawn from any address pin which is switched from low to high.

Figure 16. Read-Modify-Write Cycle

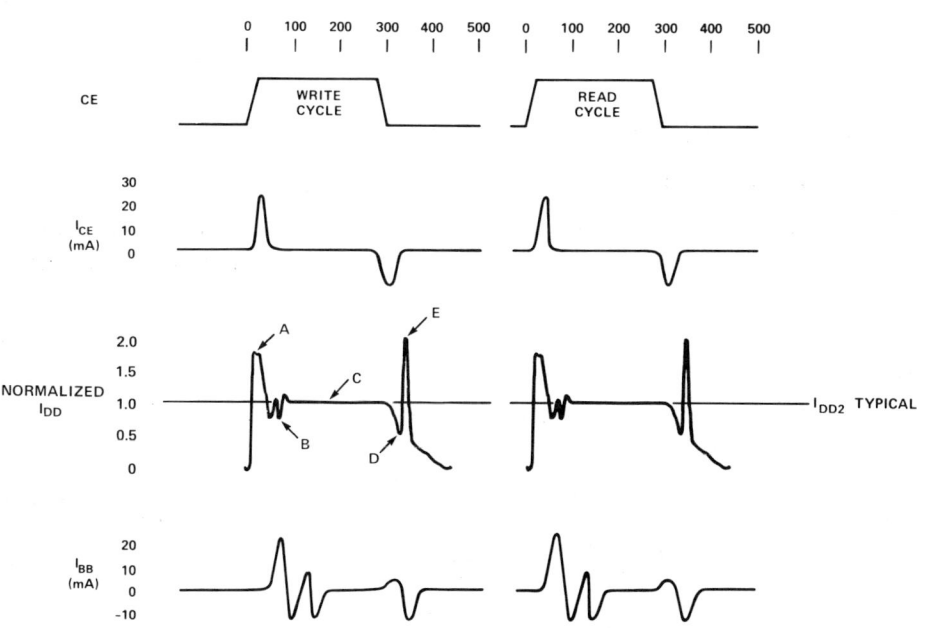

Figure 17. Typical Current Transients vs Time

These transients are characteristic of any dynamic RAM and are in part related to the density of the RAM. Again, if adequate decoupling measures are taken, very little noise will be generated on the V_{BB} system distribution.

Notice that the transient current for I_{CC1} is not shown. This is because the V_{CC} supply is connected only to the internal output device and its transient depends on the load placed on the output.

A full discussion on decoupling the power distribution in 2107B arrays appears in the decoupling section. As it is shown later, the use of a multi-layer memory board is not required by the 2107B.

SYSTEM CONSIDERATIONS

The previous sections of this application note have dealt with the characteristics of the 2107B as a stand-alone device. This section will outline the types of interface, system design considerations, power calculations and testing considerations when using the 2107B.

MOS Level Drivers

There are many types of drivers capable of driving n-channel RAMs such as the 2107B. The drivers can be used in one or more of the configurations as shown in Figure 18a, b, and c. Each of the driver types shown in Figure 18 has an optimum circuit load that it can drive and each has special design considerations. These drivers are catagorized in three general types; those which:

1. Require external drive transistors.

2. Require an additional power supply.

3. Require no special components or voltages.

In case (1) above, there is insufficient high level drive capability in the driver, hence a PNP external discrete transistor must be used to generate sufficient up-going transition (Figure 18a). Note that this transistor is driving in the saturated mode so the minimum high level criteria ($V_{DD}-1.0$) on the high level MOS clock are easily met.

Driver type (2), shown in Figure 18b, does not require external discrete transistors but does require an additional power supply. This extra supply is usually 3V higher than the V_{DD} supply for the RAM (e.g., using this type of driver with the 2107B would require $V_{DD} = 12V$ and $V_{DD1} = 15V$). The additional supply is necessary to assure that the minimum up level ($V_{DD}-1.0$) requirement of the MOS clock is met.

The 3245, shown in Figure 18c, has been designed to maintain the $V_{IH\ MIN}$ requirements of the 2107B, while some other types of drivers using a single V_{DD} supply may not maintain a sufficient $V_{IH\ MIN}$ level. The 3245 is recommended for all new designs using the 2107B.

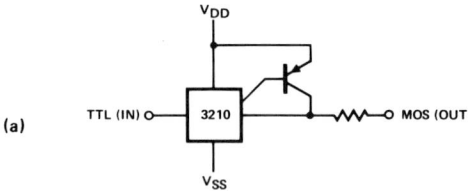

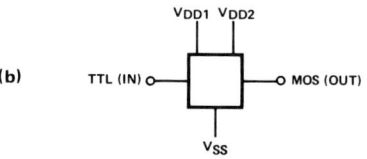

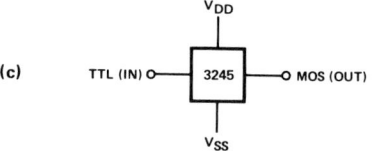

Figure 18. Three Types of MOS Level Drivers

It is important to remember to place the MOS level driver outputs physically as close as possible to the memory array. This will minimize any transmission line impedance mismatch between the unloaded stub and heavily loaded line in the memory array. The effect can most easily be seen with the aid of Figure 19. The impedance of the interconnect is:

1. $$Z_{0(1)} = \sqrt{\frac{L}{C_1}}$$

where C_1 is the capacitance per unit length of the interconnect

L is the inductance per unit length of the interconnect

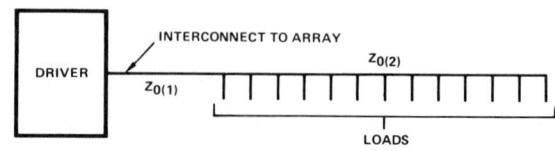

Figure 19. MOS Level Driver Loading

DESIGNING WITH 22 PIN, 4K DYNAMIC RAMS

For all practical purposes, the inductance per unit length of the printed line is independent of the externally connected loads. Therefore, the impedance of the loaded section of transmission line can be represented as:

2. $$Z_{0(2)} = \sqrt{\frac{L}{C_1 + C_2}}$$

where C_2 is the added capacitance per unit length to the printed transmission line.

For most practical systems, capacitance per unit length of an unloaded transmission line will be approximately 1–2 pF/in. (C_1). C_2 is the capacitance effect of the 2107B per unit length. Since the spacing between memory devices is approximately 0.5" the typical loading effect of C_2 is 30 pF/in. (i.e., 15 pF assumed for each chip enable input).

The ratio of the two impedances is calculated as follows:

$$\frac{Z_{0(1)}}{Z_{0(2)}} = \sqrt{\frac{C_1 + C_2}{C_1}} = \sqrt{\frac{32}{2}} = 4$$

This means that the impedance of the stub is four times the impedance of the loaded section.

If the loads are placed close to the driver output the effect of the stub will be negligible and will cause no problem.

3245 MOS Level Driver

The Intel® 3245 is a quad MOS level driver, with each driver capable of driving 250 pF load with maximum delay of 30 nsec. The 3245 requires two power supplies; V_{CC} (+5V), and V_{DD} (+12V). The pin configuration and logic diagram of the 3245 is shown in Figure 20. For reference, input/output waveforms are shown in Figure 21, with delays given in Table II for worst case conditions.

Note that Table II gives the minimum input to output delay for a lightly loaded line (C = 150 pF) and the maximum delay plus rise time for a heavier load (C = 250 pF). The minimum delay time is given so the system designer can guarantee that the chip enable driven by a particular driver does not occur *before* the address lines have stabilized. The maximum delay plus rise time is given to guarantee

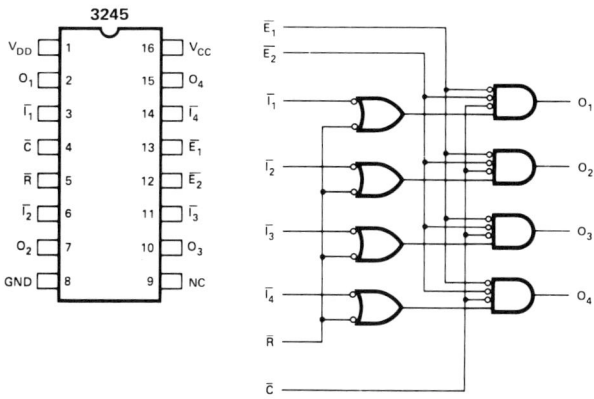

Figure 20. 3245 Pin Configuration and Logic Diagram

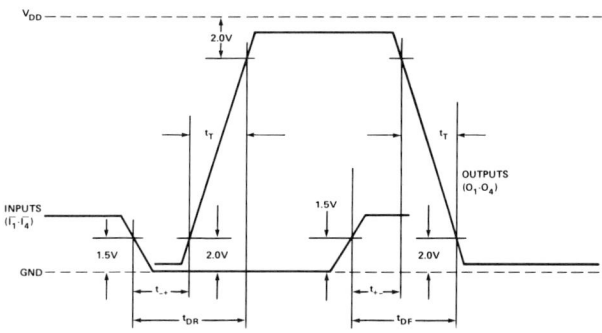

Figure 21. 3245 Input/Output Waveforms

Table II. 3245 A.C. Characteristics

$T_A = 0°C$ to $75°C$, $V_{CC} = 5.0V \pm 5\%$, $V_{DD} = 12V \pm 5\%$

Symbol	Parameter	Min.[1]	Typ.[2]	Max.[3]	Unit	Test Conditions
t_{-+}	Input to Output Delay	5	11		ns	$R_{SERIES} = 0$
t_{DR}	Delay Plus Rise Time		20	32	ns	$R_{SERIES} = 0$
t_{+-}	Input to Output Delay	3	7		ns	$R_{SERIES} = 0$
t_{DF}	Delay Plus Fall Time		18	32	ns	$R_{SERIES} = 0$
t_T	Output Transition Time	10	17	25	ns	$R_{SERIES} = 20\Omega$
t_{DR}	Delay Plus Rise Time		27	38	ns	$R_{SERIES} = 20\Omega$

NOTES: 1. $C_L = 150$ pF 2. $C_L = 200$ pF & $T_A = 25°C$ 3. $C_L = 250$ pF a required system access or cycle time can be met. The capacitance values specified for the **3245** of C = 150 pF, C = 200 pF, and C = 250 pF are representative of the minimum, typical, and maximum capacitance, respectively, of nine 2107B Chip Enable inputs plus associated stray capacitance.

Graphs showing the effect of capacitance loads on delay and rise times are shown in Figure 22a and b.

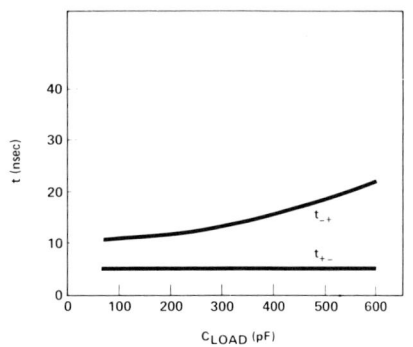

(a) INPUT TO OUTPUT DELAY

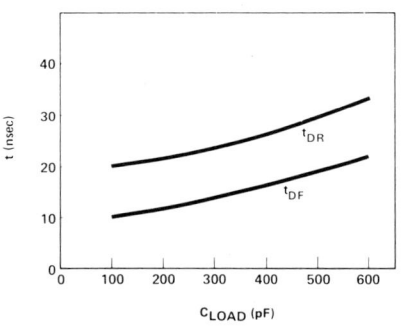

(b) DELAY PLUS TRANSITION TIME

Figure 22. 3245 Delay and Transition Time as a Function of C_{LOAD}

The **3245** offers a great deal of flexibility in driving large arrays of 2107Bs. A sample of its logic capability is shown in Figure 23. A given card is selected by $\overline{\text{Card Enable i}}$, byte control is maintained with $\overline{\text{Byte Enable}}$, and the desired row selected by $\overline{\text{Row Enable i}}$. The basic chip enable timing pulse is provided by CE timing.

At refresh time it is necessary to activate the $\overline{\text{Card Enable i}}$, $\overline{\text{Byte Enable}}$ and $\overline{\text{Refresh Enable}}$ to refresh the entire card at one time. In most systems, it is desirable to refresh all cards simultaneously. If the cards are decoupled properly (see Decoupling section), the power supply transients during refresh will be minimal and are acceptable. The basic configuration of such a card is shown in Figure 24. For

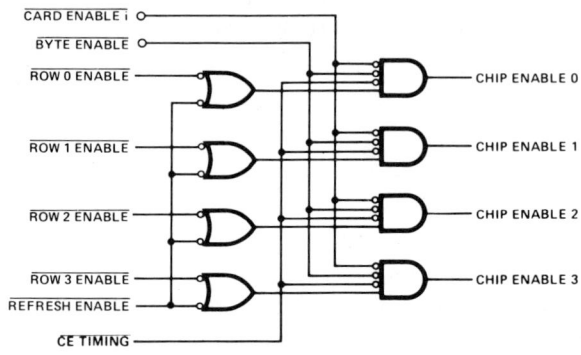

Figure 23. 3245 Enable Configuration

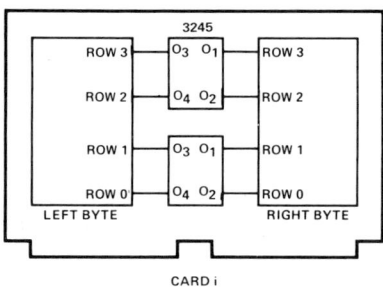

Figure 24. System Organization and Driver Placement

this system, the entire 16K × 16 memory array can be driven with two **3245**s placed as shown between the two memory arrays.

Waveforms of the 3245 driver in a system similar to that shown in Figure 24 are given in Figure 25a–d. The driver configuration used is shown in Figure 26. Figure 25a and b shows the leading and trailing edge of chip enable at both the beginning and ending of the printed line for an added series resistance R of 10Ω. Note the transition time and overshoot for each of these edges. The overshoot is worst case at the leading edge at the driver end and on the trailing edge at the end of the line. The trailing edge overshoot is 2.2V while the leading edge overshoot is 1.5V. Both values are very marginal for system operation.

The effect of increasing the series resistance to 20Ω for the above driver is shown in Figure 25c and d. Note that the transition time has increased but is

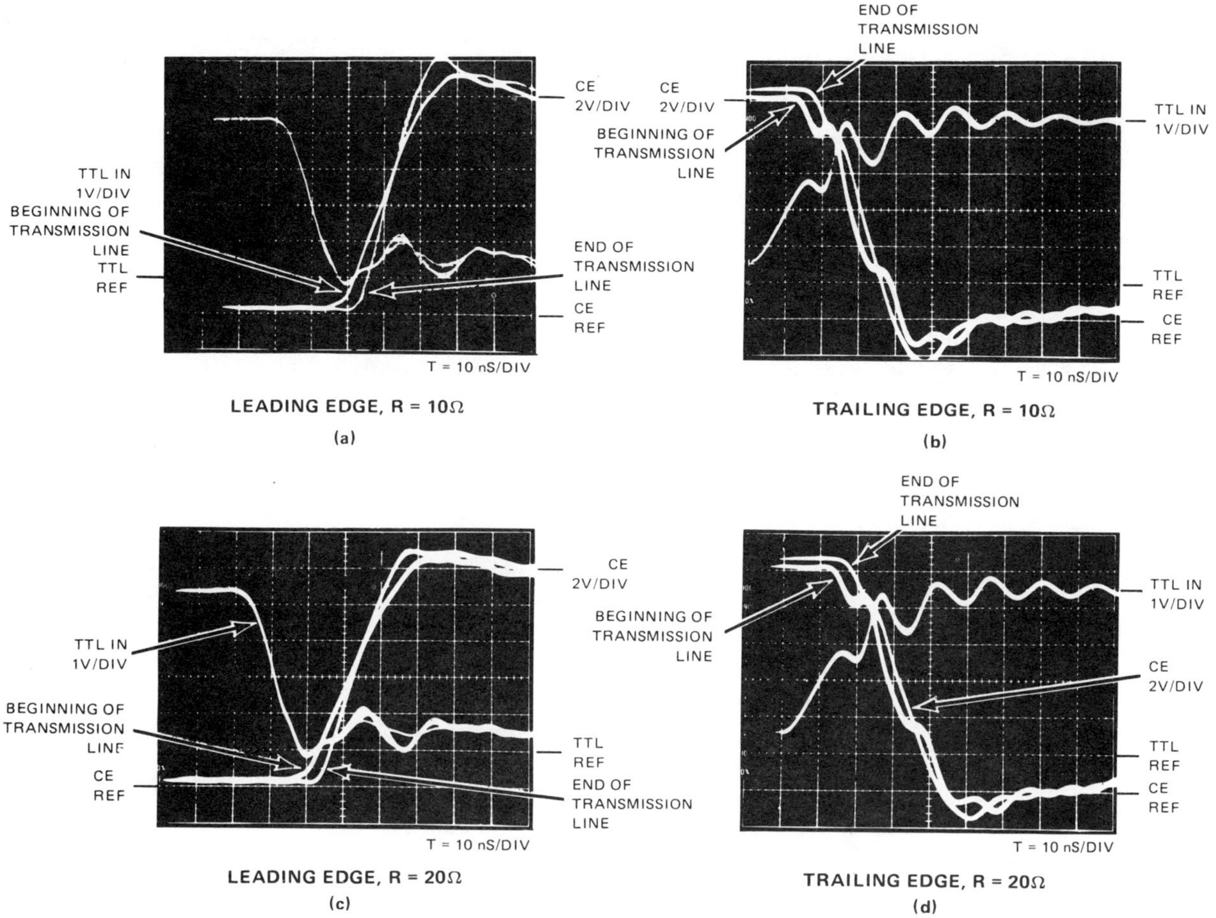

DESIGNING WITH 22 PIN, 4K DYNAMIC RAMS

Figure 25. 3245 Typical Driver Waveforms

(a) LEADING EDGE, R = 10Ω
(b) TRAILING EDGE, R = 10Ω
(c) LEADING EDGE, R = 20Ω
(d) TRAILING EDGE, R = 20Ω

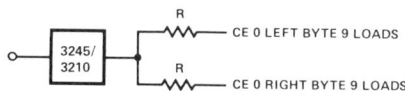

Figure 26. MOS Level Driver Configuration

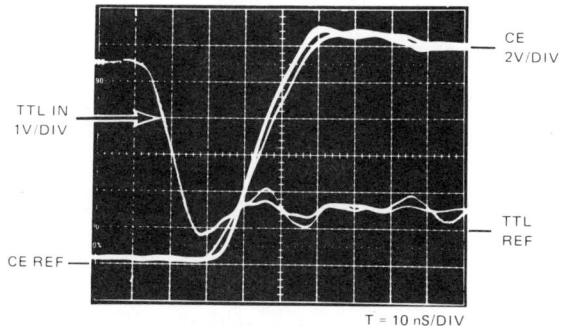

Figure 27. 3245 Driver Waveform with Temperature = 70°C

still within entirely acceptable limits and the overshoots have been cut in half. The driver is now operating in an acceptable mode with minimal overshoot.

The effect of temperatures on the 3245 is shown in Figure 27. A 20Ω series resistor is used with the driver.

The results of board measurements of a typical **3245** driver driving 18 loads and 9 loads is shown in Table III. Note that the delay does not change appreciably with temperature but the transition time increased approximately 2–3 nsec from 25°C to 70°C.

3210 MOS Level Driver

The pin configuration and logic symbol for the 3210 driver is shown in Figure 28. As shown in this figure, this driver consists of one MOS level driver and four TTL low voltage buffers. These low voltage buffers can be used to drive inputs which require a 3.5V high level (such as the 2107A address

Table III. Summary of 3245 Driver Board Delay Measurements

NUMBER 2107B LOADS AND CIRCUIT CONFIGURATION		MEASURED CONDITIONS INPUT TO OUTPUT DELAY				MEASURED DELAY[3] PLUS RISE		MEASURED DELAY[4] PLUS FALL	
		t_{-+}[1]		t_{+-}[2]					
		TYP.	WORST[5] CASE	TYP.	WORST[5] CASE	TYP.	WORST[5] CASE	TYP.	WORST[5] CASE
3245 18 LOADS[6]	R = 20Ω	12	12	10	10	34	37	33	35
3245 9 LOADS	R = 20Ω	11		10		30	33[7]	25	27[7]

NOTES:
1. TTL 1.5 to V_{SS} +1 volt
2. TTL 1.5 to V_{DD} −1 volt
3. TTL 1.5 to V_{DD} −1 volt
4. TTL 1.5 to V_{SS} +1 volt
5. Worst case driver on board at 70°C and 5% power supply variation.
6. 18 loads 20Ω split resistor (see Figure 26).
7. Projected from 18 load delay.

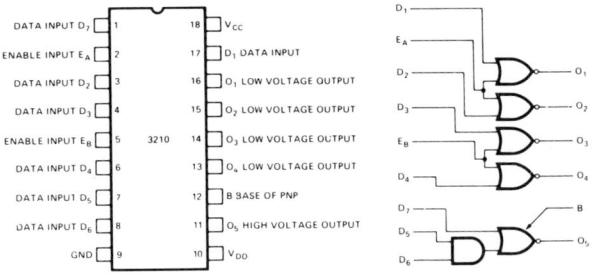

Figure 28. 3210 Pin Configuration and Logic Symbol

inputs) or they can be used to drive high capacitance loads with minimum delay. For reference, the input/output characteristics of the 3210 are shown in Figure 29 and table IV, respectively.

The driver configuration for the 3210 MOS level output is shown with the aid of photos in Figure 30a and b for series resistances of 10Ω and 20Ω. Table V summarizes the results of board measurements for the 3210 as a function of series resistance and temperature.

Low Voltage Driver/Buffers

The address, data-in, write enable, and chip select inputs on the 2107B are all low voltage TTL compatible requiring no special interface. This section will discuss the types of drivers which can be used to drive the low voltage inputs along with the advantages and disadvantages of the drivers.

The types of low level drivers capable of driving the 2107B are shown in Figure 31. Two observations are pointed out regarding the use of TTL drivers shown in Figure 31.

1. There are no pull up resistors.
2. Series 74S type gates are not recommended.

Table IV. 3210 A.C. Characteristics

T_A = 0°C to 75°C, V_{CC} = 5.0V ±5%, V_{DD} = 12V ±5%

SYMBOL	PARAMETER	MIN.	TYP.[1]	MAX.	UNITS	TEST CONDITIONS
t_{LDR}	Delay Plus Rise Time for Low Voltage Drivers		17	25	nS	C_L = 200 pF
t_{LDF}	Delay Plus Fall Time for Low Voltage Drivers		16	25	nS	C_L = 200 pF
t_{H-+}	Input to Output Delay for High Voltage Driver	9	15		nS	C_L = 175 pF
t_{HDR}	Delay Plus Rise Time for High Voltage Driver		27	40	nS	C_L = 350 pF
t_{H+-}	Input to Output Delay for High Voltage Driver	4	8		nS	C_L = 175 pF
t_{HDF}	Delay Plus Fall Time for High Voltage Driver		18	30	nS	C_L = 350 pF
t_{DB}	Delay to Base Drive to External PNP (Pin 12)	4	8	17	nS	

NOTE: 1. T_A = 25°C
A.C. CONDITIONS OF TEST:
Input Pulse Amplitudes: 3.0V Input Pulse Rise and Fall Times: 5 nS between 1 volt and 2 volts Measurement Points: See Waveforms

DESIGNING WITH 22 PIN, 4K DYNAMIC RAMS

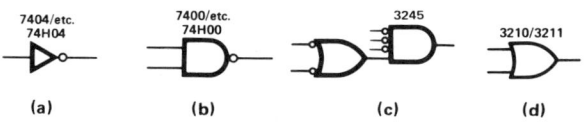

Figure 31. Low Level Drivers

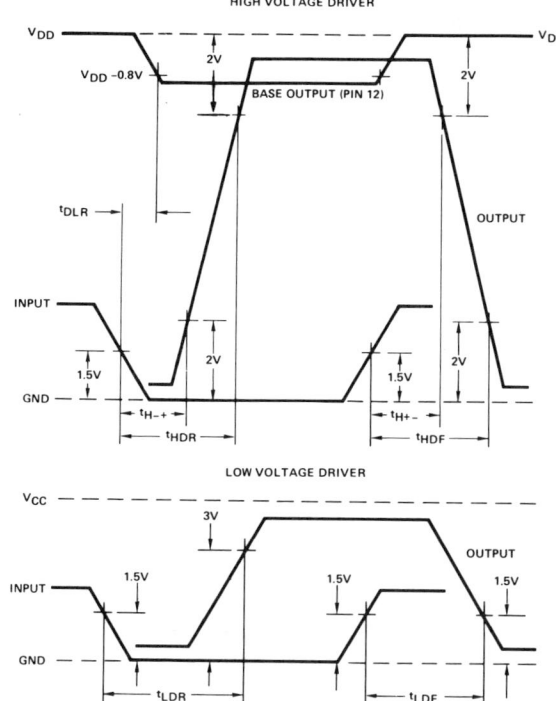

Figure 29. 3210 Input/Output Characteristics

TTL Drivers

Since TTL devices will typically pull up actively to 2.8V to 3.4V, which is well above the required minimum high level, pull up resistors are not needed. Standard Series 7400 type gates are specified to supply 400 μa up level current at 2.4V worst case. Since each address input of the 2107B has a maximum leakage current of 10 μa, this type of driver is capable of driving 40 2107B address lines. However, it should be noted that these 40 address inputs have a capacitance of 240 pF. This load will increase the delay through the series 74 gates.

When driving the 2107B address inputs with TTL gates it is advisable to use a NAND type circuit

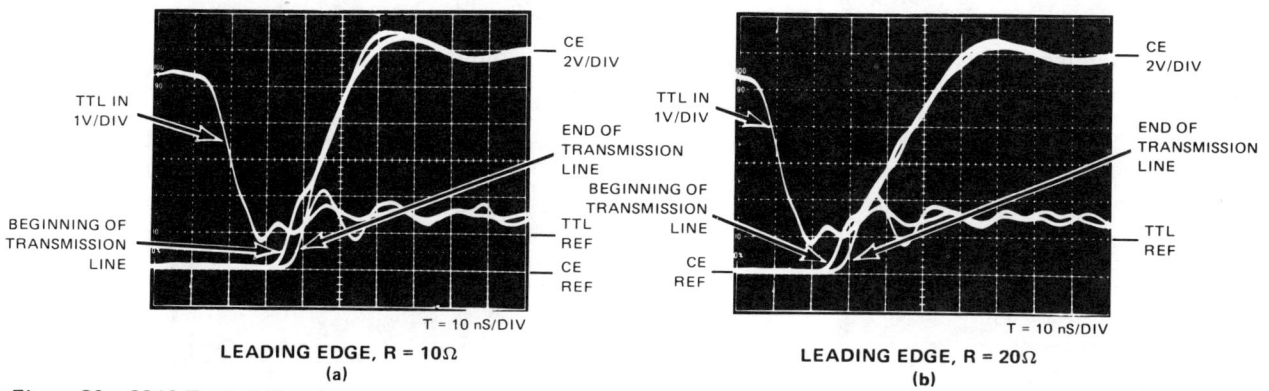

Figure 30. 3210 Typical Waveforms

Table V. Summary of 3210 Driver Board Delay Measurements

NUMBER 2107B LOADS AND CIRCUIT CONFIGURATION	MEASURED CONDITIONS INPUT TO OUTPUT DELAY				MEASURED DELAY[3] PLUS RISE		MEASURED DELAY[4] PLUS FALL	
	t_{-+}[1]		t_{+-}[2]					
	TYP.	WORST[5] CASE	TYP.	WORST[5] CASE	TYP.	WORST[5] CASE	TYP.	WORST[5] CASE
3210 18 LOADS[6] R = 20Ω	16	16	10	10	48	50	35	35
3210 9 LOADS R = 20Ω	14		8		30	32[7]	25	25[7]

NOTES:
1. TTL 1.5 to V_{SS} +1 volt
2. TTL 1.5 to V_{DD} −1 volt
3. TTL 1.5 to V_{DD} −1 volt
4. TTL 1.5 to V_{SS} +1 volt
5. Worst case driver on board at 70°C and 5% power supply variation.
6. 18 loads 20Ω split resistor (see Figure 26).
7. Projected from 18 load delay.

(such as shown in Figure 31) with an enable input. This will allow all addresses to be set up in a high state (above 2.4V) and be driven low when appropriate. Since TTL gates have much better drive capability in the high to low direction, the increase in delay due to the large capacitance is reduced.

It is not recommended that Schottky type TTL gates be used to drive the low level inputs of the 2107B. This is because under worst case conditions, the down level of the Schottky device is approximately 100 mv higher than for a regular or H series TTL gate. This higher level coupled with the address noise coupled from the 2107B (see Address Buffer/Latch section) might make some systems marginal in operation. In addition, the effect of chip enable transition on address low voltage reduces the maximum positive down level on the addresses. (See Device Specification section.)

An example of TTL circuits (7400, 74H00, 74S00) driving 36 address inputs on the 2107B at refresh time is shown in Figure 32. Figure 33 shows the same TTL gate at "Read" time. Note the high level loading effect is greatly reduced because only 9 loads (2107B) are turned on at one time. Note that a Series 74S gate was used and is shown for reference only. From these photos the amount of overshoot present in driving high-low is clearly seen. Therefore, even with TTL drivers it is desirable to use series resistors to decrease the negative overshoot. This resistor value depends on the load on the driver and 20Ω is recommended when driving 36 address loads.

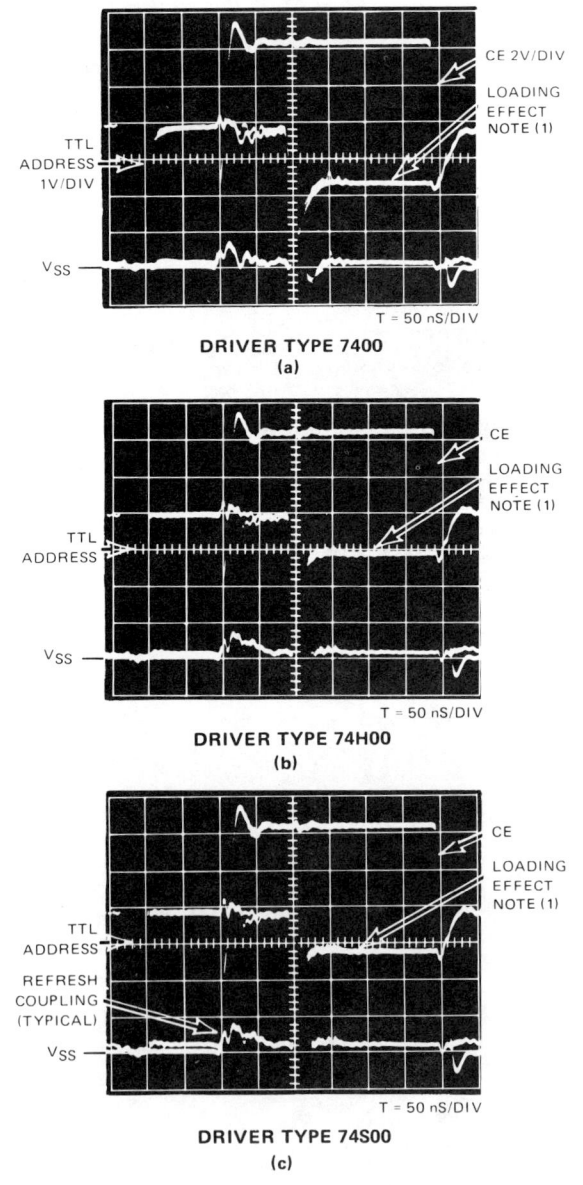

Figure 32. TTL Driver Waveforms (Refresh Cycle)

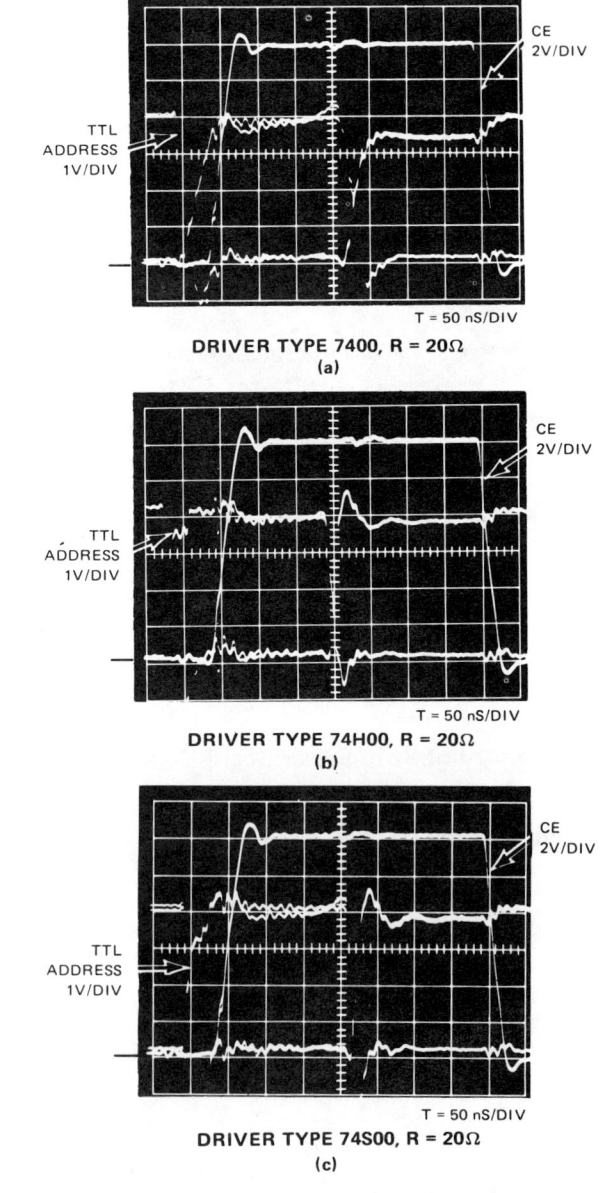

Figure 33. TTL Driver Waveforms (Read Cycle)

DESIGNING WITH 22 PIN, 4K DYNAMIC RAMS

In the discussion of the address latch circuitry (Address Buffer/Latch section) reference was made to input currents drawn by the address buffer when an address is switched from a low to high level during chip enable. The decrease in high level shown in Figure 32 is due to the 0.5 mA/2107B loading of the address line following the low to high transition while chip enable is on. (All photos from Figure 32 are taken at refresh time when all devices are on. This condition is worst case.)

Figure 32 also shows the effect of 36 memory devices coupling charge back to the address line [see Note (1) on photos]. This coupling limits the series resistance value which can be added to the address drivers to minimize overshoot. It also suggests that the address drivers be placed as close as practical to the memory array.

The photo shown in Figure 34 is the current associated with the low to high level address transition for 36 devices at refresh time. (Note the time delay of current relative to address voltage change. This is the result of delays associated with the current probe relative to the voltage probe.) For this example, the driver used is a 3210.

Other Low Voltage Driver/Buffers

When speed and high level drive capability is needed it is desirable to use drivers which are designed specifically for driving high capacitance loads with minimum delay. The 3245 and 3210 can be used to drive the 2107B low voltage inputs.

When operating the 3245 in a low voltage mode, the device is connected per schematic shown in Figure 35. As shown in this figure, the V_{DD1} pin (pin 1) is connected to V_{CC} (+5) and the V_{DD2} pin (pin 9) is connected to +12V. Photos of the waveforms of the 3245 in the low voltage drive mode are shown in Figure 36a and b. The circuit configuration is shown in Figure 35. As shown in the photo, the 3245 has very high drive capability in both the positive and negative directions.

For comparison, the low level buffer portions of the 3210 are shown in Figure 37a and b. As is shown, both the 3245 and 3210 make excellent low level buffer drivers for heavily loaded address lines.

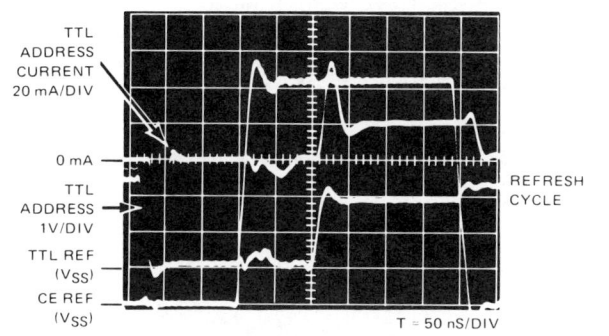

Figure 34. Typical Address Input Current

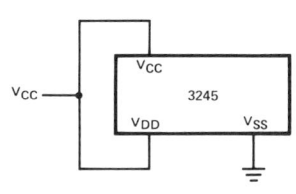

Figure 35. 3245 Connected in Low Voltage Drive Mode

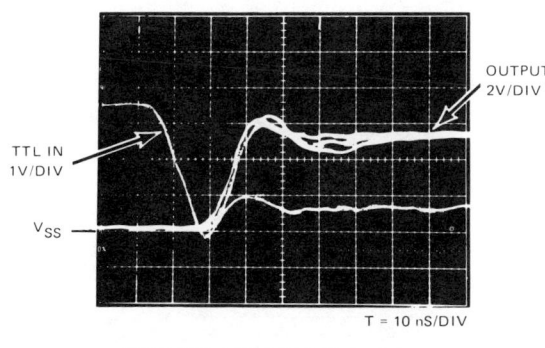

POSITIVE ADDRESS TRANSITION
(a)

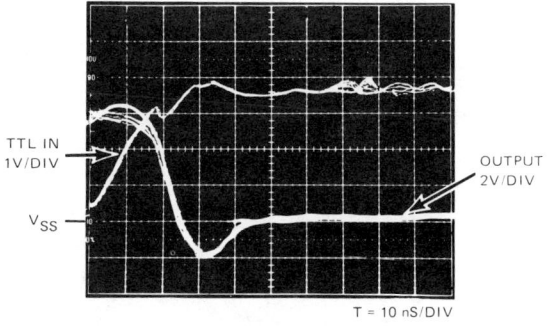

NEGATIVE ADDRESS TRANSITION
(b)

Figure 36. Typical Waveforms, 3245 Low Voltage Mode

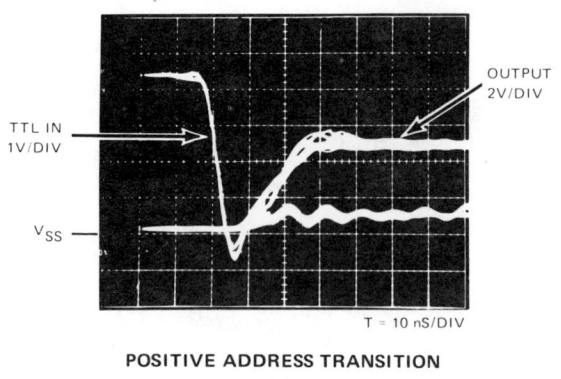
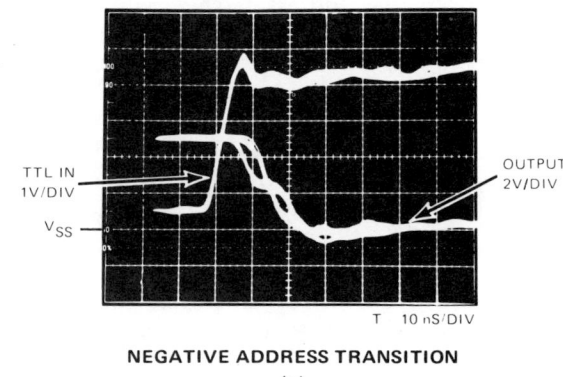

POSITIVE ADDRESS TRANSITION
(a)

NEGATIVE ADDRESS TRANSITION
(b)

Figure 37. Typical Waveforms, 3210 Low Voltage Mode

Output Sensing

The output of the 2107B can be sensed with any TTL compatible series 74, 74L, 74LS or 74S gate. In addition, Intel provides a latch (3404) which features high speed and high density in a single package. The pin configuration for the 3404 is shown in Figure 38.

The V_{CC} input to the 2107B goes only to the output buffer as shown in Figure 8. This means that other types of outputs can be used instead of standard TTL devices if so desired. However, since there are many different ways to utilize this feature, do not exceed the maximum limits on voltage when using the 2107B in a non-standard manner.

Typical curves of output current as a function of output voltage are included in Figure 39a and b to facilitate the output interface of non-TTL loads.

System Timing and Control

The simplicity of design when using the 2107B memory component is shown by the schematic given in Figures 40, 41, and 42. The basic timing for this schematic is shown in Figure 43.

The design shown is for an expandable 16K × 18 system featuring:

1. Asynchronous memory requests/multiple ports
2. Free running refresh

The timing cycle consists of a start initiated by a memory request (MREQ) which triggers the busy latch and begins chip enable. The busy signal is used to disable other ports from requesting a memory cycle while the memory is being accessed from

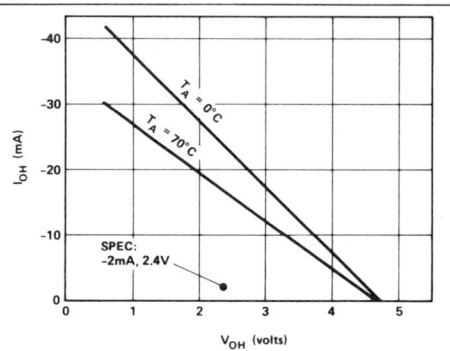

(a) HIGH LEVEL OUTPUT

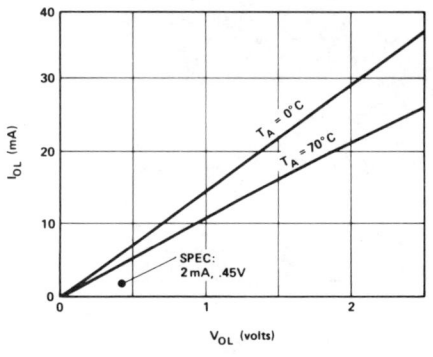

(b) LOW LEVEL OUTPUT

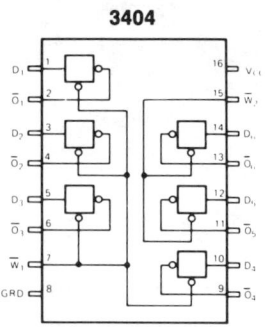

Figure 38. 3404 Pin Configuration

Figure 39. 2107B Output Characteristics

DESIGNING WITH 22 PIN, 4K DYNAMIC RAMS

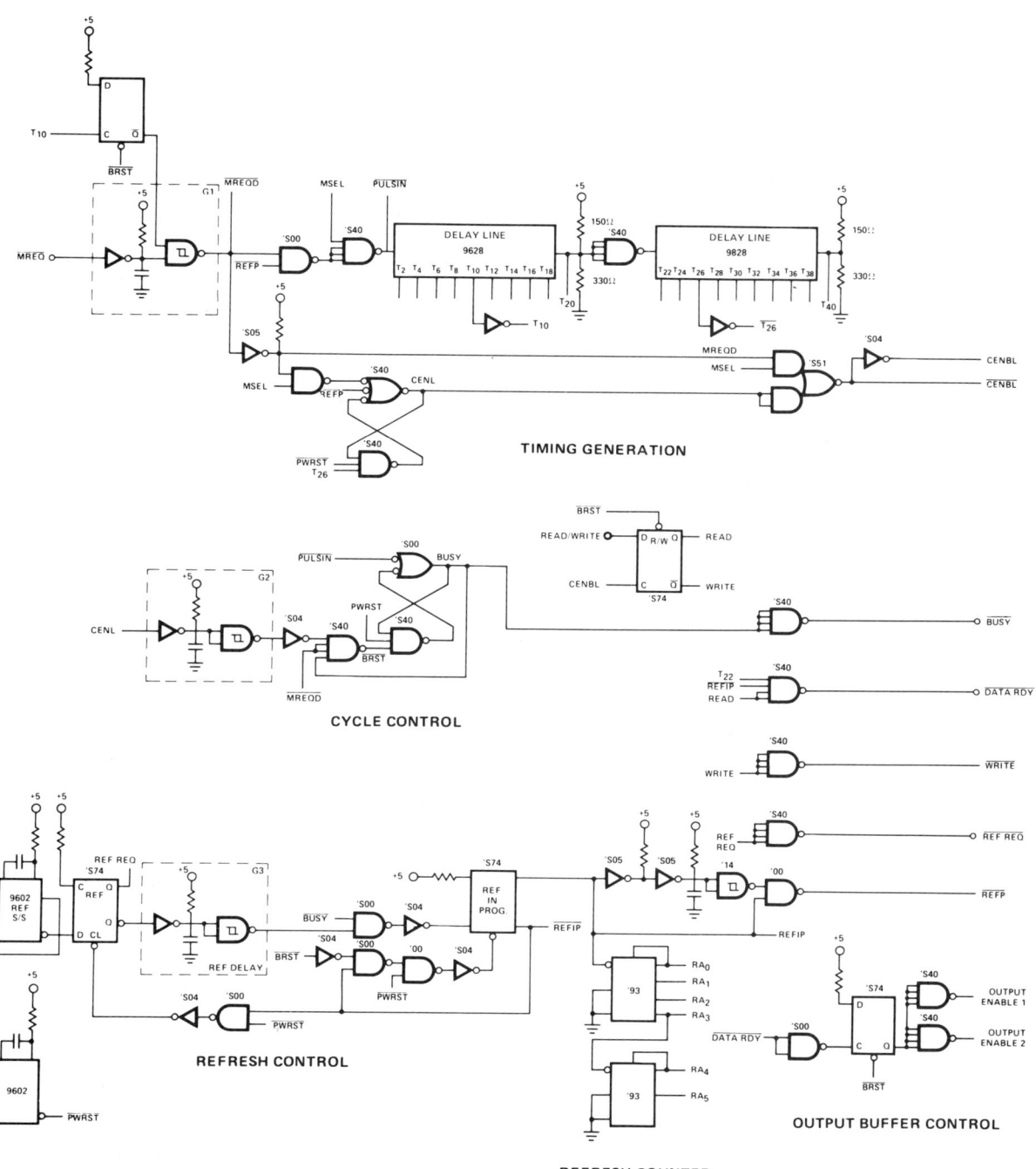

Figure 40. 16K X 18 Memory System Timing Generation

DYNAMIC RAMS

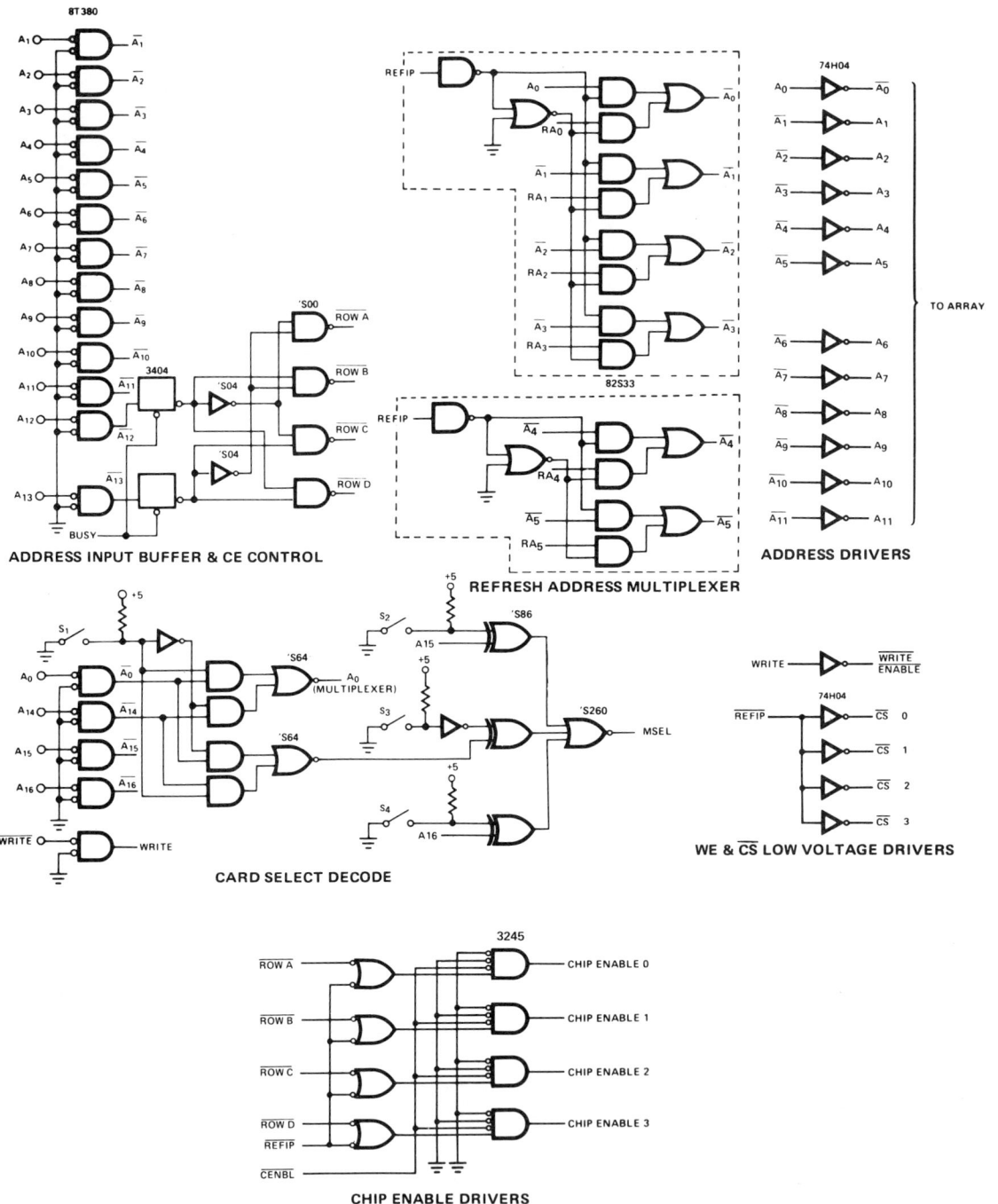

Figure 41. 16K X 18 Memory System Address Buffer Interface

DESIGNING WITH 22 PIN, 4K DYNAMIC RAMS

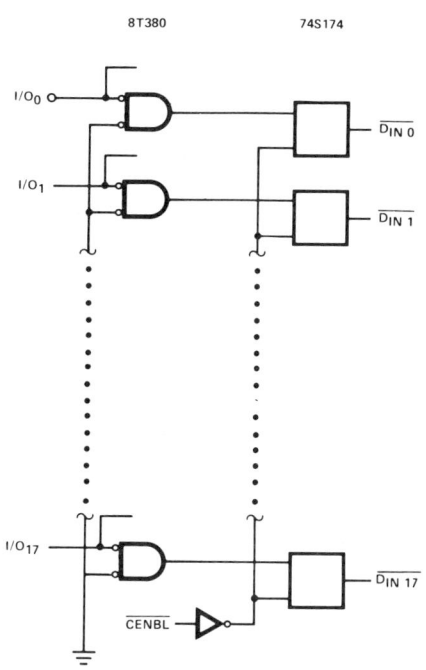
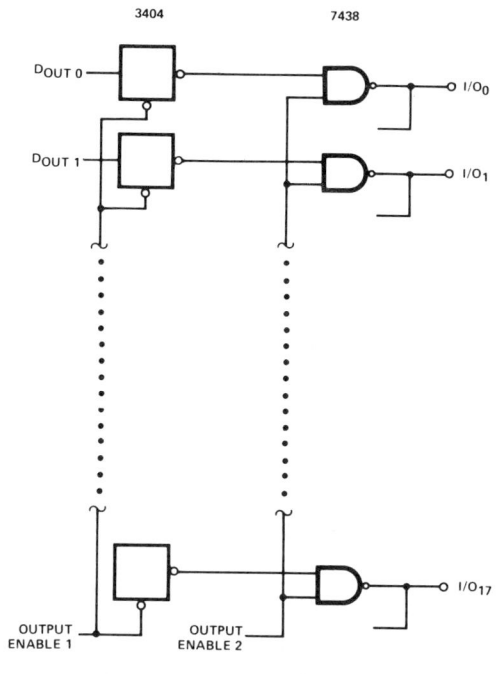

Figure 42. 16K X 18 Memory System Input/Output Interface

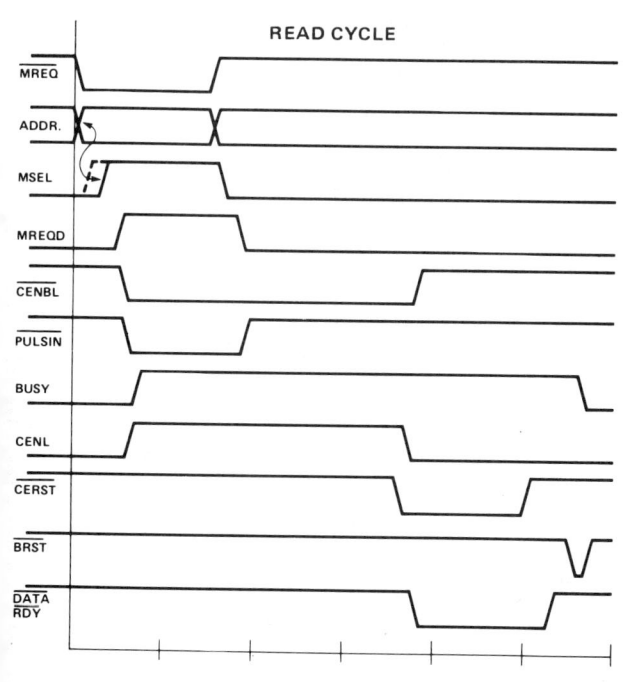

Figure 43. 16K X 18 Memory System Control Timing

another device. Since further timing signals for a read/write cycle are straightforward they will not be discussed further.

However, several things need to be said about the refresh circuitry. As many system designers know, when a memory system with asynchronous refresh runs into trouble in the checkout state, it is 99% sure to be refresh interference in one form or another in the control logic. The most likely cause of problems for asynchronous refresh is glitching between a refresh request and a normal cycle request resulting in false starts or the system not knowing whether or not it is in a refresh cycle or normal cycle.

To alleviate this problem it is necessary to determine that there are no possible requests coming from an external port to the memory when a refresh cycle is started. This will prevent the low order address line from making transitions at the wrong time (due to the multiplexer between the refresh and normal addresses) and taking excessive time to recover to the proper level.

The circuit which performs the function of delaying the onset of a refresh cycle is shown by G3 in Figure 40. Here, refresh is delayed for as long as necessary to assure that the refresh required latch (REF REQ) has had time to block further requests from all ports attached to the memory.

Attention is also called to the power on reset (PWRST) signal shown in Figure 40. This signal is necessary to assure that all latches have been reset (or set) to the proper state after power has been applied. In addition, note that the refresh/addresses RA_0 thru RA_5 are changed after the refresh cycle is complete. This assures that the address will not be changing during refresh as chip enable goes high.

Memory Array Layout

The layout for the 2107B memory array can be identical to that used for the 2107A. An example of such a layout is shown in Figure 44. The layout in this example is constructed with grided power busing which minimizes power distribution noise. When using this technique it is important to remember to bus all power lines both vertically and horizontally through every memory component.

The effect of proper power distribution in the memory array cannot be over-emphasized. It is most desirable to bus the power lines both vertically and horizontally at every memory device location (even if it means running a 15 mil wide printed line to achieve the connection). If it is not possible to make such a connection at every location, then the interconnect should be done as much as possible throughout the array.

As a general rule of thumb, power distribution can be considered adequate if the distance from each power pin (e.g., V_{DD} to capacitor and V_{SS} to capacitor) to the closest decoupling capacitor is less than or equal to 1.5 inches.

For some layouts, particularly those which have all timing and control as well as the memory on a single board, it may be desirable to build multilayer boards. Attention should be paid to the construction of the internal planes to gain maximum effectiveness from these planes. If all the required power supplies cannot be distributed on internal

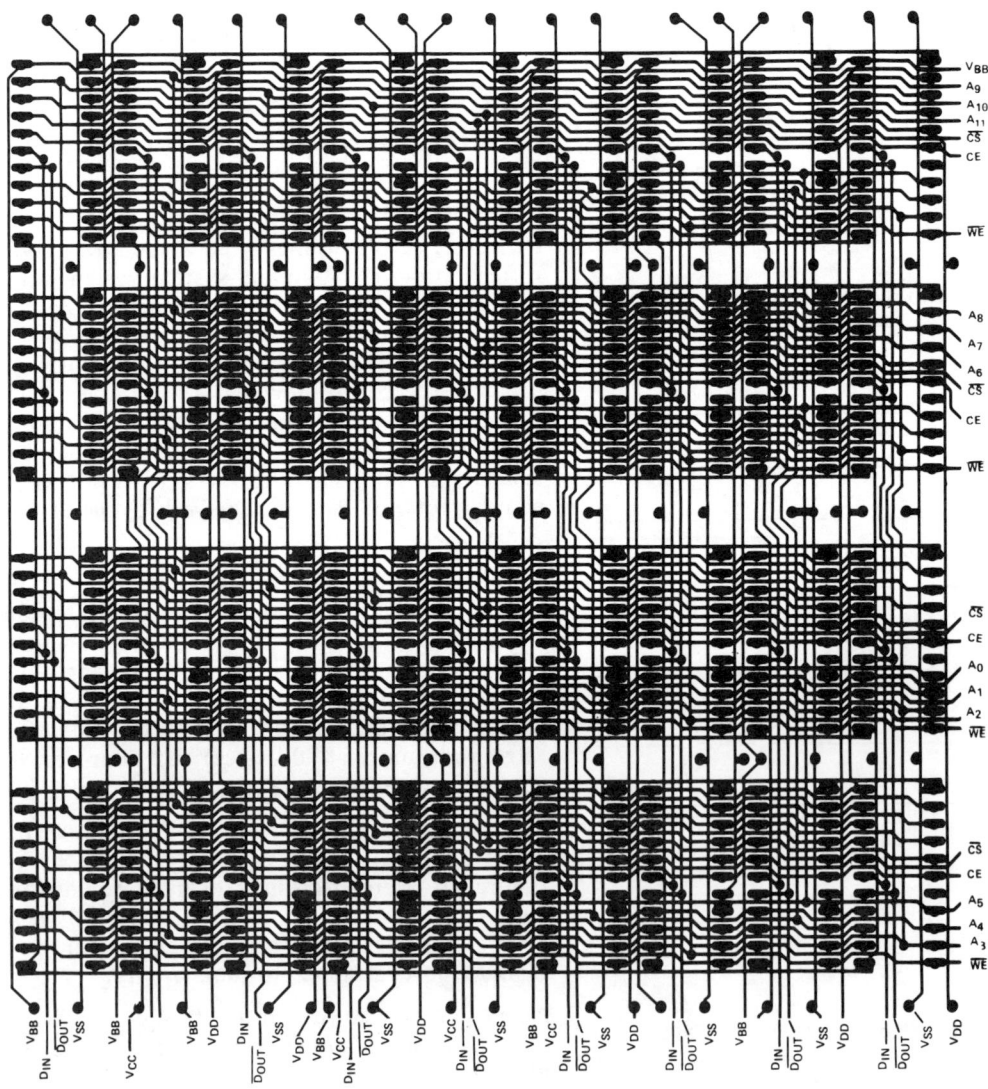

Figure 44. 2107B Memory Array Layout Using Grided Power Distribution

planes and any have to be left out and put on the upper surfaces of the board they should be removed from the internal plane in the following order:

1. V_{CC} 3. V_{DD}
2. V_{BB} 4. V_{SS}

Numbers 1 and 2 can be interchanged if there is a particularly heavy V_{CC} load due to timing, control, etc. circuitry on the board.

When constructing internal planes, care should be taken to obtain the most continuous plane possible. For example, the plane should have "fingers" between each IC feedthrough to minimize inductance.

Decoupling

As mentioned in the Transient Currents section, it is imperative to adequately decouple all supplies to the 2107B. The type and amount of decoupling recommended is most easily shown with the aid of the diagram given in Figure 45. In this figure, every other location for decoupling is V_{DD}-V_{SS} using a 1.0 μF capacitor. Alternate locations can be V_{BB}-V_{SS} or V_{CC}-V_{SS}. It is suggested that V_{BB}-V_{SS} be decoupled more heavily than V_{CC}-V_{SS} (as shown in Figure 45), because of the higher transients on V_{BB}. Noise on the V_{BB} distribution is shown in Figure 46.

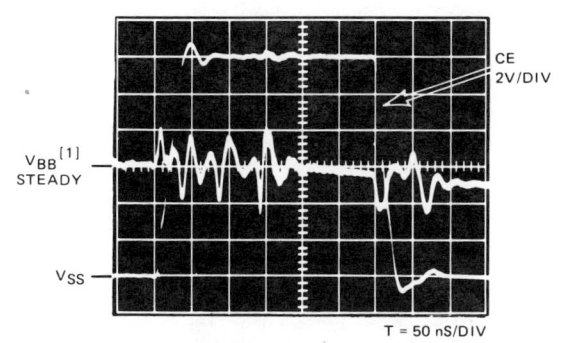

[1] V_{BB} A.C. COUPLED, 100 mV/DIV

Figure 46. Typical V_{BB} Array Noise Decoupling per Figure 45

In addition to the 1.0 μF decoupling discussed above, it is necessary to provide a bulk of ~100 μF V_{DD}-V_{SS} per 36 devices located near the memory array. Also, placing 4.7 μF capacitors between V_{DD}-V_{SS} along the end of each row as shown will eliminate noise problems during refresh time.

The effect of changing decoupling capacitance in a system is shown in Figure 47a, b, c, and d.

These photos show the effects of different decoupling schemes on the V_{DD} supply and the effect of adding a more solid power distribution bus. (In this case #22 wire was paralleled with the existing power distribution of grided 15 mil printed line.) Each of the photos shown in Figure 47 were taken at the worst case location in the memory array at refresh time.

Figure 47a shows the V_{DD} supply with 0.1 μF spaced at every third device, no additional V_{DD} busing and no bulk capacitors (4.7 μF) at the end of each row. Note that the V_{DD} supply decreases to approximately 300 mv below desired setting with spikes driving the supply down a maximum of 440 mv. This excursion is not acceptable.

Figure 47b is for the condition of decoupling with 0.1 μF every third device and adding a 4.7 μF capacitor at the end of each row. Additional power busing on the V_{DD} and V_{SS} lines was added but was observed to have little effect on the noise, shown in Figure 47b. For this case, the V_{DD} supply is observed to decrease approximately 180 mv with spikes adding a further reduction to 240 mv. The major difference between this condition and the one shown in Figure 47a is the addition of the 4.7 μF capacitors at the end of each row. However, the decrease in V_{DD} voltage is still unacceptable.

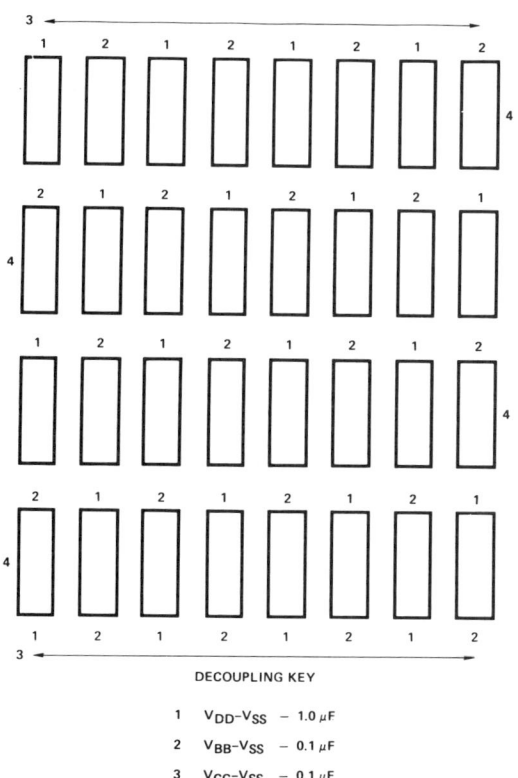

DECOUPLING KEY

1 V_{DD}-V_{SS} — 1.0 μF
2 V_{BB}-V_{SS} — 0.1 μF
3 V_{CC}-V_{SS} — 0.1 μF
4 V_{DD}-V_{SS} — 4.7 μF

Figure 45. Recommended Memory Array Decoupling

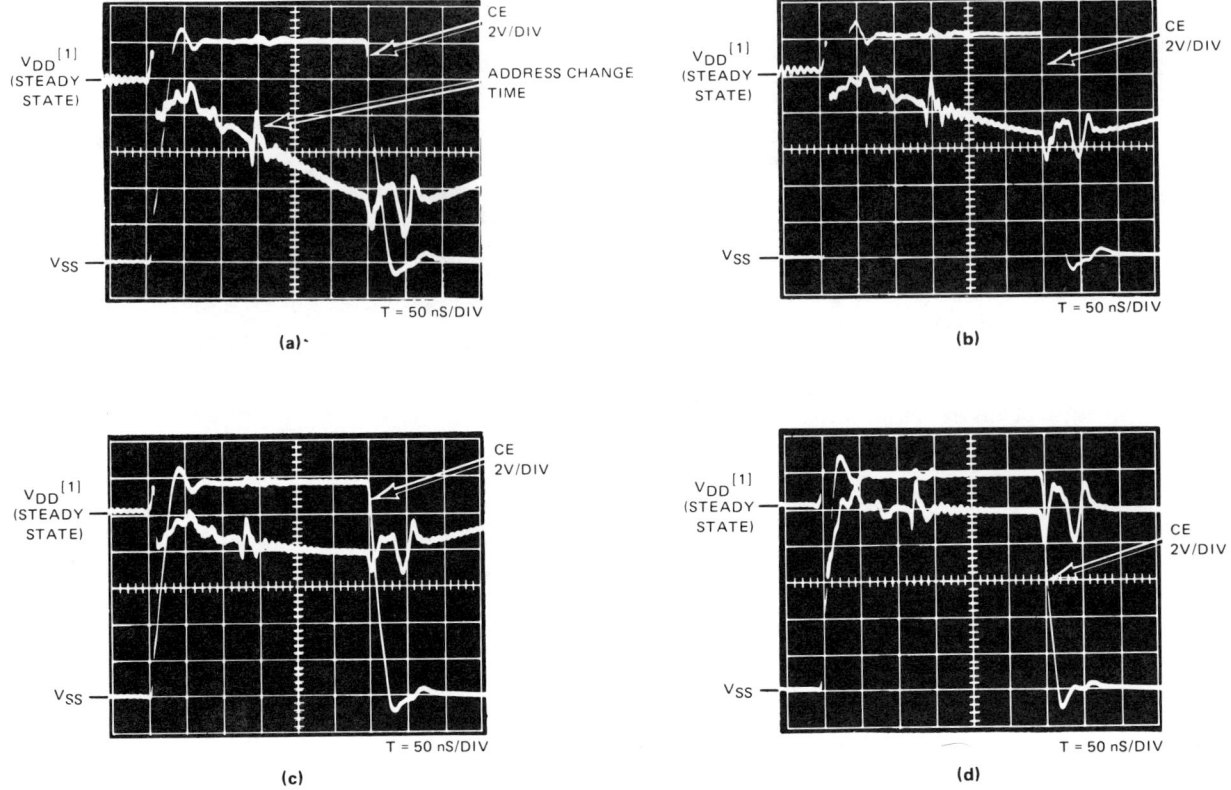

Figure 47. V_{DD} Noise as a Function of Decoupling

Figure 47c shows the V_{DD} supply where V_{DD} in the memory array is decoupled at every other memory device location. All other conditions are per Figure 47b. Note that the offset is now only 100 mv with spikes taking the supply down to 170 mv below nominal. Such decoupling results in adequate but marginally acceptable system operation.

Figure 47d shows the V_{DD} supply decoupled at every other memory device location with 1.0 μF ceramic capacitors. All other conditions are per Figure 47c. Note that the offset is approximately 20 mv with spikes lowering the V_{DD} to a maximum of 200 mv for the length of time shown.

The most desirable decoupling of V_{DD} to V_{SS} in a memory array is therefore 1.0 μF at every other device location with 4.7 μF at the end of each row.

The above recommendations on power distribution and decoupling will result in minimal memory array power noise. However, it is certainly not the only way to suppress power distribution noise. Adequate distribution and decoupling can be assumed if the following values are achieved:

1. $V_{DD}-V_{SS}$ 200 mv peak
2. $V_{BB}-V_{SS}$ 100 mv peak
3. $V_{CC}-V_{SS}$ 100 mv peak
4. $V_{SS}-V_{SS}$ 200 mv peak   (corner to diagonal corner)

Debugging A Memory System

The design and build of memory systems using the newer, easier to use dynamic RAMs, usually results in minimum system debugging time. However, when this is not the case and the control and memory are not playing together well, life can be mighty miserable for the designer while the problem is being tracked down.

This section will deal with some of the more common problems that can affect dynamic memory systems in general, their characteristics and how to better identify them. An integral part of this section is the testing of the system to identify those conditions which cause the most problems for the memory system. In the following it is assumed that power supplies and timing are set to nominal values.

In debugging a memory system the most logical place to start is to determine that all specified criteria are met. This means looking at chip enable timing during each type of cycle both high and low voltage level (read, write, read-modify-write, if used, and refresh). In all cycles make sure the addresses and chip select are set up at the proper time and are held for the minimum hold time. Check all other signals for proper levels (this especially includes the address down level at the time just after chip enable goes high). Remember that the maximum address down level is a function of the chip enable rise time. Erratic operation results from a slow transition and a marginal address down level. Next, check all voltage pins for excessive noise. It is usually desirable to check the power noise at refresh time since all memory devices will be on then and noise will be at a maximum.

After the above, sync on a read cycle and make sure that the system data strobe, if any, occurs before chip enable going low has a chance to reset the data. Also check to make sure no spurious write signals are getting through at read time. In a write cycle check the write enable waveform and make sure data-in is valid at or before write enable goes low.

In a refresh cycle, $\overline{\text{write enable}}$ should be held high unless chip select is high. Also, while in the refresh mode, make sure that all refresh addresses are being accessed. This is most easily done by syncing on the high order refresh address, A_5, and looking at the low order addresses for one cycle of A_5. Checking a read-modify-write cycle is merely a combination of the above discussion of read and write.

After confirming that the specification is met in all regards and that power supply noise is within tolerance in all cycles, the designer is probably tempted to harbor ill feelings toward the memory component and/or manufacturer of same. However, it is not yet time for such.

If the memory is failing most of the data and address patterns that are being used for the test, it is useful to inhibit refresh. When doing so, make sure that the test cycle is such that refresh is being done "automatically" by the normal cycles occurring at a fast enough rate. When inhibiting refresh it may be necessary to restrict the test addressing so that all cells can be "refreshed" by a normal cycle. If the problem goes away after inhibiting refresh, you are now in the army of people who have used dynamic RAMs to be caught with refresh interference.

The only thing that can be said about refresh interference is that refresh is coming in at the wrong time! In properly designed systems, the most likely culprit is a noise glitch getting into the refresh timing circuitry to cause the problem. One of the most common causes for other types of system design is the improper use of "D" type latches. For example, if an asynchronous input (relative to clock) is applied to the D input of a latch and clocked, there will be times where the change on the D input occurs simultaneously with the clock. In some latches this can cause an order of magnitude increase in delay instead of simply missing the D input (see Figure 48). This problem also exists when using a latch made of NAND gates (see Figure 49). The method of correcting the refresh interference problem of a system is left to the imagination and luck of the designer. If the problem is not refresh interference, do not harbor ill feelings yet!

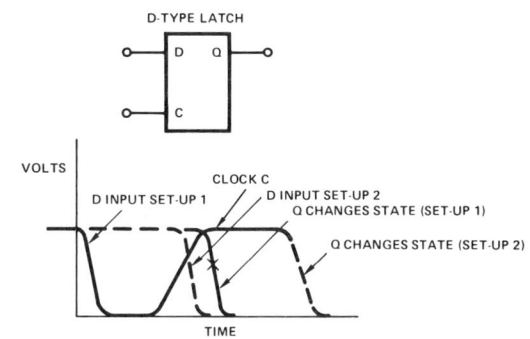

Figure 48. D-Type Latch

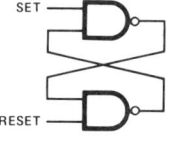

Figure 49. Cross-Coupled NAND Latch

After checking all of the above with no change in results, the next place to start is to determine whether the system is sensitive to addressing patterns. An effective test for evaluating address pattern sensitivity is Galpat. The structure of Galpat is shown in Figure 50. This test is time consuming and requires a careful monitoring of the failed data and its addresses.

Failures caused by a Galpat-only type test are most likely due to address line noise, address coupling to other signals, or refresh related. If address type noise is suspected a careful look at every point in each address path is in order. The best place to concentrate is around the address and its complement that failed.

Refresh related problems can occur during Galpat because this test takes a long time and may not "automatically" refresh the memory. (A sequential type test can refresh the memory automatically if it cycles faster than the maximum allowed

A "galloping" "1" or "0" thru memory consists of initializing the contents of memory (all "1s" or "0s") and implementing the following sequence at each successive memory location:

1. Write opposite data (from initialized state) into test address (A_{TEST})
2. Read next address ($A_{TEST} + 1$)
3. Read test address (A_{TEST})
4. Read $A_{TEST} + 2$
5. Read test address — continue read sequence for entire memory
6. Write test address back to initialize state
7. Go to next address for new A_{TEST}
8. Repeat steps 1–6 until entire memory tested
9. Complement initial data pattern and repeat steps 1–8

Figure 50. Galpat Flow Chart

refresh period.) If any address fails to get refreshed during the refresh period, Galpat will most likely pick it up.

If the above does not yield a clue, then a check of the data pattern across a word is in order. In many tests each bit in a word contains the same data. This can cause certain groups of data lines to couple into adjacent control or address lines. This problem can be tracked down by allowing only one bit in a word to change at a time.

If the memory system is having massive failures, it is very likely that the above debug procedure will reveal the problem. The second type of problem to be discussed is that of soft failures at frequent intervals. In general, these are problems caused by system noise, marginal timing, flaky peripheral device(s), or marginal memory component.

For soft failures, the first item to suspect is refresh interference. Proceed per above to isolate the problem.

A great deal of information on soft failures at nominal voltage settings can be obtained by shmooing the memory system. A shmoo consists of varying each voltage in a manner which is worst case for certain conditions.

The voltage points which emphasize certain tendencies in the memory are contained in Figure 51. The device failed address should be noted at each shmoo point to give a clue to the problem.

A broad guideline here is as follows:

Failure	Cause
1. V_{DD} low, $|V_{BB}|$ high	— timing marginal (memory tends to slow down).
2. V_{DD} high, $|V_{BB}|$ low	— noise in system. Look for V_{SS}, V_{DD}, V_{BB} noise.

Temperature variation can also reveal similar problems. For example:

Failure	Cause
1. High temperature	— timing should be suspected.
2. Low temperature	— noise should be suspected.

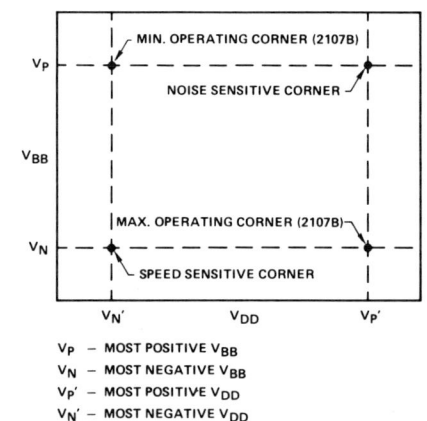

V_P — MOST POSITIVE V_{BB}
V_N — MOST NEGATIVE V_{BB}
V_P' — MOST POSITIVE V_{DD}
V_N' — MOST NEGATIVE V_{DD}

Figure 51. Example of Memory System Shmoo Plot

Power Calculations

The typical power dissipation for the 2107B with a chip enable on time of 230 nsec and a 400 nsec cycle is calculated as follows for a typical device:

Device Power

1. $P_{DOP} = \begin{bmatrix} V_{DD} \times I_{DD} \text{ AV} \\ + V_{BB} \times I_{BB} \end{bmatrix}$

$= \begin{bmatrix} 12.0 \times 38 \text{ mA} \\ + 5.0 \times 0.1 \text{ mA} \end{bmatrix} = 456.5 \text{ mw}$

Since the calculation of standby power without refresh for dynamic memory is meaningless, the following calculations are for standby with refresh:

2. $P_{DOP} = \begin{bmatrix} V_{DD} \times I_{DD1} \\ + V_{BB} \times I_{BB} \end{bmatrix}$

$= \begin{bmatrix} 12.0 \times 0.11 \text{ mA} \\ + 5.0 \times 0.1 \text{ mA} \end{bmatrix} = 1.82 \text{ mw}$

DESIGNING WITH 22 PIN, 4K DYNAMIC RAMS

3. $P_{DSB} = P_{DOP}\left(\dfrac{N\,T_{CY}}{T_{REF}}\right) + P_{NOP}\left(\dfrac{T_{REF}-N\,T_{CY}}{T_{REF}}\right)$

where:
- P_{DOP} = Operating power disspiation
- P_{NOP} = Non-operating (chip enable low) power dissipation
- P_{DSB} = Standby/Refresh power
- N = Number of refresh cycles in refresh period
- T_{REF} = Refresh period in µsec
- T_{CY} = Refresh cycle time in µsec

For the 2107B, the following values apply:
- N = 64
- T_{REF} = 2000 µsec
- T_{CY} = 0.40 µsec

4. $P_{DSB} = 456.5\left(\dfrac{64\,(0.400)}{2000}\right) + 1.82\left(\dfrac{2000-25.6}{2000}\right)$ mw

or

5. $P_{DSB} = (5.84 + 1.80)$ mw

6. $P_{DSB} = 8.6$ mw

The above calculations do not include V_{CC} power since it is dependent only upon the output load used. The output of the 2107B is in a high impedance state when chip enable is low or chip select is high and only leakage level currents flow under these conditions.

System Power

In most systems only a portion of the memory devices will be continually accessed. For example, in the system previously described (16K × 18) worst case power is a continual access of one row (the other three rows are dissipating power in the refresh only mode).

System power for the 16K × 18 system is calculated from:

1. $P_{SYS} = P_{DS} \times N + P_{DA} \times M + P_{DOP} \times D + P_{DSB} \times E$

where:
- P_{DS} = Power dissipated in drivers during standby (including refresh)
- N = Number of drivers in standby
- P_{DA} = Power dissipated in drivers during max. duty cycle operation
- M = Number of drivers in max. duty cycle operation
- P_{DOP} = Power dissipated by memory device max. duty cycle
- D = Number of devices in P_{DOP}
- P_{DSB} = Power dissipated by memory devices during standby (including refresh)
- E = Number of devices in P_{DSB}

For this example, all drivers are assumed to be 3210s. Therefore:
- P_{DS} = 387 mw, N = 6
- P_{DA} = 467 mw, M = 2
- P_{DOP} = 456.5 mw, D = 18
- P_{DSY} = 8.6 mw, E = 54

or

$P_{SYS} = \overbrace{387 \times (6) + 467\,(2)}^{\text{Driver Power}} + \overbrace{456.5\,(18) + 8.6\,(54)}^{\text{Memory Component Power}}$

2. $P_{SYS} = \overbrace{2322 + 934}^{\text{Drivers}} + \overbrace{8217 + 464}^{\text{Memory Devices}}$

$P_{DRIVERS}$ = 3256 mw

P_{MEMORY} = 8681 mw

or

Total System Power:

3. $P_{SYS} = 11.9$ watts

The power dissipated by the drivers is approximately 26% of total system power in a max.–duty cycle operating environment.

Total standby power (including refresh is calculated from equation (1) where:

$N = 8 \quad M = 0 \quad D = 0 \quad E = 72$

or:

4. $P_{SYS} = \overbrace{387\,(8)}^{\text{Driver}} + \overbrace{8.6\,(72)}^{\text{Memory}}$

$P_{SYS} = (3096 + 619)$ mw

or

$P_{DRIVERS}$ = 3092 mw

P_{MEMORY} = 619 mw

5. $P_{SYS} = 3.7$ watts

Note that in this case, driver power amounts to approximately 83% of total system power.

Power Supply Sequencing

The V_{BB} substrate bias supply must never be allowed to be more positive than 0.3V above V_{SS}, V_{DD}, or V_{CC} at any time. Catastrophic device failure can result if these criteria are not met. To minimize this problem of power sequencing and inadvertent power shorts, it is recommended that V_{BB} be referenced to V_{SS}.

ACKNOWLEDGMENT

Appreciation is extended to R. L. Papenberg of the Application Engineering Department for his work on the 3245 and 3210 driver system evaluation.

DESIGNING WITH 16K DYNAMIC RAMS

JIM COE

Photomicrograph of the Intel 16,384 × 1 Bit 2116 Dynamic RAM.

DESIGNING WITH 16K DYNAMIC RAMS

INTRODUCTION

The Intel® 2116 is a 16,384 word by 1 bit dynamic random access memory. The 2116 is fabricated using Intel's proven two-layer polysilicon, n-channel silicon gate MOS technology. The device is packaged in a standard 16-pin DIP. The pin configuration and logic symbol are shown in Figure 1.

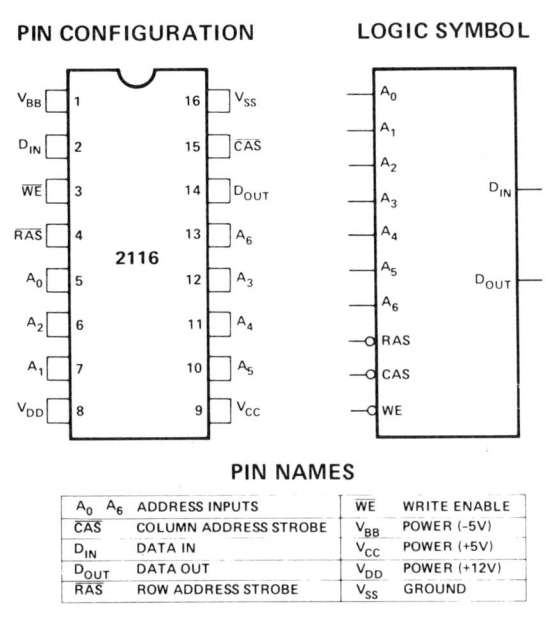

Figure 1. 2116 Pin Assignments

The 2116 operates with three power supplies relative to ground: V_{DD} (+12V), V_{BB} (-5V), and V_{CC} (+5V). The V_{CC} supply is connected only to the output buffer of the 2116 and may be turned off during power down (battery back-up) operation.

The 2116 is designed to be compatible with the industry standard 16-pin 4K RAM, the Intel® 2104A. This compatibility allows a single system design for both the 4K and 16K devices providing for memory expansion without additional engineering.

The use of the 16-pin package is made possible by multiplexing the 14 address bits (required to address 1 of 16,384 bits) into the 2116 on 7 address input pins. The two 7-bit address words are latched into the 2116 by the two TTL clocks, Row Address Strobe ($\overline{RAS}$) and Column Address Strobe ($\overline{CAS}$). Non-critical clock timing requirements allow use of the multiplexing technique while maintaining high performance.

Data is stored in the 2116 in single transistor, dynamic storage cells. The storage cells require refreshing for data retention. Refreshing is accomplished by performing a memory cycle at each of the 128 row addresses every 2 milliseconds.

The purpose of this Application Brief is to describe the basic internal operation of the 2116 and to outline the areas in design which allow a 2104A/2116 compatible memory system.

Device Internal Operation

Operation of the 2116 is most easily understood with the aid of the block diagram shown in Figure 2. As is shown in this figure, the 2116 is arranged as two 8192-bit storage arrays sharing a common set of column address decoders and a common I/O bus. Each array is arranged in a 64 row by 128 column matrix of storage cells with 128 sense amplifiers per array. Row address bit A_6 is decoded and selects one of the two arrays to be active during any given memory cycle. Thus, only one set of 128 sense amplifiers is active during a cycle maintaining low operating power.

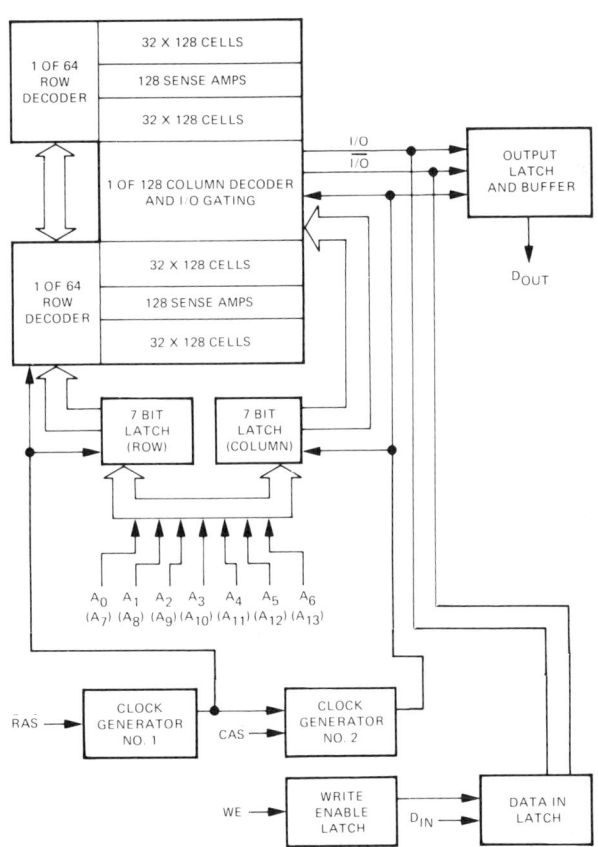

Figure 2. 2116 Block Diagram

The storage cells are implemented with a single transistor and a "storage" capacitor and are called single transistor cells. A cell is accessed by the coincidence of a row select (defined by address bits $A_0 - A_6$) and a column select (defined by address bits $A_7 - A_{13}$). On chip timing and control generators provide the internal timing signals for decoding, data sensing, read/write strobing and I/O data gating. The timing circuits in the 2116 are activated by the negative going edges of the two TTL clocks, $\overline{RAS}$ and $\overline{CAS}$.

Data Sensing

Data is stored in the 2116 storage cells as one of two discrete voltage levels on the cell capacitor; a high is $\sim V_{DD}$ (+12V) and a low is $\sim V_{SS}$ (ground). These levels must be sensed by the data sense amplifiers and propagated to the Data Output (D_{OUT}) in order to fulfill the function of a RAM device. Sensing of the stored levels is destructive and automatic restoration (rewriting) of the sensed data must also occur.

The 2116 data sensing scheme is known as the Dummy Cell Reference technique. The reference level that the sense amplifier compares the stored level to is a level stored in a special, non-accessable storage cell. The level stored in this reference or "dummy" cell is less than the minimum allowable stored high level and greater than the maximum allowable stored low. Examination of the simplified sense amplifier schematic of Figure 3 will clarify the sensing operation.

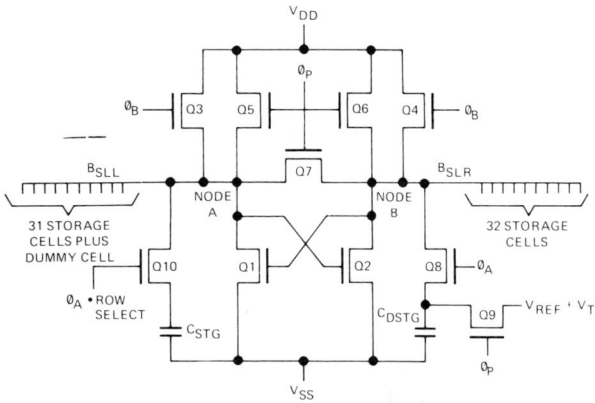

Figure 3. Simplified 2116 Data Sensing Schematic

During the $\overline{RAS}$ clock off time (high), $\emptyset_P$ turns on devices Q5 and Q6 connecting nodes A and B to V_{DD} and precharging the nodes to V_X ($\sim V_{DD} - V_T$ where V_T is the MOS device threshold voltage). Device Q7 is also turned on by $\emptyset_P$ and connects nodes A and B together assuring that they reach the same precharge level. $\emptyset_P$ also turns on device Q9 precharging the dummy storage cell capacitor (C_{DSTG}) to V_{REF}.

When $\overline{RAS}$ goes active (low), $\emptyset_P$ turns off isolating nodes A and B and the dummy cell capacitor. When the row address bits have been decoded and the row select is valid, $\emptyset_A$ turns on Q8 and Q10, the dummy cell and storage cell transistors respectively.

This connects the cell capacitors to the bit sense lines (Bit Sense Line Left [BSLL] and Bit Sense Line Right [BSLR]). If the voltage stored in C_{STG} is greater than the voltage stored in C_{DSTG} (V_{REF}), node A will be higher than node B. This voltage inequality will cause the sense amplifier (a cross-coupled latch made up of devices Q1 and Q2) to switch when load devices Q3 and Q4 are turned on by $\emptyset_B$. The latch will switch node B to V_{SS} and node A to $\sim V_{DD}$ due to the regenerative action of the latch. $\emptyset_B$ is delayed from $\emptyset_A$ sufficiently to allow the voltages on nodes A and B to stabilize prior to enabling the sense amplifier. If the voltage stored in C_{STG} had been less than that stored in C_{DSTG}, the latch would have sensed a low and switched such that node A would be at V_{SS} and node B would be at $\sim V_{DD}$.

After the stored level has been sensed against the reference level, the sense amplifier will have forced BSLL to a level corresponding to the level originally stored in the storage cell capacitor (V_{DD} if $V_{CSTG} > V_{REF}$ or V_{SS} if $V_{CSTG} < V_{REF}$). Since the storage cell transistor (Q10) is still turned on, the storage cell capacitor will be charged to the BSLL level. This effectively restores the sensed data into the cell capacitor but at full levels, not leakage or noise degraded levels. This is also what occurs when a storage cell is refreshed, the data integrity is restored through the sensing function.

Note that the stored level only has to be greater than or less than V_{REF}, not full V_{DD} or V_{SS} levels. This is important because leakage currents from the storage cell capacitor degrades a stored high level toward V_{REF} while system ground noise degrades a stored low level toward V_{REF}. Leakage degraded high levels are the most serious design problem and V_{REF} is generally set closer to V_{SS} than to V_{DD} to counteract the leakage effects. Leakage of stored high levels is also the reason dynamic storage RAMs must be periodically refreshed.

Data Storage

The block diagram in Figure 2 shows that the two 8192-bit arrays in the 2116 share a common I/O bus and common column decoders. The simplified schematic of Figure 4 shows one set of corresponding columns from the two arrays with their sense amplifiers and I/O gating. As shown, the I/O bus consists of two parallel, opposite polarity data lines which connect the column(s) to the Data In and Data Out latches. Referring to the previous discussion of the operation of the sense amplifiers and storage cells, a "stored level" or data map may be developed for the 2116.

DESIGNING WITH 16K DYNAMIC RAMS

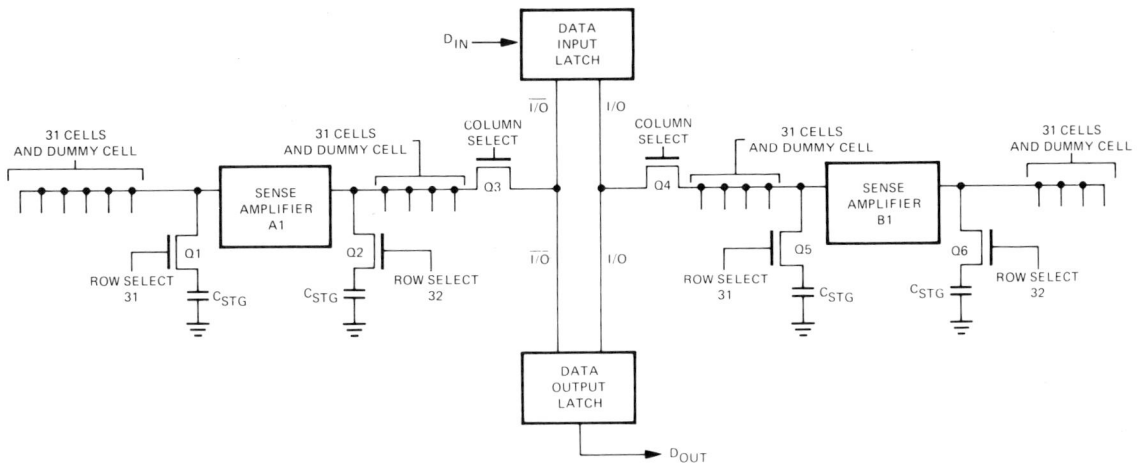

Figure 4. I/O Line and Data Column Schematic

Column select device (Q3) connects sense amplifier A1 and its related storage cells to the $\overline{I/O}$ line. Data stored in the cells on A1's BSLR will be inverted with respect to the data level at the Data Input (D_{IN}). Data in the cells on BSLL will be the same polarity as D_{IN} since the I/O bus data is inverted through the sense amplifier. Conversely, sense amplifier B1 and its related cells are connected to the I/O line by device Q4 and data on it's BSLL will be D_{IN} while data on it's BSLR will be $\overline{D_{IN}}$. These data inversions are internal to the 2116 and are invisible to the user since D_{OUT} will be the same polarity as D_{IN}. The data map for the 2116 is therefore, as shown in Figure 5. This figure also indicates the address map for the 2116.

Address Latches

The 7-bit row and column address words are latched into internal latches by $\overline{RAS}$ and $\overline{CAS}$ respectively. These latches capture the TTL level address information on the shared address input pins and convert the TTL levels to the MOS levels (12V) required internally by the 2116.

Data Latches

Both the Data Input (D_{IN}) and Data Output (D_{OUT}) information is latched by the 2116. The input data is latched by the logical AND function of $\overline{RAS}$, $\overline{CAS}$, and $\overline{WE}$. When a data cycle is being performed ($\overline{RAS}$ low), D_{IN} will be latched by the falling edge of the last of the two control signals ($\overline{CAS}$ or $\overline{WE}$) to go low. In a "fast" write cycle, i.e., $\overline{WE}$ low before $\overline{CAS}$ goes low, the $\overline{CAS}$ edge will operate the latch. In a "late" write ($\overline{CAS}$ low before $\overline{WE}$ goes low) or read-modify-write cycle, D_{IN} is latched by the falling edge of $\overline{WE}$.

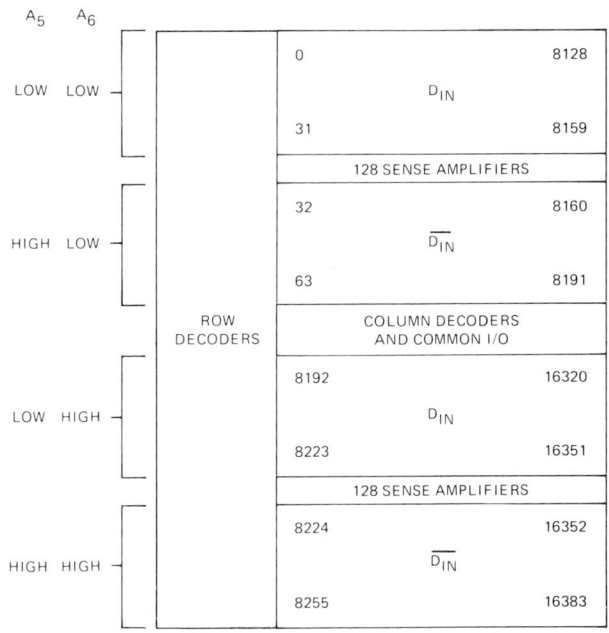

Figure 5. 2116 Address and Data Map

The Data Output (D_{OUT}) latch and buffer is controlled by $\overline{CAS}$. The leading (falling) edge of $\overline{CAS}$ in any cycle causes D_{OUT} to assume an open-circuit (HI-Z) state. At access time (t_{RAC} or t_{CAC}), D_{OUT} will assume a data state (high or low) dependent upon the type of data cyle performed or will remain in the HI-Z state if the cycle was a $\overline{CAS}$-only deselect cycle. The D_{OUT} state is latched and remains valid until the next cycle during which a $\overline{CAS}$ occurs. Table I summarizes the states the data output assumes for each type of 2116 cycle.

Refresh Modes

The data stored in the 2116 single transistor storage cells may be refreshed in any of three modes. The cells must be refreshed every 2msec.

Read Cycle Refresh: A read cycle at each of the 128 row addresses (A_0 - A_6) of the 2116 will refresh all the storage cells. This refresh mode is useful only when the memory system consists of a single row of devices (16K words X n-bits) and OR-tying of outputs is not necessary. Each device will access data during the refresh cycle and OR-tying of device outputs would result in conflict between devices for the output data bus. Write cycles also fulfill the refresh requirement but the selected cell (determined by the column address) on the row being refreshed will have new data written into it while the remaining 127 cells on the row are simply refreshed.

Table 1. Data Output Content

Type of Cycle	Data Latch Content (D_{OUT})
Read Cycle	Data from Addressed Memory Cell
Write Cycle	Input Data (D_{IN})
$\overline{RAS}$-Only Cycle	Data from previous Cycle or HI-Z if Device was Deselected in previous Cycle
$\overline{CAS}$-Only Cycle	HI-Z ($\overline{CAS}$-Only Deselects Device and Turns Output Buffer Off)
R-M-W Cycle	Data Read from Addressed Memory Cell During Read Portion of Cycle
Page Mode Entry Cycle	Data from Addressed Memory Cell
Page Mode Read Cycle	Data from Addressed Memory Cell
Page Mode Write Cycle (or Page Mode Write and Exit Cycle)	Input Data (D_{IN})

$\overline{RAS}$-Only Refresh: A cycle with only the $\overline{RAS}$ clock active, performed at each of the 128 row addresses will refresh the 2116 storage cells. This mode is useful when the memory system consists of multiple rows of devices. The data outputs of the RAMs may be OR-tied when $\overline{RAS}$-only refresh cycles are performed since the D_{OUT} line of each 2116 will remain unchanged during the refresh cycle.

$\overline{CAS}$-Before-$\overline{RAS}$-Refresh: The 2116 storage cells may be refreshed with only 64 cycles each 2msec if the $\overline{CAS}$-before-$\overline{RAS}$ mode is used. In this mode, initiated by $\overline{CAS}$ being valid (low) when $\overline{RAS}$ goes low, both 8KX1 halves (see Figure 2) of the 2116 are turned on and one row in both halves is refreshed during each cycle. Since there are 64 rows of cells in each half, only 64 cycles are required to refresh all the cells. This refresh mode is also useful in systems with multiple rows of devices since receipt of a $\overline{CAS}$ before the $\overline{RAS}$ turns off all device outputs, thereby preventing OR-tied data conflicts.

APPLICATIONS INFORMATION

The Intel® 2116 is functionally compatible with the industry standard Intel® 2104A 16-pin 4K RAM. It is pin compatible with the 2104A with the exception of the seventh address bit (A_6) input pin. The 4K RAM uses that pin as the Chip Select ($\overline{CS}$) input pin. The $\overline{CS}$ signal on the 4K RAMs was essentially treated as a seventh column address bit and, therefore, there is considerable similarity between the 16K and 4K 16-pin RAMs.

The following applications information will concentrate on designing compatible 4K/16K memory systems rather than on just using the 2116. Additional basic applications information on the use of 16-pin, multiplexed address RAMs is contained in the next section of this Handbook.

Implementing Refresh

The 2116 may be refreshed in any of three modes. Read cycles and $\overline{RAS}$-only cycles refresh the row of storage cells (1 of 128 rows) addressed by A_0 through A_6 and, therefore, require 128 cycles each 2msec to refresh the stored data. The third 2116 refresh mode, $\overline{CAS}$-before-$\overline{RAS}$, refreshes two rows of storage cells during each cycle and, therefore, only requires 64 cycles each 2 msec to refresh the stored data.

The 2104A is compatible with all three 2116 refresh modes. A very simple compatible refresh system would perform 128 $\overline{RAS}$-only refresh cycles each 2 msec on both the 2104A and 2116. The 2104A would of course be refreshed twice as often as necessary but this is not a problem. The advantage would be that no logic or timing change would be necessary to differentiate between the 4K and 16K RAMs for refreshing.

Read cycles could also be used with 128 cycles each 2 msec but the 2104A $\overline{CS}$ input would need to be driven high (deselected) during each cycle to prevent data bus conflicts between OR-tied 2104A data outputs during refresh. This requires a logic control funtion of $\overline{CS}$ during refresh (and read cycles also dissipate more power than $\overline{RAS}$-only cycles) so most systems will use $\overline{RAS}$-only refresh.

DESIGNING WITH 16K DYNAMIC RAMS

Each of the first two refresh modes require 128 refresh cycles each 2 msec. Assuming a system cycle time of 500 nsec, 3.2% of the available memory time is required for refreshing. In many systems, this loss of memory availability is of no consequence. In the high throughput memory system environments found in many large and mid-sized computer systems, however, any loss of memory availibility is undesireable.

For these systems, the 64 cycle refresh mode is advantageous since it requires only 1.6% of the available memory line, a 50% savings over 128 cycle refresh. It is also compatible with the 4K RAM systems presently in use since the the 4K RAMs require only 64 refresh cycles each 2 msec.

The 2116 automatically goes into its 64-cycle refresh mode when $\overline{CAS}$ is low (active) at the time $\overline{RAS}$ goes low (active). When this $\overline{CAS}$-before-$\overline{RAS}$ condition is satisfied, the 2116 ignores address bit A_6 and refreshes one row in each half of the device, thus refreshing all 128 rows of storage cells in only 64 cycles. Address bits A_0 through A_5 determine which rows are refreshed. The 2104A will also accept the $\overline{CAS}$-before-$\overline{RAS}$ cycle and will simply perform a READ cycle on the row addressed by address bits A_0 through A_5, thereby refreshing the row of storage cells. The 2104A $\overline{CS}$ input should be driven high (unselected) during this refresh mode to prevent conflicts between OR-tied data outputs just as with normal read cycle refreshing.

Address Multiplexing/Refresh Timing

After the refreshing mode has been selected, the address multiplexer and refresh address counter/timer must be configured to support the selected operational mode. The simplest compatible mode (128-cycle $\overline{RAS}$-only refresh) will again be developed first. Figure 6 shows the detailed block diagram of the logic required to perform the multiplexing/refresh function for the 2104A/2116 compatible system. An implementation of the required logic using the Intel® 3222 and Intel® 3242 Schottky TTL memory support devices is shown in Figure 7.

The 2104A requires 12 address bits multiplexed into 6 address input pins plus a Chip Select ($\overline{CS}$) input. The 2116 requires 14 address bits multiplexed into 7 address input pins and no $\overline{CS}$ signal. Rather than requiring jumpers or strapping at each address multiplexer input pin, the address assignments shown in Figure 1 are "scrambled" to minimize the strapping requirements as much as possible. This address scrambling effects only the column addresses to the 2116. It results in the column address to the memory devices progressing in the order 0, 2, 4, 6, ..., 124, 126, 1, 3, 5, 7, ... 125, 127 as the column address bits (A_9 through A_{13}) from the processor progress in the order 0, 1, 2, 3, 4, ..., 125, 126, 127. This does not effect refreshing since only the column address bits are scrambled. No system effects will result from this technique but it is necessary to be aware of the addressing characteristics while troubleshooting the system.

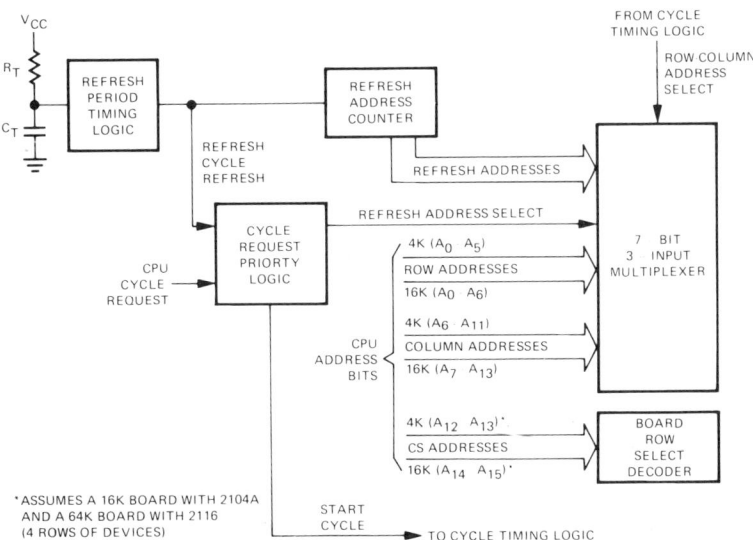

Figure 6. Detailed Block Diagram of Address and Refresh Logic

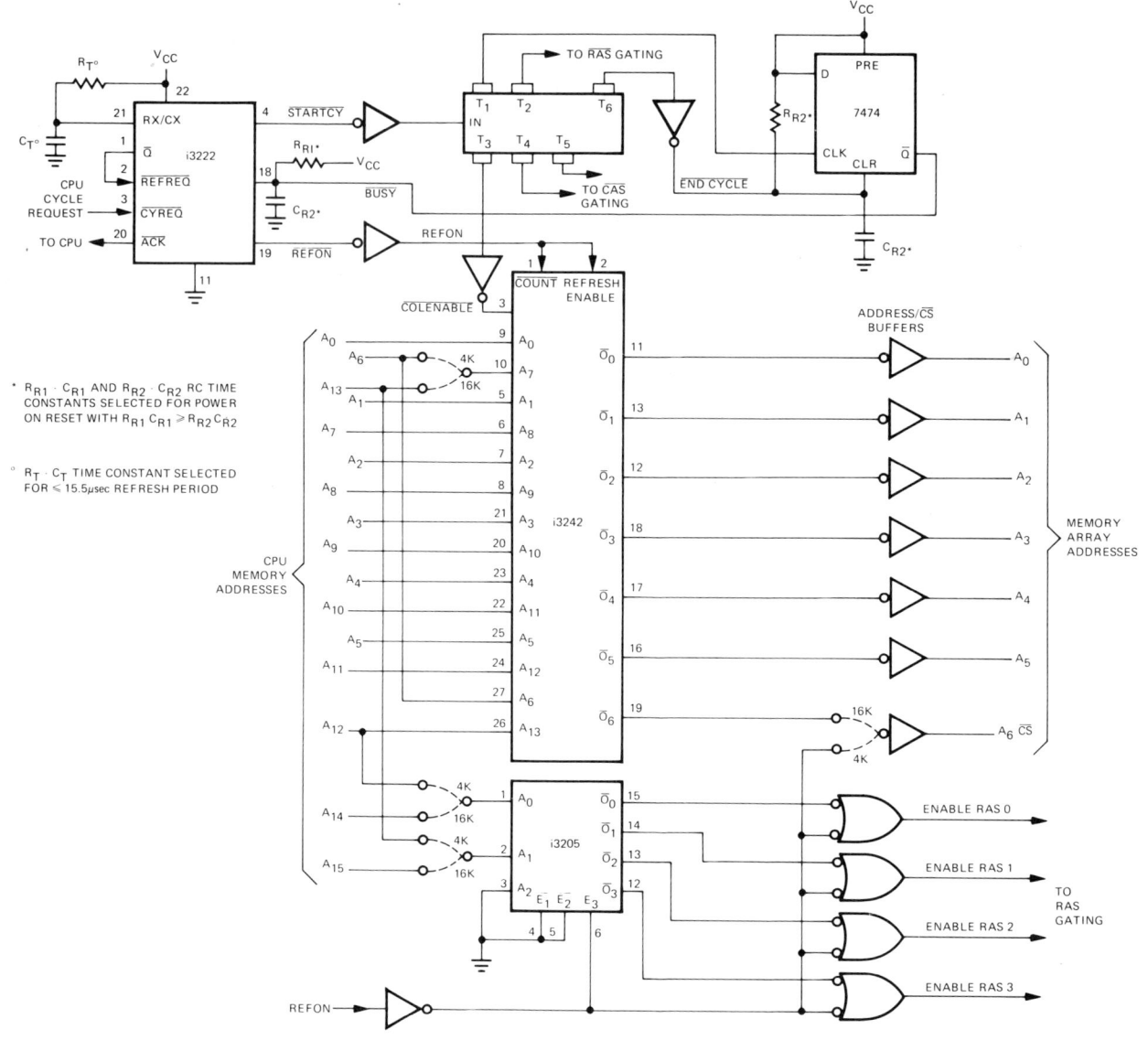

Figure 7. 4K/16K Memory System Control

The circuit of Figure 7 assumes a memory board configuration of four rows of memory devices (16K words X n-bits with the 2104A or 64K words X n-bits with the 2116). It also assumes that row selection will be via $\overline{RAS}$ gating for both the 2104A and 2116 and that the $\overline{RAS}$-only refresh mode will be used with both devices. The address decoding for row selection and $\overline{RAS}$ gating is performed by the 3205. Only address inputs A_0 and A_1 of the 3205 are used and the E_3 enable input pin is used to inhibit the address decoding during refresh cycles.

Processor address bits A_{12} and A_{13} are decoded by the 3205 when the 2104A is used (14 system address bits total) and address bits A_{14} and A_{15} are decoded for row selection with the 2116 (16 system address bits total). This requires strapping of the proper system address bits into the 3205 as indicated in Figure 7.

A 74S00 quad Nand gate is used in an inverting OR gate configuration to provide either 1-of-4 $\overline{RAS}$ enables during data cycles or 4-of-4 $\overline{RAS}$ enables during refresh cycles to refresh all rows at once.

The 3242 includes the refresh address counter and the counter is incremented following each refresh cycle by the high-to-low transition of the REFRESH ENABLE signal at the 3242 $\overline{COUNT}$ input.

An optimization of the configuration of Figure 7 would be to select between 64-cycle refresh for the

DESIGNING WITH 16K DYNAMIC RAMS

2104 A and 128-cycle refresh for the 2116. This would optimize the memory availability for each device type while using $\overline{\text{RAS}}$-only refresh cycles. Figure 8 shows the modifications required on the circuit of Figure 7 to implement the refresh switching.

If 64-cycle refresh is desired for both the 4K and 16K RAMs, clock control logic is necessary to switch $\overline{\text{CAS}}$ low prior to $\overline{\text{RAS}}$ during refresh cycles.

The $\overline{\text{CS}}$ pin of the 2104A must also be driven high during refresh cycles to prevent data output bus conflicts. One possible logic configuration to perform the switching function is shown in figure 9.

Power Distribution/Decoupling

The recommended printed circuit board layout for the memory device array is shown in Figure 10. Notice that each power supply distribution system

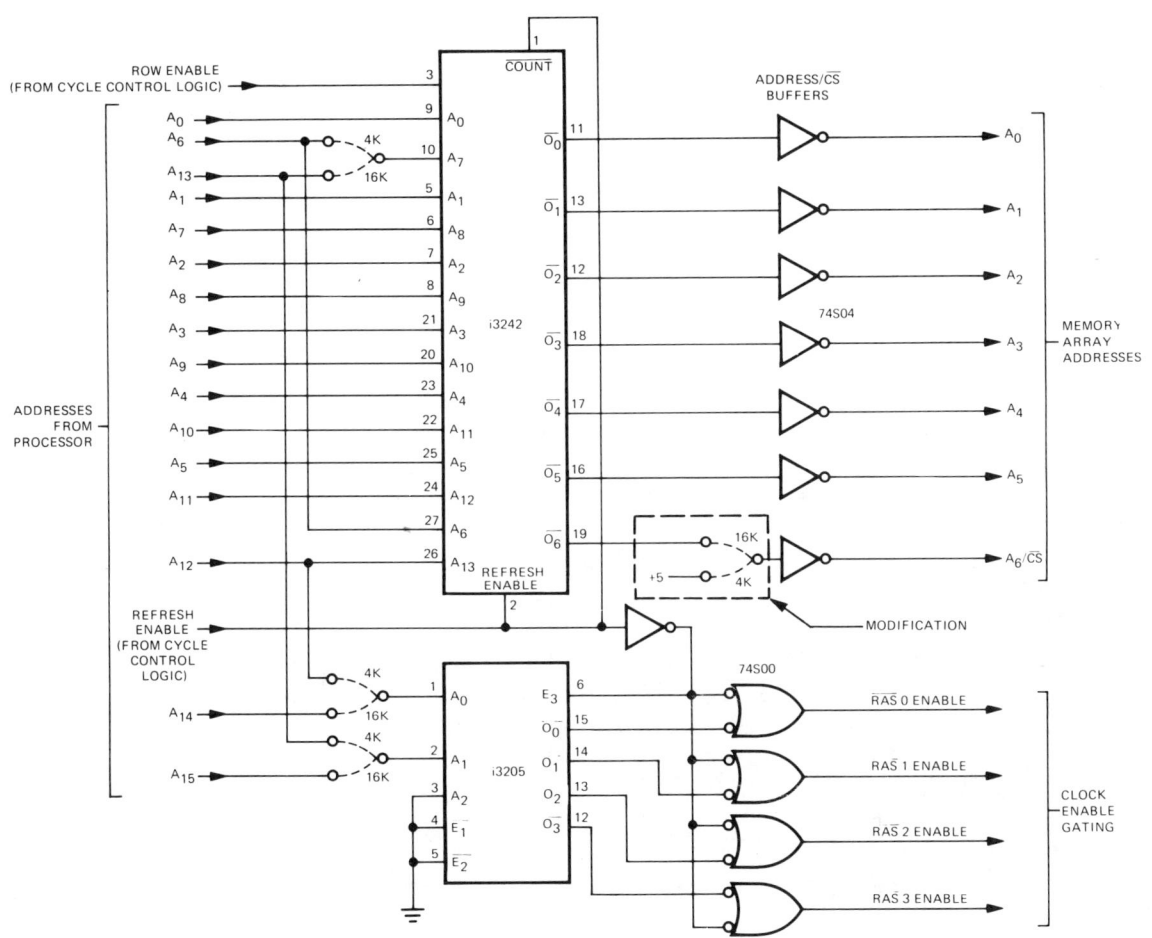

Figure 8. Refresh Timing Switching

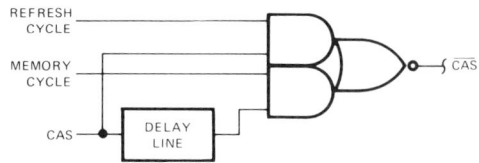

Figure 9. Multiplexed $\overline{CAS}$ for 64 Cycle Refresh

in this double-sided layout is gridded both horizontally and vertically at each device location. This technique provides a low inductance, high quality distribution system which performs as well as multi-layered layout techniques.

When the layout of Figure 2 and the following recommended decoupling capacitance values are used, power supply noise levels within the memory device matrix will be within the required operational limits for the 2104A and 2116.

Recommended decoupling for the 2104A is as Follows:

V_{DD}: A $0.1\mu F$ ceramic capacitor between V_{DD} and V_{SS} at every other device location.

A $10\mu F$ tantalum or equivalent bulk capacitor adjacent to the array for each 16 devices in the array.

V_{BB}: A $0.1\mu F$ ceramic capacitor between V_{BB} and V_{SS} at every other device location (preferably alternate devices to the V_{DD} decoupling).

A $10\mu F$ tantalum or equivalent bulk capacitor adjacent to the array for each 32 devices in the array.

V_{CC}: A $0.01\mu F$ ceramic capacitor between V_{CC} and V_{SS} for each 8 devices in the array.

Recommended decoupling for the 2116 is the same as for the 2104A with the following exceptions:

V_{DD}: Use $0.33\mu F$ ceramic capacitors rather than $0.1\mu F$.

Use a $20\mu F$ tantalum rather than a $10\mu F$.

A common configuration would be to use $0.33\mu F$ ceramics for V_{DD} decoupling with both the 2104A and 2116. Also use a $10\mu F$ tantalum on V_{DD} for each 8 devices in the array. The V_{BB} and V_{CC} decoupling would use the recommended values for the 2104A. This configuration would yield acceptable results with both the 4K and 16K devices and would most likely be more economical than using two different configurations.

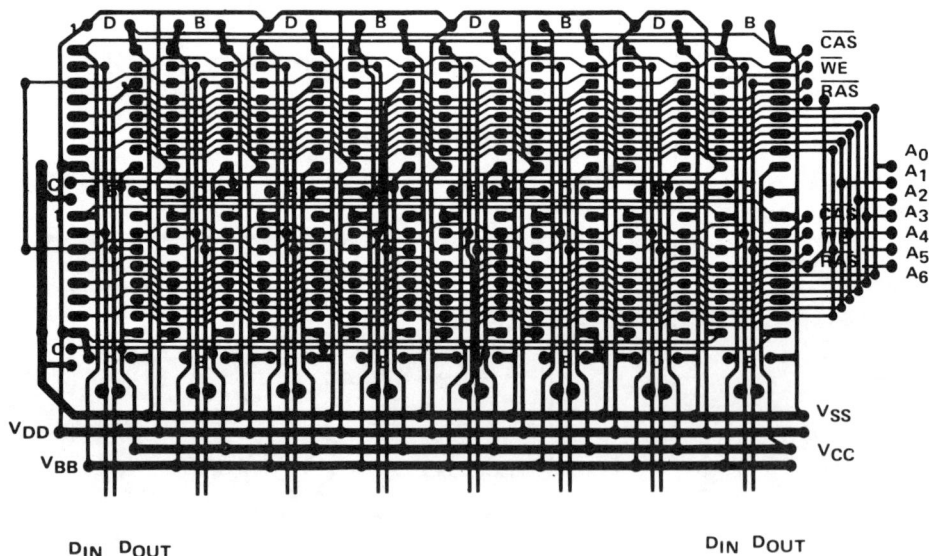

DECOUPLING CAPACITORS
D = $0.33\mu F$ to V_{DD} TO V_{SS}
B = $0.1\mu F$ V_{BB} TO V_{SS}
C = $0.01\mu F$ V_{CC} TO V_{SS}

Figure 10. Recommended two-Sided Board Layout for 2116

DYNAMIC RAMS USED WITH MICROPROCESSORS

GARY FIELLAND

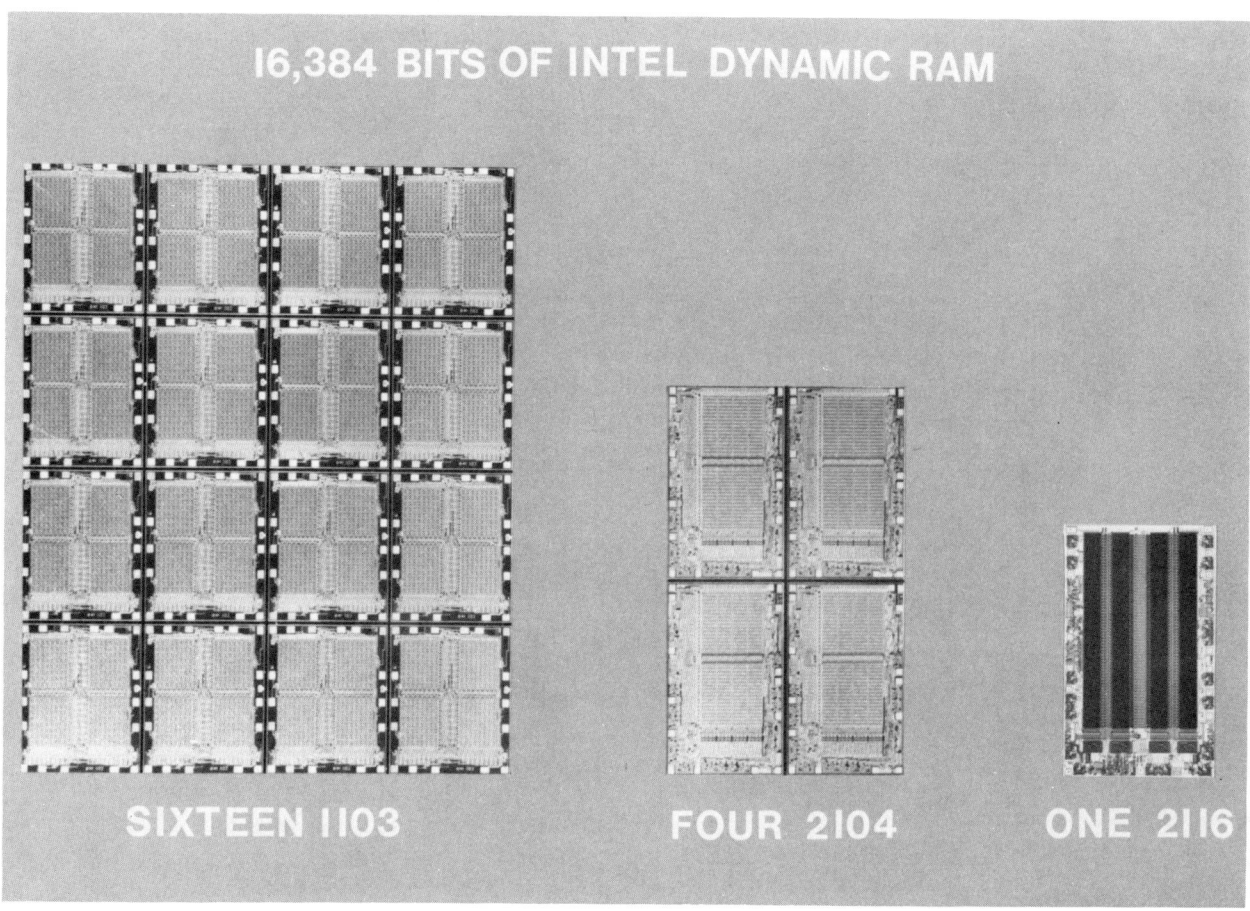

Three Generations of Intel Dynamic RAMs.

INTRODUCTION

Dynamic RAMs (Random Access Memories) have come of age in microprocessor system designs. The pervasiveness of programmed logic, the use of high-level languages and techniques, the increased power of microprocessors, and the decline of memory prices are resulting in a requirement for ever-increasing amounts of read/write memory.

Whereas the DEC® PDP-8 only provided address-ability for 4096 (4K, K = 1024) words, the Intel® 8080A microprocessor addresses 64K bytes, and some applications have added hardware to extend the addressing range even further. One can't help but agree with R. R. Everett's statement in his paper, "The Whirlwind I Computer" (AIEE-IRE Conference, pp. 70–74, 1951): "What the computer industry needs, has needed, and will probably always need is a bigger, better, and faster storage device." The dynamic RAM with its high bit density and low bit cost provides a viable solution.

This Application Note is divided into 12 sections, each dealing with some aspect of dynamic RAM/microprocessor interface design. The first three sections are fairly general and cover the differences between static and dynamic RAMs and the specific requirements of dynamic RAMs. The four following sections deal with actual design examples using Intel's 2107B, 2104A, and 2116 dynamic memories connected to an Intel® 8080A microprocessor. The four following sections provide a general look at some techniques which may be useful in dynamic RAM/microprocessor interface designs. The last section provides a summary and is followed by two appendices which include a design example and data sheets, with recommended artwork.

1. STATIC vs. DYNAMIC

Perhaps the most popular microprocessor read/write memory component is the Intel® 2102A 1024-bit static RAM. By "static" we mean that the information stored in the RAM will remain valid as long as power is provided. Contrast this with the "dynamic" RAM wherein simply providing power will not prevent the loss of the data stored within the RAM. The data storage cell is "dynamic" and must be "refreshed" (read data from cell, amplify, write back to cell) periodically to ensure the data integrity.

1.1 Density

Several observations can be made by contrasting the macroscopic characteristics of the two types of semiconductor RAM (see Table I). First we may note that dynamic RAMs maintain a factor of 4 to 1 in component density over static RAMs.

Table I. Dynamic vs. Static RAMs

DEVICE	NO. BITS	BITS/in.$^{2(1)}$	OPERATING POWER/BIT$^{(2)}$
2102A STATIC	1K	2K	0.16 mW
2114 STATIC	4K	7K	0.10 mW
2107B DYNAMIC	4K	5K	0.03 mW
2104A DYNAMIC	4K	8K	0.02 mW
2116 DYNAMIC	16K	33K	0.01 mW

NOTES: 1. board array area
2. Typical power at 2 µs system cycles. Dynamic RAMs clocked at minimum timing.

When the 1024-bit 2102A was the largest static RAM, the 4096-bit 2107B and 2104A dynamic RAMs were available. Similarly, today there is a 4096-bit 2114 static RAM and a 16,384-bit 2116 dynamic RAM. The component density advantage of dynamic RAMs is inherent to the much simpler and smaller dynamic RAM memory cell. This large density advantage means that the system designer can get more bits per square inch of board area with dynamic RAMs, even including the overhead circuitry required.

1.2 Speed

Memory speed is a fairly complex characteristic to contrast. Considering a typical microprocessor memory READ cycle (Figure 1) will help bring the term "access time" into focus. The microprocessor presents its memory address which stabilizes and this stabilization is followed some time later (typically 100 ns) by its memory read (MEMR) strobe. This strobe asks that the selected memory device provide the addressed data on the microprocessor's bidirectional data bus. Providing the address early in the cycle allows address decoding and module selection to be performed before the MEMR strobe occurs, thereby obviating any bus conflicts. Some time (typically 350 ns) after the issuance of the MEMR strobe the processor expects valid data to be available at its input port. If the data cannot be made available in the allotted time the processor can be forced to WAIT by pulling its READY input to the inactive state. As long as READY is false the processor will wait an integral number of processor clock periods (WAIT states).

Now we can say that in order to meet the "No-WAIT state" timing for the processor the memory system must be no slower than the processor

DYNAMIC RAMS USED WITH MICROPROCESSORS

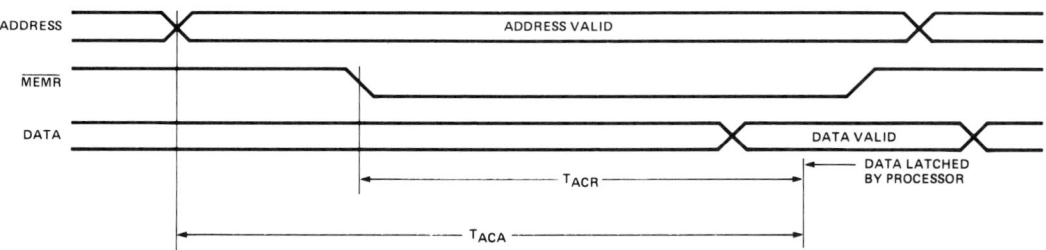

Figure 1. A typical microprocessor No-WAIT-state memory READ cycle showing the required memory system access from address (T_{ACA}) and READ strobe (T_{ACR}).

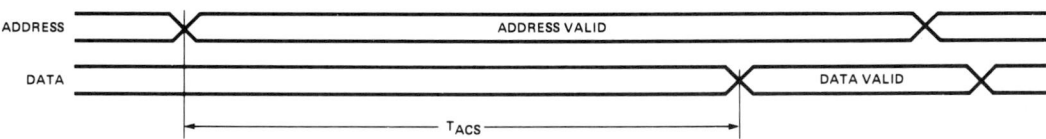

Figure 2. Static memory access time (T_{ACS}) is measured from address valid to data valid.

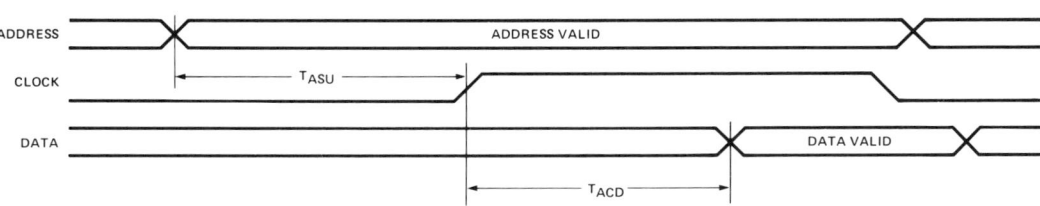

Figure 3. Dynamic memory access time (T_{ACD}) is measured from clock to data valid with an implied address set-up time (T_{ASU}).

required read access times. The first is T_{ACA} which is defined as the time from a stable processor supplied memory address until the processor expects valid data at its input port (typically 450 ns). The second time of interest is T_{ACR} which is defined as the time from the processor supplied read strobe until the processor expects valid data at its input port (typically 350 ns). These parameters are limiting factors for memory access time. Read access time T_{ACS} (Figure 2) for a static memory component is defined as the time from a stable address input until the data at the RAM output is valid. A static memory will begin accessing the addressed bit cell(s) the moment the address becomes available. If there are multiple banks of static memory, each will access the addressed bit cell(s), though only one bank will be selected by the high-order address bits decoder. Hence, ignoring bus and bus buffer delays, a static memory with $T_{ACS}=T_{ACA}$ will satisfy the No-WAIT state criterion.

Now consider the access of a dynamic RAM. the read access time T_{ACD} (Figure 3) for a dynamic memory component is defined as the time from its clock input until the data at the RAM output is valid. Note the address is implicitly assumed to be set up at the address inputs of the dynamic RAM some time (typically 0 to 10 ns) T_{ASU} before the clock input is activated. Convenience of system design, though, usually dictates that the clock input be activated by the MEMR strobe. This means the dynamic memory has less time available to access its data than would a static memory. A second consideration is that typically 50 to 100 ns are lost in the dynamic memory controller itself. Thus in order to satisfy the No-WAIT state criterion, a dynamic memory must provide a $T_{ACD} < T_{ACR}$ (typically 50 to 100 ns less). Summarizing, we note that a dynamic memory must have an access time some 150 to 200 ns faster than a static memory to satisfy a microprocessor No-

WAIT state access requirement. Figure 4 graphically depicts the memory access time requirements for an Intel® 8080A microprocessor.

1.3 Power

As Table I demonstrates, dynamic RAMs have an advantage over static RAMs in power/bit of as much

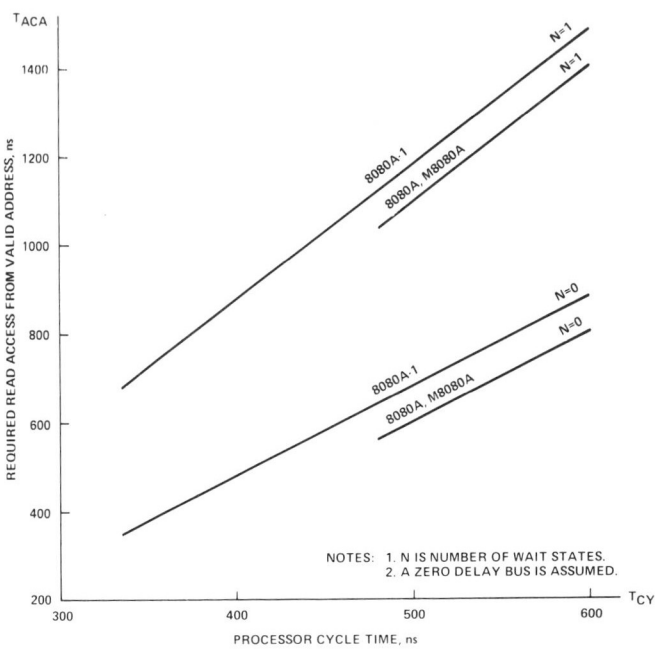

(a) READ ACCESS FROM VALID ADDRESS OUTPUT

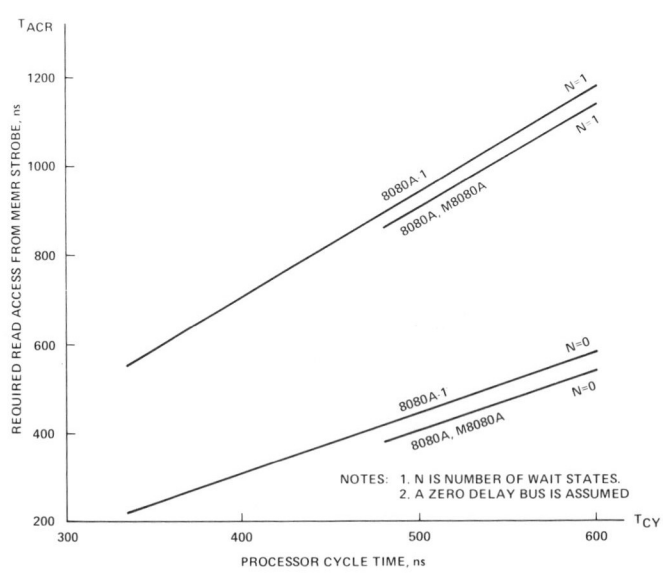

(b) READ ACCESS FROM MEMR STROBE

Figure 4. The Intel® 8080A/8224/8228/8238 processor group may operate with a wide range of memory speeds. Shown are the access time requirements from valid address and MEMR strobe outputs to valid data at the 8228/8238 data bus input assuming a zero delay bus.

DYNAMIC RAMS USED WITH MICROPROCESSORS

as 16 to 1. This is primarily due to the fact that dynamic RAMs draw nearly zero power except when being accessed or refreshed. The power figures for dynamic RAMs in Table I assume a 2-μs memory cycle time. This is typical if every microprocessor memory cycle is from the same dynamic RAM. However, most microprocessor applications use ROM (Read Only Memory) as the instruction storage, implying the effective dynamic memory cycle time will be longer than 2 μs, improving the power dissipation even more. Obviously, power is a very important consideration since the cost/watt is increasing. The nearly-zero standby power characteristic of dynamic memories is also very beneficial in battery back-up applications.

1.4 Cost

Cost is a very complex function and is discussed here only to provide a guideline. The price of a semiconductor memory is very volatile and is a function of a number of variables including access time, package, product maturity, volume, availability, competition, and special handling, among others. As a general rule one would expect a dynamic RAM to be somewhat cheaper per bit than a static RAM. Dynamic RAMs, however, require some control circuitry which must burden the system cost, but the cost of the control circuits is usually very small in comparison to the cost of the memory array. Another factor which will influence system cost is the significantly lower power required by dynamic memories. Still one additional factor which must be considered is that of memory density where the dynamic RAM can offer a factor of over 4 to 1 better than static memories. This density may or may not be significant, depending on board size and the number of bits required. In summary, we expect a dynamic memory system to usually cost less than a static memory system.

1.5 Ease of Use

There is no read/write random access memory component easier to use and more "forgiving" than a static memory. On the other hand, there is no reason that one cannot design a dynamic memory system every bit as reliable and, perhaps, more cost effective than a static memory system.

The very nature of static RAMs lends to their ease of use. Each bit is stored in a cross-coupled bistable circuit which will maintain its state (as last written) as long as power is supplied. The bistable operation of the circuit implies that quite a bit of energy must be applied differentially to cause a cell to change state, enhancing the noise immunity. The common mode rejection properties of a bistable also help reduce sensitivity to power supply variations. There is, of course, a point when the cell power supply will be so low as to defeat bistable action, giving unpredictable results. Furthermore, the bistable circuit is self-restoring. That is to say that if a transient caused the bistable to almost (but not quite) change states, the gain of the elements would quickly bring the circuit back to its original state.

Static RAMs are easy. Apply an address and read out the data, or apply address and data and write the information. The timing is noncritical, the inputs and outputs are TTL compatible, and only a single +5V supply is required.

Dynamic RAMs on the other hand, have gained somewhat of a bad reputation. It seems to be widely held that dynamic RAMs are esoteric components reserved for the main frame memory system designer. While this may have been true in the very early days of dynamic memories, it no longer holds true. Modern technology and circuit design have produced a series of dynamic RAMs which are applicable to small microprocessor systems.

The early PMOS (P-channel Metal Oxide Semiconductor) 1103 dynamic memories were difficult to use by today's standards. They required high voltage power supplies (+16V and +19V) and had tight timing requirements. They did not have TTL (Transistor-Transistor Logic) compatible inputs or outputs, and, in fact, required external sense amplifiers to sense their low-level outputs. Because of the complexity of an 1103 system design (amplified by the fact that microprocessors were still in their infancy), the component was used primarily in main frame memory applications where the overhead was justified.

However, modern dynamic memories have been greatly improved. These NMOS (N-channel) devices require microprocessor-compatible supplies (+12V, +5V, −5) and most inputs and outputs are TTL compatible. Perhaps even more beneficial to the casual user, the timing has been greatly simplified. It is now feasible to design dynamic memories into a fairly small system.

This is not to say that using dynamic memories is as easy as using static memories. Dynamic memories demand that more attention be paid to the memory system design. They also require that a dynamic memory controller be designed. The specific requirements of dynamic memories will be discussed shortly.

1.6 Static or Dynamic?

Five characteristics (density, speed, power, cost, ease of use) of semiconductor read/write random access memories have been compared; static vis-a-vis dynamic. The system designer must weigh the tradeoffs and make a decision based on the system requirements. In general, any requirement for less than 4K bytes is best handled with static memories, while any requirement for over 16K bytes is best met with dynamic memories. The intermediate memory requirements must be carefully considered and the appropriate tradeoffs made. The remainder of this Application Note will be dedicated to a treatise on effective means of implementing dynamic memories in microprocessor systems.

2. DYNAMIC MEMORY

The semiconductor dynamic RAM can be a very effective component in the implementation of reliable, high performance, low cost microprocessor memories. The device does, however, have several unique requirements which should be considered.

2.1 Bit Cell

First consider the dynamic memory bit-cell which is quite unlike the cell of a static RAM. The dynamic bit-cell conceptually consists of several switches and a capacitor (see Figure 5) implementing a sample-and-hold circuit. During a WRITE operation the datum is placed on the input and switch one (S_1) is closed, completely charging/discharging the capacitor, depending on the datum. During a READ operation switch two (S_2) is closed and the capacitor voltage is compared to a reference voltage yielding the datum on the comparator output. When in the storage mode, all switches are open with the capacitor storing the charge. The capacitor will gradually deviate from its stored voltage due to charge being removed by the leakage source. If the node is left for some time the capacitor voltage will have drifted far enough that its level no longer reflects the datum that was stored. To prevent this the cell is periodically refreshed by sequentially closing switch two (S_2), then three (S_3), then one (S_1). Thus the internally capacitor voltage is read out, amplified, and rewritten, restoring the original voltage level on the capacitor. Note the READ operation could be redefined to include this function.

2.2 Refresh

Thus, one can see that refresh is a very important requirement for a dynamic RAM. The dynamic memory controller must assure that every bit cell is refreshed often enough to maintain data integrity. The refresh interval is specified by the vendor and a typical requirement is that each bit-cell be refreshed every 2 ms. To reduce the overhead imposed by refresh, dynamic memory components have their cells organized in arrays. The Intel® 2107B and 2104A are 4096-bit memories constructed with a cell array of 64 rows and 64 columns. All columns in a single row are refreshed simultaneously so that the user has only to provide 64 refresh cycles (each with a different row address) each 2 ms. The advanced Intel® 2116 is a 16,384-bit memory component constructed with two identical cell arrays each with 64 rows and 128 columns. Again, all columns in a single row in an array are refreshed simultaneously. This means that the user must supply 128 refresh cycles (each with a different row address) each 2 ms. But the 2116, through clever circuit design, is also able to access both arrays simultaneously during refresh. This neat trick cuts the refresh overhead in half; only 64 special refresh cycles need be supplied!

In order to supply the refresh row address a counter (6 or 7 bits) is required and is incremented after each refresh cycle (Figure 6). A two-input row address multiplexer is also used to multiplex either the processor-supplied memory address or the counter-supplied refresh address onto the dynamic

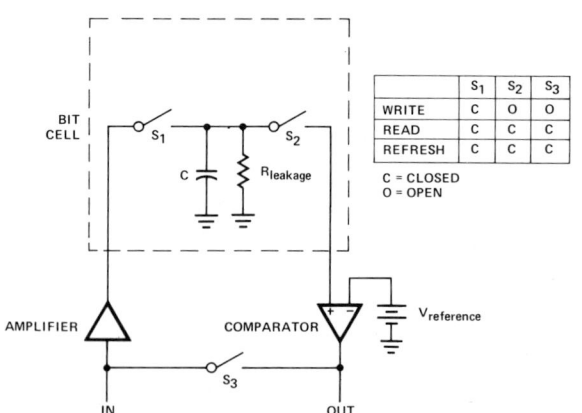

Figure 5. Conceptual dynamic memory bit cell showing sample-and-hold operation.

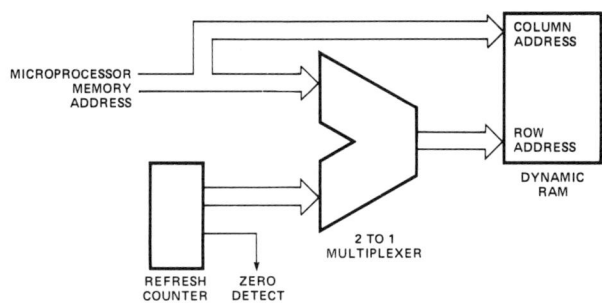

Figure 6. Dynamic RAM systems include a refresh counter, and row address multiplexer. Zero detection is useful when implementing Burst Mode Refresh systems.

memory row address inputs. These refresh cycles may be effected as a set of contiguous cycles known as *Burst Mode Refresh*, or as discrete non-adjacent cycles known as *Distributed* or *Single Cycle Refresh*. The choice between the two methods is normally made based on memory availability requirements and ease of implementation. In either case some means must be provided to arbitrate between processor memory cycles and refresh cycles. The design of this refresh arbiter can be simple or complex, dependent on the method chosen. This design will be considered in more detail.

Though supplying refresh in a timely, error-free, and cost effective manner is the dominant requirement, some dynamic RAMs have additional unique properties. The 2107B, for instance, requires a high voltage (+12V) clock input and hence a special clock driver circuit. The timing for dynamic RAMs is, in general, more critical than that for static RAMs, usually resulting in a requirement for precise time interval generation. Today's dynamic memories have nearly all inputs and outputs TTL compatible and use the microprocessor compatible supplies of +12V, +5V, and −5V.

3. REFRESH TECHNIQUES

Providing refresh for dynamic memories is of paramount importance and several different techniques have been developed to provide this refresh. Each differs slightly in complexity, generality, and memory availability. The designer should study the various approaches and pick the one right for his application.

3.1 Asynchronous Refresh

There are three common technqiues used to provide dynamic memory refresh in microprocessor systems (see Table II). The *asynchronous method* is based on the tenet that refresh is inherently a real time event (one refresh cycle every 31 μs − 2 ms/64 cycles) and is independent of the state of the processor. This popular approach yields the most flexible system since it is very loosely coupled. The asynchronous memory system normally has its own dedicated control logic and may run independently of the microprocessor. The local control logic supplies refresh as needed and couples with the processor only to provide READ or WRITE cycles. In most implementations the memory system is unaware of the processor state or any other processor particulars. For instance, an asynchronous refresh memory system may be designed such that it will operate with nearly any microprocessor or even with a powered down processor. Similarly, the memory appears no different than static memory to the microprocessor, except that the processor may occasionally have to wait for service if the memory is busy performing refresh.

While the asynchronous refresh memory system is the most modular approach to design, it frequently suffers from a high degree of complexity and consequent performance degradation. Perhaps the most significant contributing factor is that of dealing with the asynchronous requests. The design of a reliable, high-speed memory controller which must arbitrate between asynchronous refresh requests and microprocessor memory requests is somewhat tricky (Figure 7). This problem is further complicated since the beginning of the memory cycle must now be delayed until the requests have been definitively resolved and appropriate address, data, and control set-up times supplied.

Table II. Refresh Techniques

Asynchronous	—	Refresh is performed asynchronously with respect to the microprocessor.
Synchronous	—	Refresh is performed in synchronism with microprocessor events.
Semisynchronous	—	Refresh is performed in synchronism with the microprocessor clock, but asynchronously with respect to microprocessor events.

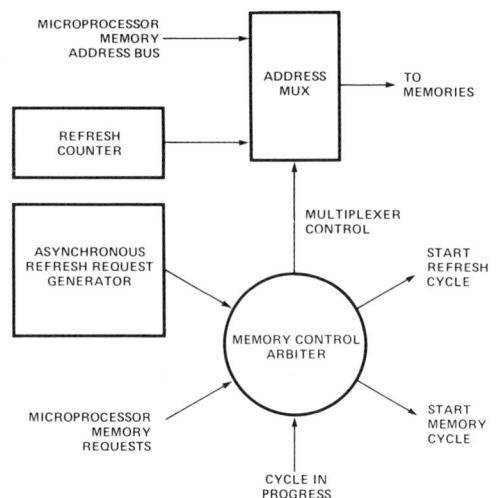

Figure 7. Asynchronous memory system. The design of a reliable memory control arbiter to resolve asynchronous requests is somewhat tricky.

Thus, this request resolution time adds directly to the system access time for each cycle. An example of an asynchronous refresh microprocessor memory system is provided in Appendix I.

3.2 Synchronous Refresh

The *synchronous method* of refresh is normally used to improve the performance and apparent availability of the memory system. With this method refresh cycles are obliged to occur in synchronism with microprocessor events and the event usually chosen is a cycle in which the microprocessor will not be using the memory. Hence, there is no contention for the memory and, furthermore, the refresh cycles do not detract from the apparent memory availability. This method of hiding the refresh cycles during times the memory would otherwise be idle is often called *Invisible Refresh*, since the processor sees no delay due to refresh. It may be noted that herein the memory is available to the processor without conflict every time a memory request is made. This absence of contention leads to a non-obvious but significant performance improvement. The memory address and data multiplexers may now select the microprocessor bus before the cycle begins. Then when the microprocessor issues a request, the cycle begins immediately. The processor normally provides sufficient address and data lead time to satisfy all address decoding, propagation delay, and memory device set-up times. Contrast this with the asynchronous approach wherein after the request is issued there may be additional delay due to a cycle already commenced and/or arbitration and multiplexer settling before the requested cycle can be started. This means the synchronous system access time is determined only by the memory and timing generation circuitry without any time required for arbitration or multiplexer settling. Thus, with today's dynamic memories, it is possible to guarantee memory cycles which require no microprocessor WAIT states. Dispensing with WAIT states yields a significant performance improvement, since every WAIT state typically consumes 500 ns with today's microprocessors.

The major design hurdle in realizing a synchronous refresh microprocessor memory system is that of the refresh scheduler design. The scheduler accepts status inputs from the refresh timer and the microprocessor and based on its knowledge of that processor, schedules the refresh cycles into idle periods. Obviously, such a scheduler is intimately linked with the particular microprocessor used and its hard-wired design is based on that processor's characteristics. For instance, it is unlikely that a synchronous memory system scheduler designed for an 8080A would be directly appropriate for another microprocessor. Further, it may be necessary to provide some type of synchronism override to guarantee refresh, should the microprocessor be detained from reaching the normal refresh event. Such might be the case in some designs if the microprocessor entered the HALT state or was held in the RESET, WAIT, or HOLD states. And this override introduces some degree of asynchronism back into the system, as the processor may begin again at any time.

One prevalent technique for implementing synchronous refresh using the Intel® 8080A microprocessor is to schedule refresh during the T4 state (see *Intel 8080A Microcomputer System User's Manual*) of an instruction fetch (M1) machine cycle. The 8080A completes a READ cycle in state T3 and does not use the memory during the T4 state, allowing more than one clock period (approximately 500 ns) for the refresh cycle. The scheduler implementation shown in Figure 8 determines the coincidence of refresh required and the beginning of a FETCH machine cycle, and sets a flip-flop. Two more flip-flops serve as counters and clock the START REFRESH flip-flop at the end of state T3. Note the implementation shown makes no provision for intervening WAIT states which migh occur if the instruction fetch were from a slow ROM. Similarly, there is no synchronous override to provide refresh for the memory in the event the 8080A stops fetching instructions. A similar

DYNAMIC RAMS USED WITH MICROPROCESSORS

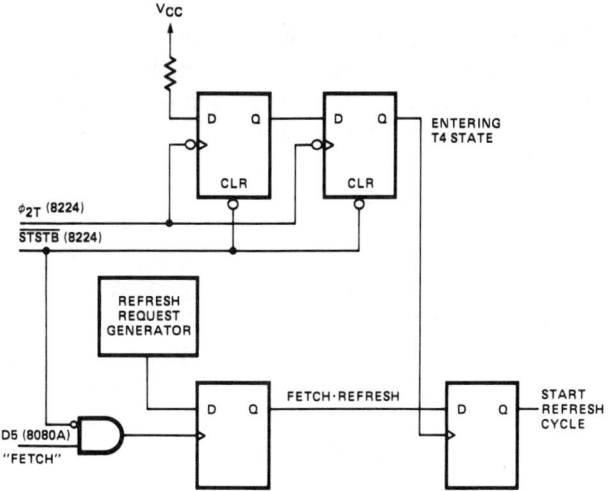

Figure 8. Synchronous Refresh Scheduler for an 8080A microprocessor memory system starts refresh cycles during the T4 state of a FETCH machine cycle. No provision is made for WAIT states in this implementation.

technique would be to perform refresh when the processor makes an instruction fetch from a ROM, leaving the dynamic RAMs otherwise idle. A special form of synchronous refresh will be explored in several of the detailed design examples given.

3.3 Semisynchronous Refresh

Semisynchronous refresh is a hybrid of asynchronous refresh. Its primary raison d'etre is to simplify the request arbiter discussed with relation to asynchronous refresh. The semisynchronous approach takes advantage of a characteristic of microprocessors in that their memory requests are initiated synchronous with a clock edge. Thus, if the refresh request is synchronous with the opposite clock edge, it can be guaranteed that the two request transitions will never occur simultaneously. This simplifies the design of the memory request arbiter. Excepting the synchronizing of refresh requests to the microprocessor clock, semisynchronous refresh is very similar to asynchronous refresh. There still must be memory cycle vs. refresh cycle arbitration since one cycle may already be in progress when the other is requested. The only advantage is that the arbitration should be simpler. The semisynchronous memory system is also quite independent of the processor state, though the processor clock must continue to run. However, incorporating a different microprocessor might require a slight modification to insure the mutual exclusion of refresh and processor request transitions. The memory controller, though, must still arbitrate between requests and lock out the tardy one. Hence, refresh is still visible and the processor may have to wait for refresh cycle completion before gaining access to the memory.

A semisynchronous system might be designed for an Intel® 8080A microprocessor system, realizing the microprocessor will never initiate a memory request on the rising edge of its $\phi 2$ clock. In the circuit shown in Figure 9, a refresh request pulse is

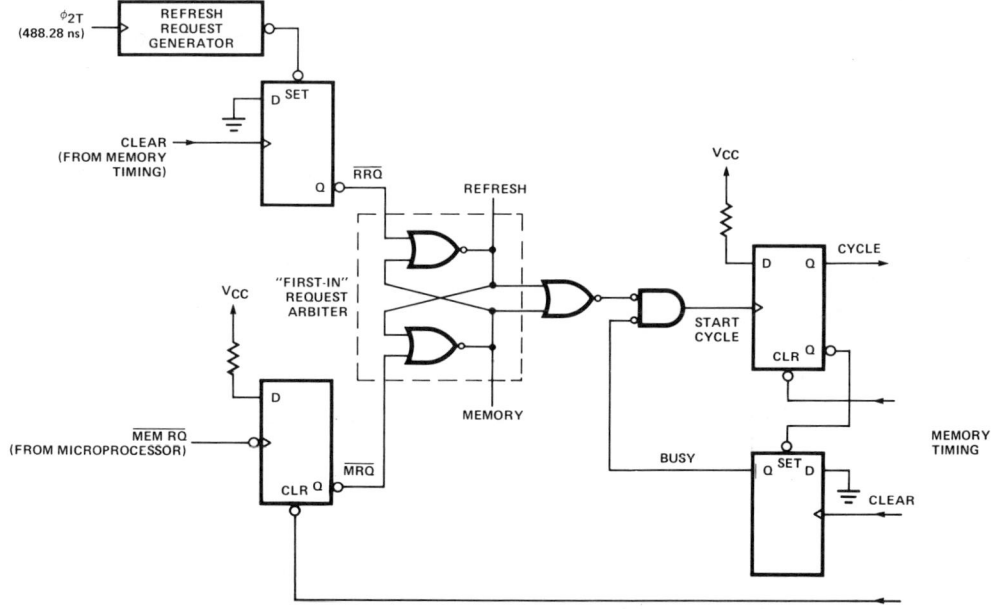

Figure 9. Semisynchronous Refresh System requests refresh cycles synchronous with the rising edge of ϕ_2.

generated periodically synchronous with the rising edge of microprocessor clock $\phi2$. This pulse sets a refresh request flip-flop. Similarly, a microprocessor memory request will set (never coincident with the rising edge of $\phi2$) a memory request flip-flop. The request which arrives first will be honored by the "first-in" request arbiter and further requests locked out until the memory is again available. Any honored request will start the memory timing circuitry and set the BUSY flip-flop. The timing circuitry should clear the flip-flops when appropriate. Note the controller will perform two consecutive cycles if both requests are pending.

3.4 Invisible or Not?

If one chooses a synchronous refresh methodology, the next question might be concerned with whether or not to try for invisible refresh. First, consider the loss due to non-invisible refresh. If one assumes a nominal memory cycle time of 500 ns, there are 4000 memory cycles available in each 2-ms refresh interval. During this interval, 64 refresh cycles must be provided and this yields a loss of less than 2% in memory availability. Now, if one looks from the processor's point of view, the situation is similar. A typical processor machine cycle is 2 μs, so assuming instant memory availability and no loss due to refresh, 1000 machine cycles can be performed each refresh interval. Now, if 64 of those machine cycles were delayed one memory cycle (500 ns) for non-invisible refresh, only 984 machine cycles could be executed each refresh interval, yielding a processing loss of less than 2%.

Thus, the designer must consider the tradeoff. If maximum performance is required, using invisible refresh can achieve a performance improvement of about 2%. The penalty is added cost and complexity of the refresh scheduler design. In some implementations though, this cost may be quite small.

3.5 LSI Refresh Controllers Help

Intel has introduced a family of refresh control circuits to simplify the design of dynamic memory controllers. Designed with high-speed Schottky bipolar logic, the 3222 was designed to complement the 2107B, while the 3242 complements the 2104A and 2116. Both the 3222 and the 3242 are able to reduce the system package count and cost by providing a major portion of the required control logic.

The 3222 is an LSI device designed to facilitate designing asynchronous refresh memory systems using dynamic RAMs. As Figure 10 shows, it contains an accurate, highly stable refresh oscillator which may be programmed by an external resistor and capacitor. This oscillator sets a latch (Q) which may be connected to the REFRESH REQUEST (REFREQ) input. Then when the refresh cycle begins, the latch will be cleared; only to be set again when the oscillator period expires. Also provided is control logic to arbitrate on a first-in basis between ordinary memory requests (CYREQ) and refresh requests (REFREQ). Once a request has been accepted the controller will ask that a cycle be started (STARTCY) and BUSY will latch the current state, locking out further requests. The controller indicates whether it is to be a refresh (REFON = TRUE) or memory (REFON = FALSE) cycle. The multiplexer will transmit the outputs of the 6-bit refresh counter during a refresh cycle, whereas the address inputs (A_5-A_0) will be transmitted otherwise. The refresh counter is incremented at the end of the refresh cycle. The 3222 accepts an input (BUSY) to determine the current status of the memory.

Though the majority of the control logic is self-contained, all memory system timing must be provided externally. To aid the designer, the 3222 data sheet specifies the requisite device timing for all request combinations, including simultaneous requests. This device is included in an asynchronous refresh memory system implementation described in Appendix I. The 3222 is also used as the controller in a synchronous system using the 2107B memory.

The Intel® 3242 (Figure 11) contains an address multiplexer and refresh counter, and is designed for use with the 2104A 4K and the 2116 16K dynamic RAMs. It is fabricated in Intel's reliable Schottky bipolar process and its high performance is specified over a 0°C to +75°C ambient temperature range. The 3242 multiplexes 14 (or 12) bits of system supplied address onto 7 (or 6) output pins. The device also contains a 7-bit refresh counter which may be multiplexed onto the output and is externally controlled so that either distributed or burst-mode refresh may be used. This part facilitates doing a memory design which would be compatible with either the 2104A or 2116 memory.

4. DESIGN EXAMPLES

The three examples that follow are all based on a special form of the synchronous refresh method. As discussed earlier, this type design is intended for

DYNAMIC RAMS USED WITH MICROPROCESSORS

PIN CONFIGURATION

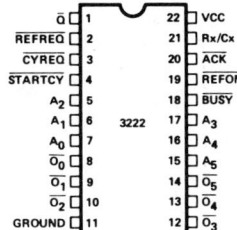

PIN NAMES

A_0 - A_5	ADDRESS INPUTS	$\overline{O}_0$ - $\overline{O}_5$	ADDRESS OUTPUTS
$\overline{ACK}$	ACKNOWLEDGE OUTPUT	$\overline{Q}$	INTERNAL REFRESH REQUEST LATCH OUTPUT
$\overline{BUSY}$	BUSY INPUT	$\overline{REFON}$	REFRESH ON OUTPUT
$\overline{CYREQ}$	CYCLE REQUEST INPUT	$\overline{REFREQ}$	REFRESH REQUEST INPUT
		RxCx	RC TIE POINT
		$\overline{STARTCY}$	START CYCLE OUTPUT
		V_{CC}	+5V SUPPLY

BLOCK DIAGRAM

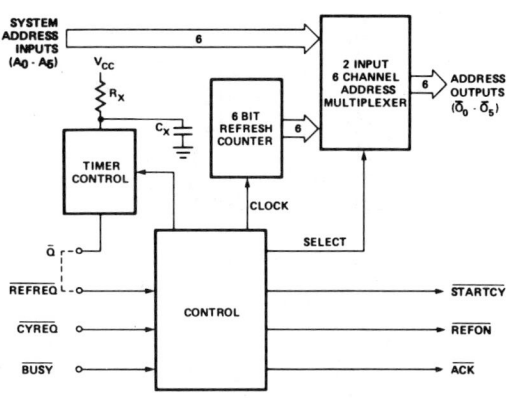

LOGIC DIAGRAM

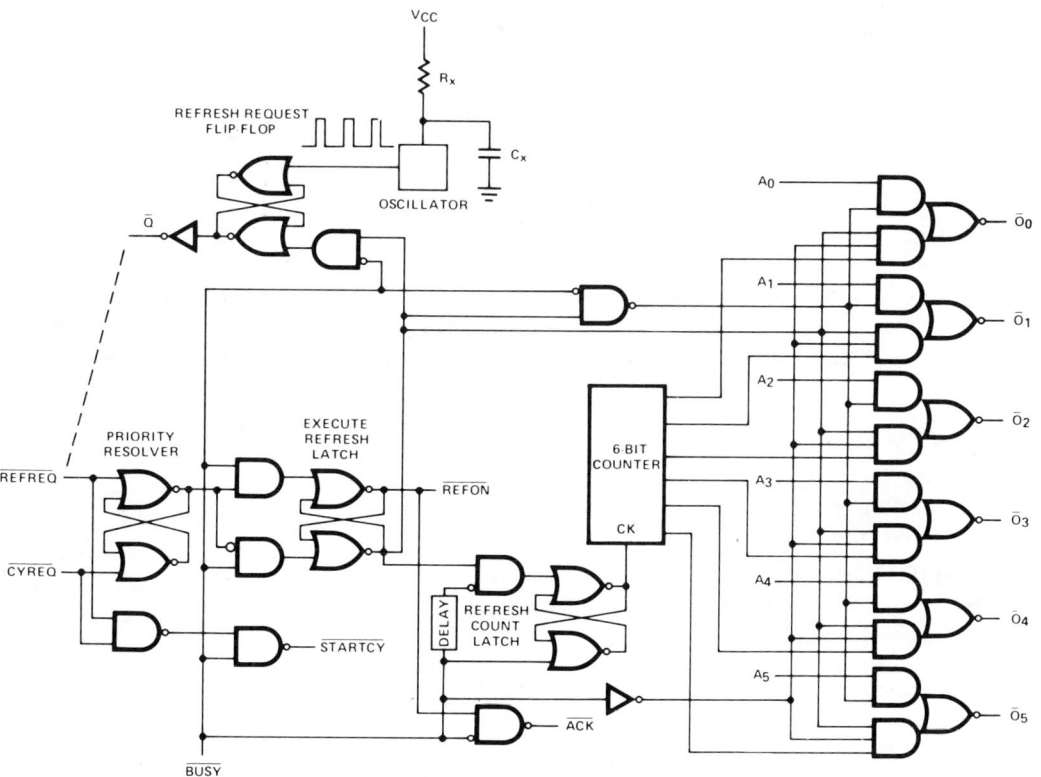

Figure 10. The Intel® 3222 is a complete refresh controller subsystem. It is designed to implement asynchronous refresh dynamic memory systems.

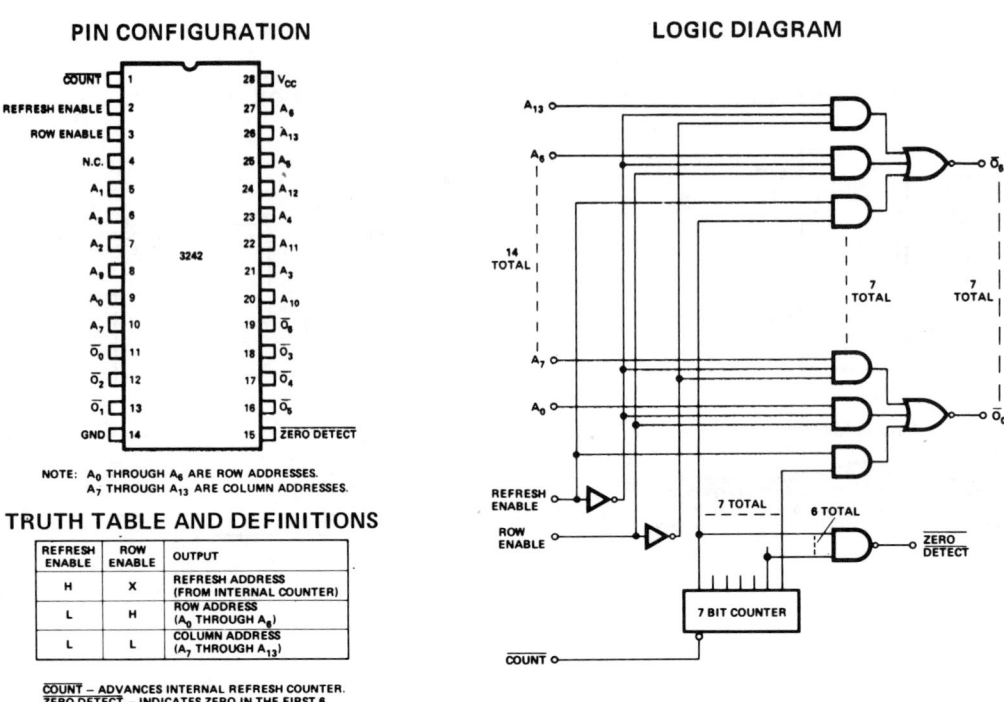

Figure 11. The Intel® 3242 is a Schottky bipolar address multiplexer and refresh counter to complement the 2104A and 2116 dynamic RAMs.

use with dynamic memories as part of a dedicated microprocessor system, as opposed to a "stand-alone", general-purpose memory system. The microprocessor used for the examples is the ubiquitous 8080A system (see *Intel 8080 Microcomputer System User's Manual*). For the examples, the 8080A processor group is configured as shown in Figure 12 and operates with an 18.432 MHz crystal.

When designing a microprocessor dynamic RAM interface, several tradeoffs face the designer. The first is a question of system requirements. The examples herein are aimed towards a dedicated system where cost is more important than generality. (For those designs where generality is critical, the example given in Appendix I provides a very versatile approach for a memory system design.) Using cost and performance as the two key objectives, the synchronous method of refresh was chosen, with no attempt made to implement invisible refresh.

4.1 Scheduler Design

At this point a decision must be made as to how the synchronous refresh scheduler should be designed. Clearly, the scheduler must have intimate knowledge of the processor state and be able to predict the appropriate moment to perform refresh. Conveniently, the 8080A already has an inherent scheduling function under another name; and who has better knowledge of the processor state than the processor itself? Two pins, HOLD and HLDA (HoLD Acknowledge), are provided for this function. HOLD is an input used to request that the 8080A suspend its use of the bus as soon as practical and HLDA is an output on which the 8080A signals that it is about to yield the bus. In 8080A systems, HLDA is commonly delayed until the trailing edge of $\phi 2$, yielding HLDAD (HoLD Acknowledge Delayed) which is used to gate off the 8080A bus drivers.

Voila! Our scheduler design is complete without any design work required. What could be better than a scheduler already designed, guaranteed, and available practically for free? This is the scheduler chosen for the examples and the signals will be given aliases to reflect our use: HOLD = RFRQ (ReFresh ReQuest), HLDA = RFAK (ReFresh AcKnowledge), and HLDAD = RFAKD (ReFresh AcKnowledge Delayed).

DYNAMIC RAMS USED WITH MICROPROCESSORS

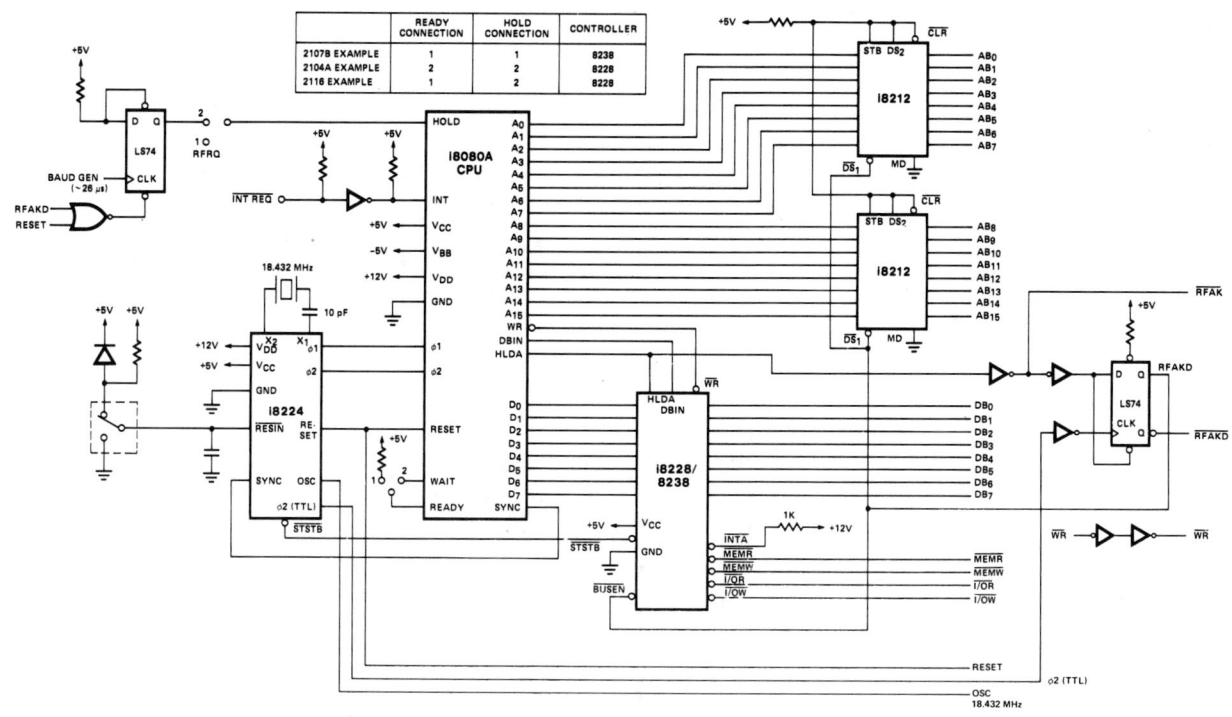

Figure 12. The Intel® 8080A processor group is configured for each of the three design examples.

4.2 Limitations

Let us now consider the limitations on our system due to our choice of scheduler. First, one notes that there are several conditions under which refresh will not be provided due to our use of the 8080A HOLD feature. Refresh will not be provided while the 8080A RESET pin is active, so RESET should not be maintained for long periods of time if the dynamic memory integrity is to be preserved. Similarly, refresh will not be provided while the 8080A is in the WAIT state; that is, READY is false. This should be no problem unless READY is being used to single-step the processor. (While this technique has been used by thousands of engineers, there are better techniques available today. One can use software single-step techniques or the very powerful ICE-80 debug tool.) Finally, refresh will not be provided while in the Interrogation Mode when using ICE-80. (See Intel *ICE-80 Hardware Reference Manual*.) During the Interrogation Mode, the user system is virtually without a processor, so it obviously will not respond to requests. Though these limitations should be kept in mind, they are of little consequence in a dedicated system. The only real drawback is during initial system debug and several comments will be made on this in the conclusion.

With the scheduler design completed, all that remains is to provide a source of refresh requests (Figure 13) and an interface to the three most popular dynamic RAMs — 2107B, 2104A, and 2116.

5. 2107B IS EASY

The Intel® 2107B is a 4096-word by 1-bit dynamic N-channel MOS RAM (see Figure 14 and Appendix II). It is fabricated using a single transistor memory cell in Intel's proven N-channel silicon gate technology to achieve high speed, high density, low power, and low cost.

This device has fairly simple timing requirements and only one high voltage input. It also provides on-board address latches and a TTL compatible three-state output. Because of the simple interface, such a device easily lends itself to a microprocessor memory system.

The Intel® 3245 (Figure 15) is a companion memory support circuit serving as a quad TTL-to-MOS driver. It provides high output current and voltage suitable for driving the clock inputs of the N-channel 2107B. It features clamping diodes to minimize line reflections and gating and selection

DYNAMIC RAMS

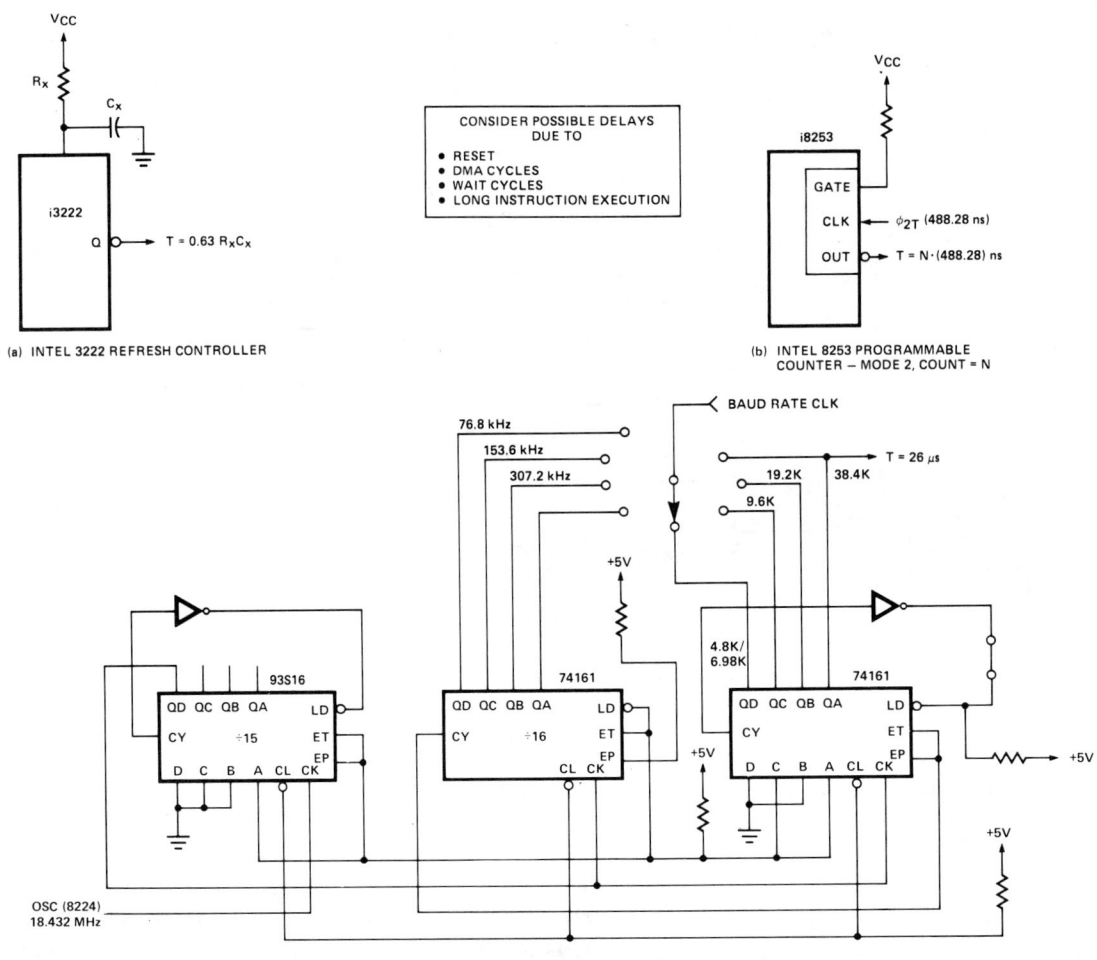

Figure 13. There are numerous sources available in a microprocessor system to generate refresh requests. The refresh request period should be chosen to insure adequate refresh keeping all possible delays in mind.

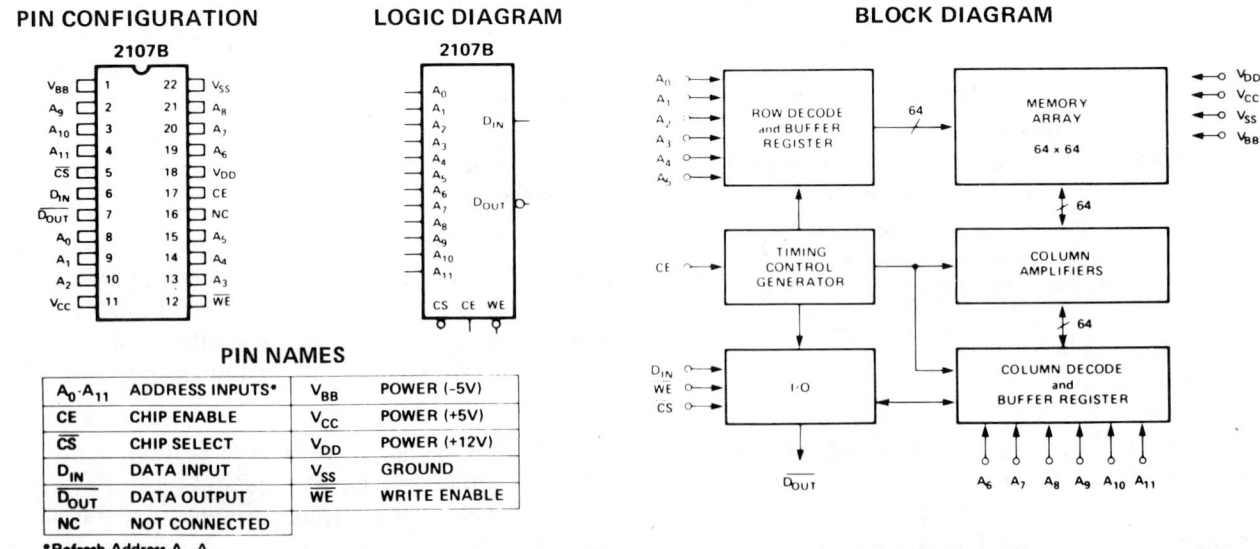

Figure 14. The Intel® 2107B is a 4096-word by 1-bit dynamic N-channel MOS RAM.

DYNAMIC RAMS USED WITH MICROPROCESSORS

PIN CONFIGURATION

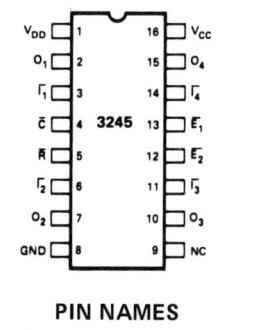

PIN NAMES

$\overline{I_1}$-$\overline{I_4}$	SELECT INPUTS	O_1-O_4	DRIVER OUTPUTS
$\overline{E_1}, \overline{E_2}$	ENABLE INPUTS	V_{CC}	+5V POWER SUPPLY
$\overline{R}$	REFRESH SELECT INPUT	V_{DD}	+12V POWER SUPPLY
$\overline{C}$	CLOCK CONTROL INPUT	NC	NOT CONNECTED

LOGIC DIAGRAM

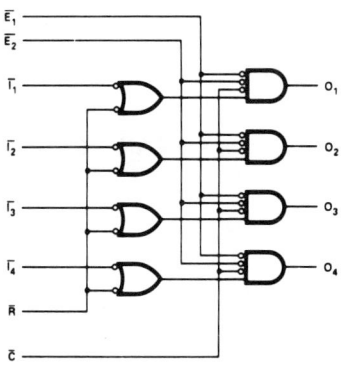

Figure 15. The Intel® 3245 provides selection logic and high voltage drive capability.

---

logic to minimize gating delays and system package count. It is fabricated by means of Intel's highly reliable Schottky bipolar process and is specified over a 0°C to +75°C ambient temperature range, driving a realistic capacitive load.

5.1 2107B Design Example

Before going into the detail of the design example shown schematically in Figure 16, consider just the 3222 refresh controller. Comparing the 3222 logic diagram of Figure 10 with the connection of the 3222 in the schematic of Figure 16, several observations can be made. It can be seen that the request inputs are not being used conventionally. Rather, the 3222 is wired with the normal cycle request (CYREQ) permanently inactive and the refresh request (REFREQ) permanently active. With this connection, the arbitration function of the 3222 has been disabled; it is being used simply for its component parts. The stable oscillator is used to generate refresh requests every 25 µs and the BUSY input is used to control the internal 6-bit multiplexer and 6-bit refresh counter. When BUSY is inactive, the outputs reflect the six address inputs. Then, when BUSY goes active, the multiplexer presents the 6-bit counter on the outputs. Finally, when BUSY goes inactive again, the counter increments. Thus, the 3222 Asynchronous Refresh Controller has been used in a non-conventional way to significantly reduce the parts count in a synchronous refresh memory system.

Now let's consider in detail the circuit design of this 16K word by 8-bit synchronous refresh memory system using the 2107B memory and designed for the 8080A microprocessor. First, consider the data path. Data from the bidirectional 8080A data bus is buffered and inverted by the Intel® 3404's and passed to the data inputs of the memories. The inverted data out of the memories is wire-ORed (selected by Chip Enable) and passed to the Intel® 8212 three-state buffer. The 8212 places the data on the 8080A bidirectional data bus when enabled by a memory read cycle on this memory module.

The four high-order address bits are decoded to generate a 16K module select and a 4K group select signal. Note the 4K group select signal can be wired to disable the module select (INHibit MEMory, INHMEM) should that 4K address space be occupied by another memory such as a PROM. The four group select signals are also applied to the linear select inputs of the 3245 clock driver. The 3245 will be enabled by either module selection or by a refresh cycle. Note a refresh cycle will also activate the 3245 refresh input, thus allowing the activation of all four outputs so that all memory devices may be refreshed simultaneously. The next 6 address bits are fed to 7437 buffers which are also used to force a stable state on the memory column addresses during refresh. The low-order 6 address bits are fed directly into the multiplexer section of the 3222, the outputs of which are fed to the memory row address inputs. Note there is no electrical requirement to maintain an exact correspondence between processor address bits and memory address bits. Thus, if it facilitates layout, one can map the address bits as desired; though the logical-to-physical mapping may not be as straightforward. The only requirement is that the six outputs of the

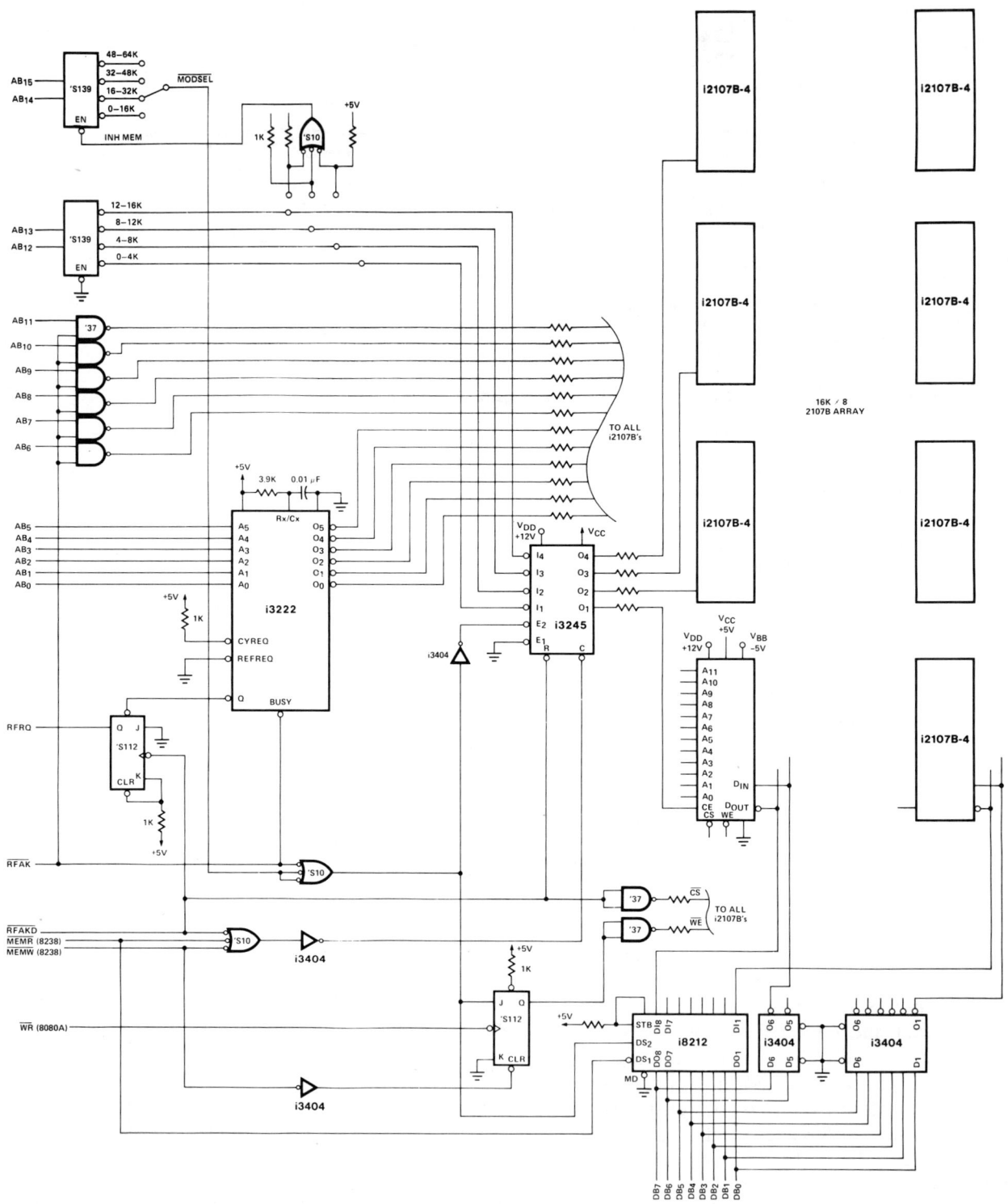

Figure 16. The Intel® 2107B facilitates the design of a 16K-byte synchronous refresh dynamic RAM system which requires no timing networks.

3222 be wired in any sequence to the six low-order address (A_5-A_0) inputs of the 2107B's. However, note that many memory test diagnostic programs rely on this exact correspondence and would have to be modified to test another mapping.

One may notice that this design does not have any delay lines, counters, or the like to generate timing. This savings is possible because of the simple timing requirements of the 2107B. In fact, in this design, the 2107B CE signal is nothing more than an inverted and amplified version of the applicable cycle request (RFAKD, MEMR, MEMW) provided by the 8080A processor group. As soon as the cycle request becomes active, CE becomes active. Because of the 8080A design, the cycle requests are spaced far enough apart in time to satisfy the 2107B cycle time restriction and the pulse widths are more than sufficient to satisfy the 2107B's CE pulse width requirement. The 2107B's Write Enable (WE) pulse is derived directly from the 8080A's WR output. (Note this example with the 2107B uses not the 8228, but rather the 8238 with advanced write timing.) Obviously, the parts count is greatly reduced since complicated timing networks need not be included; but there is one caution. Since CE is just an amplified version of the cycle request, any noise or overshoot on the cycle request will also be transmitted onto CE. The CE signal is a fairly critical one so some care must be taken to insure that it is clean. If it is not possible to insure the cycle requests will be clean, a pulse-shaping network should be added.

Consider a typical refresh cycle: the 3222 refresh oscillator Q-output sets a flip-flop which applies RFRQ to the 8080A. When the 8080A completes its use of the bus and possibly the dynamic memory module, it emits RFAK. This is used to enable the refresh address, including the contents of the 6-bit counter, to the memory array and also to enable the 3245 driver. A short time later the delayed refresh acknowledgement (RFAKD) arrives and serves several functions. It clears the RFRQ flip-flop, disables all the 2107B data outputs via the chip selects, and enables all four 3245 outputs. Finally, the signal itself is inverted, amplified and used for all the 2107B CE signals. After a minimum of one 8080A state time (500 ns), the 8080A will leave the HOLD state and remove both RFAK and RFAKD simultaneously. Several occurrences result. The 2107B CE signals will terminate, the 2107B address inputs will once again see the 8080A address bus, and the 3222 refresh counter will advance, completing the refresh cycle.

The memory cycles in this system are fairly simple. First, the 8080A provides the address which is decoded by the memory system to provide the module select, group select, and enable signals. Then the request is provided; MEMR for a memory read or MEMW for a memory write. The request is amplified and directly used for the 2107B CE inputs. If a read cycle, the 8212 will be enabled to pass the memory data onto the data bus. If a write cycle, the write enable flip-flop will be enabled. Then, when the data is available during the write cycle, the 8080A WR signal will set the WE flip-flop to generate the 2107B WE pulse. When the memory request is removed, the 2107B CE and WE terminate and the cycle is complete.

Thus, with only 10 parts, one can interface a 16K byte dynamic memory to the 8080A. This memory, realized with Intel's 2107B-4, will then be available to the 8080A with essentially no delay and hence, its access requires No WAIT states under any conditions. Obviously, this memory system could also be expanded to give a full complement of high performance, low-cost dynamic memory. So, one can see that by using the synchronous refresh method one can realize an inexpensive and high performance dynamic memory with only a small sacrifice in generality.

6. DENSITY THROUGH ADDRESS MULTIPLEXING

While the 2107B in its 22-pin package is probably the most widely used 4K bit RAM, there are some advantages gained with the 16-pin 2104A. Obviously, the 2104A's smaller package allows a greater board density within the memory array, and also the package is compatible with widely available automated handling equipment.

The Intel® 2104A (Figure 17) is a 4096-word by 1-bit dynamic MOS RAM. It is fabricated using a single transistor memory cell in Intel's proven N-channel silicon gate technology to achieve high speed, high density, low power, and low cost. The device uses the microprocessor compatible power supplies of +12V, +5V, and −5V, and includes on-chip latches for address, chip select, and data in. All inputs, including clocks, are TTL compatible, and the output data is latched and valid into the next cycle via a three-state TTL compatible output.

The use of the 16-pin package is made possible by multiplexing the 12 address bits into the 2104A

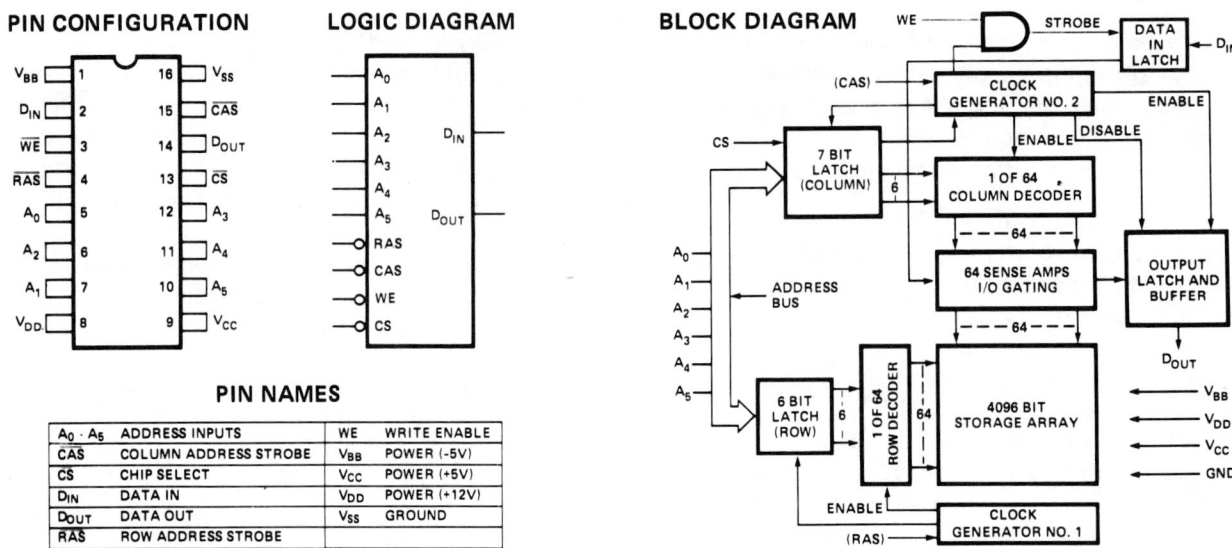

Figure 17. The Intel® 2104A is a 4096-word by 1-bit dynamic N-channel MOS RAM utilizing address multiplexing for increased density.

on 6 address input pins. The two 6-bit address words are latched into the 2104A by the two TTL clocks, Row Address Strobe (RAS) and Column Address Strobe (CAS). The noncritical timing required by the 2104A along with the existence of the Intel® 3242 address multiplexer make the 2104A a viable microprocessor memory component.

6.1 2104A Design Example

Reflecting on the characteristics of the 2104A, one might conclude the part was designed with the microprocessor user in mind. Its 16-pin package allows the high density expected in microprocessor systems, and its TTL compatibility eliminates the requirement for high voltage drivers.

An 8K word by 8-bit synchronous refresh memory system using the 2104A and 3242 has been designed for the 8080A processor group (using the 8228) discussed in Section 4. The method for providing refresh in this system is nearly identical to that used in the 2107B memory system design; the only difference is in the source of refresh requests. While the oscillator on the 3222 was used to force the refresh requests in the 2107B design, this oscillator is not available in the 3242. Fortunately, though, microprocessor systems typically have a number of clocks available, and it was just such a clock used to force refresh requests in this design. A countdown chain (Figure 13c) is commonly used to generate all the necessary communications baud rates from the 8224's 18.432 MHz oscillator output; and one of the counter outputs has a period of about 26 μs which is nearly ideal. This counter output then is used to set the refresh request (RFRQ) flip-flop every 26 μs.

For this design, cost was reduced at the expense of performance. Many applications are very cost sensitive and have more than adequate performance, so it is often a good tradeoff. The method used was to implement all memory and I/O cycles with one WAIT state always injected. This allows a less critical design, but even more importantly, it allows the use of lower performance, less expensive memories. The method used (see Figure 18) to

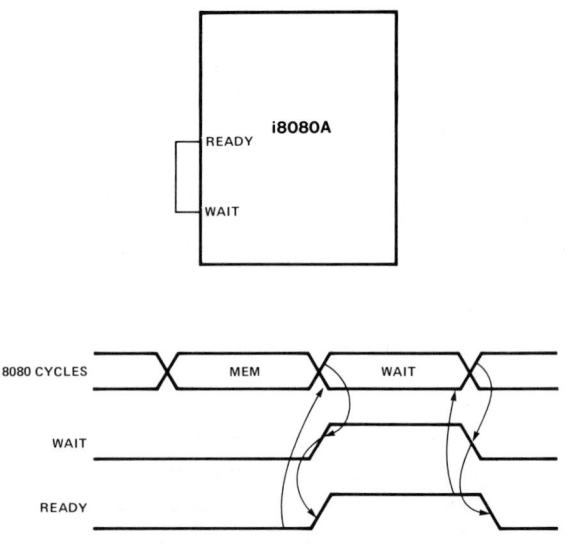

Figure 18. Single WAIT state generator for 8080A microprocessor.

DYNAMIC RAMS USED WITH MICROPROCESSORS

inject a single WAIT state on each memory and I/O cycle is quite simple. The 8080A READY input is simply strapped to the WAIT output. During a memory or I/O cycle when READY is first sampled, it is found to be inactive which then sends WAIT active and causes the processor to idle for one clock period. Then when READY is again sampled, it is found to be active. The machine cycle continues and WAIT is forced inactive in preparation for the next memory or I/O cycle. Thus, in this design a wire is used as a single WAIT state generator which facilitates the use of low-cost memory.

One requirement imposed by the use of the 2104A instead of the 2107B is the need to generate timing. Recall the 2104A uses multiplexed address inputs to reduce pin count and this technique requires the use of accurately timed row (RAS) and column (CAS) address strobes. There are many techniques which might be used to generate timing including delay lines, monostable multivibrators, counters, and shift registers. While each approach has its advantages and disadvantages, the shift register approach was used here since this approach yields a relatively low cost and highly accurate timing chain. The basis for the shift register approach is to connect the clock input of a shift register to a highly stable clock (8224 oscillator output). Then, starting with a clear register, logic ones are allowed to "walk" down the register, advancing one stage each clock period. After the register is "full of ones", logic zeros are allowed to walk down the register, giving a total time interval of 2*N*T seconds, where N is the number of register stages and T is the clock period. The real beauty of this approach is that, since only 1 bit changes at a time, decoding glitches are eliminated, thus simplifying state decode.

The design schematically represented in Figure 19 brings all these concepts together. The data path is straightforward. Data from the 8080A's bidirectional data bus is buffered by the 8216s and passed

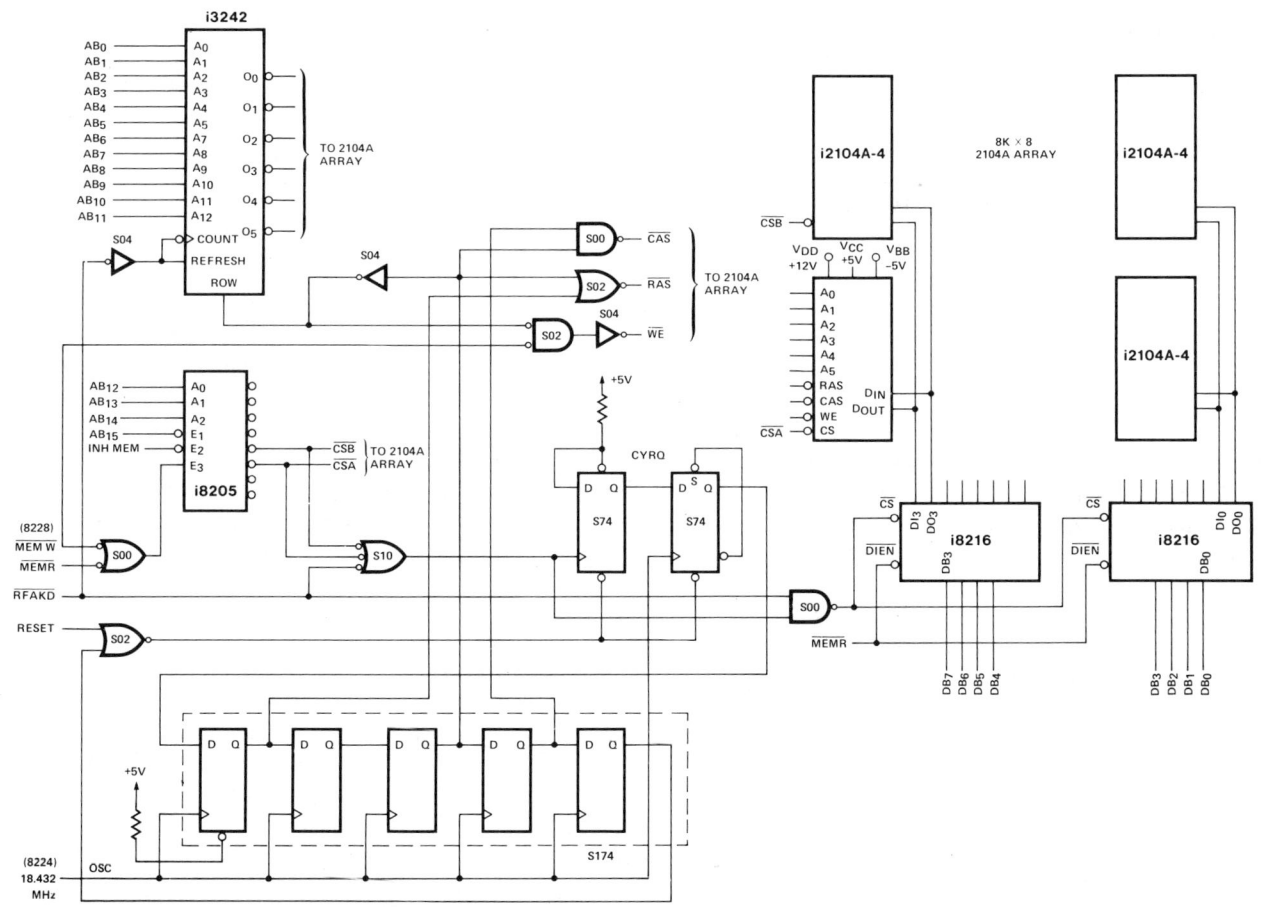

Figure 19. An 8K-word × 8-bit synchronous refresh memory system using the Intel® 2104A dynamic RAM.

to the 2104A data inputs. Data out of the 2104A (the 2104A is non-inverting) is wire-ORed (selected by Chip Select, CS) and passed to the 8216s. The three-state 8216 outputs are enabled by a memory read on this memory module and drive the 8080A's data bus.

In this design one of two banks of 2104As is selected using the Chip Select (CS) input. An alternate and perhaps better way to select is by multiplexing the RAS signal. With this method all devices would have their CS inputs active, but only one bank would receive RAS (except during refresh). This method consumes less power and promotes upward compatibility to the 2116. The 2116 design in Section 7 uses the multiplexed RAS technique.

The low-order 12 address bits are fed directly into the 3242 multiplexer, the six outputs of which are connected to the 2104A memory address inputs. Address bits A_5-A_0, $A_{12}-A_7$, or the internal refresh counter may be presented on these six outputs, dependent on the state of the ROW input and the overriding REFRESH input. The REFRESH input is activated upon a refresh cycle (RFAKD); the trailing edge of which advances the internal refresh counter.

The 4 high-order address bits are decoded and enabled by a memory read (MEMR) or memory write (MEMW) cycle. The decoded outputs are used to select one of two 4K memory groups (CSA, CSB) and to set a cycle request flip-flop, starting the timing chain. Note this flip-flop will also be set by a refresh cycle (RFAKD). Once started, the cycle request will be synchronized with the 18.432 MHz oscillator (from 8224) and propagated down the shift register (74S174) in 54-ns steps. When a logic one reaches the final shift register stage the request and synchronizing flip-flops are cleared, allowing logic zeros to then propagate down the register. All the required timing, including the generation of the row (RAS) and column (CAS) address strobes, the 3242 multiplexer switching (ROW), and the write enable (WE) pulse are derived from three taps on this shift register.

While there is some ambiguity in the delay from cycle request to a logic one reaching the first shift register stage, this delay is not critical. Since the cycle request is not synchronized to the oscillator, the output of the synchronizing flip-flop may become active anytime from zero to one clock period after the cycle request. There is also a possibility that this flip-flop could hang due to a set-up time violation, but the probability of it remaining hung for one clock period is quite small. The important aspect of this design is that all critical times are relative, not to the cycle request, but rather to the moment the first register stage is set. Thereby all critical times are simply measured in integral numbers of crystal controlled clock periods, with the normal accounting for propagation delays. Thus the synchronizing delay ambiguity does not affect most of the system timing parameters. This ambiguity however must be considered when calculating the memory response time to a request; but this time is non-critical due to the WAIT state insertion.

With this background, consider a typical processor memory cycle. The processor sends out its 16-bit address and the low-order 6 bits propagate through the 3242 to the 2104A address inputs. The 4 high-order address bits are decoded and enabled when the memory request (MEMR) or (MEMW) arrives from the 8228. The decoded and enabled output activates the appropriate 2104A group's chip select (CS) and sets the cycle request (CYRQ) flip-flop. It also enables the 8216's onto the 8080A data bus if a MEMR cycle is in progress. After synchronization is achieved, the first register stage will go active, initiating the row address strobe (RAS).

Two clock periods later the 3242 multiplexer is switched so that address bits $A_{12}-A_7$ are presented to the 2104A address inputs. Simultaneously, the 2104A's will be presented with write enable (WE) if a MEMW cycle is in progress. One clock period later the addresses have settled and the column address strobe (CAS) is initiated. After another clock period the cycle request and synchronizing flip-flop are cleared. Three clock periods later RAS and CAS are removed, WE is removed (if applicable), and the 3242 multiplexer is returned to its normal state. When the memory request is removed the 8216's are disabled (if applicable) and the memory cycle is complete.

Consider a refresh cycle. The refresh request flip-flop (RFRQ) will be set after a 26-μs period by a transition on the output of a counter in the baud rate generation chain. After the 8080A completes its use of the bus, it will issue a refresh acknowledge (RFAKD). This signal will enable the refresh counter onto the 3242 outputs, and hence the 2104A address inputs. It will clear the refresh request flip-flop and set the cycle request flip-flop, starting the timing chain. The timing for refresh is

DYNAMIC RAMS USED WITH MICROPROCESSORS

identical to that for a memory read cycle. The only differences are that the refresh counter will be maintained as the address throughout the cycle, and that none of the 2104A chip select (CS) inputs will be active. A minimum of one 8080A state time after the refresh request is removed, the 8080A will remove the acknowledge (RFAKD), advancing the 3242 refresh counter, re-enabling address multiplexing, and terminating the refresh cycle.

Operating on the premis that cost is more important then speed it is possible to build an inexpensive 8K word by 8-bit dynamic memory system. The circuitry required to interface this 2104A memory to an 8080A processor totals fewer than 10 TTL packages. Furthermore, using the single WAIT state generation techniques it is possible to use the slowest and least expensive memories while still achieving acceptable system performance. The addition of a wait state to every machine cycle only costs about 20% in system performance, an acceptable degradation in many applications. Thus it can be seen that interfacing dynamic RAMs to microprocessors can be not only easy, but inexpensive as well.

7. 64K IN 16 SQUARE INCHES!

While the 2104A was seen to be somewhat more dense than the 2107B, there is no comparison in density with Intel's new 2116 16K bit dynamic MOS RAM in a 16-pin package. With this exciting new memory component it is possible to build a 64K byte (8 bits/byte) memory array in a printed circuit board area of 16 square inches! That's 4 inches on a side!

The Intel® 2116 (Figure 20) is a 16,384-word by 1-bit dynamic MOS RAM. It is fabricated using a single transistor dynamic storage cell in Intel's production-proven, two-layer polysilicon N-channel technology to achieve high reliability, high density, high speed, and low power dissipation. The device uses the microprocessor compatible supplies of +12V, +5V, and −5V, and includes on-chip latches for address and data in. All inputs including the clocks are TTL compatible, and the output data is latched and valid into the next cycle via a three-state TTL compatible output, just as in the 2104A.

The use of the 16-pin package is made possible by multiplexing the 14 address bits into the 2116 on 7 address input pins. The two 7-bit address words are latched into the 2116 by the two TTL clocks, Row Address Strobe (RAS) and Column Address Strobe (CAS). No chip select is provided on the 2116, however, the output is brought to a high impedance state by a CAS-only deselect cycle or by a CAS-beforeRAS refresh cycle. Refreshing can

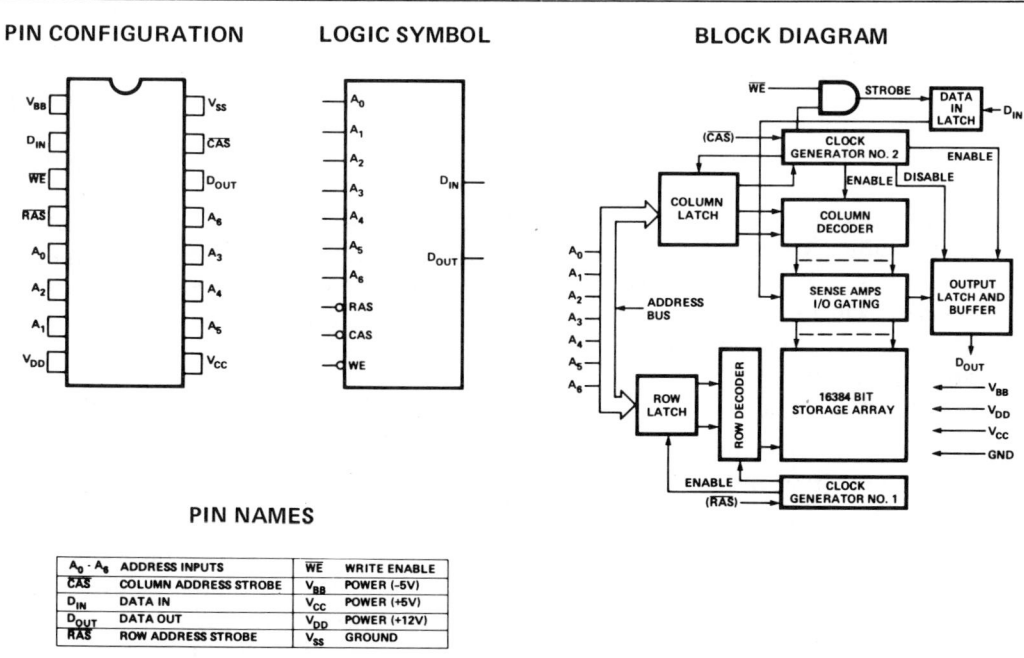

Figure 20. The Intel® 2116 is a 16K-word by 1-bit dynamic N-channel MOS RAM utilizing address multiplexing for maximum density.

be accomplished every 2 ms by any one of the three following methods: (1) CAS-before-RAS cycles on 64 addresses, A_0-A_5, (2) RAS-only cycles on 128 addresses, A_0-A_6, or (3) normal read cycles on 128 addresses, A_0-A_6. The non-critical timing required by the 2116 along with the existence of the Intel® 3242 address multiplexer make the 2116 a viable microprocessor memory component.

7.1 2116 Design Example

Since the 2116 is quite similar in application to the 2104A, it should be fairly straightforward to construct a 2116 microprocessor memory after having done so with the 2104A. The multiplexing and most of the timing is much the same. The only real difference is in the method of chip selection (the 2116 has no CS input) and possibly the detail of refresh.

A 64K byte synchronous refresh memory system using the 2116 and 3242 has been designed for the 8080A processor group (using the 8228) discussed in Section 4. Though the refresh cycles themselves are implemented differently than with the 2104A, the method for providing synchronous refresh in this system is just as it is in the 2104A example of Section 6. An output (26 μs period) from a counter in the baud rate generation chain is used to set the refresh request (RFRQ) flip-flop. And just as before this RFRQ is applied to the 8080A's HOLD input, with hold acknowledge (HLDA) being used as refresh acknowledge (RFAK).

For this design, maximizing performance was taken to be a precept in spite of the consequent increase in cost. This system provides 64K bytes available to an 8080A (488 ns clock) system without a WAIT state required. The primary cost penalty is that the design requires the use of the high-speed 2116-3 memory component. Obviously, the design could be slightly modified and made to operate with lower-speed parts and use the single WAIT state injection technique discussed in Section 6.

In this design the shift register technique is also used to generate highly accurate time intervals just as it is in the 2104A design. One notable difference in this design, though, is encountered because the 2116 does not have a chip select pin. Rather, herein the devices are selected by decoding the addresses and using the RAS input for selection, while all devices receive CAS. Thus, a device receiving CAS with no RAS is deselected, performs no memory operations, and its output remains in the high impedance state. Only the group of devices receiving both RAS then CAS are enabled to perform memory operations and have their outputs enabled.

Another difference is in the CAS-before-RAS refresh technique unique to the 2116. Since the 2116 has 128 rows, one might expect being required to perform 128 refresh cycles to refresh the entire memory. However, the 2116 is implemented as two arrays of 64 rows each and, through clever circuit design, a row in both arrays can be refreshed simultaneously. The CAS-before-RAS cycle is the technique used to inform the 2116 that a refresh is to occur and therefore allow access to both arrays simultaneously. This 2116 microprocessor memory system is designed to take advantage of that feature, reducing the memory "dead-time" by a factor of two.

The 64K byte 2116 microprocessor memory system design is shown schematically in Figure 21. The data path is straightforward. Data from the 8080A's bidirectional data bus is buffered by the 8216s and passed to the 2116 data inputs. Data out of the 2116s (the 2116 is non-inverting) is wire-ORed (selected by RAS) and passed to the 8216s. The three-state 8216 outputs are enabled and drive the 8080A data bus during all memory read cycles so long as the inhibit is not active.

The 14 low-order address bits are fed directly into the 3242 multiplexer, the seven outputs of which are connected to the 2116 memory address inputs. Address bits $A_{13}-A_7$, A_6-A_0, or the internal 7-bit counter may be presented on these seven outputs, dependent on the state of the ROW input and the overriding REFRESH input. The REFRESH input is activated upon the start of a refresh cycle (REF), the trailing edge of which advances the internal refresh counter. The two high-order addresses are decoded and used to select one of four banks of 16K byte memories.

Any non-inhibited memory cycle request (including refresh) will set the cycle request (CYRQ) flip-flop, starting the timing chain. Note that for a refresh cycle, RFAKD, the refresh (REF) flip-flop will have already been set by RFAK. Setting the refresh flip-flop early allows time for the multiplexer outputs to settle on the refresh counter, enables all four RAS drivers, and generates the early CAS for CAS-before-RAS refresh.

Once set, the cycle request will be immediately applied as RAS and also synchronized with the

DYNAMIC RAMS USED WITH MICROPROCESSORS

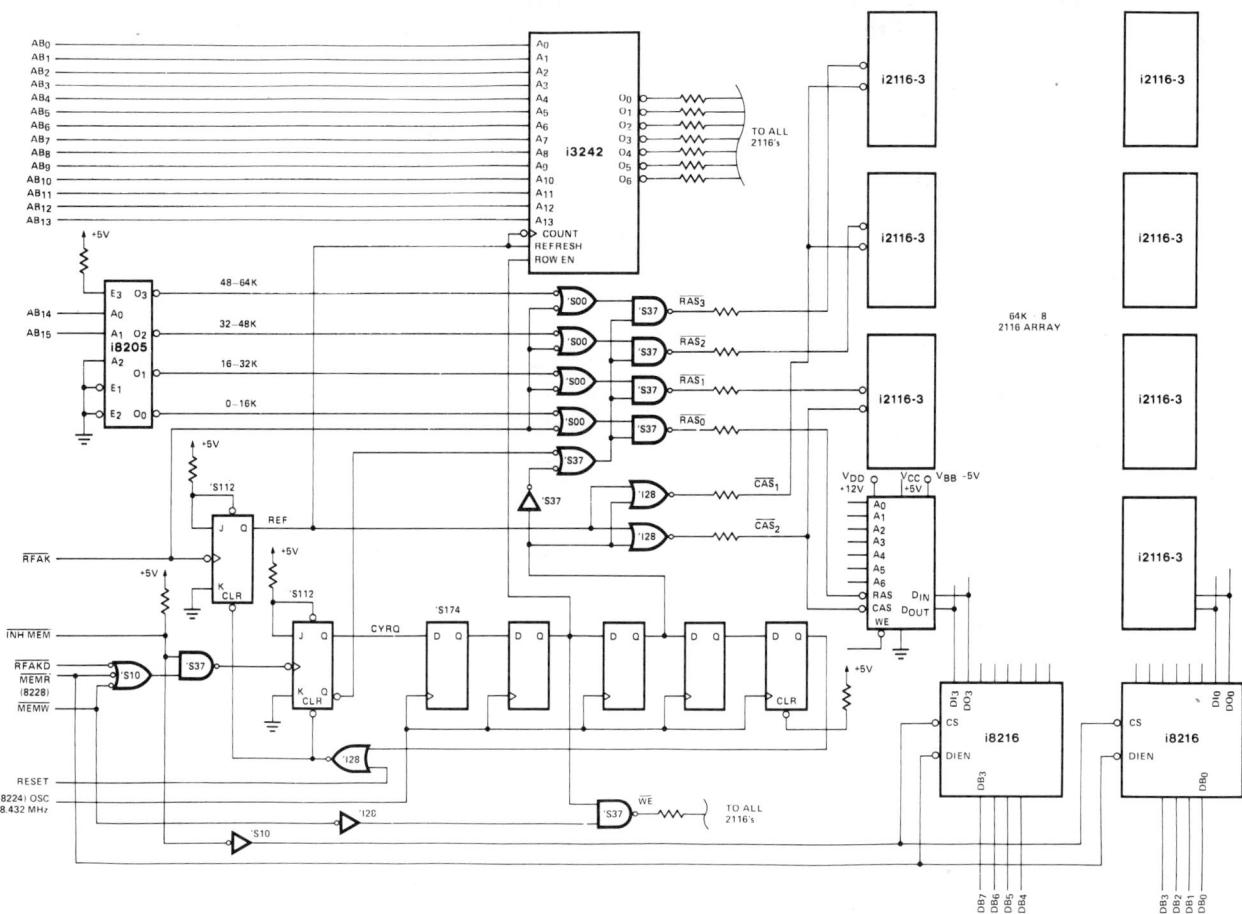

Figure 21. A 64K-byte synchronous refresh dynamic memory system using the 2116 memory.

18.432 MHz oscillator (from 8224) and propagated down the shift register (74S174) in 54-ns steps. This is the earliest that RAS can be applied and this must be done in order to meet the 8080A No-WAIT-state access requirement. When a logic one reaches the final shift register stage the cycle request, and refresh (if applicable) flip-flops are cleared, allowing logic zeros to then propagate down the register. Clearing the refresh flip-flop also advances the refresh counter and returns the multiplexer to its system address multiplexing function. All the critical timing, including the generation of the row (RAS) and column (CAS) address strobe intervals, the 3242 multiplexer switching (ROW), and the write enable (WE) pulse are derived from two taps on the shift register.

Because of the speed requirement, this design has the synchronizing delay inside the timing chain.

As mentioned, upon a valid cycle request the cycle request flip-flop is set, immediately applying the row address strobe (RAS). Note this is different than in the 2104A design where the request was first synchronized to the clock and then RAS applied. The advantage of the approach taken herein is that the memory access may start as soon as possible, rather than after a synchronizing delay. The disadvantage is that now the time ambiguity is a part of the RAS pulse. This is inconvenient but not insurmountable. One simply insures that all memory set-up and hold times will be met for either extreme of the ambiguous synchronizing interval. This ambiguity must also be kept in mind when calculating the worst case system access time.

It might be instructive to now consider a typical processor memory cycle. The processor sends out its 16-bit address and bits A_{13}–A_7 propagate through the 3242 to the 2116 address inputs. The 2 high-order address bits are decoded to select one of four 16K banks by enabling the corresponding RAS driver. Simultaneously any inhibit circuitry which may reside in the system is decoding the address and applying the inhibit, if applicable. When the memory request (MEMR or MEMW) arrives from the 8228 the cycle request (CYRQ) flip-flop will be set (assuming the memory inhibit, INH MEM, was inactive), activating the selected RAS driver and starting the cycle. If a memory read (MEMR) is requested the 8216 bus drivers will be enabled onto the 8080A system data bus. After the synchronizing delay of from zero to one clock period the first register stage will set, and after another clock period the second stage will set. At this time the 3242 multiplexer is switched so that address bits A_6-A_0 are presented to the 2116 address inputs. Simultaneously the 2116's will be presented with write enable (WE) if a memory write (MEMW) cycle is in progress. One clock period later the addresses have settled and the column address strobe (CAS) is initiated. After two more clock periods the cycle request flip-flop is cleared, allowing the second shift register stage to be clear after two additional clock periods. At this time the WE pulse is removed (if applicable) and the 3242 multiplexer is returned to its normal state. One clock period later RAS and CAS are simultaneously removed. When the memory request is removed the 8216's are disabled (if applicable) and the memory cycle is complete.

Looking at refresh, the refresh request (RFRQ) flip-flop will be set after a 26-μs period by a transition on the output of a counter in the baud rate generation chain. After the 8080A completes its use of the bus it will issue a refresh acknowledge (RFAK). This signal enables all four RAS drivers and sets the refresh (REF) flip-flop. The setting of REF enables the refresh counter onto the 3242 outputs, and also activates the CAS driver. Recall CAS-only will deselect all 2116 outputs. When the delayed refresh acknowledge (RFAKD) is issued it will clear the refresh request flip-flop and set the cycle request flip-flop, activating all four RAS drivers and starting the timing chain. The rest of the timing for refresh is identical to that for a memory read cycle. The only differences are that the refresh counter will be maintained as the address throughout most of the cycle, and that all the RAS drivers will be active simultaneously. When the REF flip-flop is cleared (at the same time CYRQ is cleared), the 3242 refresh counter ad-

vances and the address multiplexing function is re-enabled. A minimum of one 8080A state time after the refresh request is removed, the 8080A will remove both acknowledges (RFAK and RFAKD), disabling the four RAS drivers and terminating the refresh cycle.

The 2116 is a real technological breakthrough and a tremendous benefit for microprocessor systems! Think of it, a 64K byte array in only 16 square inches. And, further, the entire 64K byte memory interface requires only 11 packages and provides memory accesses without a single WAIT state, implementing an 8080A system which fulfills the requirements of even the highest performance microprocessor applications.

8. MEMORY OVERLAP

A few moments thought should be sufficient to realize that an 8080A system which has 64K of dynamic RAM, as its only memory is useless. When the power is first applied, the content of the RAM is indeterminate and since 64K encompasses the entire memory address space, there is no provision for a non-volatile memory such as a ROM or PROM. Thus the 8080A will have no meaningful instructions to execute and the system is useless. There are several techniques which can be used to circumvent this problem. Both involve overlapping a ROM or PROM into the memory address space and require an inhibit feature on the dynamic RAM system.

Because semiconductor memories are available in discrete sizes, it may be necessary to have two memories overlap in address space. For instance, it might be desired to have a system with 4K bytes of ROM in addresses 0 to 4095 and 60K bytes of RAM in addresses 4096 to 65,353. Such a system could be implemented most economically with 16K RAMs but then this 64K byte RAM memory would also occupy the first 4K bytes. In order to map the ROM into the first 4K bytes, the first 4K of RAM would essentially be discarded by forever inhibiting the RAM from functioning while in that address range. This could be implemented with an address decoding network and memory inhibit (INH MEM) line as shown in Figure 22a.

Another possibility might be that it is desirable to have the ROM in the address space and the RAM inhibited only at system start-up ("bootstrap" or "shadow" ROM). Then once the system has reached steady state the bootstrap program could

DYNAMIC RAMS USED WITH MICROPROCESSORS

perform an OUTPUT, inhibiting the bootstrap ROM and enabling the RAM it overlaps (shadows). Thus, the overlapped RAM is not forever discarded; it is simply switched out while the ROM is being used to bootstrap the system and possibly load new programs into the remaining RAM. This allows an easy startup while still maintaining the flexibility to later exchange the ROM for RAM. Such a system could be implemented with an address decoding network, an output port, and RAM memory inhibit line as shown in Figure 22b.

A slight modification of the above technique is sometimes convenient. Rather than completely inhibiting the RAM during start-up, it is only inhibited for READ cycles. Then all READs in that address range will be from ROM, but all WRITEs will be to both. This effectively just directs the WRITEs to RAM since a WRITE to a ROM is a no-op. This permits a bootstrap ROM to load a program into RAM even in locations the bootstrap ROM currently occupies. In fact, a bootstrap ROM could even copy itself into the corresponding RAM locations. Of course, after steady state has been achieved, the ROM may be completely disabled, allowing both READs and WRITEs from the RAM. The implementation of this technique is quite similar to that discussed previously. The only difference is that the read inhibit (INH RD) in the RAM module simply keeps the data bus drivers in their high impedance state (Figure 22c).

9. THINGS TO WATCH OUT FOR

It should always be remembered that dynamic RAMs, while appearing to be rather simple digital devices, are in fact highly complex analog systems. They include differential sensing amplifiers which must detect decivolt signals buried in noise and which must operate in tens of nanoseconds. Also, typically included are timing circuits and power supply sequencing circuits used to dynamically allocate power to subsystems only as required. While a large power savings is achieved, this technique combined with capacitive charging transients generate significant, and quite fast, transients in the power supply current of the device.

Because of this highly complex analog nature one should consult the vendor's memory applications literature (Intel's *Memory Design Handbook*) before proceeding with a design. From the beginning the designer should respect the complexity involved and take steps necessary to insure a

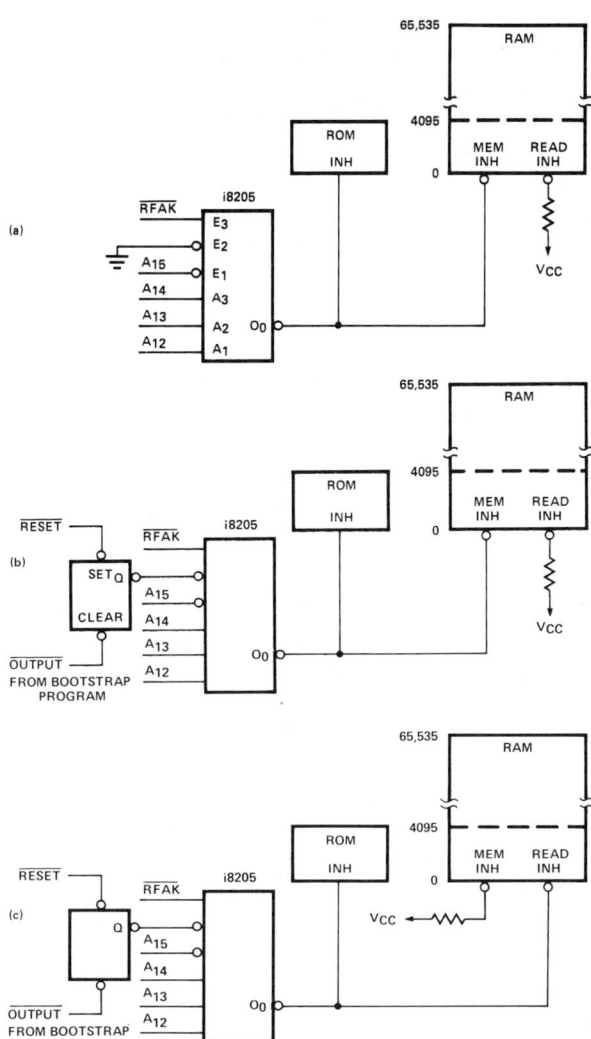

Figure 22. Memory overlap provided by (a) discarding 4K bytes of RAM, (b) inhibiting 4K bytes of RAM during startup, and (c) inhibiting only RAM Reads during startup.

trouble-free design. Be warned that it is much easier to properly design a dynamic memory system than it is to make a hastily designed system function reliably.

9.1 Layout

A proper printed circut layout can contribute much towards achieving a reliable dynamic memory system design. It should have an effective gridded power supply distribution network to adequately supply the current and to minimize the inductive effects. The distribution of circuit ground is of paramount importance to reduce ground noise and inductive offsets, and to provide a ground plane for the signal lines.

Another consideration during layout is finding a geometry that minimizes the length of the signal and clock lines. This may require breaking the array in two and driving it from the middle or some similar technique.

Included in Appendix II are recommended two-sided printed circuit layouts for the three most popular dynamic RAMs — the 2107B, 2104A, and 2116.

9.2 Bypass

The importance of bypassing the power supplies cannot be overstressed. Recalling that large current spikes are inevitable, capacitors must be provided to supply these transients. The capacitors required fall into two categories. First are the capacitors of small physical size and low inherent inductance, such as the monolithic and other ceramic capacitors. These should be used quite liberally and intimately integrated throughout the array. The second type is the larger bulk capacitor used to prevent supply droop. These also should be included within the array for good distribution. The vendor's literature will normally make specific recommendations for capacitive bypass. It would be unwise to provide anything less than recommended, or less than top quality components in a prototype design. If it is later desired to cut corners by reducing the quality or quantity of capacitors, let that be a separate exercise.

Recommended bypass networks are provided in Appendix II.

9.3 Transmission Line Effects

By carefully laying out the circuit to minimize signal path length, one can reduce the effect due to the transmission line properties of a printed circuit trace. However, this may not be sufficient and then one must consider other technqiues. Most clock drivers include clamps which help minimize over- and undershoot. Another frequently used technique is to put a series resistor in the line to help match impedances and damp out reflections. The optimum value for this resistor is very dependent on layout and driver and receiver characteristics, and is normally determined empirically. A value of 20 to 30 ohms is a fairly good starting value. When using a series resistor one must also insure that the voltage drop across it does not severely degrade the logic level noise margins.

9.4 Crosstalk

While crosstalk is usually not a severe problem, it should be considered during layout. One must avoid the temptation of running two or more signals very close together for long lengths when the signals can adversely affect one another.

9.5 Refresh Interference

If it is decided to design an asynchronous refresh memory system, be very careful with the refresh arbiter. It is quite a tricky design that can very rapidly arbitrate between asynchronous requests (see box). Most circuit design problems in asynchronous refresh memory systems are found to be linked with refresh interference. And, this can be a most insidious problem since it is of an asynchronous nature and may only manifest itself under infrequent circumstances.

10. IT ALMOST WORKS!

This is a situation which most of us have faced. The system is showing signs of life but is not operating as desired. Let us consider some debugging strategies applicable to a microprocessor system which includes dynamic RAM for all or part of its memory.

The initial debug of such a system should be done with the dynamic RAM subsystem totally disabled, maybe even removed. In this way the microprocessor system, sans the dynamic RAM, can be effectively debugged. You say you were going to use 64K of dynamic RAM as your only memory, save a bootstrap ROM? How can you debug your microprocessor without any RAM? Intel's ICE-80 saves the day! It lets you borrow memory from your MDS-800 with just a few keystrokes required.

The debugging effort at this stage should be quite thorough. One should ascertain that all the addresses and control lines reach the RAM system, and that the data paths are as designed. A logic pulser and probe are invaluable in this effort.

After convincing oneself that the entire system, except the dynamic RAMs, has been thoroughly tested and proven correct, the fun may begin. Try to operate the dynamic RAM system statically forcing all inputs as required. While the dynamic memory system cannot be totally debugged statically, a great deal can be accomplished. All the interface circuitry can be shown to be functional, DC levels and power supplies can be checked, and the many interconnections can be verified.

Now that the design works statically it remains to

DYNAMIC RAMS USED WITH MICROPROCESSORS

Asynchronism and Bistable Hang-Up

Though most activities of a computing system are synchronous, there are occasionally times when asynchronous events must be introduced. This introduction must be handled with great care. Bistable devices are normally designed with an input set-up and hold time requirement with respect to the "clock" input (see Figure). This timing requirement insures the bistable circuit designer that the bistable will not be forced into its astable state. Hence, knowing the normal delay paths the designer may specify the delay time required to change state. This delay time, however, is only valid so long as the required set-up and hold times are honored.

Whenever the data input occurs asynchronously to the clock input, there is the possibility that the set-up time will be violated. If this is the case, the bistable may "hang" in its astable state for an indeterminate amount of time, extending the specified delay. This hang-up, under forced conditions, has been observed to last for as long as 20 ns with a 74S74. Furthermore, during this hang-up time, the outputs are undefined. Depending on the circuit design, they may do nothing, exhibit a slow transition, or even oscillate. Obviously, care should be taken when using the output of a bistable device operating in the asynchronous mode. In fact, improper handling of asynchronism is one of the most common circuit design problems when designing dynamic RAM systems.

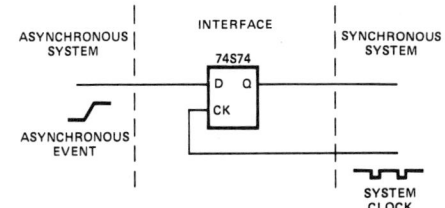

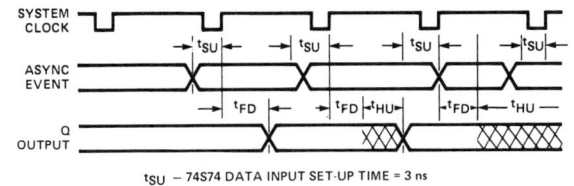

t_{SU} – 74S74 DATA INPUT SET-UP TIME = 3 ns
t_{FD} – 74S74 DELAY FROM CLOCK ↑ = 9 ns
t_{HU} – HANG UP DUE TO SET-UP TIME VIOLATION

SYNCHRONIZING CIRCUIT EXHIBITING HANG-UP

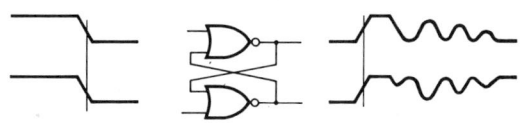

"FIRST IN" ARBITER EXHIBITING HANG-UP

---

prove it in actual operation. For these dynamic tests the microprocessor and dynamic RAM subsystems will work together. Refresh should be disabled to avoid any problems thereto related. At room temperature today's dynamic RAMs will retain their data without refresh for over 2 ms, and 10 ms is probably typical. The tests will be so programmed that any location of interest will be accessed frequently enough to maintain the data. A very simple program is written and executed which simply writes a location, reads the value written, and loops back to the beginning. With this simple program one can get a very stable oscilloscope display and examine all the parameters of interest; e.g., address and data set-up and hold, noise, signal quality and levels, access times, pulse widths, and so on. Before continuing one should ascertain that the system operates perfectly to this point. Once again it should be pointed out that ICE-80 is invaluable in this debug work. With this tool one can very rapidly write, assemble, load and execute a program to test the memories. It also gives breakpoint and trace capability to help uncover the source of a problem.

At this point most of the problems will have been found. So the procedure is to write/execute programs which subject the RAM subsystem to progressively more difficult tests. If at any point an error is found and the cause cannot be easily deduced, write/execute a very specific test program to prove/disprove one's hypotheses. The 8080A microprocessor is a very powerful ally; it can be easily and quickly programmed to generate nearly any desired diagnostic.

Only when one has gained some confidence that the RAM system operates flawlessly should refresh be introduced. Then any problems which arise should be attributable to some type of refresh interference. Finding the cause of refresh interference is not always easy and every available aid (storage scope, logic analyzer, ICE-80, etc.) should be used. Once again, if difficulties arise, simplify the test program as much as possible to gain a repetitive program (for stable viewing) and to test a specific hypothesis. Using the synchronous refresh approach, though, should reduce the probability of encountering problems with refresh interference.

When the system works correctly for all test programs even when refresh is enabled, one might like to employ a very comprehensive diagnostic program. This type program is designed to test not only the system but the RAMs themselves, for such subtle failures as pattern sensitivity and adjacent bit alteration. A test such as this could be used not only in development but in production as well. Figure 23 provides a debug flowchart summarizing the above discussion.

11. TWICE THE FUN

There are some applications in which there is a need for two or more processors. Such a need can arise for any of several reasons such as increased throughput, division of labor, specialized tasks, modularity, redundancy and so on. While the implementation of multiprocessor systems can be a very complex problem, only a small, well defined subset of the problem will be treated herein. For purposes of discussion a multiprocessor system will be defined as any system where there may exist more than one "master" which may take control of the system bus, and thereby control the "slave" resources, such as a memory, on that bus.

Clearly, this definition will then subsume systems with both a single central processing unit and a Direct Memory Access (DMA) unit as multiprocessor systems. A DMA controller has the ability to gain control of the system bus and effect transfers to and from slave resources on that bus. Hence, the DMA controller is also a master per the above definition.

The purpose of this section, however, is not to classify multiprocessor systems or components, but rather to discuss the relationship of dynamic memory systems as discussed earlier with respect to multiple microprocessor systems. In a multiprocessor system it is perhaps easiest to envision the role of a slave device. The slave simply accepts or supplies data as commanded on a set of common control lines, heedless of which master is currently controlling the bus. Thus the autonomous asynchronous refresh dynamic memory system fits quite naturally into a multiprocessor environment. Since the asynchronous refresh memory system can stand alone without requiring any outside assistance, it serves well as a slave. It accepts requests over its control lines and supplies/accepts data as required. When refresh is required it provides same for itself, delaying memory request acknowledges as necessary. The design in Appendix I is perfectly ammenable to use in a multiprocessor system.

The asynchronous refresh dynamic memory system provides facility in multiple microprocessor systems, but one should consider performance as well. In every multiprocessor system there must be arbitration for the system bus to assure that at any given moment only a single master has control of that bus. This arbitration may be quite complex and time consuming in a large system and may have to precede each system bus cycle. The additional time delay overhead then imposed on each memory cycle by an asynchronous refresh memory may be insigificant. On the other hand, a very small system (8080A and 8257 DMA controller) may have quite simple and efficient bus arbitration. The 8080A assumes control of the bus and only yields it when requested to do so. There is no visible arbitration overhead on each 8080A cycle and the time delay imposed on each memory cycle by an asynchronous refresh memory system could be quite significant.

A valid question then might be "Well, how would a synchronous refresh memory fit into a multi-processor environment?" The underlying concept is to elevate the role of the memory so that it is no longer simply a slave. A request for refresh is actually a request for a system resource, namely the memory, and can be handled via the same arbitration mechanism as every other system resource request. Refresh would then be performed only after a request had been made and the system bus granted. The rest of the time the memory would act as a slave, responding to memory requests instantly without any additional delay due to refresh arbitration.

Consider a multiprocessor system composed of a synchronous refresh memory, an 8080A microprocessor, and an 8257 DMA controller (see the

DYNAMIC RAMS USED WITH MICROPROCESSORS

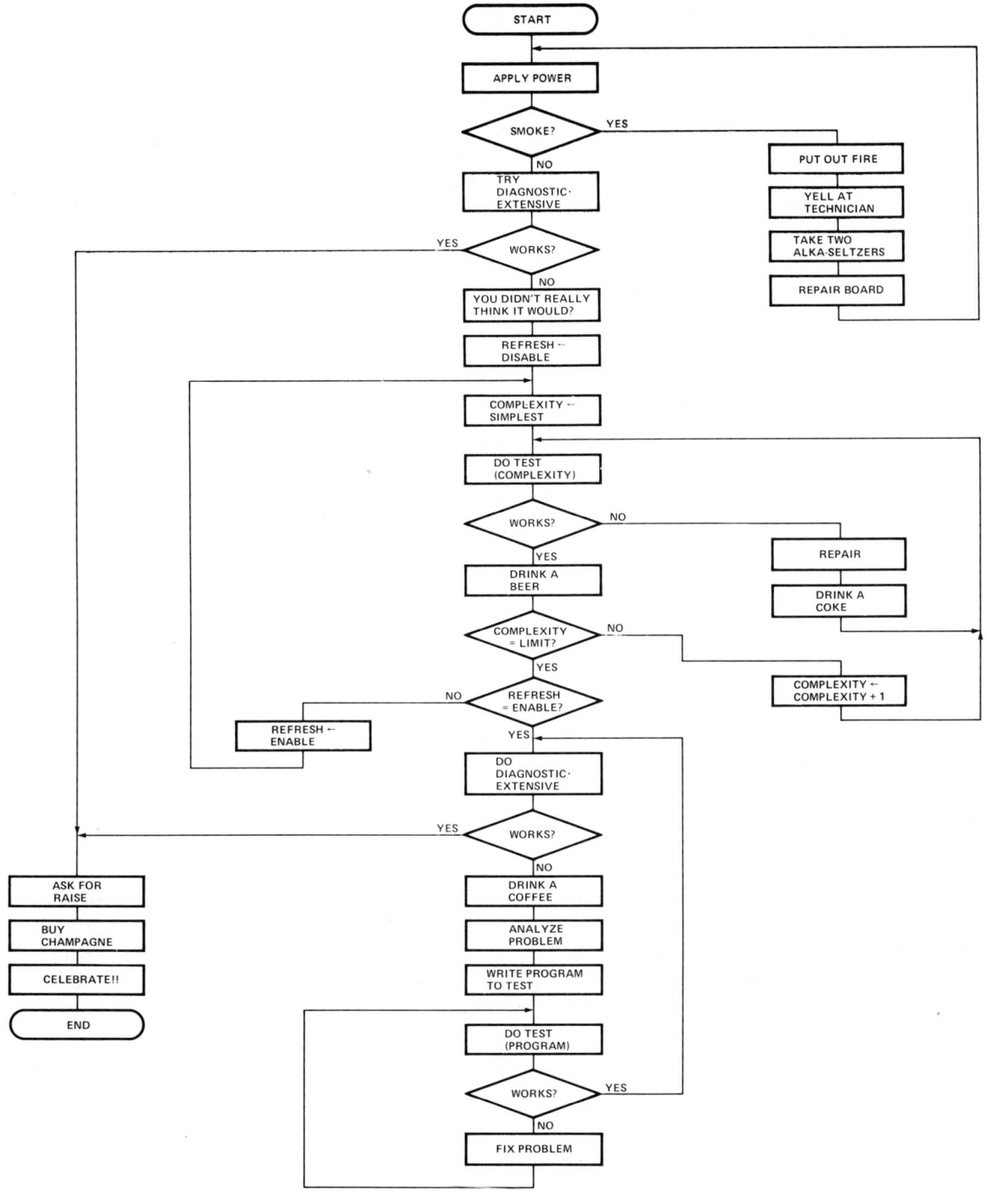

Figure 23. A dynamic RAM/microprocessor interface debugging flowchart.

Intel *8080 Microcomputer Peripherals User's Manual*). Such a system might be more correctly called a multi-master system. Normally, the 8080A assumes control of the system bus and the 8257 requests control by activating the HOLD input of the 8080A (see Figure 24). When the 8080A has completed its use of the bus it releases control to the 8257 by emitting Hold Acknowledge (HLDA). The 8257 then commands the bus for as long as necessary and then releases the bus by removing its request. However, recall that it was this same HOLD and HLDA mechanism which was used to provide synchronous refresh. What shall we do?

There are two different approaches one might take. One possibility would be to use one of the channels (the highest priority) of the DMA controller (see Figure 25). This channel would be programmed to be in the VERIFY mode (no memory or I/O commands generated). The address (refresh controller supplies address) and count (STOP on Terminal Count mode not used) programmed would be irrelevant. Then a refresh request would be applied to the channel DMA request input, and the channel DMA acknowledge would be treated as refresh acknowledge. Refresh now looks much the same as it did before to the memory, though the refresh cycle time and system idle time will be quite a bit longer than before.

Another alternative would be to design a HOLD input arbiter. One such mechanism (Figure 26) would arbitrate HOLD requests on a "first-in" priority basis, that is the first request would receive the acknowledge. With the design as shown the 8080A would be allowed to complete another bus cycle after the release of a request even if the other request were pending. This is useful in that the acknowledgement procedure to each requestor is exactly as it would be if the other were not present. The only difference is in the delay a requestor sees from request to acknowledgement.

Each of these approaches introduces further constraints to assure adequate refresh. The latter approach would not prevent the DMA controller from hogging every bus cycle, thus preventing refresh. The system should be designed to avoid this condition. (The 8257 has a control override feature allowing one to gracefully take control of the bus by inhibiting HLDA to the 8257 and using AEN as a signal indicating the 8257 has relinquished control.) Likewise, the approach using a channel of the 8257 is vulnerable to a programmer error. The programmer could inadvertently disable the refresh channel (perhaps intermittently) leading to some very strange symptoms.

Neither the asynchronous nor synchronous refresh method will be right for every multimicroprocessor system. There are a large number of tradeoffs which must be made by the system designer. In a given application the instant response of the synchronous refresh system may allow the removal of a WAIT state from every memory cycle. This could significantly enhance the overall system performance even though refresh now requires system bus cycles. In another application there may already be so much bus arbitration overhead that the additional delay due to refresh arbitration in an asynchronous refresh memory is insignificant. If most memory requests were to a ROM memory the asynchronous refresh memory would usually refresh itself without degrading system performance, whereas the synchronous refresh memory would still require system bus cycles. The system designer should carefully weigh the contributing factors to optimize system cost and performance.

12. "MEMORY IS CHEAP"

"The programmer who tries to save a byte usually gets bitten." How many times have you heard comments similar to these? It is inevitable that memory usage will continue to increase, calling for larger and larger microprocessor memories. While many factors (such as faster processors, increased features, high level languages, structured programming and declining prices) contribute to this trend, it nevertheless gives dynamic memories increased importance as elements for microprocessor system designs. There will continue to be microprocessor applications with RAM requirements under 4K bytes, where perhaps static memories fit best. However, there will also surely be an increasing number of microprocessor applications with RAM requirements of 16K bytes and over, where dynamic memories offer the only viable solution.

Fortunately for the microprocessor system designer, dynamic RAMs are becoming increasingly facile in application. Though correct usage still requires a solicitous design, many of the more difficult memory characteristics have been removed. The required timing and array drivers and receivers for today's dynamic memories are much simpler than those required with the early PMOS memories. Refresh is one requirement which remains, but Intel has a growing family of memory

DYNAMIC RAMS USED WITH MICROPROCESSORS

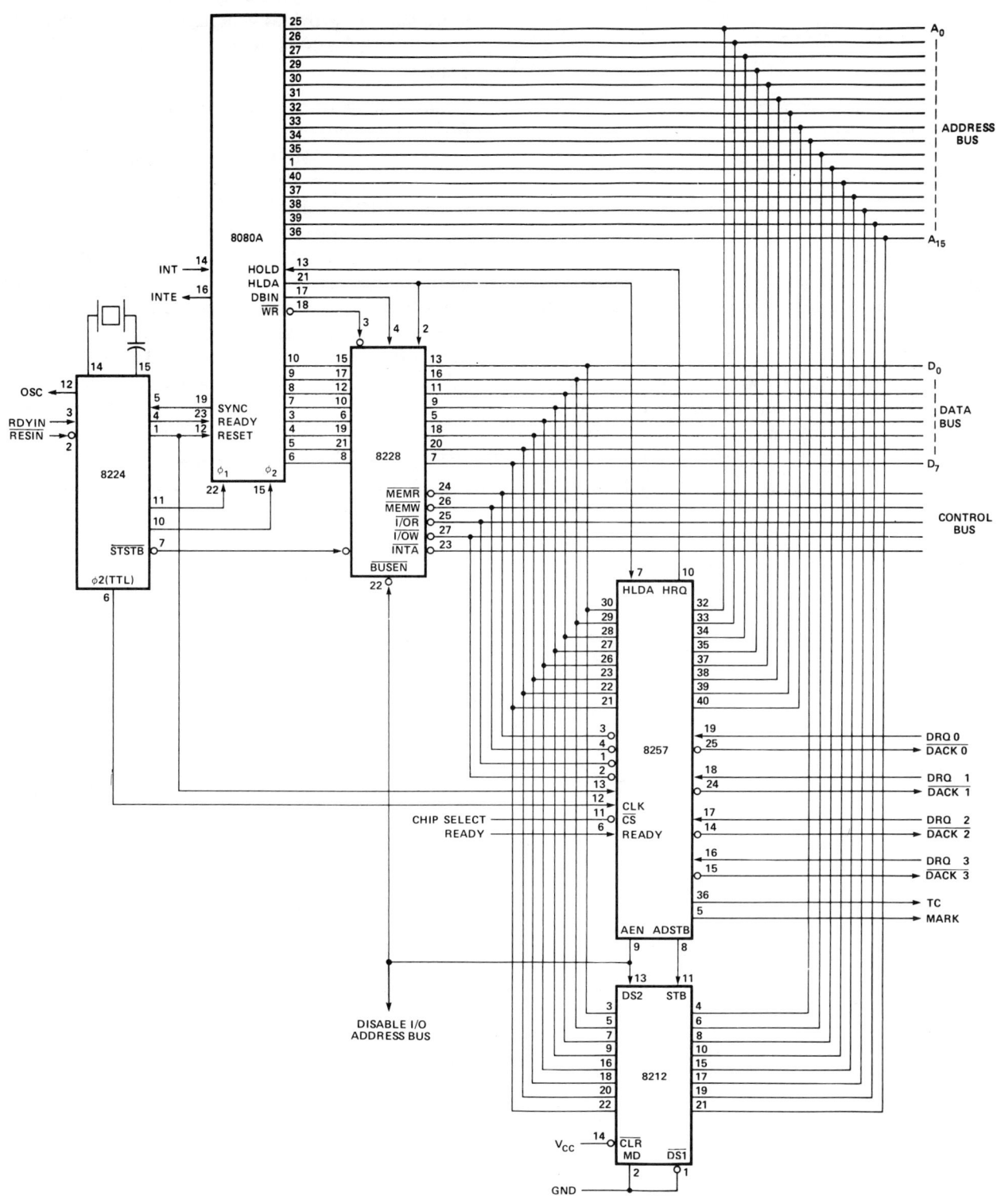

Figure 24. Normal 8257 DMA/8080 System Interface.

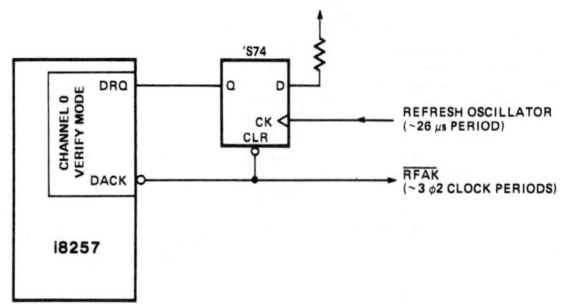

Figure 25. Using the highest priority DMA channel to provide synchronous refresh for dynamic memories.

support circuits which facilitate dynamic RAM designs. These components can be used in a microprocessor system to implement refresh with the traditional asynchronous refresh method or with the synchronous refresh method. The asynchronous method provides generality and the synchronous method offers high performance and low cost.

Today's designer of a microprocessor system has a tremendous number of memory options. If the RAM requirement is small enough, use static RAMs; they're easiest. If dynamic RAMs are dictated by the requirements, don't dispair. There are an infinite number of implementations from which to choose, each with its own particular advantages and disadvantages. Furthermore, today's memories are easier to use and a knowledgeable, prudent designer should bear a reliable, cost-effective microprocessor memory system. But, if you find yourself working late one night trying vehemently to debug your microprocessor-dynamic RAM design, just remember two things, "Memory is cheap," and "They're only capacitors."

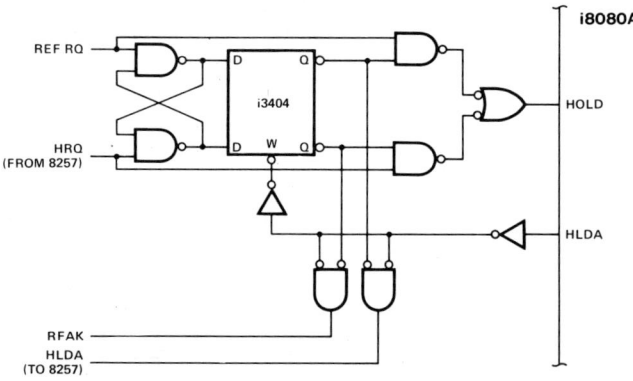

Figure 26. Arbitration between refresh and DMA requests for 8080A's HOLD input.

APPENDIX I

A VERSATILE ASYNCHRONOUS REFRESH DYNAMIC MEMORY SYSTEM

The Intel® SBC 104/108 Combination Memory and I/O Expansion Board is a good example of a very versatile asynchronous refresh dynamic memory system. It uses the Intel® 2104/2108 dynamic RAM and uses the 3222 as an asynchronous refresh memory controller. This general-purpose system is capable of operating with any of the SBC 80 Single Board Computers. In fact, it may also operate in a system with multiple SBC 80s. Similarly, multiple SBC 104/108s can be used with a single SBC 80.

This Appendix represents an edited version of the SBC 104/108 Hardware Reference Manual, along with appropriate schematic diagrams. Further information may be obtained from the source document.

CHAPTER 1
INTRODUCTION

The SBC 104/108 Combination Memory and I/O Expansion Boards are members of Intel's complete line of SBC 80 memory and I/O expansion boards. They interface directly (via the system bus) with any SBC 80 Single Board Computer to expand system interrupt levels, RAM and ROM memory capacity, and serial and parallel I/O lines. The SBC 108 is functionally identical to the SBC 104 except in RAM memory capacity. The SBC 104 has 4K of RAM implemented with Intel® 2104's while the SBC 108 has 8K of RAM implemented with Intel® 2108's. This manual presents a complete functional and physical description of the SBC 104. Except where otherwise noted, or reference is made to RAM capacity, this description is also true of the SBC 108.

The RAM memory expansion feature consists of eight Intel dynamic RAM memory components, providing 4K/8K bytes of read/write memory. Refresh control is provided on-board using an Intel® 3222 refresh controller. Refresh cycles are initiated asynchronously with respect to system memory (read/write) cycles. Any resulting competition between refresh and system memory cycle requests is arbitrated by the refresh control logic. All cycle requests are handled on a first-received, first-serviced basis. Pending cycle requests are serviced as soon as the previous cycle is complete.

CHAPTER 2
FUNCTIONAL/PROGRAMMING CHARACTERISTICS

This chapter briefly describes the organization of the SBC 104 from two points of view. The principal functions performed by the hardware are identified and the general data flow is illustrated in Section 2.1. This section is intended as an introduction to the detailed information provided in Chapter 3, Theory of Operations.

2.1 FUNCTIONAL DESCRIPTION

To facilitate the following description, the SBC 104 is divided into seven functional blocks, as shown in Figure 2-1.

1. Bus Interface
2. Random Access Memory (RAM)
3. Read Only Memory (ROM/PROM)
4. Parallel I/O Interface
5. Serial I/O Interface
6. Interrupt Status/Mask Registers
7. Memory Enhancements

The *Bus Interface* logic consists of those circuit elements most directly involved with communication between the bus master and the SBC 104. These include bus address/control line receivers, bidirectional data buffer, memory request decode logic, I/O port select decode logic and transfer acknowledge generation and line driver circuits. Also included is the memory and I/O address assignment hardware, consisting of two rocker switch packages and various sets of wire-wrap jumper pins.

The *Random Access Memory (RAM)* section provides 4096 × 8 bits of read/write storage. Eight Intel® 2104 dynamic memory chips (4096 × 1 bits each) are used for the memory elements. The other circuit functions directly related to RAM functions include the following:

Intel® 8222 refresh controller, which coordinates RAM cycle requests from the system bus with internally generated refresh requests. Sixty-four refresh cycles are required to refresh all memory cells in the array.

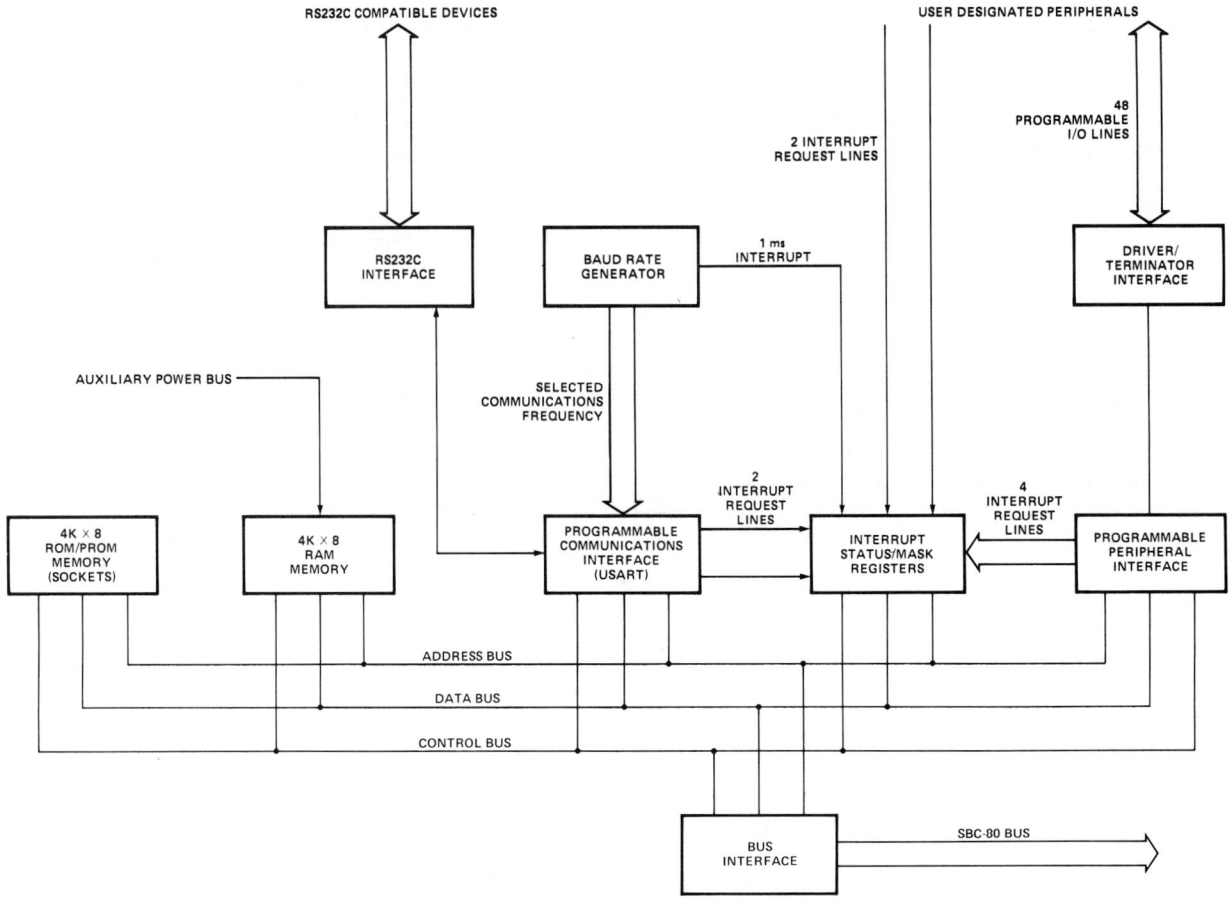

Figure 2-1. SBC 104 Functional Block Diagram.

- Address multiplexer, which multiplexes the 6 or 7 low-order address bits (row) with the 6 or 7 high-order address bits (column). During refresh cycles, only the row address bits are used.
- Intel® 8212 latched buffer, which stores the read data that is present at the RAM array's data out pins.
- Various flip-flops, shift registers and combinatorial logic elements that, together, control access to the RAM and its output buffer. These elements also control the generation of RAM access acknowledge signals.
- Intel® 8224 crystal-stabilized clock generator which provides the fundamental timing reference for the RAM access control logic.

CHAPTER 3

THEORY OF OPERATION

In the preceding chapter, each of the SBC 104 functional blocks was identified and briefly defined. This chapter explains how these functions are implemented.

NOTE: Both active-high (positive true) and active-low (negative true) signals appear in the SBC 104 schematics. To avoid confusion when referring to these signals in this chapter, the following convention is used. The mnemonic (signal label) for each active-low signal is terminated by a slash; e.g., IOW/ means that the signal level on that line will be low when the I/O write command is true (active). A mnemonic without the slash refers to an active-high signal; e.g., the line labeled MEM W is at the high logic level when the memory write signal is true.

DYNAMIC RAMS USED WITH MICROPROCESSORS

3.1 BUS INTERFACE

The Bus Interface refers to those logic elements that participate directly in the following types of system bus activity:

1. System address, control and data buffering
2. System address decoding
3. System control signal propagation
4. Transfer acknowledge generation.

The four groups of Bus Interface logic responsible for these tasks are described in the following paragraphs.

3.1.1 BUS ADDRESS, CONTROL, DATA BUFFERS

The bus address and control signal buffer circuits consist of inverting line receivers of the 74LS04 (address) and 74S04 (control) types. These circuits restore the signals on the system bus lines to their proper logic levels with very high switching speed.

The data buffers are formed by two Intel® 8226 inverting bidirectional driver/receiver chips (A74 and A75). The system data bus is connected to the devices' DB pins. The DO and DI pins of each chip are connected, via printed wires, to the interrupt mask register, RAM data inputs and outputs, PROM data inputs, parallel I/O interface ports and serial I/O interface USART chip.

Directional control (DIEN/) for A74 and A75 is exercised by the memory read and I/O read commands (MRDC/ and IORC/) through NOR gate A41-1. If either read command is asserted by the bus master, the data buffer's driver mode is selected. At all other times, the data buffer's receiver circuits are enabled.

Control of chip select (CS/) for the data buffer is exercised through a set of combinatorial logic, whose final stage is NOR gate A41-4. The data buffer is selected if the following set of conditions is decoded:

RAMAD · INH 1/(NOT) · (MWTC/ + MRDC/) – If a RAM address is decoded AND the memory space is not shared by the PROM AND either a memory write or memory read command is received, a chip select is generated for A74 and A75.

3.1.2 SYSTEM ADDRESS DECODE LOGIC

This logic decodes the appropriate system bus address bits into a RAM request, PROM request or an I/O select. Associated with the decode logic are switches and jumper connections that permit field-modifications of base address assignments.

NOTE: Row and column address decoding for memory accessing is a separate function, which is performed in the selected memory chips.

Memory address decoding is carried out by a pair of 74S151 devices, designated A56 (RAM address decoder) and A57 (PROM address decoder). These are 8-into-1 multiplexers, with true and complementary outputs.

Device A56 tests bus address lines ABC, ABD and ABE for the base address assigned to the RAM expansion memory. This base address is assigned by *opening* the appropriate switch on S3, which applies a high (true) logic level at the corresponding data input of the address decoder.

NOTE: The address decoder's data inputs are high-true, so that an OPEN switch is required to activate an input.

Address bit F is applied to the RAM address decoder's strobe input in either its true or complementary form (ABF or ABF/). This selection determines the bit significance of the S3 switches. If ABF is used, the switch values range from 0 to 7. ABF/ increases the switch values to the range 8 to F.

Jumper pins 90, 89, and 91 are provided for this selection (90–89 selects ABF; 90–91 selects ABF/).

The 3-bit address (ABC, ABD, ABE) enables one of the switch controlled data inputs through to the RAM address decoder's Y (true) output. When that address value corresponds to the switch-selected base address, the address decoder's output will go high (true).

This output, designated RAMAD, is ANDed with various other logic conditions to provide a chip select to the bidirectional data buffer (A74, A75) and to produce the RAM logic enable RAM REQ. The conditions required to generate RAM REQ are:

RAMAD · $\overline{\text{INH 1/}}$ · (MWRC/ + MRDC/)

INH 1/ is a RAM inhibit that is part of the system bus. It is used to prevent RAM devices in the system from responding whenever system memory space used by a PROM is addressed. On the SBC 104, it is used to prevent the generation of RAM REQ. INH 1/ must be false for RAM REQ and the bidirectional data buffer chip select to be generated.

RAM REQ is used to enable various elements of the RAM control logic for operation. Details regarding its implementation are provided in Section 3.2, Random Access Memory (RAM).

Additional address space can be assigned to the SBC 104 RAM elements by opening (selecting) more than one S3 switch. Each open switch specifies a base address for a separate 4K block of RAM.

3.1.3 SYSTEM CONTROL SIGNAL PROPAGATION

These are the circuits that forward the I/O and memory read/write commands to their respective destinations.

The memory write command, MWTC/, is received by the high-speed hex inverter A69-6 to become MEM W. MEM W is sent to the RAM control logic where it is used with timing control circuits to provide a write enable to the RAM. It is also combined with various conditions at the interface to produce a chip select for the bidirectional data buffer (A74 and A75) and to generate the signal RAM REQ. See Sections 3.1.1 and 3.1.2 for details.

The memory read command, MRDC/, is received by the high-speed hex inverter A69-4 to become MEM R. MEM R is forwarded to the RAM output buffer via a second hex inverter. There, it gates the contents of the buffer onto the internal (SBC 104) data bus. MEM R is also combined with other interface conditions to generate either a PROM REQ/ or a chip select for A74, A75 and a RAM REQ. See Sections 3.1.1 and 3.1.2 for details.

3.1.4 TRANSFER ACKNOWLEDGE GENERATION

This logic provides a transfer acknowledge response, XACK/, to notify the bus master that write data provided by the bus master has been accepted or that read data it has requested is available on the system bus. XACK/ allows the bus master to conclude the current input or output instruction cycle. Timing for RAM transfer acknowledge is shown in Figure 3-2.

The SBC 104/108 also provides timing control for an earlier acknowledge signal, labeled AACK/ (advanced acknowledge). AACK/ can be used in some 8080-based systems, but not by the SBC

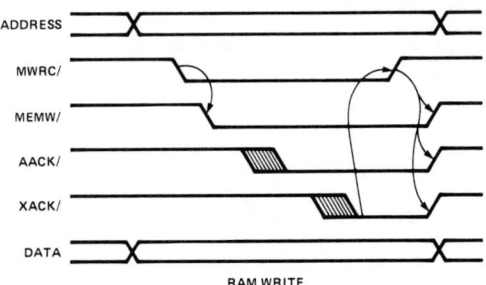

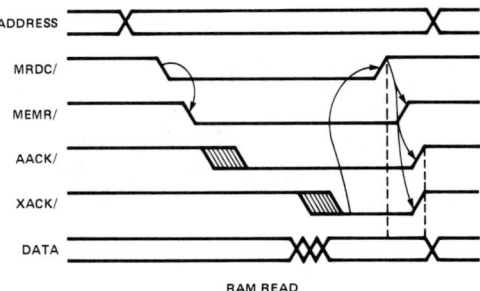

Figure 3-2. RAM Transfer Acknowledge Timing.

80/10, as advance notification that requested data will be valid when the bus master is ready to use it.

This early acknowledge avoids a carry-over into another WAIT state, as would be required if XACK/ were used. Provision is also made for generating AACK/ in response to RAM requests. Timing for this path is determined by the RAM control logic.

3.2 RANDOM ACCESS MEMORY (RAM)

The RAM section of the SBC 104 is illustrated on sheet 3 of the schematics. It consists of eight Intel® 2104 dynamic RAM memory chips, a pair of 74157 devices for multiplexing row and column address bits, an Intel® 8222 refresh controller chip, an Intel® 8212 chip for read data buffering, an Intel® 8224 clock generator and a pair of 74S195 shift registers for timing control and various gates and flip-flops for general sequence control.

RAM transfer acknowledge (RAM XACK) circuits are included as part of the general RAM control logic.

A battery backup circuit is also provided to assure full refresh services to the RAM in the event of primary power loss. This circuit supplies auxiliary

DYNAMIC RAMS USED WITH MICROPROCESSORS

+5V, −5V and +12V power to critical circuts in the RAM section and prevents any access to the RAM except for refresh cycles.

There are three types of operations associated with the dynamic RAM elements.

1. Memory Write — In this operation, the contents of the internal data bus are written into an addressed location in the RAM array (1 bit per 2104 chip) and a RAM transfer acknowledge signal is forwarded to the XACK/ circuits at the interface. Row and column address information is taken from address bits AB0 to AB5 and AB6 to ABB, respectively. These are applied to the 2104 address inputs in multiplexed fashion via devices A65 and A81. RAM control logic provides multiplex control of the row and column addresses, timing control of the 2104 chip select, write enable and address strobe inputs and timing control of the RAM transfer acknowledge signal RAM XAK.

2. Memory Read — In this operation, the contents of an addressed RAM location are loaded into the 8212 data buffer and then transferred to the bidirectional data buffer at the interface. RAM XAK is also forwarded to the XACK/ circuits at the interface. As with memory write operations, all sequences are coordinated by the RAM control logic.

3. Refresh Cycle — For this cycle, an SBC 104 performs a normal read operation (with all chips deselected) while the SBC 108 performs a "CAS before RAS" operation at a row address specified by the 8222 refresh controller. For each new refresh cycle, the row address is increased by one; 64 such cycles refresh the entire memory array. RAM control logic coordinates read/write requests with refresh requests so that competing requests are honored on a first received, first serviced basis. A timing control circuit that is part of the 8222 assures that the frequency of refresh requests is sufficient to maintain full data retention in the array.

Two of these memory operations, memory read and memory write, are initiated by the system and are referred to as system memory cycles. Details regarding these operations are presented in Section 3.2.2. Refresh cycles are discussed further in Section 3.2.3. Section 3.2.1 provides general information.

3.2.1 GENERAL RAM CONTROL

This logic coordinates memory read/write requests with refresh requests. It also provides sequence and timing control of the row/column address multiplexing, of the 2104 control inputs (RAS, CAS, CS and WE) and of the RAM transfer acknowledge circuits.

Requests for access to the RAM array are made through the Intel® 8222 refresh controller A48.

The two types of system requests, memory read and memory write, are consolidated into a single request input to the 8222. This low-true input is designated C REQ/. Requests for a refresh cycle, which are designated R REQ/, are generated by a refresh timing circuit within A48.

A third input, BUSY/, informs the refresh controller whether or not a memory cycle (system or refresh) is in progress. If BUSY/ is low (memory is busy), neither C REQ/ nor R REQ/ will be accepted.

When BUSY/ is high, the refresh controller honors the next memory request it receives. If a request is received while BUSY/ is low, the controller services that pending request after BUSY/ goes high.

The refresh controller initiates a memory control sequence in the RAM control circuits by bringing its SC/ (start cycle) output low. A second refresh controller output RO/ (refresh on) determines the nature of the memory control sequence. If RO/ is low when SC/ goes low, a refresh control sequence is executed. A high level at RO/ causes a memory read or memory write control sequence.

When a system memory cycle is specified (SC/ without RO/), two inputs from the interface logic, MEM R and MEM W, determine which type of system memory cycle is performed.

The RAM control sequence for system memory cycles is described further in the next section. Refer to Figure 3-3 for timing control information.

3.2.2 SYSTEM MEMORY CYCLES (C REQ/)

A single request is made to the refresh controller for either a memory read or memory write cycle. This request, C REQ/, is made by NAND gate A51-12 as a result of successfully decoding the following three general conditions:

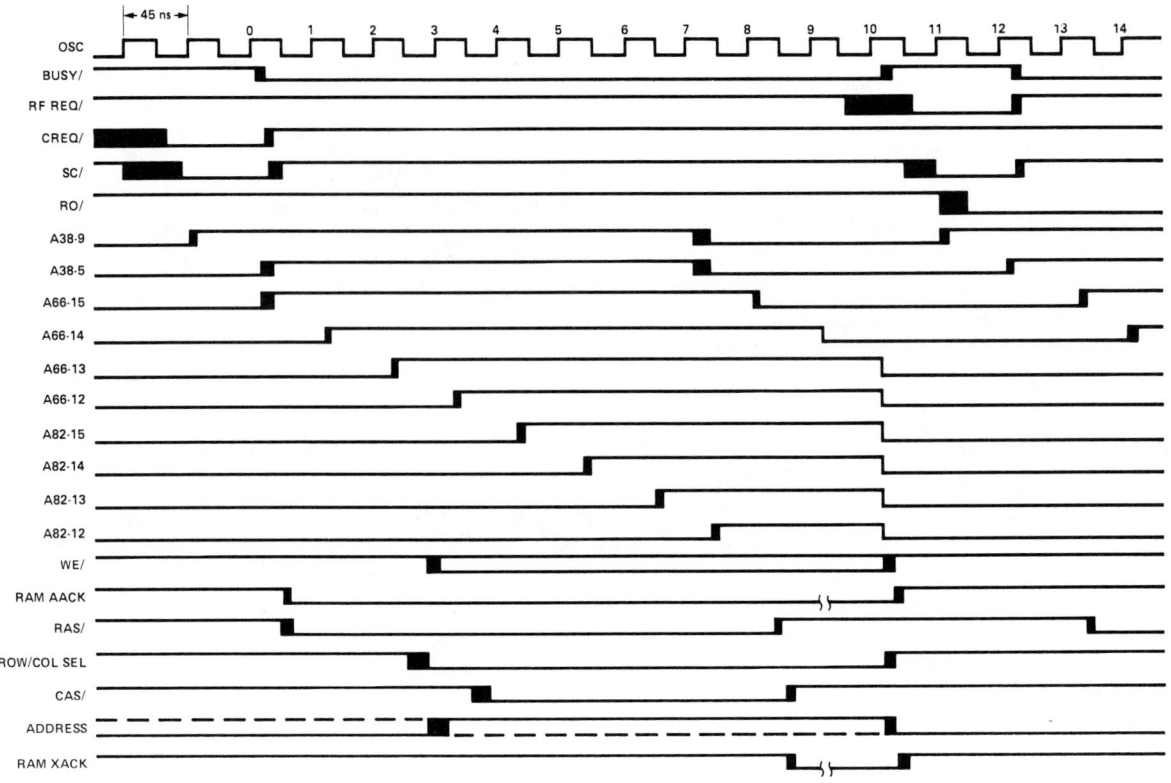

Figure 3-3. RAM Control Timing Summary.

1. The RAM REQ line from the Bus Interface must be high (true).

2. Flip-flop A67-9/8 must be reset. This condition indicates that the current RAM REQ is a new request. When it sets later in the cycle, it disables REQ/, thereby preventing the 8222 from responding to the same RAM REQ twice.

3. The memory protect latch, A25-9/8, is in the reset state. This circuit is described in Section 3.2.4.

When these three conditions are satisfied, A51-12 applies a cycle request to the refresh controller A48. If neither the R REQ/ (refresh request) nor BUSY/ input to the refresh controller is active (low), the cycle request will be accepted and a start cycle (SC/) signal will be applied to FF1 (A38-9/8) of the RAM control logic.

FF1 will be clocked set by the next rising edge of the OSC clock. This 22.1184 MHz square-wave is generated by the Intel® 8224 clock generator circuit. It serves as the primary reference clock for synchronizing the various RAM control functions as well as the XACK control circuits in the interface. Other clock generator outputs are the baud clock (BD CLK) and $\phi 2$ TTL clock, which are forwarded to the serial interface. Their uses are discussed in Section 3.5.

Since a refresh operation is not in progress at this point, the refresh controller's refresh on (RO/) output is high (false). This level allows the set state of FF1 to be propagated through gates A50-11 and A51-8 to the J and K inputs of shift register A66.

With the J and $\overline{K}$ inputs of A66 both set to 1, that device begins shifting 1's through, starting with the next positive edge of OSC. The A66 outputs then provide a sequence of transitions that are used to synchronously step the various RAM control functions through a read or write operation. One A66 output is also applied to a second shift register, A82, to extend the transition sequence. The full set of shift register outputs are referred to as OA through OH. OA corresponds to the Q0 output of A66 and OH corresponds to the Q3 output of A82. See Figure 3-3.

The next OSC edge after FF1 sets causes FF2

(A38-5/6) to set. FF2 brings the refresh controller's BUSY/ input low (true) via NOR gate A83-4. This in turn disables the refresh controller's SC/ (start cycle) output.

At the same time BUSY/ goes true, the OA output of A66 goes high, with the following consequences.

- The RAM array's RAS/ input goes true, which loads the row address (AB0–AB5) into the chip's internal address register. These address bits are propagated through the 8222 refresh controller because a refresh cycle is not in effect and through the row/column multiplexer (A65, A81) because the multiplexer's SEL input is high. SEL does not go low until later in the control sequence, just before CAS/ is asserted.

- The RAM AAK (Advance Acknowledge) latch A67-9/8 is clocked set. Since the advance acknowledge signal is not used in SBC 80/10 systems, this flip-flop's chief function in this case is to disable the cycle request gate, A51-12.

- One input to the CAS control gate A50-3 is satisfied. However, this gate will not provide CAS/ to the RAM array until the OD output of A66 goes high.

The OA output of A66 is also tied to a jumper set, W7, to allow its use as an alternative clock for the RAM AAK latch. It has no significance in this application.

The OC output causes the row/column address multiplexer's SEL iput to go low. This level selects the column address bits (AB6–ABB).

If the RAM REQ is for a write cycle (MEM W input is high), OC also activates the RAM array's WE (write enable) input. This strobes the contents of the SBC 104 data bus into the RAM's data in register.

Next, the OD output of A66 enables gate A50-3. As a result, the RAM array's CAS/ input goes low, which strobes the column address into the internal address regsiter.

The OD output also provides 1's to the J and $\overline{K}$ inputs of the second shift register, A82. This causes A82 to begin contributing sequence control outputs to the operation.

If a memory read cycle has been requested, the RAM array's WE/ input will be high (false) during CAS/. This combination causes the array's data out pins to go to their high impedance states during cell access time. At the end of the period, valid data is presented at the 8212 data in pins. This timing is controlled within the 2104.

So long as the RAM XAK latch A67-5/6 remains reset, the STB (strobe) input to the 8212 will be high and the data latch in the 8212 will reflect the contents of the data in lines. This condition changes as a result of the next event described.

The fifth positive OSC edge after CAS/ goes low causes the shift register's OA output to go low, with the following consequences:

- The RAM array's RAS/ input goes high.
- A50-3 is disabled, causing the array's CAS/ input to go high.
- A50-3 also clocks the RAM XAK latch set, which causes the 8212 STB input to go low. This latches the RAM array's output data into the 8212's data latch. RAM XAK is also forwarded to the XACK/ drivers at the interface.

The 8212's tri-state output buffer is enabled by the chip's DS1/ and DS2 inputs. The low-true DS1/ condition is satisfied by MEM R, inverted. DS2 is satisfied by RAM REQ. When enabled, these drivers forward the contents of the 8212's data latch to the bidirectional data buffer at the interface.

3.2.3 REFRESH CYCLE (R REQ/)

Whenever the refresh controller's BUSY/ and C REQ/ (cycle request) inputs are not active, the controller is able to honor a refresh request.

Refresh requests are made approximately every 14 μs by a timing control circuit that is part of the 8222 device. An external RC network determines the duration of the refresh request intervals.

Each request is made by bringing the 8222's R REQ/ input low.

R REQ/ is provided by a latch in the 8222 so that if the BUSY/ input is low at the time R REQ/ is asserted, the refresh request will be maintained until it is serviced.

When the 8222 accepts R REQ/, the controller's SC/ (start cycle) output goes low, followed by the RO/ (refresh on) output. As in a read/write cycle, SC/ allows OSC to clock FF1 (A38-9/8) set. However, RO/ disables the shift register until FF2

(A38-5/6) sets one OSC cycle later. Then, the shift register begins its output sequence.

When FF2 sets, the refresh controller's BUSY/ input goes low to clear SC/ and to block any cycle request that might occur during refresh.

Next, the 8222 gates the 6-bit refresh address out to the row/column address multiplexer (A65, A81) in place of system address bits AB0–AB5.

This refresh address is supplied by a 6-bit counter in the 8222 device, which is incremented by one count after each refresh cycle. The counter's output is multiplexed within the 8222 with the row address bits supplied by the system (AB0–AB5). Internal address select logic selects the system address bits during read or write and the refresh address bits during refresh cycles.

NAND gate A50-8 prevents the row/column address multiplexer from selecting its column address inputs so long as RO/ is low.

The shift register's OA output enables the RAM array's RAS/ input and then, eight OSC cycles later, it disables RAS/.

Two clocks later, BUSY/ is held up by A26-4. The trailing edge of BUSY/ causes the 8222's RO/ output to clear, which marks the end of the refresh cycle.

The refresh timer begins a new timing cycle and, after approximately 14 μs, generates a new R REQ/.

The SBC 108 performs a special "CAS before RAS" refresh cycle. When jumper W8 is removed the normal shift register sequence is modified such that the CAS input to the memory array goes low one clock cycle before RAS. This sequence informs the RAMs that a refresh cycle is being performed.

3.2.4 MEMORY PROTECT

The memory protect feature consists of a set of auxiliary power busses, which can supply power to critical RAM and RAM control circuits when main power is lost. It also includes a flip-flop, A25-9/8, which prevents new cycle requests from being initiated once the memory protect feature is invoked.

Before power is lost, a low-true signal labeled MEMORY PROTECT must be asserted at P2-20. This signal must be generated in such a fashion that it predicts imminent loss of main power. Its low level causes A25-9/8 to reset. The true-to-false transition of BUSY/ is used as the source for the A25-9/8 clock. This synchronizes the resetting of the memory protect latch with the conclusion of a memory cycle. Memory protect must thus precede loss of power by 16 μs.

When A25-9/8 resets, its Q output disables the C REQ/ generator A51-12. No new cycle requests can be issued until MEMORY PROTECT/ goes false, allowing A25-9/8 to set again. Refresh requests are not affected.

CHAPTER 6
COMPATIBLE EQUIPMENT

The SBC 104 is designed to interface directly with any SBC 80 Single Board Computer via the system bus.

6.1 SBC 80/10

The SBC 80/10 is completely compatible with the SBC 104 OEM Combination Memory and I/O

Board module. The SBC 80/10 can be interfaced with up to 10 combination modules. Table 6-1 summarizes access characteristics of the SBC 104.

6.2 MASTER MODULES

The SBC 104 can operate in systems containing more than one master module.

APPENDIX A
SBC 104/108 SCHEMATICS

Schematic drawings for the SBC 104 and SBC 108 are provided in this Appendix. Information and diagrams in this section are subject to change without notice. References should be made to schematics shipped with this module.

DYNAMIC RAMS USED WITH MICROPROCESSORS

TABLE 6-1. SBC 104 ACCESS CHARACTERISTICS WHEN USED WITH SBC 80/10 CPU

MODULE	INSTRUCTION	CPU CYCLES		CPU WAIT STATES		CYCLE TIME (μs)		REFRESH DEGRADATION RAM	
		MIN	MAX	MIN	MAX	MIN	MAX*	MIN	MAX
COMBINATION	MR	8	8	1	1	3.9	4.4	0	+1
	MW	9	9	2	2	4.4	4.9	0	+1
	IOR	11	11	1	1	5.4	––		
	IOW	12	12	2	2	5.9	––		

*Includes maximum refresh degradation.

REFERENCE:

   MR – MEMORY READ: MOV A,M – 7 CYCLES
   MW – MEMORY WRITE: MOV M,A – 7 CYCLES
   IOR – I/O READ: IN Addr   – 10 CYCLES
   IOW – I/O WRITE: OUT Addr – 10 CYCLES

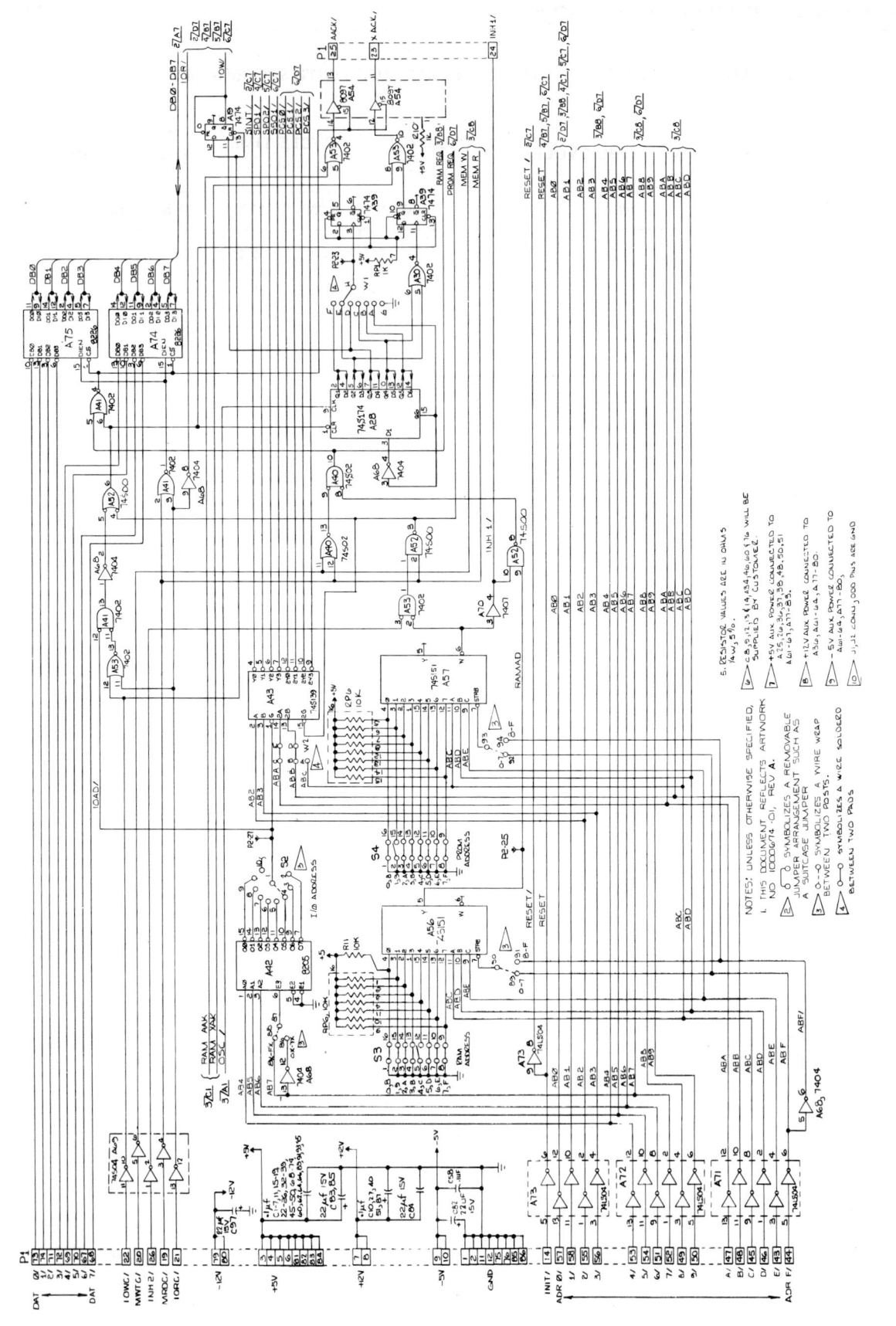

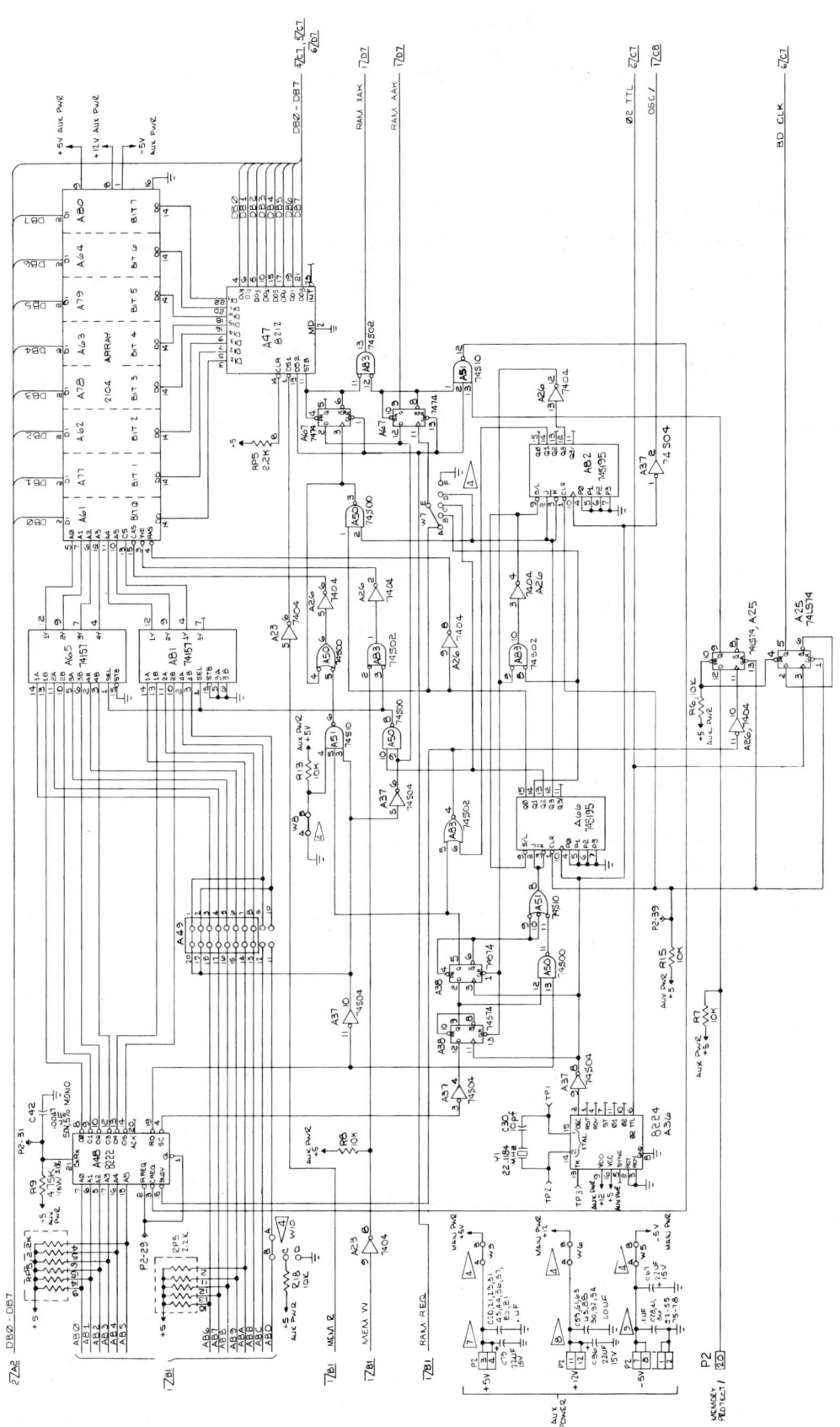

1103
1024 X 1 BIT DYNAMIC RAM

- **Low Power Dissipation** — Dissipates Power Primarily on Selected Chips
- **Access Time** — 300 nsec
- **Cycle Time** — 580 nsec
- **Refresh Period** ... 2 milliseconds for 0–70°C Ambient
- **OR-Tie Capability**
- **Simple Memory Expansion** — Chip Enable Input Lead
- **Fully Decoded** — on Chip Address Decode
- **Inputs Protected** — All Inputs Have Protection Against Static Charge
- **Ceramic and Plastic Package** -- 18 Pin Dual In-Line Configuration.

The Intel 1103 is designed primarily for main memory applications where high performance, low cost, and large bit storage are important design objectives.

It is a 1024 word by 1 bit random access memory element using normally off *P*-channel MOS devices integrated on a monolithic array. It is fully decoded, permitting the use of an 18 pin dual in-line package. It uses dynamic circuitry and primarily dissipates power only during precharge.

Information stored in the memory is non-destructively read. Refreshing of all 1024 bits is accomplished in 32 read cycles and is required every two milliseconds.

A separate **cenable** (chip enable) lead allows easy selection of an individual package when outputs are OR-tied.

The Intel 1103 is fabricated with **silicon gate technology.** This **low threshold** technology allows the design and production of higher performance MOS circuits and provides a higher functional density on a monolithic chip than conventional MOS technologies.

Intel's silicon gate technology also provides excellent protection against contamination. This permits the use of low cost plastic packaging.

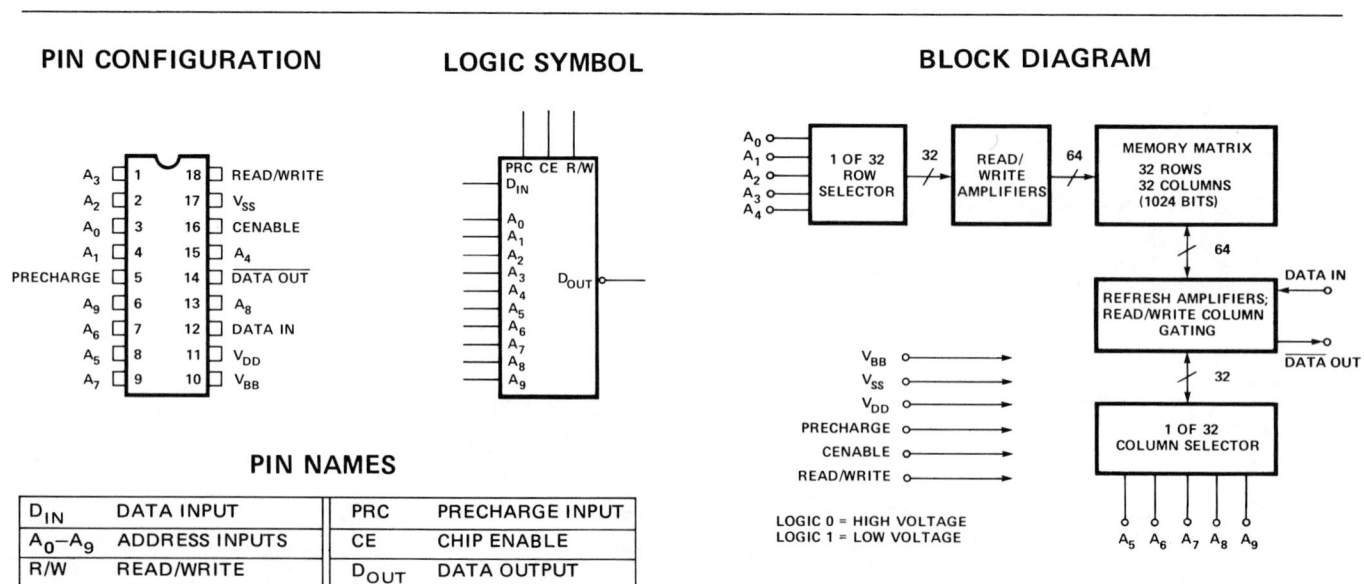

Maximum Guaranteed Ratings*

Temperature Under Bias	0°C to 70°C
Storage Temperature	−65°C to +150°C
All Input or Output Voltages with Respect to the Most Positive Supply Voltage, V_{BB}	−25V to 0.3V
Supply Voltages V_{DD} and V_{SS} with Respect to V_{BB}	−25V to 0.3V
Power Dissipation	1.0 W

*COMMENT:

Stresses above those listed under "Maximum Guaranteed Rating" may cause permanent damage to the device. This is a stress rating only and functional operation of the device at these or at any other condition above those indicated in the operational sections of this specification is not implied. Exposure to absolute maximum rating conditions for extended periods may affect device reliability.

D.C. and Operating Characteristics

$T_A = 0°C$ to $+70°C$, $V_{SS}^{(1)} = 16V \pm 5\%$, $(V_{BB} - V_{SS})^{(6)} = 3V$ to $4V$, $V_{DD} = 0V$ unless otherwise specified

SYMBOL	TEST	MIN.	TYP.	MAX.	UNIT	CONDITIONS
I_{LI}	INPUT LOAD CURRENT (ALL INPUT PINS)			1	µA	$V_{IN} = 0V$
I_{LO}	OUTPUT LEAKAGE CURRENT			1	µA	$V_{OUT} = 0V$
I_{BB}	V_{BB} SUPPLY CURRENT			100	µA	
$I_{DD1}^{(2)}$	SUPPLY CURRENT DURING T_{PC}		37	56	mA	ALL ADDRESSES = 0V; PRECHARGE = 0V; CENABLE = V_{SS}; $T_A = 25°C$
$I_{DD2}^{(2)}$	SUPPLY CURRENT DURING T_{OV}		38	59	mA	ALL ADDRESSES = 0V; PRECHARGE = 0V; CENABLE = 0V; $T_A = 25°C$
$I_{DD3}^{(2)}$	SUPPLY CURRENT DURING T_{POV}		5.5	11	mA	PRECHARGE = V_{SS}; CENABLE = 0V; $T_A = 25°C$
$I_{DD4}^{(2)}$	SUPPLY CURRENT DURING T_{CP}		3	4	mA	PRECHARGE = V_{SS}; CENABLE = V_{SS}; $T_A = 25°C$
$I_{DDAV}^{(5)}$	AVERAGE SUPPLY CURRENT		17	25	mA	CYCLE TIME = 580 ns; PRECHARGE WIDTH = 190 ns; $T_A = 25°C$
$V_{IL1}^{(7)}$	INPUT LOW VOLTAGE (ALL ADDRESS & DATA-IN LINES)	$V_{SS}-17$		$V_{SS}-14.2$	V	$T_A = 0°C$
$V_{IL2}^{(7)}$	INPUT LOW VOLTAGE (ALL ADDRESS & DATA-IN LINES)	$V_{SS}-17$		$V_{SS}-14.5$	V	$T_A = 70°C$
$V_{IL3}^{(7,8)}$	INPUT LOW VOLTAGE (PRECHARGE CENABLE & READ/WRITE INPUTS)	$V_{SS}-17$		$V_{SS}-14.7$	V	$T_A = 0°C$
$V_{IL4}^{(7,8)}$	INPUT LOW VOLTAGE (PRECHARGE CENABLE & READ/WRITE INPUTS)	$V_{SS}-17$		$V_{SS}-15.0$	V	$T_A = 70°C$
$V_{IH1}^{(7)}$	INPUT HIGH VOLTAGE (ALL INPUTS)	$V_{SS}-1$		$V_{SS}+1$	V	$T_A = 0°C$
$V_{IH2}^{(7)}$	INPUT HIGH VOLTAGE (ALL INPUTS)	$V_{SS}-0.7$		$V_{SS}+1$	V	$T_A = 70°C$
I_{OH1}	OUTPUT HIGH CURRENT	600	900	4000	µA	$T_A = 25°C$; $R_{LOAD} = 100\Omega$ (4)
I_{OH2}	OUTPUT HIGH CURRENT	500	800	4000	µA	$T_A = 70°C$; $R_{LOAD} = 100\Omega$ (4)
I_{OL}	OUTPUT LOW CURRENT		See Note 3			$R_{LOAD} = 100\Omega$ (4)
V_{OH1}	OUTPUT HIGH VOLTAGE	60	90	400	mV	$T_A = 25°C$
V_{OH2}	OUTPUT HIGH VOLTAGE	50	80	400	mV	$T_A = 70°C$
V_{OL}	OUTPUT LOW VOLTAGE		See Note 3			

Note 1: The V_{SS} current drain is equal to ($I_{DD} + I_{OH}$) or ($I_{DD} + I_{OL}$).
Note 2: See Supply Current vs. Temperature (p. 3) for guaranteed current at the temperature extremes. These values are taken from a single pulse measurement.
Note 3: The output current when reading a low output is the leakage current of the 1103 plus external noise coupled into the output line from the clocks. V_{OL} equals I_{OL} across the load resistor.
Note 4: This value of load resistance is used for measurement purposes. In applications the resistance may range from 100Ω to 1 kΩ.
Note 5: This parameter is periodically sampled and is not 100% tested.
Note 6: ($V_{BB} - V_{SS}$) supply should be applied at or before V_{SS}.
Note 7: The maximum values for V_{IL} and the minimum values for V_{IH} are linearly related to temperature between 0°C and 70°C. Thus any value in between 0°C and 70°C can be calculated by using a straight-line relationship.
Note 8: The maximum values for V_{IL} (for precharge, cenable & read/write) may be increased to $V_{SS}-14.2$ @ 0°C and $V_{SS}-14.5$ @ 70°C (same values as those specified for the address & data-in lines) with a 40ns degradation (worst case) in t_{AC}, t_{PC}, t_{RC}, t_{WC}, t_{RWC}, t_{ACC1} and t_{ACC2}.

Supply Current vs Temperature

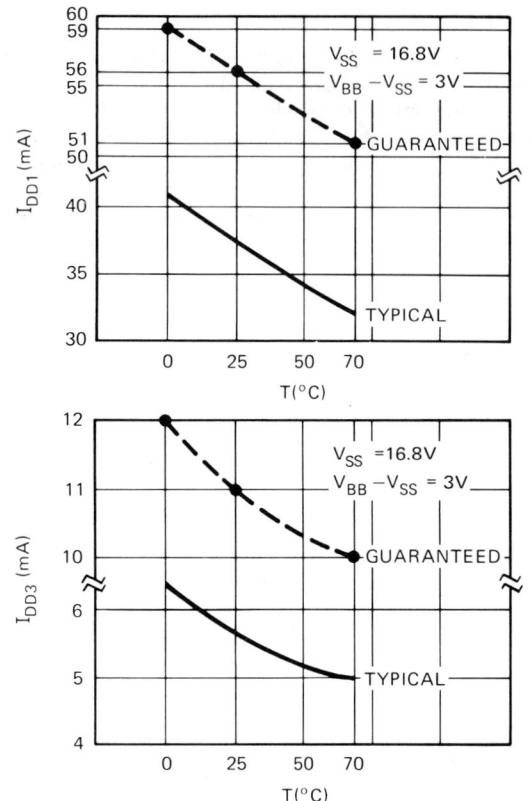

Typical Characteristics

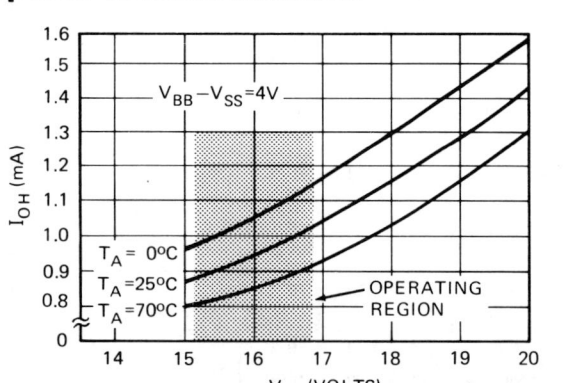

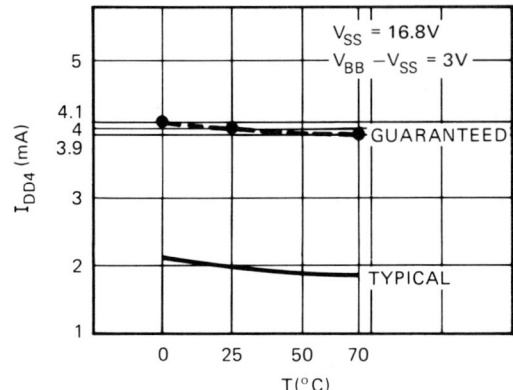

Note 1. ΔI_{DD} is due to charging of internal device node capacitance at precharge

Note 2. These values are taken from a single pulse measurement

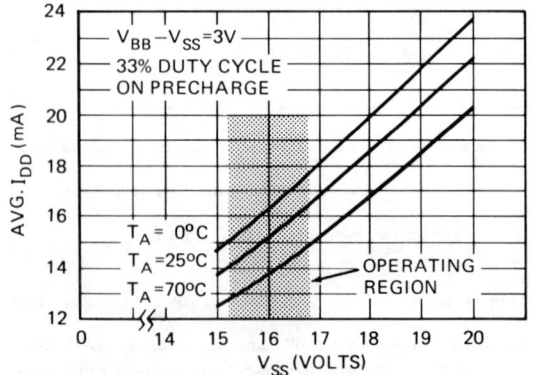

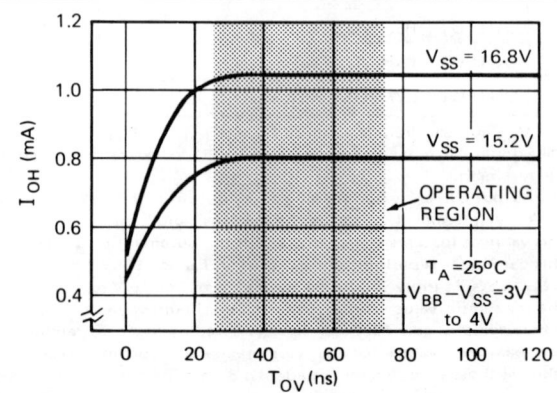

AC Characteristics $T_A = 0°C$ to $70°C$, $V_{SS} = 16 \pm 5\%$, $(V_{BB} - V_{SS}) = 3.0V$ to $4.0V$, $V_{DD} = 0V$

READ, WRITE, AND READ/WRITE CYCLE

SYMBOL	TEST	MIN.	TYP.	MAX.	UNIT	CONDITIONS
t_{REF}	TIME BETWEEN REFRESH			2	ms	
t_{AC} [1]	ADDRESS TO CENABLE SET UP TIME	115			ns	
t_{CA}	CENABLE TO ADDRESS HOLD TIME	20			ns	
t_{PC} [1]	PRECHARGE TO CENABLE DELAY	125			ns	
t_{CP}	CENABLE TO PRECHARGE DELAY	85			ns	
t_{OVL}	PRECHARGE & CENABLE OVERLAP, LOW	25		75	ns	$t_T = 20$ ns
t_{OVH}	PRECHARGE & CENABLE OVERLAP, HIGH			140	ns	$t_T = 20$ ns
t_{OVM}	PRECHARGE & CENABLE OVERLAP, 50% POINTS	45		95	ns	

READ CYCLE

SYMBOL	TEST	MIN.	TYP.	MAX.	UNIT	CONDITIONS
t_{RC} [1]	READ CYCLE	480			ns	
t_{POV}	PRECHARGE TO END OF CENABLE	165		500	ns	
t_{PO}	END OF PRECHARGE TO OUTPUT DELAY			120	ns	
t_{ACC1} [1]	ADDRESS TO OUTPUT ACCESS	300			ns	$t_{ACmin} + t_{OVLmin} + t_{POmax} + 2t_T$
t_{ACC2} [1]	PRECHARGE TO OUTPUT ACCESS	310			ns	$t_{PCmin} + t_{OVLmin} + t_{POmax} + 2t_T$

Conditions: $t_T = 20$ ns, $C_{LOAD} = 100$ pF, $R_{LOAD} = 100\Omega$, $V_{REF} = 40$ mV

WRITE OR READ/WRITE CYCLE

SYMBOL	TEST	MIN.	TYP.	MAX.	UNIT	CONDITIONS
t_{WC} [1]	WRITE CYCLE	580			ns	$t_T = 20$ ns
t_{RWC} [1]	READ/WRITE CYCLE	580			ns	
t_{PW}	PRECHARGE TO READ/WRITE DELAY	165		500	ns	
t_{WP}	READ/WRITE PULSE WIDTH	50			ns	
t_W	READ/WRITE SET UP TIME	80			ns	
t_{DW}	DATA SET UP TIME	105			ns	
t_{DH}	DATA HOLD TIME	10			ns	
t_{PO}	END OF PRECHARGE TO OUTPUT DELAY			120	ns	$C_{LOAD} = 100$ pF, $R_{LOAD} = 100\Omega$, $V_{REF} = 40$ mV
t_{CW}	RELATIONSHIP BETWEEN CENABLE AND READ/WRITE			0	ns	

Note 1: These times will degrade by 40 ns (worst case) if the maximum values for V_{IL} (for precharge, cenable and read/write inputs) go to $V_{SS}-14.2V$ @ $0°C$ and $V_{SS}-14.5V$ @ $70°C$ as defined on page 2.

*CAPACITANCE $T_A = 25$ C

SYMBOL	TEST	TYP.	PLASTIC PKG. MAX.	CERAMIC PKG. MAX.	UNIT	CONDITIONS
C_{AD}	ADDRESS CAPACITANCE	5	7	12	pF	$V_{IN} = V_{SS}$
C_{PR}	PRECHARGE CAPACITANCE	15	18	19.5	pF	$V_{IN} = V_{SS}$
C_{CE}	CENABLE CAPACITANCE	15	18	21	pF	$V_{IN} = V_{SS}$
C_{RW}	READ/WRITE CAPACITANCE	11	15	19.5	pF	$V_{IN} = V_{SS}$
C_{IN1}	DATA INPUT CAPACITANCE	4	5	7.5	pF	CENABLE = 0V, $V_{IN} = V_{SS}$
C_{IN2}	DATA INPUT CAPACITANCE	2	4	6.5	pF	CENABLE = V_{SS}, $V_{IN} = V_{SS}$
C_{OUT}	DATA OUTPUT CAPACITANCE	2	3	7	pF	$V_{OUT} = 0V$

Conditions: $f = 1$ MHz, All Unused Pins Are At A.C. Ground

*This parameter is periodically sampled and is not 100% tested. They are measured at worst case operating conditions.

WRITE CYCLE OR READ/WRITE CYCLE

Timing illustrated for minimum cycle.

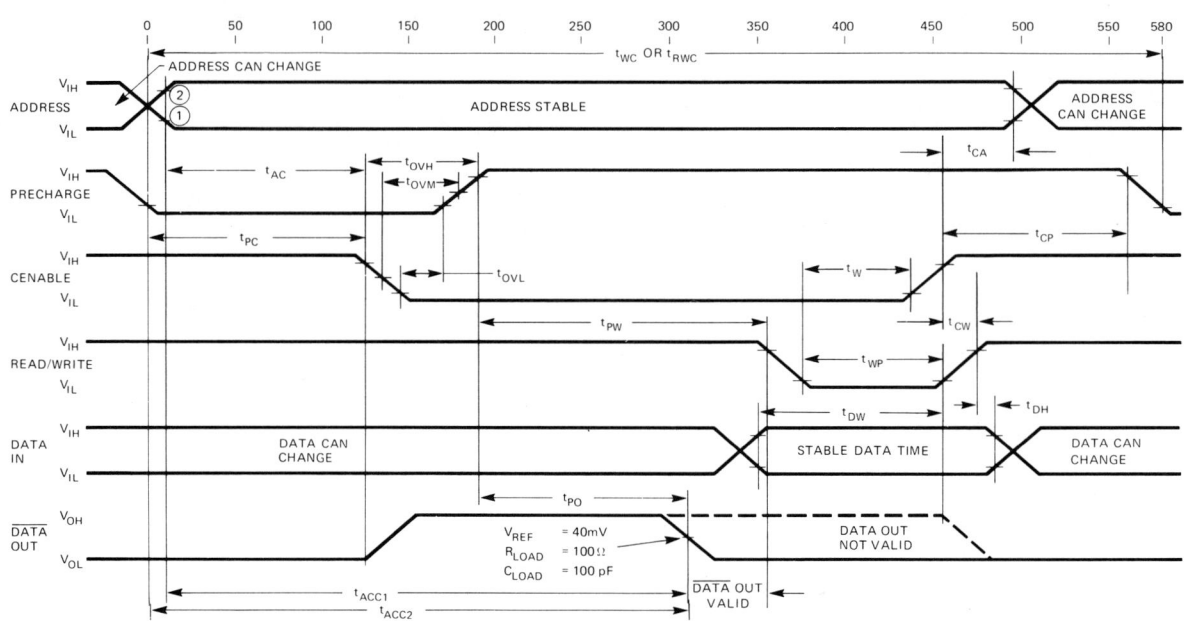

READ CYCLE

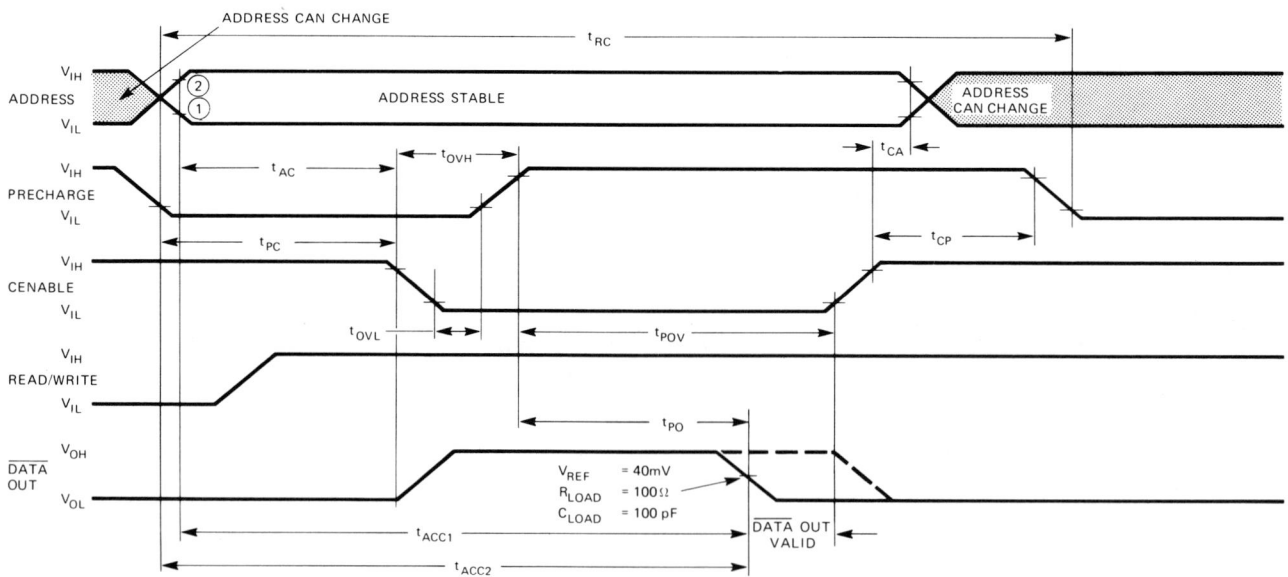

NOTE ① $V_{DD} + 2V$
NOTE ② $V_{SS} - 2V$ ⎦ t_T IS DEFINED AS THE TRANSITIONS BETWEEN THESE TWO POINTS
NOTE 3 t_{DW} IS REFERENCED TO POINT ① OF THE RISING EDGE OF CENABLE OR READ/WRITE WHICHEVER OCCURS FIRST
NOTE 4 t_{DH} IS REFERENCED TO POINT ② OF THE RISING EDGE OF CENABLE OR READ/WRITE WHICHEVER OCCURS FIRST

1103-1
1024 × 1 BIT DYNAMIC RAM

The Intel® 1103-1 is a high speed 1024 bit dynamic random access memory and is the high speed version of the standard 1103. The DC and AC Characteristics for the 1103-1 are given in the following three pages. The absolute maximum ratings for the 1103-1 are the same as for the 1103 on page 2-8.

- **Access Time — 150 nsec**
- **Cycle Time — 340 nsec**

D.C. and Operating Characteristics

($T_A = 0°C$ to $+55°C$, $V_{SS}^1 = 19V \pm 5\%$ $(V_{BB} - V_{SS})^6 = 3V$ to $4V$, $V_{DD} = 0V$ unless otherwise specified)

SYMBOL	TEST	MIN.	TYP.	MAX.	UNIT	CONDITIONS
I_{LI}	INPUT LOAD CURRENT (ALL INPUT PINS)			10	μA	$V_{IN} = 0V$
I_{LO}	OUTPUT LEAKAGE CURRENT			10	μA	$V_{OUT} = 0V$
I_{BB}	V_{BB} SUPPLY CURRENT			100	μA	
I_{DD1}^2	SUPPLY CURRENT DURING T_{PC}		45	60	mA	ALL ADDRESSES = 0V, PRECHARGE = 0V, CENABLE = V_{SS}, $T_A = 25°C$
I_{DD2}^2	SUPPLY CURRENT DURING T_{OV}		50	68.5	mA	ALL ADDRESSES = 0V, PRECHARGE = 0V, CENABLE = 0V, $T_A = 25°C$
I_{DD3}^2	SUPPLY CURRENT DURING T_{POV}		8.5	11	mA	PRECHARGE = V_{SS}, CENABLE = 0V, $T_A = 25°C$
I_{DD4}^2	SUPPLY CURRENT DURING T_{CP}		3.0	4	mA	PRECHARGE = V_{SS}, CENABLE = V_{SS}, $T_A = 25°C$
$I_{DD\ AVG}^5$	AVERAGE SUPPLY CURRENT		20	23	mA	CYCLE TIME = 340 ns, PRECHARGE WIDTH @50% 105ns, $T_A = 25°C$
V_{IL}	INPUT LOW VOLTAGE	$V_{SS} - 20$		$V_{SS} - 18$	V	
V_{IH}	INPUT HIGH VOLTAGE	$V_{SS} - 1$		$V_{SS} + 1$	V	
I_{OH1}	OUTPUT HIGH CURRENT	1150	1300	7000	μA	$T_A = 25°C$
I_{OH2}	OUTPUT HIGH CURRENT	900	1150	7000	μA	$T_A = 55°C$
I_{OL}^3	OUTPUT LOW CURRENT		See Note 3			$R_{LOAD}^4 = 100\ \Omega$
V_{OH1}	OUTPUT HIGH VOLTAGE	115	130	700	mV	$T_A = 25°C$,
V_{OH2}	OUTPUT HIGH VOLTAGE	90	115	700	mV	$T_A = 55°C$,
V_{OL}^3	OUTPUT LOW VOLTAGE		See Note 3			

Note 1: The V_{SS} current drain is equal to ($I_{DD} + I_{OH}$) or ($I_{DD} + I_{OL}$).
Note 2: See Supply Current vs. Temperature (p. 2-9) for guaranteed current at the temperature extremes. These values are taken from a single pulse measurement.
Note 3: The output current when reading a low output is the leakage current of the 1103 plus external noise coupled into the output line from the clocks. V_{OL} equals I_{OL} across the load resistor.
Note 4: This value of load resistance is used for measurement purposes. In applications the resistance may range from 100 Ω to 1 kΩ.
Note 5: This parameter is periodically sampled and is not 100% tested.
Note 6: ($V_{BB} - V_{SS}$) supply should be applied at or before V_{SS}.

AC Characteristics ($T_A = 0°C$ to $55°C$, $V_{SS} = 19 \pm 5\%$, $V_{BB} - V_{SS} = 3.0V$ to $4.0V$, $V_{DD} = 0V$)

READ, WRITE, AND READ/WRITE CYCLE

SYMBOL	TEST	MIN.	TYP.	MAX.	UNIT	CONDITIONS
t_{REF}	TIME BETWEEN REFRESH			1	ms	
t_{AC}	ADDRESS TO CENABLE SET UP TIME	30			ns	
t_{CA}	CENABLE TO ADDRESS HOLD TIME	10			ns	
t_{PC}	PRECHARGE TO CENABLE DELAY	60			ns	
t_{CP}	CENABLE TO PRECHARGE DELAY	40			ns	
t_{OVL}	PRECHARGE & CENABLE OVERLAP, LOW	5		30	ns	$t_T = 20$ ns
t_{OVH}	PRECHARGE & CENABLE OVERLAP, HIGH			85	ns	$t_T = 20$ ns
t_{OVM}	PRECHARGE & CENABLE OVERLAP, 50% POINTS	25		50	ns	

READ CYCLE

SYMBOL	TEST	MIN.	TYP.	MAX.	UNIT	CONDITIONS
t_{RC} [1]	READ CYCLE	300			ns	$t_T = 20$ ns
t_{POV}	PRECHARGE TO END OF CENABLE	115		500	ns	
t_{PO} [1]	END OF PRECHARGE TO OUTPUT DELAY			75	ns	$C_{LOAD} = 50$ pF, $R_{LOAD} = 100\Omega$, $V_{REF} = 80$ mV
t_{ACC1} [1]	ADDRESS TO OUTPUT ACCESS	150			ns	$t_{ACmin} + t_{OVLmin} + t_{POmax} + 2 t_T$, $C_{LOAD} = 50$ pF, $R_{LOAD} = 100\Omega$, $V_{REF} = 80$ mV
t_{ACC2} [1]	PRECHARGE TO OUTPUT ACCESS	180			ns	$t_{PCmin} + t_{OVLmin} + t_{POmax} + 2 t_T$, $C_{LOAD} = 50$ pF, $R_{LOAD} = 100\Omega$, $V_{REF} = 80$ mV

WRITE OR READ/WRITE CYCLE

SYMBOL	TEST	MIN.	TYP.	MAX.	UNIT	CONDITIONS
t_{WC}	WRITE CYCLE	340			ns	$t_T = 20$ ns
t_{RWC} [1]	READ/WRITE CYCLE	340			ns	
t_{PW}	PRECHARGE TO READ/WRITE DELAY	115		500	ns	
t_{WP}	READ/WRITE PULSE WIDTH	20			ns	
t_W	READ/WRITE SET UP TIME	20			ns	
t_{DW}	DATA SET UP TIME	40			ns	
t_{DH}	DATA HOLD TIME	10			ns	
t_{PO} [1]	END OF PRECHARGE TO OUTPUT DELAY			75	ns	$C_{LOAD} = 50$ pF, $R_{LOAD} = 100\Omega$, $V_{REF} = 80$ mV
t_{CW}	RELATIONSHIP BETWEEN CENABLE AND READ/WRITE			0	ns	

NOTE 1: These times will degrade by 35 nsec if a V_{REF} point of 40 mV is chosen instead of the 80 mV point defined in the spec.

*CAPACITANCE $T_A = 25°C$

SYMBOL	TEST	TYP.	PLASTIC PKG. MAX.	CERAMIC PKG. MAX.	UNIT	CONDITIONS
C_{AD}	ADDRESS CAPACITANCE	5	7	12	pF	$V_{IN} = V_{SS}$
C_{PR}	PRECHARGE CAPACITANCE	15	18	19.5	pF	$V_{IN} = V_{SS}$
C_{CE}	CENABLE CAPACITANCE	15	18	21	pF	$V_{IN} = V_{SS}$
C_{RW}	READ/WRITE CAPACITANCE	11	15	19.5	pF	$V_{IN} = V_{SS}$
C_{IN1}	DATA INPUT CAPACITANCE	4	5	7.5	pF	CENABLE = 0V, $V_{IN} = V_{SS}$
C_{IN2}	DATA INPUT CAPACITANCE	2	4	6.5	pF	CENABLE = V_{SS}, $V_{IN} = V_{SS}$
C_{OUT}	DATA OUTPUT CAPACITANCE	2	3	7	pF	$V_{OUT} = 0V$

$f = 1$ MHz, All Unused Pins Are At A.C. Ground

*This parameter is periodically sampled and is not 100% tested. They are measured at worst case operating conditions.

WRITE OR READ/WRITE CYCLE

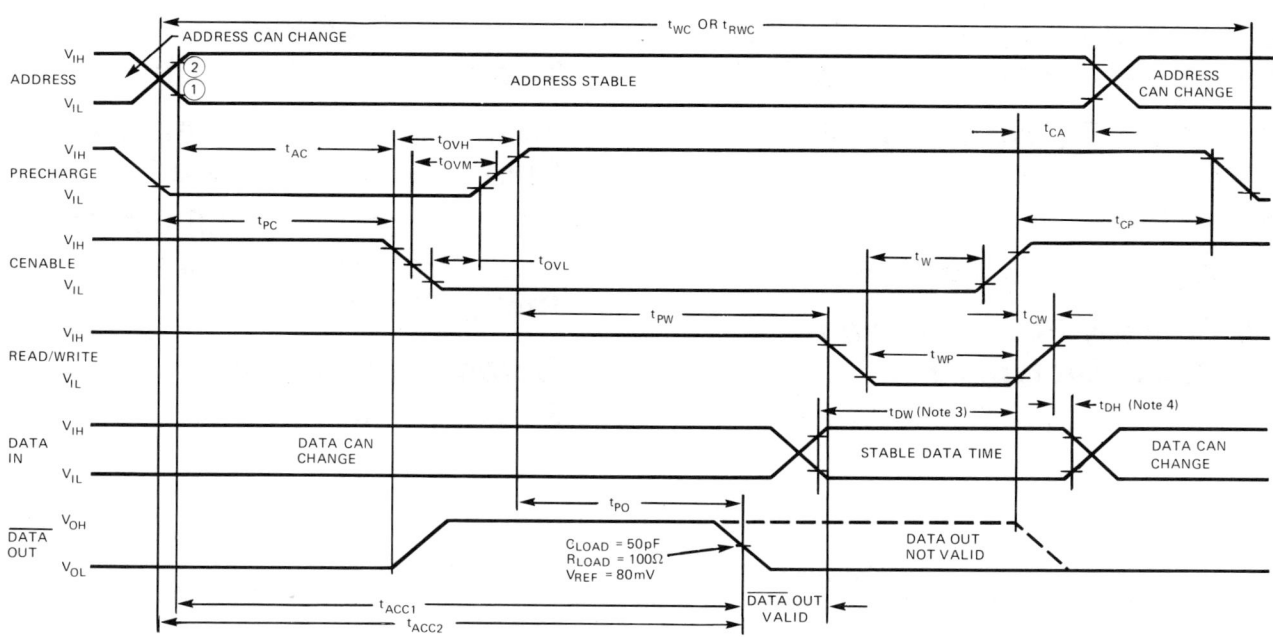

READ CYCLE

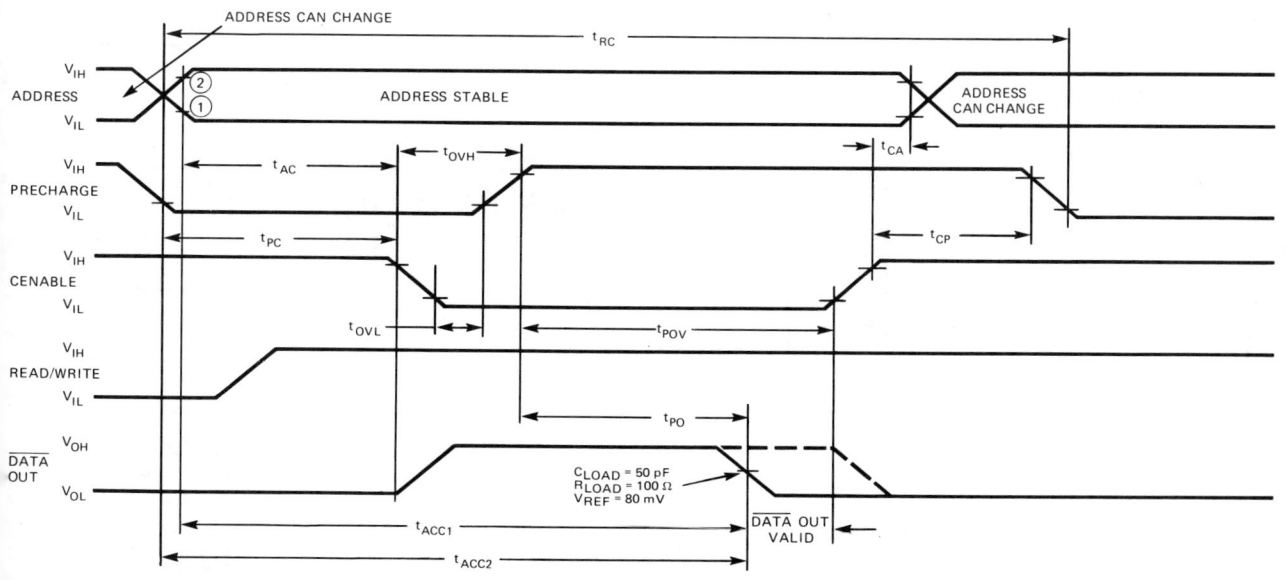

NOTE ① $V_{DD} + 2V$
NOTE ② $V_{SS} - 2V$] t_T IS DEFINED AS THE TRANSITIONS BETWEEN THESE TWO POINTS
NOTE 3 t_{DW} IS REFERENCED TO POINT ① OF THE RISING EDGE OF CHIP ENABLE OR READ/WRITE WHICHEVER OCCURS FIRST
NOTE 4 t_{DH} IS REFERENCED TO POINT ② OF THE RISING EDGE OF CHIP ENABLE OR READ/WRITE WHICHEVER OCCURS FIRST

1103A
1024 X 1 BIT DYNAMIC RAM

* **No Precharge Required -- Critical Precharge Timing is Eliminated**
- **Electrically Equivalent to 1103 -- Pin-for-Pin/Functionally Compatible**
- **Fast Access Time -- 205ns max.**
- **Low Standby Power Dissipation -- 2 µW/Bit typical**
- **Address Registers Incorporated on the Chip**
- **Simple Memory Expansion -- Chip Enable Input Lead**
- **Inputs Protected -- All Inputs Have Protection Against Static Charge**
- **Ceramic and Plastic Package -- 18-Pin DIP**

The 1103A is a 1024 word by 1 bit dynamic RAM. It is designed primarily for main memory applications where high performance, low cost, and large bit storage are important design objectives. The 1103A is electrically equivalent to the 1103.

1103A systems may be simplified due to the elimination of the precharge clock, its associated circuitry, and critical overlap timing. Only one external clock, CENABLE, is required.

Information stored in the memory is non-destructively read. Refreshing of all 1024 bits is accomplished in 32 read cycles (addressing A_0 to A_4) and is required every two milliseconds. The memory may be used in a low power standby mode by having cenable at V_{SS} potential.

The 1103A is fabricated with silicon gate technology. This low threshold technology allows the design and production of higher performance MOS circuits and provides a higher functional density on a monolithic chip than conventional MOS technologies.

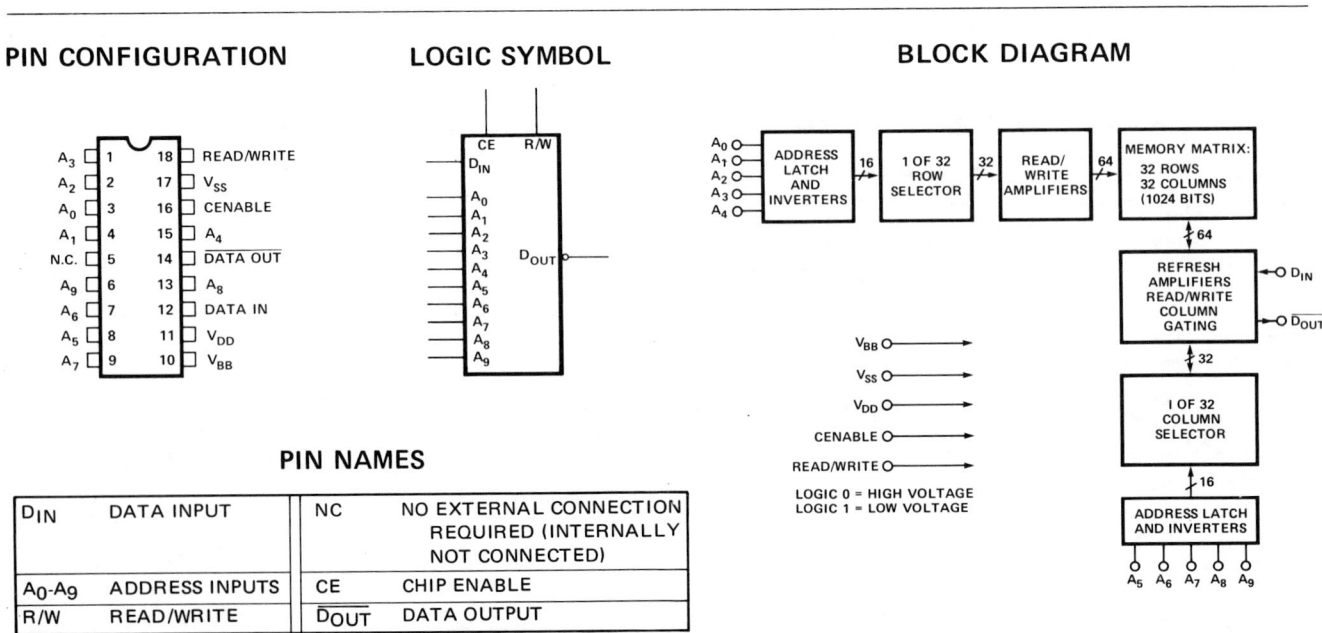

PIN NAMES

D_{IN}	DATA INPUT	NC	NO EXTERNAL CONNECTION REQUIRED (INTERNALLY NOT CONNECTED)
A_0-A_9	ADDRESS INPUTS	CE	CHIP ENABLE
R/W	READ/WRITE	$\overline{D}_{OUT}$	DATA OUTPUT

Absolute Maximum Ratings*

Temperature Under Bias	0°C to 70°C
Storage Temperature	−65°C to +150°C
All Input or Output Voltages with Respect to the most Positive Supply Voltage, V_{BB}	−25V to 0.3V
Supply Voltages V_{DD} and V_{SS} with Respect to V_{BB}	−25V to 0.3V
Power Dissipation	1.0W

*COMMENT:
Stresses above those listed under "Absolute Maximum Ratings" may cause permanent damage to the device. This is a stress rating only and functional operation of the device at these or any other conditions above those indicated in the operational sections of this specification is not implied. Exposure to absolute maximum rating conditions for extended periods may affect device reliability.

D. C. and Operating Characteristics

T_A = 0°C to +70°C, V_{SS}[1] = 16V ± 5%, $(V_{BB} - V_{SS})$[2] = 3V to 4V, V_{DD} = 0V unless otherwise specified.

Symbol	Test	Min.	Typ.	Max.	Unit	Conditions	
I_{LI}	Input Load Current (All Input Pins)			1	μA	V_{IN} = 0V	
I_{LO}	Output Leakage Current			1	μA	V_{OUT} = 0V	
I_{BB}	V_{BB} Supply Current			100	μA		
I_{DD1}	Supply Current During Cenable On		4	11	mA	Cenable = 0V; T_A = 25°C	
I_{DD2}	Supply Current During Cenable Off		0.1	4	mA	Cenable = V_{SS}; T_A = 25°C	
I_{DDAV}	Average Supply Current		17	25	mA	Cycle Time = 580ns; T_A = 25°C	
V_{IL}	Input Low Voltage	V_{DD} − 1		V_{DD} + 1	V		
V_{IH}	Input High Voltage	V_{SS} − 1		V_{SS} + 1	V		
I_{OH1}	Output High Current	600	1800	4000	μA	T_A = 25°C	
I_{OH2}	Output High Current	500	1500	4000	μA	T_A = 70°C	R_{LOAD}[4] = 100Ω
I_{OL}	Output Low Current		See Note Three				
V_{OH1}	Output High Voltage	60	180	400	mV	T_A = 25°C	
V_{OH2}	Output High Voltage	50	150	400	mV	T_A = 70°C	
V_{OL}	Output Low Voltage		See Note Three				

NOTES:
1. The V_{SS} current drain is equal to (I_{DD} + I_{OH}) or (I_{DD} + I_{OL}).
2. (V_{BB} − V_{SS}) supply should be applied at or before V_{SS}.
3. The output current when reading a low output is the leakage current of the 1103 plus external noise coupled into the output line from the clocks. V_{OL} equals I_{OL} across the load resistor.
4. This value of load resistance is used for measurement purposes. In applications the resistance may range from 100Ω to 1 kΩ.

Supply Current vs Temperature

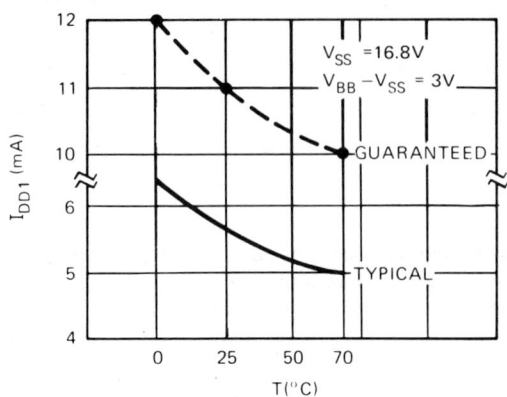
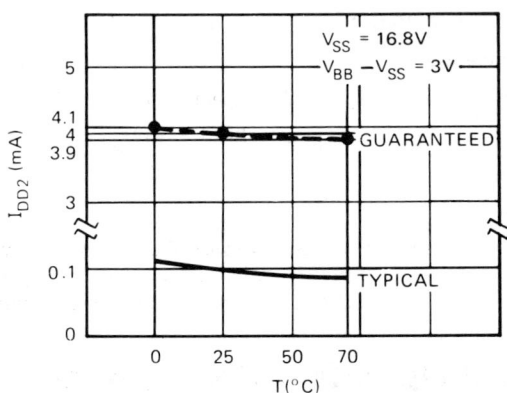

Typical Characteristics

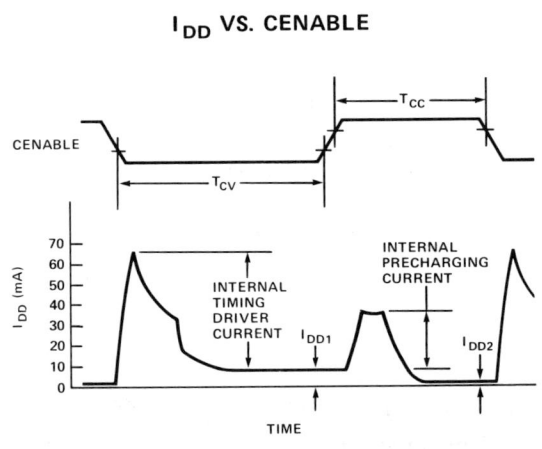

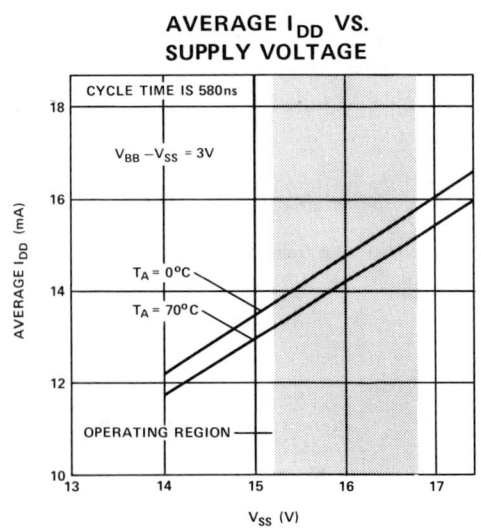
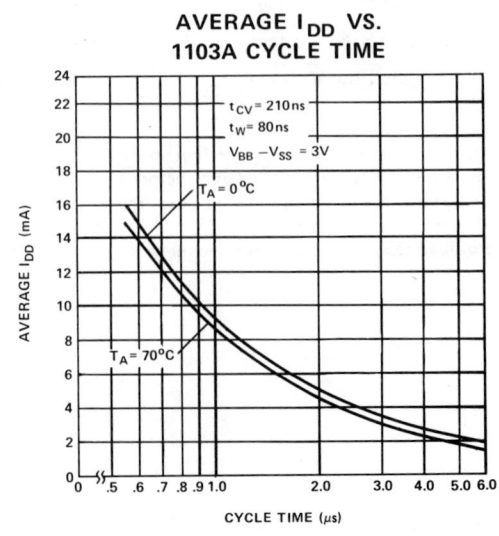

A.C. Characteristics
$T_A = 0°C$ to $70°C$, $V_{SS} = 16V \pm 5\%$, $(V_{BB} - V_{SS}) = 3.0V$ to $4.0V$, $V_{DD} = 0V$

READ, WRITE, AND READ/WRITE CYCLE

Symbol	Test	Min.	Max.	Unit	Conditions
t_{REF}	Time Between Refresh		2	ms	
t_{AC}	Address to Cenable Set Up Time	0		ns	
t_{AH}	Address Hold Time	100		ns	
t_{CC}	Cenable Off Time	230		ns	

READ CYCLE

Symbol	Test	Min.	Max.	Unit	Conditions
t_{RC}	Read Cycle	480		ns	$t_T = 20$ ns
t_{CV}	Cenable on Time	210	500	ns	$C_{LOAD} = 100$ pF
t_{CO}	Cenable Output Delay		185	ns	$R_{LOAD} = 100\Omega$
t_{ACC}	ADDRESS TO OUTPUT ACCESS		205	ns	$t_{ACC} = t_{AC\,MIN} + t_{CO} + t_T$; $V_{REF} = 40$ mV
t_{WH}	Read/Write Hold Time	30		ns	

WRITE OP. READ/WRITE CYCLE

Symbol	Test	Min.	Max.	Unit	Conditions
t_{WCY}	Write Cycle	580		ns	$t_T = 20$ ns
t_{RWC}	Read/Write Cycle	580		ns	
t_{CW}	Cenable to Read/Write Delay	210	500	ns	
t_{WP}	Read/Write Pulse Width	50		ns	
t_W	Read/Write Set Up Time	80		ns	
t_{DW}	Data Set Up Time	105		ns	
t_{DH}	Data Hold Time	10		ns	
t_{CO}	Output Delay		185	ns	$C_{LOAD} = 100$ pF; $R_{LOAD} = 100\Omega$; $V_{REF} = 40$ mV
t_{WC}	Read/Write to Cenable	0		ns	

CAPACITANCE[1] $T_A = 25°C$

Symbol	Test	Typ. Plastic	Plastic Pkg. Max.	Ceramic Pkg. Max.	Unit	Conditions	
C_{AD}	Address Capacitance	5	7	12	pF	$V_{IN} = V_{SS}$	
C_{CE}	Cenable Capacitance	22	25	28	pF	$V_{IN} = V_{SS}$	
C_{RW}	Read/Write Capacitance	11	15	19.5	pF	$V_{IN} = V_{SS}$	$f = 1$ MHz. All unused pins are at A.C. ground.
C_{IN1}	Data Input Capacitance	4	5	7.5	pF	Cenable = 0V; $V_{IN} = V_{SS}$	
C_{IN2}	Data Input Capacitance	2	4	6.5	pF	Cenable = V_{SS}	
C_{OUT}	Data Output Capacitance	2	3	7.0	pF	$V_{IN} = V_{SS}$; $V_{OUT} = 0V$	

NOTES: 1. These parameters are periodically sampled and are not 100% tested. They are measured at worst case operating conditions.

WRITE CYCLE OR READ/WRITE CYCLE
Timing illustrated for minimum cycle.

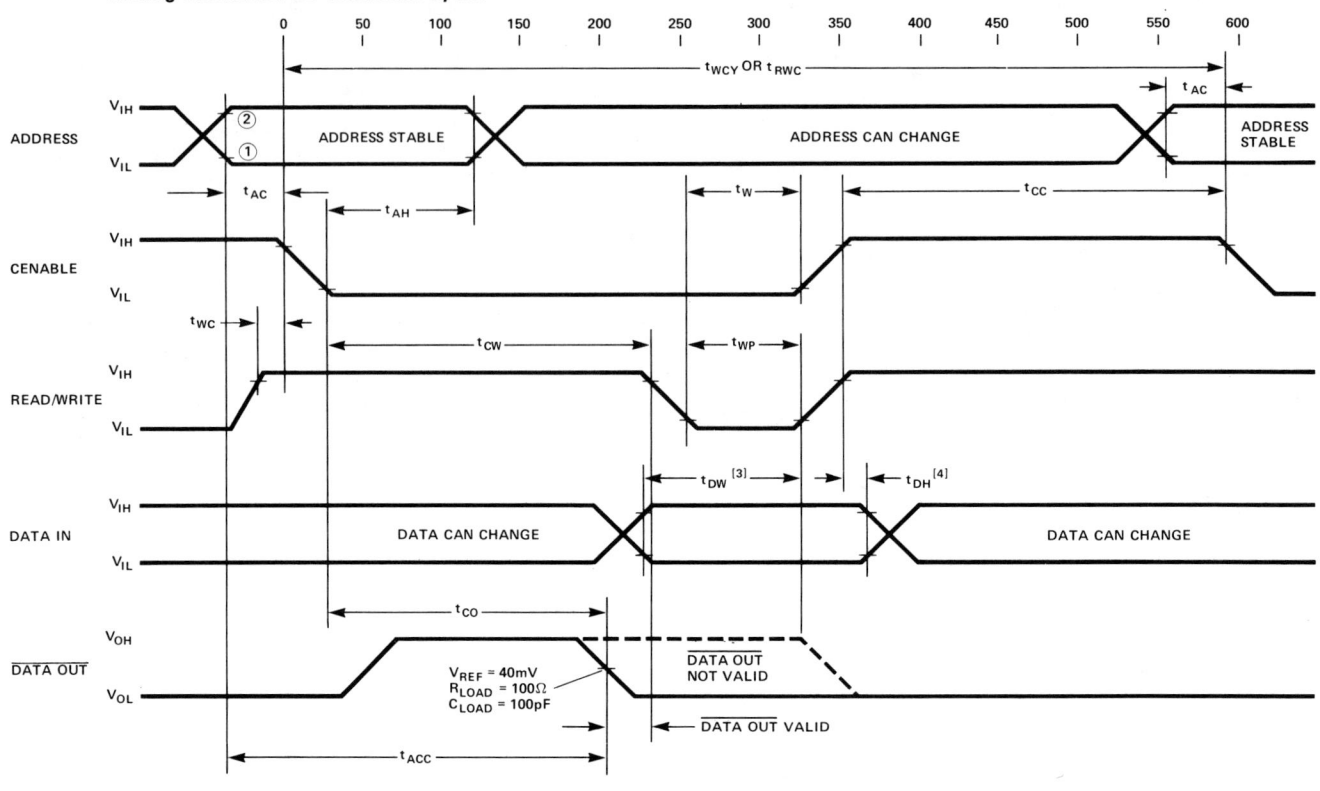

READ CYCLE

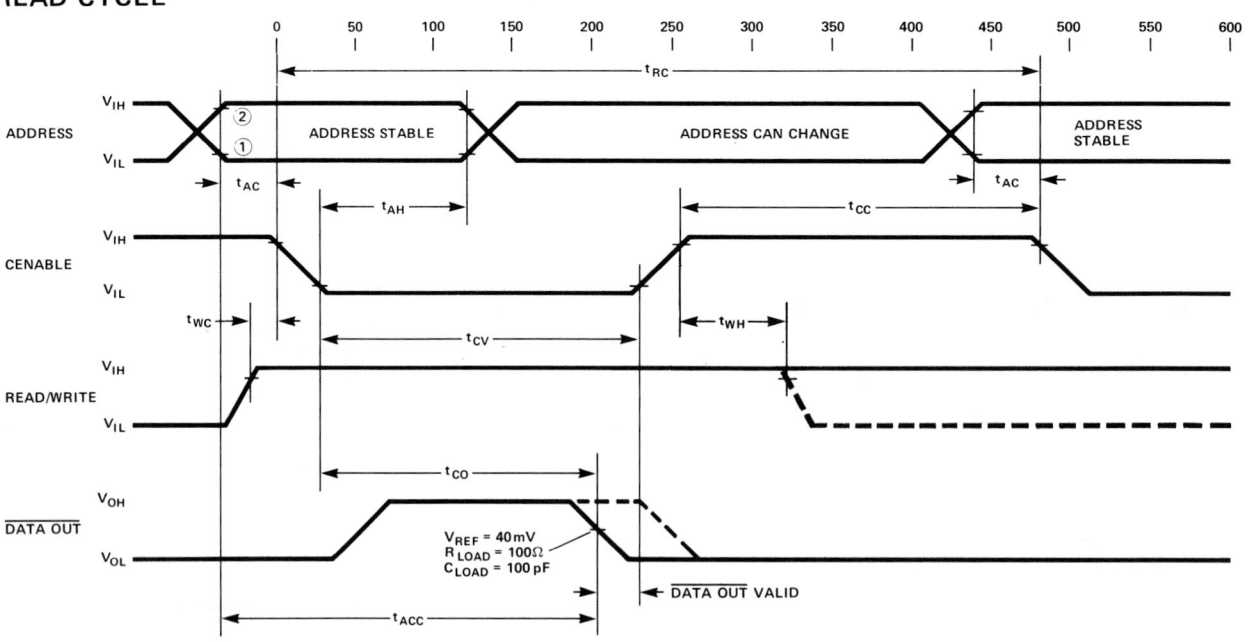

NOTES:
1. $V_{DD} + 2V$
2. $V_{SS} - 2V$ — t_T is defined as the transition between these two points.
3. t_{DW} is referenced to point 1 of the rising edge of cenable or Read/Write, whichever occurs first.
4. t_{DH} is referenced to point 2 of the rising edge of Read/Write.

1103A-1
1024 x 1 BIT DYNAMIC RAM

- **High Speed 1103A — Access Time — 145 ns / Cycle Time — 340 ns**

* **No Precharge Required -- Critical Precharge Timing is Eliminated**

- **Low Standby Power Dissipation -- 0.2 µW/Bit Typical**

- **Address Registers Incorporated on the Chip**

- **Simple Memory Expansion -- Chip Enable Input Lead**

- **Inputs Protected -- All Inputs Have Protection Against Static Charge**

- **Standard 18-Pin Dual In-Line Packages**

The Intel® 1103A-1 is a high speed 1024 bit dynamic random access memory and is the fastest version of the standard 1103A. It is designed primarily for main memory applications where high performance, low cost, and large bit storage are important design objectives.

1103A-1 systems may be simplified due to the elimination of the precharge clock, its associated circuitry, and critical overlap timing. Only one external clock, CENABLE, is required.

Information stored in the memory is non-destructively read. Refreshing of all 1024 bits is accomplished in 32 read cycles (addressing A_0 to A_4) and is required every one millisecond. The memory may be used in a low power standby mode by having cenable at V_{SS} potential.

The 1103A-1 is fabricated with silicon gate technology. This low threshold technology allows the design and production of higher performance MOS circuits and provides a higher functional density on a monolithic chip than conventional MOS technologies.

PIN CONFIGURATION   LOGIC SYMBOL

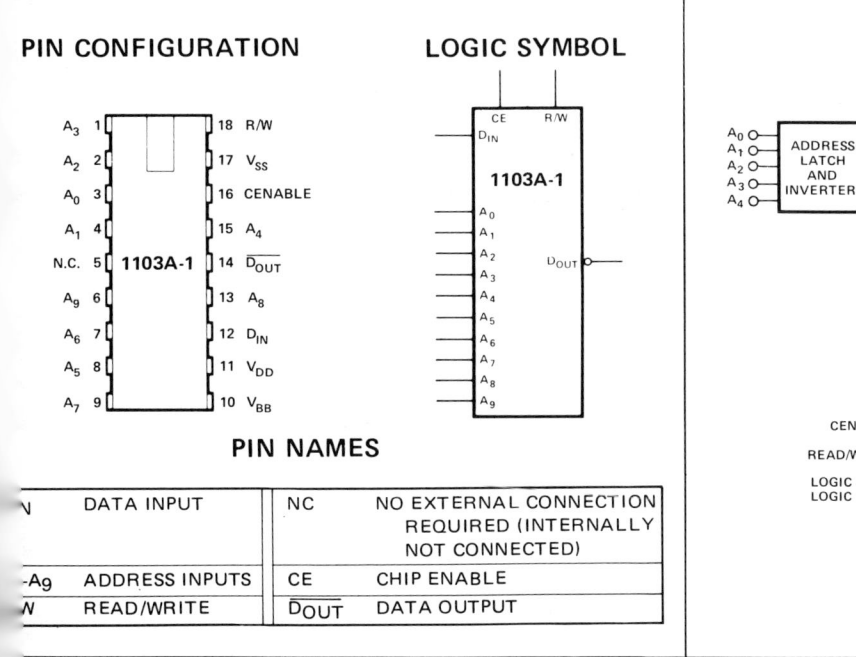

PIN NAMES

D_{IN}	DATA INPUT	NC	NO EXTERNAL CONNECTION REQUIRED (INTERNALLY NOT CONNECTED)
A_0-A_9	ADDRESS INPUTS	CE	CHIP ENABLE
R/W	READ/WRITE	$\overline{D}_{OUT}$	DATA OUTPUT

BLOCK DIAGRAM

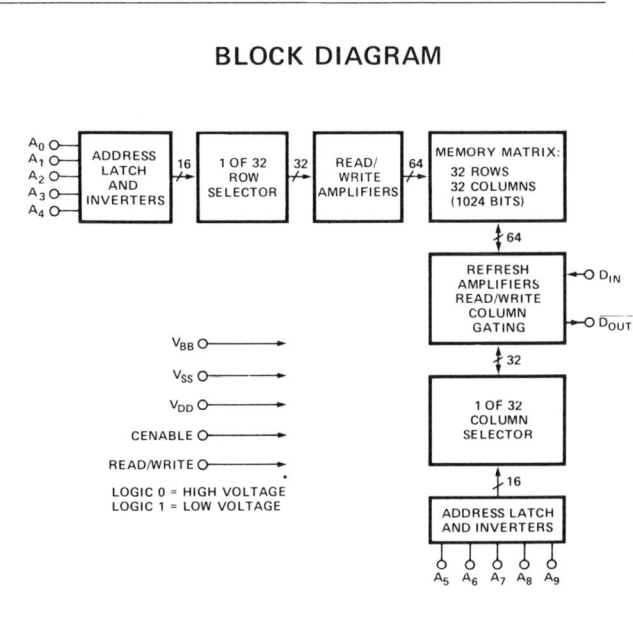

Absolute Maximum Ratings*

Temperature Under Bias	0°C to 70°C
Storage Temperature	−65°C to +150°C
All Input or Output Voltages with Respect to the most Positive Supply Voltage, V_{BB}	−25V to 0.3V
Supply Voltages V_{DD} and V_{SS} with Respect to V_{BB}	−25V to 0.3V
Power Dissipation	1.0W

*COMMENT:
Stresses above those listed under "Absolute Maximum Ratings" may cause permanent damage to the device. This is a stress rating only and functional operation of the device at these or any other conditions above those indicated in the operational sections of this specification is not implied. Exposure to absolute maximum rating conditions for extended periods may affect device reliability.

D. C. and Operating Characteristics

T_A = 0°C to +55°C, V_{SS}[1] = 19V ± 5%, ($V_{BB} - V_{SS}$)[2] = 3V to 4V, V_{DD} = 0V unless otherwise specified.

Symbol	Test	Min.	Typ.	Max.	Unit	Conditions
I_{LI}	Input Load Current (All Input Pins)			10	μA	V_{IN} = 0V
I_{LO}	Output Leakage Current			10	μA	V_{OUT} = 0V
I_{BB}	V_{BB} Supply Current			100	μA	
I_{DD1}	Supply Current During Cenable On		7	11	mA	Cenable = 0V; T_A = 25°C
I_{DD2}	Supply Current During Cenable Off		0.01	0.5	mA	Cenable = V_{SS}; T_A = 25°C
I_{DDAV}	Average Supply Current		25	33	mA	Cycle Time = 340ns; T_A = 25°C
V_{IL}	Input Low Voltage	V_{DD} −1		V_{DD} +1	V	
V_{IH}	Input High Voltage	V_{SS} −1		V_{SS} +1	V	
I_{OH1}	Output High Current	1150	1800	7000	μA	T_A = 25°C
I_{OH2}	Output High Current	900	1600	7000	μA	T_A = 55°C
I_{OL}	Output Low Current		See Note Three			R_{LOAD}[4] = 100Ω
V_{OH1}	Output High Voltage	115	180	700	mV	T_A = 25°C
V_{OH2}	Output High Voltage	90	160	700	mV	T_A = 55°C
V_{OL}	Output Low Voltage		See Note Three			

NOTES:
1. The V_{SS} current drain is equal to ($I_{DD} + I_{OH}$) or ($I_{DD} + I_{OL}$).
2. ($V_{BB} - V_{SS}$) supply should be applied at or before V_{SS}.
3. The output current when reading a low output is the leakage current of the 1103 plus external noise coupled into the output line from the clocks. V_{OL} equals I_{OL} across the load resistor.
4. This value of load resistance is used for measurement purposes. In applications the resistance may range from 100Ω to 1 kΩ.

A.C. Characteristics
$T_A = 0°C$ to $55°C$, $V_{SS} = 19V \pm 5\%$, $(V_{BB} - V_{SS}) = 3.0V$ to $4.0V$, $V_{DD} = 0V$.

READ, WRITE, AND READ/WRITE CYCLE

Symbol	Test	Min.	Max.	Unit	Conditions
t_{REF}	Time Between Refresh		1	ms	
t_{AC}	Address to Cenable Set Up Time	0		ns	
t_{AH}	Address Hold Time	100		ns	
t_{CC}	Cenable Off Time	120		ns	

READ CYCLE

Symbol	Test	Min.	Max.	Unit	Conditions
t_{RC}	Read Cycle	300		ns	$t_T = 20\,ns$
t_{CV}	Cenable on Time	140	500	ns	$C_{LOAD} = 50pF$
t_{CO}	Cenable Output Delay		125	ns	$R_{LOAD} = 100\Omega$
t_{ACC}	ADDRESS TO OUTPUT ACCESS		145	ns	$t_{ACC} = t_{AC\,MIN} + t_{CO} + t_T$, $V_{REF} = 80mV$
t_{WH}	Read/Write Hold Time	30		ns	

WRITE OR READ/WRITE CYCLE

Symbol	Test	Min.	Max.	Unit	Conditions
t_{WCY}	Write Cycle	340		ns	$t_T = 20\,ns$
t_{RWC}	Read/Write Cycle	340		ns	
t_{CW}	Cenable to Read/Write Delay	140	500	ns	
t_{WP}	Read/Write Pulse Width	20		ns	
t_W	Read/Write Set Up Time	20		ns	
t_{DW}	Data Set Up Time	40		ns	
t_{DH}	Data Hold Time	10		ns	
t_{CO}	Output Delay		125	ns	$C_{LOAD} = 50pF$; $R_{LOAD} = 100\Omega$, $V_{REF} = 80mV$
t_{WC}	Read/Write to Cenable	0		ns	

CAPACITANCE[1] $T_A = 25°C$

Symbol	Test	Typ. Plastic	Plastic Pkg. Max.	Ceramic Pkg. Max.	Unit	Conditions
C_{AD}	Address Capacitance	5	7	12	pF	$V_{IN} = V_{SS}$
C_{CE}	Cenable Capacitance	22	25	28	pF	$V_{IN} = V_{SS}$
C_{RW}	Read/Write Capacitance	11	15	19.5	pF	$V_{IN} = V_{SS}$
C_{IN1}	Data Input Capacitance	4	5	7.5	pF	Cenable = 0V, $V_{IN} = V_{SS}$
C_{IN2}	Data Input Capacitance	2	4	6.5	pF	Cenable = V_{SS}
C_{OUT}	Data Output Capacitance	2	3	7.0	pF	$V_{IN} = V_{SS}$, $V_{OUT} = 0V$

$f = 1\,MHz$. All unused pins are at A.C. ground.

NOTES: 1. These parameters are periodically sampled and are not 100% tested. They are measured at worst case operating conditions.

WRITE CYCLE OR READ/WRITE CYCLE

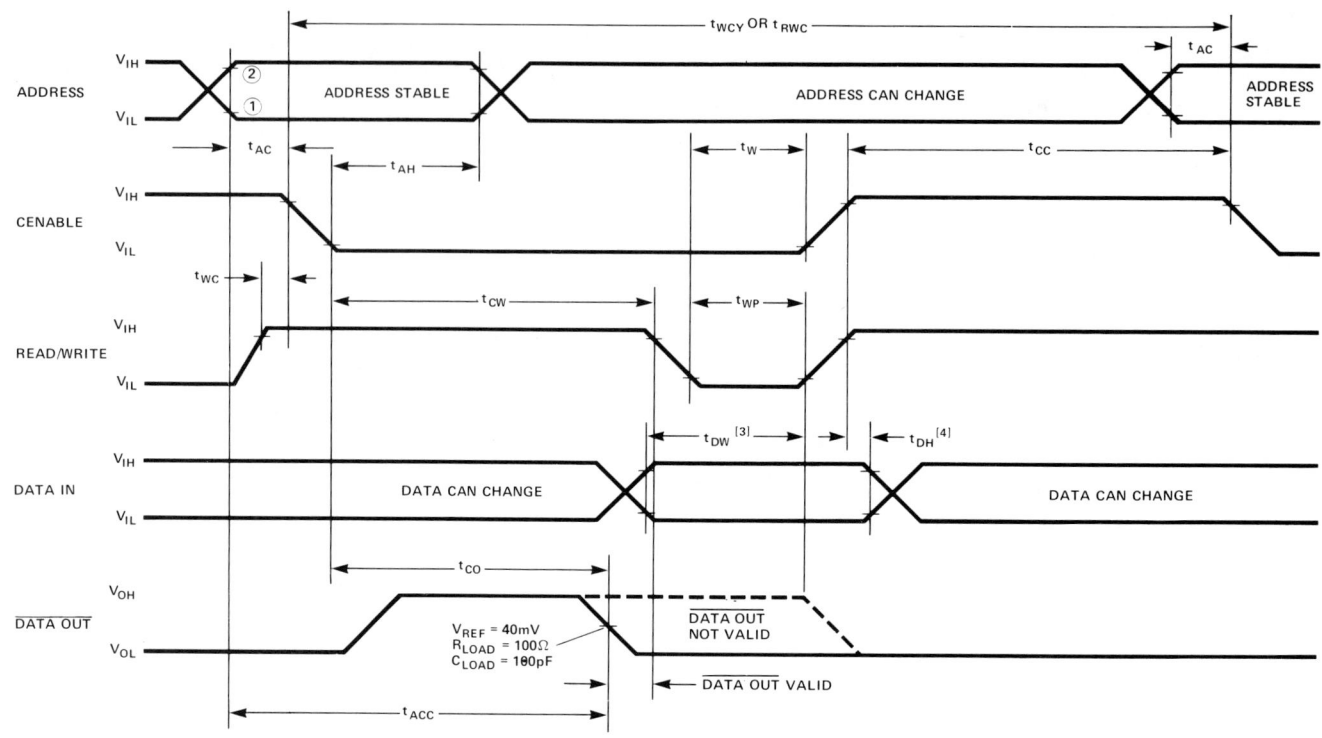

READ CYCLE

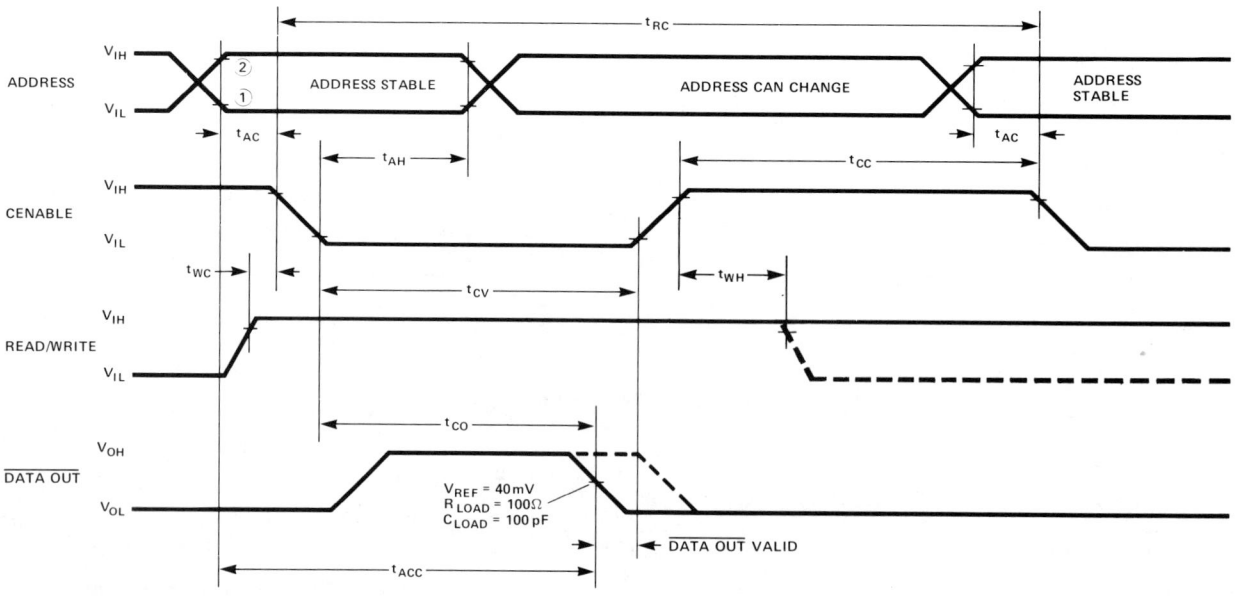

NOTES:
1. $V_{DD} + 2V$
2. $V_{SS} - 2V$ ⎬ t_T is defined as the transition between these two points.
3. t_{DW} is referenced to point 1 of the rising edge of cenable or Read/Write, whichever occurs first.
4. t_{DH} is referenced to point 2 of the rising edge of Read/Write.

Supply Current vs Temperature

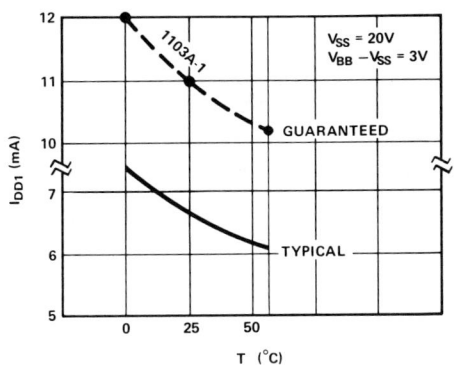

Typical Characteristics

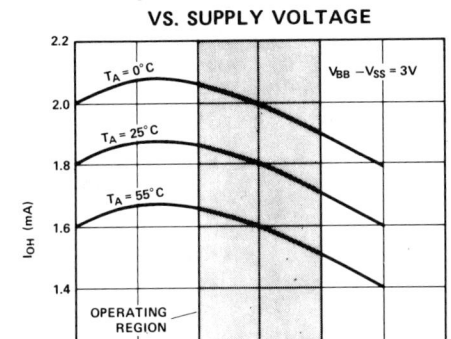

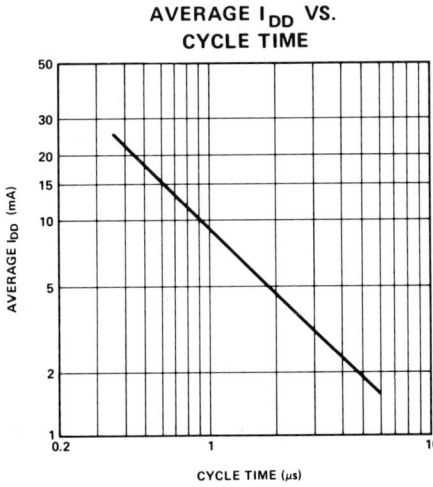

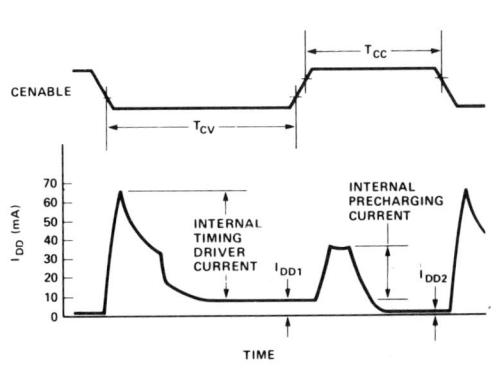

1103A-2
1024 x 1 BIT DYNAMIC RAM

- **High Speed 1103A — Access Time — 145 ns / Cycle Time — 400 ns**

- **✱ No Precharge Required -- Critical Precharge Timing is Eliminated**

- **Low Standby Power Dissipation -- 0.2 µW/Bit Typical**

- **Address Registers Incorporated on the Chip**

- **Simple Memory Expansion -- Chip Enable Input Lead**

- **Inputs Protected -- All Inputs Have Protection Against Static Charge**

- **Standard 18-Pin Dual In-Line Packages**

The Intel® 1130A-2 is a high speed 1024 bit dynamic random access memory and is the 400 ns cycle time version of the standard 1103A. It is designed primarily for main memory applications where high performance, low cost, and large bit storage are important design objectives.

1103A-2 systems may be simplified due to the elimination of the precharge clock, its associated circuitry, and critical overlap timing. Only one external clock, CENABLE, is required.

Information stored in the memory is non-destructively read. Refreshing of all 1024 bits is accomplished in 32 read cycles (addressing A_0 to A_4) and is required every one millisecond. The memory may be used in a low power standby mode by having cenable at V_{SS} potential.

The 1103A-2 is fabricated with silicon gate technology. This low threshold technology allows the design and production of higher performance MOS circuits and provides a higher functional density on a monolithic chip than conventional MOS technologies.

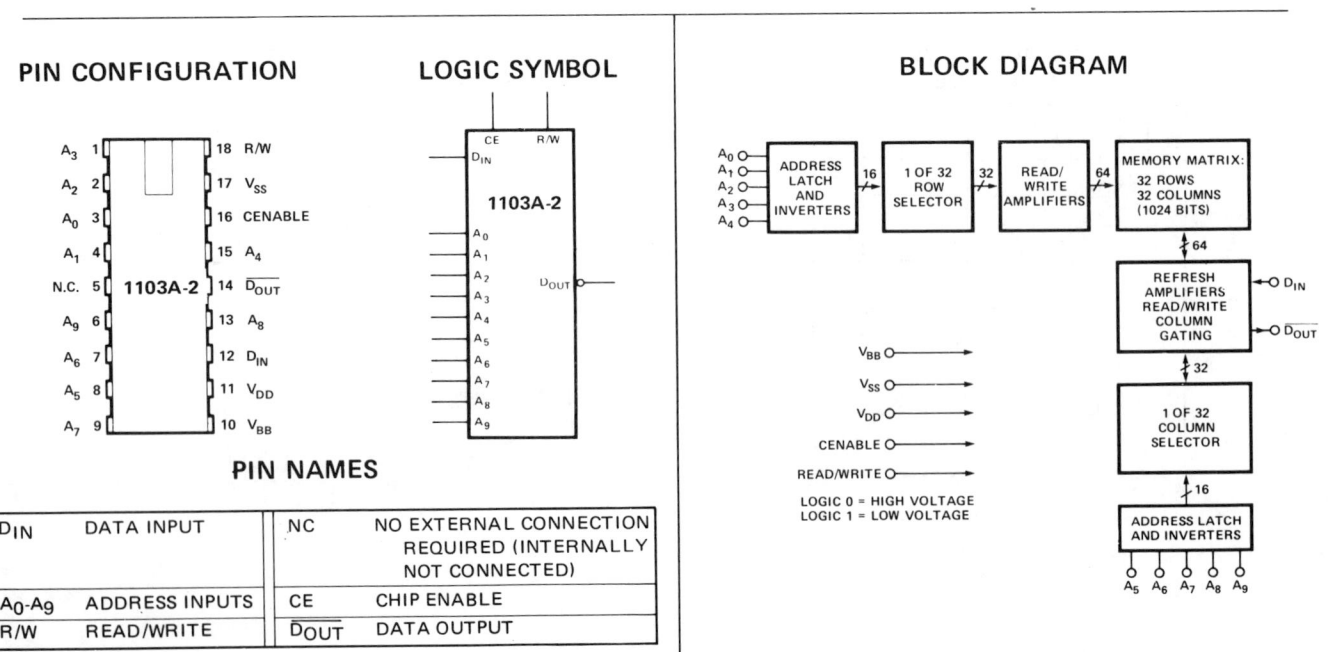

PIN CONFIGURATION   LOGIC SYMBOL   BLOCK DIAGRAM

PIN NAMES

D_{IN}	DATA INPUT	NC	NO EXTERNAL CONNECTION REQUIRED (INTERNALLY NOT CONNECTED)
A_0-A_9	ADDRESS INPUTS	CE	CHIP ENABLE
R/W	READ/WRITE	$\overline{D}_{OUT}$	DATA OUTPUT

Absolute Maximum Ratings*

Temperature Under Bias	0°C to 70°C
Storage Temperature	−65°C to +150°C
All Input or Output Voltages with Respect to the most Positive Supply Voltage, V_{BB}	−25V to 0.3V
Supply Voltages V_{DD} and V_{SS} with Respect to V_{BB}	−25V to 0.3V
Power Dissipation	1.0W

*COMMENT:
Stresses above those listed under "Absolute Maximum Ratings" may cause permanent damage to the device. This is a stress rating only and functional operation of the device at these or any other conditions above those indicated in the operational sections of this specification is not implied. Exposure to absolute maximum rating conditions for extended periods may affect device reliability.

D. C. and Operating Characteristics

$T_A = 0°C$ to $+55°C$, V_{SS}[1] $= 19V \pm 5\%$, $(V_{BB} - V_{SS})$[2] $= 3V$ to $4V$, $V_{DD} = 0V$ unless otherwise specified.

Symbol	Test	Min.	Typ.	Max.	Unit	Conditions
I_{LI}	Input Load Current (All Input Pins)			10	µA	$V_{IN} = 0V$
I_{LO}	Output Leakage Current			10	µA	$V_{OUT} = 0V$
I_{BB}	V_{BB} Supply Current			100	µA	
I_{DD1}	Supply Current During Cenable On		7	11	mA	Cenable = 0V; $T_A = 25°C$
I_{DD2}	Supply Current During Cenable Off		0.01	0.5	mA	Cenable = V_{SS}; $T_A = 25°C$
I_{DDAV}	Average Supply Current		22	30	mA	Cycle Time = 400 ns; $T_A = 25°C$
V_{IL}	Input Low Voltage	$V_{DD} - 1$		$V_{DD} + 1$	V	
V_{IH}	Input High Voltage	$V_{SS} - 1$		$V_{SS} + 1$	V	
I_{OH1}	Output High Current	1150	1800	7000	µA	$T_A = 25°C$
I_{OH2}	Output High Current	900	1600	7000	µA	$T_A = 55°C$
I_{OL}	Output Low Current		See Note Three			R_{LOAD}[4] $= 100\Omega$
V_{OH1}	Output High Voltage	115	180	700	mV	$T_A = 25°C$
V_{OH2}	Output High Voltage	90	160	700	mV	$T_A = 55°C$
V_{OL}	Output Low Voltage		See Note Three			

NOTES:
1. The V_{SS} current drain is equal to $(I_{DD} + I_{OH})$ or $(I_{DD} + I_{OL})$.
2. $(V_{BB} - V_{SS})$ supply should be applied at or before V_{SS}.
3. The output current when reading a low output is the leakage current of the 1103 plus external noise coupled into the output line from the clocks. V_{OL} equals I_{OL} across the load resistor.
4. This value of load resistance is used for measurement purposes. In applications the resistance may range from 100Ω to 1 kΩ.

A.C. Characteristics $T_A = 0°C$ to $55°C$, $V_{SS} = 19V \pm 5\%$, $(V_{BB} - V_{SS}) = 3.0V$ to $4.0V$, $V_{DD} = 0V$.

READ, WRITE, AND READ/WRITE CYCLE

Refer to page 2-23 for definitions.

Symbol	Test	Min.	Max.	Unit	Conditions
t_{REF}	Time Between Refresh		1	ms	
t_{AC}	Address to Cenable Set Up Time	0		ns	
t_{AH}	Address Hold Time	100		ns	
t_{CC}	Cenable Off Time	180		ns	

READ CYCLE

Symbol	Test	Min.	Max.	Unit	Conditions
t_{RC}	Read Cycle	360		ns	$t_T = 20$ ns
t_{CV}	Cenable on Time	140	500	ns	$C_{LOAD} = 50pF$
t_{CO}	Cenable Output Delay		125	ns	$R_{LOAD} = 100\Omega$
t_{ACC}	ADDRESS TO OUTPUT ACCESS		145	ns	$t_{ACC} = t_{AC\,MIN} + t_{CO} + t_T$, $V_{REF} = 80mV$
t_{WH}	Read/Write Hold Time	30		ns	

WRITE OR READ/WRITE CYCLE

Symbol	Test	Min.	Max.	Unit	Conditions
t_{WCY}	Write Cycle	400		ns	$t_T = 20$ ns
t_{RWC}	Read/Write Cycle	400		ns	
t_{CW}	Cenable to Read/Write Delay	140	500	ns	
t_{WP}	Read/Write Pulse Width	20		ns	
t_W	Read/Write Set Up Time	20		ns	
t_{DW}	Data Set Up Time	40		ns	
t_{DH}	Data Hold Time	10		ns	
t_{CO}	Output Delay		125	ns	$C_{LOAD} = 50pF$; $R_{LOAD} = 100\Omega$, $V_{REF} = 80mV$
t_{WC}	Read/Write to Cenable	0		ns	

CAPACITANCE[1] $T_A = 25°C$

Symbol	Test	Typ. Plastic	Plastic Pkg. Max.	Ceramic Pkg. Max.	Unit	Conditions	
C_{AD}	Address Capacitance	5	7	12	pF	$V_{IN} = V_{SS}$	
C_{CE}	Cenable Capacitance	22	25	28	pF	$V_{IN} = V_{SS}$	
C_{RW}	Read/Write Capacitance	11	15	19.5	pF	$V_{IN} = V_{SS}$	$f = 1$MHz. All unused pins are at A.C. ground.
C_{IN1}	Data Input Capacitance	4	5	7.5	pF	Cenable = 0V, $V_{IN} = V_{SS}$	
C_{IN2}	Data Input Capacitance	2	4	6.5	pF	Cenable = V_{SS}	
C_{OUT}	Data Output Capacitance	2	3	7.0	pF	$V_{IN} = V_{SS}$, $V_{OUT} = 0V$	

NOTES: 1. These parameters are periodically sampled and are not 100% tested. They are measured at worst case operating conditions.

Supply Current vs Temperature

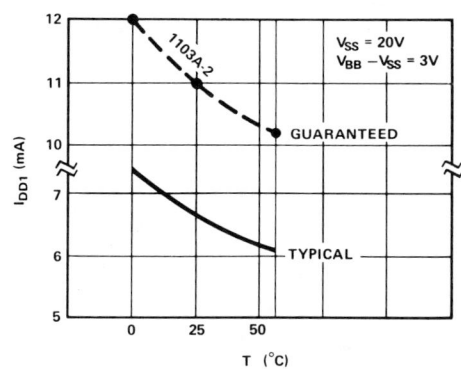

Typical Characteristics

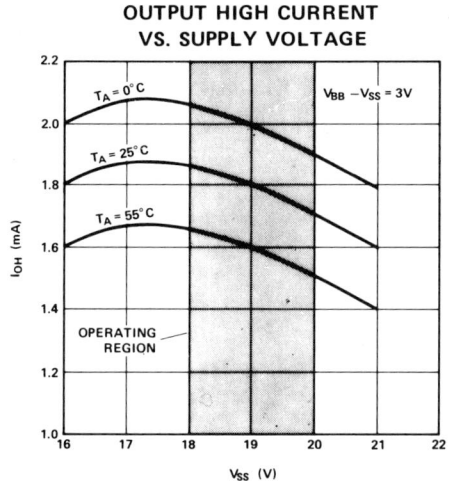

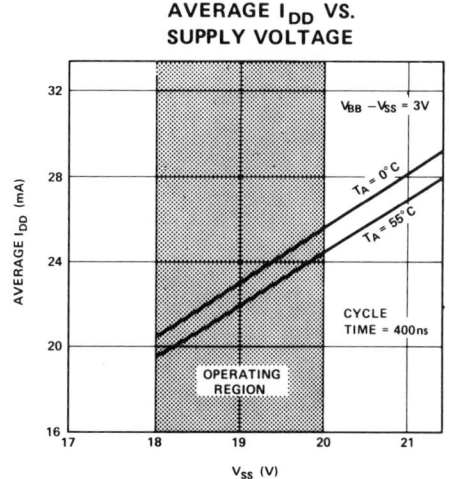

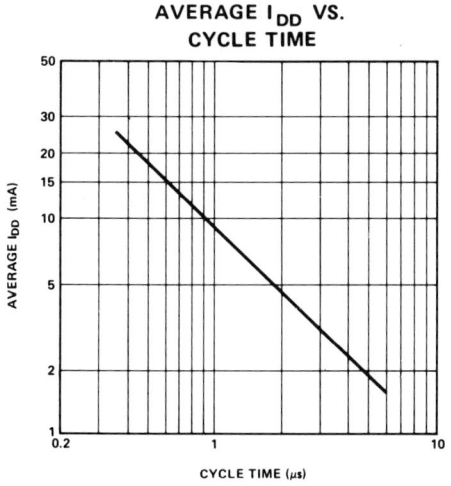

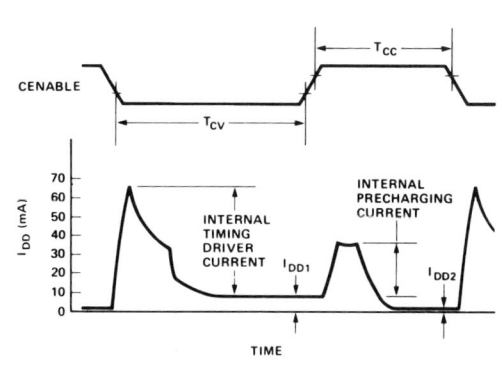

2104A
4096 x 1 BIT DYNAMIC RAM

	2104A-1	2104A-2	2104A-3	2104A-4
Max. Access Time (ns)	150	200	250	300
Read, Write Cycle (ns)	320	320	375	425
Max. IDD (mA)	35	32	30	30

- **Highest Density 4K RAM Industry Standard 16 Pin Package**
- **Low Power 4K RAM**
- **All Inputs Including Clocks TTL Compatible**
- **±10% Tolerance on All Power Supplies +12V, +5V, -5V**
- **Refresh Period: 2 ms**
- **On-Chip Latches for Addresses, Chip Select and Data In**
- **Simple Memory Expansion: Chip Select**
- **Output is Three-State, TTL Compatible; Data is Latched and Valid into Next Cycle**
- **Compatible with Intel® 2116 16K RAM**

The Intel® 2104A is a 4096 word by 1 bit MOS RAM fabricated with N-channel silicon gate technology for high performance and high functional density.

The efficient design of the 2104A allows it to be packaged in the industry standard 16 pin dual-in-line package. The 16 pin package provides the highest system bit densities and is compatible with widely available automated handling equipment.

The use of the 16 pin package is made possible by multiplexing the 12 address bits (required to address 1 of 4096 bits) into the 2104A on 6 address input pins. The two 6 bit address words are latched into the 2104A by the two TTL clocks, Row Address Strobe ($\overline{RAS}$) and Column Address Strobe ($\overline{CAS}$). Non-critical clock timing requirements allow use of the multiplexing technique while maintaining high performance.

A new unique dynamic storage cell provides high speed along with low power dissipation and wide voltage margins. The memory cell requires refreshing for data retention. Refreshing is most easily accomplished by performing a read cycle at each of the 64 row addresses every 2 milliseconds.

The 2104A is designed for page mode operation, "$\overline{RAS}$-only refreshing," and "$\overline{CAS}$-only deselection." Thus it is compatible with the Intel® 2116, 16K RAM.

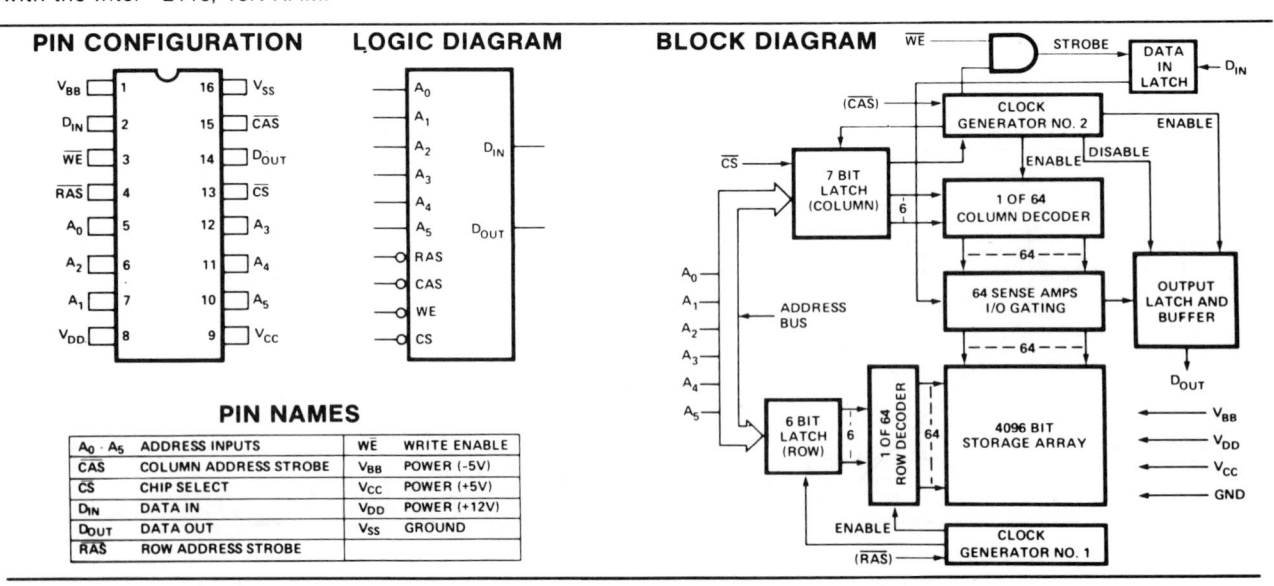

ABSOLUTE MAXIMUM RATINGS*

Ambient Temperature Under Bias ..... $-10°C$ to $+80°C$
Storage Temperature .... $-65°C$ to $+150°C$
Voltage on any Pin Relative to V_{BB}
  ($V_{SS} - V_{BB} \geq 4.5V$) .... $-0.3V$ to $+20V$
Power Dissipation 1.0W
Data Out Current .... 50 mA

*COMMENT:

Stresses above those listed under "Absolute Maximum Ratings" may cause permanent damage to the device. This is a stress rating only and functional operation of the device at these or any other conditions above those indicated in the operational sections of this specification is not implied. Exposure to absolute maximum rating conditions for extended periods may affect device reliability.

D.C. AND OPERATING CHARACTERISTICS[1]

$T_A = 0°$ to $70°C$, $V_{DD} = +12V \pm 10\%$, $V_{CC} = +5V \pm 10\%$, $V_{BB} = -5V \pm 10\%$, $V_{SS} = 0V$, unless otherwise noted.

Symbol	Parameter	Limits			Unit	Conditions	
		Min.	Typ.(2)	Max.			
I_{LI}	Input Load Current (any input)			10	µA	$V_{IN} = V_{IL\ MIN}$ to $V_{IH\ MAX}$	
$|I_{LO}|$	Output Leakage Current for High Impedance State			10	µA	Chip deselected: $\overline{RAS}$ and $\overline{CAS}$ at V_{IH}, $V_{OUT} = 0$ to $5.5V$	
I_{DD1}[3]	V_{DD} Standby Current		0.7	2	mA	$V_{DD} = 13.2V$	$\overline{CAS}$ and $\overline{RAS}$ at V_{IH}. Chip deselected prior to measurement. See Note 5.
			0.7	1.5	mA	$V_{DD} = 12.6V$	
I_{BB1}	V_{BB} Standby Current		5	50	µA	$V_{DD} = 13.2V$	
I_{DD2}[3]	Operating V_{DD} Current (Device Selected)		24	35	mA	2104A-1	$t_{CYC} = 320$ ns
			22	32	mA	2104A-2	$t_{CYC} = 320$ ns
			20	30	mA	2104A-3, 4	$t_{CYC} = 375$ ns
I_{BB2}	Operating V_{BB} Current		160	400	µA	Device Selected. Min cycle time.	
I_{CC1}[4]	V_{CC} Supply Current when Deselected			10	µA		
I_{DD3}	Operating V_{DD} Current ($\overline{RAS}$-only cycle)		12	25	mA	2104A-1, 2	$t_{CYC} = 320$ ns
			10	22	mA	2104A-3, 4	$t_{CYC} = 375$ ns
V_{IL}	Input Low Voltage (any input)	-1.0		0.8	V		
V_{IH}	Input High Voltage	2.4		7.0	V		
V_{OL}	Output Low Voltage	0.0		0.4	V	$I_{OL} = 3.2$ mA	
V_{OH}	Output High Voltage	2.4		V_{CC}	V	$I_{OH} = -5$ mA	

CAPACITANCE[6] $T_A = 25°C$

Symbol	Test	Typ.	Max.	Unit	Conditions
C_{I1}	Input Capacitance (A_0-A_5), D_{IN}, $\overline{CS}$	3	7	pF	$V_{IN} = V_{SS}$
C_{I2}	Input Capacitance $\overline{RAS}$, $\overline{WRITE}$	3	7	pF	$V_{IN} = V_{SS}$
C_O	Output Capacitance (D_{OUT})	4	7	pF	$V_{OUT} = 0V$
C_{I3}	Input Capacitance $\overline{CAS}$	6	7	pF	$V_{IN} = V_{SS}$

Notes:
1. All voltages referenced to V_{SS}. The only requirement for the sequence of applying voltages to the device is that V_{DD}, V_{CC}, and V_{SS} should never be 0.3V or more negative than V_{BB}. After the application of supply voltages or after extended periods of operation without clocks, the device must perform a minimum of one initialization cycle (any valid memory cycles containing both $\overline{RAS}$ and $\overline{CAS}$) prior to normal operation.
2. Typical values are for $T_A = 25°C$ and nominal power supply voltages.
3. The I_{DD} current flows to V_{SS}.
4. When chip is selected V_{CC} supply current is dependent on output loading. V_{CC} is connected to output buffer only.
5. The chip is deselected; i.e., output is brought to high impedance state by $\overline{CAS}$-only cycle or by a read cycle with $\overline{CS}$ at V_{IH}.
6. Capacitance measured with Boonton Meter.

DYNAMIC RAMS

A.C. CHARACTERISTICS[1]

$T_A = 0°C$ to $70°C$, $V_{DD} = 12V \pm 10\%$, $V_{CC} = 5V \pm 10\%$, $V_{BB} = -5V \pm 10\%$, $V_{SS} = 0V$, unless otherwise noted.

READ, WRITE, AND READ MODIFY WRITE CYCLES

Symbol	Parameter	2104A-1 Min.	2104A-1 Max.	2104A-2 Min.	2104A-2 Max.	2104A-3 Min.	2104A-3 Max.	2104A-4 Min.	2104A-4 Max.	Unit
t_{REF}	Time Between Refresh		2		2		2		2	ms
t_{RP}	$\overline{RAS}$ Precharge Time	100		115		115		125		ns
t_{CP}	$\overline{CAS}$ Precharge Time	60		80		110		110		ns
t_{RCL}[2]	$\overline{RAS}$ to $\overline{CAS}$ Leading Edge Lead Time	20	50	25	70	35	110	80	135	ns
t_{CRP}	$\overline{CAS}$ to $\overline{RAS}$ Precharge Time	0		0		0		0		ns
t_{RSH}	$\overline{RAS}$ Hold Time	100		130		140		165		ns
t_{CSH}	$\overline{CAS}$ Hold Time	150		200		250		300		ns
t_{AR}	$\overline{RAS}$ to Address or $\overline{CS}$ Hold Time	95		120		160		215		ns
t_{ASR}	Row Address Set-Up Time	0		0		0		0		ns
t_{ASC}	Column Address or $\overline{CS}$ Set-Up Time	-5		0		0		0		ns
t_{RAH}	Row Address Hold Time	20		25		35		80		ns
t_{CAH}	Column Address or $\overline{CS}$ Hold Time	45		50		50		80		ns
t_T	Rise or Fall Time		50		50		50		50	ns
t_{OFF}	Output Buffer Turn-Off Delay	0	50	0	60	0	60	0	80	ns
t_{CAC}[3]	Access Time From $\overline{CAS}$		100		130		140		165	ns
t_{RAC}[3]	Access Time From $\overline{RAS}$		150		200		250		300	ns

READ CYCLE

Symbol	Parameter	2104A-1 Min.	2104A-1 Max.	2104A-2 Min.	2104A-2 Max.	2104A-3 Min.	2104A-3 Max.	2104A-4 Min.	2104A-4 Max.	Unit
t_{RC}	Random Read or Write Cycle Time	320		320		375		425		ns
t_{RAS}	$\overline{RAS}$ Pulse Width	150	32000	200	32000	250	32000	300	32000	ns
t_{CAS}	$\overline{CAS}$ Pulse Width	100		130		140		165		ns
t_{RCS}	Read Command Set-Up Time	0		0		0		0		ns
t_{RCH}	Read Command Hold Time	0		0		0		0		ns
t_{DOH}	Data Out Hold Time	32		32		32		32		μs

WRITE CYCLE[4]

Symbol	Parameter	2104A-1 Min.	2104A-1 Max.	2104A-2 Min.	2104A-2 Max.	2104A-3 Min.	2104A-3 Max.	2104A-4 Min.	2104A-4 Max.	Unit
t_{RC}	Random Read or Write Cycle Time	320		320		375		425		ns
t_{RAS}	$\overline{RAS}$ Pulse Width	150	32000	200	32000	250	32000	300	32000	ns
t_{CAS}	$\overline{CAS}$ Pulse Width	100		130		140		165		ns
t_{WCS}	Write Command Set-Up Time	0		0		0		0		ns
t_{WCH}	Write Command Hold Time	55		75		75		80		ns
t_{WCR}	Write Command Hold Time Referenced to $\overline{RAS}$	105		145		185		215		ns
t_{WP}	Write Command Pulse Width	45		55		75		80		ns
t_{RWL}	Write Command to $\overline{RAS}$ Lead Time	100		130		140		150		ns
t_{CWL}	Write Command to $\overline{CAS}$ Lead Time	100		130		140		150		ns
t_{DS}	Data-In Set-Up Time	0		0		0		0		ns
t_{DH}	Data-In Hold Time	55		75		75		80		ns
t_{DHR}	Data-In Hold Time Referenced to $\overline{RAS}$	105		145		185		215		ns

Notes:
1. All voltages referenced to V_{SS}. Minimum timings do not allow for t_T or skews.
2. $\overline{CAS}$ must remain at V_{IH} a minimum of t_{RCL} MIN after RAS switches to V_{IL}. To achieve the minimum guaranteed access time (t_{RAC}), $\overline{CAS}$ must switch to V_{IL} at or before t_{RCL} of $t_{RAC} - t_T - t_{CAC}$ as described in the Applications Information on page 6. t_{RCL} MAX is given for reference only as $t_{RAC} - t_{CAC}$.
3. Load = 2 TTL and 100 pF. See Applications Information.
4. In a write cycle D_{OUT} latch will contain data written into cell. In a read-modify-write cycle D_{OUT} latch will contain data read from cell. If $\overline{WE}$ goes low after $\overline{CAS}$ and prior to t_{CAC}, D_{OUT} is indeterminate.

WAVEFORMS

READ CYCLE

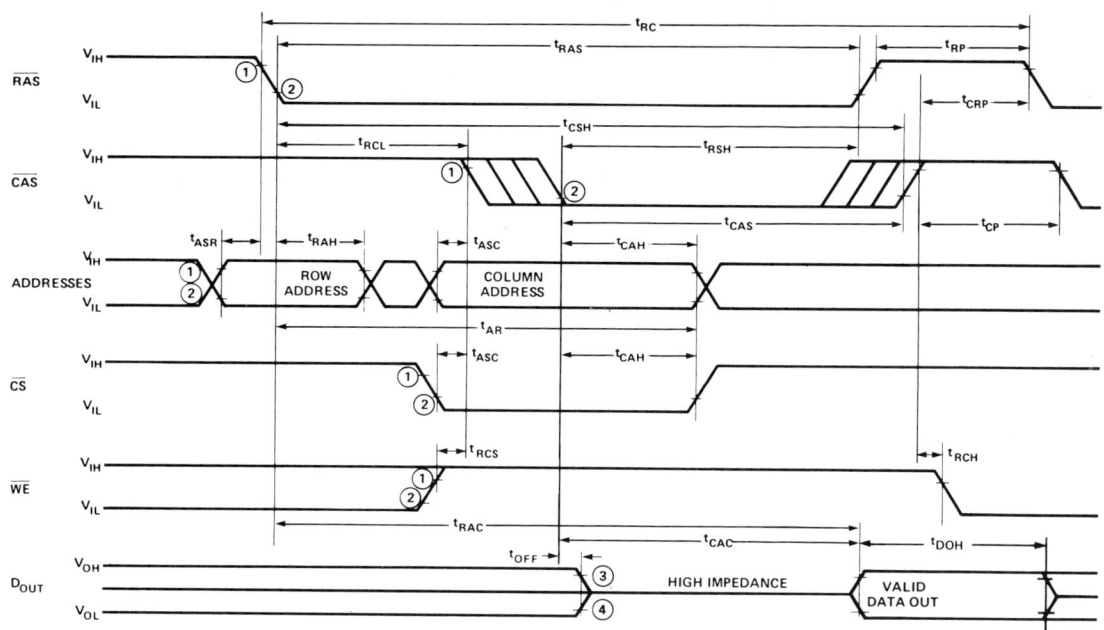

WRITE CYCLE

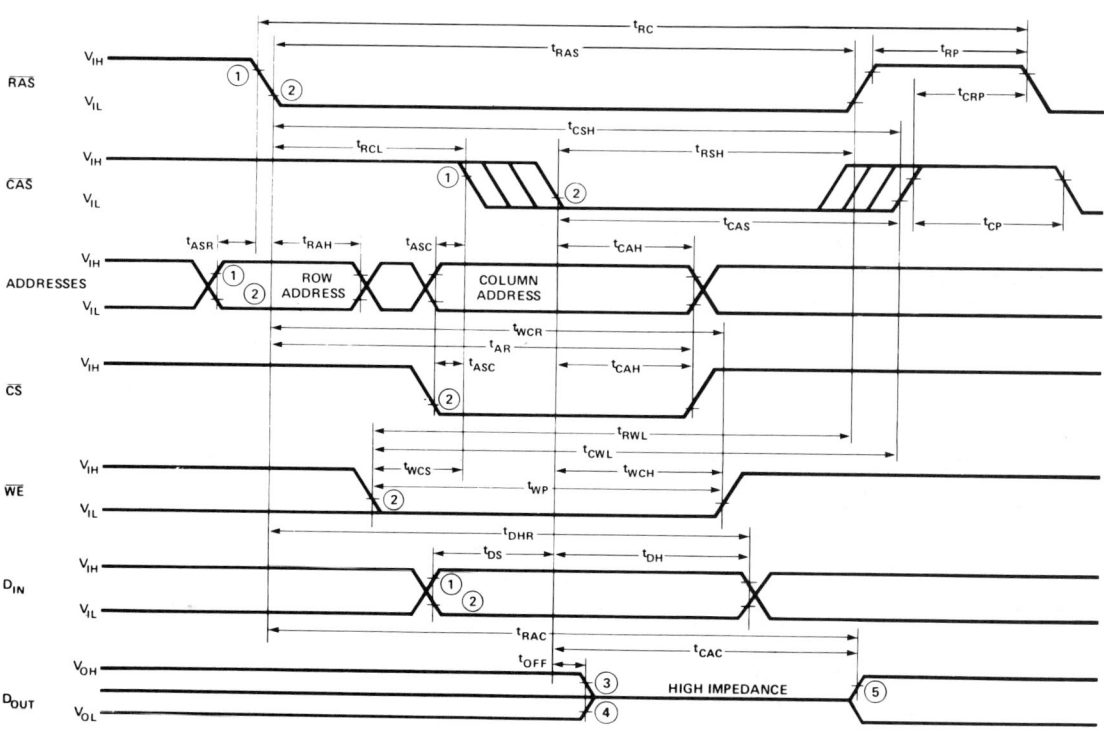

DYNAMIC RAMS

A.C. CHARACTERISTICS[1]

$T_A = 0°$ to $70°C$, $V_{DD}=12V \pm10\%$, $V_{CC}=5V \pm10\%$, $V_{BB}=-5V \pm10\%$, $V_{SS}=0V$, unless otherwise noted.

READ-MODIFY-WRITE CYCLE

Symbol	Parameter	2104A-1 Min.	2104A-1 Max.	2104A-2 Min.	2104A-2 Max.	2104A-3 Min.	2104A-3 Max.	2104A-4 Min.	2104A-4 Max.	Unit
t_{RWC}	Read Modify Write Cycle Time[2]	350		445		505		575		ns
t_{CRW}	RMW Cycle $\overline{CAS}$ Width	200		260		280		315		ns
t_{RRW}	RMW Cycle $\overline{RAS}$ Width	250		330		390		450		ns
t_{RWL}	RMW Cycle $\overline{RAS}$ Lead Time	100		130		140		150		ns
t_{CWH}	RMW Cycle $\overline{CAS}$ Hold Time	250		330		390		450		ns
t_{CWL}	Write Command to $\overline{CAS}$ Lead Time	100		130		140		150		ns
t_{WP}	Write Command Pulse Width	45		55		75		80		ns
t_{RCS}	Read Command Set-Up Time	0		0		0		0		ns
t_{MOD}	Modify Time	0	10	0	10	0	10	0	10	µs
t_{DS}	Data-In Set-Up Time	0		0		0		0		ns
t_{DH}	Data-In Hold Time	55		75		75		80		ns

Notes: 1. All voltages referenced to V_{SS}.
2. The minimum cycle timing does not allow for t_T or skews.

WAVEFORMS

READ-MODIFY-WRITE CYCLE

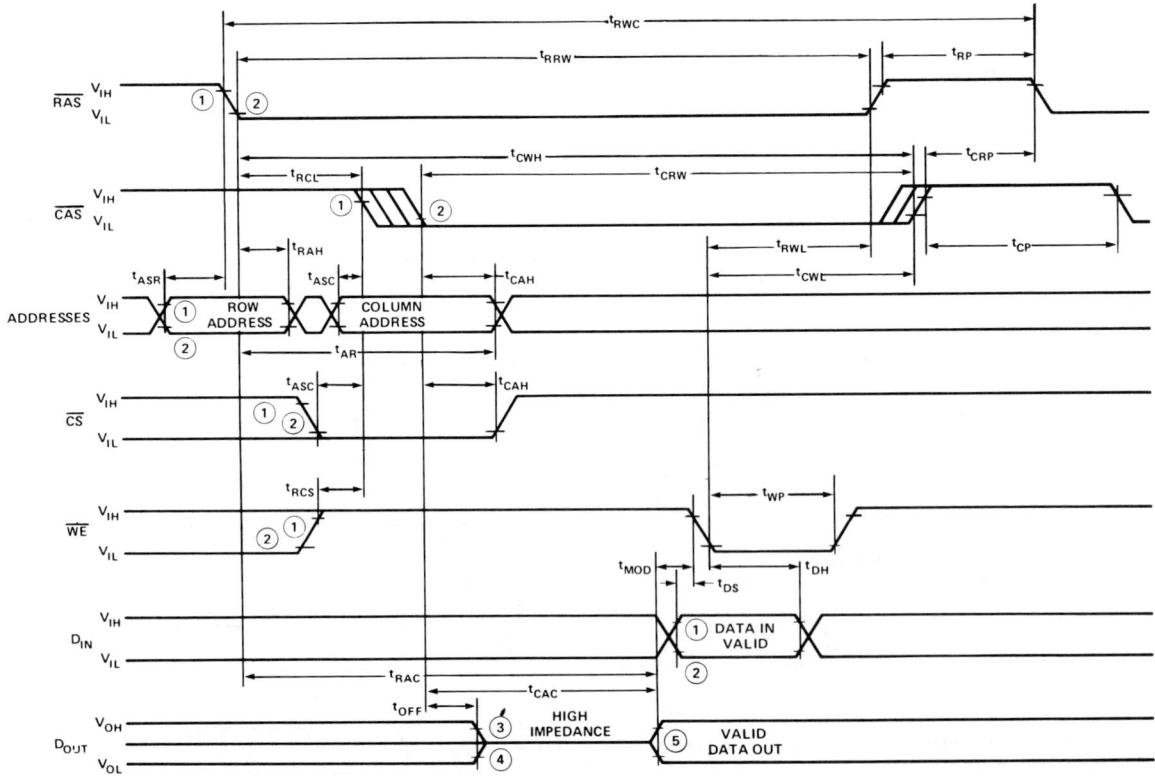

Notes: 1,2. V_{IHMIN} and V_{ILMAX} are reference levels for measuring timing of input signals.
3,4. V_{OHMIN} and V_{OLMAX} are reference levels for measuring timing of D_{OUT}.
5. In a write cycle D_{OUT} latch will contain data written into cell. In a read-modify-write cycle D_{OUT} latch will contain data read from cell. If $\overline{WE}$ goes low after $\overline{CAS}$ and prior to t_{CAC}, D_{OUT} is indeterminate.

TYPICAL CHARACTERISTICS

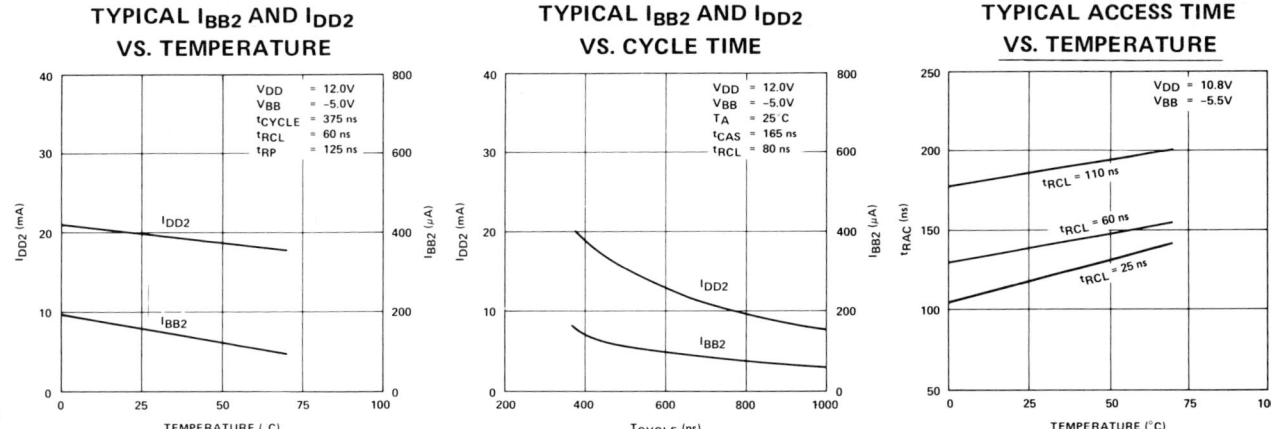

APPLICATIONS

ADDRESSING

Two externally applied negative going TTL clocks, Row Address Strobe ($\overline{RAS}$), and Column Address Strobe ($\overline{CAS}$), are used to strobe the two sets of 6 addresses into internal address buffer registers. The first clock, $\overline{RAS}$, strobes in the six low order addresses (A_0-A_5) which selects one of 64 rows and begins the timing which enables the column sense amplifiers. The second clock, $\overline{CAS}$, strobes in the six high order addresses (A_6-A_{11}) to select one of 64 column sense amplifiers and Chip Select ($\overline{CS}$) which enables the data out buffer.

An address map of the 2104A is shown below. Address "0" corresponds to all addresses at V_{IL}. All addresses are sequentially located on the chip.

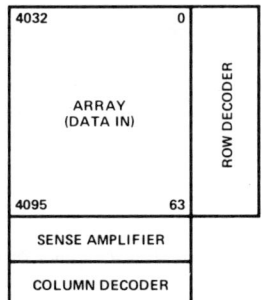

2104A Address Map

DATA CYCLES/TIMING

A memory cycle begins with addresses stable and a negative transition of $\overline{RAS}$. See the waveforms on page 4. It is not necessary to know whether a Read or Write cycle is to be performed until $\overline{CAS}$ becomes valid.

Note that Chip Select ($\overline{CS}$) does not have to be valid until the second clock, $\overline{CAS}$. It is, therefore, possible to start a memory cycle before it is known which device must be selected. This can result in a significant improvement in system access time since the decode time for chip select does not enter into the calculation for access time.

Both the $\overline{RAS}$ and $\overline{CAS}$ clocks are TTL compatible and do not require level shifting and driving at high voltage MOS levels. Buffers internal to the 2104A convert the TTL level signals to MOS levels inside the device. Therefore, the delay associated with external TTL-MOS level converters is not added to the 2104A system access time.

READ CYCLE

A Read cycle is performed by maintaining Write Enable ($\overline{WE}$) high during $\overline{CAS}$. The output pin of a selected device will unconditionally go to a high impedance state immediately following the leading edge of $\overline{CAS}$ and remain in this state until valid data appears at the output at access time. The selected output data is internally latched and will remain valid until a subsequent $\overline{CAS}$ is given to the device by a Read, Write, Read-Modify-Write, $\overline{CAS}$ only or Refresh cycle. Data-out goes to a high impedance state for all non-selected devices.

Device access time, t_{ACC}, is the longer of two calculated intervals:

1. $t_{ACC} = t_{RAC}$ OR   2. $t_{ACC} = t_{RCL} + t_T + t_{CAC}$

Access time from $\overline{RAS}$, t_{RAC}, and access time from $\overline{CAS}$, t_{CAC}, are device parameters. Row to column address strobe lead time, t_{RCL}, and transition time, t_T, are system dependent timing parameters. For example, substituting the device parameters of the S1648 and assuming a TTL level transition time of 5 ns yields:

3. $t_{ACC} = t_{RAC} = 300$ns for 80 nsec $\leq t_{RCL} \leq 130$nsec

OR

4. $t_{ACC} = t_{RCL} + t_T + t_{CAC} = t_{RCL} + 170$ns for $t_{RCL} > 130$ns.

Note that if 80 nsec $\leq t_{RCL} \leq$ 130 nsec, device access time is determined by equation 3 and is equal to t_{RAC}. If $t_{RCL} >$ 130 nsec, access time is determined by equation 4. This 50ns interval (shown in the t_{RCL} inequality in equation 3) in which the falling edge of $\overline{CAS}$ can occur without affecting access time is provided to allow for system timing skew in the generation of $\overline{CAS}$. This allowance for a t_{RCL} skew is designed in at the device level to allow minimum access times to be achieved in practical system designs.

WRITE CYCLE

A Write Cycle is generally performed by bringing Write Enable ($\overline{WE}$) low before $\overline{CAS}$. D_{OUT} will be the data written into the cell addressed. If $\overline{WE}$ goes low after $\overline{CAS}$ and prior to t_{CAC}, D_{OUT} will be indeterminate.

READ-MODIFY-WRITE CYCLE

A Read-Modify-Write Cycle is performed by bringing Write Enable ($\overline{WE}$) low after access time, t_{RAC}, with $\overline{RAS}$ and $\overline{CAS}$ low. Data in must be valid at or before the falling edge of $\overline{WE}$. In a read-modify-write cycle D_{OUT} is data read and does not change during the modify-write portion of the cycle.

$\overline{CAS}$ ONLY (DESELECT) CYCLE

In some applications, it is desirable to be able to deselect all memory devices without running a regular memory cycle. This may be accomplished with the 2104A by performing a $\overline{CAS}$-Only Cycle. Receipt of a $\overline{CAS}$ without $\overline{RAS}$ deselects the 2104A and forces the Data Output to the high-impedance state. This places the 2104A in its lowest power, standby condition. I_{DD} will be about twice I_{DD1} for the first cycle of $\overline{CAS}$-only deselection and I_{DD1} for any additional $\overline{CAS}$-only cycles. The cycle timing and $\overline{CAS}$ timing should be just as if a normal $\overline{RAS}/\overline{CAS}$ cycle was being performed.

CHIP SELECTION/DESELECTION

The 2104A is selected by driving $\overline{CS}$ low during a Read, Write, or Read-Modify-Write cycle. A device is deselected by 1) driving $\overline{CS}$ high during a Read, Write, or Read-Modify-Write cycle or 2) performing a $\overline{CAS}$ Only cycle independent of the state of $\overline{CS}$.

REFRESH CYCLES

Each of the 64 rows internal to the 2104A must be refreshed every 2 msec to maintain data. Any data cycle (Read, Write, Read-Modify-Write) refreshes the entire selected row (defined by the low order row addresses). The refresh operation is independent of the state of chip select. It is evident, of course, that if a Write or Read-Modify-Write cycle is used to refresh a row, the device should be deselected ($\overline{CS}$ high) if it is desired not to change the state of the selected cell.

$\overline{RAS}/\overline{CAS}$ TIMING

The device clocks, $\overline{RAS}$ and $\overline{CAS}$, control operation of the 2104A. The timing of each clock and the timing relationships of the two clocks must be understood by the user in order to obtain maximum performance in a memory system.

The $\overline{RAS}$ and $\overline{CAS}$ have minimum pulse widths as defined by t_{RAS} and t_{CAS} respectively. These minimum pulse widths must be maintained for proper device operation and data integrity. A cycle, once begun by driving $\overline{RAS}$ and/or $\overline{CAS}$ low must not be ended or aborted prior to fulfilling the minimum clock signal pulse width(s). A new cycle must not begin until the minimum precharge time, t_{RP}, has been met.

PAGE MODE OPERATION

The 2104A is designed for page mode operation and is presently being characterized for that mode. Specifications will be available at a later date.

POWER SUPPLY

Typical power supply current waveforms versus time are shown below for both a $\overline{RAS}/\overline{CAS}$ cycle and a $\overline{CAS}$ only cycle. I_{DD} and I_{BB} current surges at $\overline{RAS}$ and $\overline{CAS}$ edges make adequate decoupling of these supplies important. Due to the high frequency noise component content of the current waveforms, the decoupling capacitors should be low inductance, ceramic units selected for their high frequency performance.

It is recommended that a 0.1 μF ceramic capacitor be connected between V_{DD} and V_{SS} at every other device in the memory array. A 0.1 μF ceramic capacitor should also be connected between V_{BB} and V_{SS} at every other device (preferably the alternate devices to the V_{DD} decoupling). For each 16 devices, a 10 μF tantalum or equivalent capacitor should be connected between V_{DD} and V_{SS} near the array. An equal or slightly smaller bulk capacitor is also recommended between V_{BB} and V_{SS} for every 32 devices.

A 0.01 μF ceramic capacitor is recommended between V_{CC} and V_{SS} at every eighth device to prevent noise coupling to the V_{CC} line which may affect the TTL peripheral logic in the system.

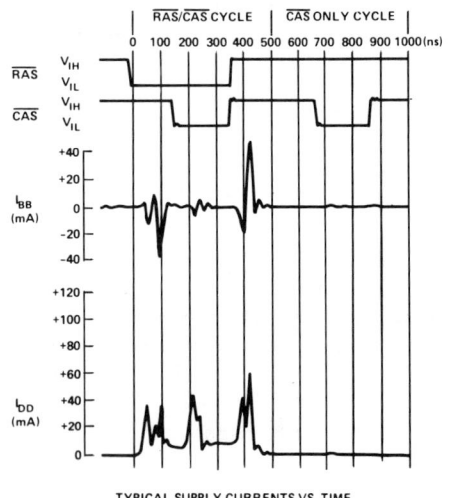

TYPICAL SUPPLY CURRENTS VS. TIME

Due to the high frequency characteristics of the current waveforms, the inductance of the power supply distribution system on the array board should be minimized. It is recommended that the V_{DD}, V_{BB}, and V_{SS} supply lines be gridded both horizontally and vertically at each device in the array. This technique allows use of double-sided circuit boards with noise performance equal to or better than multi-layered circuit boards.

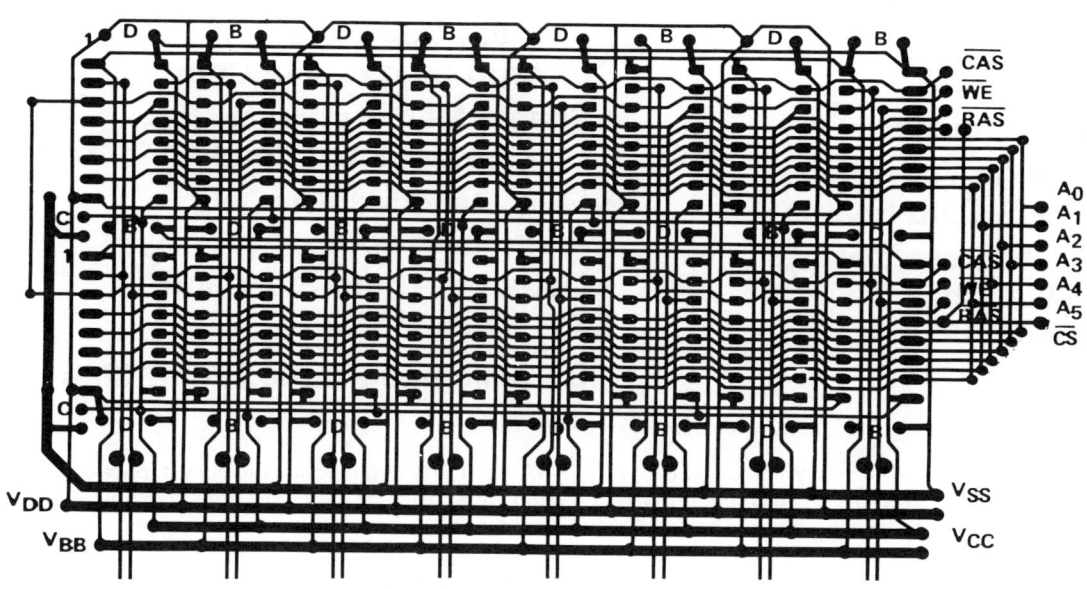

DECOUPLING CAPACITORS
D = 0.1 μF to V_{DD} TO V_{SS}
B = 0.1 μF V_{BB} TO V_{SS}
C = 0.01 μF V_{CC} TO V_{SS}

2107A
4096 x 1 BIT DYNAMIC RAM

Product	2107A-1	2107A	2107A-4	2107A-5
Access Time	280 ns	300 ns	350 ns	420 ns

- **Low Cost Per Bit**
- **Low Standby Power**
- **Easy System Interface**
- **Only One High Voltage Input Signal — Chip Enable**
- **Low Level Address, Data, Write Enable, Chip Select Inputs**
- **Address Registers Incorporated on the Chip**
- **Simple Memory Expansion: Chip Select Input Lead**
- **Fully Decoded: On Chip Address Decode**
- **Output is Three State and TTL Compatible**
- **Ceramic and Plastic 22-Pin DIPs**

The Intel 2107A is a 4096 word by 1 bit dynamic n-channel MOS RAM. It was designed for memory applications where very low cost and large bit storage are important design objectives. The 2107A uses dynamic circuitry which reduces the operation and standby power dissipation.

Reading information from the memory is non-destructive. Refreshing is accomplished by performing one read cycle on each of the 64 row addresses. Each row address must be refreshed every two milliseconds. The memory is refreshed whether Chip Select is a logic one or a logic zero.

The 2107A is fabricated with n-channel silicon gate technology. This technology allows the design and production of high performance, easy to use MOS circuits and provides a higher functional density on a monolithic chip than other MOS technologies.

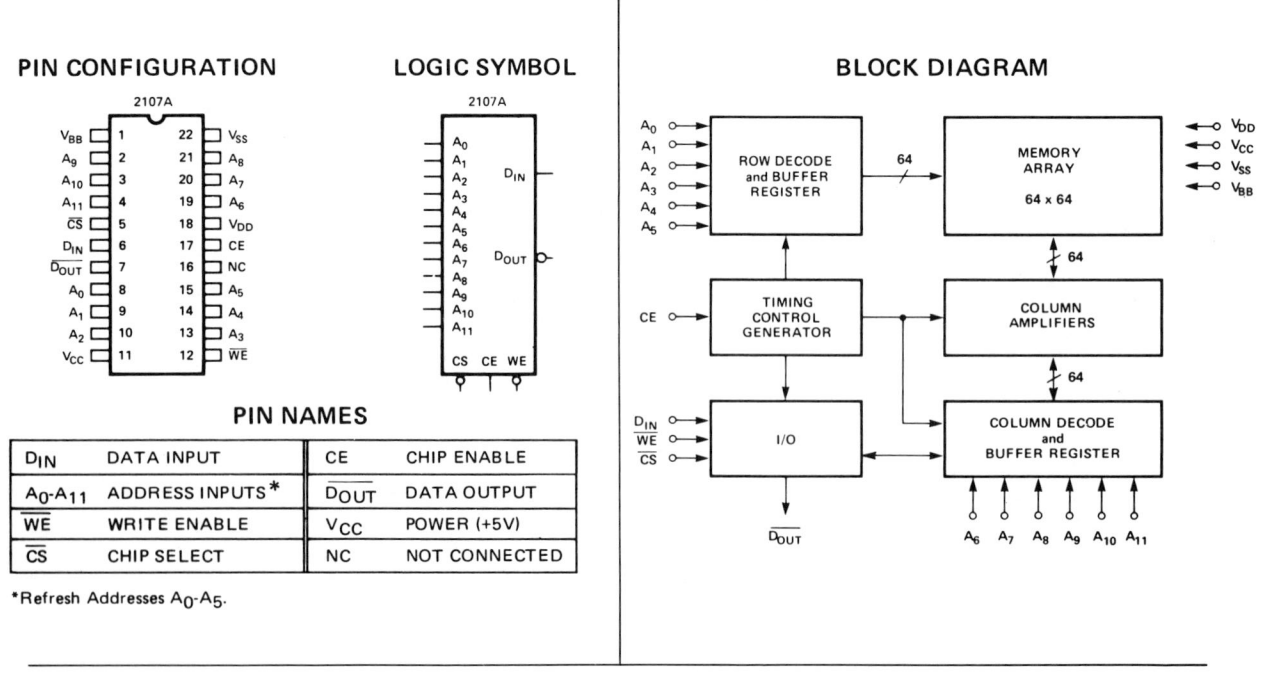

*Refresh Addresses A_0-A_5.

212

Absolute Maximum Ratings*

Temperature Under Bias	0°C to 70°C
Storage Temperature	−65°C to +150°C
All Input or Output Voltages with Respect to the most Negative Supply Voltage, V_{BB}	+25V to −0.3V
Supply Voltages V_{DD}, V_{CC}, and V_{SS} with Respect to V_{BB}	+20V to −0.3V
Power Dissipation	1.0W

*COMMENT:
Stresses above those listed under "Absolute Maximum Ratings" may cause permanent damage to the device. This is a stress rating only and functional operation of the device at these or any other conditions above those indicated in the operational sections of this specification is not implied. Exposure to absolute maximum rating conditions for extended periods may affect device reliability.

D.C. and Operating Characteristics

T_A = 0°C to 70°C, V_{DD} = +12V ± 5%, V_{CC} = +5V ± 5%, V_{BB} [1] = −5V ± 5%, V_{SS} = 0V, unless otherwise notes.

Symbol	Parameter	Min.	Typ.[2]	Max.	Unit	Conditions
I_{LI}	Input Load Current (all inputs except CE)		.01	10	μA	V_{IN} = $V_{IL\ MIN}$ to $V_{IH\ MAX}$
I_{LC}	Input Load Current		.01	10	μA	V_{IN} = $V_{IL\ MIN}$ to $V_{IH\ MAX}$
$|I_{LO}|$	Output Leakage Current for high impedance state		.01	10	μA	CE = −1V to +.8V or $\overline{CS}$ = 3.5V, V_O = 0V to 5.25V
I_{DD1}	V_{DD} Supply Current during CE off[3]		.1	100	μA	CE = −1V to +.8V
I_{DD2}	V_{DD} Supply Current during CE on[5]		14	22	mA	CE = V_{IHC}, T_A = 25°C
$I_{DD\ AV}$	Average V_{DD} Supply Current	(See Table 1)				T_A = 25°C, Fig. 1,3
I_{CC1}	V_{CC} Supply Current during CE off		.01	10	μA	CE = −1V to +.8V
I_{CC2}	V_{CC} Supply Current during CE on		5	10	mA	CE = V_{IHC}, T_A = 25°C
$I_{CC\ AV}$	Average V_{CC} Supply Current	(See Table 1)				T_A = 25°C, Fig. 2,4
I_{BB}	V_{BB} Supply Current		1	100	μA	
V_{IL}	Input Low Voltage[4]	−1.0		0.8	V	
V_{IH}	Input High Voltage[4]	3.5		V_{CC} +1	V	
V_{ILC}	CE Input Low Voltage[4]	−1.0		+1.0	V	
V_{IHC}	CE Input High Voltage	V_{DD} −1		V_{DD} +1	V	
V_{OL}	Output Low Voltage[4]	0.0		0.45	V	I_{OL} = 1.7mA, Fig. 6
V_{OH}	Output High Voltage[4]	2.4		V_{CC}	V	I_{OH} = −100μA, Fig. 5

NOTES:
1. The only requirement for the sequence of applying voltage to the device is that V_{DD}, V_{CC}, and V_{SS} should never be .3V or more negative than V_{BB}.
2. Typical values are for T_A = 25°C and nominal power supply voltages.
3. The I_{DD} and I_{CC} currents flow to V_{SS}. The I_{BB} current is the sum of all leakage currents.
4. Referenced to V_{SS} unless otherwise noted.
5. For 2107A-4 and 2107A-5 I_{DD2} is 25mA max.

A.C. Characteristics

$T_A = 0°C$ to $70°C$, $V_{DD} = 12V \pm 5\%$, $V_{CC} = 5V \pm 5\%$, $V_{BB} = -5V \pm 5\%$,
READ, WRITE, AND READ MODIFY/WRITE CYCLE $V_{SS} = 0V$, unless otherwise noted.

Symbol	Parameters	2107A Min.	2107A Max.	2107A-1 Min.	2107A-1 Max.	2107A-4 Min.	2107A-4 Max.	2107A-5 Min.	2107A-5 Max.	Units
t_{REF} [1]	Time Between Refresh		2		1		2		2	ms
t_{AC}	Address to CE Set Up Time	0		0		0		0		ns
t_{AH}	Address Hold Time	100		100		100		100		ns
t_{CC}	CE Off Time	180		100		200		250		ns
t_T	CE Transition Time		50		50		50		50	ns
t_{CF}	CE Off to Output High Impedance State	0		0		0		0		ns

READ CYCLE

Symbol	Parameters	Min.	Max.	Min.	Max.	Min.	Max.	Min.	Max.	Units
t_{RCY} [2]	Read Cycle Time	500		420		570		690		ns
t_{CER}	CE On Time During Read	280	3000	280	3000	330	3000	400	300	ns
t_{CO}	CE Output Delay		280		260		330		400	ns
t_{ACC} [3]	Address to Output Access		300		280		350		420	ns
t_{WL}	CE to $\overline{WE}$ Low	0		0		0		0		ns
t_{WC}	$\overline{WE}$ to CE on	0		0		0		0		ns

WRITE CYCLE

Symbol	Parameters	Min.	Max.	Min.	Max.	Min.	Max.	Min.	Max.	Units
t_{WCY} [2]	Write Cycle Time	700		550		840		970		ns
t_{CEW}	CE Width During Write	480	3000	410	3000	600	3000	680	3000	ns
t_W	$\overline{WE}$ to CE Off	340		250		400		450		ns
t_{CW}	CE to $\overline{WE}$ High	300		250		—		—		ns
t_{DW}	D_{IN} to $\overline{WE}$ Set Up	0		0		0		0		ns
t_{CD} [4]	CE to D_{IN} Set Up		50		50		50		50	ns
t_{DH}	D_{IN} Hold Time	0		0		0		0		ns
t_{WP}	$\overline{WE}$ Pulse Width	150		150		200		200		ns
t_{WW} [5]	$\overline{WE}$ Wait	0		0		170		200		ns
t_{WC}	$\overline{WE}$ to CE On	0		0		0		0		ns

Capacitance [6] $T_A = 25°C$

Symbol	Test	Plastic And Ceramic Pkg. Typ.	Plastic And Ceramic Pkg. Max.	Unit	Conditions
C_{AD}	Address Capacitance, $\overline{CS}$, $\overline{WE}$, D_{IN}	3	6	pF	$V_{IN} = V_{SS}$
C_{CE}	CE Capacitance	17	25	pF	$V_{IN} = V_{SS}$
C_{OUT}	Data Output Capacitance	3	6	pF	$V_{OUT} = 0V$

Notes:
1. For plastic 2107A-4 and 2107A-5 $t_{REF} = 1mS$.
2. $t_T = 20ns$.
3. $C_{LOAD} = 50$ pf; Load = 1 TTL; Ref = 2.0V for high, 0.8V for low; $t_{ACC} = t_{AC} + t_{CO} + 1\ t_T$.
4. t_{CD} applies only when $t_W > t_{CEW} - 50ns$.
5. The 2107A and 2107A-1 should not be operated with t_{WW} in the 50 to 170 ns range.
6. Capacitance measured with Boonton Meter or effective capacitance calculated from the equation
$$C = \frac{I \Delta t}{\Delta V}$$
with the current equal to a constant 20mA.

D.C. Characteristics

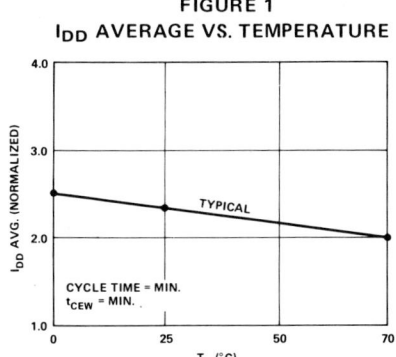

FIGURE 1
I_{DD} AVERAGE VS. TEMPERATURE

FIGURE 2
I_{CC} AVERAGE VS. TEMPERATURE

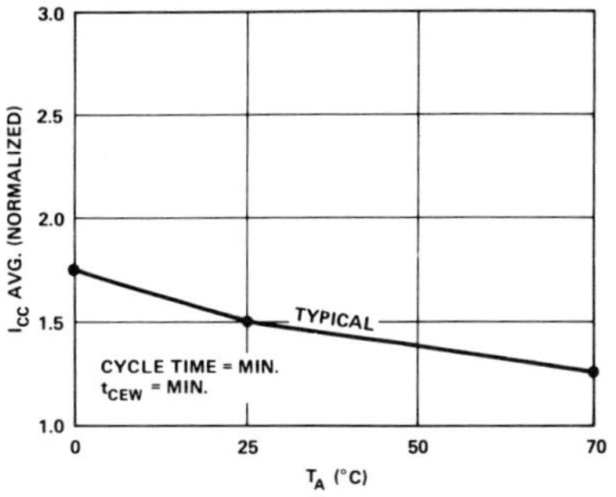

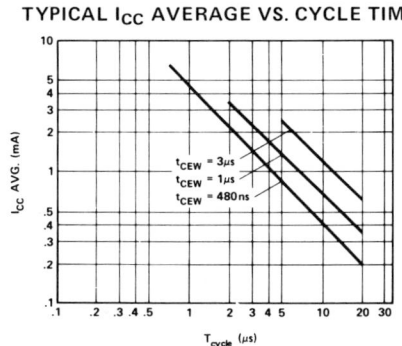

FIGURE 3
TYPICAL I_{DD} AVERAGE VS. CYCLE TIME

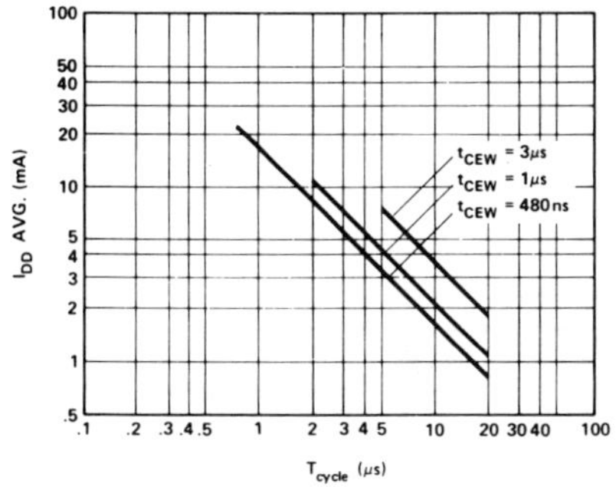

FIGURE 4
TYPICAL I_{CC} AVERAGE VS. CYCLE TIME

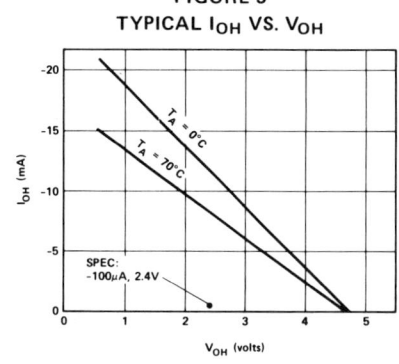

FIGURE 5
TYPICAL I_{OH} VS. V_{OH}

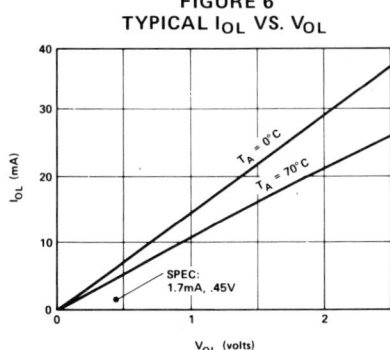

FIGURE 6
TYPICAL I_{OL} VS. V_{OL}

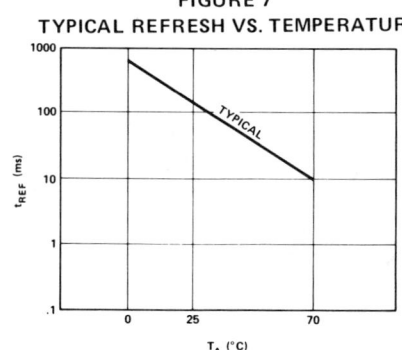

FIGURE 7
TYPICAL REFRESH VS. TEMPERATURE

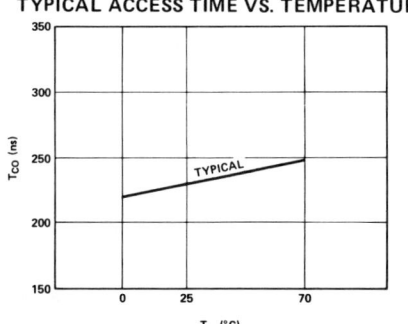

FIGURE 8
TYPICAL ACCESS TIME VS. TEMPERATURE

Table 1. I_{DDAV} and I_{CCAV} Characteristics.

Product	I_{DDAV} (Typ)	I_{DDAV} (Max)	I_{CCAV} (Typ)	I_{CCAV} (Max)	Cycle	t_{CEW}
2107A	23mA	34mA	6mA	10mA	700ns	480ns
2107A-1	28mA	38mA	8mA	12mA	550ns	410ns
2107A-4	22mA	33mA	5mA	9mA	840ns	600ns
2107A-5	18mA	28mA	4mA	8mA	970ns	680ns

Read Modify Write Cycle

Symbol	Parameters	2107A Min.	2107A Max.	2107A-1 Min.	2107A-1 Max.	2107A-4 Min.	2107A-4 Max.	2107A-5 Min.	2107A-5 Max.	Units
t_{RWC}[1]	Read Modify Write (RMW) Cycle Time	840		670		970		1140		ns
t_{CRW}[2]	CE Width During RMW	620	3000	530	3000	730	3000	850	3000	ns
t_{WC}	$\overline{WE}$ to CE on	0		0		0		0		ns
t_W	$\overline{WE}$ to CE off	340		250		400		450		ns
t_{WP}	$\overline{WE}$ Pulse Width	150		150		200		200		ns
t_{DW}	D_{IN} to $\overline{WE}$ Set Up	0		0		0		0		ns
t_{DH}	D_{IN} Hold Time	0		0		0		0		ns
t_{CO}	CE to Output Delay		280		260		330		400	ns
t_{ACC}[3]	Access Time		300		280		350		420	ns
t_{WD}	D_{OUT} Valid After $\overline{WE}$	0		0		0		0		ns

Notes: 1. t_T = 20ns

2. $t_{CRW} - t_W = t_{CO}$

3. C_{LOAD} = 50 pf; Load = One TTL Gate; Ref = 2.0V for High, 0.8V for low; $t_{ACC} = t_{AC} + t_{CO}$ + 1 TTL

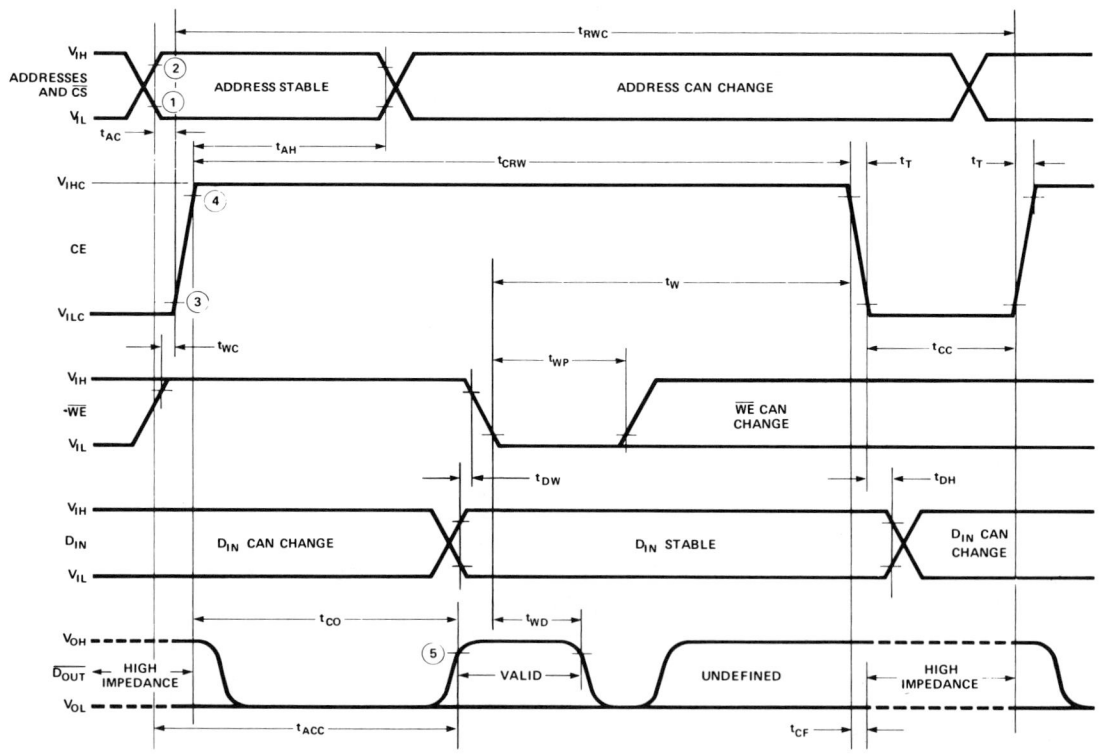

NOTES:
1. V_{SS} + 1.5V is the reference level for measuring timing of the address CS, WE, and D_{IN}.
2. V_{SS} + 3.0V is the reference level for measuring timing of the address, CS, WE, and D_{IN}.
3. V_{SS} + 2.0V is the reference level for measuring timing of CE.
4. V_{DD} −2V is the reference level for measuring timing of CE.
5. V_{SS} + 2.0V is the reference level for measuring the timing of D_{OUT}.

Read and Refresh Cycle [1]

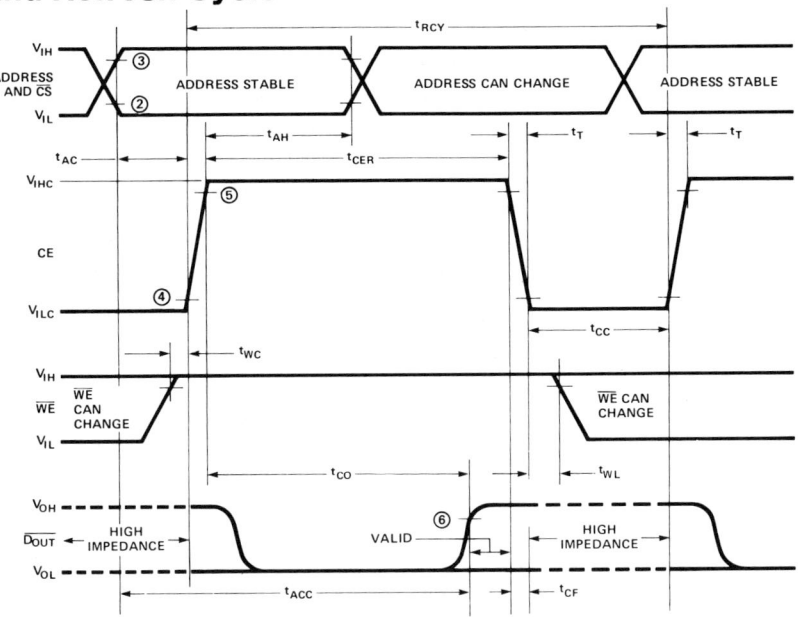

Write Cycle

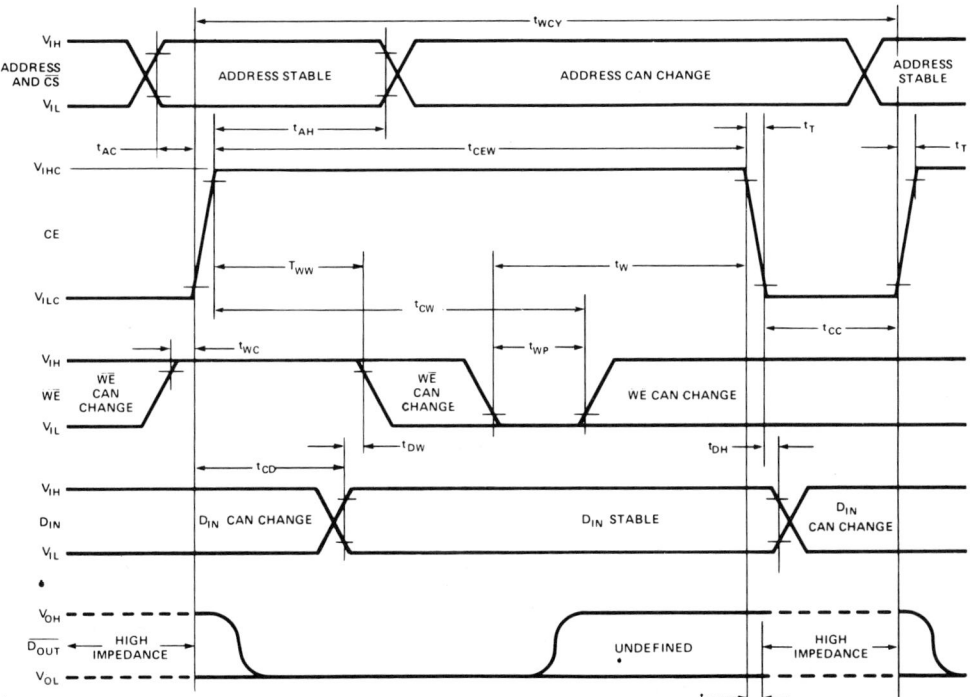

NOTES:
1. For Refresh cycle row and column addresses must be stable before t_{AC} and remain stable for entire t_{AH} period.
2. V_{SS} + 1.5V is the reference level for measuring timing of the addresses, $\overline{CS}$, $\overline{WE}$, and D_{IN}.
3. V_{SS} + 3.0V is the reference level for measuring timing of the addresses, $\overline{CS}$, $\overline{WE}$, and D_{IN}.
4. V_{SS} + 2.0V is the reference level for measuring timing of CE.
5. V_{DD} −2V is the reference level for measuring timing of CE.
6. V_{SS} + 2.0V is the reference level for measuring the timing of $\overline{D_{OUT}}$.

2107B
4096 BIT DYNAMIC RAM

	2107B	2107B-4	2107B-5
Access Time	200ns	270ns	300ns
Read, Write Cycle	400ns	470ns	590ns
RMW Cycle	520ns	590ns	750ns

- Low Cost Per Bit
- Low Standby Power
- Easy System Interface
- Only One High Voltage Input Signal — Chip Enable
- TTL Compatible — All Address, Data, Write Enable, Chip Select Inputs
- Refresh Period—2ms for 2107B, 2107B-4, 1ms for 2107B-5 @70°C
- Address Registers Incorporated on the Chip
- Simple Memory Expansion — Chip Select Input Lead
- Fully Decoded — On Chip Address Decode
- Output is Three State and TTL Compatible
- Industry Standard 22-Pin Configuration

The Intel® 2107B is a 4096 word by 1 bit dynamic n-channel MOS RAM. It was designed for memory applications where very low cost and large bit storage are important design objectives. The 2107B uses dynamic circuitry which reduces the standby power dissipation.

Reading information from the memory is non-destructive. Refreshing is most easily accomplished by performing one read cycle on each of the 64 row addresses. Each row address must be refreshed every two milliseconds. The memory is refreshed whether Chip Select is a logic one or a logic zero.

The 2107B is fabricated with n-channel silicon gate technology. This technology allows the design and production of high performance, easy to use MOS circuits and provides a higher functional density on a monolithic chip than other MOS technologies. The 2107B uses a single transistor cell to achieve high speed and low cost. It is a replacement for the 2107A.

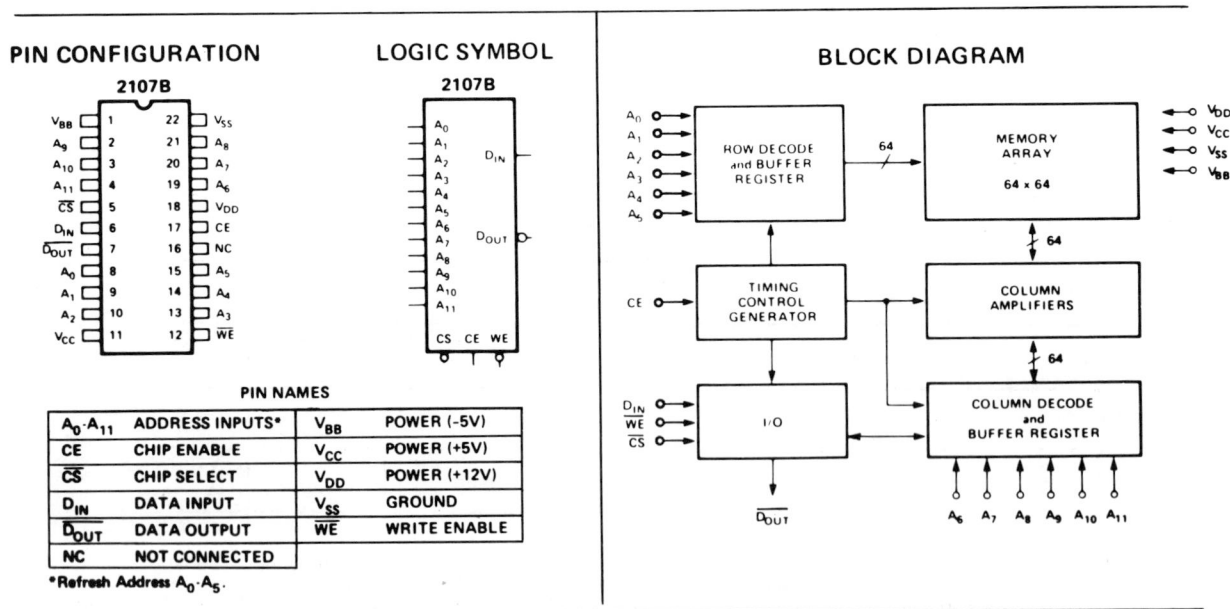

Absolute Maximum Ratings*

Temperature Under Bias	0°C to 70°C
Storage Temperature	−65°C to +150°C
All Input or Output Voltages with Respect to the most Negative Supply Voltage, V_{BB}	+25V to −0.3V
Supply Voltages V_{DD}, V_{CC}, and V_{SS} with Respect to V_{BB}	+20V to −0.3V
Power Dissipation	1.25W

*COMMENT:
Stresses above those listed under "Absolute Maximum Ratings" may cause permanent damage to the device. This is a stress rating only and functional operation of the device at these or any other conditions above those indicated in the operational sections of this specification is not implied. Exposure to absolute maximum rating conditions for extended periods may affect device reliability.

D.C. and Operating Characteristics

T_A = 0°C to 70°C, V_{DD} = +12V ±5%, V_{CC} = +5V ±10%, V_{BB}[1] = −5V ±5%, V_{SS} = 0V, unless otherwise noted.

Symbol	Parameter	Min.	Typ.[2]	Max.	Unit	Conditions
I_{LI}[6]	Input Load Current (all inputs except CE)		.01	50	µA	V_{IN} = $V_{IL\,MIN}$ to $V_{IH\,MAX}$ CE = V_{ILC} or V_{IHC}
I_{LC}	Input Load Current		.01	2	µA	V_{IN} = $V_{IL\,MIN}$ to $V_{IH\,MAX}$
$|I_{LO}|$	Output Leakage Current for high impedance state		.01	10	µA	CE = V_{ILC} or $\overline{CS}$ = V_{IH} V_O = 0V to 5.25V
I_{DD1}	V_{DD} Supply Current during CE off[3]		110	200[5]	µA	CE = −1V to +.6V
I_{DD2}	V_{DD} Supply Current during CE on			60	mA	CE = V_{IHC}, $\overline{CS}$ = V_{IL}
$I_{DD\,AV}$	Average V_{DD} Current		38	54	mA	$\overline{CS}$ = V_{IL}; T_A = 25°C; Min cycle time, Min t_{CE}
I_{CC1}[4]	V_{CC} Supply Current during CE off		.01	10	µA	CE = V_{ILC} or $\overline{CS}$ = V_{IH}
I_{BB}	V_{BB} Supply Current		5	400	µA	
V_{IL}	Input Low Voltage	−1.0		0.6	V	t_T = 20ns, V_{ILC} = +1.0V
V_{IH}	Input High Voltage	2.4		V_{CC}+1	V	t_T = 20ns
V_{ILC}	CE Input Low Voltage	−1.0		+1.0	V	
V_{IHC}	CE Input High Voltage	V_{DD}−1		V_{DD}+1	V	
V_{OL}	Output Low Voltage	0.0		0.45	V	I_{OL} = 2.0mA
V_{OH}	Output High Voltage	2.4		V_{CC}	V	I_{OH} = −2.0mA

NOTES:
1. The only requirement for the sequence of applying voltage to the device is that V_{DD}, V_{CC}, and V_{SS} should never be .3V or more negative than V_{BB}.
2. Typical values are for T_A = 25°C and nominal power supply voltages.
3. The I_{DD} and I_{CC} currents flow to V_{SS}. The I_{BB} current is the sum of all leakage currents.
4. During CE on V_{CC} supply current is dependent on output loading, V_{CC} is connected to output buffer only.
5. Maximum I_{DD1} for 2107B-5 is 250 µA.
6. During CE high a current of 0.5mA typical, 1.5mA maximum will be drawn from any address pin which is switched from low to high.

DYNAMIC RAMS

A.C. Characteristics
$T_A = 0°C$ to $70°C$, $V_{DD} = 12V \pm 5\%$, $V_{CC} = 5V \pm 10\%$, $V_{BB} = -5V \pm 5\%$,
READ, WRITE, AND READ MODIFY/WRITE CYCLE $V_{SS} = 0V$, unless otherwise noted.

Symbol	Parameter	2107B Min.	2107B Max.	2107B-4 Min.	2107B-4 Max.	2107B-5 Min.	2107B-5 Max.	Units	Note
t_{REF}	Time Between Refresh		2		2		1	ms	7
t_{AC}	Address to CE Set Up Time	0		0		10		ns	3
t_{AH}	Address Hold Time	100		100		100		ns	
t_{CC}	CE Off Time	130		130		200		ns	
t_T	CE Transition Time	10	40	10	40	10	40	ns	
t_{CF}	CE Off to Output High Impedance State	0		0		0		ns	

READ CYCLE

Symbol	Parameter	2107B Min.	2107B Max.	2107B-4 Min.	2107B-4 Max.	2107B-5 Min.	2107B-5 Max.	Units	Note
t_{CY}	Cycle Time	400		470		590		ns	4
t_{CE}	CE On Time	230	4000	300	4000	350	3000	ns	
t_{CO}	CE Output Delay		180		250		280	ns	5
t_{ACC}	Address to Output Access		200		270		300	ns	6
t_{WL}	CE to $\overline{WE}$	0		0		0		ns	
t_{WC}	$\overline{WE}$ to CE On	0		0		0		ns	

WRITE CYCLE

Symbol	Parameter	2107B Min.	2107B Max.	2107B-4 Min.	2107B-4 Max.	2107B-5 Min.	2107B-5 Max.	Units	Note
t_{CY}	Cycle Time	400		470		590		ns	4
t_{CE}	CE On Time	230	4000	300	4000	350	3000	ns	
t_W	$\overline{WE}$ to CE Off	125		150		200		ns	
t_{CW}	CE to $\overline{WE}$	150		150		150		ns	
t_{DW}	D_{IN} to $\overline{WE}$ Set Up	0		0		0		ns	1
t_{DH}	D_{IN} Hold Time	0		0		0		ns	
t_{WP}	$\overline{WE}$ Pulse Width	50		50		75		ns	
t_{WW}	$\overline{WE}$ Delay	75		75		75		ns	

Capacitance[2] $T_A = 25°C$

Symbol	Test	Plastic And Ceramic Pkg. Typ.	Plastic And Ceramic Pkg. Max.	Unit	Conditions
C_{AD}	Address Capacitance, $\overline{CS}$	4	6	pF	$V_{IN} = V_{SS}$
C_{CE}	CE Capacitance	17	25	pF	$V_{IN} = V_{SS}$
C_{OUT}	Data Output Capacitance	5	7	pF	$V_{OUT} = 0V$
C_{IN}	D_{IN} and $\overline{WE}$ Capacitance	8	10	pF	$V_{IN} = V_{SS}$

Notes:
1. If $\overline{WE}$ is low before CE goes high then D_{IN} must be valid when CE goes high.
2. Capacitance measured with Boonton Meter or effective capacitance calculated from the equation.
$C = \frac{I \Delta t}{\Delta V}$ with the current equal to a constant 20mA.
3. t_{AC} is measured from end of address transition.
4. $t_T = 20ns$
5. $C_{LOAD} = 50pF$, Load = One TTL Gate, Ref = 2.0V.
6. $t_{ACC} = t_{AC} + t_{CO} + 1 t_T$
7. $t_{REF} = 2ms$ at $T_A = 55°C$ for the 2107B-5.

Read and Refresh Cycle [1]

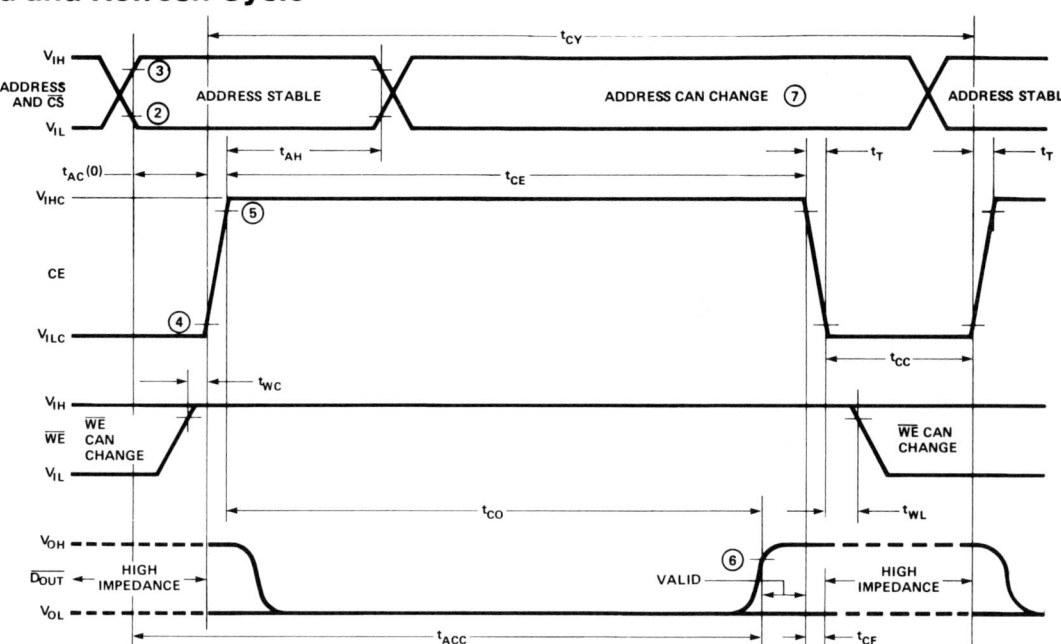

Write Cycle

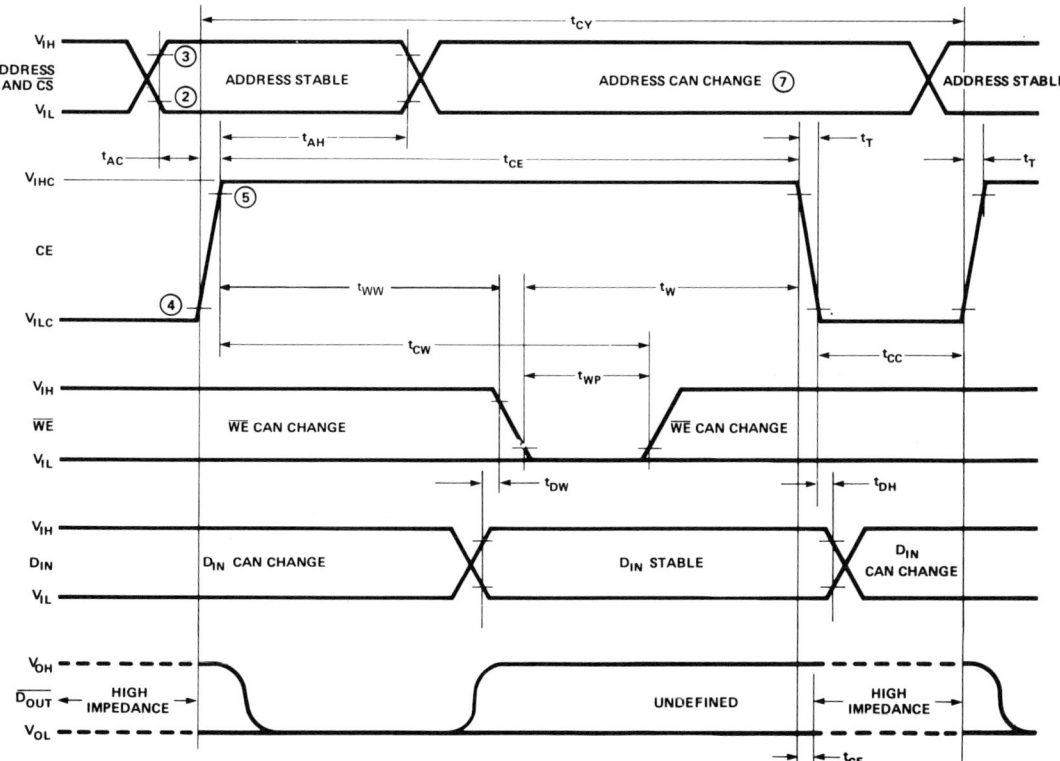

NOTES:
1. For Refresh cycle row and column addresses must be stable before t_{AC} and remain stable for entire t_{AH} period.
2. V_{IL} MAX is the reference level for measuring timing of the addresses, $\overline{CS}$, $\overline{WE}$, and D_{IN}.
3. V_{IH} MIN is the reference level for measuring timing of the addresses, $\overline{CS}$, $\overline{WE}$, and D_{IN}.
4. V_{SS} +2.0V is the reference level for measuring timing of CE.
5. V_{DD} -2V is the reference level for measuring timing of CE.
6. V_{SS} +2.0V is the reference level for measuring the timing of $\overline{D_{OUT}}$.
7. During CE high typically 0.5mA will be drawn from any address pin which is switched from low to high.

Read Modify Write Cycle [1]

Symbol	Parameter	2107B Min.	2107B Max.	2107B-4 Min.	2107B-4 Max.	2107B-5 Min.	2107B-5 Max.	Unit
t_{RWC}	Read Modify Write (RMW) Cycle Time	520		590		750		ns
t_{CRW}	CE Width During RMW	350	4000	420	4000	510	3000	ns
t_{WC}	$\overline{WE}$ to CE on	0		0		0		ns
t_W	$\overline{WE}$ to CE off	150		150		200		ns
t_{WP}	$\overline{WE}$ Pulse Width	50		50		100		ns
t_{DW}	D_{IN} to $\overline{WE}$ Set Up	0		0		0		ns
t_{DH}	D_{IN} Hold Time	0		0		0		ns
t_{CO}	CE to Output Delay		180		250		280	ns
t_{ACC}	Access Time ($t_{ACC} = t_{AC} + t_{CO} + 1t_T$)		200		270		300	ns

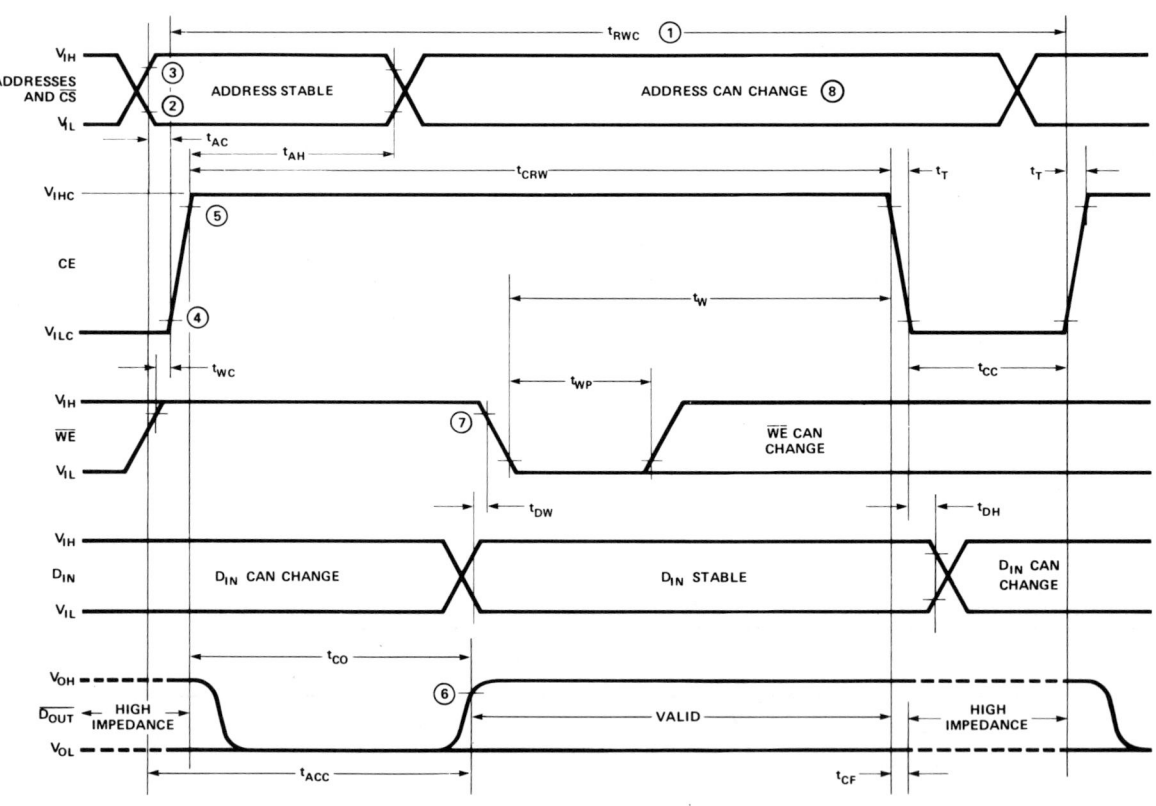

NOTES:
1. Minimum cycle timing is based on t_T of 20ns.
2. V_{IL} MAX is the reference level for measuring timing of the addresses, $\overline{CS}$, $\overline{WE}$, and D_{IN}.
3. V_{IH} MIN is the reference level for measuring timing of the addresses, $\overline{CS}$, $\overline{WE}$, and D_{IN}.
4. V_{SS} +2.0V is the reference level for measuring timing of CE.
5. V_{DD} -2V is the reference level for measuring timing of CE.
6. V_{SS} +2.0V is the reference level for measuring the timing of $\overline{D_{OUT}}$. C_{LOAD} = 50pF. Load = One TTL Gate.
7. $\overline{WE}$ must be at V_{IH} until end of t_{CO}.
8. During CE high typically 0.5mA will be drawn from any address pin which is switched from low to high.

Typical Characteristics

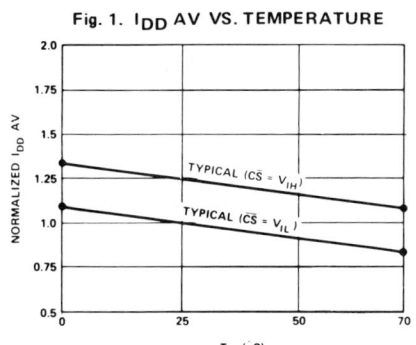

Fig. 1. I_{DD} AV VS. TEMPERATURE

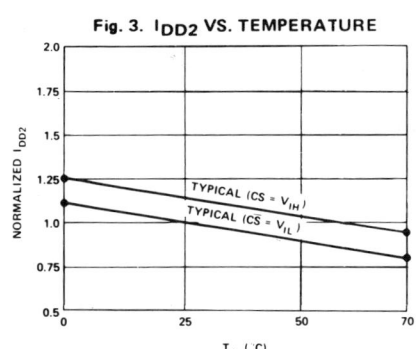

Fig. 3. I_{DD2} VS. TEMPERATURE

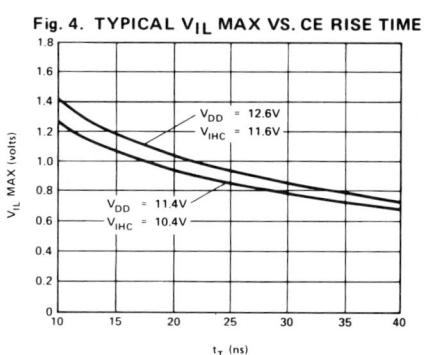

Fig. 2. TYPICAL I_{DD} AVERAGE VS. CYCLE TIME

Fig. 4. TYPICAL V_{IL} MAX VS. CE RISE TIME

Typical Current Transients vs. Time

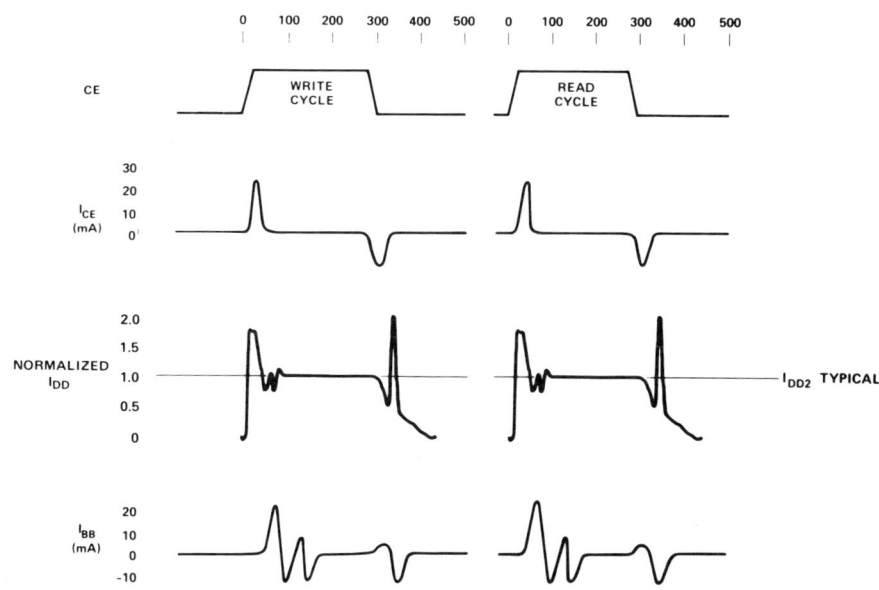

For additional typical characteristics and applications information please refer to Intel Application Note AP-10, "Memory System Design With the Intel 2107B 4K RAM" or Intel's Memory Design Handbook.

2107C
4096-BIT DYNAMIC RAM

	2107C-1	2107C-2	2107C	2107C-4
Access Time (ns)	150	200	250	300
Read, Write Cycle (ns)	380	400	430	470
RMW Cycle (ns)	450	500	550	590
Max $I_{DD\ AV}$ (mA)	35	33	30	30

- **Direct Replacement for Industry Standard 22-Pin 4K RAMs**
- **Low Operating Power**
- **Low Standby Power**
- **Only One High Voltage Input Signal– Chip Enable**
- **150 ns Access Time**
- **± 10% Tolerance on all Power Supplies**
- **Output is Three-State and TTL Compatible**
- **TTL Compatible — All Address, Data, Write Enable, Chip Select Inputs**
- **Refresh Period 2 ms**

The Intel® 2107C is a 4096-word by 1-bit dynamic n-channel MOS RAM. It was designed for memory applications where very low cost and large bit storage are important design objectives. A new unique dynamic storage cell provides high speed and wide operating margins. The 2107C uses dynamic circuitry which reduces the standby power dissipation.

Reading information from the memory is non-destructive. Refreshing is most easily accomplished by performing one read cycle on each of the 64 row addresses. Each row address must be refreshed every two milliseconds. The memory is refreshed whether Chip Select is a logic one or a logic zero.

The 2107C is fabricated with n-channel silicon gate technology. This technology allows the design and production of high performance, easy to use MOS circuits and provides a higher functional density on a monolithic chip than other MOS technologies. The 2107C is a replacement for the 2107A, 2107B and other industry standard 22-pin 4K RAMs.

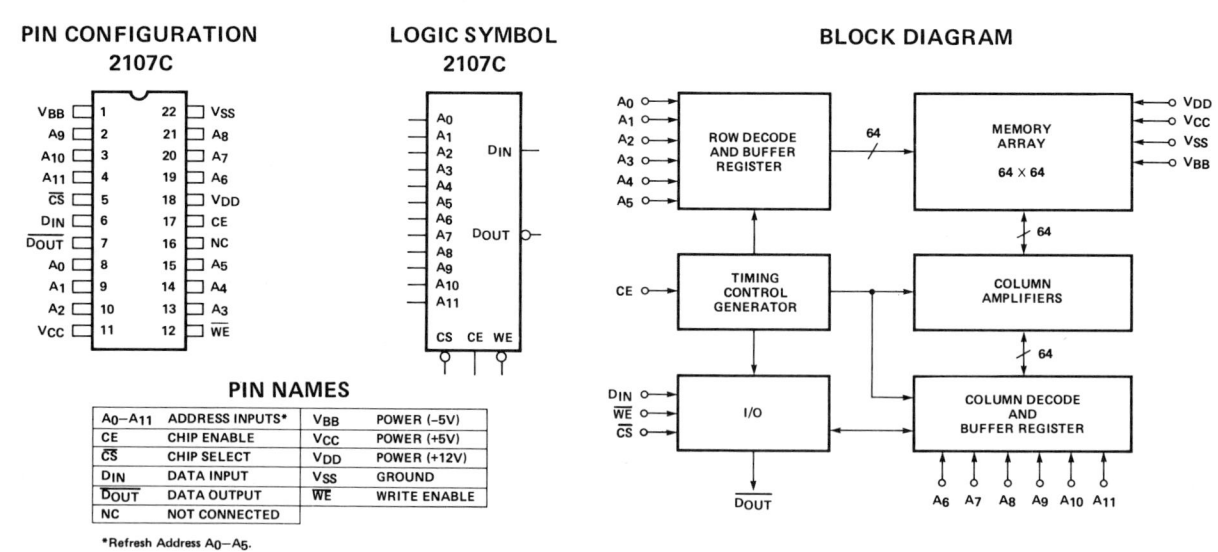

Absolute Maximum Ratings*

Temperature Under Bias .... -10°C to 80°C
Storage Temperature .... -65°C to +150°C
Voltage on any Pin Relative to V_{BB} ($V_{SS} - V_{BB} \geqslant 4.5$) .... -0.3V to +20V
Power Dissipation 1.00W

*COMMENT:
Stresses above those listed under "Absolute Maximum Ratings" may cause permanent damage to the device. This is a stress rating only and functional operation of the device at these or any other conditions above those indicated in the operational sections of this specification is not implied. Exposure to absolute maximum rating conditions for extended periods may affect device reliability.

D.C. and Operating Characteristics

$T_A = 0°C$ to $70°C$, $V_{DD} = +12V \pm 10\%$, $V_{CC} = +5V \pm 10\%$, V_{BB}[1] $= -5V \pm 10\%$, $V_{SS} = 0V$, unless otherwise noted.

Symbol	Parameter	Min.	Typ.[2]	Max.	Unit	Conditions
I_{LI}	Input Load Current (all inputs except CE)			10	μA	V_{IN} = 0V to $V_{IH\ MAX}$ CE = V_{ILC} or V_{IHC}
I_{LC}	Input Load Current, CE			2	μA	V_{IN} = 0V to $V_{IHC\ MAX}$
$|I_{LO}|$	Output Leakage Current for high impedance state			10	μA	CE = V_{ILC} or $\overline{CS}$ = V_{IH} V_O = 0V to 5.5V
I_{DD1}[3]	V_{DD} Supply Current — standby [3]		20	200	μA	CE = -1V to +0.6V
$I_{DD\ AV}$	Average V_{DD} Current — operating		24	35	mA	2107C-1, t_{CYC} = 380
			22	33	mA	2107C-2, t_{CYC} = 400
			20	30	mA	2107C, t_{CYC} = 430
			20	30	mA	2107C-4, t_{CYC} = 470
I_{CC1}[3,4]	V_{CC} Supply Current — standby			10	μA	CE = V_{ILC} or $\overline{CS}$ = V_{IH}
I_{BB1}	V_{BB} Supply Current — standby		5	50	μA	CE = -1V to +0.6V
$I_{BB\ AV}$	Average V_{BB} Current — operating		100	400	μA	Min. cycle time, Min. t_{CE}
V_{IL}	Input Low Voltage	-1.0		0.8	V	
V_{IH}	Input High Voltage	2.4		V_{CC}+1	V	
V_{ILC}	CE Input Low Voltage	-1.0		+1.0	V	
V_{IHC}	CE Input High Voltage	V_{DD}-1		V_{DD}+1	V	
V_{OL}	Output Low Voltage	0.0		0.40	V	I_{OL} = 3.2 mA
V_{OH}	Output High Voltage	2.4		V_{CC}	V	I_{OH} = -2.0 mA

NOTES:
1. The only requirement for the sequence of applying voltage to the device is that V_{DD}, V_{CC}, and V_{SS} should never be 0.3V or more negative than V_{BB}.
2. Typical values are for $T_A = 25°C$ and nominal power supply voltages.
3. The I_{DD} and I_{CC} currents flow to V_{SS}.
4. During CE on V_{CC} supply current is dependent on output loading. V_{CC} is connected to output buffer only.

A.C. Characteristics [1]

T_A = 0°C to 70°C, V_{DD} = 12V ±10%, V_{CC} = 5V ±10%, V_{BB} = –5V ±10%, V_{SS} = 0V, unless otherwise noted.

READ, WRITE, AND READ MODIFY/WRITE CYCLE

Symbol	Parameter	2107C-1 Min.	2107C-1 Max	2107C-2 Min.	2107C-2 Max.	2107C Min.	2107C Max.	2107C-4 Min.	2107C-4 Max.	Units	Note
t_{REF}	Time Between Refresh		2		2		2		2	ms	
t_{AC}	Address to CE Set-Up Time	0		0		0		0		ns	2
t_{AH}	Address Hold Time	50		50		100		100		ns	
t_{CC}	CE Off Time	130		130		130		130		ns	
t_T	CE Transition Time		40		40		40		40	ns	
t_{CD}	CE Off to Output Disable Time		30		30		30		30	ns	3

READ CYCLE

Symbol	Parameter	2107C-1 Min.	2107C-1 Max.	2107C-2 Min.	2107C-2 Max.	2107C Min.	2107C Max.	2107C-4 Min.	2107C-4 Max.	Units	Note
t_{CY}	Cycle Time	380		400		430		470		ns	3
t_{CE}	CE On Time	210	4000	230	4000	260	4000	300	4000	ns	
t_{CO}	CE Output Delay		130		180		230		280	ns	4
t_{ACC}	Address to Output Access		150		200		250		300	ns	5
t_{WL}	CE to $\overline{WE}$	0		0		0		0		ns	
t_{WC}	$\overline{WE}$ to CE On	0		0		0		0		ns	

WRITE CYCLE

Symbol	Parameter	2107C-1 Min.	2107C-1 Max.	2107C-2 Min.	2107C-2 Max.	2107C Min.	2107C Max.	2107C-4 Min.	2107C-4 Max.	Units	Note
t_{CY}	Cycle Time	380		400		430		470		ns	3
t_{CE}	CE On Time	210	4000	230	4000	260	4000	300	4000	ns	
t_W	$\overline{WE}$ to CE Off	125		125		125		175		ns	
t_{CW}	CE to $\overline{WE}$	150		150		150		200		ns	
t_{DW}	D_{IN} to $\overline{WE}$ Set-Up	0		0		0		0		ns	6
t_{DH}	D_{IN} Hold Time	0		0		0		0		ns	
t_{WP}	$\overline{WE}$ Pulse Width	50		50		50		100		ns	
t_{WD}	$\overline{WE}$ to Output Disable Time		15		15		15		15	ns	

Capacitance [7] T_A = 25°C

Symbol	Test	Plastic and Ceramic Package Typ.	Plastic and Ceramic Package Max.	Unit	Conditions
C_{AD}	Address Capacitance, $\overline{CS}$, D_{IN}	5	7	pF	$V_{IN} = V_{SS}$
C_{CE}	CE Capacitance	10	15	pF	$V_{IN} = V_{SS}$
C_{OUT}	Data Output Capacitance	5	7	pF	$V_{OUT} = 0V$
$C_{\overline{WE}}$	$\overline{WE}$ Capacitance	6	8	pF	$V_{IN} = V_{SS}$

NOTES:
1. After the application of supply voltages or after extended periods of operation without CE, the device must perform a minimum of one initialization cycle (any valid memory cycle or refresh cycle) prior to normal operation.
2. t_{AC} is measured from end of address transition.
3. t_T = 20 ns.
4. C_{LOAD} = 50 pF, Load = One TTL Gate, Ref = 2.0V.
5. $t_{ACC} = t_{AC} + t_{CO} + 1 t_T$.
6. If WE is low before CE goes high then D_{IN} must be valid when CE goes high.
7. Capacitance measured with Boonton Meter or effective capacitance calculated from the equation:

$$C = \frac{I \Delta t}{\Delta V}$$ with the current equal to a constant 20 mA.

Read and Refresh Cycle [1]

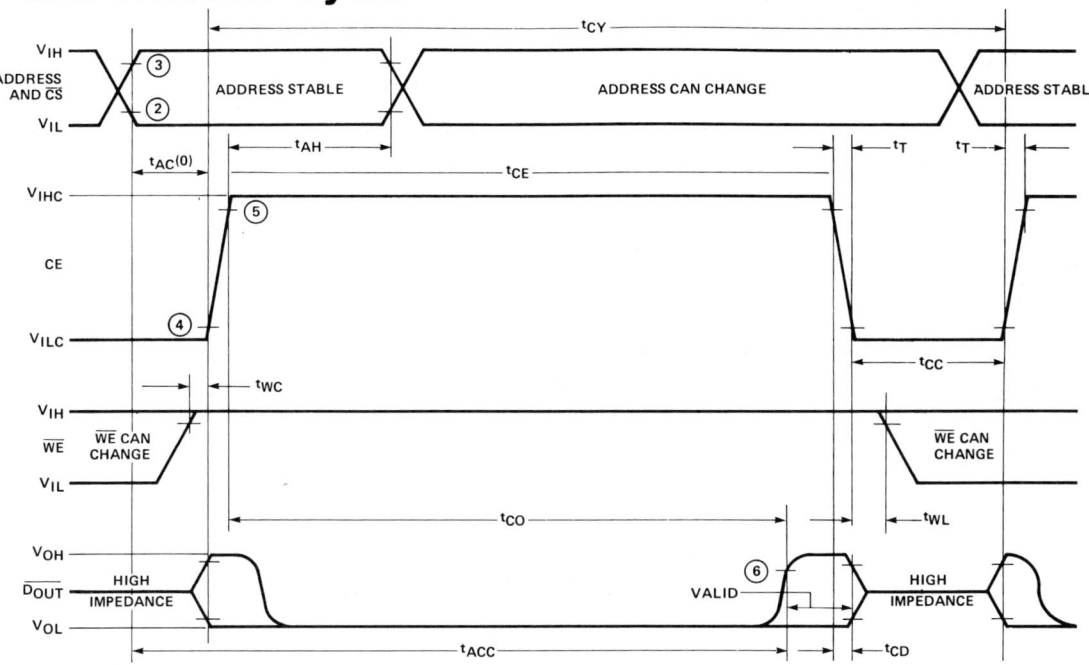

Write Cycle

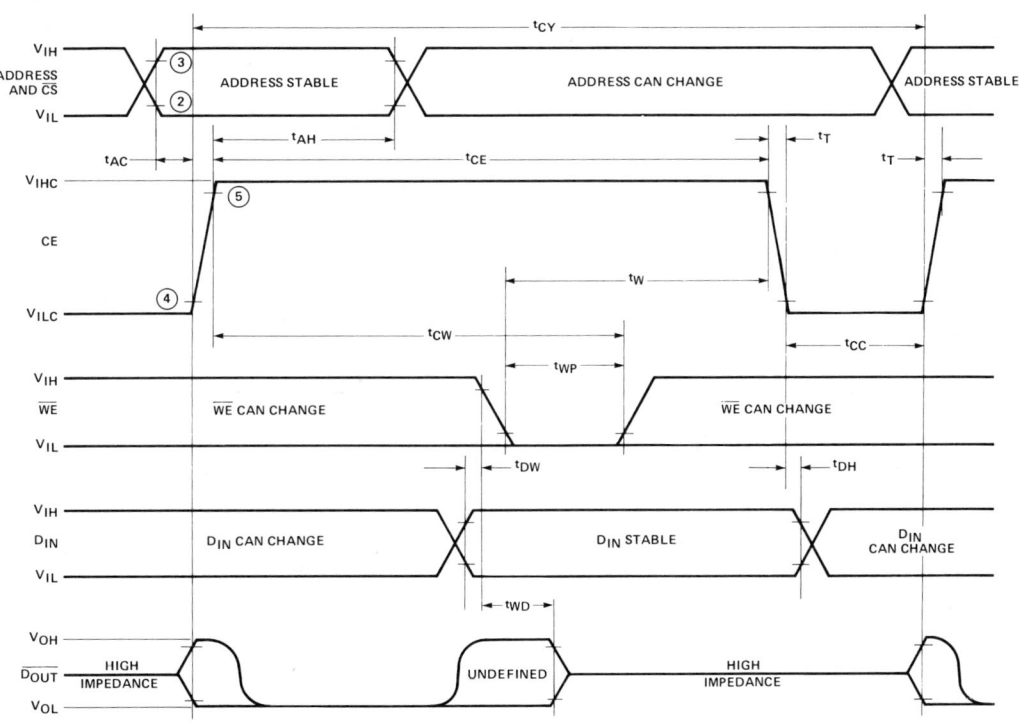

NOTES:
1. For Refresh cycle, row and column addresses must be stable before t_{AC} and remain stable for entire t_{AH} period.
2. V_{IL} MAX is the reference level for measuring timing of the addresses, $\overline{CS}$, $\overline{WE}$, and D_{IN}.
3. V_{IN} MIN is the reference level for measuring timing of the addresses, $\overline{CS}$, $\overline{WE}$, and D_{IN}.
4. V_{SS} +2.0V is the reference level for measuring timing of CE.
5. V_{DD} −2V is the reference level for measuring timing of CE.
6. V_{SS} +2.0V is the reference level for measuring the timing of $\overline{D_{OUT}}$.

Read Modify Write Cycle

Symbol	Parameter	2107C-1		2107C-2		2107C		2107C-4		Units	Note
		Min.	Max.	Min.	Max.	Min.	Max.	Min.	Max.		
t_{RWC}	Read Modify Write (RMW) Cycle	450		500		550		590		ns	1
t_{CRW}	CE Width During RMW	280	4000	330	4000	380	4000	420	4000	ns	
t_{WC}	$\overline{WE}$ to CE On	0		0		0		0		ns	
t_W	$\overline{WE}$ to CE Off	125		125		125		175		ns	
t_{WP}	$\overline{WE}$ Pulse Width	50		50		50		100		ns	
t_{DW}	D_{IN} to $\overline{WE}$ Setup	0		0		0		0		ns	
t_{DH}	D_{IN} Hold Time	0		0		0		0		ns	
t_{CO}	CE to Output Delay		130		180		230		280	ns	
t_{ACC}	Access Time		150		200		250		300	ns	
t_{WD}	$\overline{WE}$ to Output Disable Time	15		15		15		15		ns	

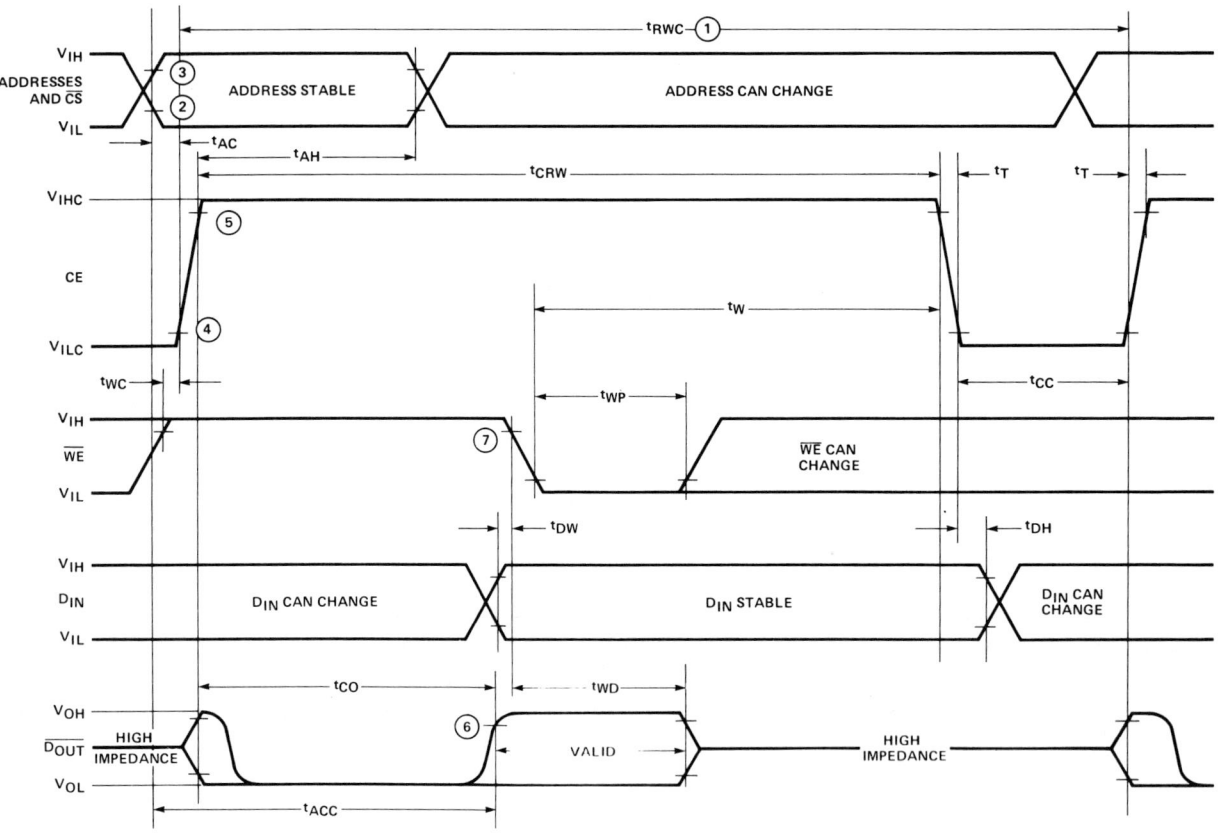

NOTES:
1. t_T of 20 ns.
2. V_{IL} MAX is the reference level for measuring timing of the addresses, $\overline{CS}$, $\overline{WE}$, and D_{IN}.
3. V_{IH} MIN is the reference level for measuring timing of the addresses, $\overline{CS}$, $\overline{WE}$, and D_{IN}.
4. V_{SS} +2.0V is the reference level for measuring timing of CE.
5. V_{DD} −2V is the reference level for measuring timing of CE.
6. V_{SS} +2.0V is the reference level for measuring the timing of $\overline{D_{OUT}}$. C_{LOAD} = 50 pF. Load = One TTL Gate.
7. $\overline{WE}$ must be at V_{IH} until end of t_{CO}.

2108-2 AND 2108-4
8192 X 1 BIT DYNAMIC RAM

	2108-2	2108-4
	S1572, S1573	S1626, S1627
Max. Access Time (ns)	200	300
Read, Write Cycle (ns)	350	425
Read-Modify-Write Cycle (ns)	400	595

- 8K RAM in Industry Standard 16 Pin Package
- Low Standby Power
- All Inputs Including Clocks TTL Compatible
- Standard Power Supplies +12V, +5V, -5V
- Only 64 Refresh Cycles Required Every 2 ms
- On-Chip Input Latches
- Output is Three-State, TTL Compatible; Data is Latched and Valid into Next Cycle
- Fully Compatible with 4K and 16K Dynamic RAMs

The Intel® 2108 is a 8K Dynamic MOS RAM organized as 8192 words by 1 bit. The 2108 employs the same masks and highly reliable, production-proven two layer polysilicon N-MOS technology as the Intel® 2116 16K RAM. As shown in the block diagram below, the 2116 is organized as two 8K RAMs on a single silicon die. Each of these 8K RAMs contains its own row decoders, sense amplifiers, and storage cells. The 2108 is fully tested to insure that one 8K RAM meets all AC and DC specifications.

The 2108 is available as either the upper or lower half of the 2116. Address A_6 selects the operating half. For S1572 or S1627, A_6 should be high (V_{IH}) during row address strobe (RAS). For S1573 or S1626, A_6 should be low (V_{IL}) during RAS. The use of the Intel® 3242 Address Multiplexer/Refresh Counter with a 2108 is described on page 2-66. The 2108 is packaged in the industry standard 16-pin DIP which is compatible with widely available automated handling equipment and facilitates easy upgrading from 2104A-type 4K RAM Systems and up to 2116-type 16K RAM Systems.

As in the 2104A-type 4K RAM and 2116-type 16K RAM, the 2108 has non-critical clock timing requirements which allow use of addressing multiplexing while maintaining high performance. Three methods of refreshing are permissable; they are described in the applications section of this data sheet.

The 2108 will provide the same reliable operation in its system usage as any Intel product. Information on the details of reliability tests performed on the 2108 and field data on the use of partial devices are available from Intel Corporation.

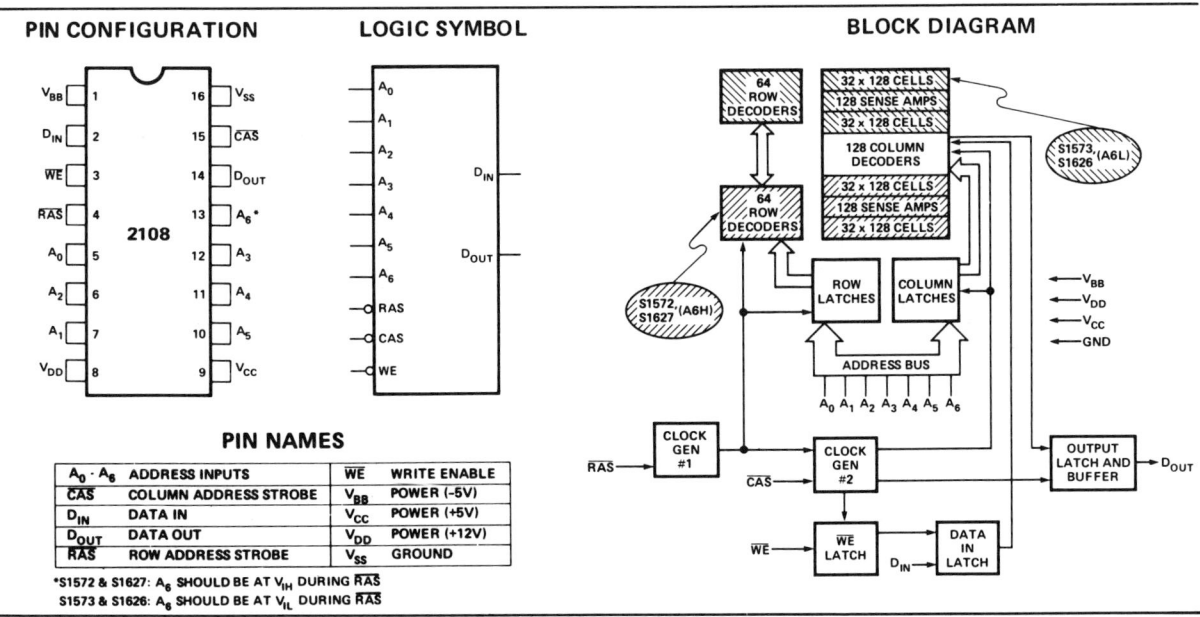

PIN NAMES

A_0 - A_6	ADDRESS INPUTS	WE	WRITE ENABLE
CAS	COLUMN ADDRESS STROBE	V_{BB}	POWER (-5V)
D_{IN}	DATA IN	V_{CC}	POWER (+5V)
D_{OUT}	DATA OUT	V_{DD}	POWER (+12V)
RAS	ROW ADDRESS STROBE	V_{SS}	GROUND

*S1572 & S1627: A_6 SHOULD BE AT V_{IH} DURING RAS
S1573 & S1626: A_6 SHOULD BE AT V_{IL} DURING RAS

Absolute Maximum Ratings*

Ambient Temperature Under Bias	$-10°C$ to $+80°C$
Storage Temperature	$-65°C$ to $+150°C$
Voltage on any Pin Relative to V_{BB} ($V_{SS} - V_{BB} \geq 4V$)	$-0.3V$ to $+20V$
Power Dissipation	1.25W

*COMMENT:
Stresses above those listed under "Absolute Maximum Ratings" may cause permanent damage to the device. This is a stress rating only and functional operation of the device at these or any other conditions above those indicated in the operational sections of this specification is not implied. Exposure to absolute maximum rating conditions for extended periods may affect device reliability.

D.C. and Operating Characteristics [1],[2]

$T_A = 0°C$ to $70°C$, $V_{DD} = +12V \pm 10\%$, $V_{CC} = +5V \pm 10\%$, $V_{BB} = -5V \pm 10\%$, $V_{SS} = 0V$, unless otherwise noted.

Symbol	Parameter	Min.	Typ.[3]	Max.	Unit	Conditions	
I_{LI}	Input Load Current (any input)			10	μA	$V_{IN} = V_{IL}$ MIN to V_{IH} MAX	
$|I_{LO}|$	Output Leakage Current for high impedance state		0.1	10	μA	Chip deselected: $\overline{RAS}$ and $\overline{CAS}$ at V_{IH} $V_{OUT} = 0$ to $5.5V$	
I_{DD1}	V_{DD} Supply Current		1.2	2	mA	$\overline{CAS}$ and $\overline{RAS}$ at V_{IH} or $\overline{CAS}$-only cycle. Chip deselected prior to measurement. See Note 5.	
I_{BB1}	V_{BB} Supply Current		1	50	μA		
I_{DD2}[4]	Operating V_{DD} Current		53	69	mA	2108-2 t_{CYC} = 350 ns	$T_A = 25°C$ Device selected. See Note 6.
			49	65	mA	2108-4 t_{CYC} = 425 ns	
I_{BB2}	Operating V_{BB} Current		120	400	μA	Device selected	
I_{CC1}[7]	V_{CC} Supply Current when deselected			10	μA		
V_{IL}	Input Low Voltage (any input)	-1.0		0.8	V		
V_{IH}	Input High Voltage (any input)	2.4		$V_{CC}+1$	V		
V_{OL}	Output Low Voltage	0.0		0.4	V	I_{OL} = 4.1 mA (Read Cycle Only)	
V_{OH}	Output High Voltage	2.4		V_{CC}	V	I_{OH} = -5 mA (Read Cycle Only)	

Capacitance [8]

$T_A = 25°C$, $V_{DD} = 12V \pm 10\%$, $V_{CC} = 5V \pm 10\%$, $V_{BB} = -5V \pm 10\%$, $V_{SS} = 0V$, unless otherwise noted.

Symbol	Parameter	Typ.	Max.	Unit	Conditions
C_{I1}	Address, Data In & $\overline{WE}$ Capacitance	4	7	pF	$V_{IN} = V_{SS}$
C_{I2}	$\overline{RAS}$ Capacitance	3	5	pF	$V_{IN} = V_{SS}$
C_{I3}	$\overline{CAS}$ Capacitance	6	10	pF	$V_{IN} = V_{SS}$
C_O	Data Output Capacitance	3	7	pF	$V_{OUT} = 0V$

Notes:
1. All voltages referenced to V_{SS}. No power supply sequencing is required but V_{DD}, V_{CC}, and V_{SS} should never be 0.3V or more negative than V_{BB}.
2. To avoid self-clocking, $\overline{RAS}$ should not be allowed to float.
3. Typical values are for $T_A = 25°C$ and nominal power supply voltages.
4. For $\overline{RAS}$-only refresh $I_{DD} = 0.78\ I_{DD2}$. For $\overline{CAS}$-before-$\overline{RAS}$ (64 cycle refresh) $I_{DD} = 0.96\ I_{DD2}$.
5. The chip is deselected (i.e., output is brought to high impedance state) by $\overline{CAS}$-only cycle or by $\overline{CAS}$-before-$\overline{RAS}$ cycle. The current flowing in a selected (i.e., output on) chip with $\overline{RAS}$ and $\overline{CAS}$ at V_{IH} is approximately twice I_{DD1}.
6. See Page 2-62 for typical I_{DD} characteristics under other conditions.
7. When chip is selected V_{CC} supply current is dependent on output loading; V_{CC} is connected to output buffer only.
8. Capacitance measured with Boonton Meter.

TYPICAL CHARACTERISTICS

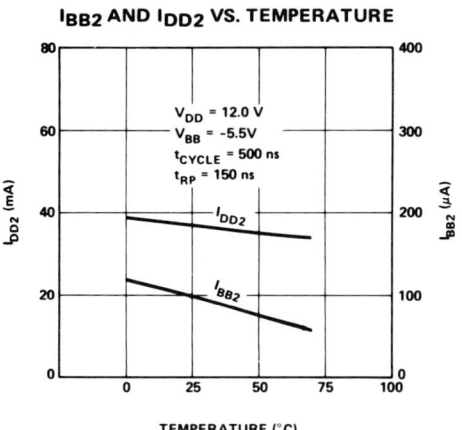

Figure 1.

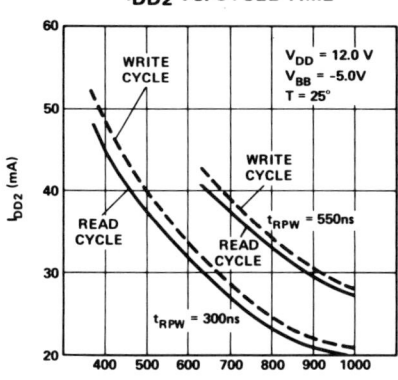

Figure 2.

Standby Power Calculations:

$$P_{REF} = P_{OP}\left(N\frac{t_{CYC}}{t_{REF}}\right) + P_{SB}\left(1 - N\frac{t_{CYC}}{t_{REF}}\right) \text{ where}$$

P_{OP} = Power dissipation — continuous operation = $V_{DD} \times I_{DD2}$.

N = Number of refresh cycles (64).

t_{CYC} = Cycle time for a refresh cycle.

t_{REF} = Time between refreshes

P_{SB} = Standby power dissipation = $V_{DD} \times I_{DD1} + |V_{BB}| \times I_{BB}$

Note that I_{DD2} depends upon refresh as follows:

1. For ($\overline{RAS}$ before $\overline{CAS}$) use I_{DD2} from Figures 1 and 2.
2. For ($\overline{CAS}$ before $\overline{RAS}$) multiply I_{DD2} determined in (1) by 0.96.
3. For ($\overline{RAS}$ only) multiply I_{DD2} determined in (1) by 0.78.

Examples of typical calculations for V_{BB} = −5.0V, V_{DD} = 12.0V, T_A = 25°C, t_{CYC} = 0.425 μs, t_{RAS} = 0.3 μs. t_{REF} = 2000 μs:

1. 128 cycle ($\overline{RAS}$ before $\overline{CAS}$): P_{OP} = 12.0V × 43 mA = 516 mW

$$P_{REF} = 516\left(128\frac{0.425}{2000}\right) + (12 \times 1.2 + 5 \times 0.001)\left(1 - 128\frac{0.425}{2000}\right)$$

P_{REF} = 28.0 mW

2. 64 cycle ($\overline{CAS}$ before $\overline{RAS}$); P_{OP} = 12.0V × 43 (0.96) mA = 495 mW.

$$P_{REF} = 495\left(64\frac{0.425}{2000}\right) + (12 \times 1.2 + 5 \times 0.001)\left(1 - 64\frac{0.425}{2000}\right) =$$

P_{REF} = 20.9 mW

3. 128 cycle ($\overline{RAS}$ only): P_{OP} = 12.0V × 43 (0.78) mA = 402 mW

P_{REF} = 25.0 mW

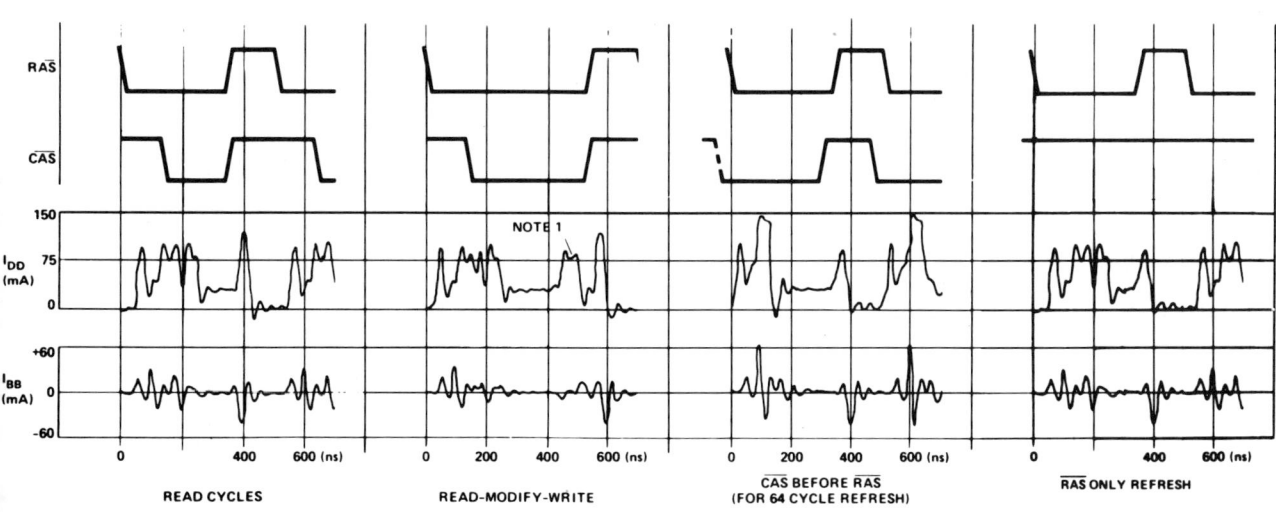

Note 1: Increase in current due to $\overline{WE}$ going low. Width of this current pulse is independent of $\overline{WE}$ pulse width.

Figure 3. Supply Current Waveforms.

A.C. Characteristics [1]

$T_A=0°C$ to $70°C$, $V_{DD}=12V \pm 10\%$, $V_{CC}=5V \pm 10\%$, $V_{BB}=-5V \pm 10\%$, $V_{SS}=0V$, unless otherwise noted.

READ, WRITE, READ-MODIFY-WRITE AND REFRESH CYCLES

Symbol	Parameter	2108-2 Min.	2108-2 Max.	2108-4 Min.	2108-4 Max.	Unit
t_{REF}	Time Between Refresh		2		2	ms
t_{RP}	$\overline{RAS}$ Precharge Time	75		95		ns
t_{CP}	$\overline{CAS}$ Precharge Time	100		125		ns
t_{RCL} [2]	$\overline{RAS}$ to $\overline{CAS}$ Leading Edge Lead Time	45	75	60	110	ns
t_{CRP} [3]	$\overline{CAS}$ to $\overline{RAS}$ Precharge Time	0		0		ns
t_{RSH}	$\overline{RAS}$ Hold Time	160		220		ns
t_{CSH}	$\overline{CAS}$ Hold Time	200		300		ns
t_{ASR}	Row Address Set-Up Time	0		0		ns
t_{ASC}	Column Address Set-Up Time	−10		−10		ns
t_{AH}	Address Hold Time	45		60		ns
t_T	Transition Time (Rise and Fall)		50		50	ns
t_{OFF}	Output Buffer Turn Off Delay	0	60	0	80	ns
t_{CAC} [4]	Access Time From $\overline{CAS}$		125		190	ns
t_{RAC} [4]	Access Time From $\overline{RAS}$		200		300	ns

READ AND REFRESH CYCLES

Symbol	Parameter	2108-2 Min.	2108-2 Max.	2108-4 Min.	2108-4 Max.	Unit
t_{CYC} [5]	Random Read Cycle Time	350		425		ns
t_{RAS}	$\overline{RAS}$ Pulse Width	275	32000	330	32000	ns
t_{CAS}	$\overline{CAS}$ Pulse Width	125	3000	190	3000	ns
t_{CH}	$\overline{CAS}$ Hold Time for $\overline{RAS}$-Only Refresh	30		30		ns
t_{CPR}	$\overline{CAS}$ Precharge for 64 Cycle Refresh	30		30		ns
t_{RCH}	Read Command Hold Time	20		20		ns
t_{RCS}	Read Command Set-Up Time	0		0		ns
t_{DOH}	Data-Out Hold Time	32		32		μs

WRITE CYCLE

Symbol	Parameter	2108-2 Min.	2108-2 Max.	2108-4 Min.	2108-4 Max.	Unit
t_{CYC} [5]	Random Write Cycle Time	350		425		ns
t_{RAS}	$\overline{RAS}$ Pulse Width	275	32000	330	32000	ns
t_{CAS}	$\overline{CAS}$ Pulse Width	125	10000	190	10000	ns
t_{WCH}	Write Command Hold Time	75		100		ns
t_{WP}	Write Command Pulse Width	50		100		ns
t_{RWL}	Write Command to $\overline{RAS}$ Lead Time	125		200		ns
t_{CWL}	Write Command to $\overline{CAS}$ Lead Time	100		160		ns
t_{DS} [6]	Data-In Set-Up Time	0		0		ns
t_{DH} [6]	Data-In Hold Time	100		125		ns

Notes:
1. All voltages referenced to V_{SS}.
2. $\overline{CAS}$ must remain at V_{IH} a minimum of t_{RCL} MIN after $\overline{RAS}$ switches to V_{IL}. To achieve the minimum guaranteed access time (t_{RAC}), $\overline{CAS}$ must switch to V_{IL} at or before t_{RCL} (MAX) = t_{RAC} − t_{CAC}. Device operation is not guaranteed for t_{RCL}>2 μs.
3. The t_{CRP} specification is less restrictive than the t_{CRL} range which was specified in the 2108 preliminary data sheet.
4. Load = 1 TTL and 50 pF.
5. The minimum cycle timing does not allow for t_T or skews.
6. Referenced to $\overline{CAS}$ or $\overline{WE}$, whichever occurs last.

Waveforms

READ CYCLE

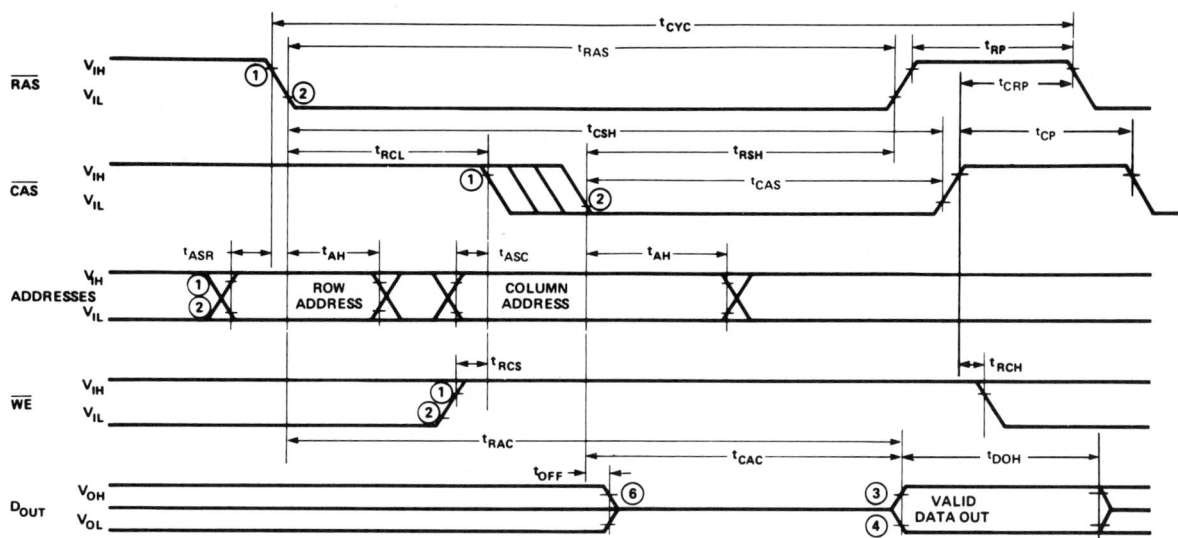

WRITE CYCLE

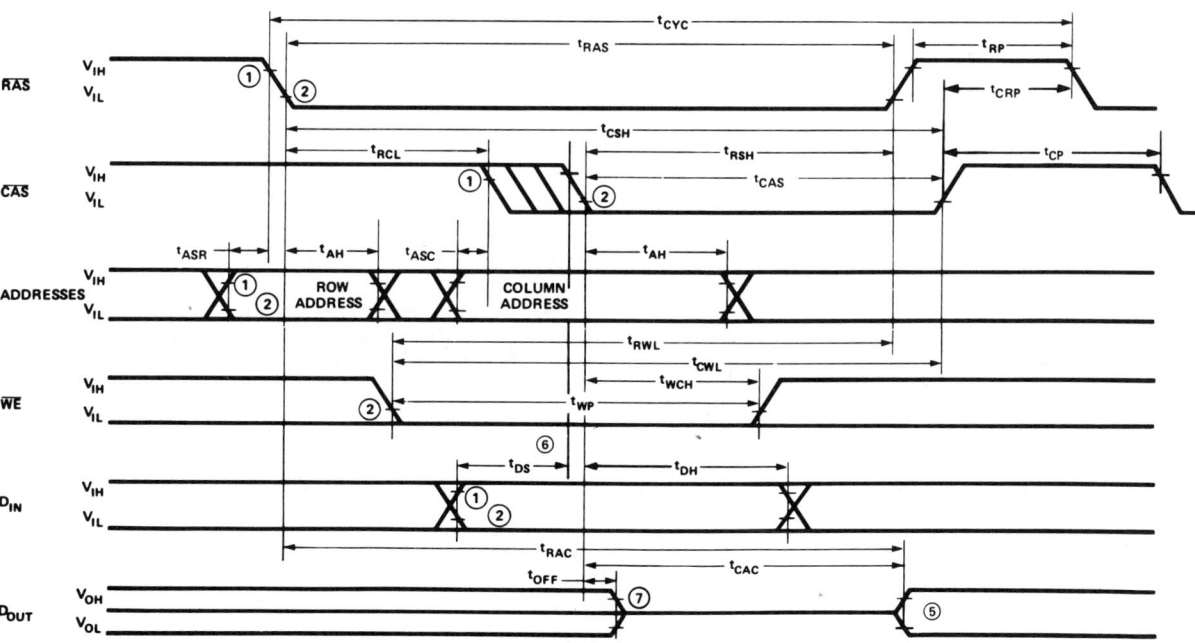

Notes:
1,2. $V_{IH\ MIN}$ and $V_{IL\ MAX}$ are reference levels for measuring timing of input signals.
3,4. $V_{OH\ MIN}$ and $V_{OL\ MAX}$ are reference levels for measuring timing of D_{OUT}.
5. D_{OUT} follows D_{IN} when writing, with $\overline{WE}$ before $\overline{CAS}$.
6. Referenced to $\overline{CAS}$ or $\overline{WE}$, whichever occurs last.
7. t_{OFF} is measured to $I_{OUT} \leq |I_{LO}|$.

DYNAMIC RAMS

A.C. Characteristics

T_A = 0°C to 70°C, V_{DD} = 12V ±10%, V_{CC} = 5V ±10%, V_{BB} = –5V ±10%, V_{SS} = 0V, unless otherwise noted.

READ-MODIFY-WRITE CYCLE

Symbol	Parameter	2108-2 Min.	2108-2 Max.	2108-4 Min.	2108-4 Max.	Unit
t_{RMW}	Read-Modify-Write Cycle Time	400		595		ns
t_{CRW}	RMW Cycle $\overline{CAS}$ Width	225	3000	350	3000	ns
t_{RRW}	RMW Cycle $\overline{RAS}$ Width	325	32000	500	32000	ns
t_{RWH}	RMW Cycle $\overline{RAS}$ Hold Time	250		390		ns
t_{CWH}	RMW Cycle $\overline{CAS}$ Hold Time	300		460		ns
t_{RWL}	Write Command to $\overline{RAS}$ Lead Time	125		200		ns
t_{CWL}	Write Command to $\overline{CAS}$ Lead Time	100		160		ns
t_{WP}	Write Command Pulse Width	50		100		ns
t_{RCS}	Read Command Set-Up Time	0		0		ns
t_{MOD}	Modify Time	0	10	0	10	μs
t_{DS}	Data-In Set-Up Time	0		0		ns
t_{DHM}	Data-In Hold Time (RMW Cycle)	50		125		ns

Waveforms

READ MODIFY WRITE CYCLE

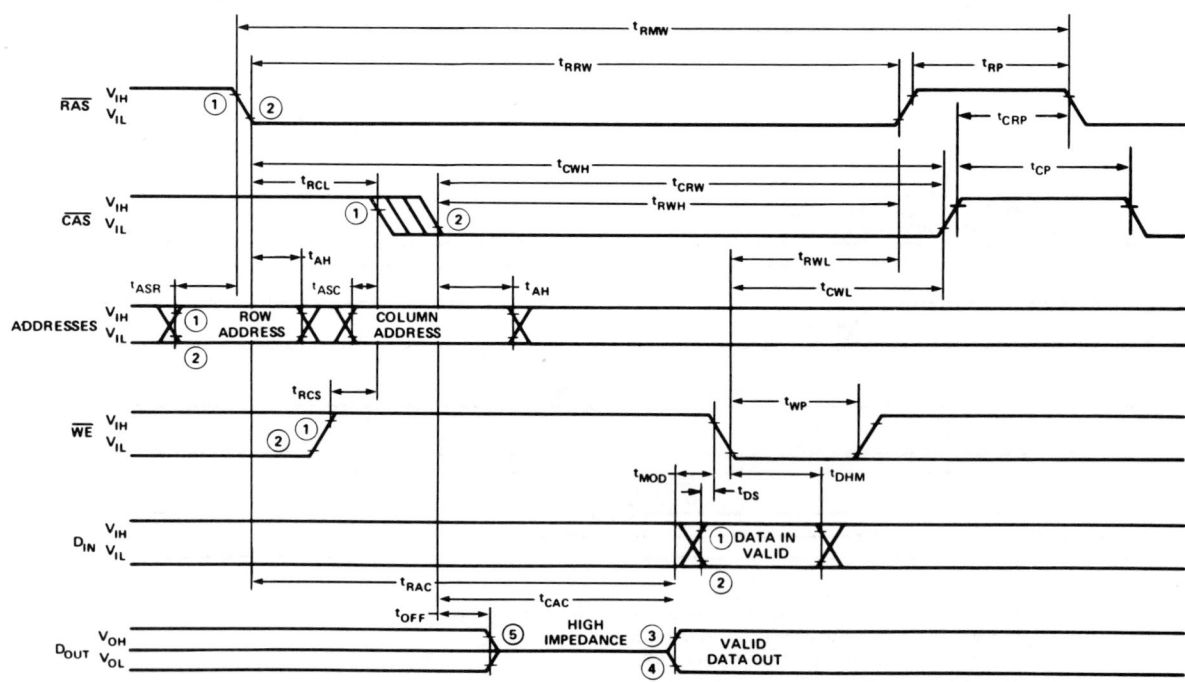

Notes: 1,2. V_{IHMIN} and V_{ILMAX} are reference levels for measuring timing of input signals.
       3,4. V_{OHMIN} and V_{OLMAX} are reference levels for measuring timing of D_{OUT}.
        5. t_{OFF} is measured to $I_{OUT} \le |I_{LO}|$.

Refresh Cycle Waveforms

$\overline{CAS}$ BEFORE $\overline{RAS}$ CYCLES. (64 CYCLE REFRESH)

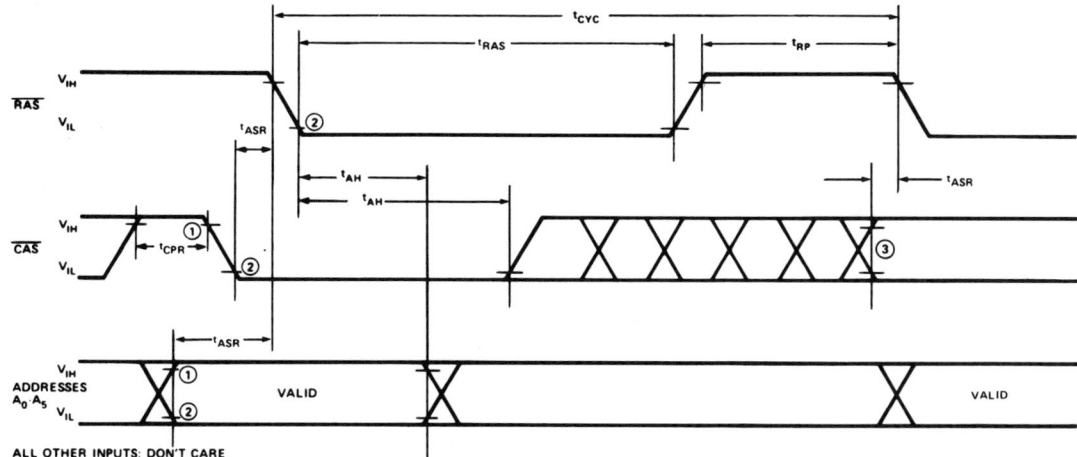

$\overline{RAS}$ ONLY CYCLES (128 CYCLE REFRESH)

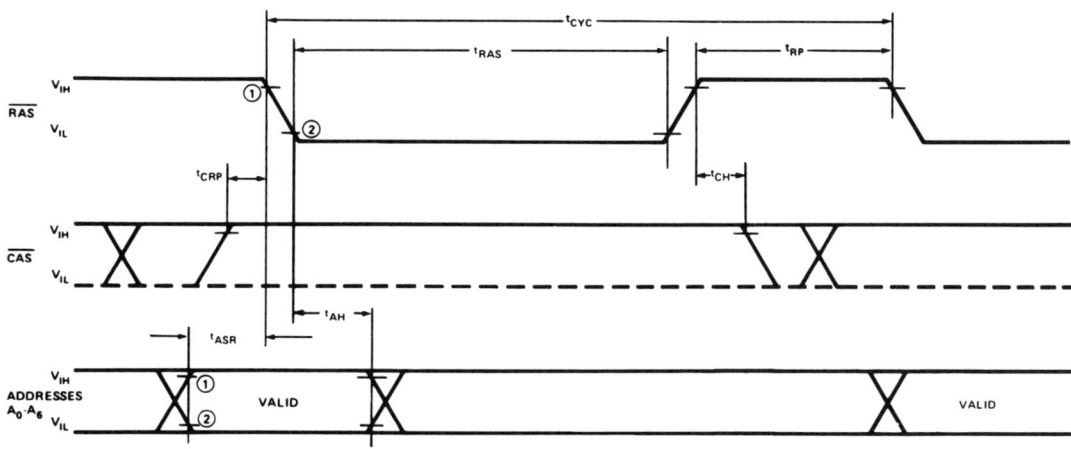

Notes: 1,2. V_{IHMIN} and V_{ILMAX} are reference levels for measuring timing of input signals.
3. $\overline{CAS}$ must be high or low as appropriate for the next cycle.

Applications Information

The 2108 may be refreshed in any of three modes: read cycles with $\overline{RAS}$ before $\overline{CAS}$ timing as shown on page 5, $\overline{RAS}$ only cycles (page 7), or $\overline{CAS}$ before $\overline{RAS}$ cycles (page 7). In all three modes A_6 must be held high for the S1572 and S1627 or low for the S1573 and S1626. The row addressed by A_0 through A_5 is refreshed. Therefore, 64 cycles are required to refresh the stored data.

The $\overline{CAS}$-before-$\overline{RAS}$ mode is useful in the 2116 as a technique for increasing memory availability and minimizing standby power dissipation by requiring only 64 refresh cycles every 2 ms. Systems employing the 2108 in a $\overline{CAS}$-before-$\overline{RAS}$ refresh mode can be easily upgraded to the most efficient 16K RAM capability.

Since the 2108 input pin A_6 supplies two system addresses (A_6 and A_{13}) to the internal memory array, it is not possible to simply tie this input high or low. The 2108 Input A_6 must be tied to the appropriate level only during row address strobe ($\overline{RAS}$) and then used to supply the high order system address A_{13} during column address strobe ($\overline{CAS}$). Control of A_6 in a system may be implemented as shown at right. In this circuit the output A_6 of multiplexer M supplies the appropriate high or low level (determined by S1572, S1627, S1573, or S1626) during $\overline{RAS}$ for both a memory cycle and refresh cycle. During $\overline{CAS}$, system address A_{13} is multiplexed on A_6 as shown. See the 2116 section for additional applications information.

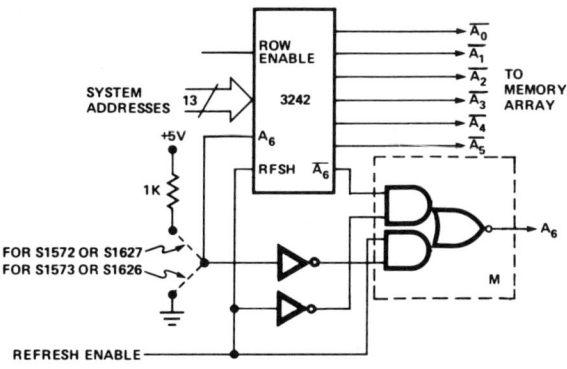

Power Supply Decoupling/Distribution

Power supply current waveforms for the 2108 are shown in Figure 3. The V_{DD} supply provides virtually all of the operating current for the 2108. The V_{DD} supply current, I_{DD}, has two components: transient current peaks when the clocks change state and a DC component while the clocks are active (low). When selecting the decoupling capacitors for the V_{DD} supply, the characteristics of capacitors as well as the current waveform must be considered. Suppression of transient or pulse currents require capacitors with small physical size and low inherent inductance. Monolithic and other ceramic capacitors exhibit these desirable characteristics. When the current waveform indicates a DC component, bulk capacity must be located near the current load to supply the load power. Inductive effects of PC board traces and bus bars preclude supplying the DC component from bulk capacitors at the periphery of a memory matrix without voltage droop during the active portion of a memory cycle. This means that some bulk capacity in the form of electrolytic or large ceramic capacitors should be distributed around or within the memory matrix.

The V_{BB} supply current, I_{BB}, has high transient current peaks, with essentially no DC component (less than 400 microamperes). The V_{BB} capacitors should be selected for transient suppression characteristics. The following capacitance values and locations are recommended for the 2108:

1. A 0.33 μF ceramic capacitor between V_{DD} and V_{SS} (ground) at every other device.

2. A 0.1 μF ceramic capacitor between V_{BB} and V_{SS} at every other device (preferably alternate devices to the V_{DD} decoupling above).

3. A 4.7 μF electrolytic capacitor between V_{DD} and V_{SS} for each eight devices and located adjacent to the devices.

The V_{CC} supply is connected only to the 2108 output buffer and is not used internally. The load current from the V_{CC} supply is dependent only upon the output loading and is usually only the input high level current to a TTL gate and the output leakage currents of any OR-tied 2108 (typically 100 μA or less total). Intel recommends that a 0.1 or 0.01 μF ceramic capacitor be connected between V_{CC} and V_{SS} for every eight devices to preclude coupled noise from affecting the TTL devices in the system.

Intel recommends a power supply distribution system such that each power supply is grided both horizontally and vertically at each memory device. This technique minimizes the power distribution system impedance and enhances the effect of the decoupling capacitors.

Output Data Latch

The 2108 contains an output data latch eliminating the need for an external system data latch and the timing circuitry required to strobe an external latch. The output latch operates identically to the 16-pin 4K RAM (Intel 2104) output latch enhancing the system compatibility of the 16K and 4K devices.

Operation of the output latch is controlled by $\overline{CAS}$. The data output will go to the high-impedance state immediately following the $\overline{CAS}$ leading edge during each data cycle and will either go to valid data at access time on selected devices (devices receiving both $\overline{RAS}$ and $\overline{CAS}$) or will remain in the high impedance state on unselected devices (devices receiving only $\overline{CAS}$). During $\overline{RAS}$-only refresh cycles, the data output remains in the state it was prior to the $\overline{RAS}$-only cycle. This unique feature of latched output RAMs allows a refresh cycle to be hidden among data cycles without impacting data availability. For instance, a $\overline{RAS}$-only refresh cycle could follow each data cycle in a microprocessor system but the accessed data would remain at the device output and the microprocessor could take the data at any time within the cycle. Non-latched output devices do not provide this type of hidden refresh capability since their data output would go to the high impedance state at the end of the data cycle.

Page Mode Operation

The 2108 is designed for page mode operation and is presently being characterized for that mode. Specifications will be available at a later date.

Packaging Information

16-LEAD HERMETIC DUAL IN-LINE PACKAGE

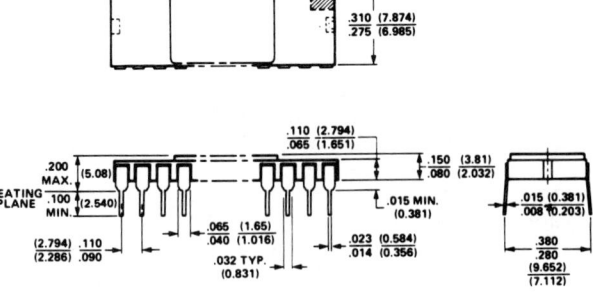

2116

16,384 X 1 BIT DYNAMIC RAM

	2116-2	2116-3	2116-4
Max. Access Time (ns)	200	250	300
Read, Write Cycle (ns)	350	375	425
Read-Modify-Write Cycle (ns)	400	525	595

- **Highest Density 16K RAM: Industry Standard 16 Pin Package**
- **Low Standby Power**
- **All Inputs Including Clocks TTL Compatible**
- **±10% Tolerance on all Power Supplies +12V, +5V, −5V**
- **On-Chip Latches for Address and Data In**
- **Only 64 Refresh Cycles Required Every 2 ms**
- **Output is Three-State, TTL Compatible; Data is Latched and Valid into Next Cycle**

The Intel® 2116 is a 16,384 word by 1 bit MOS RAM fabricated with two layer polysilicon N-MOS technology — a production-proven process for high performance, high reliability, and high functional density. The 2116 uses a single transistor dynamic storage cell and dynamic circuitry to achieve high speed and low power dissipation.

The unique design of the 2116 allows it to be packaged in the industry standard 16 pin dual-in-line package. The 16 pin package provides the highest system bit densities and is compatible with widely available automated handling equipment. The 2116 is designed to facilitate upgrading of 2104A-type 4K RAM systems to 16K capabilities.

The use of the 16 pin package is made possible by multiplexing the 14 address bits (required to address 1 of 16,384 bits) into the 2116 on 7 address input pins. The two 7 bit address words are latched into the 2116 by the two TTL clocks, Row Address Strobe ($\overline{RAS}$) and Column Address Strobe ($\overline{CAS}$). Non-critical clock timing requirements allow use of the multiplexing technique while maintaining high performance.

The single transistor dynamic storage cell provides high speed along with low power dissipation. The memory cell requires refreshing for data retention. Refreshing can be accomplished every 2 ms by any one of the three following methods: (1) $\overline{CAS}$ before $\overline{RAS}$ cycles on 64 addresses, A_0–A_5, (2) $\overline{RAS}$-only cycles on 128 address, A_0–A_6, or (3) normal read or write cycles on 128 addresses, A_0–A_6. A write cycle will refresh stored data on all bits of the selected row except the bit which is addressed. The output is brought to a high impedance state by a $\overline{CAS}$-only cycle or by a $\overline{CAS}$-before-$\overline{RAS}$ refresh cycle.

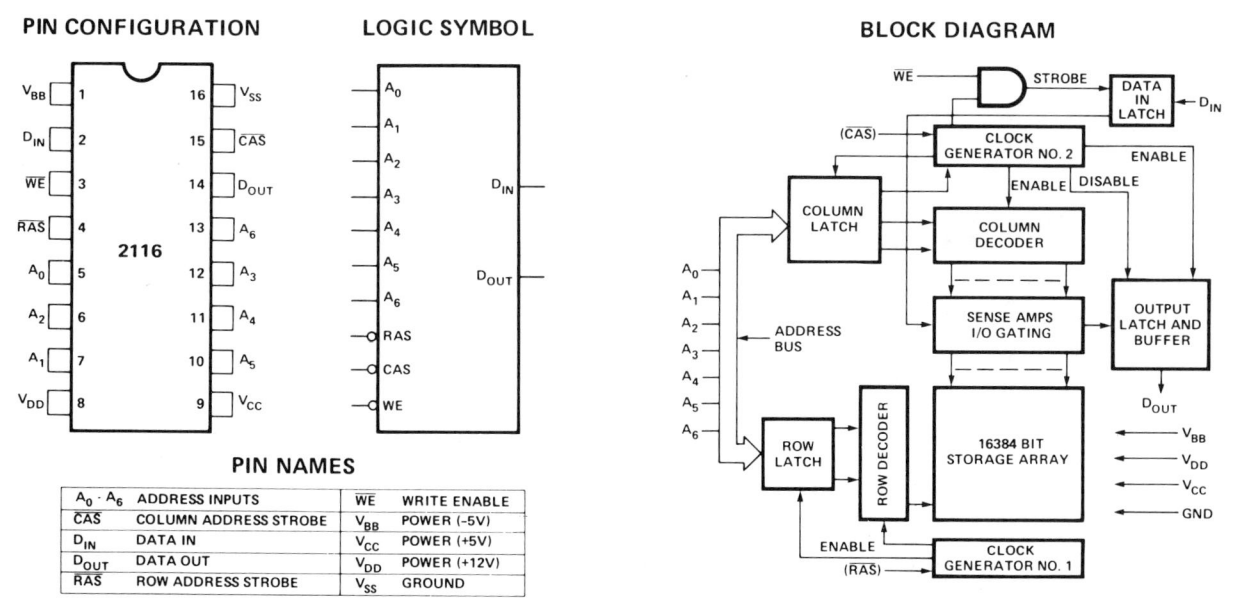

Absolute Maximum Ratings*

Ambient Temperature Under Bias	−10°C to +80°C
Storage Temperature	−65°C to +150°C
Voltage on any Pin Relative to V_{BB} ($V_{SS} − V_{BB} \geq 4V$)	−0.3V to +20V
Power Dissipation	1.25W

*COMMENT:
Stresses above those listed under "Absolute Maximum Ratings" may cause permanent damage to the device. This is a stress rating only and functional operation of the device at these or any other conditions above those indicated in the operational sections of this specification is not implied. Exposure to absolute maximum rating conditions for extended periods may affect device reliability.

D.C. and Operating Characteristics [1], [2]

$T_A = 0°C$ to $70°C$, $V_{DD} = +12V \pm 10\%$, $V_{CC} = +5V \pm 10\%$, $V_{BB} = -5V \pm 10\%$, $V_{SS} = 0V$, unless otherwise noted.

Symbol	Parameter	Min.	Typ.(3)	Max.	Unit	Conditions
I_{LI}	Input Load Current (any input)			10	µA	$V_{IN} = V_{IL\,MIN}$ to $V_{IH\,MAX}$
$|I_{LO}|$	Output Leakage Current for high impedance state		0.1	10	µA	Chip deselected: $\overline{RAS}$ and $\overline{CAS}$ at V_{IH} $V_{OUT} = 0$ to 5.5V
I_{DD1}	V_{DD} Supply Current		1.2	2	mA	$\overline{CAS}$ and $\overline{RAS}$ at V_{IH} or $\overline{CAS}$-only cycle. Chip deselected prior to measurement. See Note 5.
I_{BB1}	V_{BB} Supply Current		1	50	µA	
I_{DD2}[4]	Operating V_{DD} Current		53	69	mA	2116-2 t_{CYC} = 350 ns — $T_A = 25°C$ Device selected. See Note 6.
			51	68	mA	2116-3 t_{CYC} = 375 ns
			49	65	mA	2116-4 t_{CYC} = 425 ns
I_{BB2}	Operating V_{BB} Current		120	400	µA	Device selected
I_{CC1}[7]	V_{CC} Supply Current when deselected			10	µA	
V_{IL}	Input Low Voltage (any input)	−1.0		0.8	V	
V_{IH}	Input High Voltage (any input)	2.4		$V_{CC}+1$	V	
V_{OL}	Output Low Voltage	0.0		0.4	V	I_{OL} = 4.1 mA (Read Cycle Only)
V_{OH}	Output High Voltage	2.4		V_{CC}	V	I_{OH} = −5 mA (Read Cycle Only)

Capacitance [8]

$T_A = 25°C$, $V_{DD} = 12V \pm 10\%$, $V_{CC} = 5V \pm 10\%$, $V_{BB} = -5V \pm 10\%$, $V_{SS} = 0V$, unless otherwise noted.

Symbol	Parameter	Typ.	Max.	Unit	Conditions
C_{I1}	Address, Data In & $\overline{WE}$ Capacitance	4	7	pF	$V_{IN} = V_{SS}$
C_{I2}	$\overline{RAS}$ Capacitance	3	5	pF	$V_{IN} = V_{SS}$
C_{I3}	$\overline{CAS}$ Capacitance	6	10	pF	$V_{IN} = V_{SS}$
C_O	Data Output Capacitance	3	7	pF	$V_{OUT} = 0V$

Notes:
1. All voltages referenced to V_{SS}. No power supply sequencing is required but V_{DD}, V_{CC}, and V_{SS} should never be 0.3V or more negative than V_{BB}.
2. To avoid self-clocking, $\overline{RAS}$ should not be allowed to float.
3. Typical values are for $T_A = 25°C$ and nominal power supply voltages.
4. For $\overline{RAS}$-only refresh $I_{DD} = 0.78\ I_{DD2}$. For $\overline{CAS}$-before-$\overline{RAS}$ (64 cycle refresh) $I_{DD} = 0.96\ I_{DD2}$.
5. The chip is deselected (i.e., output is brought to high impedance state) by $\overline{CAS}$-only cycle or by $\overline{CAS}$-before-$\overline{RAS}$ cycle. The current flowing in a selected (i.e., output on) chip with $\overline{RAS}$ and $\overline{CAS}$ at V_{IH} is approximately twice I_{DD1}.
6. See Page 3 for typical I_{DD} characteristics under other conditions.
7. When chip is selected V_{CC} supply current is dependent on output loading; V_{CC} is connected to output buffer only.
8. Capacitance measured with Boonton Meter.

Typical Characteristics

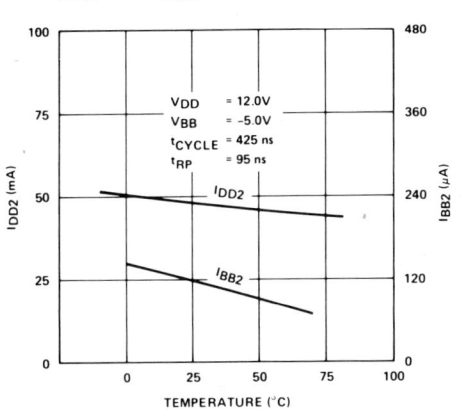

Figure 1.

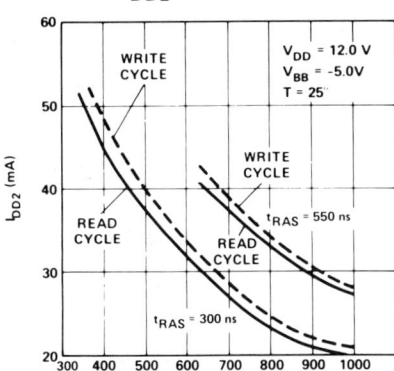

Figure 2.

Standby Power Calculations:

$$P_{REF} = P_{OP} \left(N \frac{t_{CYC}}{t_{REF}}\right) + P_{SB} \left(1 - N \frac{t_{CYC}}{t_{REF}}\right) \text{ where}$$

P_{OP} = Power dissipation (continuous operation) $\cong V_{DD} \times I_{DD2}$.

N = Number of refresh cycles (64 or 128)

t_{CYC} = Cycle time for a refresh cycle.

t_{REF} = Time between refreshes

P_{SB} = Standby power dissipation = $V_{DD} \times I_{DD1} + |V_{BB}| \times I_{BB}$

Note that I_{DD2} depends upon refresh as follows:

1. For 128 cycle ($\overline{RAS}$ before $\overline{CAS}$) use I_{DD2} from Figures 1 and 2.
2. For 64 cycle ($\overline{CAS}$ before $\overline{RAS}$) multiply I_{DD2} determined in (1) by 0.96.
3. For 128 cycle ($\overline{RAS}$ only) multiply I_{DD2} determined in (1) by 0.78.

Examples of typical calculations for V_{BB} = -5.0V, V_{DD} = 12.0V, T_A = 25°C, t_{CYC} = 0.425 μs, t_{RAS} = 0.3 μs, t_{REF} = 2000 μs:

1. 128 cycle ($\overline{RAS}$ before $\overline{CAS}$): P_{OP} = 12.0V x 43 mA = 516 mW

$$P_{REF} = 516 \left(128 \frac{0.425}{2000}\right) + (12 \times 1.2 + 5 \times 0.001)\left(1 - 128 \frac{0.425}{2000}\right)$$

P_{REF} = 28.0 mW

2. 64 cycle ($\overline{CAS}$ before $\overline{RAS}$); P_{OP} = 12.0V x 43 (0.96) mA = 495 mW.

$$P_{REF} = 495 \left(64 \frac{0.425}{2000}\right) + (12 \times 1.2 + 5 \times 0.001)\left(1 - 64 \frac{0.425}{2000}\right) =$$

P_{REF} = 20.9 mW

3. 128 cycle ($\overline{RAS}$ only): P_{OP} = 12.0V x 43 (0.78) mA = 402 mW

P_{REF} = 25.0 mW

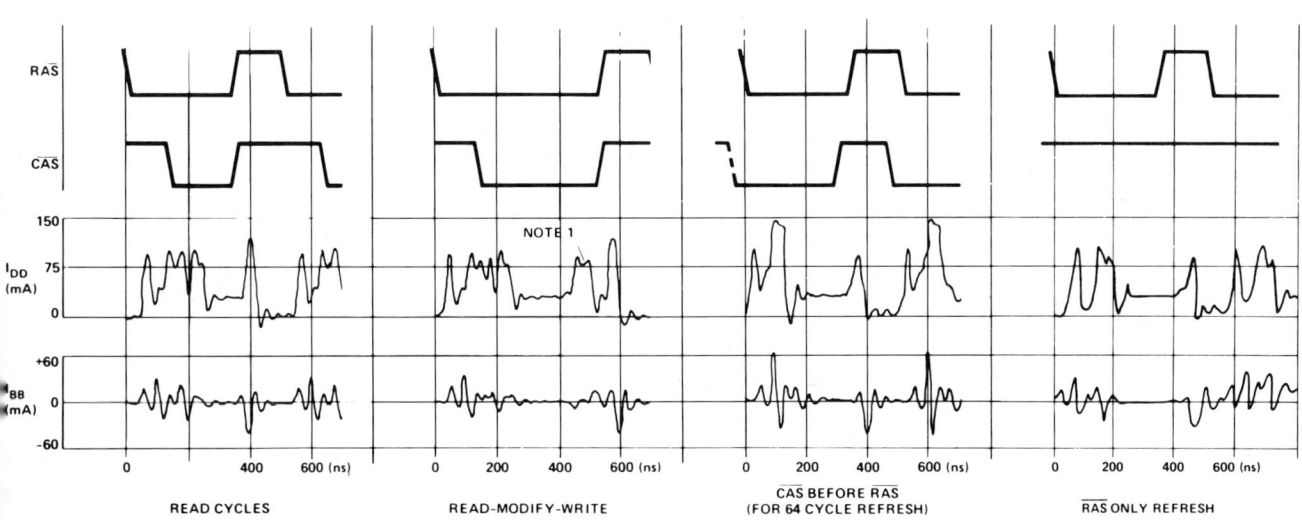

Note 1: Increase in current due to $\overline{WE}$ going low. Width of this current pulse is independent of $\overline{WE}$ pulse width.

Figure 3. Supply Current Waveforms.

A.C. Characteristics[1]

T_A=0°C to 70°C, V_{DD}=12V ±10%, V_{CC}=5V ±10%, V_{BB}=−5V ±10%, V_{SS}=0V, unless otherwise noted.

READ, WRITE, READ-MODIFY-WRITE AND REFRESH CYCLES

Symbol	Parameter	2116-2 Min.	2116-2 Max.	2116-3 Min.	2116-3 Max.	2116-4 Min.	2116-4 Max.	Unit
t_{REF}	Time Between Refresh		2		2		2	ms
t_{RP}	RAS Precharge Time	75		75		95		ns
t_{CP}	CAS Precharge Time	100		125		125		ns
t_{RCL}[2]	RAS to CAS Leading Edge Lead Time	45	75	50	110	60	110	ns
t_{CRP}[3]	CAS to RAS Precharge Time	0		0		0		ns
t_{RSH}	RAS Hold Time	160		200		220		ns
t_{CSH}	CAS Hold Time	200		250		300		ns
t_{ASR}	Row Address Set-Up Time	0		0		0		ns
t_{ASC}	Column Address Set-Up Time	−10		−10		−10		ns
t_{AH}	Address Hold Time	45		50		60		ns
t_T	Transition Time (Rise and Fall)		50		50		50	ns
t_{OFF}	Output Buffer Turn Off Delay	0	60	0	60	0	80	ns
t_{CAC}[4]	Access Time From CAS		125		150		190	ns
t_{RAC}[4]	Access Time From RAS		200		250		300	ns

READ AND REFRESH CYCLES

Symbol	Parameter	2116-2 Min.	2116-2 Max.	2116-3 Min.	2116-3 Max.	2116-4 Min.	2116-4 Max.	Unit
t_{CYC}[5]	Random Read Cycle Time	350		375		425		ns
t_{RAS}	RAS Pulse Width	275	32000	300	32000	330	32000	ns
t_{CAS}	CAS Pulse Width	125	10000	150	10000	190	10000	ns
t_{CH}	CAS Hold Time for RAS-Only Refresh	30		30		30		ns
t_{CPR}	CAS Precharge for 64 Cycle Refresh	30		30		30		ns
t_{RCH}	Read Command Hold Time	20		20		20		ns
t_{RCS}	Read Command Set-Up Time	0		0		0		ns
t_{DOH}	Data-Out Hold Time	32		32		32		μs

WRITE CYCLE

Symbol	Parameter	2116-2 Min.	2116-2 Max.	2116-3 Min.	2116-3 Max.	2116-4 Min.	2116-4 Max.	Unit
t_{CYC}[5]	Random Write Cycle Time	350		375		425		ns
t_{RAS}	RAS Pulse Width	275	32000	300	32000	330	32000	ns
t_{CAS}	CAS Pulse Width	125	10000	150	10000	190	10000	ns
t_{WCH}	Write Command Hold Time	75		100		100		ns
t_{WP}	Write Command Pulse Width	50		100		100		ns
t_{RWL}	Write Command to RAS Lead Time	125		200		200		ns
t_{CWL}	Write Command to CAS Lead Time	100		150		160		ns
t_{DS}[6]	Data-In Set-Up Time	0		0		0		ns
t_{DH}[6]	Data-In Hold Time	100		100		125		ns

Notes:
1. All voltages referenced to V_{SS}.
2. CAS must remain at V_{IH} a minimum of t_{RCL} MIN after RAS switches to V_{IL}. To achieve the minimum guaranteed access time (t_{RAC}), CAS must switch to V_{IL} at or before t_{RCL} (MAX) = t_{RAC} −t_{CAC}. Device operation is not guaranteed for t_{RCL}>2 μs.
3. The t_{CRP} specification is less restrictive than the t_{CRL} range which was specified in the 2116 preliminary data sheet.
4. Load = 1 TTL and 50 pF.
5. The minimum cycle timing does not allow for t_T or skews.
6. Referenced to CAS or WE, whichever occurs last.

Waveforms

READ CYCLE

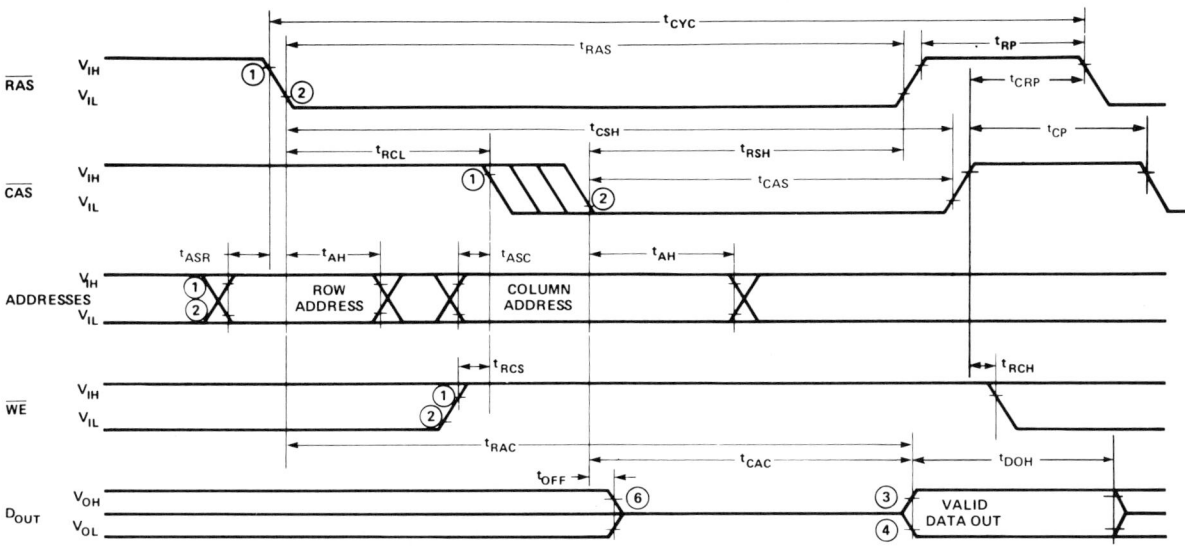

WRITE CYCLE

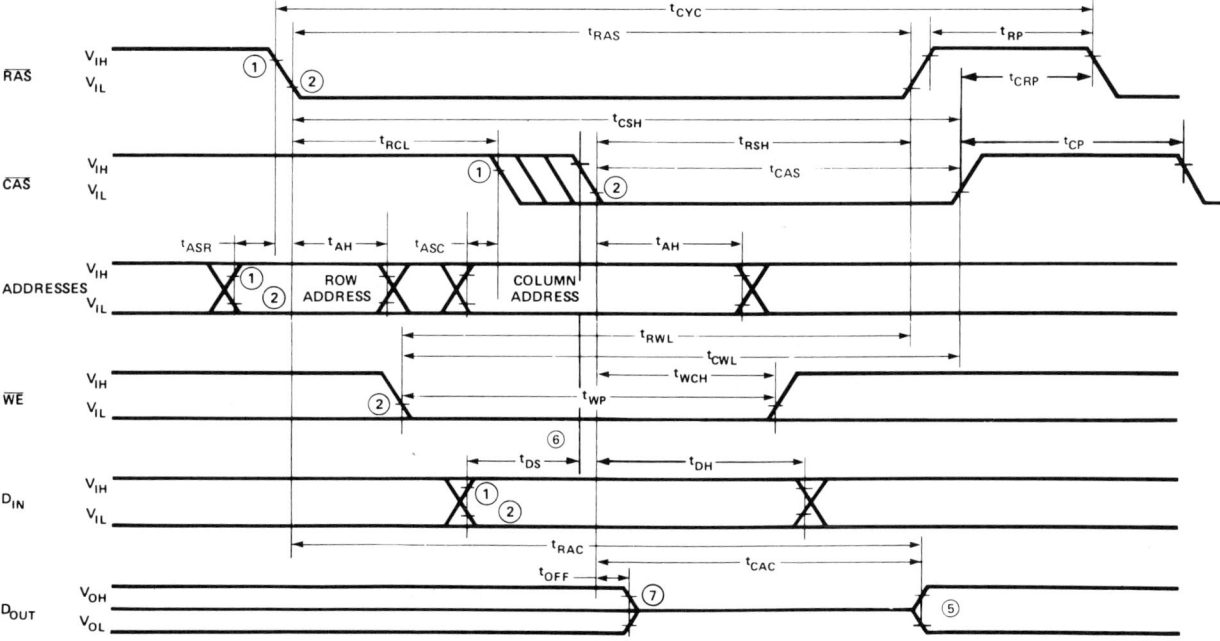

Notes: 1,2. V_{IH} MIN and V_{IL} MAX are reference levels for measuring timing of input signals.
3,4. V_{OH} MIN and V_{OL} MAX are reference levels for measuring timing of D_{OUT}.
5. D_{OUT} follows D_{IN} when writing, with $\overline{WE}$ before $\overline{CAS}$.
6. Referenced to $\overline{CAS}$ or $\overline{WE}$, whichever occurs last.
7. t_{OFF} is measured to $I_{OUT} \leq |I_{LO}|$.

DYNAMIC RAMS

A.C. Characteristics

$T_A = 0°C$ to $70°C$, $V_{DD} = 12V \pm 10\%$, $V_{CC} = 5V \pm 10\%$, $V_{BB} = -5V \pm 10\%$, $V_{SS} = 0V$, unless otherwise noted.

READ-MODIFY-WRITE CYCLE

Symbol	Parameter	2116-2 Min.	2116-2 Max.	2116-3 Min.	2116-3 Max.	2116-4 Min.	2116-4 Max.	Unit
t_{RMW}	Read-Modify-Write Cycle Time	400		525		595		ns
t_{CRW}	RMW Cycle $\overline{CAS}$ Width	225	10000	310	10000	350	10000	ns
t_{RRW}	RMW Cycle $\overline{RAS}$ Width	325	32000	450	32000	500	32000	ns
t_{RWH}	RMW Cycle $\overline{RAS}$ Hold Time	250		350		390		ns
t_{CWH}	RMW Cycle $\overline{CAS}$ Hold Time	300		410		460		ns
t_{RWL}	Write Command to $\overline{RAS}$ Lead Time	125		200		200		ns
t_{CWL}	Write Command to $\overline{CAS}$ Lead Time	100		160		160		ns
t_{WP}	Write Command Pulse Width	50		100		100		ns
t_{RCS}	Read Command Set-Up Time	0		0		0		ns
t_{MOD}	Modify Time	0	10	0	10	0	10	μs
t_{DS}	Data-In Set-Up Time	0		0		0		ns
t_{DHM}	Data-In Hold Time (RMW Cycle)	50		100		125		ns

Waveforms

READ MODIFY WRITE CYCLE

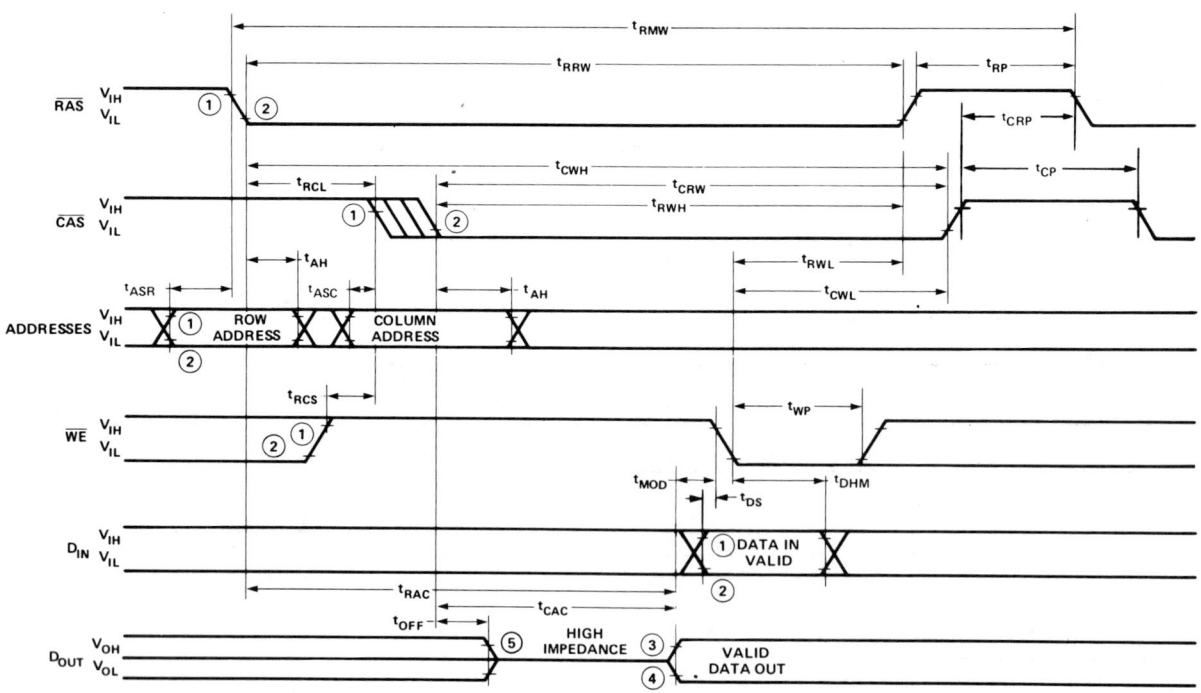

Notes: 1,2. V_{IHMIN} and V_{ILMAX} are reference levels for measuring timing of input signals.
3,4. V_{OHMIN} and V_{OLMAX} are reference levels for measuring timing of D_{OUT}.
5. t_{OFF} is measured to $I_{OUT} \leq |I_{LO}|$.

Refresh Cycle Waveforms

$\overline{CAS}$ BEFORE $\overline{RAS}$ CYCLES. (64 CYCLE REFRESH)

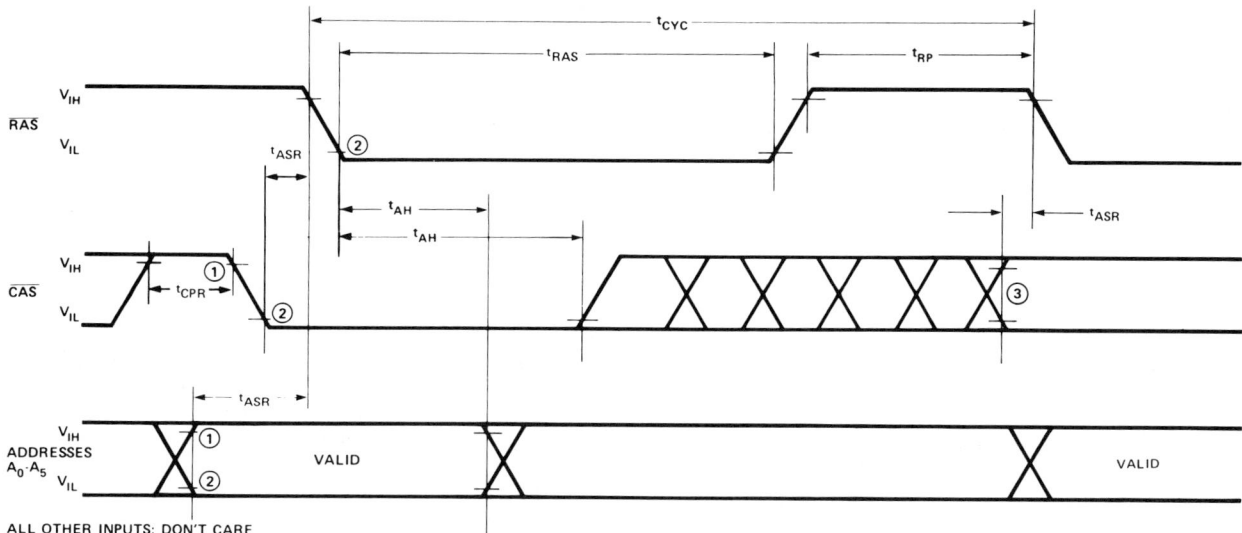

$\overline{RAS}$ ONLY CYCLES (128 CYCLE REFRESH)

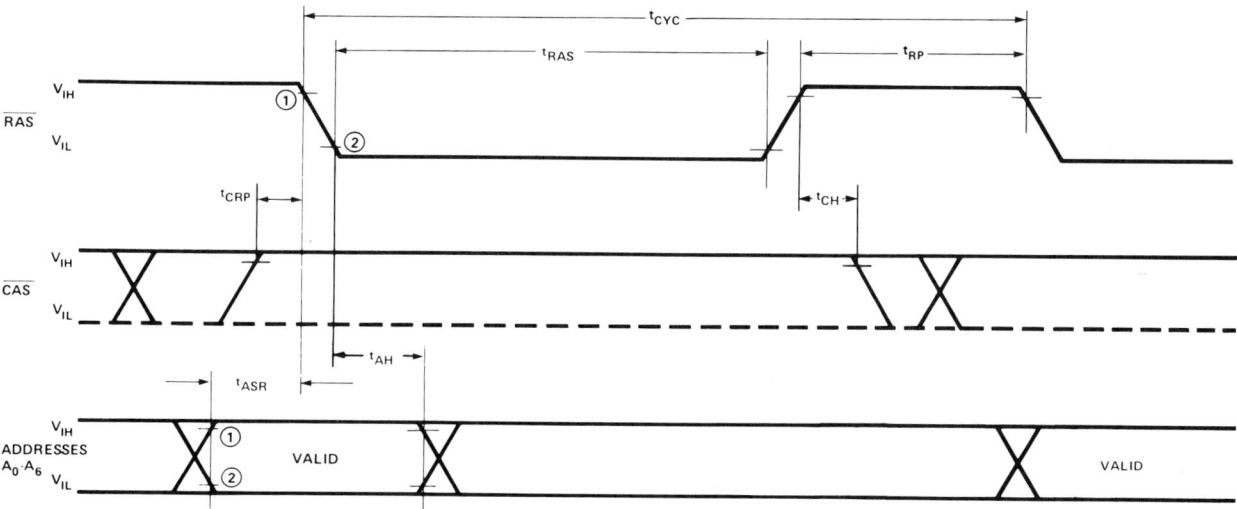

Notes: 1,2. V_{IHMIN} and V_{ILMAX} are reference levels for measuring timing of input signals.
3. $\overline{CAS}$ must be high or low as appropriate for the next cycle.

Applications Information

REFRESH MODES

The 2116 may be refreshed in any of three modes. Read/Refresh cycles and $\overline{RAS}$-only cycles refresh the row addressed by A_0 through A_6 and therefore require 128 cycles to refresh the stored data. Assuming a 500 nsec system cycle time, the refresh operations require 64 μsec out of each 2.0 msec refresh period or 3.2% of the available memory time. The third 2116 refresh mode, $\overline{CAS}$-before-$\overline{RAS}$, allows refresh of the stored data in only 64 cycles and requires only 32 μsec or 1.6% of the available memory time (equal to the 64-cycle refresh 4K RAMs). While some 2116 aplications would not be impacted by the 3.2% memory lockout time using 128 cycle refresh, most large mainframe memory applications would suffer throughput degradation in that refresh mode. Intel designed the 2116 to allow either 128-cycle or 64-cycle refresh, allowing the system designer to choose the refresh mode which fits his system needs. In addition to allowing higher memory throughput, the $\overline{CAS}$-before-RAS 64-cycle refresh mode dissipates approximately 14% less power than the 128-cycle $\overline{RAS}$-only mode and 23% less power than the 128-cycle Read/Refresh mode (refer to the Standby Power Calculation section).

POWER SUPPLY DECOUPLING/DISTRIBUTION

Power supply current waveforms for the 2116 are shown in Figure 3. The V_{DD} supply provides virtually all of the operating current for the 2116. The V_{DD} supply current, I_{DD}, has two components: transient current peaks when the clocks change state and a DC component while the clocks are active (low). When selecting the decoupling capacitors for the V_{DD} supply, the characteristics of capacitors as well as the current waveform must be considered. Suppression of transient or pulse currents require capacitors with small physical size and low inherent inductance. Monolithic and other ceramic capacitors exhibit these desirable characteristics. When the current waveform indicates a DC component, bulk capacity must be located near the current load to supply the load power. Inductive effects of PC board traces and bus bars preclude supplying the DC component from bulk capacitors at the periphery of a memory matrix without voltage droop during the active portion of a memory cycle. This means that some bulk capacity in the form of electrolytic or large ceramic capacitors should be distributed around or within the memory matrix.

The V_{BB} supply current, I_{BB}, has high transient current peaks, with essentially no DC component (less than 400 microamperes). The V_{BB} capacitors should be selected for transient suppression characteristics. The following capacitance values and locations are recommended for the 2116:

1. A 0.33 μF ceramic capacitor between V_{DD} and V_{SS} (ground) at every other device.

2. A 0.1 μF ceramic capacitor between V_{BB} and V_{SS} at every other device (preferably alternate devices to the V_{DD} decoupling above).

3. A 4.7 μF electrolytic capacitor between V_{DD} and V_{SS} for each eight devices and located adjacent to the devices.

The V_{CC} supply is connected only to the 2116 output buffer and is not used internally. The load current from the V_{CC} supply is dependent only upon the output loading and is usually only the input high level current to a TTL gate and the output leakage currents of any OR-tied 2116s (typically 100 μA or less total). Intel recommends that a 0.1 or 0.01 μF ceramic capacitor be connected between V_{CC} and V_{SS} for every eight devices to preclude coupled noise from affecting the TTL devices in the system.

Intel recommends a power supply distribution system such that each power supply is grided both horizontally and vertically at each memory device. This technique minimizes the power distribution system impedance and enhances the effect of the decoupling capacitors.

OUTPUT DATA LATCH

The 2116 contains an output data latch eliminating the need for an external system data latch and the timing circuitry required to strobe an external latch. The 2116 output latch operates identically to the output latch found on all industry standard 16-pin, 4K RAMs and enhances the system compatibility of the 16K and 4K devices.

Operation of the output latch is controlled by $\overline{CAS}$. The data output will go to the high-impedance state immediately following the $\overline{CAS}$ leading edge during each data cycle and will either go to valid data at access time on selected devices (devices receiving both $\overline{RAS}$ and $\overline{CAS}$) or will remain in the high impedance state on unselected devices (devices receiving only $\overline{CAS}$). During $\overline{RAS}$-only refresh cycles, the data output remains in the state it was prior to the $\overline{RAS}$-only cycle. This unique feature of latched output RAMs allows a refresh cycle to be hidden among data cycles without impacting data availability. For instance, a $\overline{RAS}$-only refresh cycle could follow each data cycle in a microprocessor system but the accessed data would remain at the device output and the microprocessor could take the data at any time within the cycle. Non-latched output devices do not provide this type of hidden refresh capability since their data output would go to the high impedance state at the end of the data cycle.

PAGE MODE OPERATION

The 2116 is designed for page mode operation and is presently being characterized for that mode. Specifications will be available at a later date.

Packaging Information

16-LEAD HERMETIC DUAL IN-LINE PACKAGE

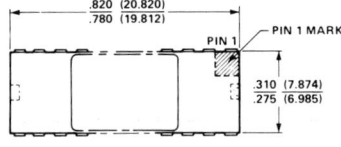

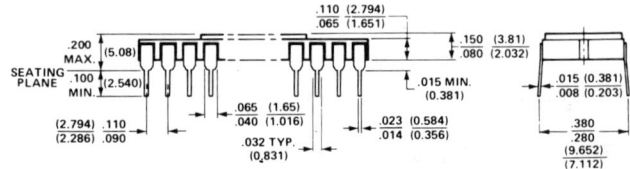

2117
16,384 x 1 BIT DYNAMIC RAM

	2117-2	2117-3	2117-4
Maximum Access Time (ns)	150	200	250
Read, Write Cycle (ns)	320	375	410
Read-Modify-Write Cycle (ns)	330	375	475

- Industry Standard 16-Pin Configuration
- ±10% Tolerance on All Power Supplies: +12V, +5V, -5V
- Low Power: 462mW Max. Operating, 20mW Max. Standby
- Low I_{DD} Current Transients
- All Inputs, Including Clocks, TTL Compatible
- Non-Latched Output is Three-State, TTL Compatible
- $\overline{RAS}$ Only Refresh
- 128 Refresh Cycles Required Every 2ms
- Page Mode Capability
- $\overline{CAS}$ Controlled Output Allows Hidden Refresh

The Intel® 2117 is a 16,384 word by 1-bit Dynamic MOS RAM fabricated with Intel's standard two layer polysilicon NMOS technology — a production proven process for high performance, high reliability, and high storage density.

The 2117 uses a single transistor dynamic storage cell and advanced dynamic circuitry to achieve high speed with low power dissipation. The circuit design minimizes the current transients typical of dynamic RAM operation. These low current transients and ±10% tolerance on all power supplies contribute to the high noise immunity of the 2117 in a system environment.

Multiplexing the 14 address bits into the 7 address input pins allows the 2117 to be packaged in the industry standard 16-pin DIP. The two 7-bit address words are latched into the 2117 by the two TTL clocks, Row Address Strobe ($\overline{RAS}$) and Column Address Strobe ($\overline{CAS}$). Non-critical timing requirements for $\overline{RAS}$ and $\overline{CAS}$ allow use of the address multiplexing technique while maintaining high performance.

The 2117 three-state output is controlled by $\overline{CAS}$, independent of $\overline{RAS}$. After a valid read or read-modify-write cycle, data is latched on the output by holding $\overline{CAS}$ low. The data out pin is returned to the high impedance state by returning $\overline{CAS}$ to a high state. The 2117 hidden refresh feature allows $\overline{CAS}$ to be held low to maintain latched data while $\overline{RAS}$ is used to execute $\overline{RAS}$-only refresh cycles.

The single transistor storage cell requires refreshing for data retention. Refreshing is accomplished by performing $\overline{RAS}$-only refresh cycles, hidden refresh cycles, or normal read or write cycles on the 128 address combinations of A_0 through A_6 during a 2ms period. A write cycle will refresh stored data on all bits of the selected row except the bit which is addressed.

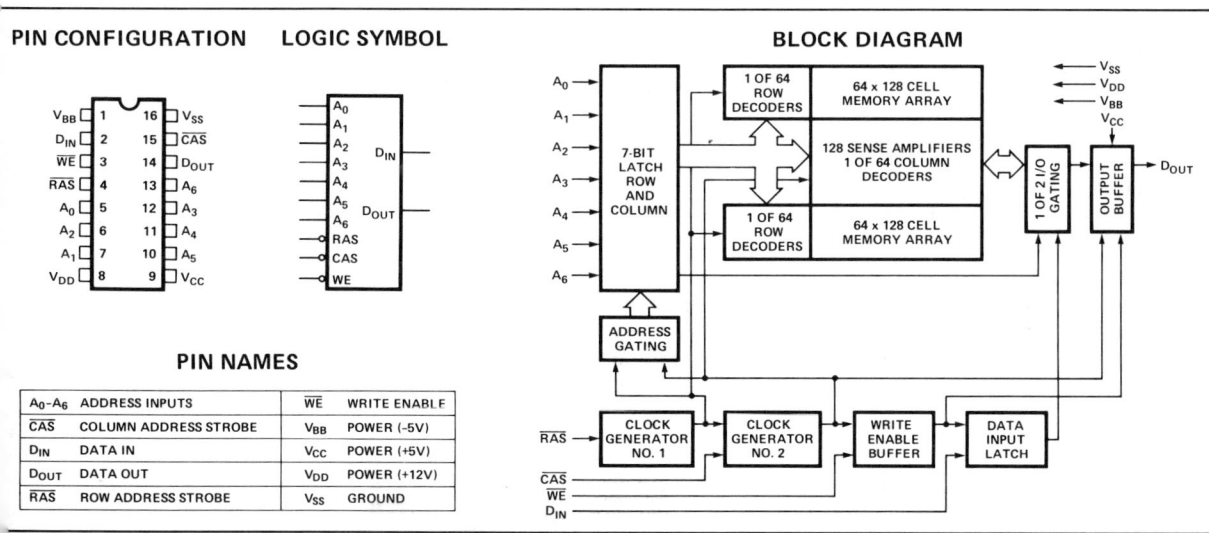

ABSOLUTE MAXIMUM RATINGS*

Ambient Temperature Under Bias ... -10°C to +80°C
Storage Temperature .... -65°C to +150°C
Voltage on Any Pin Relative to V_{BB}
  ($V_{SS} - V_{BB} \geq 4V$) .... -0.3V to +20V
Data Out Current 50mA
Power Dissipation 1.0W

*COMMENT:

Stresses above those listed under "Absolute Maximum Rating" may cause permanent damage to the device. This is a stress rating only and functional operation of the device at these or at any other condition above those indicated in the operational sections of this specification is not implied. Exposure to absolute maximum rating conditions for extended periods may affect device reliability.

D.C. AND OPERATING CHARACTERISTICS[1,2]

$T_A = 0°C$ to $70°C$, $V_{DD} = 12V \pm 10\%$, $V_{CC} = 5V \pm 10\%$, $V_{BB} = -5V \pm 10\%$, $V_{SS} = 0V$, unless otherwise noted.

Symbol	Parameter	Min.	Typ.[3]	Max.	Unit	Test Conditions	Notes
$\|I_{LI}\|$	Input Load Current (any input)		0.1	10	μA	$V_{IN} = V_{SS}$ to 7.0V, $V_{BB} = -5.0V$	
$\|I_{LO}\|$	Output Leakage Current for High Impedance State		0.1	10	μA	Chip Deselected: $\overline{CAS}$ at V_{IH}, V_{OUT} = 0 to 5.5V	
I_{DD1}	V_{DD} Supply Current, Standby			1.5	mA	$\overline{CAS}$ and $\overline{RAS}$ at V_{IH}	4
I_{BB1}	V_{BB} Supply Current, Standby		1.0	50	μA		
I_{CC1}	V_{CC} Supply Current, Output Deselected		0.1	10	μA	$\overline{CAS}$ at V_{IH}	5
I_{DD2}	V_{DD} Supply Current, Operating			35	mA	2117-2, t_{RC} = 375ns, t_{RAS} = 150ns	4,6
				35	mA	2117-3, t_{RC} = 375ns, t_{RAS} = 200ns	4
				33	mA	2117-4, t_{RC} = 410ns, t_{RAS} = 250ns	4
I_{BB2}	V_{BB} Supply Current, Operating, $\overline{RAS}$-Only Refresh, Page Mode		150	300	μA	$T_A = 0°C$	
I_{DD3}	V_{DD} Supply Current, $\overline{RAS}$-Only Refresh			27	mA	2117-2, t_{RC} = 375ns, t_{RAS} = 150ns	4,6
				27	mA	2117-3, t_{RC} = 375ns, t_{RAS} = 200ns	4
				26	mA	2117-4, t_{RC} = 410ns, t_{RAS} = 250ns	4
I_{DD5}	V_{DD} Supply Current, Standby, Output Enabled		1.5	3	mA	$\overline{CAS}$ at V_{IL}, $\overline{RAS}$ at V_{IH}	
V_{IL}	Input Low Voltage (all inputs)	-1.0		0.8	V		
V_{IH}	Input High Voltage (all inputs)	2.4		6.0	V		
V_{OL}	Output Low Voltage			0.4	V	I_{OL} = 4.2mA	4
V_{OH}	Output High Voltage	2.4			V	I_{OH} = -5mA	4

NOTES:
1. All voltages referenced to V_{SS}.
2. No power supply sequencing is required. However, V_{DD}, V_{CC} and V_{SS} should never be more negative than -0.3V with respect to V_{BB} as required by the absolute maximum ratings.
3. Typical values are for T_A = 25°C and nominal supply voltages.
4. See the Typical Characteristics Section for values of this parameter under alternate conditions.
5. I_{CC} is dependent on output loading when the device output is selected. V_{CC} is connected to the output buffer only. V_{CC} may be reduced to V_{SS} without affecting refresh operation or maintenance of internal device data.
6. For the 2117-2 at t_{RC} = 320ns, t_{RAS} = 150ns, I_{DD2} max. is 45mA and I_{DD3} max. is 31mA.

TYPICAL SUPPLY CURRENT WAVEFORMS

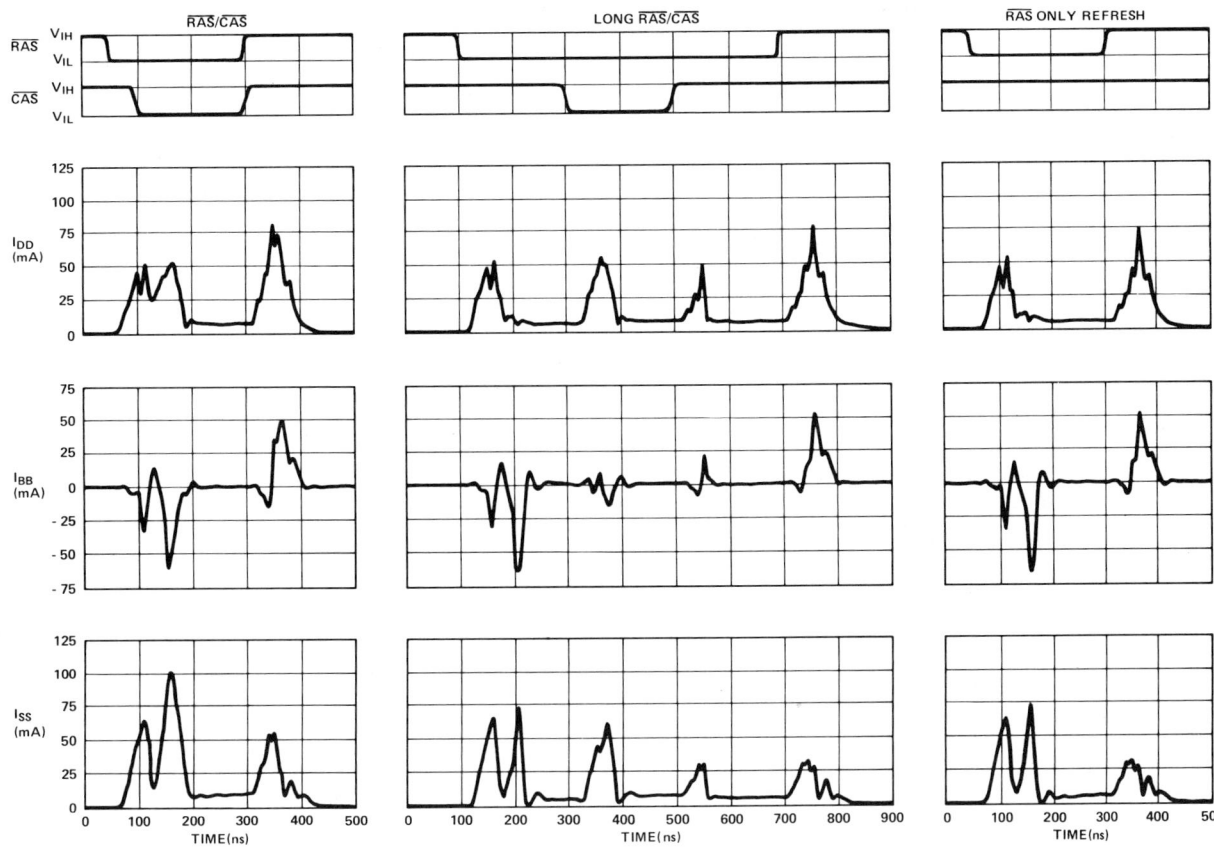

Typical power supply current waveforms vs. time are shown for the $\overline{RAS}/\overline{CAS}$ timings of Read/Write, Read/Write (Long $\overline{RAS}/\overline{CAS}$), and $\overline{RAS}$-only refresh cycles. I_{DD} and I_{BB} current transients at the $\overline{RAS}$ and $\overline{CAS}$ edges require adequate decoupling of these supplies. Decoupling recommendations are provided in the Applications section.

The effects of cycle time, V_{DD} supply voltage and ambient temperature on the I_{DD} current are shown in graphs included in the Typical Characteristics Section. Each family of curves for I_{DD1}, I_{DD2}, and I_{DD3} is related by a common point at $V_{DD} = 12.0V$ and $T_A = 25°C$ for two given t_{RAS} pulse widths. The typical I_{DD} current for a given condition of cycle time, V_{DD} and T_A can be determined by combining the effects of the appropriate family of curves.

CAPACITANCE[1]

$T_A = 25°C$, $V_{DD} = 12V \pm 10\%$, $V_{CC} = 5V \pm 10\%$, $V_{BB} = -5V \pm 10\%$, $V_{SS} = 0V$, unless otherwise specified.

Symbol	Parameter	Typ.	Max.	Unit
C_{I1}	Address, Data In	3	5	pF
C_{I2}	$\overline{RAS}$ Capacitance, $\overline{WE}$ Capacitance	4	7	pF
C_{I3}	$\overline{CAS}$ Capacitance	6	10	pF
C_O	Data Output Capacitance	4	7	pF

NOTES:
1. Capacitance measured with Boonton Meter or effective capacitance calculated from the equation:
$C = \frac{I \Delta t}{\Delta V}$ with ΔV equal to 3 volts and power supplies at nominal levels.

A.C. CHARACTERISTICS[1,2,3]

$T_A = 0°C$ to $70°C$, $V_{DD} = 12V \pm 10\%$, $V_{CC} = 5V \pm 10\%$, $V_{BB} = -5V \pm 10\%$, $V_{SS} = 0V$, unless otherwise noted.

READ, WRITE, READ-MODIFY-WRITE AND REFRESH CYCLES

Symbol	Parameter	2117-2 Min.	2117-2 Max.	2117-3 Min.	2117-3 Max.	2117-4 Min.	2117-4 Max.	Unit	Notes
t_{RAC}	Access Time From RAS		150		200		250	ns	4,5
t_{CAC}	Access Time From CAS		100		135		165	ns	4,5,6
t_{REF}	Time Between Refresh		2		2		2	ms	
t_{RP}	RAS Precharge Time	100		120		150		ns	
t_{CPN}	CAS Precharge Time (non-page cycles)	25		25		25		ns	
t_{CRP}	CAS to RAS Precharge Time	-20		-20		-20		ns	
t_{RCD}	RAS to CAS Delay Time	20	50	25	65	35	85	ns	7
t_{RSH}	RAS Hold Time	100		135		165		ns	
t_{CSH}	CAS Hold Time	150		200		250		ns	
t_{ASR}	Row Address Set-Up Time	0		0		0		ns	
t_{RAH}	Row Address Hold Time	20		25		35		ns	
t_{ASC}	Column Address Set-Up Time	-10		-10		-10		ns	
t_{CAH}	Column Address Hold Time	45		55		75		ns	
t_{AR}	Column Address Hold Time, to RAS	95		120		160		ns	
t_T	Transition Time (Rise and Fall)	3	50	3	50	3	50	ns	8
t_{OFF}	Output Buffer Turn Off Delay	0	50	0	60	0	70	ns	

READ AND REFRESH CYCLES

Symbol	Parameter	2117-2 Min.	2117-2 Max.	2117-3 Min.	2117-3 Max.	2117-4 Min.	2117-4 Max.	Unit	Notes
t_{RC}	Random Read Cycle Time	320		375		410		ns	
t_{RAS}	RAS Pulse Width	150	10000	200	10000	250	10000	ns	
t_{CAS}	CAS Pulse Width	100	10000	135	10000	165	10000	ns	
t_{RCS}	Read Command Set-Up Time	0		0		0		ns	
t_{RCH}	Read Command Hold Time	0		0		0		ns	

WRITE CYCLE

Symbol	Parameter	2117-2 Min.	2117-2 Max.	2117-3 Min.	2117-3 Max.	2117-4 Min.	2117-4 Max.	Unit	Notes
t_{RC}	Random Write Cycle Time	320		375		410		ns	
t_{RAS}	RAS Pulse Width	150	10000	200	10000	250	10000	ns	
t_{CAS}	CAS Pulse Width	100	10000	135	10000	165	10000	ns	
t_{WCS}	Write Command Set-Up Time	-20		-20		-20		ns	9
t_{WCH}	Write Command Hold Time	45		55		75		ns	
t_{WCR}	Write Command Hold Time, to RAS	95		120		160		ns	
t_{WP}	Write Command Pulse Width	45		55		75		ns	
t_{RWL}	Write Command to RAS Lead Time	60		80		100		ns	
t_{CWL}	Write Command to CAS Lead Time	60		80		100		ns	
t_{DS}	Data-In Set-Up Time	0		0		0		ns	
t_{DH}	Data-In Hold Time	45		55		75		ns	
t_{DHR}	Data-In Hold Time, to RAS	95		120		160		ns	

READ-MODIFY-WRITE CYCLE

Symbol	Parameter	2117-2 Min.	2117-2 Max.	2117-3 Min.	2117-3 Max.	2117-4 Min.	2117-4 Max.	Unit	Notes
t_{RWC}	Read-Modify-Write Cycle Time	330		375		475		ns	
t_{RRW}	RMW Cycle RAS Pulse Width	185	10000	245	10000	305	10000	ns	
t_{CRW}	RMW Cycle CAS Pulse Width	135	10000	180	10000	230	10000	ns	
t_{RWD}	RAS to WE Delay	120		160		200		ns	9
t_{CWD}	CAS to WE Delay	70		95		125		ns	9

Notes: See following page for A.C. Characteristics Notes.

WAVEFORMS

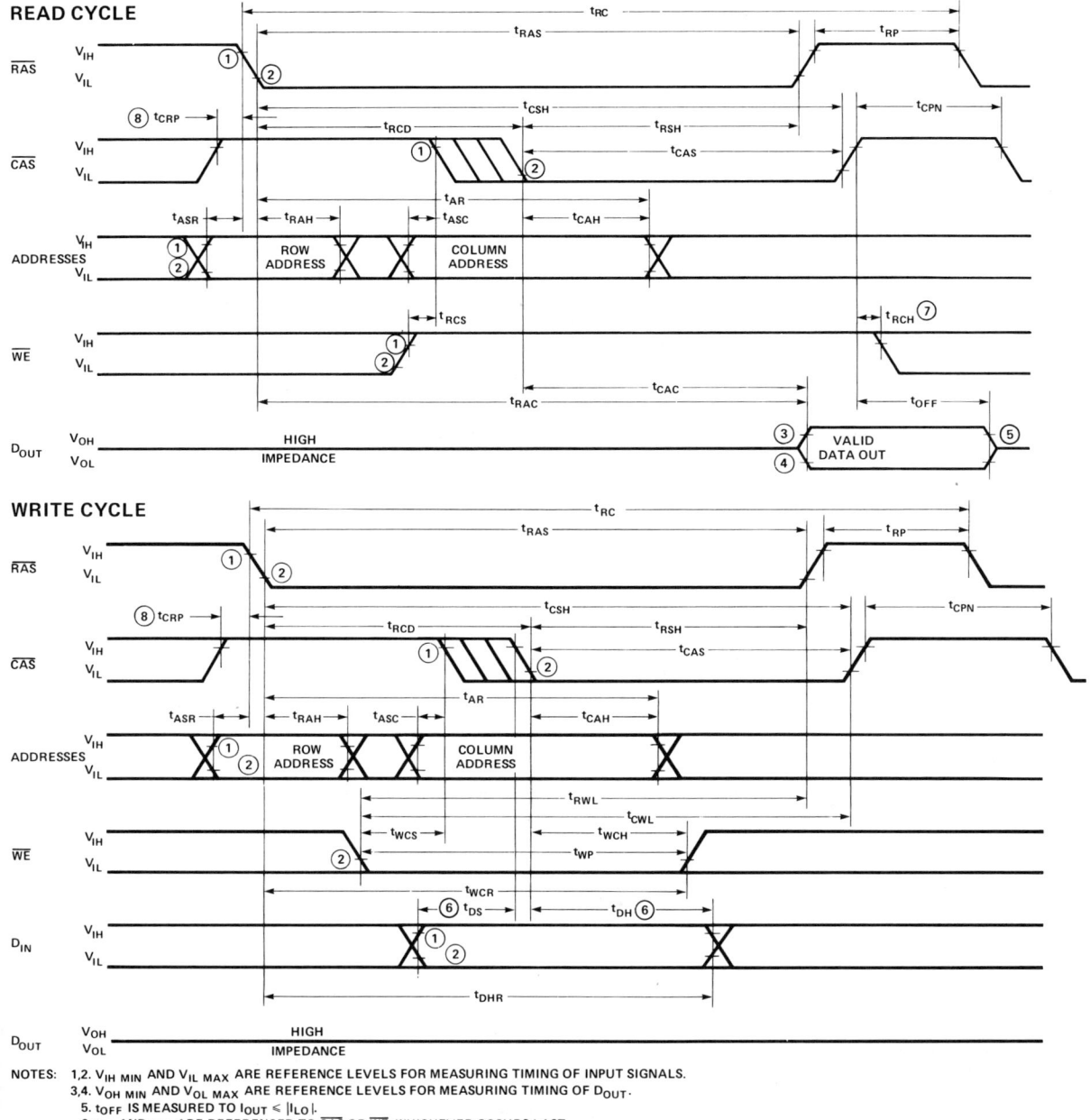

NOTES:
1,2. $V_{IH\ MIN}$ AND $V_{IL\ MAX}$ ARE REFERENCE LEVELS FOR MEASURING TIMING OF INPUT SIGNALS.
3,4. $V_{OH\ MIN}$ AND $V_{OL\ MAX}$ ARE REFERENCE LEVELS FOR MEASURING TIMING OF D_{OUT}.
5. t_{OFF} IS MEASURED TO $I_{OUT} \leq |I_{LO}|$.
6. t_{DS} AND t_{DH} ARE REFERENCED TO $\overline{CAS}$ OR $\overline{WE}$, WHICHEVER OCCURS LAST.
7. t_{RCH} IS REFERENCED TO THE TRAILING EDGE OF $\overline{CAS}$ OR $\overline{RAS}$, WHICHEVER OCCURS FIRST.
8. t_{CRP} REQUIREMENT IS ONLY APPLICABLE FOR $\overline{RAS}/\overline{CAS}$ CYCLES PRECEEDED BY A $\overline{CAS}$-ONLY CYCLE (i.e., FOR SYSTEMS WHERE $\overline{CAS}$ HAS NOT BEEN DECODED WITH $\overline{RAS}$).

A.C. CHARACTERISTICS NOTES (From Previous Page)

1. All voltages referenced to V_{SS}.
2. Eight cycles are required after power-up or prolonged periods (greater than 2ms) of $\overline{RAS}$ inactivity before proper device operation is achieved. Any 8 cycles which perform refresh are adequate for this purpose.
3. A.C. Characteristics assume $t_T = 5ns$.
4. Assume that $t_{RCD} \leq t_{RCD}$ (max.). If t_{RCD} is greater than t_{RCD} (max.) then t_{RAC} will increase by the amount that t_{RCD} exceeds t_{RCD} (max.).
5. Load = 2 TTL loads and 100pF.
6. Assumes $t_{RCD} \geq t_{RCD}$ (max.).
7. t_{RCD} (max.) is specified as a reference point only; if t_{RCD} is less than t_{RCD} (max.) access time is t_{RAC}, if t_{RCD} is greater than t_{RCD} (max.) access time is $t_{RCD} + t_{CAC}$.
8. t_T is measured between V_{IH} (min.) and V_{IL} (max.).
9. t_{WCS}, t_{CWD} and t_{RWD} are specified as reference points only. If $t_{WCS} \geq t_{WCS}$ (min.) the cycle is an early write cycle and the data out pin will remain high impedance throughout the entire cycle. If $t_{CWD} \geq t_{CWD}$ (min.) and $t_{RWD} \geq t_{RWD}$ (min.), the cycle is a read-modify-write cycle and the data out will contain the data read from the selected address. If neither of the above conditions is satisfied, the condition of the data out is indeterminate.

WAVEFORMS

READ-MODIFY-WRITE CYCLE

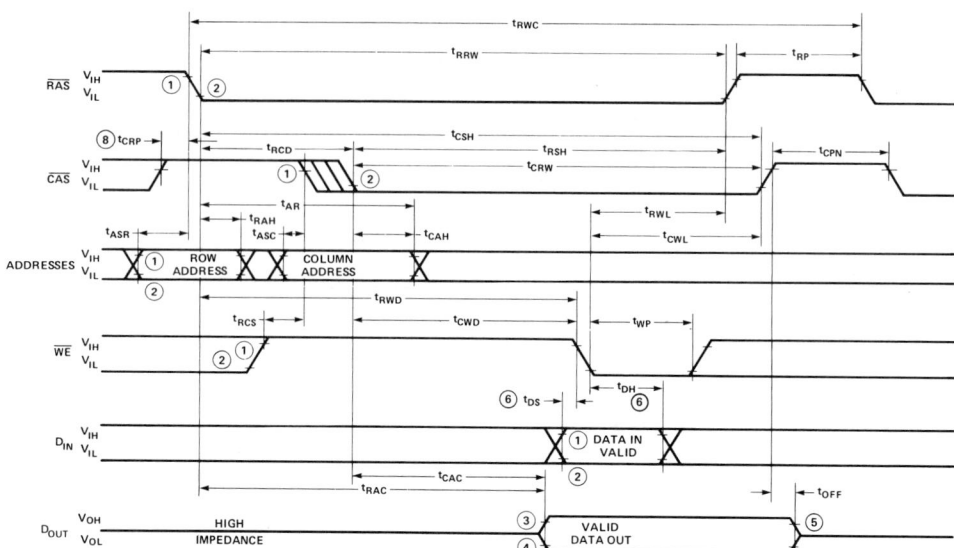

RAS-ONLY REFRESH CYCLE

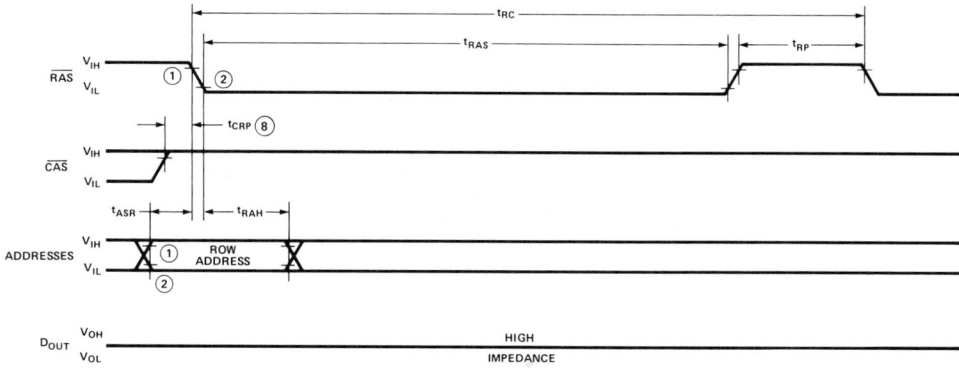

HIDDEN REFRESH CYCLE

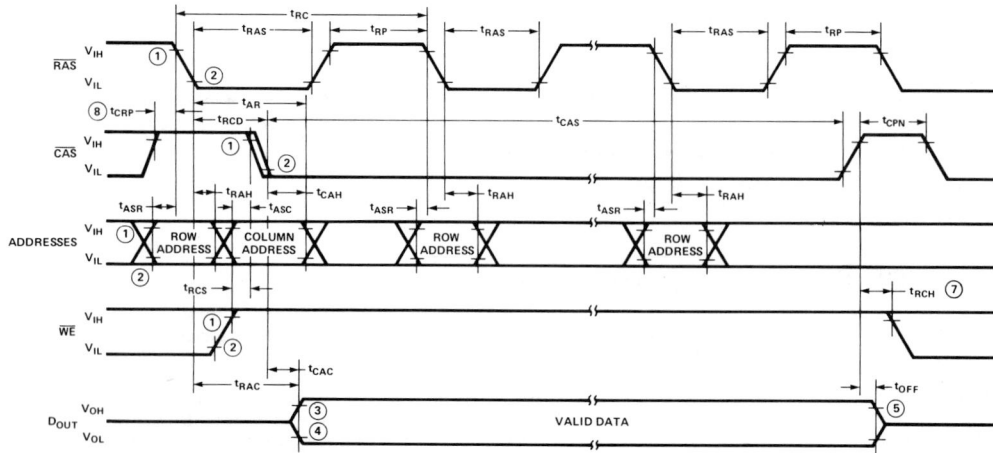

NOTES: 1,2. V_{IH} MIN AND V_{IL} MAX ARE REFERENCE LEVELS FOR MEASURING TIMING OF INPUT SIGNALS.
3,4. V_{OH} MIN AND V_{OL} MAX ARE REFERENCE LEVELS FOR MEASURING TIMING OF D_{OUT}.
5. t_{OFF} IS MEASURED TO $I_{OUT} \leq |I_{LO}|$.
6. t_{DS} AND t_{DH} ARE REFERENCED TO $\overline{CAS}$ OR $\overline{WE}$, WHICHEVER OCCURS LAST.
7. t_{RCH} IS REFERENCED TO THE TRAILING EDGE OF $\overline{CAS}$ OR $\overline{RAS}$, WHICHEVER OCCURS FIRST.
8. t_{CRP} REQUIREMENT IS ONLY APPLICABLE FOR $\overline{RAS}/\overline{CAS}$ CYCLES PRECEEDED BY A $\overline{CAS}$-ONLY CYCLE (i.e., FOR SYSTEMS WHERE $\overline{CAS}$ HAS NOT BEEN DECODED WITH $\overline{RAS}$).

TYPICAL CHARACTERISTICS[1]

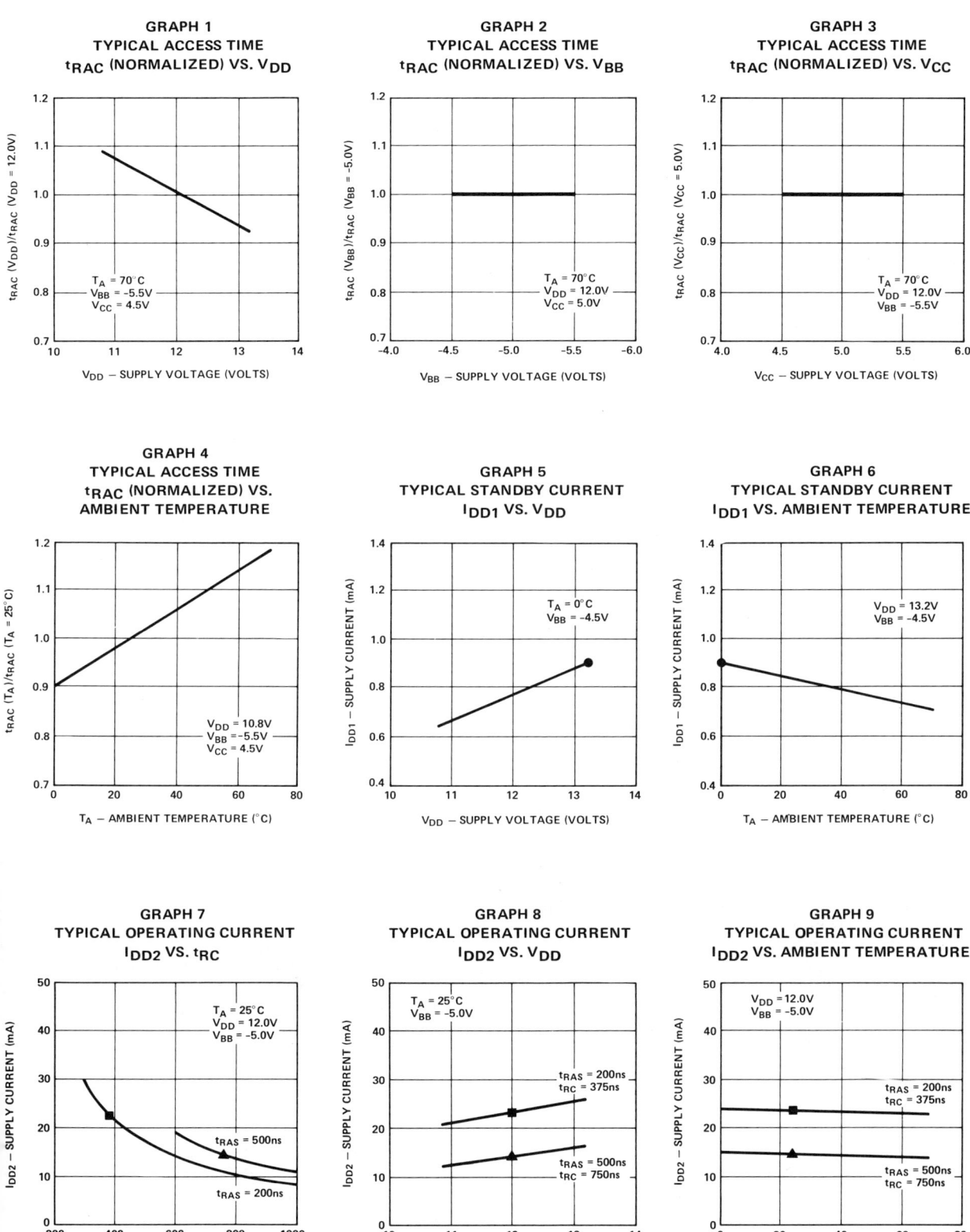

NOTES: See following page for Typical Characteristics Notes.

TYPICAL CHARACTERISTICS [1]

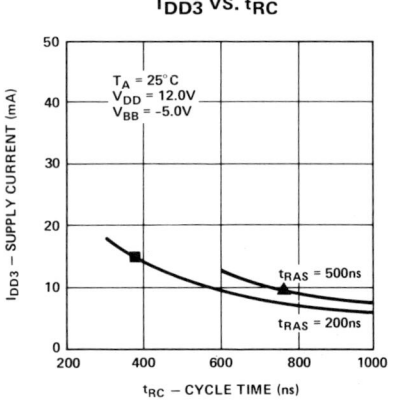

GRAPH 10
TYPICAL RAS ONLY
REFRESH CURRENT
I_{DD3} VS. t_{RC}

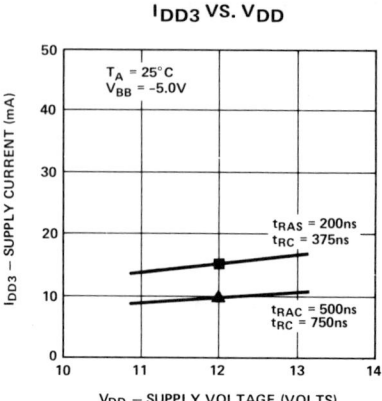

GRAPH 11
TYPICAL RAS ONLY
REFRESH CURRENT
I_{DD3} VS. V_{DD}

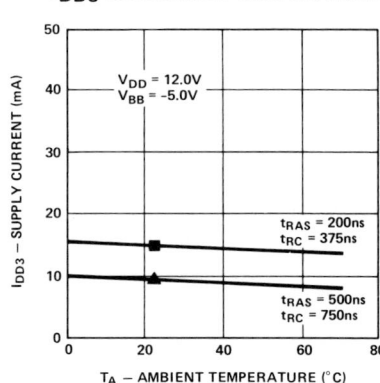

GRAPH 12
TYPICAL RAS ONLY
REFRESH CURRENT
I_{DD3} VS. AMBIENT TEMPERATURE

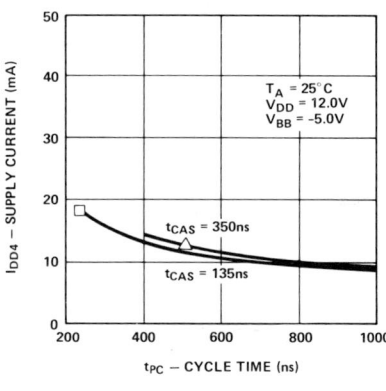

GRAPH 13
TYPICAL PAGE MODE CURRENT
I_{DD4} VS. t_{PC}

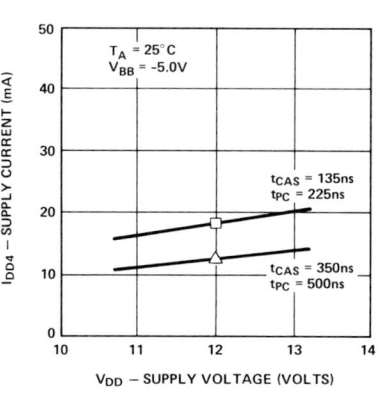

GRAPH 14
TYPICAL PAGE MODE CURRENT
I_{DD4} VS. V_{DD}

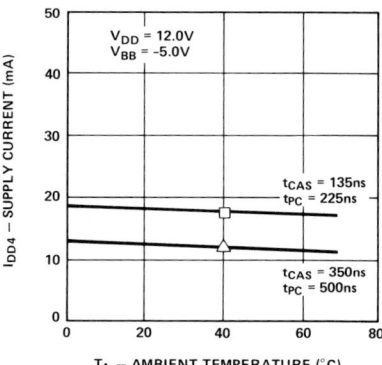

GRAPH 15
TYPICAL PAGE MODE CURRENT
I_{DD4} VS. AMBIENT TEMPERATURE

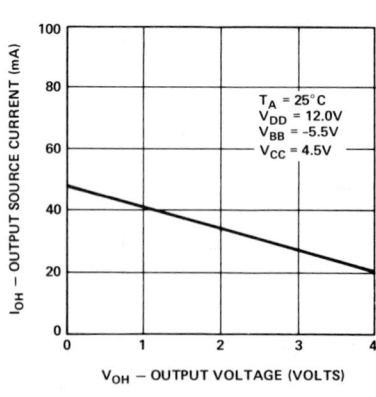

GRAPH 16
TYPICAL OUTPUT SOURCE CURRENT
I_{OH} VS. OUTPUT VOLTAGE V_{OH}

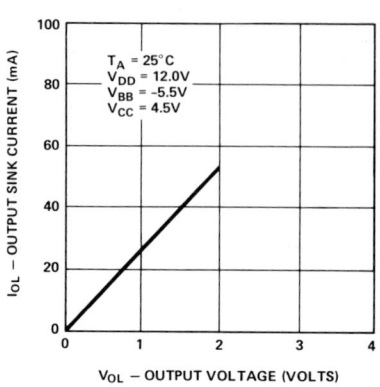

GRAPH 17
TYPICAL OUTPUT SINK CURRENT
I_{OL} VS. OUTPUT VOLTAGE V_{OL}

NOTES:

1. The cycle time, V_{DD} supply voltage, and ambient temperature dependence of I_{DD1}, I_{DD2}, I_{DD3} and I_{DD4} is shown in related graphs. Common points of related curves are indicated:

   ● I_{DD1} @ V_{DD} = 13.2V, T_A = 0°C
   ■ I_{DD2} or I_{DD3} @ t_{RAS} = 200ns, t_{RC} = 375ns, V_{DD} = 12.0V, T_A = 25°C
   ▲ I_{DD2} or I_{DD3} @ t_{RAS} = 500ns, t_{RC} = 750ns, V_{DD} = 12.0V, T_A = 25°C
   □ I_{DD4} @ t_{CAS} = 135ns, t_{PC} = 225ns, V_{DD} = 12.0V, T_A = 25°C
   △ I_{DD4} @ t_{CAS} = 350ns, t_{PC} = 500ns, V_{DD} = 12.0V, T_A = 25°C

   The typical I_{DD} current for a given combination of cycle time, V_{DD} supply voltage and ambient temperature may be determined by combining the effects of the appropriate family of curves.

D.C. AND A.C. CHARACTERISTICS, PAGE MODE[7,8,11]

$T_A = 0°C$ to $70°C$, $V_{DD} = 12V \pm 10\%$, $V_{CC} = 5V \pm 10\%$, $V_{BB} = -5V \pm 10\%$, $V_{SS} = 0V$, unless otherwise noted.
For Page Mode Operation order 2117-2 S6053, 2117-3 S6054, or 2117-4 S6055.

Symbol	Parameter	2117-2 S6053 Min.	Max.	2117-3 S6054 Min.	Max.	2117-4 S6055 Min.	Max.	Unit	Notes
t_{PC}	Page Mode Read or Write Cycle	170		225		275		ns	
t_{PCM}	Page Mode Read Modify Write	205		270		340		ns	
t_{CP}	$\overline{CAS}$ Precharge Time, Page Cycle	60		80		100		ns	
t_{RPM}	$\overline{RAS}$ Pulse Width, Page Mode	150	10,000	200	10,000	250	10,000	ns	
t_{CAS}	$\overline{CAS}$ Pulse Width	100	10,000	135	10,000	165	10,000	ns	
I_{DD4}	V_{DD} Supply Current Page Mode, Minimum t_{PC}, Minimum t_{CAS}		38		30		26	mA	9

WAVEFORMS

PAGE MODE READ CYCLE

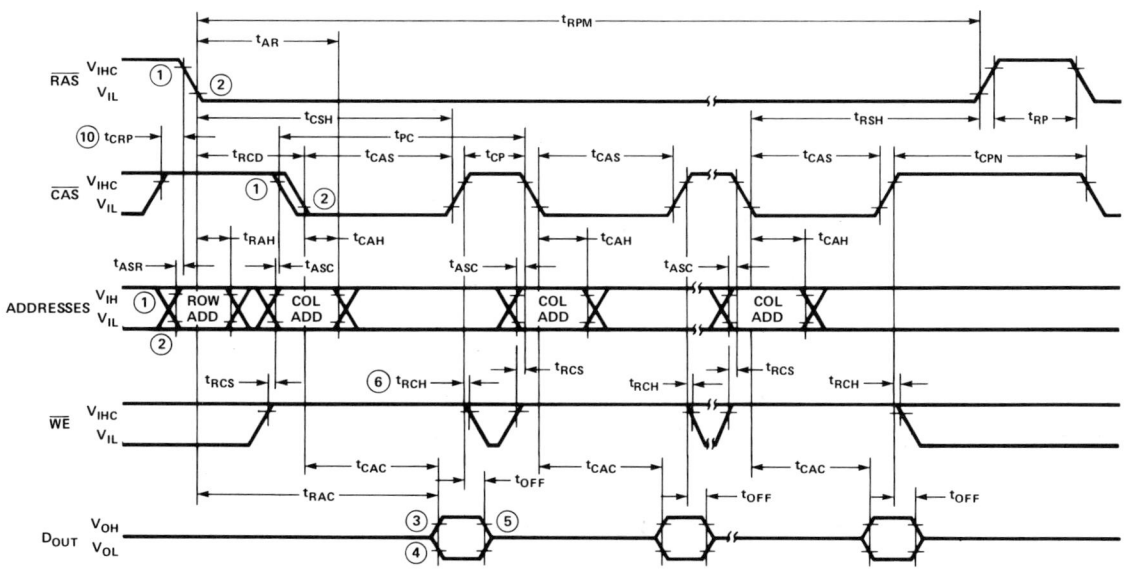

NOTES:
1,2. $V_{IH\ MIN}$ AND $V_{IL\ MAX}$ ARE REFERENCE LEVELS FOR MEASURING TIMING OF INPUT SIGNALS.
3,4. $V_{OH\ MIN}$ AND $V_{OL\ MAX}$ ARE REFERENCE LEVELS FOR MEASURING TIMING OF D_{OUT}.
5. t_{OFF} IS MEASURED TO $I_{OUT} \leq |I_{LO}|$.
6. t_{RCH} IS REFERENCED TO THE TRAILING EDGE OF $\overline{CAS}$ OR $\overline{RAS}$, WHICHEVER OCCURS FIRST.
7. ALL VOLTAGES REFERENCED TO V_{SS}.
8. AC CHARACTERISTIC ASSUME $t_T = 5ns$.
9. SEE THE TYPICAL CHARACTERISTICS SECTION FOR VALUES OF THIS PARAMETER UNDER ALTERNATE CONDITIONS.
10. t_{CRP} REQUIREMENT IS ONLY APPLICABLE FOR $\overline{RAS}/\overline{CAS}$ CYCLES PRECEEDED BY A $\overline{CAS}$-ONLY CYCLE (i.e., FOR SYSTEMS WHERE $\overline{CAS}$ HAS NOT BEEN DECODED WITH $\overline{RAS}$).
11. ALL PREVIOUSLY SPECIFIED A.C. AND D.C. CHARACTERISTICS ARE APPLICABLE TO THEIR RESPECTIVE PAGE MODE DEVICE (i.e., 2117-3, S6054 WILL OPERATE AS A 2117-3).

PAGE MODE WRITE CYCLE

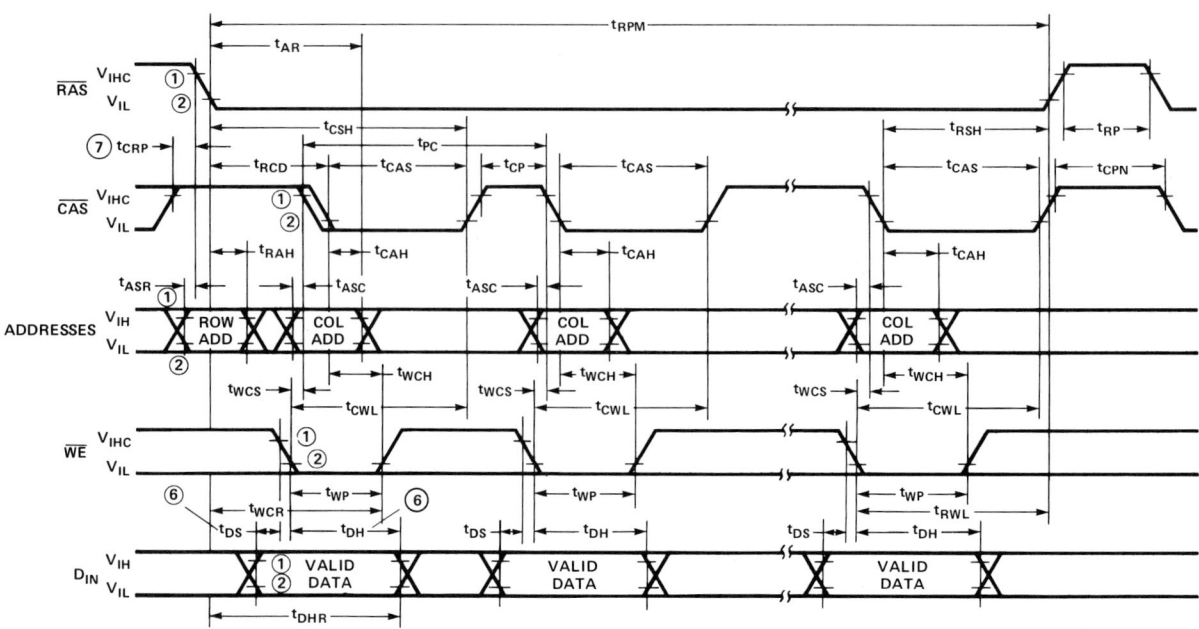

PAGE MODE READ-MODIFY-WRITE CYCLE

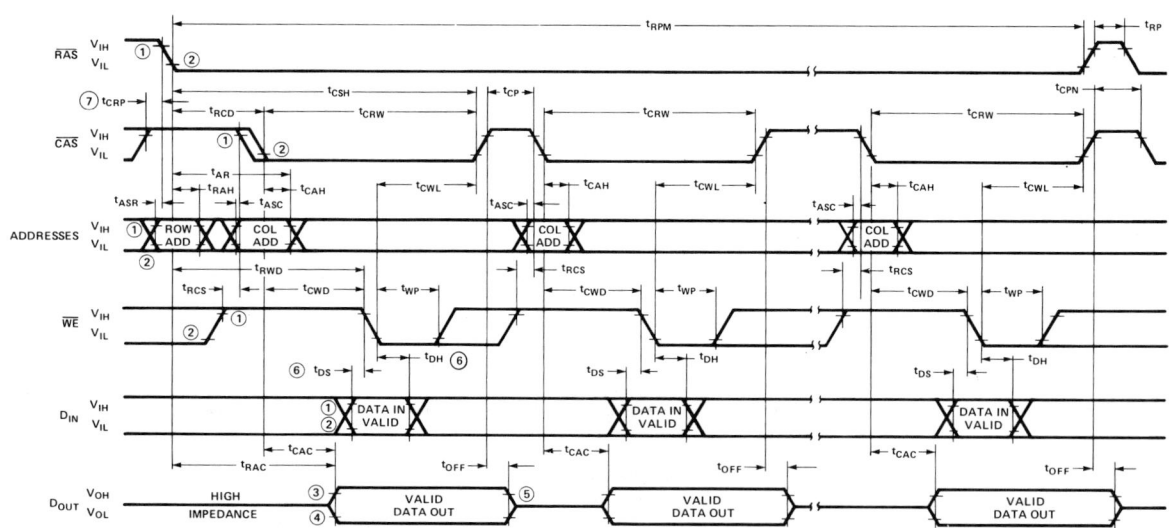

NOTES:
1,2. $V_{IH\,MIN}$ AND $V_{IL\,MAX}$ ARE REFERENCE LEVELS FOR MEASURING TIMING OF INPUT SIGNALS.
3,4. $V_{OH\,MIN}$ AND $V_{OL\,MAX}$ ARE REFERENCE LEVELS FOR MEASURING TIMING OF D_{OUT}.
5. t_{OFF} IS MEASURED TO $I_{OUT} \leq |I_{LO}|$.
6. t_{DS} AND t_{DH} ARE REFERENCED TO $\overline{CAS}$ OR $\overline{WE}$, WHICHEVER OCCURS LAST.
7. t_{CRP} REQUIREMENT IS ONLY APPLICABLE FOR $\overline{RAS}/\overline{CAS}$ CYCLES PRECEEDED BY A $\overline{CAS}$-ONLY CYCLE (i.e., FOR SYSTEMS WHERE $\overline{CAS}$ HAS NOT BEEN DECODED WITH $\overline{RAS}$).

APPLICATIONS

READ CYCLE

A Read cycle is performed by maintaining Write Enable ($\overline{WE}$) high during a $\overline{RAS}/\overline{CAS}$ operation. The output pin of a selected device will remain in a high impedance state until valid data appears at the output at access time.

Device access time, t_{ACC}, is the longer of the two calculated intervals:

1. $t_{ACC} = t_{RAC}$ OR 2. $t_{ACC} = t_{RCD} + t_{CAC}$

Access time from $\overline{RAS}$, t_{RAC}, and access time from $\overline{CAS}$, t_{CAC}, are device parameters. Row to column address strobe delay time, t_{RCD}, are system dependent timing parameters. For example, substituting the device parameters of the 2117-3 yields:

3. $t_{ACC} = t_{RAC} = 200$nsec for 25nsec $\leq t_{RCL} \leq 65$ nsec
   OR
4. $t_{ACC} = t_{RCD} + t_{CAC} = t_{RCD} + 135$ for $t_{RCD} > 65$nsec

Note that if 25nsec $\leq t_{RCD} \leq 65$nsec device access time is determined by equation 3 and is equal to t_{RAC}. If $t_{RCL} > 65$nsec, access time is determined by equation 4. This 40nsec interval (shown in the t_{RCD} inequality in equation 3) in which the falling edge of $\overline{CAS}$ can occur without affecting access time is provided to allow for system timing skew in the generation of $\overline{CAS}$.

REFRESH CYCLES

Each of the 128 rows of the 2117 must be refreshed every 2 milliseconds to maintain data. Any memory cycle:

1. Read Cycle
2. Write Cycle (Early Write, Delayed Write or Read-Modify-Write)
3. $\overline{RAS}$-only Cycle

refreshes the selected row as defined by the low order ($\overline{RAS}$) addresses. Any Write cycle, of course, may change the state of the selected cell. Using a Read, Write, or Read-Modify-Write cycle for refresh is not recommended for systems which utilize "wire-OR" outputs since output bus contention will occur.

A $\overline{RAS}$-only refresh cycle is the recommended technique for most applications to provide for data retention. A $\overline{RAS}$-only refresh cycle maintains the D_{OUT} in the high impedance state with a typical power reduction of 20% over a Read or Write cycle.

$\overline{RAS}/\overline{CAS}$ TIMING

$\overline{RAS}$ and $\overline{CAS}$ have minimum pulse widths as defined by t_{RAS} and t_{CAS} respectively. These minimum pulse widths must be maintained for proper device operation and data integrity. A cycle, once begun by driving $\overline{RAS}$ and/or $\overline{CAS}$ low must not be ended or aborted prior to fulfilling the minimum clock signal pulse width(s). A new cycle can not begin until the minimum precharge time, t_{RP}, has been met.

DATA OUTPUT OPERATION

The 2117 Data Output (D_{OUT}), which has three-state capability, is controlled by $\overline{CAS}$. During $\overline{CAS}$ high state ($\overline{CAS}$ at V_{IH}) the output is in the high impedance state. The following table summarizes the D_{OUT} state for various types of cycles.

Intel 2117 Data Output Operation for Various Types of Cycles

Type of Cycle	D_{OUT} State
Read Cycle	Data From Addressed Memory Cell
Fast Write Cycle	HI-Z
$\overline{RAS}$-Only Refresh Cycle	HI-Z
$\overline{CAS}$-Only Cycle	HI-Z
Read/Modify/Write Cycle	Data From Addressed Memory Cell
Delayed Write Cycle	Indeterminate

HIDDEN REFRESH

A feature of the 2117 is that refresh cycles may be performed while maintaining valid data at the output pin. This feature is referred to as Hidden Refresh. Hidden Refresh is performed by holding $\overline{CAS}$ at V_{IL} and taking $\overline{RAS}$ high and after a specified precharge period (t_{RP}), executing a "$\overline{RAS}$-Only" refresh cycle, but with $\overline{CAS}$ held low (see Figure below).

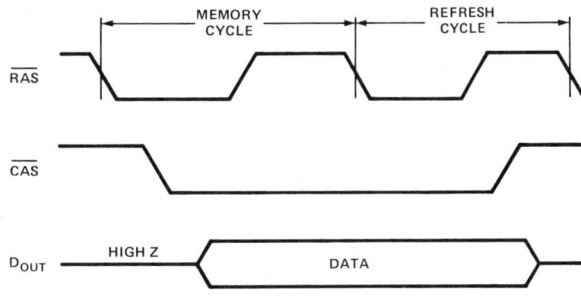

This feature allows a refresh cycle to be "hidden" among data cycles without affecting the data availability.

POWER ON

The 2117 requires no power on sequence providing absolute maximum ratings are not exceeded. After the application of supply voltages or after extended periods of bias (greater than 2 milliseconds) without clocks, the device must perform a minimum of eight initialization cycles (any combination of cycles containing a $\overline{RAS}$ clock, such as $\overline{RAS}$-Only refresh) prior to normal operation.

DYNAMIC RAMS

POWER SUPPLY DECOUPLING/DISTRIBUTION

It is recommended that a 0.1μF ceramic capacitor be connected between V_{DD} and V_{SS} at every other device in the memory array. A 0.1μF ceramic capacitor should also be connected between V_{BB} and V_{SS} at every other device (preferably the alternate devices to the V_{DD} decoupling). For each 16 devices, a 10μF tantalum or equivalent capacitor should be connected between V_{DD} and V_{SS} near the array. An equal or slightly smaller bulk capacitor is also recommended between V_{BB} and V_{SS} for every 32 devices.

The V_{CC} supply is connected only to the 2117 output buffer and is not used internally. The load current from the V_{CC} supply is dependent only upon the output loading and is associated with the input high level current to a TTL gate and the output leakage currents of any OR-tied 2117's (typically 100μA or less total). Intel recommends that a 0.1 or 0.01μF ceramic capacitor be connected between V_{CC} and V_{SS} for every eight memory devices.

Due to the high frequency characteristics of the current waveforms, the inductance of the power supply distribution system on the array board should be minimized. It is recommended that the V_{DD}, V_{BB}, and V_{SS} supply lines be gridded both horizontally and vertically at each device in the array. This technique allows use of double sided circuit boards with noise performance equal to or better than multi-layered circuit boards.

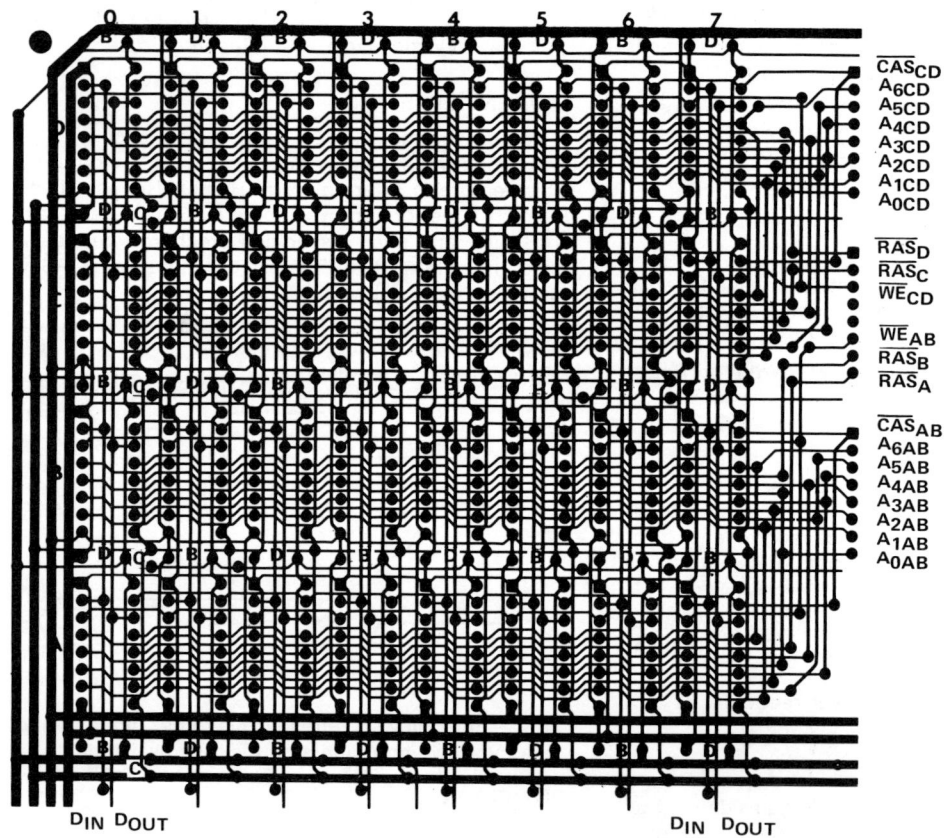

DECOUPLING CAPACITORS
D = 0.1μF TO V_{DD} TO V_{SS}
B = 0.1μF V_{BB} TO V_{SS}
C = 0.01μF V_{CC} TO V_{SS}

SAMPLE P.C. BOARD LAYOUT EMPLOYING VERTICAL AND HORIZONTAL GRIDDING ON ALL POWER SUPPLIES.
BOARD ORGANIZATION: 64K WORDS BY 8-BITS.

64K BYTE STORAGE ARRAY LAYOUT

CHAPTER THREE
PROMs and ROMs

Designing with PROMs and ROMs	**258**
Designing with 2708 EPROM	**294**
Designing with 2716 EPROM	**308**
1702A 2K (256 × 8) UV EPROM	322
1702AL 2K (256 × 8) Low Power UV EPROM	326
2308 8192 Bit Static MOS ROM	329
2316E 16,384 Bit Static ROM	333
2708 8K and 4K UV EPROM	336
2716 16K (2K × 8) UV EPROM	341
2758 8K (1K × 8) Low Power +5V UV EPROM	345
3602A, 3622A 3602, 3622 2048 Bit (512 × 4) High Speed PROM	351
3604A, 3624A and 3604, 3624 4096 Bit (512 × 8) High Speed PROM	354
3605, 3625 4K Bipolar PROM	357
3608, 3628 8K (1K × 8) Bipolar PROM	360
ROM/PROM Programming Instructions	**363**

DESIGNING WITH PROMS AND ROMS

BOB GREENE AND DAVE HOUSE

Photomicrograph at the Intel 3604 4K (512 × 8) Bipolar PROM.

DESIGNING WITH PROMS AND ROMS

INTRODUCTION

The combination of low cost, high speed, system design flexibility and data non-volatility has made read only memories an important part of many digital systems in production today. The rapid development of semiconductor read only memories has produced a succession of faster, larger, and more flexible devices.

Today, Intel combines the best of Schottky bipolar and P- and N-channel MOS semiconductor processing technologies to manufacture the fastest and largest line of read only memory products available anywhere in the world.

This chapter is divided into three sections. In the first section, *Understanding the Technology*, the various technologies used to produce read only memories and programmable read only memories are discussed in order to achieve a good understanding of how these devices operate and how one technology differs from another.

The second section describes the device from an operational and programming point of view, and presents Intel's extensive line of read only memories. In the third section, *System Applications*, the system aspects of address driving, output ORing, array configuration, printed circuit board layout, and power supply decoupling are presented.

Read Only Memories

A read only memory is an array of selectively open and closed unidirectional contacts. In the 16-bit array example shown in Figure 1, half of the address lines are decoded and used to energize one of the four row lines. This, in turn, activates those column lines which have a closed contact to the one selected row line. The remaining address lines are decoded and enable one of the column sense amplifiers. If chip select is true, the data is gated to the output pin by the output driver.

The primary differences in read only memories is in the forming of the open or closed contact; that is, in the design of the cell. In mask programmable read only memories (ROMs) the contact is made to selectively including or excluding a small conducting jumper during the final phase of semiconductor manufacture. In bipolar programmable read only memories (PROMs) the contact is made with a fusible material such that the contact can later be opened, allowing the data pattern to be configured by the user after the device has been manufactured.

Once programmed, Erasable Programmable Read Only Memories (EPROMs) allow the programmed

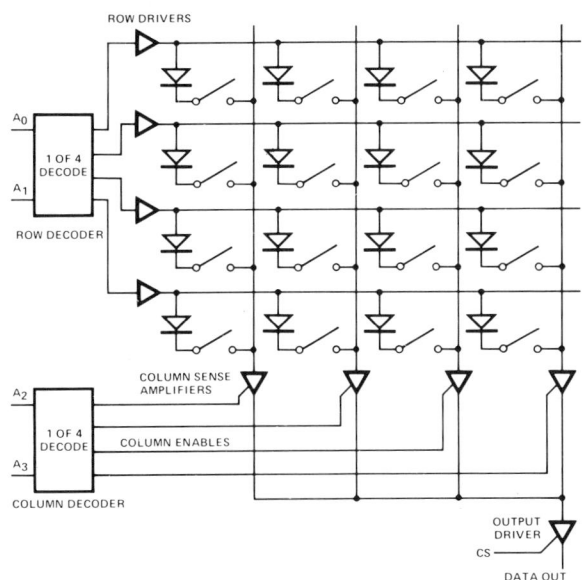

Figure 1. 16-Bit Simplified Array.

contacts to be restored to their initial state, such that they can be re-programmed as often as desired.

UNDERSTANDING THE TECHNOLOGY

As shown in Figure 2, there are two basic PROM/ROM technologies — bipolar and MOS. Their primary difference is in access time; bipolar access times are approximately 50–90 nS, and MOS access times are about an order of magnitude higher. Bipolar read only memories are available in 1K, 2K, and 4K bit sizes, while MOS read only memories are available in 2K through 16K bit sizes. Although PROMs and ROMs are available from both technologies, EPROMs are available only with MOS technology.

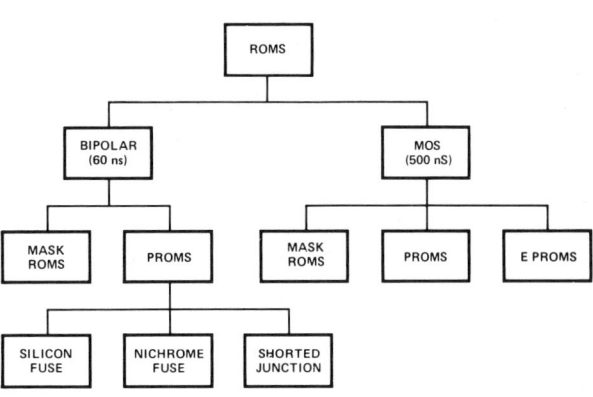

Figure 2. PROM/ROM Technology Family Tree.

Bipolar Technologies

As previously mentioned, bipolar devices offer higher speeds than MOS devices. For very high volume usage with those devices whose data pattern never change, mask programmable read only memories, commonly called ROMs, provide the lowest cost.

Electrically programmable read only memories, or PROMs, allow the data pattern to be defined when the device is used rather than when the device is manufactured.

MASK PROGRAMMABLE READ ONLY MEMORIES

Integrated circuit devices are fabricated from a wafer of silicon through a number of processing steps, including photo masking, etching, and diffusing in order to create a pattern of junctions and interconnections across the surface of the wafer. One of the final steps in the manufacturing process is to coat the entire surface of the silicon wafer with a layer of aluminum, and then to selectively etch away portions of the aluminum, leaving the desired interconnecting pattern. In the manufacture of mask programmed read only memories, the row-to-column contacts are selectively made by the inclusion or exclusion of aluminum connections in the final aluminum etch process.

The normal lead time required for fabrication of a new integrated circuit can be foreshortened from 9 to 10 weeks to about 4 to 6 weeks because the wafers can be manufactured through the point of metalization and held in storage until the data pattern is defined. By this method, the lead time required for delivery of a particular ROM pattern is only the time required to produce the mask and etch the final metal pattern on the wafer.

ELECTRICALLY PROGRAMMABLE READ ONLY MEMORIES

Electrically programmable read only memories allow the data pattern to be defined after final packaging rather than when the device is manufactured.

Three types of electrically programmable read only memories, commonly called PROMs, will be discussed here.

The Nichrome Fuse

The first PROMs were made with a nichrome fuse technology. Nichrome, an alloy of nickel and chrome, is deposited as a very thin film link to the column lines of the PROM. Heavy currents cause this film to "blow", opening the connection between the row and column lines. The cell is actually constructed of a transistor switch and the nichrome fuse, as shown in Figure 3. When the row is selected, the transistor, Q_{XY}, is turned on, and, if the fuse is intact, the column bus is pulled towards V_{CC} (+5V). If the fuse is "blown" or open, the column bus is left floating.

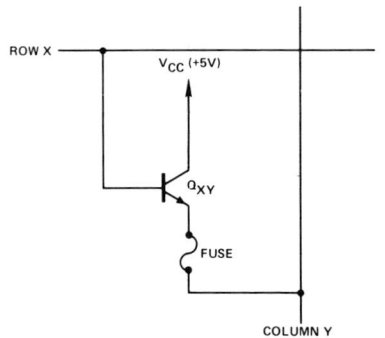

Figure 3. Typical Fuse Cell.

Nichrome Problems

Problems with nichrome fuses are all related to the technology. The selection of aluminum as the conductive material in integrated circuits and transistors did involve some serious metallurgical considerations. Of major importance is the fact that aluminum readily adheres to silicon dioxide, but does not rapidly diffuse through it. In addition, aluminum forms non-rectifying (ohmic) contacts with silicon.[1] Still, the formation of good silicon-to-aluminum contacts has always been a problem; the formation of good, reliable nichrome contact is a greater problem.

In addition, nichrome is not the easiest material to work with, especially considering the extremely thin layer (about 200 Angstroms) that must be deposited in order to achieve the desired resistance in the fuse. This deposition is very hard to control and the nichrome is additionally subject to corrosion.[2]

---

[1] Parker, G. H., J. C. Cornet, and W. S. Pinter. "Reliability Considerations in the Design and Fabrication of Polysilicon Fusible Link PROMs." A lecture to the IEEE 12th Annual Proceedings on Reliability Physics, 1974.

[2] Bauer, Joseph B. "Military Microcircuit Packaging," *The Electronic Engineer*, July 1972.

The most serious problem associated with nichrome fuse technology is probably the phenomenon commonly referred to as "growback",[3] the reversal of the programming process such that a single bit will, after some time, go from the programmed state back to the unprogrammed state.[4] Considerable analysis has been done to investigate this growback phenomenon[2-7] in nichrome fuse PROMs to understand how the nichrome fuse blows,[3,5,6] to determine the location and movement of the metals before and after fusion,[5,6] and to determine why a small number of these fuses (once blown) appear to reconnect.[2,5,6]

Fusion occurs under a layer of glass which has been added to the entire wafer to provide scratch protection and to minimize electron migration in the metal. Since fusion takes place without oxygen, or any other atmosphere, oxidation cannot play an important part in the fusing. It appears, rather, that the nichrome heats up under heavy current and becomes molten, forming a very narrow gap. Figure 4 is a picture taken with a scanning electron microscope of a blown nichrome fuse. Notice the fingers, or dendrites, of nichrome. Studies indicate that it is dendritic relinking that causes the fuse to begin to reconduct after some period of time.[5]

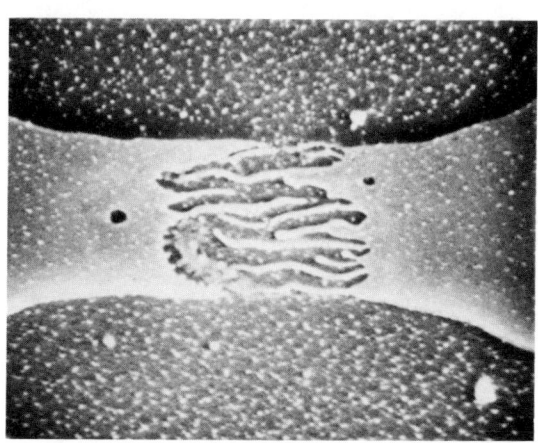

Photo courtesy of HI-REL Laboratories
San Marina, California

Figure 4. Blown Nichrome Fuse.

Electron microscope investigations reveal that random concentrations of nickel appear around the fused links,[6] and that the nickel reacts with the underlying SiO_2 in the gap,[5] forming a nickel-glass structure that resists chemical etching.[3] Also, chromium was found present in the gap.[5]

A CASE STUDY OF NICHROME GROWBACK

In one particular study of nichrome fuse failures performed by Litton,[5] the PROMs "were random samples from PROMs supplied by four manufacturers representing a buy of about twenty thousand 1024 PROMs over a three-year period. These PROMs were purchased to a high reliability full-temperature range specification reflecting a Mil Std-883 Class B screening requirement."

From this same study . . .

"It was found that there appeared to be a glass nichrome reaction which resulted in the formation of a glass structure which resisted the etch. These resistors had exhibited the reappearing bit phenomena after being in the computer in service in the field for some length of time. Further analysis of these resistors by the electron beam microprobe provided information needed to achieve a conceptual understanding of this growback reaction."

"Hard to program bits were associated with process control and PROM design. It was concluded that better than state of the art process controls were required. Therefore, additional screens were needed to insure reliability. The limitation of the number of program pulses and the energy to program was of extreme importance."

The reliability problems with nichrome fuse PROMs all relate to nichrome fuse processing technology; "growback" is inherent in the use of this technology. Some efforts have been made to find tests that will isolate PROM fuses that have a higher probability of relinking. These testing techniques include temperature stressing, temperature cycling, high temperature burn-in, and testing at reduced voltages. No test has been devised which will eliminate the relinking problem.

---

[3] Barnes, D. E. and J. E. Thomas. "Reliability Assessment of a Semiconductor Memory by Design Analysis." A lecture to the IEEE 12th Annual Proceedings on Reliability Physics, 1974.

[4] Devaney, John R. and A. M. Sheble, III. "Plasma Etching PROMs and Other Problems." A lecture to the IEEE 12th Annual Proceedings on Reliability Physics, 1974.

[5] Eisenberg, P. H. and R. Nosler. "Nichrome Resistors in Programmable Read Only Memory Integrated Circuits." A lecture to the IEEE 12th Annual Proceedings on Reliability Physics, 1974.

[6] Franklin, Paul and David Burgess. "Reliability Aspects of Nichrome Fusible Link PROMs (Programmable Read Only Memories)." A lecture to the IEEE 12th Annual Proceedings on Reliability Physics, 1974.

[7] Baitinger, W. E., N. Winograd, J. W. Amy, and J. A. Munarin. "Nichrome Resistor Failures as Studied by X-Ray Photoelectron Spectroscopy (XPS or ESCA)." A lecture to the IEEE 12th Annual Proceedings on Reliability Physics, 1974.

The Silicon Fuse

Intel bipolar PROMs, a typical cell of which is shown in Figure 3, work in the same manner as do the nichrome fuses, with the exception that the fuse material is polycrystalline silicon, which is deposited in a thick layer at the appropriate stage in the manufacturing process. This is the same standard, reliable technique that has been used by Intel in producing millions of MOS LSI circuits every week using polycrystalline silicon.

All of the Intel bipolar PROMs have included on the die a test row and column which are blown at wafer sort. The extra row and column are incorporated primarily to improve the programming yield of the final end product. By addressing this test row, the functionality of the decoders and the programmability of the fuses can be verified. The test fuse circuitry is designed such that arrays with unusual fuses that could cause programming yield problems can be screened at electrical test.

The fuse, shown in Figure 5, is a notched strip of polycrystalline silicon. Figure 6 shows an array of 12 cells. Each cell consists of a single transistor in an emitter-follower configuration with the silicon fuse connecting to the column line as shown in Figure 3. A cross section of the cell is shown in Figure 7.

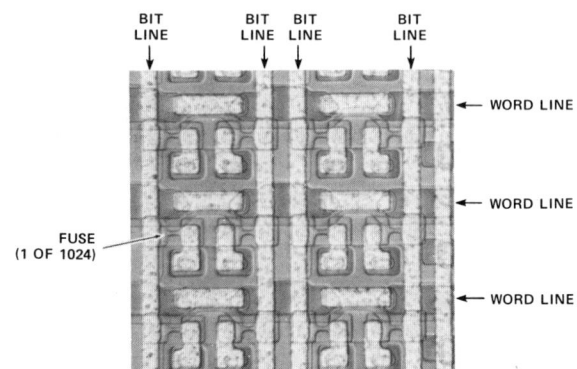

Figure 6. Polysilicon Cell Array.

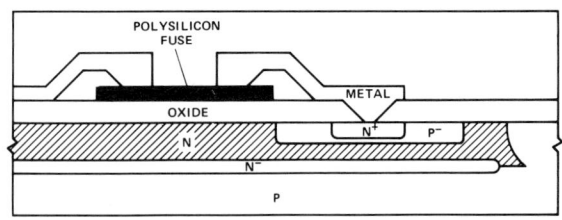

Figure 7. Polysilicon Fuse Cross Section.

The fuse is blown with a pulse train of successively wider pulses, with a current of 20–30 mA typically needed to blow the fuse. During this "blowing" operation, temperatures estimated at 1400°C are reached in the notch of the polysilicon fuse. At these temperatures, the silicon oxidizes and forms an insulating material. Figure 8 shows a blown and unblown fuse. *The use of silicon eliminates conductive dendrites and the existence of conductive materials in the fused gap.*

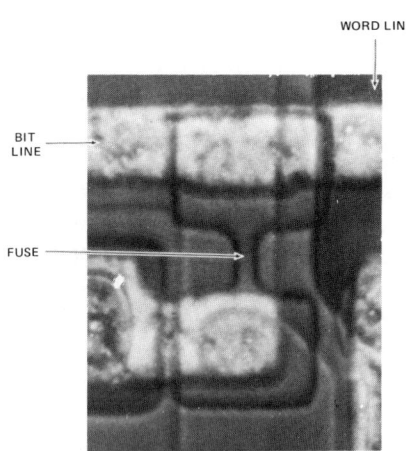

Figure 5. Unblown Polysilicon Fuse.

The thickness of the silicon fuse is nominally 3000 Angstroms, 15 times the thickness of the nichrome fuse. Resistivity of the fuse is controlled by doping, as in standard integrated circuits.

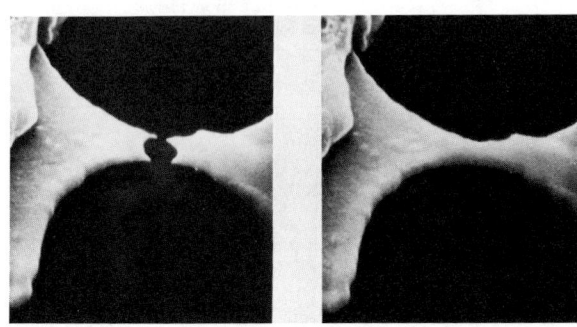

Figure 8. Blown and Unblown Polysilicon Fuse.

DESIGNING WITH PROMS AND ROMS

Since silicon is a standard integrated circuit material, no new contact problems or problems with dissimilar materials are encountered. Growback does not exist with the silicon fuse. *After 3 billion fuse hours of system life testing at 85°C, zero failures have been found.*

The Shorted Junction

A third type of bipolar PROM implementation is the shorted junction. The shorted junction cell is shown in Figures 9 and 10. In this cell, diode Q_1 is reverse-biased and the heavy flow of electrons in the reverse direction causes aluminum atoms from the emitter contact to migrate through the emitter to the base, causing an emitter-to-base short. Extreme care must be taken such that sufficient contact is made to the base without actually puncturing and shorting through the base.

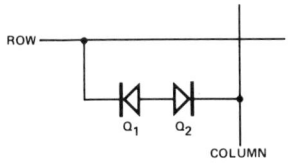

Figure 9. Schematic of Shorted Junction Cells.

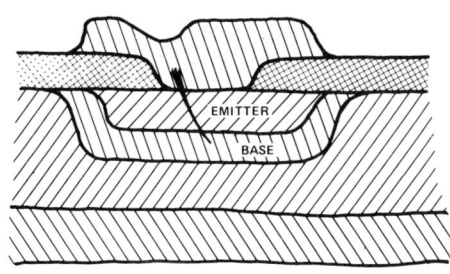

Figure 10. Cross Section of Shorted Junction Cell.

Although the shorted junction PROM does not have the reliability problems associated with the nichrome fuse, programming is greatly complicated by the fact that underprogramming results in insufficient or intermittent contact with the base and overprogramming results in possible internal shorts. The problem of distributing heavy currents around the chip requires the use of multiple-layer metalization, and, as a result, most major semiconductor companies have not committed to the shorted junction technology.

MOS Technology

FAMOS IMPLEMENTATION

As mentioned earlier, it is possible to produce PROMs and ROMs using MOS technology. In 1971 Intel introduced a unique erasable PROM that allows the programmed information to be erased by exposure to ultraviolet light of the correct wavelength and intensity.

The storage element is the Floating gate Avalanche-injection MOS (FAMOS) charge storage device, a cross section of which is shown in Figure 11. The

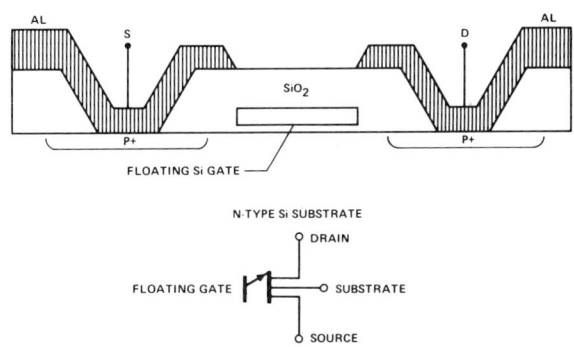

Figure 11. FAMOS Storage Cell.

operation of the cell depends on charge transport to the floating gate by avalanche injection of electrons. The device is essentially a silicon gate MOS field effect transistor in which no connection is made to the silicon gate. Operation of the FAMOS memory structure depends on charge transport to the floating gate by avalanche injection of electrons from either the source or drain. A junction voltage in excess of −30V applied to a p-channel FAMOS device will result in the injection of high-energy electrons from the p–n junction surface avalanche region to the floating silicon gate. The amount of charge transferred to the floating gate is a function of amplitude and duration of the applied junction voltage, as shown in Figure 12. The presence or absence of charge can be sensed by measuring the conductance between the source and drain.

Once the applied junction voltage is removed, no discharge path is available for the accumulated electrons since the gate is surrounded by thermal oxide, which is a very low conductivity dielectric. The electric field in the structure after the removal of junction voltage is due only to the accumulated

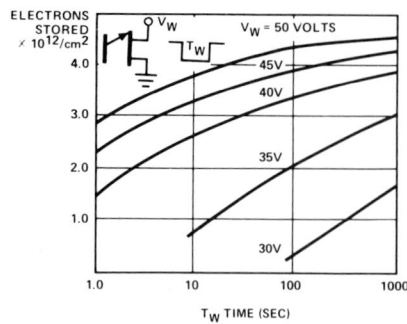

Figure 12. Charge Transfer vs. Programming Pulse Width.

electron charge and is not sufficient to cause charge transport across the polysilicon-thermal-oxide energy barrier.

Charge decay plots as a function of time at 125°C and 300°C are shown in Figure 13. An extrapolation of the 300°C charge decay results indicates that 70% of the initial induced charge will be retained for as long as 10 years at 125°C.

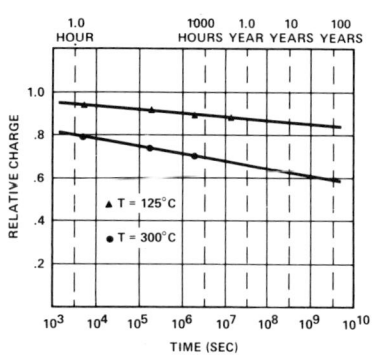

Figure 13. Charge Decay vs. Time.

Since the gate electrode is not electrically accessible, the charge cannot be removed by an electrical pulse. However, the initial condition of no electronic charge on the gate can be restored by illuminating the FAMOS device with ultraviolet light, which results in the flow of a photocurrent from the floating gate back to the silicon substrate, thereby discharging the gate to its initial condition. This erase method allows complete testing of a complex programmable read only memory array.

SPECIFIC DEVICE DESCRIPTIONS

Bipolar Devices

Intel manufacturers a complete line of bipolar PROMs and ROMs as shown in Table I, and in the Product Selection Guide, page PSG-2.

Table I. The Intel PROM/ROM Family.

	1K (256 x 4) 16-Pin		2K (512 x 4) 16-Pin		4K (512 x 8) 24-Pin	
	OC[1]	TS[2]	OC[1]	TS[2]	OC[1]	TS[2]
PROMs	3601	3621	3602	3622	3604	3624
ROMs	3301A	— —	3302	3322	3304A	3324A

NOTES: 1. Open-collector output
2. Three-state output.

Each PROM is pin-for-pin compatible with its mask ROM counterpart. Programming is accomplished by "blowing" a polysilicon fuse in the emitter leg of the bipolar transistor that serves as a data storage cell. Because of the internal circuitry, the initial (unprogrammed) state of the output of the 3601 PROM is low, while the 3602/3622 and 3604/3624 devices have an initial state that produces a high output.

In the 2K and 4K sizes, the part can be ordered with either an open-collector or three-state output.

3601/3621 AND 3301A

The 3601/3621 and 3301A 1K PROM and ROM pin configuration and logic symbol are shown in

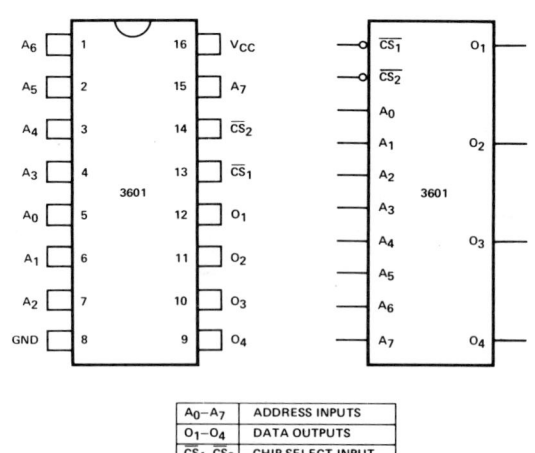

Figure 14. 3601/3621 and 3301A Pin Configuration and Logic Symbol.

Figure 14, and the address and data waveforms are shown in Figure 15. The device is organized as 256 4-bit words. The AC characteristics are summarized in Table II, and the DC characteristics in Table III. Capacitance is shown in Table IV.

DESIGNING WITH PROMS AND ROMS

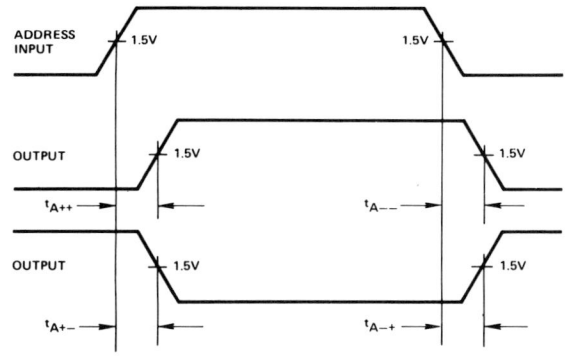
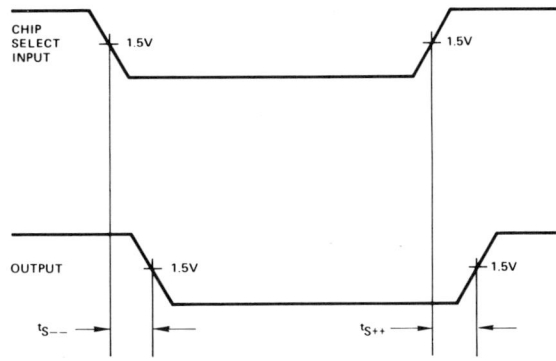

Figure 15. 3601/3301A Address and Data Waveforms.

Table II. 3601/3301A A.C. Characteristics.

SYMBOL	PARAMETER	DEVICE TYPE	LIMITS 0°C	LIMITS 25°C	LIMITS 75°C	UNIT	TEST CONDITIONS
t_{A++}, t_{A--} t_{A++}, t_{A-+}	Address to Output Delay	3301A	45	45	45	nS	$\overline{CS}_1$ and $\overline{CS}_2$ must be at V_{IL} to activate the PROM.
		3601	70	60	70	nS	
		3601-1	50	50	50	nS	
t_{S++}, t_{S--}	Chip Select to Output Delay	3301A	20	20	20	nS	
		3601	25	25	25	nS	
		3601-1	25	25	25	nS	

A simplified block diagram is shown in Figure 16, with a typical 1-bit schematic shown in Figure 17.

Addresses A_3-A_7 select 1 of 32 rows by activating 1 of 32 decoders. Each row consists of 32 cells. Addresses A_0-A_2 enable the 1 of 8 decoders, multiplexing 1 of 8 bits to the appropriate sense amplifier as shown in Figure 17. The logical AND of $\overline{CS}_2 \cdot \overline{CS}_1$ energizes all the columns in the array and provides a programming path as will be described later. $\overline{CS}_1$, which is also active low, enables each of the four output buffers.

The transistors are Schottky barrier diode clamped to allow faster switching speeds than devices fabricated with a conventional gold diffusion process.

Each of the address lines has a low voltage diode input clamp to minimize line reflections.

The outputs are open-collector, which allows them to be OR-connected for memory expansion. The capacitance of the data out pins is typically 7 pF, as shown in Table IV, or 56 pF for eight devices OR-tied together.

Programming the 3601 is accomplished by pulsing V_{CC} and $\overline{CS}_2$ with waveforms as described in the programming section. The initial (unprogrammed) state of the device is with all outputs low; that is, a bit is considered programmed when the output is high.

3602/3622 AND 3302/3322

The 3602/3622 and 3302/3322 pin configuration and logic symbol are shown in Figure 18.

The 3602/3302 has an open-collector output, while the 3622/3322 is a three-state output. A simplified block diagram of the part is shown in Figure 16. The schematic is shown in Figure 17. As indicated in Figure 16, the organization is 512 X 4 bits. Operation is analogous to the 1K PROM described earlier.

Table III. 3601/3301A D.C. Characteristics.

All limits apply for V_{CC} = +5.0V ±5%, T_A = 0°C to +75°C

SYMBOL	PARAMETER		MIN	TYP[1]	MAX	UNIT	TEST CONDITIONS
I_{FA}	Address Input Load Current			−0.05	−0.25	mA	V_{CC} = 5.25V, V_A = 0.45V
I_{FS}	Chip Select Input Load Current			−0.05	−0.25	mA	V_{CC} = 5.25V, V_S = 0.45V
I_{RA}	Address Input Leakage Current				40	μA	V_{CC} = 5.25V, V_A = 4.0V
I_{RS}	Chip Select Input Leakage Current				40	μA	V_{CC} 5.25V, V_S = 4.0V
V_{CA}	Address Input Clamp Voltage			−0.7	−1.0	V	V_{CC} = 4.75V, I_A = −5.0 mA
V_{CS}	Chip Select Input Clamp Voltage			−0.7	−1.0	V	V_{CC} = 4.75V, I_S = −5.0 mA
V_{OL}	Output Low Voltage			0.3	0.45	V	V_{CC} = 4.75V, I_{OL} = 15 mA
I_{CEX}	Output Leakage Current				100	μA	V_{CC} = 5.25V, V_{CE} = 5.25V
I_{CC}	Power Supply Current	3601		90	130	mA	V_{CC} = 5.25V, $V_{A0} \rightarrow V_{A7}$ = 0V, $V_{S0} = V_{S1}$ = 0V
		3301A		90	125	mA	
V_{IL}	Input "Low" Voltage				0.85	V	V_{CC} = 5.0V
V_{IH}	Input "High" Voltage		2.0			V	V_{CC} = 5.0V

NOTE: 1. Typical values are at 25°C and at nominal voltage.

Table IV. 3601/3301A Capacitance[1].

SYMBOL	PARAMETER	TYP	MAX	UNIT	TEST CONDITIONS
C_{INA}	Address Input Capacitance	4	10	pF	V_{CC} = 5V V_{IN} = 2.5V
C_{INS}	Chip Select Input Capacitance	6	10	pF	V_{CC} = 5V V_{IN} = 2.5V
C_{OUT}	Output Capacitance	7	12	pF	V_{CC} = 5V V_{OUT} = 2.5V

NOTE: 1. This parameter is only periodically samples and is not 100% tested.

Referring to Figure 17, addresses A_3–A_8 select 1 of 64 rows, each row consisting of 32 cells. Addresses A_0–A_2 enable the 1 of 8 decoders, multiplexing 1 of 8 bits to the appropriate sense amplifier.

Chip select, $\overline{CS}$, enables the output buffer, and provides the programming path.

The 3302 and 3322 provide ROM capability for applications that have matured sufficiently to allow use of a fixed data pattern.

The 3602L-6 and 3622L-6 have the additional capability of reducing power whenever the chip is deselected; i.e., $\overline{CS}$ high. The standby power is 236mW, compared to 685mW for the 3602 and 3622.

3604/3624 AND 3304A/3324A

The 3604/3304A pin configuration and logic symbol are shown in Figure 19, and typical waveforms

DESIGNING WITH PROMS AND ROMS

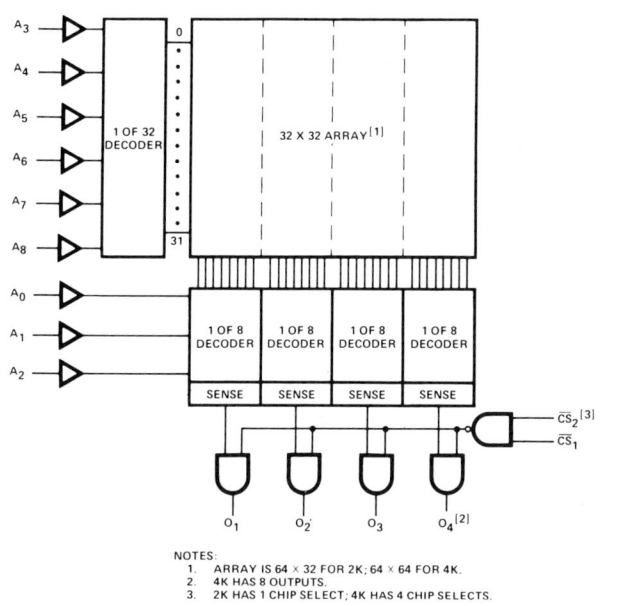

NOTES:
1. ARRAY IS 64 × 32 FOR 2K; 64 × 64 FOR 4K.
2. 4K HAS 8 OUTPUTS.
3. 2K HAS 1 CHIP SELECT; 4K HAS 4 CHIP SELECTS.

Figure 16. Simplified Bipolar PROM Block Diagram.

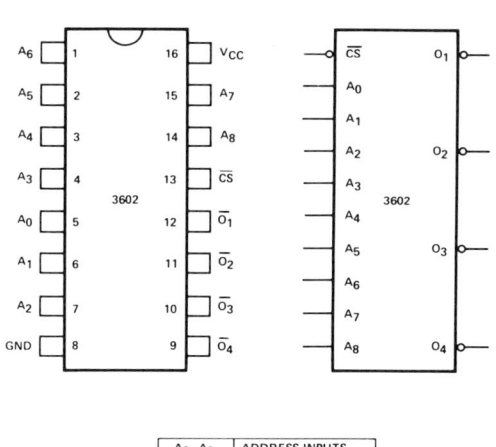

A_0–A_8	ADDRESS INPUTS
O_1–O_4	DATA OUTPUTS
$\overline{CS}$	CHIP SELECT INPUT

Figure 18. 3602/3622 and 3302/3322 Pin Configuration and Logic Symbol.

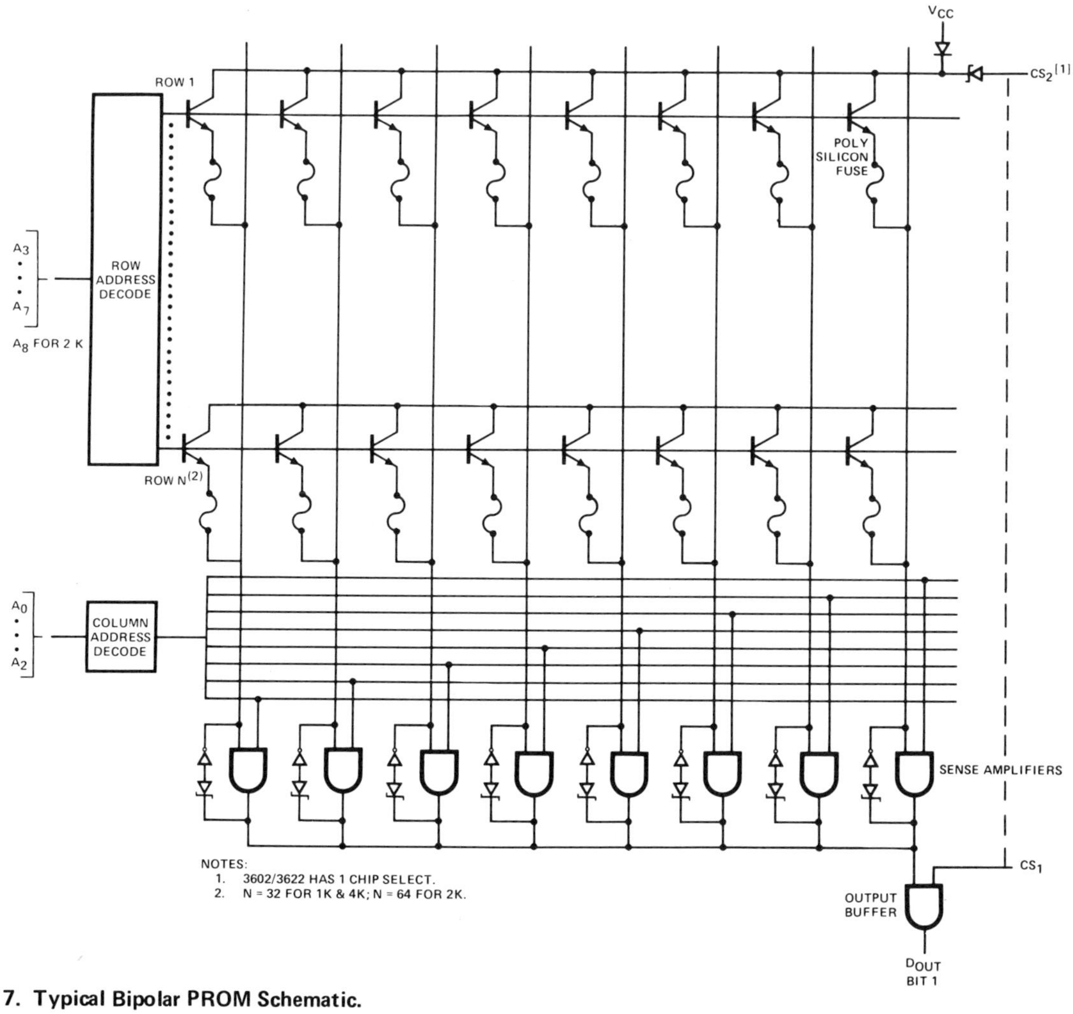

NOTES:
1. 3602/3622 HAS 1 CHIP SELECT.
2. N = 32 FOR 1K & 4K; N = 64 FOR 2K.

Figure 17. Typical Bipolar PROM Schematic.

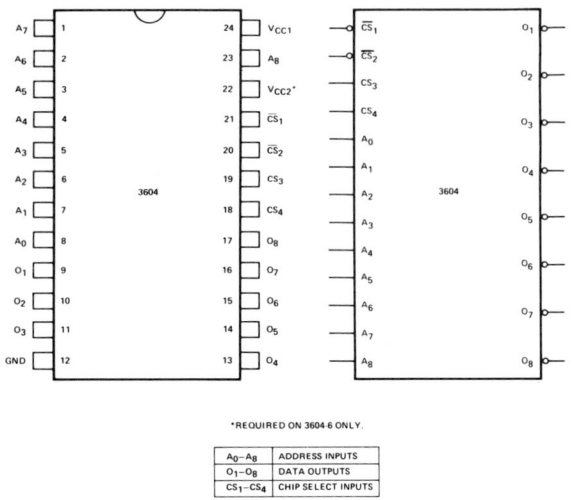

Figure 19. 3604/3304A Pin Configuration and Logic Symbol.

are shown in Figure 20. Table V summarizes the AC characteristics, with capacitance and DC characteristics summarized in Tables VI and VII, respectively.

The organization of the device is 512 × 8 bits. The basic operation of the 4K device is directly analogous to the 1K PROM as shown in Figure 17. Addresses A_3–A_8 select 1 of 64 rows, each row consisting of 64 cells. Addresses A_0–A_2 enable the decoders, multiplexing 1 of 8 bits to the appropriate sense amplifier. $\overline{CS}_1$ provides the programming path, while $\overline{CS}_1 \cdot \overline{CS}_2 \cdot CS_3 \cdot CS_4$ provide an enable to the output. The 4-chip select terms may facilitate decoding when working with large arrays.

The 3604L-6 has the additional feature that when the chip is selected (i.e., $\overline{CS}_1$ or $\overline{CS}_2$ high), the power is reduced by approximately 70%. To utilize this feature, pins 22 and 24 must be connected as shown in Table VIII, which compares the 3604 and 3604L-6 V_{CC} connections.

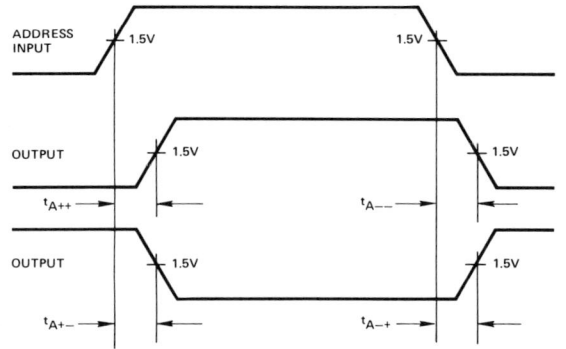
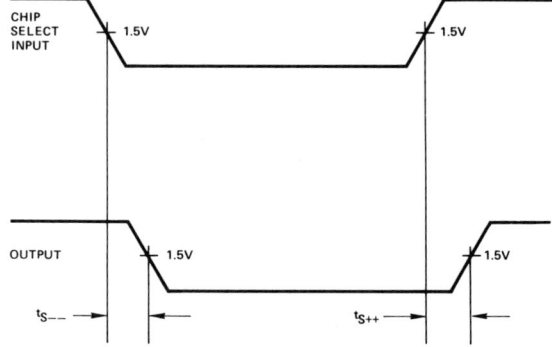

Figure 20. 3304A/3604 Address and Data Waveforms.

Table V. 3304A/3604 A.C. Characteristics.

V_{CC} = +5V ±5%, T_A = 0°C to +75°C

SYMBOL	PARAMETER		MAX	UNIT	TEST CONDITIONS
t_{A++}, t_{A--} t_{A+-}, t_{A-+}	Address to Output Delay	3304A 3604 3604L-6	70 70 90	nS nS nS	$\overline{CS}_1 = \overline{CS}_2 = V_{IL}$ and $CS_3 = CS_4 = V_{IH}$ to select the PROM.
t_{S++}	Chip Select to Output Delay	3304A 3604 3604L-6	30 30 30	nS nS nS	
t_{S--}	Chip Select to Output Delay	3304A 3604 3604L-6	30 30 120	nS nS nS	

DESIGNING WITH PROMS AND ROMS

Table VI. 3304A/3604 Capacitance[1].

SYMBOL	PARAMETER	TYP	MAX	UNIT	TEST CONDITIONS	
C_{INA}	Address Input Capacitance	4	10	pF	$V_{CC} = 5V$	$V_{IN} = 2.5V$
C_{INS}	Chip Select Input Capacitance	6	10	pF	$V_{CC} = 5V$	$V_{IN} = 2.5V$
C_{OUT}	Output Capacitance	7	12	pF	$V_{CC} = 5V$	$V_{OUT} = 2.5V$

NOTE: 1. This parameter is only periodically samples and is not 100% tested.

Table VII. 3304A/3604 D.C. Characteristics.

All limits apply for $V_{CC} = +5.0V \pm 5\%$, $T_A = 0°C$ to $+75°C$

SYMBOL	PARAMETER	MIN	TYP[1]	MAX	UNIT	TEST CONDITIONS
I_{FA}	Address Input Load Current		−0.05	−0.25	mA	$V_{CC} = 5.25V$, $V_A = 0.45V$
I_{FS}	Chip Select Input Load Current		−0.05	−0.50	mA	$V_{CC} = 5.25V$, $V_S = 0.45V$
I_{RA}	Address Input Leakage Current			40	μA	$V_{CC} = 5.25V$, $V_A = 5.25V$
I_{RS}	Chip Select Input Leakage Current			40	μA	$V_{CC} = 5.25V$, $V_S = 4.0V$
V_{CA}	Address Input Clamp Voltage		−0.7	−1.0	V	$V_{CC} = 4.75V$, $I_A = -5.0$ mA
V_{CS}	Chip Select Input Clamp Voltage		−0.7	−1.0	V	$V_{CC} = 4.75V$, $I_S = -5.0$ mA
V_{OL}	Output Low Voltage		0.3	0.45	V	$V_{CC} = 4.75V$, $I_{OL} = 15$ mA
I_{CEX}	Output Leakage Current			100	μA	$V_{CC} = 5.25V$, $V_{CE} = 5.25V$
I_{CC1}	Power Supply Current 3304A and 3604			190	mA	$V_{CC1} = 5.25V$, $V_{A0} \rightarrow V_{A7} = 0V$ $CS_1 = CS_2 = 0V$ $CS_3 = CS_4 = 5.25V$
I_{CC2}	Power Supply Current (3604L-6) Active			140	mA	$V_{CC2} = 5.25V$, $V_{CC1} = $ Open Chip Selected
	Standby			45	mA	Chip Deselected
V_{IL}	Input "Low" Voltage			0.85	V	$V_{CC} = 5.0V$
V_{IH}	Input "High" Voltage	2.0			V	$V_{CC} = 5.0V$

NOTE: 1. Typical values are at 25°C and at nominal voltage.

Table VIII. 3604/3604L-6 Connections and Power Consumption.

DEVICE TYPE	CONNECTION	READ	PROGRAM	POWER
3604	Pin 22	+5V or No Connect	Pulsed 12.5V	998 mW maximum
	Pin 24	+5V	Pulsed 12.5V	
3604L-6	Pin 22	+5V	Pulsed 12.5V	735 mW maximum with chip selected
	Pin 24	No Connect [1]	Pulsed 12.5V	236 mW maximum with chip deselected

NOTE: 1. Do not connect pin 24 of the 3604L-6 to any other pin

The 3304AL6 mask ROM is available for ROM users who wish to take advantage of the power reduction feature of the reduced standby power.

MOS Devices

The Intel family of MOS PROMs and ROMs can also be divided into two groups. The 1602A/1702A, and its mask programmable counterpart, the 1302, are 2048 bit MOS devices, organized as 256 x 8 bits. The 2704 and 2708 are, respectively, 4K (512 x 8) and 8K (1024 x 8). The 2308 is the mask counterpart of the 2708. The 1702A, the 2704 and the 2708 EPROMs are all implemented with the Intel FAMOS technology.

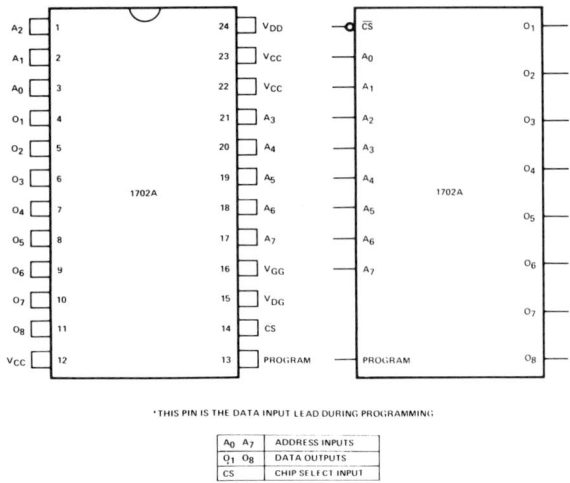

Figure 21. 1302/1602A/1702A Pin Configuration and Logic Symbol.

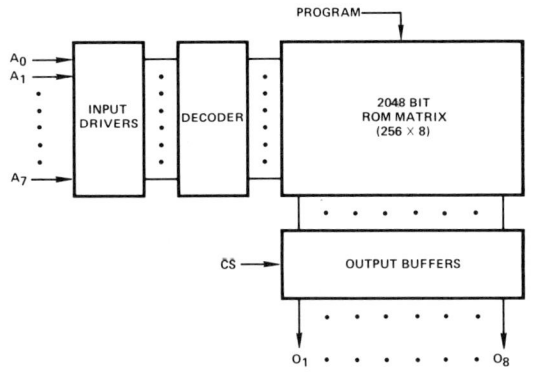

Figure 22. 1602A/1702A Block Diagram.

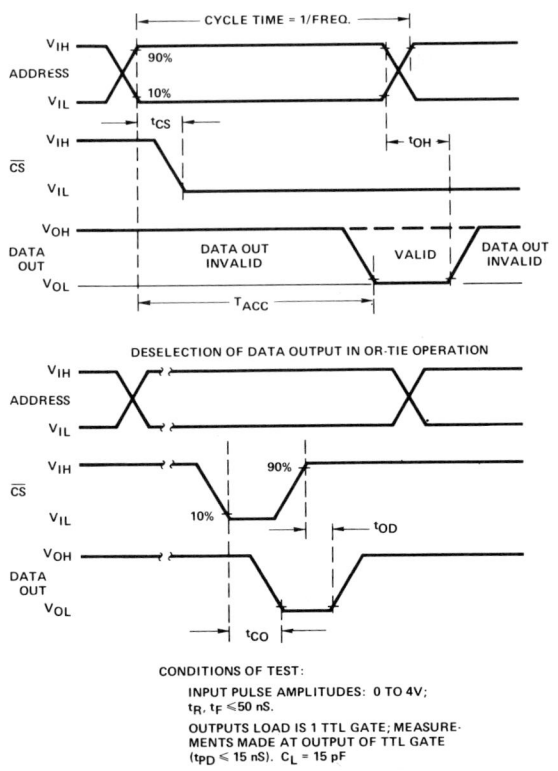

Figure 23. 1302/1602A/1702A Waveforms.

1602A/1702A/1302

The logic symbol and pin configuration for these devices is shown in Figure 21, and the block diagram is shown in Figure 22. The 1302 mask programmable device is pin for pin compatible with the electrically programmable devices.

The FAMOS data storage cell used in the 1602A/1702A is described in the technology section.

The operation and electrical characteristics of the 1602A and the 1702A are identical; the 1702A is packaged with a quartz lid to allow erasure by high intensity ultraviolet light as described in the technology chapter. The 1602A/1702A switching characteristics are shown in Figure 23, and AC and DC characteristics are summarized in Tables IX and X, respectively. Capacitance is shown in Table XI.

The operation of the 1602A/1702A is similar to the bipolar PROMs described earlier. The higher order address bits A_5-A_7 perform the row decode function, while the low order address bits provide the column decode. Chip select, $\overline{CS}$, is active low and enables the eight output buffers.

For low power applications, with the 1602AL/1702AL it is possible to clock the V_{GG} (–9V) supply, resulting in a decrease of power proportional to the V_{GG} duty cycle.

As with the bipolar PROMs, care should be taken with the number of devices that are OR-tied together such that access time is not compromised.

The initial (unprogrammed) state of the 1602A/1702A is all "0's" (output low). Programming is accomplished by writing "1's" (output high) in the proper bit locations.

Figure 24 presents various parametric curves that will assist the designer in determining worst case conditions when using the device.

DESIGNING WITH PROMS AND ROMS

Table IX. 1302/1602A/1702A A.C. Characteristics.

$T_A = 0°C$ to $+70°C$, $V_{CC} = +5V \pm 5\%$, $V_{DD} = -9V \pm 5\%$, $V_{GG} = -9V \pm 5\%$, unless otherwise specified.

SYMBOL	PARAMETER		MIN	TYP[1]	MAX	UNIT
Frequency	Repetition Rate				1	MHz
t_{OH}	Previous Read Data Valid				100	nS
t_{ACC}	Address to Output Delay			0.7	1	μS
t_{DVGG}	Clocked V_{GG} Set Up	1602AL/1702AL	1			μS
		1302	1			μS
t_{CS}	Chip Select Delay	1602A/1702A			100	nS
		1302			200	nS
t_{CO}	Output Delay from CS	1602A/1702A			900	nS
		1302			500	nS
t_{OD}	Output Deselect				300	nS
t_{OHC}	Data Out Hold In Clocked V_{GG} Mode[2]				5	μS

NOTES: 1. Typical values are at 25°C and at nominal voltage.

2. The outputs will remain valid for t_{OHC} as long as clocked V_{GG} is at V_{CC}. An address change may occur as soon as the output is sensed (clocked V_{GG} may still be at V_{CC}). Data becomes invalid for the old address when clocked V_{GG} is returned to V_{GG}.

Table X. 1302/1602A/1702A D.C. and Operating Characteristics[1].

$T_A = 0°C$ to $+70°C$, $V_{CC} = +5V \pm 5\%$, $V_{DD} = -9V \pm 5\%$, $V_{GG}^{[2]} = -9V \pm 5\%$, unless otherwise specified.

SYMBOL	PARAMETER	MIN	TYP[3]	MAX	UNIT	TEST CONDITIONS	
I_{LI}	Address and Chip Select Input Load Current			1	μA	$V_{IN} = 0V$	
I_{LO}	Output Leakage Current			1	μA	$V_{OUT} = 0.0V$, $\overline{CS} = V_{CC}-2$	
I_{DD0}	Power Supply Current		5	10	mA	$V_{GG} = V_{CC}$, $\overline{CS} = V_{CC}-2$ $I_{OL} = 0.0$ mA, $T_A = 25°C$	
I_{DD1}	Power Supply Current		35	50	mA	$\overline{CS} = V_{CC}-2$ $I_{OL} = 0.0$ mA, $T_A = 25°C$	
I_{DD2}	Power Supply Current		32	46	mA	$\overline{CS} = 0.0$ $I_{OL} = 0.0$ mA, $T_A = 25°C$	Continuous Operation
I_{DD3}	Power Supply Current		38.5	60	mA	$\overline{CS} = V_{CC}-2$ $I_{OL} = 0.0$ mA, $T_A = 0°C$	
I_{CF1}	Output Clamp Current		8	14	mA	$V_{OUT} = -1.0V$, $T_A = 0°C$	
I_{CF2}	Output Clamp Current			13	mA	$V_{OUT} = -1.0V$, $T_A = 25°C$	
I_{GG}	Gate Supply Current			1	μA		
V_{IL1}	Input Low Voltage for TTL Interface	-1.0		0.65	V		
V_{IL2}	Input Low Voltage for MOS Interface	V_{DD}		$V_{CC}-6$	V		
V_{IH}	Address and Chip Select Input High Voltage	$V_{CC}-2$		$V_{CC}+0.3$	V		
I_{OL}	Output Sink Current	1.6	4		mA	$V_{OUT} = 0.45V$	
I_{OH}	Output Source Current	-2.0			mA	$V_{OUT} = 0.0V$	
V_{OL}	Output Low Voltage		-0.7	0.45	V	$I_{OL} = 1.6$ mA	
V_{OH}	Output High Voltage	3.5	4.5		V	$I_{OH} = -100$ μA	

NOTES: 1. In the programming mode, data inputs 1–8 are pins 4–11, respectively; $\overline{CS}$ = GND.

2. V_{GG} may be clocked to reduce power dissipation. In this mode average I_{DD} decreases in proportion to V_{GG} duty cycle.

3. Typical values are at 25°C and at nominal voltage.

Table XI. 1302/1602A Capacitance.

$T_A = 25°C$

SYMBOL	PARAMETER		TYP	MAX	UNIT	TEST CONDITIONS	
C_{IN}	Input Capacitance	1302	5	10	pF	$V_{IN} = V_{CC}$	
		1602A/1702A	8	15	pF	$\overline{CS} = V_{CC}$	All unused pins are at AC ground.
C_{OUT}	Output Capacitance	1302	5	10	pF	$V_{OUT} = V_{CC}$	
		1602A/1702A	10	15	pF	$V_{GG} = V_{CC}$	
C_{VGG}	V_{GG} Capacitance (Clocked V_{GG} Mode)			30	pF		

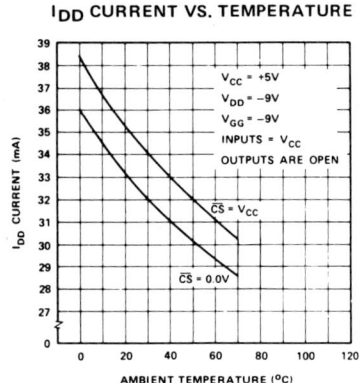
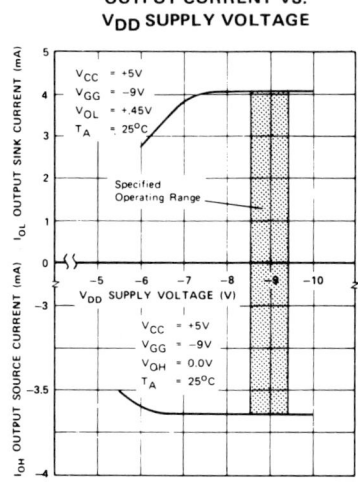
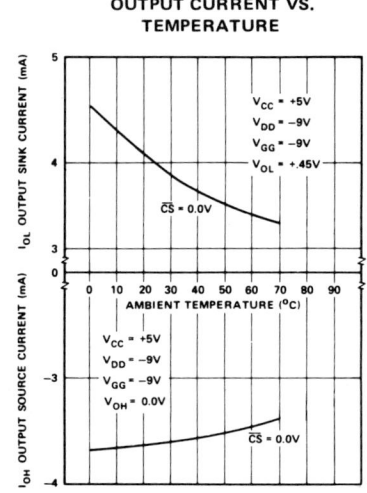
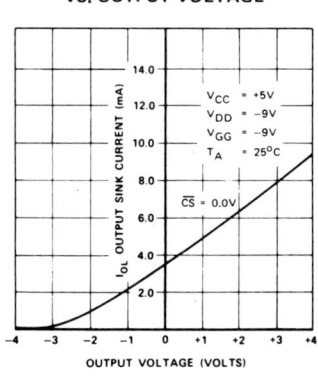
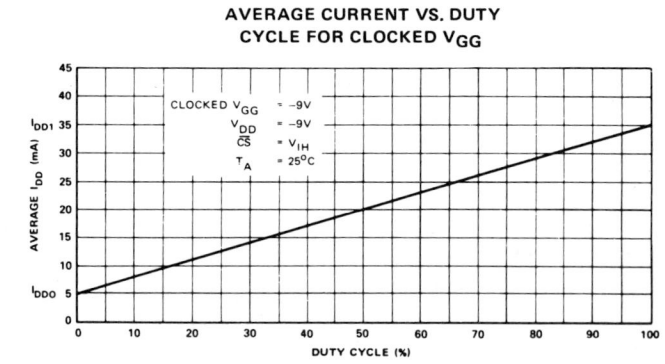
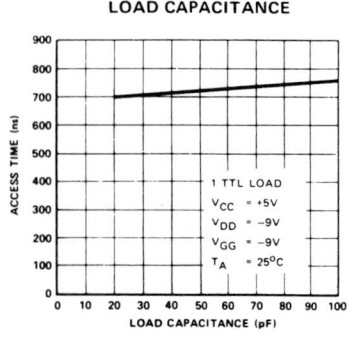
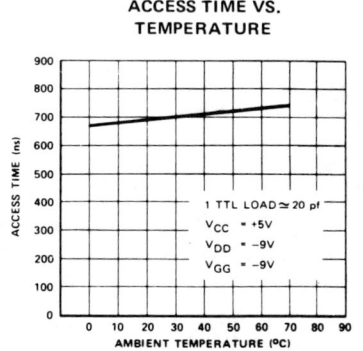

Figure 24. 1602A/1702A Parametric Curves.

DESIGNING WITH PROMS AND ROMS

PROGRAMMING

All of the PROMs described earlier require that data be entered by a technique different from that required to read. There are two ways of programming a PROM; one is to satisfy the control requirements for a particular address, apply some sort of pulsed voltage to the appropriate connection, and proceed to the next location. The other is to put the information on a mark/sense card or paper tape and give it to somebody. Both methods are presented here.

PROGRAMMING THE 1K BIPOLAR PROM (3601)

The 3601 may be programmed using the basic circuit of Figure 25. Address inputs are at standard TTL levels. Only one output may be programmed at a time. The output to be programmed must be connected to V_{CC} through a 300Ω resistor. This will force the proper programming current (3-6mA) into the output when the V_{CC} supply is later raised to 10V. All other outputs must be held at a TTL low level (0.4V maximum).

The programming pulse generator produces a series of pulses to the 3601 V_{CC} and $\overline{CS_2}$ leads as shown in Figure 26 V_{CC} is pulsed from a low of 4.5V ±0.25V to a high of 10V ±0.25V, while $\overline{CS_2}$ is pulsed from a low of ground (TTL logic 0) to a high of 15V ±0.25V. It is important to accurately maintain these voltage levels; otherwise, improper programming may result.

The pulses applied must maintain a duty cycle of 50% ±10%, and start with an initial width of 1 µS ±10%, *and increase linearly over a period of approximately 100 mS to a maximum width of 8 µS* ±10%. Typical devices have their fuse blown within 1 mS, but occasionally a fuse may take up to 400 mS to blow.

During the application of the program pulse, current to $\overline{CS_2}$ must be limited to 100 mA. The output of the 3601 is sensed when $\overline{CS_2}$ is at a TTL low level output. A programmed bit will have a TTL high output. After a fuse is blown, the V_{CC} and $\overline{CS_2}$ pulse trains must be applied for another 500 µS.

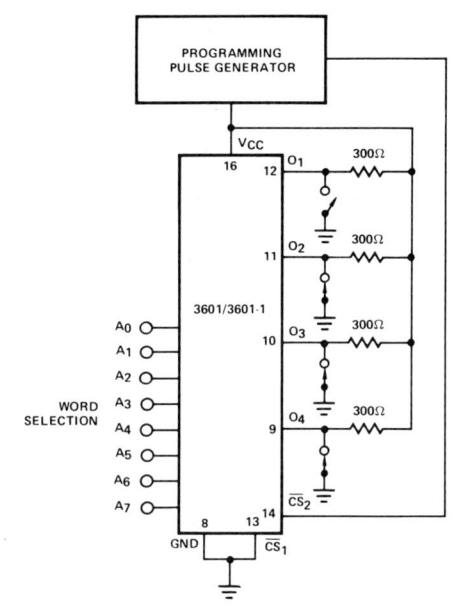

Figure 25. 3601 Programming Connections.

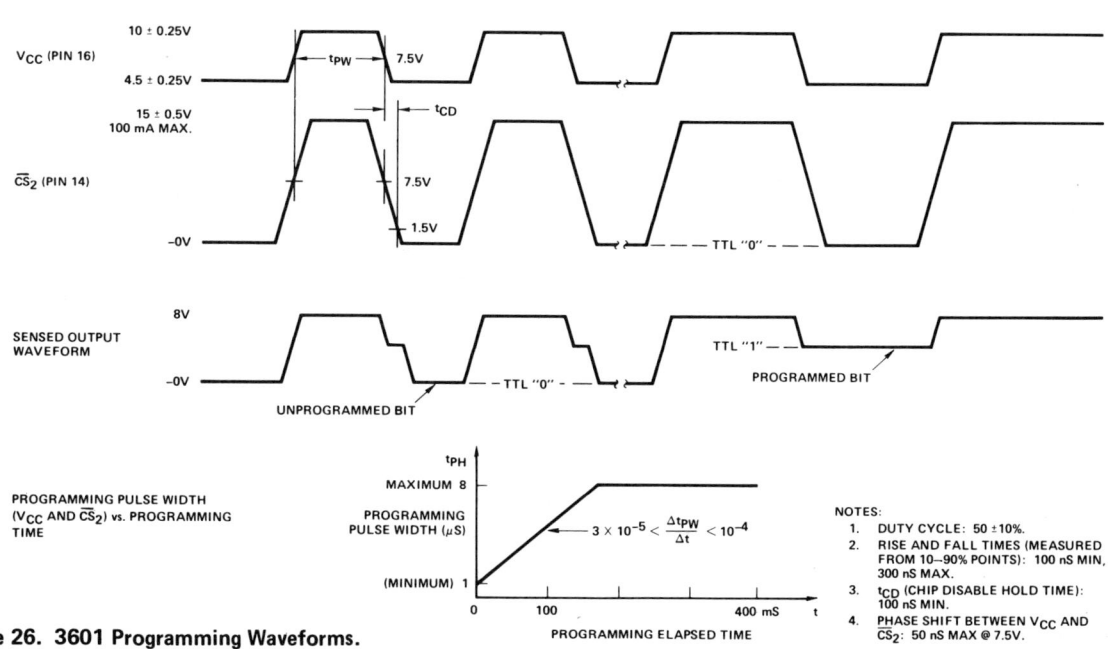

Figure 26. 3601 Programming Waveforms.

PROMS AND ROMS

PROGRAMMING THE 2K AND 4K BIPOLAR PROMs (3602/3622 AND 3604/3624) [1]

The 3602/3622 and 3604/3624 parts are also programmed by forcing current into the output, but with the 3622 and 3624, the three-state outputs that are not being programmed must be allowed to float. Figure 27 shows the basic circuit for programming the 2K and 4K family of PROMs.

The programming current that must be provided to each output is 5 mA ±10% for the 3602/3622 and 3604/3624.

The low standby power devices can be programmed in the same way, the only differences being that V_{CC1}, V_{CC2}, and $\overline{CS}_1$ must be connected and pulsed in accordance with Table XII and Figure 28. Note that pin 24 of the 3604L-6 must not be connected to any other pin, or the power down circuit will not operate.

Note that the V_{CC1} and V_{CC2} programming levels are 12.5V ±0.5V for the 3602, 3604, 3622, and 3624.

1. The 3621 is also programmed by this technique.

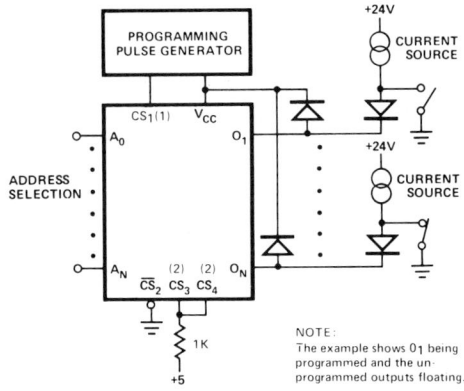

NOTES:
1. For the 3621 and 3605/3625 family only the program pulse may be applied to either CS_1 or CS_2.
2. CS_3, CS_4 are only for the 3604/3624 PROM family.

Figure 27. 3621, 2K and 4K Bipolar PROM Programming Connections.

Table XII. 3604/3604AL6 Programming Connections.

MODE	PIN	22 V_{CC2}	24 V_{CC1}
READ	3604	No Connect or +5V	+5V
	3604L-6	+5V	No Connect [1]
PROGRAM	3604	Pulsed 12.5V	Pulsed 12.5V
	3604L-6	Pulsed 12.5V	Pulsed 12.5V
STANDBY POWER	3604L-6	Power dissipation is automatically reduced whenever the 3604L-6 is deselected.	

NOTE: 1. Do not connect pin 24 of the 3604L-6 to any other pin

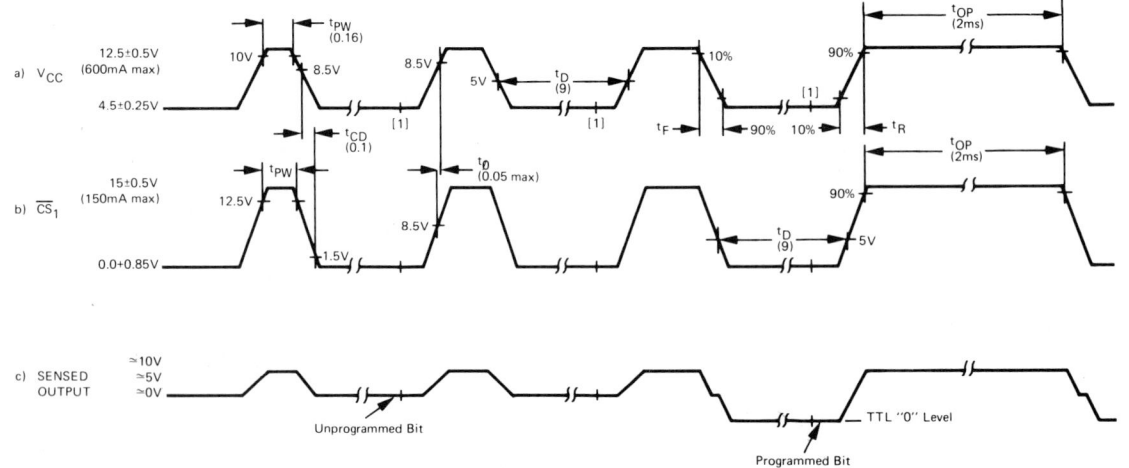

[1] Data Sense Time should be at 90% (or greater) of t_D. A bit is considered programmed after 128 successful verifications. The program pulses should continue to increase in accordance with the ramp shown above. After 128 successful verifications, the DC over program time can start.

NOTE: All times in parenthesis are in microseconds and are in minimum times unless otherwise specified.

Figure 28. 3602/3622 and 3604/3624 Programming Waveforms.

DESIGNING WITH PROMS AND ROMS

Programming the 1602A/1702A

In its initial state the 1602A/1702A array will have all outputs low. Programming is accomplished by writing highs in the proper bit locations. The peak I_{DD} current that must be provided for programming the 1602A/1702A is approximately 200 mA, and the entire device can be programmed in about 2 minutes. Figure 29 shows the waveforms required for programming, while Table XIII shows the connections used. Table XIV and XV show the A.C. and D.C. characteristics for programming.

During programming, V_{CC} should be held at ground and V_{BB} should be held at +12V. Address levels are approximately –40V for a logic "0" (output low), and approximately 0V for a logic "1" (output high). Note that these levels are larger in magnitude but in the same polarity sense as those used for reading from the memory:

logic "0": $-1V \leqslant$ logic "0" $\leqslant .65V$,

logic "1": $\geqslant V_{CC} - 2$

where $V_{CC} = 5V \pm 5\%$.

When programming, the negative-going power supplies (V_{DD}) must be pulsed. V_{DD} is pulsed to –47V ±1V. V_{GG} is brought to –35 to –40V, and the complement of the address to be programmed is applied. After the power has been applied for at least 25µS, the address must be returned to its true form 10µS or more after the address has reached its true state, and at least 100 µS after turning on power, the 3 mS program pulse (pin 13) at –47V ±1V may be applied. During the interval when V_{DD} is applied, data signals must be applied to the data output lines. A data level of approximately 0V will result in the location remaining unchanged, while a level of –47V ±1V will program a logic "1" (output high in read mode). After the program pulse is turned off, the V_{DD} and V_{GG} voltages should be turned off. This turn-off should occur from 10–100µS after removal of the program pulse.

For best results, the 1602A/1702A should be programmed by scanning through the addresses in binary sequence 32 times. Each pass repeats the same series of programming pulses. The duty cycle for applied power must not exceed 20%. As a re-

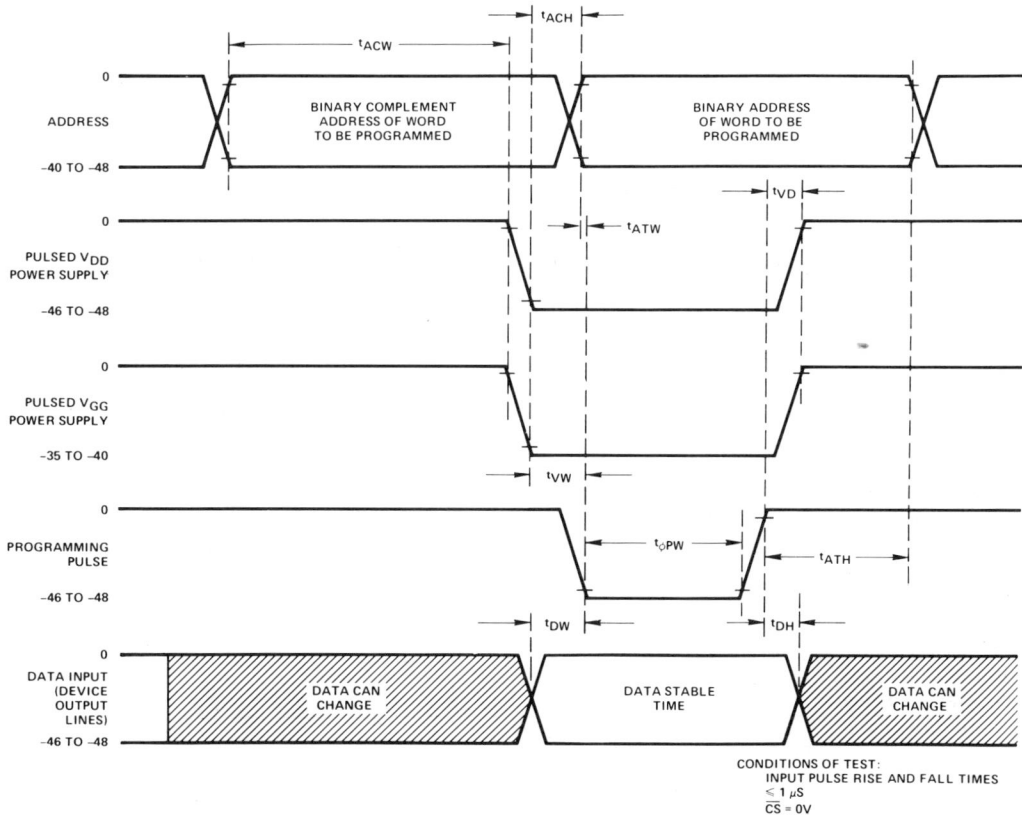

Figure 29. 1602A/1702A Programming Waveforms.

sult, each pass takes about 4 seconds, with the 32 passes taking just over 2 minutes.

ERASING THE 1702A

The 1702A EPROM may be erased by exposure to high intensity, short-wave ultraviolet light at a wavelength of 2537 Angstroms. The recommended integrated dose (i.e., UV intensity × intensity time) is 6W-sec/cm^2. The devices are made with a transparent quartz lid covering the silicon die. Conventional room light, fluorescent light, or sunlight has no measurable effect on stored data, even after years of exposure. However, after 10–20 minutes under a suitable source, the device is erased to a state of all "0's" (outputs low). To prevent damage to the device, it is recommended that no more ultraviolet light exposure be used than that necessary to erase the 1702A.

CAUTION

When using an ultraviolet source of this type, care should be taken not to expose the eyes or skin to the ultraviolet rays, as damage to vision or burns may occur. Also, these shortwave rays may generate considerable amounts of ozone which is potentially hazardous.

Table XIII. 1602A/1702A Programming Connections.

PIN MODE	12 (V_{CC})	13 (Program)	14 (CS)	15 (V_{BB})	16 (V_{GG})	22 (V_{CC})	23 (V_{CC})
Read	V_{CC}	V_{CC}	GND	V_{CC}	V_{GG}	V_{CC}	V_{CC}
Programming	GND	Program Pulse	GND	V_{BB}	Pulsed V_{GG} (V_{IL4P})	GND	GND

Table XIV. 1602A/1702A D.C. and Operating Characteristics for Programming Operation.

$T_A = 25°C$, $V_{CC} = 0V$, $V_{BB} = +12V \pm 10\%$, $\overline{CS} = 0V$ unless otherwise noted

SYMBOL	TEST	MIN.	TYP.	MAX.	UNIT	CONDITIONS
I_{LI1P}	Address and Data Input Load Current			10	mA	$V_{IN} = -48V$
I_{LI2P}	Program and V_{GG} Load Current			10	mA	$V_{IN} = -48V$
I_{BB}	V_{BB} Supply Load Current		10		mA	
I_{DDP}(1)	Peak I_{DD} Supply Load Current		200		mA	$V_{DD} = V_{prog} = -48V$ $V_{GG} = -35V$
V_{IHP}	Input High Voltage			0.3	V	
V_{IL1P}	Pulsed Data Input Low Voltage	−46		−48	V	
V_{IL2P}	Address Input Low Voltage	−40		−48	V	
V_{IL3P}	Pulsed Input Low V_{DD} and Program Voltage	−46		−48	V	
V_{IL4P}	Pulsed Input Low V_{GG} Voltage	−35		−40	V	

Note 1: I_{DDP} flows only during V_{DD}, V_{GG} on time. I_{DDP} should not be allowed to exceed 300 mA for greater than 100 μsec. Average power supply current I_{DDP} is typically 40 mA at 20% duty cycle.

2. The V_{BB} supply must be limited to 100 mA max current to prevent damage to the device.

DESIGNING WITH PROMS AND ROMS

Table XV. 1602A/1702A A.C. Characteristics for Programming Operation.
$T_A = 25°C$, $V_{CC} = 0V$, $V_{BB} = +12V \pm 10\%$, $\overline{CS} = 0V$ unless otherwise noted

SYMBOL	TEST	MIN.	TYP.	MAX.	UNIT	CONDITIONS
	Duty Cycle			20	%	
$t_{\phi PW}$	Program Pulse Width			3	ms	$V_{GG} = -35V$, $V_{DD} = V_{prog} = -48V$
t_{DW}	Data Set Up Time	25			µs	
t_{DH}	Data Hold Time	10			µs	
t_{VW}	V_{DD}, V_{GG} Set Up	100			µs	
t_{VD}	V_{DD}, V_{GG} Hold	10		100	µs	
t_{ACW} [3]	Address Complement Set Up	25			µs	
t_{ACH} [3]	Address Complement Hold	25			µs	
t_{ATW}	Address True Set Up	10			µs	
t_{ATH}	Address True Hold	10			µs	

Note 3. All 8 address bits must be in the complement state when pulsed V_{DD} and V_{GG} move to their negative levels. The addresses (0 through 255) must be programmed as shown in the timing diagram for a minimum of 32 times.

Programmers

Table XVI summarized some of the available programmers that support Intel PROMs. Specific questions regarding prices, availability, and options should be directed to the particular manufacturer.

Programmed Parts

All of the electrically programmable parts manufactured by Intel can be programmed by the end user with Intel approved equipment, or can be ordered from local distributors who are equipped with programmers compatible with each device type. In general, orders for less than 1000 pieces of programmed PROMs should be handled by local distributors, while orders for greater than that quantity should be referred to the factory. In either case, the data must be prepared in accordance with the following paragraphs.

Programming information should be sent in the form of computer punched cards or punched paper tape. In all cases, the order should be accompanied by a printout of the truth table.

The following general format is applicable to the programming information sent to Intel:

1. A data field should start with the most significant bit (O_8) and end with the least significant bit (O_1).

2. The data field should consist of P's and N's. P indicates a high level output (most positive), and N a low level output (most negative). If the programming information is sent on a punched paper tape, a start character (B) and an end character (F) must be used in the data field.

Table XVI. Approved Programmers.

	1602A/1702A Family	2704 Family	2708 Family	3601 Family	3602/3622 Family	3604/3624 Family
Intel MDS-UPP-100 Santa Clara, Calif.	X	X	X	X	X	X
Data I/O Model V Issaquah, Wash.	X	X	X	X	X	X
Prolog Series 90 Monterey, Calif.	X	Note 1	Note 1	X	X	X
Spectrum Dynamics Series 550 Burlington, Mass.	X	Note 1	Note 1	Note 1	Note 1	Note 1

Note 1. This programming card is pending Intel approval.

PROMS AND ROMS

PUNCHED CARD FORMAT

1. An 80-column Hollerith card (preferably interpreted at time of punching) punched by an IBM 026 or 029 keypunch should be submitted. The first card will be a title card, formatted as shown in Figure 30.

2. For a N words × 4-bit organization only, card 2 and the following cards should be punched as shown in Figure 31. Each card specified the 4-bit output of 14 words.

3. For a N words × 8-bit organization only, card 2 and the following cards should be punched as shown in Figure 32. Each card specifies the 8-bit output of eight words.

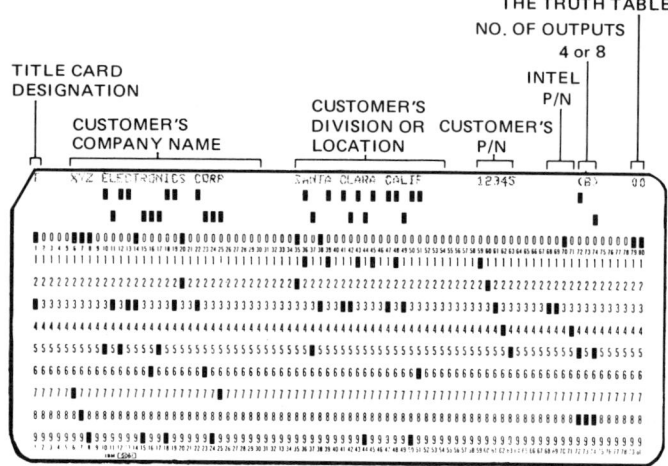

Column	Data
1	Punch a T
2-5	Blank
6-30	Customer Company Name
31-34	Blank
35-54	Customer's Company Division or location
55-58	Blank
59-63	Customer Part Number
64-67	Blank
68-74	Punch the Intel® 4-digit basic part number and in ( ) the number of output bits; e.g., 1702 (8), 3304 (8), 3301 (4), or 3601 (4).
75-78	Blank
79-80	Punch a 2 digit decimal number to identify the truth table number. The first truth table will be ØØ, second Ø1, third Ø3, etc.

Figure 30. Title Card Format.

For a N words × 4-bit organization only, cards 2 and those following should be punched as shown. Each card specifies the 4-bit output of 14 words.

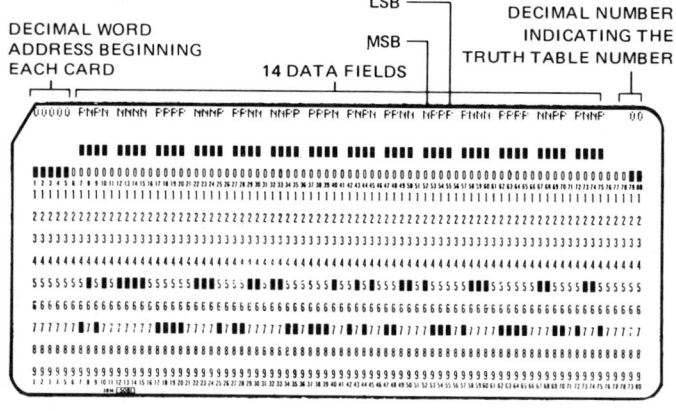

Column	Data
1-5	Punch the 5 digit decimal equivalent of the binary coded location which begins each card. The address is right justified, i.e., ØØØØØ, ØØØ14, ØØØ28, etc.
6	Blank
7-10	Data Field
11	Blank
12-15	Data Field
16	Blank
17-20	Data Field
21	Blank
22-25	Data Field
26	Blank
27-30	Data Field
31	Blank
32-35	Data Field
36	Blank
37-40	Data Field
41	Blank
42-45	Data Field
46	Blank
47-50	Data Field
51	Blank
52-55	Data Field
56	Blank
57-60	Data Field
61	Blank
62-65	Data Field
66	Blank
67-70	Data Field
71	Blank
72-75	Data Field
76-78	Blank
79-80	Punch same 2 digit decimal number as in title card.

Figure 31. 4-Bit Data Card Format.

DESIGNING WITH PROMS AND ROMS

For a N words × 8-bit organization only, cards 2 and those following should be punched as shown. Each card specifies the 8-bit output of 8 words.

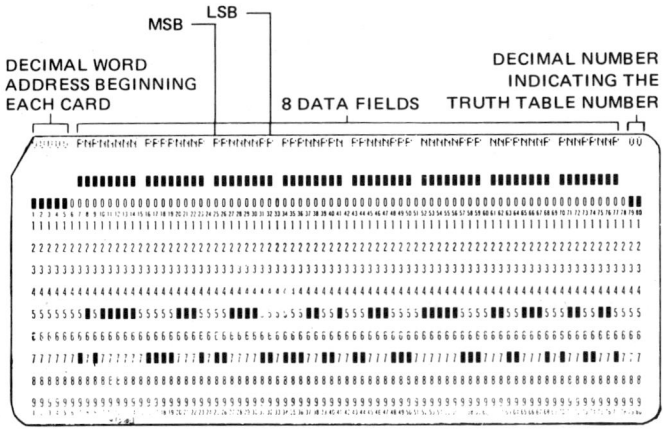

Column	Data
1-5	Punch the 5 digit decimal equivalent of the binary coded location which begins each card. The address is right justified, i.e., 00000, 00008, 00016, etc.
6	Blank
7-14	Data Field
15	Blank
16-23	Data Field
24	Blank
25-32	Data Field
33	Blank
34-41	Data Field
42	Blank
43-50	Data Field
51	Blank
52-59	Data Field
60	Blank
61-68	Data Field
69	Blank
70-77	Data Field
78	Blank
79-80	Punch same 2 digit decimal number as in title card.

Figure 32. 8-Bit Data Card Format.

PAPER TAPE FORMAT (Figure 33)

1. 1 inch-wide paper tape using 7- or 8-bit ASCII code, such as a model 33 ASR Teletype produces or
2. 11/16-inch-wide paper tape using a 5-bit Baudot code, such as a Telex produces.

The format requirements are as follows:

1. All word fields are to be punched in consecutive order, starting with word field 0 (all addresses low). There must be exactly N word fields for the N × 8 or N × 4 ROM organization.
2. Each word field must begin with the start character B and end with the stop character F. There must be exactly 8 or 4 data characters between the B and F for the N × 8 or N × 4 organization, respectively.

   No other characters, such as rubouts, are allowed anywhere in a word field. If in preparing the tape an error is made, the entire word field, including the B and F must be rubbed out. Within the word field, a P results in a high level output, and an N results in a low level output.

3. Preceding the first word field and following the last word field, there must be a leader/trailer length of at least 25 characters. This should consist of rubout punches (letter key for Telex tapes).

4. Between word fields, comments not containing B's or F's may be inserted. Carriage return and line feed characters should be inserted (as a "comment") just before each word field (or at least between every four word fields). When these carriage returns, etc., are inserted, the tape may be easily listed on the Teletype for purposes of error checking. The customer may also find it helpful to insert the word number (as a comment) at least every four word fields.

5. Included in the tape before the leader should be the customer's complete Telex or TWX number, and, if more than one pattern is being transmitted, the ROM pattern number.

6. MSB and LSB are the most and least significant bits of the device outputs. Refer to the data sheet for the pin numbers.

Example of 256 × 8 format (N = 256):

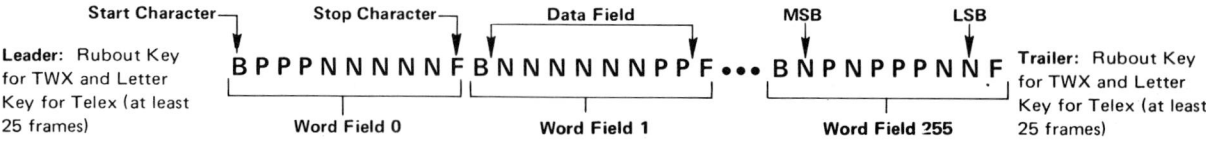

Example of 512 × 4 format (N = 2048):

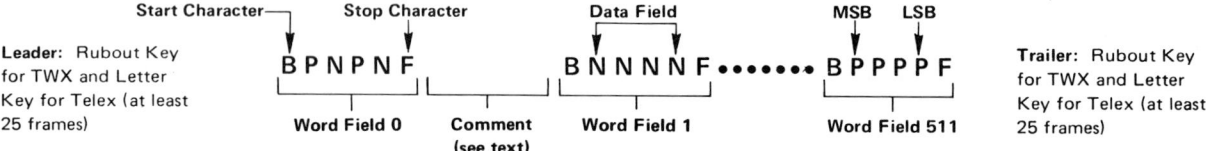

Figure 33. Paper Tape Format.

SYSTEM APPLICATIONS

System Organization

Most PROM/ROM devices contain 1K, 2K, or 4K bits and are organized as 256 or 512 locations with 4 or 8 bits per word, as shown in Figure 34. The implementation of most PROM/ROM systems requires that one or more of these devices be interconnected to provide the required number of locations and the number of bits per location.

NUMBER OF LOCATIONS	BITS PER LOCATION	
	4	8
256	3301 3601/21	1302 1602A 1702A
512	3302/22 3602/22	3304A/24 3604/24

Figure 34. PROM/ROM Device Organizations.

WORD EXPANSION

The simplest type of expansion involves the paralleling of devices to increase the number of bits per word, otherwise known as word expansion.

This type of expansion is illustrated in Figure 35, where two 3601's are paralleled to produce a 256 × 8-bit memory. The number of parallel devices can easily be calculated from the following formula:

$$\text{No. of Devices} = \frac{\text{No. of Bits per Word of the System}}{\text{No. of Bits per Word of the Device}}$$

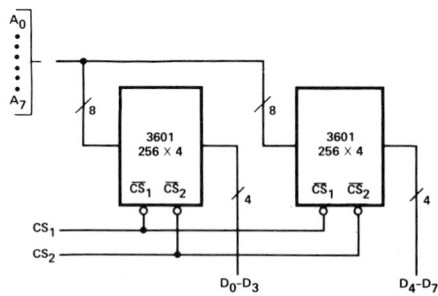

Figure 35. Word Expansion.

Therefore, a system employing a 32-bit word would require the paralleling of eight 3601's or 3602's, or four 3604's.

Word expansion requires nothing more than the parallel connection (i.e., tying together) of each individual address and chip select input; outputs remain separate.

DESIGNING WITH PROMS AND ROMS

ADDRESS EXPANSION

Just as inputs are OR-tied to obtain more bits per word, outputs can be OR-tied to obtain more words of memory (see Figure 36). When OR-tying outputs, it is necessary to select only one chip at a time to insure that the correct data is accessed. This requires the addition of logic to decode addresses and to activate the chip select for a single memory address.

ARRAY CONFIGURATIONS

Word expansion and address expansion can obviously be combined to produce a memory array of any size, provided the array size is an integral multiple of the device size.

Figure 37 shows a memory array configured by OR-tying inputs to obtain word expansion and OR-tying outputs to expand the number of words.

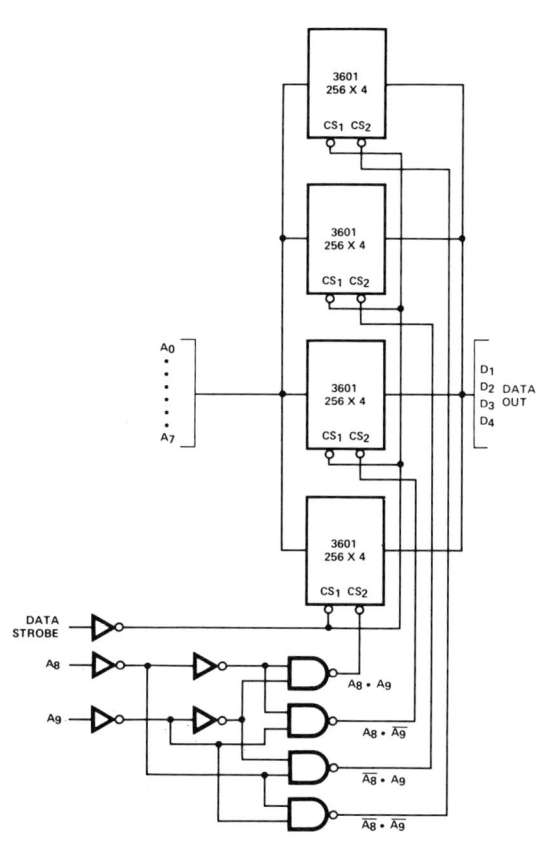

Figure 36. Address Expansion.

The number of devices required to obtain a given number of words of memory can be calculated from the following formula:

$$\text{No. of Devices} = \frac{\text{No. of Words Required in the System}}{\text{No. of Words in the Device}}$$

A system requiring 1024 words would require four 3601's, or two 3602's or 3604's.

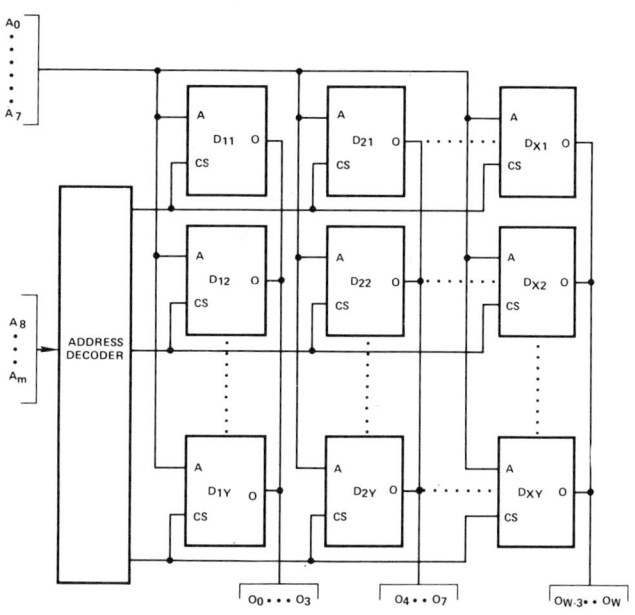

Figure 37. Combined Word and Address Expansion.

System Performance

The paralleling of inputs and outputs in memory array configurations affects capacitive loading and, therefore, system performance. Analysis of these loading effects requires consideration of buffers for driving the PROM/ROM inputs, as well as the output drive characteristics of the memory devices themselves.

BUFFERS

Buffers for driving address and chip select inputs are generally TTL devices. The effect of capacitive loading on standard, high speed, and Schottky TTL devices is shown in Figure 38.

The degradation in buffer propagation delay is directly due to increased transition time under increased capacitive loads. Figure 39 consists of

PROMS AND ROMS

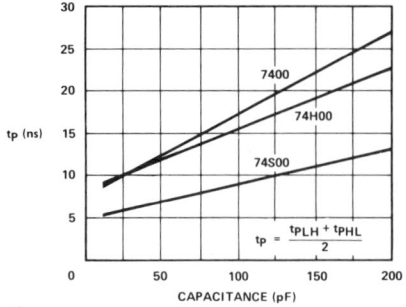

Figure 38. Capacitance vs. Propagation Time (tp).

multiple exposed photographs showing the effects of increased capacitive loads on different families of TTL gates. Figure 40 shows the same results for an increased number of chip select loads.

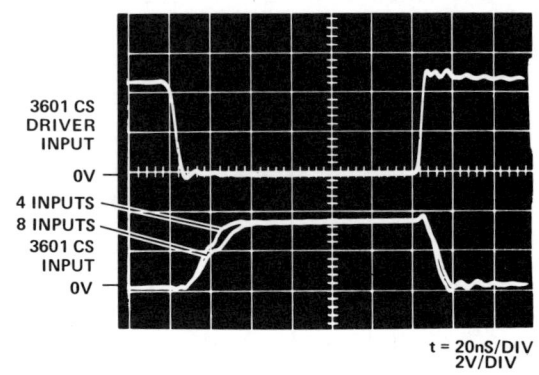

Figure 40. Input and Output of 7400 Driving 4 and 8 3601 $\overline{CS}$ Inputs.

OUTPUT LOADING

Address expansion by PROM/ROM output OR-tying increases the capacitive load on each PROM/ROM output, and results in some reduction in device access time. Figure 41 shows that going from two outputs to four outputs OR-tied increases access typically by 4 nS. The access times of the Intel bipolar PROM/ROMs are specified with a capacitive load of 30 pF, which is equivalent to the typical capacitive of output OR-tying four devices. The OR-connection of any fewer devices can reduce access time.

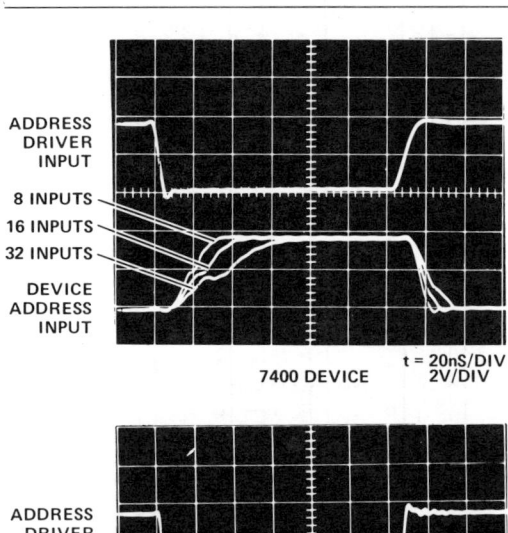

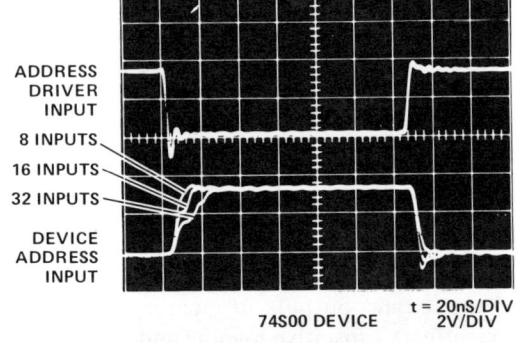

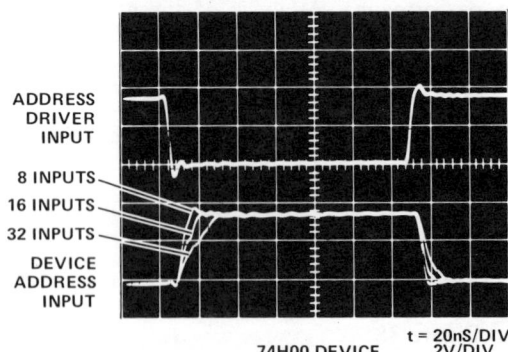

Figure 39. Various Standard TTL Devices Driving 8, 16 and 32 3601 Address Inputs.

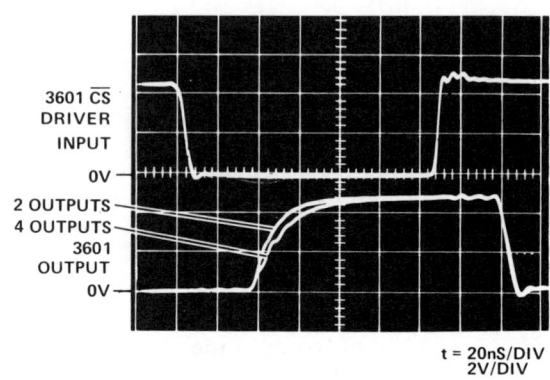

Figure 41. 2 and 4 3601 Outputs OR-Tied.

DESIGNING WITH PROMS AND ROMS

Another consideration when OR-tying outputs of PROMs and ROMs is the case where n-channel and p-channel devices must be tied together. Consider the microprocessor system shown in Figure 55. If the ROM were a p-channel 1702A and the RAM an Intel n-channel 2102A, a problem could occur. The 1702A has negative supply of –9V, and if its output pulls the 2102A output below $V_{SS}-0.8V$, the output circuit of the 2102A can be destroyed because of forward biasing the drain–substrate junction. A method of providing protection is to use an exclusive OR gate as shown in Figure 42. In this case, the value of R_1, the current limiting resistor, is determined by the maximum sink current drawn by the ROM and the maximum acceptable (most positive) low level required for the input of the exclusive OR gate.

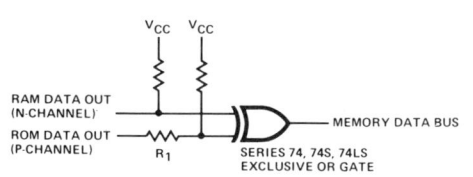

Figure 42. OR-Connecting P-Channel and N-Channel Devices.

Case Study

The selection of a memory device for a system implementation can be illustrated by the consideration of a hypothetical 1K × 32 PROM system, such as would be used for a computer microprogram control memory. Access time requirements for this system are assumed to be less than 100 nS.

The number of words required is specified as 1K, which allows the use of any device that is organized such that its number of words in the device divided into the required number of words for the system is an integer. For this application, 256 or 512 word organization would be possible devices. In a similar manner, a 4-bit or 8-bit device will allow an integral number of devices to be used to form the required 32-bit word.

Referring to the product selection guide, the required speed of 100 nS dictates the use of a bipolar device, eliminating the MOS devices. The 3601, 3602/3622, and 3604/3624 are all possible candidates for use in this system implementation. The system design resulting from the use of these three product types will be compared.

The system must be implemented using combinations of word expansion and address expansion as shown in Figure 37. Array layouts for these three parts are compared in Figures 43, 44, and 45. Each layout was done using the same rules: 0.200-inch adjacent spacing and 0.300-inch end-to-end spacing.

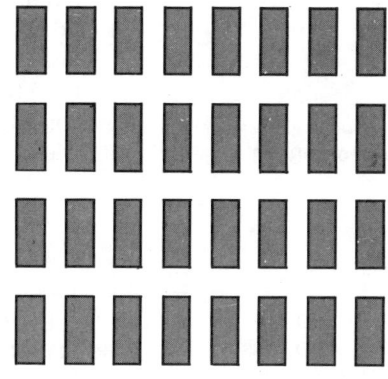

32 DEVICE ARRAY = 3.7 × 3.8 = 14.06 SQUARE INCHES
OR 2276 BITS/SQUARE INCH

Figure 43. 1K × 32-Bit System with 1K PROM (16-Pin Package).

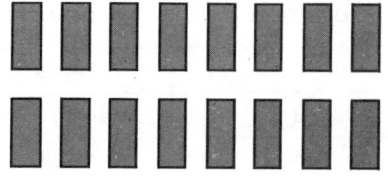

16 DEVICE ARRAY = 1.7 × 3.8 = 6.46 SQUARE INCHES
OR 4953 BITS/SQUARE INCH

Figure 44. 1K × 32-Bit System with 2K PROM (16-Pin Package).

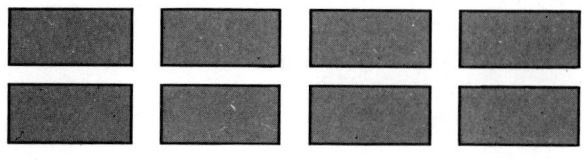

8 DEVICE ARRAY = 5.9 × 1.4 = 8.26 SQUARE INCHES

OR 3874 BITS/SQUARE INCH

Figure 45. 1K x 32-Bit System with 4K PROM (22-Pin Package).

Performance comparisons require the investigation of capacitive loading effects and device access times. Table XVII compares the number of address lines tied together, the resulting capacitance, and the resulting TTL delays for the three devices under consideration. Table XIX makes the same comparison for the chip select inputs, and Table XIX compares output loading effects. Total array power dissipation is summarized in Table XX and the complete results of the three different designs are summarized in Table XXI.

As can be seen in Table XXI, fastest system access time is achieved with the 3601, smallest printed circuit board layout area is realized with the 3602/3622, and the 3604 provides lowest system power dissipation. Selection of the optimum device is therefore left to other system development or cost considerations since all three parts are more than adequate for the stated system requirements.

Table XVII. Address Input Loading.

DEVICE SIZE	NO. OF DEVICES REQUIRED	NO. OF ADDRESS INPUTS DRIVEN	TYPICAL INPUT CAPACITANCE/DEVICE	TOTAL CAPACITANCE	MEASURED ADDRESS DRIVER DELAY
1K	32	32	4 pF	128 pF	22 nS
2K	16	16	4 pF	64 pF	16 nS
4K	8	8	4 pF	32 pF	12 nS

Table XVIII. Chip Select Input Loading.

DEVICE SIZE	NO. OF DEVICES REQUIRED	NO. OF CHIP SELECT LINES OR–TIED	TYPICAL INPUT CAPACITANCE/DEVICE	TOTAL CAPACITANCE	MEASURED CHIP SELECT DRIVER DELAY
1K	32	8	6 pF	48 pF	16 nS
2K	16	8	6 pF	48 pF	16 nS
4K	8	4	6 pF	24 pF	12 nS

Table XIX. Output Loading.

DEVICE SIZE	NO. OF DEVICES REQUIRED	NO. OF OUTPUTS OR-TIED	TYPICAL OUTPUT CAPACITANCE/DEVICE	TOTAL CAPACITANCE	MEASURED OUTPUT DELAY	OUTPUT OR-ING DELAY
1K	32	4	7 pF	28 pF	32 nS	4 nS
2K	16	2	7 pF	14 pF	28 nS	0 nS
4K	8	2	7 pF	14 pF	28 nS	0 nS

DESIGNING WITH PROMS AND ROMS

Table XX. Power Dissipation.

DEVICE SIZE	NO. OF DEVICES REQUIRED	POWER DISSIPATION (TYPICAL)		REDUCED POWER MODE (MAXIMUM)	
		mW PER DEVICE	mW PER BIT (SYSTEM)	mW ACTIVE/STANDBY PER DEVICE	mW PER BIT (SYSTEM)
1K	32	450	0.45	[1]	[1]
2K	16	700	0.35	550/225	0.19
4K	8	950	0.24	700/225	0.11

NOTE: 1. Power reduction not offered.

Table XXI. Device Comparison for 1K x 32 Memory System.

DEVICE SIZE	NO. OF DEVICES	ARRAY SIZE	MEASURED ADDRESS DRIVER TIME[1]	OUTPUT OR-ING DELAY[1]	MAXIMUM DEVICE ACCESS TIME[1]	SYSTEM ACCESS TIME[1,2]
1K	32	14.06 sq in.	22 nS	0 nS	50 nS	72 nS
2K	16	6.46 sq in.	16 nS	−4 nS	70 nS	82 nS
4K	8	8.26 sq in.	12 nS	−4 nS	70 nS	78 nS

NOTES: 1. All times taken at 1.5V points.
2. This time is the sum of device maximum and measured buffer delays.

Printed Circuit Layout Considerations

PROMs and ROMs can easily be used in much the same manner as other types of TTL design elements. The usual attention should be paid to ground distribution and decoupling.

Ideally, the circuit board ground system should consist of a ground plane on one side of the board and all signal and power distribution on the other. In reality, this is very difficult to achieve because of component densities, but the concept should be carried out as far as possible. The ground distribution should be as wide as possible everywhere, even if it means large variations in the width of the conductor.

To further approach a ground plane or mesh, horizontal and vertical power and ground traces on opposite sides of the board should be tied together at each DIP site, or as often as possible. The tying of the horizontal and vertical traces is important because long "floating" distribution lines can easily act as an antenna or a noise distribution system, allowing noise to propagate and exceed device thresholds.

In addition to reducing ground noise, an effective ground grid can serve to reduce cross-talk between address and data lines.

As can be seen in Figure 41 (previous section) the high to low transition can be very rapid, and if proper attention is not paid to the ground and power distribution, the noise resulting from these transitions can couple throughout the board and into the system. The memory devices should be decoupled at approximately every other DIP site with high frequency disc ceramic capacitors. There also should be bulk decoupling at the point where the V_{CC} line enters the board, usually one or more tantalum capacitors in the 10–50 μF range.

The layouts shown in Figure 46 and 47 are a good example of proper ground distribution. Notice that the ground plane forms a complete loop around the array (board) on both sides and that the two sides are connected periodically by horizontal and vertical traces.

The decoupling for the 1K/2K array consists of eight 0.1 μF high frequency capacitors, or one capacitor for each four devices, distributed through the array. This decoupling is adequate, but a better arrangement would be 0.1 μF located at every other device site, similar to the scheme used on the 4K layout.

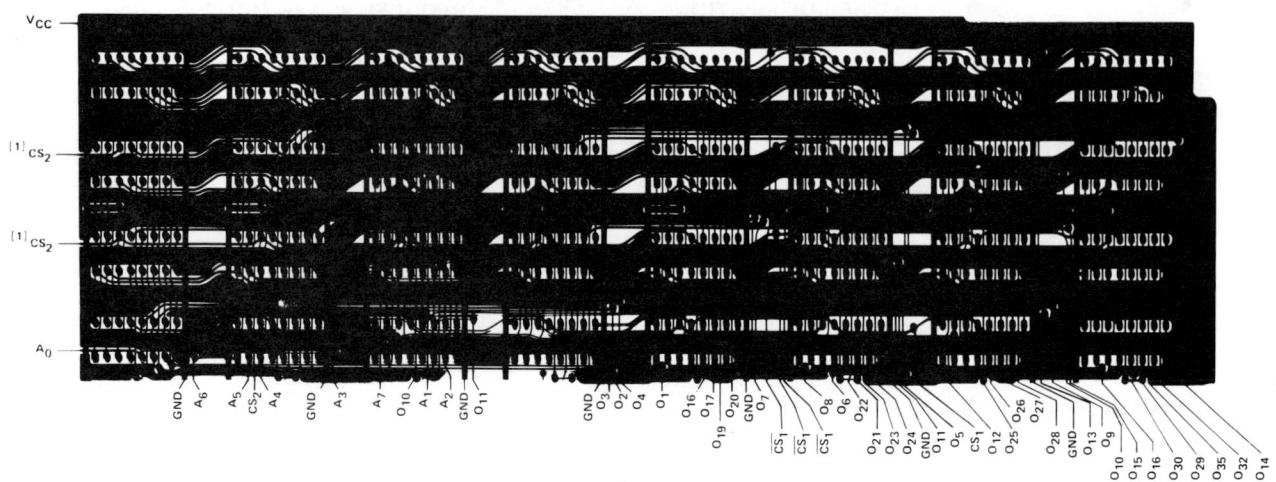

NOTE:
1. CS_2 UPPER & CS_2 LOWER CAN BE CONNECTED TOGETHER AND USED AS A_8, THUS ALLOWING THIS ARRAY TO BE 32 × 4K OR 32K × 8K, DEPENDING ON THE DEVICE USED (3601 OR 3602).

This array illustrates the layout for a 1K × 32-bit PROM memory. Power has been distributed with a grid system, and decoupling is located in the center of the array. By routing all the $\overline{CS_2}$ lines to a common point, the memory can be expanded to use 2K parts simply by supplying an additional address, A_8, and connecting it to all the former $\overline{CS_2}$ inputs.

Array layout courtesy of
MICRODATA INC.,
Irvine, California

Figure 46. 1K/2K PROM Array.

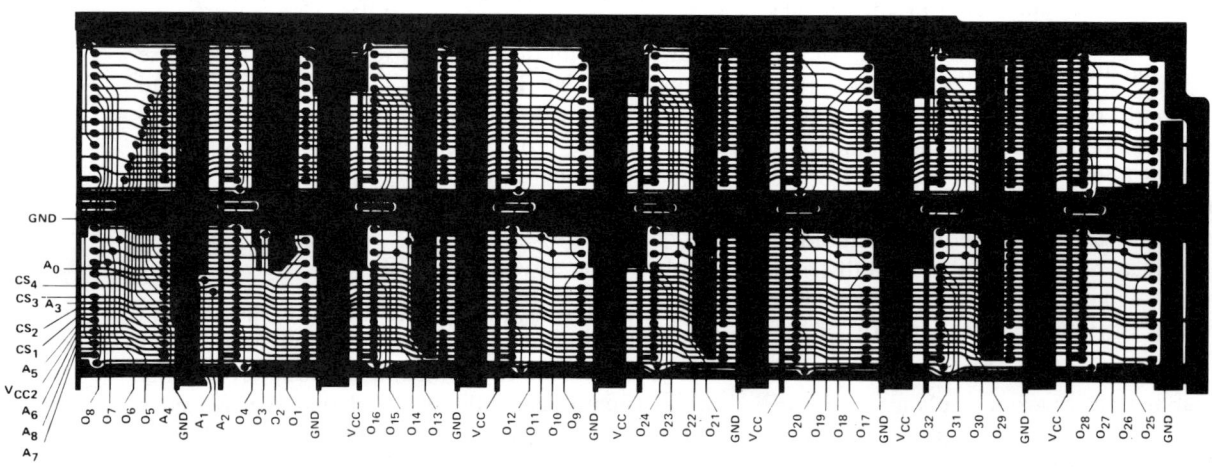

This array illustrates the use of a 24-pin 4K PROM to implement a 1K × 32-bit control store memory. As with the 1K/2K array, gridded power distribution has been used, with decoupling capacitors located in the center of the array.

Array layout courtesy of
MICRODATA INC.,
Irvine, California

Figure 47. 4K PROM Array.

DESIGNING WITH PROMS AND ROMS

Uses of PROM/ROMs

CODE CONVERSION

Read only memories lend themselves readily to converting from one binary code to another (such as from binary to Gray). This conversion is particularly useful in electromechanical systems controlled by a computer.

For example, consider a computer-controlled electromechanical encoder system. The computer performs data operation in binary form and outputs the x–y coordinates in the same form. If the stepping motor has a binary data input, erratic movement of the motor will be observed as the motor moves sequentially from one set of coordinates to another, because, as many data bits change and their exact switching relationship is not fixed, the motor will receive multiple codes until the data stabilizes. Consider the case of changing from decimal 7 to decimal 8 as shown in the truth table (Figure 48a), where 4 binary bits will change state. This transition will generate several random binary codes (up to 8) until the data stabilizes causing the stepping motor to move erratically.

It would be highly desirable to have a code where sequential motor stepping could be accomplished by changing only one bit per word between adjacent steps. The Gray code is such a code. By using the Gray code for the above example in moving from decimal 7 to 8 requires the change of only one bit (0100 to 1100). The stepping motor now moves smoothly without jitter or ambiguity as one Gray code bit changes after another. (Note that since the bit positions are not numerically weighted in Gray code, it is not possible to perform conventional binary arithmetic on the word. Therefore, the computer does not operate with such a code internally).

The code conversion from binary to Gray code for communication between the computer and the system motor becomes a simple matter if a read only memory is used. To use the truth table to convert from binary to Gray code it is merely necessary to use the binary data as the address to a ROM and read the corresponding Gray code at the output of the ROM. The example presented here is a 4-bit code but can be expanded to provide the desired resolution.

The conversion from binary to Gray code is only one of many code conversions possible. PROM/ROMs can be used to encode data for secured data transmission systems. These types of codes can be as simple or as complex as desired. A terminal attached to a central computer can "talk" to the computer over a secured line if both the terminal and computer have the proper encoding/decoding PROM/ROMs. Multiple terminals, each with its separate code, can likewise be connected to the computer. Of course for multiple terminals the computer must have the proper encoding/decoding PROM/ROM circuitry.

An example of such an encoding/decoding scheme is shown in the truth table (Figure 48b). To encode for data transmission, a standard binary code is presented to the address inputs of a ROM. The output of the ROM contains the code for the particular character to be sent. At the receiving end the order is reversed with the encoded data presented to another ROM address input whose output corresponds to the original character or data sent by the terminal. Data transmission in the reverse direction is handled in an identical manner.

DECIMAL	BINARY	GRAY
0	0 0 0 0	0 0 0 0
1	0 0 0 1	0 0 0 1
2	0 0 1 0	0 0 1 1
3	0 0 1 1	0 0 1 0
4	0 1 0 0	0 1 1 0
5	0 1 0 1	0 1 1 1
6	0 1 1 0	0 1 0 1
7	0 1 1 1	0 1 0 0
8	1 0 0 0	1 1 0 0
9	1 0 0 1	1 1 0 1
10	1 0 1 0	1 1 1 1
11	1 0 1 1	1 1 1 0
12	1 1 0 0	1 0 1 0
13	1 1 0 1	1 0 1 1
14	1 1 1 0	1 0 0 1
15	1 1 1 1	1 0 0 0

(a)

TRANSMISSION SCRAMBLER		RECEIVING SCRAMBLER	
ADDRESS	DATA	ADDRESS	DATA
BINARY CODE	SCRAMBLER CODE	SCRAMBLER CODE	BINARY CODE
0 0 0 0	0 0 0 1	0 0 0 1	0 0 0 0
0 0 0 1	1 1 0 1	1 1 0 1	0 0 0 1
0 0 1 0	1 1 1 1	1 1 1 1	0 0 1 0
0 0 1 1	1 0 1 0	1 0 1 0	0 0 1 1
0 1 0 0	0 1 1 0	0 1 1 0	0 1 0 0
0 1 0 1	0 0 1 1	0 0 1 1	0 1 0 1
0 1 1 0	0 0 1 0	0 0 1 0	0 1 1 0
0 1 1 1	0 0 0 0	0 0 0 0	0 1 1 1
1 0 0 0	1 0 0 0	1 0 0 0	1 0 0 0
1 0 0 1	1 0 1 1	1 0 1 1	1 0 0 1
1 0 1 0	1 1 1 0	1 1 1 0	1 0 1 0
1 0 1 1	1 0 0 1	1 0 0 1	1 0 1 1
1 1 0 0	1 1 0 0	1 1 0 0	1 1 0 0
1 1 0 1	0 1 1 1	0 1 1 1	1 1 0 1
1 1 1 0	0 1 1 0	0 1 1 0	1 1 1 0
1 1 1 1	0 1 0 0	0 1 0 0	1 1 1 1

(b)

Figure 48. Code Conversion Truth Tables.

COMBINATORIAL CIRCUITS

Digital circuits are often divided into two categories: combinatorial and sequential. Combinatorial circuits have no internal storage elements. As a result, the output signals are functions only of the inputs supplied at the time the output is measured (neglecting propagation delays). A ROM may be used to generate combinatorial functions when the number of input signals is not excessive. For example, a 256 word by 4 bit ROM has 8 input leads (addresses) and 4 output leads and so can be used to generate any 4 combinatorial functions of 8 variables. Additional functions may be generated by adding more ROMs — doubling the number of ROMs doubles the number of functions which can be generated.

Expanding the number of input variables is much more costly, however, Additional input variable may be decoded to operate chip selects just as additional addresses inputs are decoded in a memory array. However, each additional input variable doubles the amount of ROM required.

Various authors have expressed the option that 8 to 16 bits of ROM are equivalent to one logic gate. However, this ratio does not apply to all designs. For example, to make a quad full adder (5 outputs, 9 inputs) would require 5×2^9 or 2560 bits of ROM, but can be realized with less than 40 gates — for ratio greater than 64 bits/gate.

When using ROM to replace wired logic gates, the designer should remember that the ROM is not guaranteed to give a single output transition for a single input transition. Figure 49 illustrates the way the designer should view the ROMs behavior. In Figure 49, after a short hold time, the outputs are undefined until a period equal to the ROMs access time has elapsed. During this undefined interval, the ROM outputs may show noise or extra transitions. Not all ROMs specify a hold time. Even when a hold time is specified, it is valid only when access to a location has been made, and is measured from the first address transition.

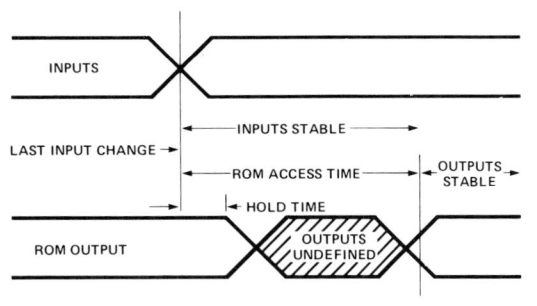

Figure 49. ROM Behavior for Combinatorial Logic.

SEQUENTIAL CIRCUITS

Sequential circuits are logic circuits with internal storage. As a result, outputs are a function of past as well as present inputs. Seqential circuits are often realized by a collection of storage elements (flip-flops) together with combinatorial logic. Outputs of the sequential network are combinatorial functions of the inputs to the network and the flip-flop outputs. The inputs to the flip-flops are combinatorial functions of network inputs and flip-flop outputs.

When a sequential digital system is described in the above manner, the state of the circuit is determined by the contents of the flip-flops. Therefore, a machine with n flip-flops can have at most 2^n internal states. To describe the circuit behavior, two sets of information must be known:

1. The outputs as function of inputs and internal states; and

2. The next states as functions of inputs and internal states.

This information may be presented via tables or graphically in the form of a state sequence diagram, such as that shown in Figure 50a. The state sequence diagram is usually drawn as a collection of circles, each labelled to correspond to one state of the machine. The circles (states) are connected by directed lines (arrows) indicating which state transitions may take place. Each such transition line is labelled with the values of the input variables for which the transition takes place, unless the input variables have no effect. In that case, the state transition always takes place and the arrow is unlabelled.

Some digital circuits are clocked, i.e., state transitions take place only upon occurrence of a clock pulse. If for some input conditions no state transition takes place at a clock time, it is indicated on the diagram as an arrow which leaves and re-enters the same circle. This arrow is labelled, like any other, with the corresponding input conditions. Clocked sequential circuits are readily designed using clocked flip-flops of the JK or D variety such as those shown in Figures 50b and 50c.

State Assignment

The state-sequence diagram describes the digital circuit behavior independent of the assignment of states to the circles of the diagram. Each circle in the diagram must be assigned a unique set of values for the state variables. Each state variable can take on the value of 1 or 0, so that n state variables can provide values for up to 2^n circles in the diagram. However, the way the values are assigned to the circles can make a significant difference in the ease of realization when JK or D flip-flops are used. At present, no known technique, other than repeated trails, exists for determining the minimum cost state assignment. The designer's insight and experience contribute significantly to the design efficiency. However, when ROMs are used, state assignments are less critical than for realization with wired logic gates.

DESIGNING WITH PROMS AND ROMS

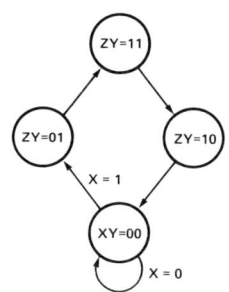

a. State Diagram

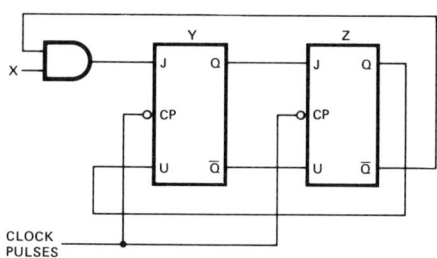

b. Corresponding Sequential Circuit — JK Realization

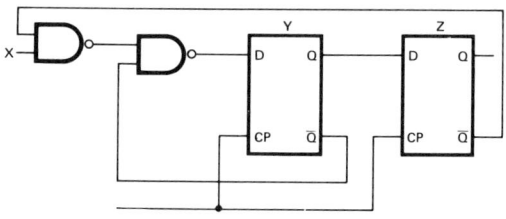

c. Corresponding Sequential Circuit — D Realization

Figure 50. Sequential Circuit State Diagram Realizations.

Asynchronous Input to Clocked System

When a clocked system has asynchronous input variables, i.e., variables which can change at other than clock times, proper behavior may depend upon the state assignment used. For example, if the values of a given asynchronous input variable can affect the values of two state variables in a given state transition, differential delays in the logic may allow 4 rather than 2 possible state changes to take place: neither, either, or both of the variables may change. To avoid this situation, state assignments should be such that only one state variable is a function of each asynchronous input variable or the asynchronous input variable should be made synchronous by clocking it into a flip-flop. Of course, the latter procedure increases the response time of the system to the input signal.

These considerations also apply to the asynchronous flip-flop forcing inputs. In general, these inputs can force the network into one or more of a subset of the states where it will remain until the forcing input is removed. If the network clocked transitions attempt to change more than one forced state variable, asynchronous removal of the forcing signal may result in any of several state transitions: any or all of the variables attempting to change may do so, depending upon differential delays in flip-flop responses, clock distribution, and distribution of the forcing signal.

Realizations with D Flip-Flops

Having assigned state variable values for each state, realization with D flip-flops is very straightforward.* First, a truth table or set of Karanaugh maps is prepared. The source variables include all state variables and all input variables. The functions to be generated involve all state variables (next state value) and all output functions. Those functions representing the next state values are used as the data inputs to the corresponding D flip-flops.

Figure 51 shows a symbolic diagram of such a network. The "clocked register" is an array of n D-type flip-flops.

A read only memory array with p address inputs and q outputs (2^P x q bits) can generate a total of q output functions of p inputs. Thus for Figure 51, if n state variables are required, p-n input variables may be used and q-n output signals may be generated.

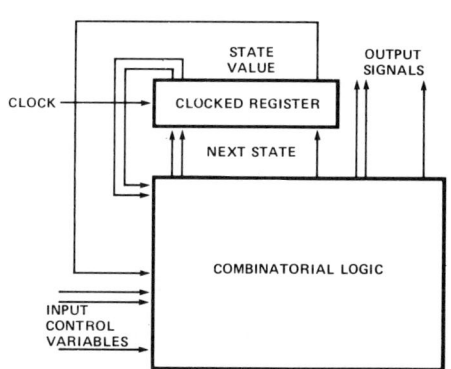

Figure 51. Realization of Digital Machine.

*For sequential networks wired with logic gates, JK flip-flops may reduce the gate count as in Figure 50b and 50c. However, ROM realizations are more economical when D flip-flops are used, because fewer functions need be generated.

Because a ROMs internal realization is quite different than that of a conventional combinatorial logic network, different considerations apply to ROM designs than for conventional designs. For example:

1. State variable assignment has little or no effect on circuit complexity when ROM realization is used. Therefore, the designer may use state variables to form output functions directly with greater ease than for conventional designs. If, however, additional logic circuits are added to reduce total ROM requirements or allow asynchronous input variables (see next paragraph), some of this design freedom may be removed.

2. All outputs of a ROM must be considered functions of all inputs. Therefore, asynchronous inputs to the ROM should not be permitted to change within an access time prior to clocking the output register, or the contents of the output register may be completely unpredictable. Additional latches or separate logic between the ROM output and D flip-flop inputs should be used so that the conditions described above (under Asynchronous Inputs to Clocked Systems) can be met. Additional ROM outputs may be used to enable or disable this logic.

Methods of Reducing ROM Size

If the number of input or output variables is large, a straightforward realization with ROM may not be practical. However, it may be possible in certain areas to reduce the amount of ROMs by adding a small amount of additional logic. Several methods for reducing the size of a ROM needed to perform a given function are described below. The use of these techniques when appropriate may permit a ROM approach to be used in a situation where it would normally be impractical to do so. Most of these techniques are illustrated in Figure 52.

Multiplexing Input Variable

Instead of using all input control variables at all times, many digital machines have only a few states where the next state decision is affected by the input variables. Therefore, a multiplexer may be used to select the input variables which are active for each given state of the machine. The effective number of input control variables at the ROM may be reduced to a number equivalent to the largest number active at any one time.

The control signals for the multiplexer may be generated by logic circuits which decode the state information or by extra output variables from the ROM. In general, these extra ROM output variables are far less expensive than the extra ROM inputs that would otherwise be necessary.

Bypassing the ROM for Input Control Variables

If the state assignments are made so that the next state is a simple function of the input variables, a small amount of logic may be placed between the ROM output and the clocked register. Some of the input control variables are then brought into the system via this logic rather than through the ROM. As in the case with input multiplexing, additional output signals may be used to enable this logic.

One simple form of this method uses a multiplexer between the ROM output and the clocked register. Certain of the

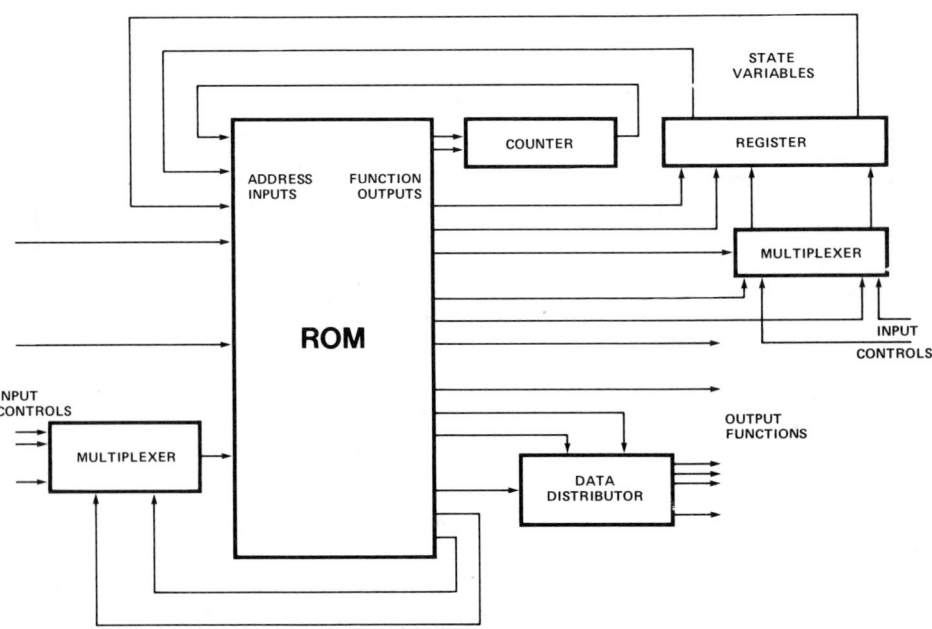

Figure 52. ROM Realization of Sequential Machine.

DESIGNING WITH PROMS AND ROMS

state variables take on the values of the input variables whenever the multiplexer is set to accept these inputs. This method places restrictions on state assignment.

A similar technique is usually necessary for use with input control variables which are asynchronous.

Output Function Distribution

When a large number of control functions must be generated, but only one or two are active at one time, data distributors may be used to generate a large number of control functions from a few ROM outputs. As an example of the type of coding which might be used, 8 non-simultaneous control functions might be generated using one data bit and 3 selection bits. The Intel® 3205 decoder may also find use in ROM output expansion. Eight selection signals can be generated from three ROM function outputs.

External State Generators/Partitioning/Factoring

When a large number of states of a state diagram fall in an easy to generate sequence, the number of state variables generated by the ROM may in some cases be reduced by generating the additional states with external circuits such as counters or shift registers. Functions of these separately realized state variables may be used as equivalent state variable inputs to the ROM.

As an example of this technique, consider a binary counter connected to a ROM such that the ROM can generate a preset or count enable variable and accept a carry output as equivalent to a state input variable. The ROM may be programmed so that for some states of the conventional state variables, the counter counts from its preset values until it overflows with the ROM staying the same state throughout the counting sequence. In this example, one input (equivalent state) variable replaces all of the state variables in the counter.

The example above is a special case of a more general technique which may be called partitioning. Instead of using an external counter with the ROM system, another ROM/register sequential machine might have been connected. The net result is a ROM/register realization of a sequential digital machine in which not all state variables are used as inputs to all of the ROM. In effect, the machine is partitioned into a number of smaller, but interactive, machines.

To partition a circuit, the state variables must be isolated into two or more groups. A new state diagram can be generated for each group. In these partitioned state diagrams, the state variables for one state diagram are treated as if they were input control variables in the other. In general, for partitioning to be effective, the state variables must be such that they can be divided into relatively independent groups.

These examples are but a few of those available to the designer wishing to take advantage of ROM. As the design complexity progresses, the structure approaches the complexity of a microprogrammed processor — one application where ROM is extensively applied.

ROM, even in complicated networks like that of Figure 52 or a microprogrammed processor, offers much easier modification of machine structure than wired logic. With the availability of programmable ROMs, ROM approaches to sequential circuit design merit serious consideration.

Even when a prototype system has been developed using the 3601 or 1702A, once ROM patterns have been fixed, the prototypes can be easily converted to use the 3301A or 1302 for production for these mask programmed devices are pin and signal compatible with their field programmable counterparts.

WAVEFORM GENERATOR

To show one simple application of ROM, consider the signal generator shown in Figure 53. An 8-bit counter driven by an oscillator drives a 2048-bit ROM (256 words of 8 bits). The ROM outputs are converted to an analog voltage by a digital to analog converter (DAC). By properly coding the ROM, a wide variety of waveforms may be generated.

For the system shown in Figure 53, each step of the 8-bit counter represents $\frac{360}{256}$ degrees of phase angle. The value at each address in ROM is the digital number representing the signal output for that phase angle. Multiple ROM/DAC combinations might be used to generate several simultaneous waveforms, or multiple phases of a signal, for example. The output of the DACs will change in small discrete steps, each less than 1% of full scale. Where this might be a problem, filtering might be used. However, undesired harmonic content of the signal will be limited to the upper harmonics.

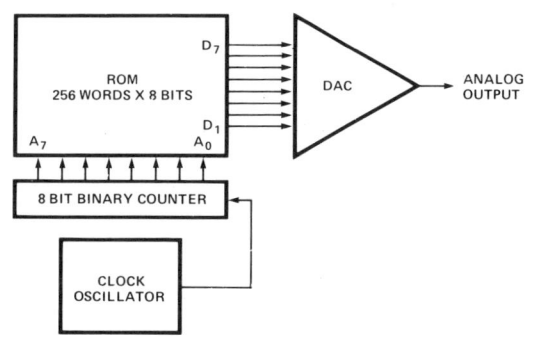

Figure 53. Digitally Controlled Waveform Generator.

CLOCK GENERATION

ROMs can be used to generate clock phases for use in multi-clock systems (such as driving a 16K CCD, Intel's 2416). An example of the generation of a general clock generator is shown in (a). The desired timing is shown in (b).

Basic operation is as follows: A clock input is applied to the input of a 74161 TTL counter as shown in (a). The counter sequentially counts six cycles of the input as shown in (b). The counter output is presented to the PROM as an address. The output of the PROM as a function of the address is shown in truth table (c). The outputs of the PROM are latched in the 74174 by the input clock. This prevents unwanted transients from occurring on the four-phase clock lines.

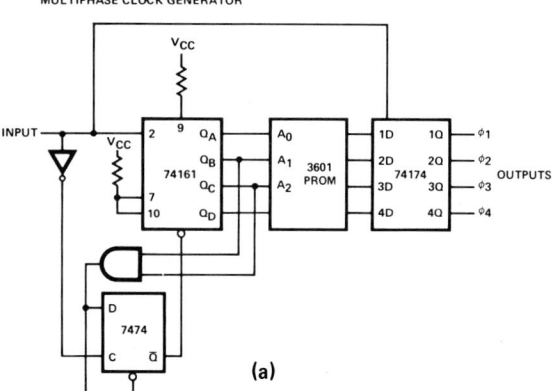

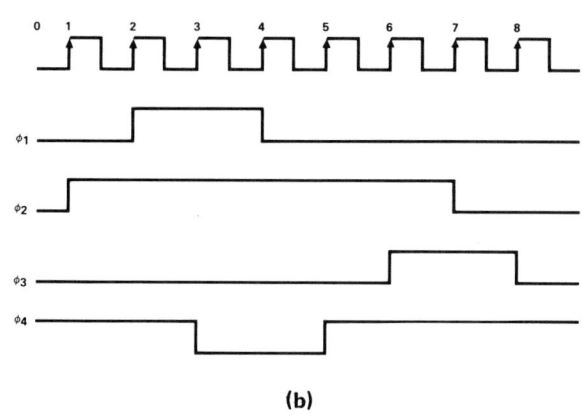

(b)

ADDRESS	O_1	O_2	O_3	O_4
0	0	0	0	1
1	0	1	0	1
2	1	1	0	1
3	1	1	0	0
4	0	1	0	0
5	0	1	0	1
6	0	1	1	0
7	0	0	1	0

(c)

Figure 54. Clock Generation.

MICROPROCESSOR SYSTEM

Illustrated here is a typical microprocessor system, based on the Intel® 8080. The 2708 provides 8K (1024 × 8) of PROM storage, while the combination of two 8101/5101's provides 2K (256 × 4) of RAM storage. A_{10} is used to select the desired type of storage access; A_{10} high selects RAM and A_{10} low selects the PROM storage. The other address bits, A_{11}–A_{16}, are generated by the 8080 and can be used for system expansion. The ROM storage can be modified to use a 1702A for 2K (256 × 8) of EPROM storage, or a 2316 for 16K (2K × 8) of ROM storage. In the case of the 2316, 11 of the 8080 address lines would be used.

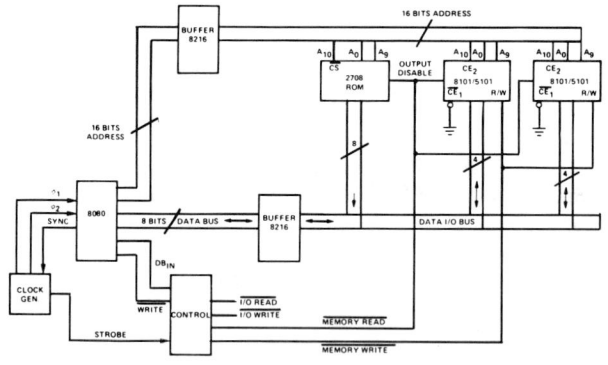

Figure 55. Microprocessor System.

4K BYTE SYSTEM

A 32 kilo-bit PROM memory organized as 4K × 8 utilizing the 1702A is shown in Figures 55 and 56. Each of the 1702A's is accessed individually by means of a 3-to-8 encoder (Intel® 3205) with decoder enables connected as shown. The three low-order addresses (A_0–A_2) are decoded simultaneously by two 3205's and the proper 1702A selected by address A_{11} activating either the right or left 3205. Since all unselected 1702A's have high impedance outputs, the selected module controls the internal data bus with its output gated through the three-state output buffers shown in the schematic.

The 4K × 8 board shown is expandable with a given PROM board identified by the PROM Board Resident Select Switches.

DESIGNING WITH PROMS AND ROMS

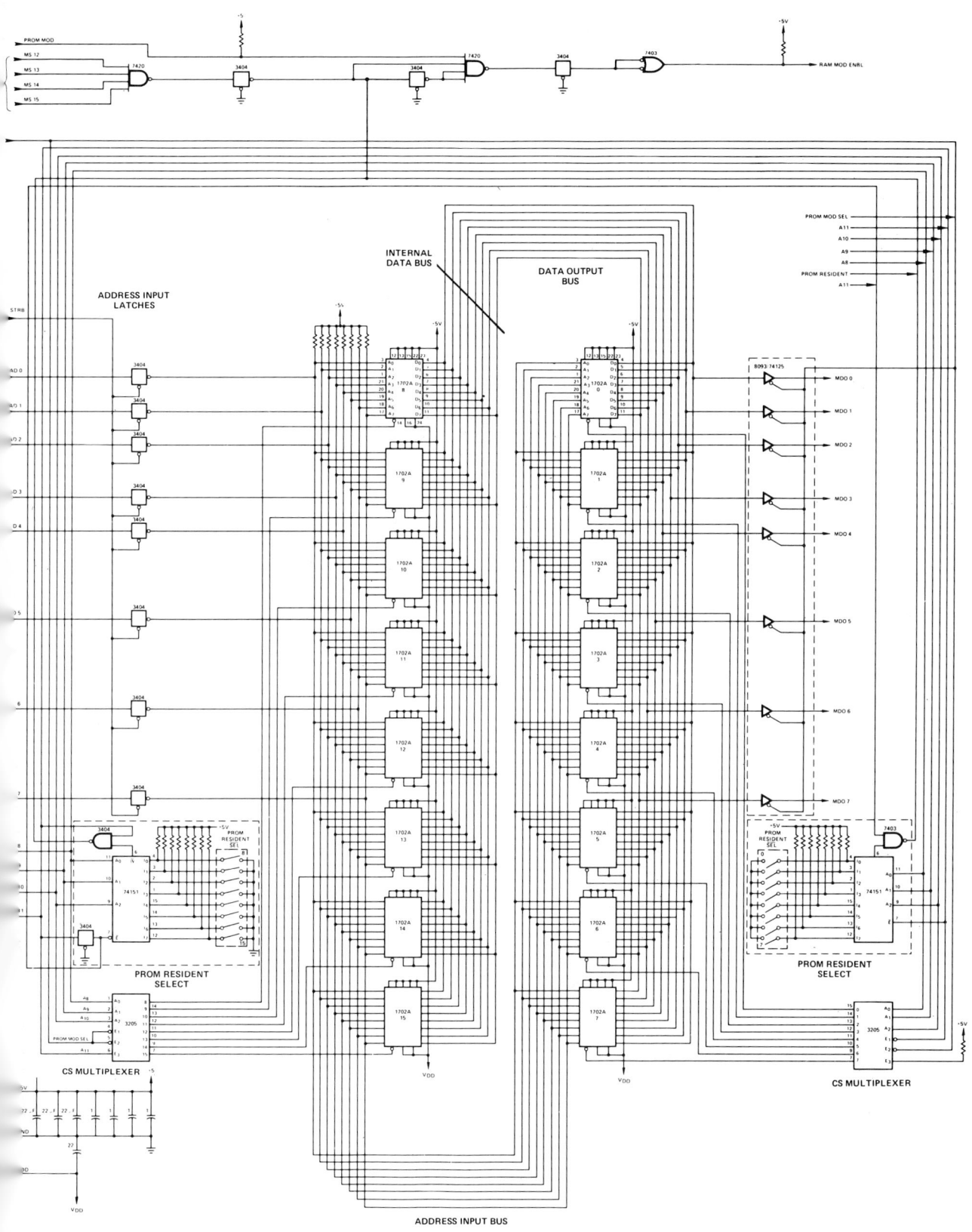

Figure 56. 4K Byte PROM System.

DESIGNING WITH 2708 EPROM

BOB GREENE

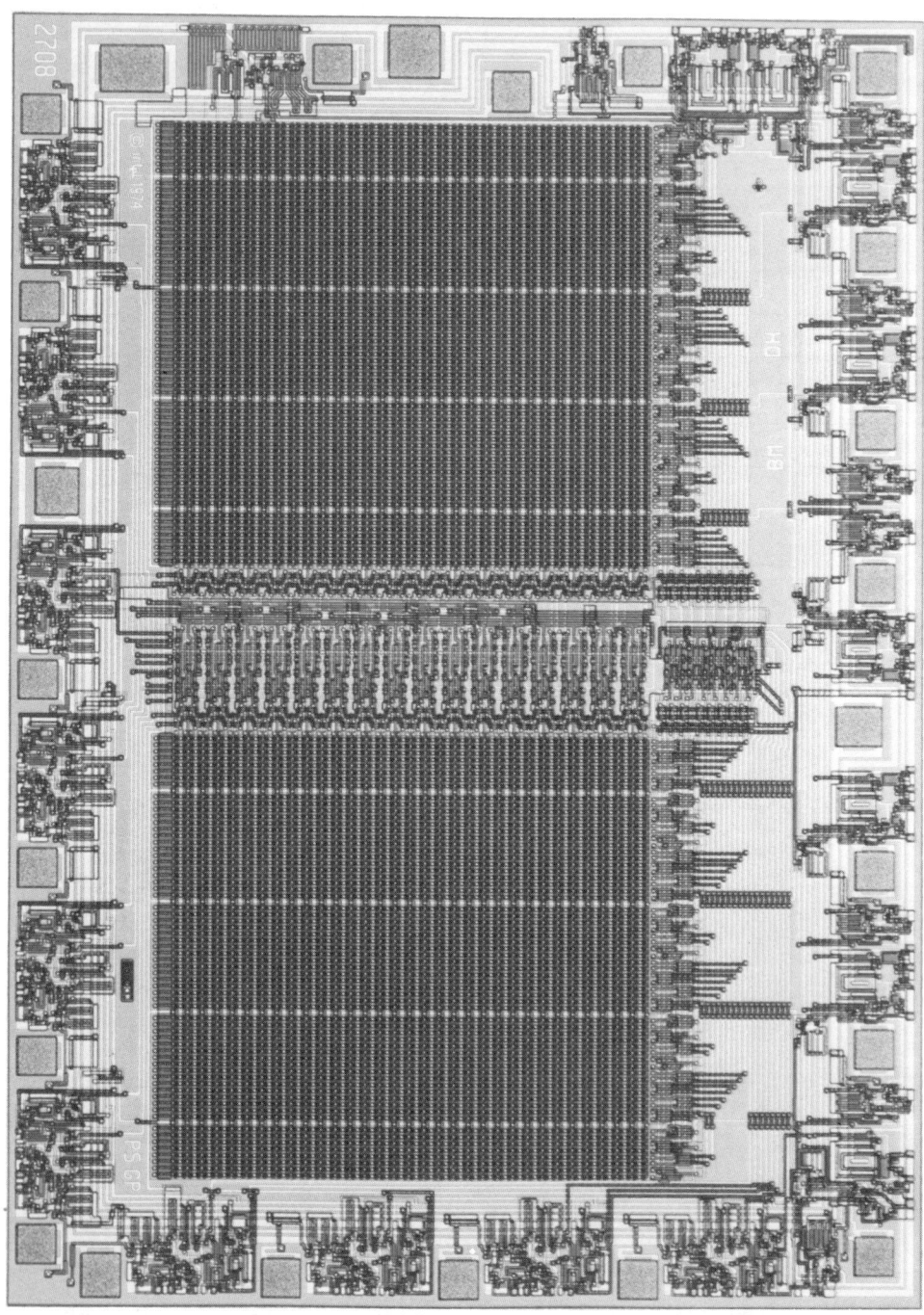

Photomicrograph of the Intel 2708 8K UV EPROM.

DESIGNING WITH 2708 EPROM

INTRODUCTION

The Intel® 2708 is a static 8192-bit (1024 × 8) Erasable Programmable Read Only Memory, or EPROM. The device is packaged in a standard 24-pin package, which has a transparent lid to allow erasure in a manner similar to the Intel® 1702A. Maximum access time is 450 ns. The device requires three power supplies, ±5V and +12V, for normal read cycles; while for programming a 26V pulse is required on the Program pin.

The address inputs and data I/Os are TTL compatible during read and programming. The data outputs are three state to facilitate memory expansion by OR-tying. Initially, and after each erasure, the device contains all "1's". Programming, or introducing "0's", is accomplished by: applying TTL level addresses and TTL level data; a +12V Write Enable signal; then sequencing through all addresses consecutively a minimum of 100 times, applying a 26V program pulse at each address. ALL ADDRESSES MUST BE PROGRAMMED DURING EACH PROGRAMMING SESSION; PROGRAMMING OF SINGLE WORDS OR SMALL BLOCKS OF WORDS IS NOT ALLOWED. As discussed in detail in the PROGRAMMING section, approximately 100 seconds are required to program the entire device.

DEVICE DESCRIPTION

The device is packaged in an industry standard 24-pin package as shown in Figure 1. The Program pin (18) receives 26V pulses during programming; during read operations it must be connected to V_{SS} (GND) or held at V_{IL}.

Pin 20, the $\overline{CS}$/WE connection, serves three functions. When at V_{IL} (0V) the device is selected for normal read operation; when at V_{IH} (3.0V min) the device is deselected and the outputs are placed in the high impedance state; and when at V_{IHW} (11.4V min) the device is Write Enabled and ready to receive program pulses.

A block diagram of the 2708 is shown in Figure 2. The low order address bits (A_0–A_3) perform column (or Y) selection, while the high order address

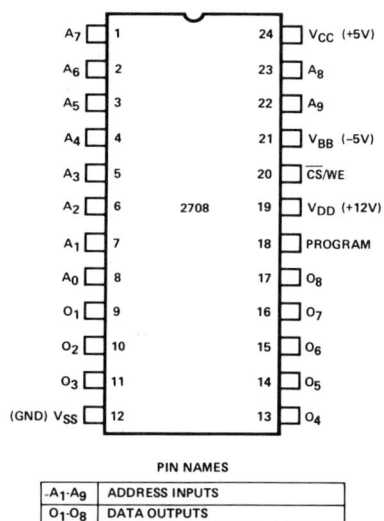

Figure 1. 2708 Pin Configuration

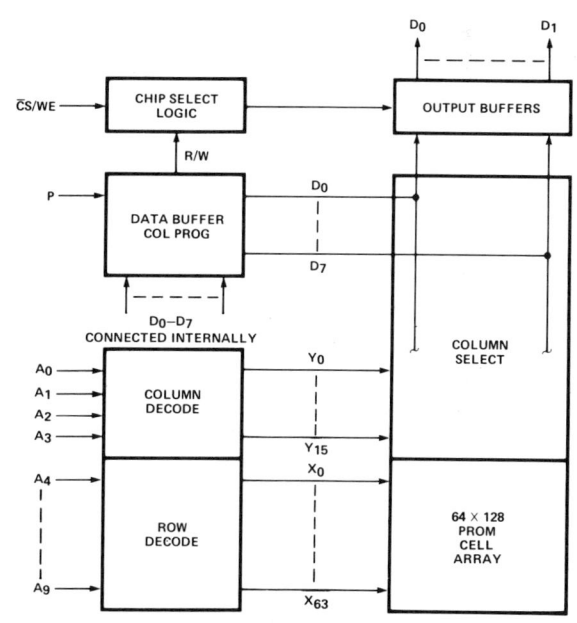

Figure 2. Detailed Block Diagram

Table I. 2708 Pin Connections and Functions

Function Pin Number Mode	Data I/O 9–11, 13–17	Address Inputs 1–7, 23, 22	V_{SS} (GND) 12	Program 18	V_{DD} Supply 19	$\overline{CS}$/WE 20	V_{BB} Supply 21	V_{CC} Supply 24
Read	D_{OUT}	A_{IN}	GND	GND	+12V	V_{IL}	–5V	+5V
Deselect	High Impedance	Don't Care	GND	GND	+12V	V_{IH}	–5V	+5V
Program	D_{IN}	A_{IN}	GND	Pulsed +26V	+12V	V_{IHW}	–5V	+5V

bits (A_4-A_9) perform the row (or X) selection. Table I assists in determining the proper voltage connections for the three modes of operation; Read, Deselect and Program.

Cell Description

The heart of the 2708 is the single transistor stacked gate cell, implemented with two layer polysilicon. The cell consists of a bottom floating gate and a top select gate, as shown in Figure 3. The top gate is connected to the row decoder, while the floating gate is used for charge storage. The cell is programmed by injection of high energy electrons through the oxide and onto the floating gate. Once there the charge is trapped, as there are no electrical connections to this floating gate. The presence of charge on the floating gate causes a shift to the cell threshold, as shown in Figure 4. In the initial state the cell has a very low threshold and selection of the cell, by way of the top select gate, will cause the transistor to turn on. Programming shifts the threshold to a higher level and selection of the cell will not allow it to turn on. The status of the cell is determined by examining its state at the sense threshold, also indicated in Figure 4. If a "1" is programmed into the cell, selection will allow a higher current to flow between the source and drain than if a "0" is programmed into the cell.

As there are no electrical connections to the floating gate, erasure must be accomplished by non-electrical means. Illumination of the cell with ultraviolet light of the correct frequency (2537Å) and duration will impart sufficient photon energy to the trapped electrons to allow them to overcome the inherent energy barrier and be transported through the oxide to the substrate.

Memory Array Operation

The cells described in the previous paragraph are interconnected to form a 64 × 128 matrix, or array, as shown in Figure 5. Access to a particular cell is described as follows: When the Row Address is stable, one row is selected, turning on the row line to all 128 cells in the row. The Column Address connects 8 of the 128 column lines to their respective sense amplifiers. The row line provides bias to all the top gates in a particular selected row, and, depending on the state of the cells, the column lines will be left at the precharged level (for a programmed "0") or will be discharged, pulling the column lines down to a low level (for a programmed "1"). To provide the very fast Chip Select to Output Delay time (t_{CO}) of 120 ns, all of the sense amplifiers are turned on when the device is deselected, and, when $\overline{CS}/WE$ reaches V_{IL}, those which are not selected are turned off, and the remaining eight amplifiers convert the charge on the column lines to TTL output levels by way of the output buffers.

During programming the selected row and column lines are pulsed to approximately 26 volts and the floating gate is charged as was described in the previous section. It is the presence of these 26V pulses on the interconnected top gates that lead to the requirement that ALL ADDRESSES MUST BE PROGRAMMED SEQUENTIALLY; PROGRAMMING OF SINGLE WORDS OR SMALL BLOCKS OF WORDS IS NOT ALLOWED, as transients may be generated that could partially alter the charge state of the cell.

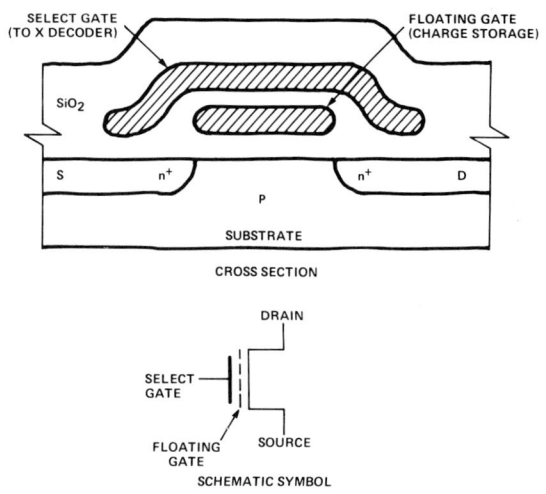

Figure 3. 2708 Storage Cell

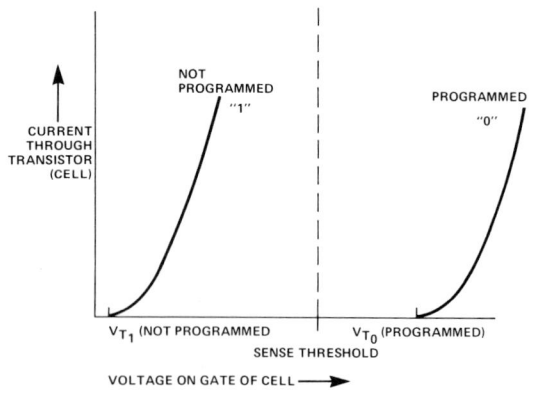

Figure 4. Storage Cell Threshold Shift

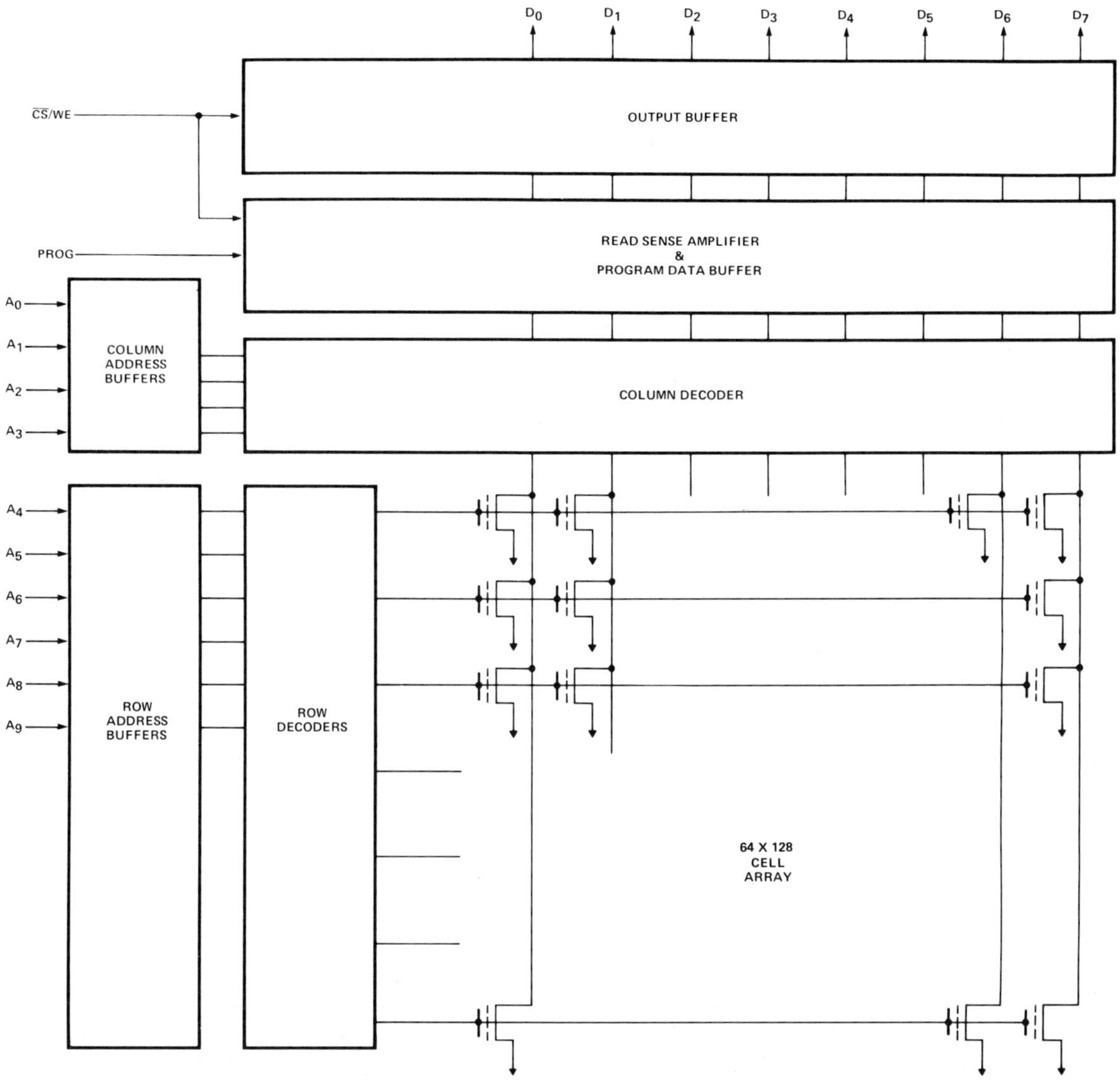

DESIGNING WITH 2708 EPROM

Figure 5. Expanded Block Diagram

Output Buffer

The equivalent schematic of the Output Buffer is shown in Figure 6. As is shown, the output buffer consists of a pair of MOS transistors, connected in a push-pull configuration. $\overline{CS}$ enables both transistors when true, while when $\overline{CS}$ is false both output devices are turned off, providing three state output operation. The output buffer will provide a V_{OL} of 0.45V at an I_{OL} of 1.6 mA, and a V_{OH} of 2.4V at −1.0 mA.

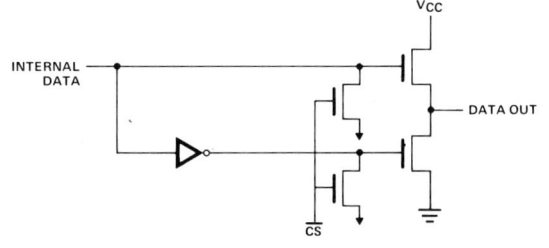

Figure 6. 2708 Output Buffer

Table II. D.C. Read Mode Characteristics

Symbol	Parameter	Min.	Typ.[1]	Max.	Unit	Conditions
I_{LI}	Address and Chip Select Input Sink Current		1	10	μA	V_{IN} = 5.25 V or V_{IN} = V_{IL}
I_{LO}	Output Leakage Current		1	10	μA	V_{OUT} = 5.25V, $\overline{CS}$/WE = 5V
I_{DD}[2]	V_{DD} Supply Current		50	65	mA	Worst Case Supply Currents:
I_{CC}[2]	V_{CC} Supply Current		6	10	mA	All Inputs High
I_{BB}[2]	V_{BB} Supply Current		30	45	mA	$\overline{CS}$/WE = 5V; T_A = 0°C
V_{IL}	Input Low Voltage	V_{SS}		0.65	V	
V_{IH}	Input High Voltage	3.0		V_{CC}+1	V	
V_{OL}	Output Low Voltage			0.45	V	I_{OL} = 1.6mA
V_{OH1}	Output High Voltage	3.7			V	I_{OH} = -100μA
V_{OH2}	Output High Voltage	2.4			V	I_{OH} = -1mA
P_D	Power Dissipation			800	mW	T_A = 70°C

NOTES:
1. Typical values are for T_A = 25°C and nominal supply voltages.
2. The total power dissipation of the 2708 is specified at 800 mW. It is not calculable by summing the various currents (I_{DD}, I_{CC}, and I_{BB}) multiplied by their respective voltages since current paths exist between the various power supplies and V_{SS}. The I_{DD}, I_{CC}, and I_{BB} currents should be used to determine power supply capacity only.

D.C. DEVICE CHARACTERISTICS

Only those D.C. Characteristics that require special attention by the user are presented in this section. The reader is referred to the 2708 device data sheet for further details. The pertinent D.C. device specifications are tabulated in Table II.

The 2708 requires three power supplies, +12V and ±5V. The device is rated to meet all applicable specifications with these supplies held within ±5% of their normal value. The Absolute Maximum Ratings in the data sheet are the maximum that the various device parameters can withstand and should not be exceeded during any phase of device operation, including programming.

Read Mode

The range of values of currents from the three power supplies, V_{DD} (+12V), V_{CC} (+5V) and V_{BB} (-5V) are shown in Table II, presented for the worst case conditions; i.e., $\overline{CS}$/WE = 5V and T_A = 0°C. The I_{DD}, I_{CC} and I_{BB} data presented indicates the maximum current drawn by the respective power input. These inputs cannot simultaneously draw maximum current. Refer to the APPLICATIONS SECTION for measured laboratory data of the interactive effects of switching the various supplies off to conserve power.

The addresses are TTL compatible, requiring V_{IL} between V_{SS} and 0.65V and V_{IH} between 3V and V_{CC} + 1. Care should be exercised in selecting address buffers to insure the minimum V_{IH} level is met by use of appropriate TTL circuit elements or pull-up resistors to V_{CC}.

During the Read mode, the $\overline{CS}$/WE input (pin 20) is also TTL compatible; however, the V_{IL} and V_{IH} requirements for the address inputs are still applicable.

The outputs are also TTL compatible, producing a V_{OL} of 0.45V maximum @ 1.6 mA and a V_{OH} of 3.7V with -100 μA capability, or 2.4V with -1 mA capability. Typical output sink current is plotted in Figure 7 as a function of the output voltage and temperature for applications requiring higher than normal I_{OL} currents.

Figure 8 illustrates several points regarding the 2708 power supply currents. First of all, as with all MOS devices, the power supply currents will decrease as a function of increasing temperature. The second point is that the current requirements of the device increase when it is deselected, i.e., when $\overline{CS}$/WE is at V_{IH}. The reason for this is that in order to meet the very fast t_{CO} time of 120 ns, all of the decoders and output stages are turned on when $\overline{CS}$/WE is at V_{IH}, and the decoders deselect those that are not required for the given data cycle. The graph also illustrates that the V_{DD} power supply is the most logical supply to be selected for switching to reduce power. Of course, if the system configuration permits, $\overline{CS}$/WE can be tied to V_{SS} to reduce power.

DESIGNING WITH 2708 EPROM

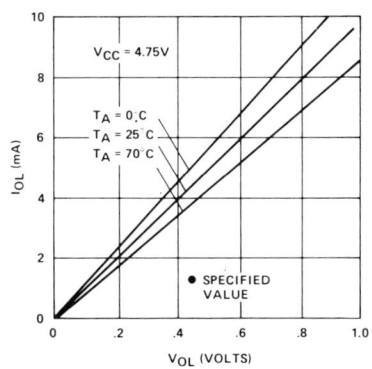

Figure 7. 2708 Typical Output Sink Current vs. Output Voltage

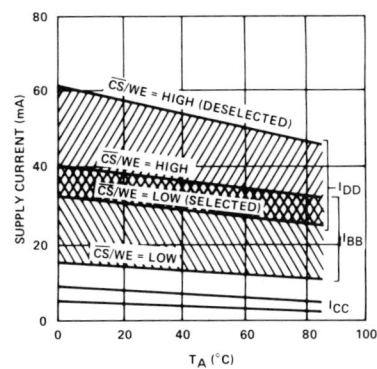

Figure 8. 2708 Power Supply Currents

Table III. D.C. Programming Characteristics

Symbol	Parameter	Min.	Typ.	Max.	Units	Test Conditions
I_{LI}	Address and $\overline{CS}/WE$ Input Sink Current			10	µA	$V_{IN} = 5.25V$
I_{IPL}	Program Pulse Source Current			3	mA	
I_{IPH}	Program Pulse Sink Current			20	mA	
I_{DD}	V_{DD} Supply Current		50	65	mA	Worst Case Supply Currents:
I_{CC}	V_{CC} Supply Current		6	10	mA	All Inputs High
I_{BB}	V_{BB} Supply Current		30	45	mA	$\overline{CS}/WE = 5V; T_A = 0°C$
V_{IL}	Input Low Level (except Program)	V_{SS}		0.65	V	
V_{IH}	Input High Level for all Addresses and Data	3.0		$V_{CC}+1$	V	
V_{IHW}	$\overline{CS}/WE$ Input High Level	11.4		12.6	V	Referenced to V_{SS}
V_{IHP}	Program Pulse High Level	25		27	V	Referenced to V_{SS}
V_{ILP}	Program Pulse Low Level	V_{SS}		1	V	$V_{IHP} - V_{ILP} = 25V$ min.

Program Mode

The address and data inputs are low level compatible during programming, with the same requirements of V_{IL} and V_{IH} as for the Read mode. The D.C. characteristics for programming are shown in Table III. To enable the device for programming, the $\overline{CS}/WE$ pin is raised to V_{IHW}, (11.4V). If the system requirements dictate that the device stay in the same socket or location for both reading and programming, it should be recalled that this pin will require three input levels: V_{IL} of V_{SS} to 0.65V to select the device for a read operation, a V_{IH} of 3V to V_{CC} +1 to deselct the device and place the output in the high impedance state, and a V_{IHW} of 11.4 to 12.6V to Write Enable, or allow programming of the device. Several circuits for generating these three active levels (V_{IL}, V_{IH} and V_{IHW}) are shown in the PROGRAMMING section (page 7).

During program operation, the outputs become the data inputs and should be treated as a three state bus. The same V_{IL} and V_{IH} levels apply to the data I/O pins as apply to the address pins.

The program pulse, which is applied to pin 18 during programming, must meet a V_{ILP} requirement (V_{SS} to 1V) and a V_{IHP} requirement (26V ± 1V).

The program pulse source must be capable of supplying a maximum of 20 mA per device when high (V_{IHP}), and be able to withstand the Program Pulse Sink current of 3 mA (I_{IPL}). This sink current should be considered when designing the program pulse driver, as, if a resistive pull-down is used, the voltage drop across the resistor can violate the V_{ILP} max requirement of 1V. It also should be noted that the program pulse will not meet specification if V_{IHP} is taken at its minimum value (25V) and V_{ILP} is taken at its maximum value (1V), as $V_{IHP}-V_{ILP}$ must equal 25 volts minimum. Several circuits are presented in the PROGRAMMING section to provide program pulses which easily meet the 25V minimum requirement for $V_{IHP}-V_{ILP}$.

A.C. DEVICE CHARACTERISTICS

Read Mode

Figure 9, the Read mode timing, indicates the maximum or minimum timing for the various timing parameters. Particular attention should be paid to t_{DF}, chip deselect to output float time. This indicates that the output buffers of the 2708 are not guaranteed to reach the high impedance state until 120 ns after $\overline{CS}/WE$ reaches the 2.8V point. If another device attempts to take control of the output node during this time, very high I_{CC} current will be drawn, generating noise on the supply lines and possibly reducing the V_{CC} level such that other devices may become inoperative. t_{DF} is also a factor to consider when switched V_{DD} is used. See the APPLICATIONS section for further discussion.

Program Mode

Figure 10 indicates the Program mode timing, while Table IV tabulates the various programming A.C. parameters.

Several options are available to the user when programming the 2708, as shown in the data sheet. The waveforms shown in Figure 10 represent the most efficient method. The various parameters are self-explanatory; two will be discussed here. The program pulse rise and fall times, t_{PR} and t_{PF}, must be held within the range of 0.5 and 2 μs to minimize the transient coupling effects discussed in the memory array section. This usually requires a series RC network on the output of the program pulse driver to slow down the rise time. Exotic waveform generators are not required. Refer to the PROGRAMMING section for circuit recommendations.

The other parameter of concern to the user is the transition from Program mode to Read mode. If the $\overline{CS}/WE$ transition does not occur after the final program pulse transition and before the address transition, as shown in Figure 10, nodes internal to the device will not discharge, causing the output buffers to indicate false data for several milliseconds.

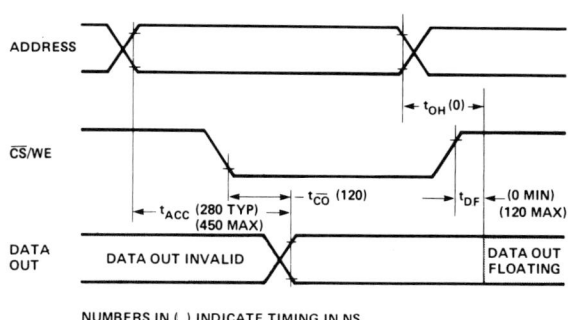

Figure 9. 2708 Read Cycle Waveforms

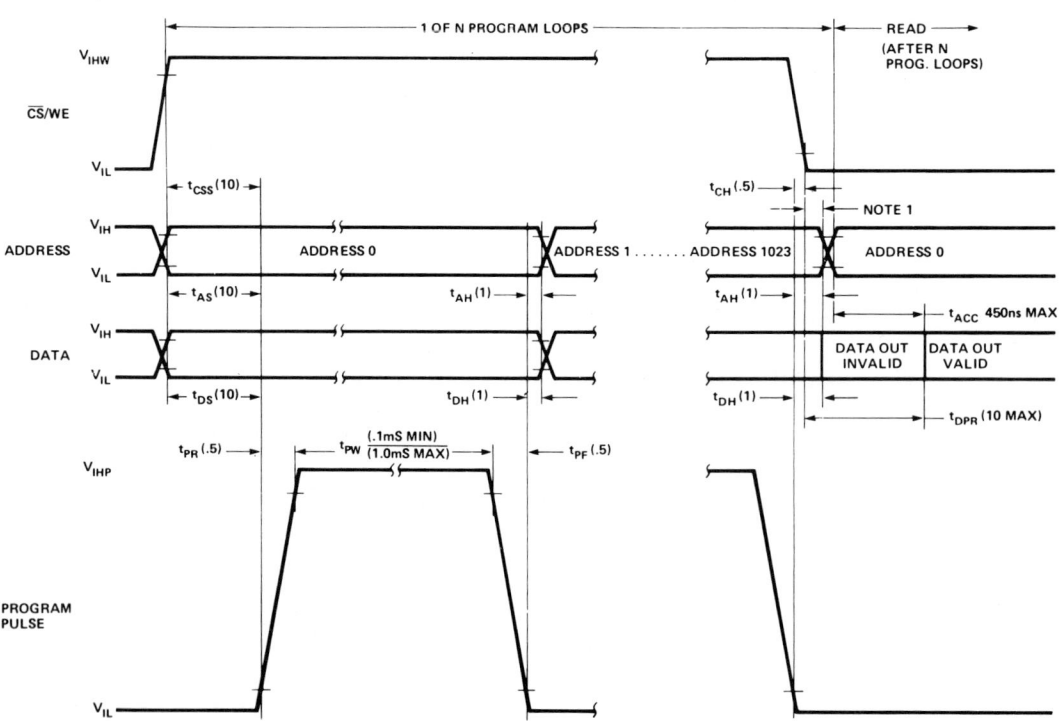

NOTE 1. THE $\overline{CS}/WE$ TRANSITION MUST OCCUR AFTER THE PROGRAM PULSE TRANSITION AND BEFORE THE ADDRESS TRANSITION.

NOTE 2. NUMBERS IN ( ) INDICATE MINIMUM TIMING IN μS UNLESS OTHERWISE SPECIFIED.

Figure 10. 2708 Programming Waveforms

Table IV. A.C. Programming Characteristics

Symbol	Parameter	Min.	Typ.	Max.	Units
t_{AS}	Address Setup Time	10			µs
t_{CSS}	$\overline{CS}$/WE Setup Time	10			µs
t_{DS}	Data Setup Time	10			µs
t_{AH}	Address Hold Time	1			µs
t_{CH}	$\overline{CS}$/WE Hold Time	.5			µs
t_{DH}	Data Hold Time	1			µs
t_{DF}	Chip Deselect to Output Float Delay	0		120	ns
t_{DPR}	Program To Read Delay			10	µs
t_{PW}	Program Pulse Width	.1		1.0	ms
t_{PR}	Program Pulse Rise Time	.5		2.0	µs
t_{PF}	Program Pulse Fall Time	.5		2.0	µs

NOTE: Intel's standard product warranty applies only to devices programmed to specifications described herein.

This will appear as an excessively long t_{DPR}, Program to Read Delay. If the $\overline{CS}$/WE timing is difficult to adjust, providing the binary complement of the first address to be verified before actually verifying will also discharge the internal nodes.

PROGRAMMING

A number of programmers are commercially available that will properly program the 2708. Intel maintains a service whereby commercial programmer manufacturers obtain design approval prior to marketing their device, in order to assure compatibility with Intel specifications. This approval should be verified with the particular programmer manufacturer prior to purchase.

It is also possible to build a programmer as part of the user's system, by adhering to the following description: The device is set up for programming operation by raising the $\overline{CS}$/WE input (pin 20) to V_{IHW} (+12V). The word address is then selected in the same manner as in the Read mode. Data to be programmed are presented, 8 bits in parallel, to the data output pins (O_1–O_8). Logic levels for address and data lines and the supply voltages are the same as for the Read mode. After address and data set up times (t_{AS} and t_{DS}, Fig. 10), one program pulse of width t_{PW} is applied to the program pin (pin 18). This sequence is then repeated for the next address. One pass through all 1024 addresses is defined as a program loop. The number of program loops (N) required is a function of the program pulse width (t_{PW}) according to the formula:

$$N \times t_{PW} \geq 100 \text{ ms}$$

where

N is the number of program loops

t_{PW} is the program pulse width.

The width of the program pulse can vary from 0.1 to 1.0 ms. The number of loops (N) can vary from a minimum of 100 (t_{PW} = 1.0 ms) to greater than 1000 (t_{PW} = 0.1 ms), depending on the value chosen for t_{PW}. IT IS NOT PERMITTED TO APPLY N PROGRAM PULSES TO AN ADDRESS AND THEN CHANGE TO THE NEXT ADDRESS AND APPLY N PROGRAM PULSES. THERE MUST BE N SUCCESSIVE LOOPS THROUGH ALL 1024 ADDRESSES.

Referring to the timing diagram, Figure 10, optimum or most efficient programming is achieved when:

$$t_{CSS} = t_{AS} = t_{DS} = 10 \text{ µs}$$
$$t_{PW} = 1.0 \text{ ms}$$
$$t_{AH} = t_{DH} = 1.0 \text{ µs}$$
$$t_{PR} = t_{PF} = 0.5 \text{ µs}$$

and the time for 1 address becomes:

$$t_{AS} + t_{PR} + t_{PW} + t_{PF} + t_{AH} = 1.012 \text{ ms}$$

or, for 100 loops and 1024 addresses, the total time to program an entire device will be 1.012 ms/address × 100 loops × 1024 addresses, or 103.6 sec. Note that the program pulse duty cycle is approximately 99%. Whatever the length of the program pulse, the requirement for making successive passes through all addresses cannot be eliminated.

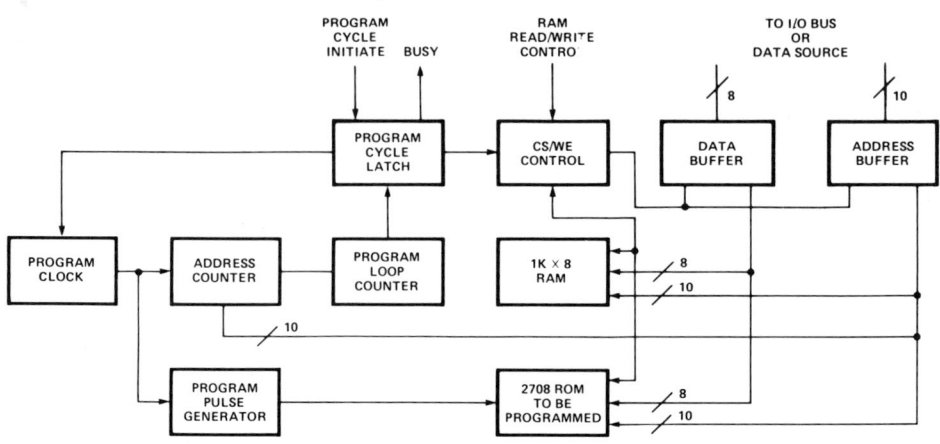

Figure 11. Typical Programming Block Diagram

Typical Programmer

Figure 11 illustrates a block diagram of a typical programmer that meets all the requirements for programming the 2708, as well as facilitating interface to a microcomputer I/O bus if it is desired to use the microprocessor system as a data source. Keyboard entry is also possible, although it does become tedious to manually enter data for 1024 PROM locations.

Operation of the programmer is as follows: While the data is being generated, the RAM Read/Write Control line allows information to pass through the Data Buffer and Address Buffer as in normal microcomputer memory operation. When the data is finalized in the 1K by 8 RAM, a Program Cycle Initiate command is generated, which responds, via the Program Cycle Latch, by generating a Busy signal back to the processor, and disenables the Data and Address Buffers, inhibiting further communication with the I/O bus until the program cycle is complete. The Program Cycle Latch also starts the Program Clock, enables the RAM, and Write Enables the PROM. It also initializes the Address and Program Loop Counters. The Program Clock activates the Program Pulse Generator, causing the information from RAM address A_0 to be programmed into the PROM. The next clock pulse increments the address counter and when the RAM data corresponding to that address is presented to the PROM inputs (outputs during read), it again increments the address counter and continues until the Address Counter overflows on the 1024th pulse, at which time the Program Loop Counter is incremented.

The entire process is then repeated until the required number of program pulses has been received by each PROM location, and the Program Loop Counter overflows, resetting the Program Cycle Latch. The PROM can now be read or verified by way of the PROM cycle request.

To modify data in a partially programmed PROM it is only necessary to read the PROM into the RAM, enter the new data pattern, and check to be sure that no bits will be attempting to program 0's to 1's, and reprogram the PROM as described above. The only method of programming a "1" where there is a "0" is to erase the entire device and reprogram. This process is illustrated graphically in Figure 12.

STATUS	PROM OUTPUTS							
	O_1	O_2	O_3	O_4	O_5	O_6	O_7	O_8
INITIAL STATE	1	1	1	1	1	1	1	1
FIRST PROGRAMMING	1	1	0	0	1	0	1	0
FIRST REPROGRAMMING	0	1	0	0	0	0	1	0
SECOND REPROGRAMMING	0	0	0	0	0	0	1	0
FINAL REPROGRAMMING	0	0	0	0	0	0	0	0
ERASURE	1	1	1	1	1	1	1	1

Figure 12. Reprogramming 2708 Outputs

DESIGNING WITH 2708 EPROM

Program Pulse Driver Circuits

Figure 13 presents several circuits which have been successfully used to generate the required 26V pulse for programming, and one circuit which should not be used.

The circuit shown in Figure 13a should not be used, as the resistive pull-down will not meet the V_{ILP} requirement of 1V max, thus not allowing $V_{IHP} - V_{ILP}$ to be equal to or greater than 25V. As was mentioned earlier, the reason for this is that the Program pin, Pin 18, sources I_{ILP} of about 2 mA when the program pulse is low and $\overline{CS}/WE$ is at +12V. The other circuits, b and c, do meet all the A.C. and D.C. specifications associated with the program pulse.

$\overline{CS}$/WE Driver Circuits

Figure 14 presents several circuits for generating the $\overline{CS}$/WE signal. Circuit a is very simple, providing the three necessary levels for on board programming. Circuit b has increased driving capability and isolation over circuit a, and will allow more noise margin. In addition, the inclusion of the two 100Ω resistors provide short circuit protection in case of socketing or soldering errors. A truth table is included with circuits a and b to indicate the various input/output conditions. Circuit c provides only two levels, V_{IL} (0V) and V_{IHW} (+12V), for use in "program and verify only" circuits; the PROM cannot be deselected using this circuit. Another way of generating the 0 and +12V signals would be to use a TTL to MOS driver, such as the Intel® 3245.

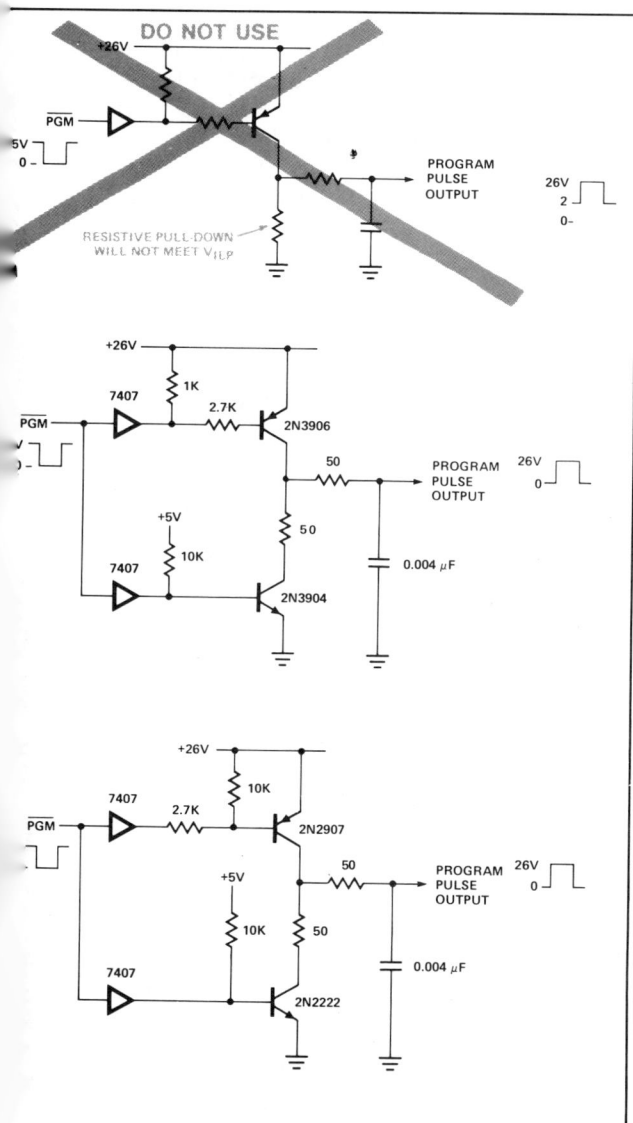

Figure 13. Program Pulse Driver Circuits

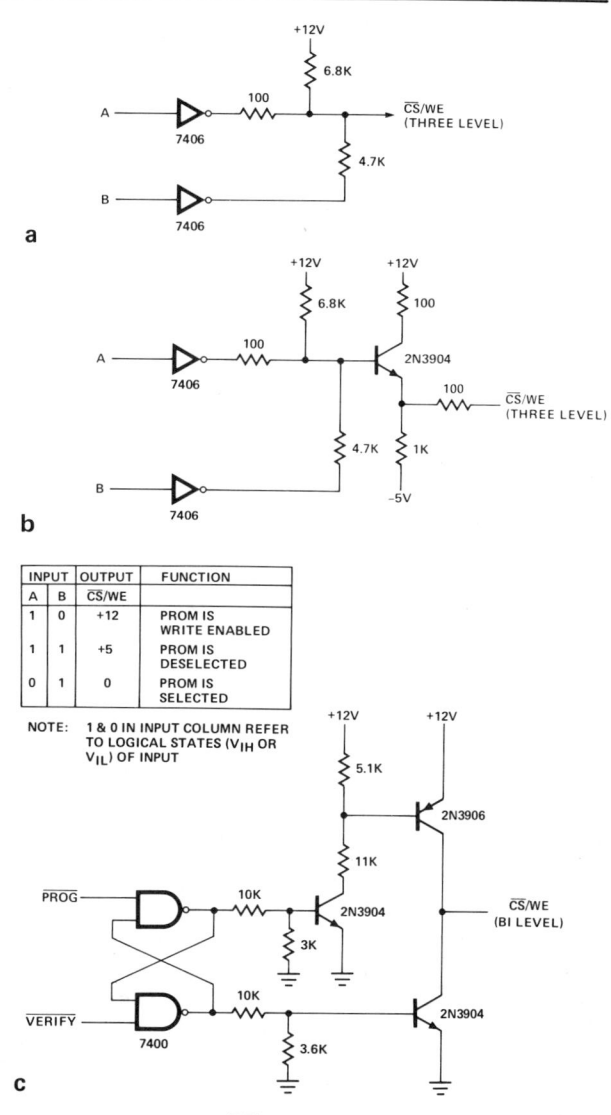

Figure 14. $\overline{CS}$/WE Driver Circuits

On Board Programming

Unlike many other erasable and programmable Read Only Memories, the 2708 can be soldered directly into a printed circuit board and programmed while "in circuit", as the inputs and outputs stay low level compatible during both read and program modes of operation. When erasure is required, the circuit board is unplugged and placed under a UV lamp for the required period of time.

In many microprocessor systems, it is quite easy to implement the RAM storage required for a data base when programming by using available RAM storage. Be sure to observe all the required setup times if the address and data bus will be performing non-programming related functions while the PROM is being programmed.

Figure 15 illustrates a possible scheme for implementing a data output/input buffer.

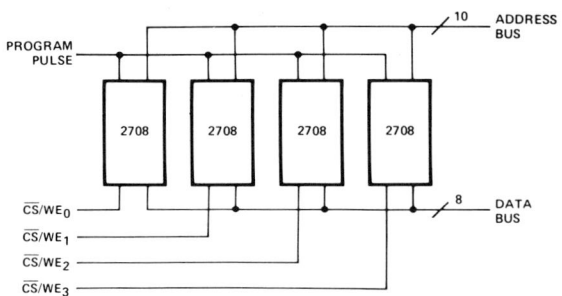

Figure 16. Circuit Implementation for On-Board Programming

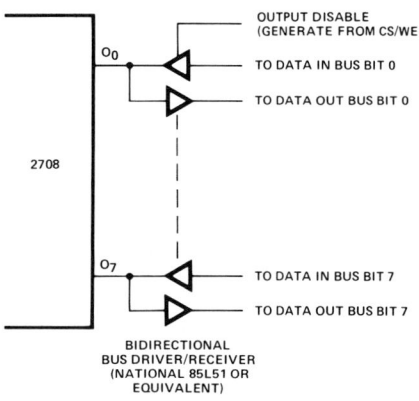

Figure 15. Data Output/Input Buffer

To take advantage of this feature, which is not tested or included as part of the device specifications, the program pulse should be applied to all devices as shown in Figure 16. Program decode is then accomplished by way of the $\overline{CS}/WE$ pin. PROM's to be programmed have this pin raised to the V_{IHW} level (+12V), while it is left at V_{IH} ($\overline{CS}$=3V) for those parts which are not to be programmed. Reserve should be built into the program pulse power supply when operation in this mode is planned, but in no case will it exceed the maximum of 20 mA per 2708 as specified in Table III.

ERASING

The 2708 is erased by exposure to ultra-violet light at a wavelength of 2537Å. The recommended integrated dosage (i.e. UV intensity × exposure time) is 15 W-sec/cm^2. In order to insure that all bits are erased, this dosage includes a guard band and is not equal to the dosage required to see the last bits return to the initial state. A guard band of 3 to 4 times the initial period (that time which appears to erase all bits) is suggested so that the device will appear erased at extremes of temperature and voltage.

Table V. UV Sources for Erasing the 2708

Model	Power Rating	Typical Time to Erase a 2708 Device
S-68	12000 uW/cm^2	15 minutes
S-52	12000 uW/cm^2	15 minutes
UVS-54	5700 uW/cm^2	45 minutes
R-52	13000 uW/cm^2	15 minutes
UVS-11	5500 uW/cm^2	45 minutes

Table V lists several UV sources for erasing the 2708. The model numbers referred to are manufactured by Ultra-Violet Products, Inc. (5114 Walnut Grove Avenue, San Gabriel, CA).

The times indicated are for the lamps placed about 1 inch away from the parts to be erased and without shortwave filters installed. For lamps other than those listed, the required times can be determined empirically or by means of an ultra-violet intensity meter, such as the UV Products Model J-225. When a meter is used, the intensity should be determined at the same location (distance from UV tube) as the PROM will be placed; this will require careful measurement to insure that the sensor is receiving exactly the same amount of UV light that the PROMs will receive.

DESIGNING WITH 2708 EPROM

APPLICATIONS

Switched Power Supplies

Although not specified in the D.C. and A.C. DEVICE SPECIFICATIONS sections, the 2708 can be operated in a power down mode by switching off the V_{DD} power supply. This is advantageous in many applications where power dissipation is a critical factor, such as battery operated or battery backed-up systems. Referring to Table II, the maximum I_{DD} power that can be saved by switching the V_{DD} power supply is 780 mW. Two factors should be noted, however. First of all, the access time will increase somewhat, as shown in Figure 17.

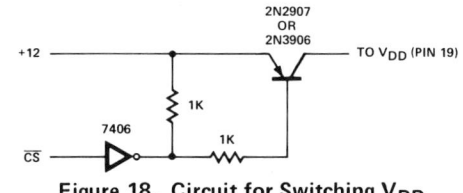

Figure 18. Circuit for Switching V_{DD}

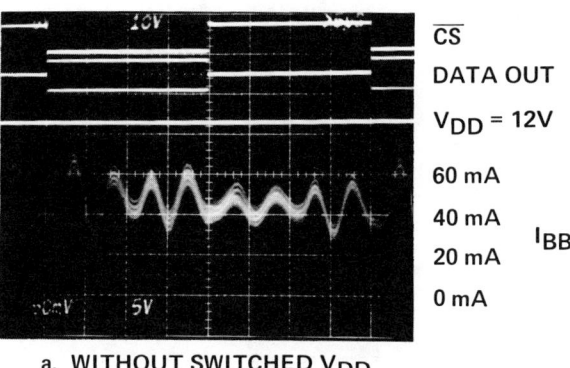

a. WITHOUT SWITCHED V_{DD}

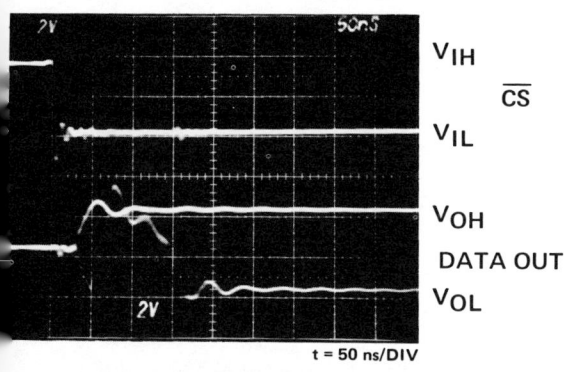

a. WITHOUT SWITCHED V_{DD}

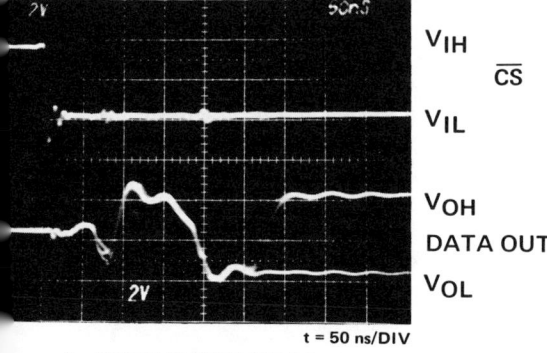

b. WITH SWITCHED V_{DD}

Figure 17. 2708 Access Time

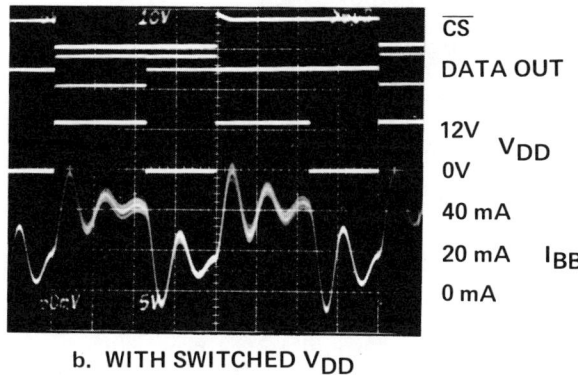

b. WITH SWITCHED V_{DD}

Figure 19. 2708 I_{BB} Current

The photos were taken using the circuit of Figure 8, at room temperature and with a small sample of parts. Based on this information, the PROM data strobe should be moved out approximately 50 ns to allow a guard band for the system. The second point related to the switching of V_{DD} is the reduction of V_{BB} current (I_{BB}). Figure 19 indicates that I_{BB} decreases to an average of approximately [?] mA when V_{DD} is off.

As shown in Figure 20, output deselection occurs within t_{DF} (Chip Select to Output Float Delay) when $\overline{CS}$ is held low and V_{DD} is switched.

Switching off V_{CC} will save some power, but the maximum value is so low (10 mA) that the extra components required for switching are probably not justified. Typical values of I_{CC} decrease about 50% when the V_{DD} supply is switched off.

The V_{BB} supply could also be switched, but, considering the reduction when the V_{DD} supply is switched off, the additional components required to switch this supply would probably not be justified, either. In addition, unless an extra power supply of -10 to -15 volts is available for a driver circuit, access time would be significantly degraded (laboratory data indicates about 50 μs).

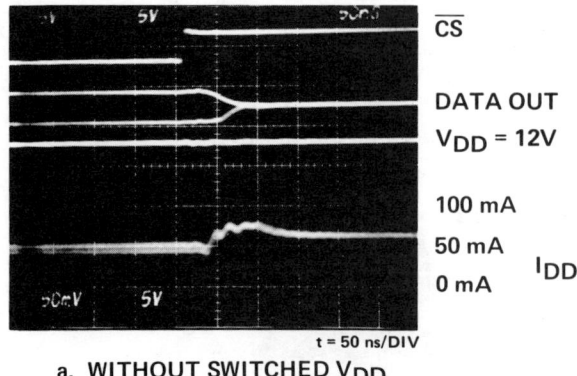

a. WITHOUT SWITCHED V_{DD}

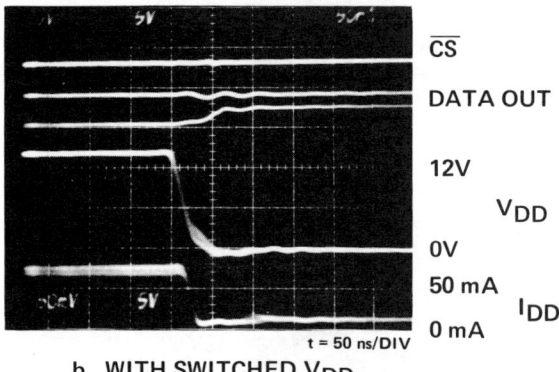

b. WITH SWITCHED V_{DD}

Figure 20. 2708 Output Deselection and I_{DD} Current

Another way of reducing power is to leave the device continuously selected and control the output by way of an enable signal on a latch or gate. Referring to Figure 8, this method would reduce power dissipation nearly 50%, as the device does dissipate less power when $\overline{CS}$/WE is low.

OR Tie Considerations

When two or more 2708's are wire ORed together, care should be exercised to see that valid data will be obtained. Referring back to Figure 7 and Figure 21, if two devices are selected at the same time, a current path can exist from Q_1 to Q_4 is shown in Figure 21. This current can be destructive to the output stage of one of the devices, or, the transistor with greater current sourcing or sinking capability can cause false data to be read from the output bus. In addition, the very high V_{CC} current drawn while both Q_1 and Q_4 are on will generate noise on the V_{CC} power supply lines, and possibly reduce the V_{CC} that is connected to other TTL control circuits, causing momentary false indications. If the maximum chip deselect to output float delay (t_{DF}) is observed, there will be no problem. The same type of situation can occur when the 2708 is used in conjunction with other memory devices, such as the RAM portion of the programmer shown in Figure 11. Careful analysis of the system timing requirements and maximum delay paths can eliminate these problems before they occur at the final checkout of a system.

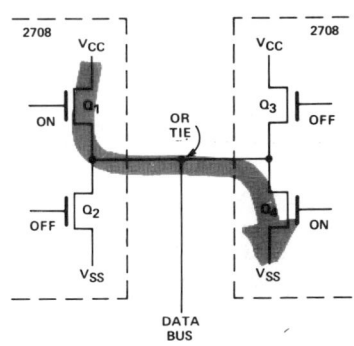

Figure 21. Results of Improper Timing when OR Tying 2708's

High Voltage CMOS Interface

Because the 2708 is erased by the same technique as the Intel® 1702A, some users have assumed that the various techniques for interfacing to high voltage CMOS circuits are similar. In fact, they are not. The 1702A is a p-channel device, requiring two power supplies (+5V and −9V), while the 2708 is a n-channel device and requires three power supplies (+12V, +5V and −5V). It is permissible to assign the ground (0V) to the most negative supply and reference all the other supplies to it; however, suitable level shifters must be used to provide the 2708 with suitable input level signals, and to convert the output signals back to the system reference levels. Figure 22 shows a possible voltage translation.

SUPPLY	2708 VOLTAGE	SYSTEM VOLTAGE
V_{DD}	+12V	+17V
V_{CC}	+ 5V	+10V
V_{SS}	0V	+ 5V
V_{BB}	− 5V	0V
V_{IL}	0 to +0.65	+5.0 to +5.65
V_{IH}	+3.0 to +6.0	+8.0 to +11.0
V_{OL}	+0.45	+5.45
V_{OH}	+2.4 @ −1 mA	+7.4

Figure 22. 2708 Voltage Translation

DESIGNING WITH 2708 EPROM

Some suitable translator circuits are: RCA CD4009/4010 or National F/4104/34104. The use of these circuits also allows some high voltage CMOS logic to be implemented, such as address and data clocks, at the CMOS levels, rather than convert them to TTL levels for operation of the 2708.

Another incorrect method of attempting to interface directly to CMOS circuits is to change the V_{CC} supply to the new interface voltage. In devices such as the Intel® 2107B this is permissible, as the V_{CC} supply is connected to the output buffer stage, but in the 2708, the +5V is used in the sense amplifier and other internal circuitry, so this should not be done.

Under Programming and Under Erasing

It is possible to "under program" the 2708, such that the cell characteristic crosses the sense threshold. The result of this is that the cell apparently drops or picks up bits. As can be seen in Figure 23, the threshold characteristic has been shifted such that small changes in voltage or temperature will cause a "1" or a "0" to be sensed. This is always the result of insufficient erasing or programming. For erasure to cause this problem, the device has only been partially programmed, and the characteristic curve has only been shifted to the sense threshold point and the device will again seem to either pick up or drop bits. The cure, in either case,

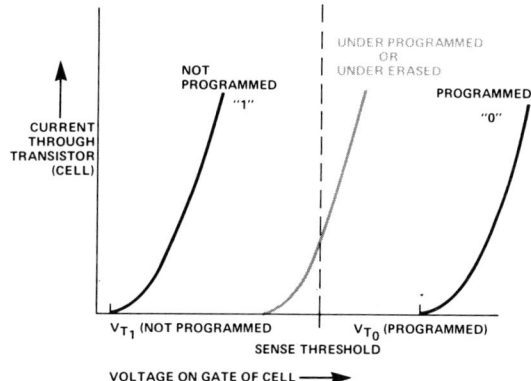

Figure 23. Effect of Under Programming or Under Erasure

is to 1) adequately erase by providing the required 10 W-sec/cm^2 of UV light at a frequency of 2537Å, or 2) program in accordance with the specifications.

ACKNOWLEDGEMENT

I would like to extend my thanks and appreciation to the Technology Development Group for their assistance in preparing the background material for this Application Note.

DESIGNING WITH 2716 EPROM

BOB GREENE

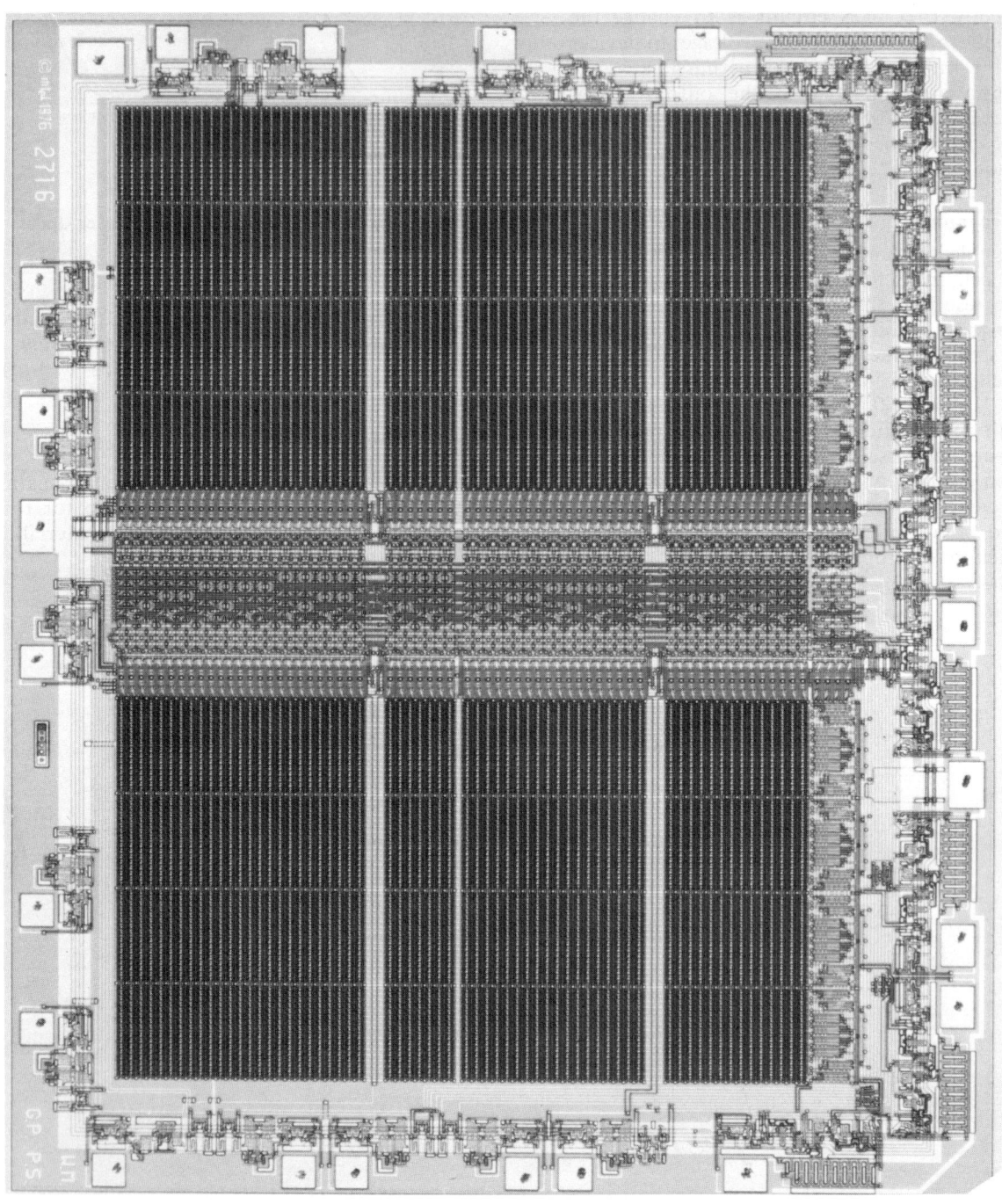

Photomicrograph of Intel 2716 16K EPROM.

DESIGNING WITH 2716 EPROM

INTRODUCTION

The INTEL® 2716 is a fully static 16,384-bit (2048 x 8) Eraseable Programmable Read Only Memory, or EPROM. The device is packaged in a standard 24-pin DIP, which has a transparent lid to allow erasure in a manner similar to that of the INTEL® 1702A and 2708. Maximum access time is 450ns. The device requires a single power supply (V_{CC} = 5V ±5%) for normal read cycles; during programming the program power supply (V_{PP}) must be raised to +25V to program each location, a single TTL level pulse is required; one 50ms pulse per address programs 8 bits in parallel. The addresses can be randomly programmed.

All input signals are fully TTL compatible during both the read and program modes. The data outputs are three state to facilitate memory expansion by OR tying. Initially and after each erasure the 2716 contains all TTL highs ("1"s); programming or introducing TTL lows ("0"s) is accomplished by: 1) raising the V_{PP} pin from +5V to +25V, 2) applying TTL level addresses and TTL level data, 3) raising the $\overline{CS}$ pin to a TTL high, and 4) applying a single 50ms TTL level pulse to the PD/PGM input.

The V_{PP} supply may be left at the +25V level for program verification, but should be returned to +5V level during normal read cycles to reduce power dissipation.

DEVICE DESCRIPTION

The 2716 is packaged in an industry standard 24 pin DIP as shown in Figure 1. The functions of the various control pins are shown in Table I.

During read operation $\overline{CS}$ is used to select and deselect the 2716. The PD/PGM pin is maintained at V_{IL}, while V_{PP}, the program power supply, is maintained at +5V. As shown in the D.C. Device Characteristics Section, I_{PP1} (the current required by pin 21) is 5mA maximum during read mode, so pin 20 should be kept at V_{CC} except when programming. As a convenience to users, it is allowable to keep the V_{PP} pin at +25 volts for program verification, but it must be returned to +5V upon completing program verification. This is easily accomplished by connecting a diode from pin 24 to pin 21 as shown in Figure 2. The tolerance on V_{PP} allows for a diode drop as discussed in the D.C. Operating Characteristics section. For read only applications, the V_{PP} pin may be tied directly to the V_{CC} pin.

Table I. 2716 Pin Connections and Functions.

PINS MODE	PD/PGM (18)	$\overline{CS}$ (20)	V_{PP} (21)	V_{CC} (24)	OUTPUTS (9-11, 13-17)
Read	V_{IL}	V_{IL}	+5	+5	D_{OUT}
Deselect	Don't Care	V_{IH}	+5	+5	High Z
Power Down	V_{IH}	Don't Care	+5	+5	High Z
Program	Pulsed V_{IL} to V_{IH}	V_{IH}	+25	+5	D_{IN}
Program Verify	V_{IL}	V_{IL}	+25	+5	D_{OUT}
Program Inhibit	V_{IL}	V_{IH}	+25	+5	High Z

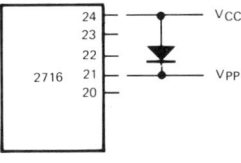

Figure 2. 2716 Power Supply Connections.

The PD/PGM input serves several functions. When low this signal enables the address, data and $\overline{CS}$ input buffers, whether V_{PP} is at +25V or +5V. When high with V_{PP} at +5V, the 2716 is powered down and the outputs are deselected without regard for the state of $\overline{CS}$. In this mode the maximum I_{CC} current is reduced from 100mA to 25mA. When $\overline{CS}$ is high and V_{PP} is at 25V, the data present on the output will be programmed into the selected address when PD/PGM is pulsed high (from V_{IL} to V_{IH}) for 50ms.

A block diagram for the 2716 is shown in Figure 3. The array of stacked gate cells is arranged as two 64 x 128 matrices, each of which is split into four 16 x 128 segments. The high order address bits (A_4-A_{10}) determine which of the 128 rows is to be accessed by way of the top select gate, while the low order address bits (A_0-A_3) perform the column decode function by activating the 1 of 16 decoders which are associated with each output bit.

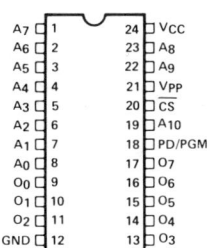

PIN NAMES

A_0–A_{10}	ADDRESSES
PD/PGM	POWER DOWN/PROGRAM
$\overline{CS}$	CHIP SELECT
O_0–O_7	OUTPUTS

Figure 1. 2716 Pin Configuration.

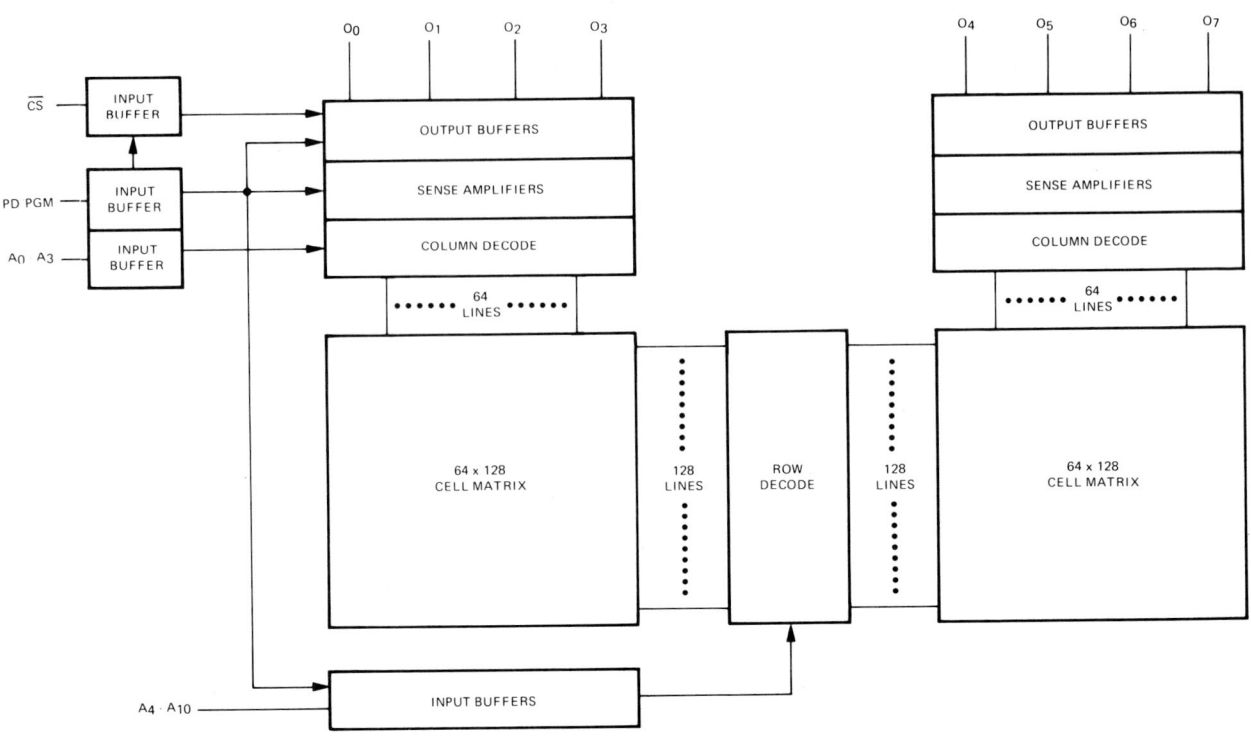

Figure 3. Detailed Block Diagram.

Cell Description

The heart of the 2716 is the single transistor stacked gate cell, which is similar to the cell used in the INTEL® 2708. The cell consists of a floating gate, used to store charge, and a top select gate which is connected to the output of the row decoder. The cell is programmed by injection of high energy electrons through the isolating oxide and onto the floating gate. Once there, the charge is trapped, as there are no electrical connections to the floating gate. The presence of electrons on the floating gate causes a shift in cell threshold, as shown in Figure 4. In the initial or erased state the threshold of the cell is low, selection via the top gate will cause the column line to discharge, which is sensed as a "HIGH" by the sense amplifier. Programming shifts the threshold to a higher level, and selection of the cell will not turn it on, the column line will not discharge, and a low will be sensed by the sense amplifier. The status of the cell is determined by examining its state at the sense threshold; if the cell is erased (HIGH data) selection will cause a higher current to flow between the source and drain than if the cell is programmed (LOW data).

Memory Array Operation

The cells described in the previous paragraph are interconnected to form a split 128 x 128 cell matrix, as shown in Figure 3. This array is divided into 8 sections organized as 16 x 128 cells. Each of these sections is connected to a column decoder, which selects one of 16 columns, connecting it to the sense amplifier which is associated with the particular bit. The sense amplifier is directly connected to the output buffer associated with the same bit. This data flow is illustrated graphically in Figure 5.

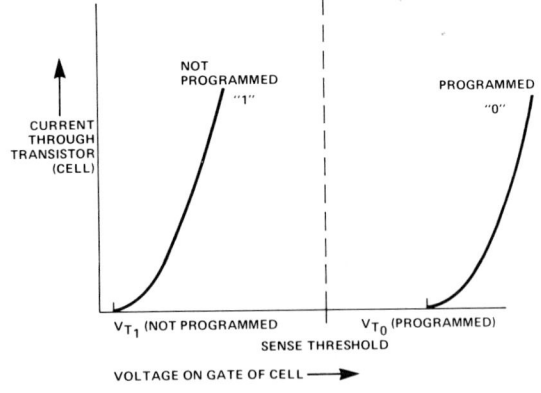

Figure 4. Storage Cell Threshold Shift

DESIGNING WITH 2716 EPROM

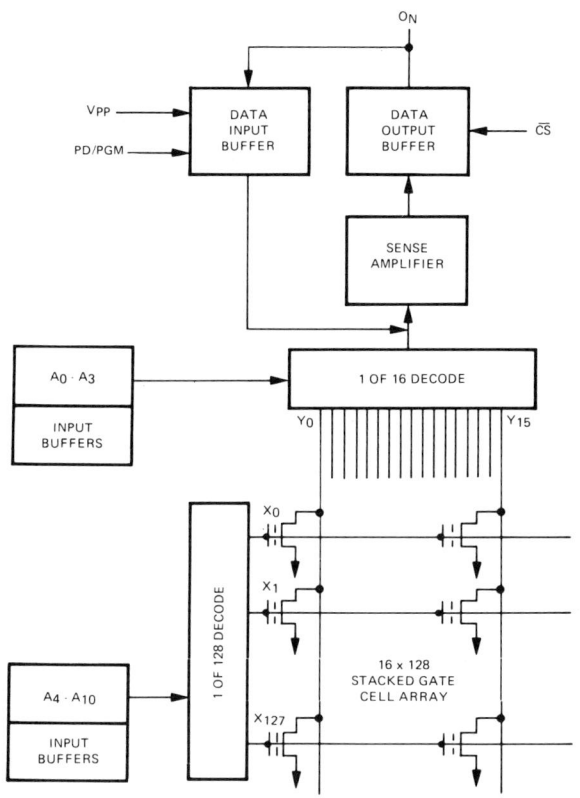

Figure 5. 2716 Single Bit Data Flow.

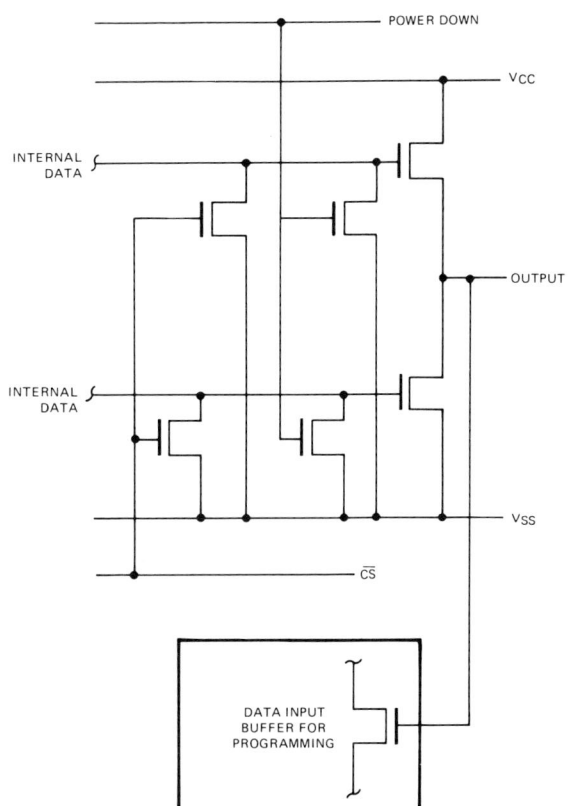

Figure 6. 2716 Output Buffer.

Output Buffer

An equivalent schematic of the output buffer is shown in Figure 6. As is shown, the output buffer consists of a pair of MOS transistors, connected in a push-pull configuration. $\overline{CS}$ enables both transistors when true; when $\overline{CS}$ is false both output devices are turned off. The PD/PGM input also is related to the output buffer and does place the output buffer in the high impedance state when the internal signal Power Down is high. This signal is normally low for regular read operations, and functions as an output deselect when high. Remember that if V_{PP} is at +25V and $\overline{CS}$ is high, raising PD/PGM high will cause a program cycle on the selected address.

READ MODE

The 2716 requires only one power supply, +5V. The device is rated to meet all applicable specifications with this supply held within ±5% of its nominal value. The Absolute Maximum Ratings in the data sheet are the maximum that the various device parameters can withstand and should not be exceeded during any phase of device operation, including programming.

D.C. Characteristics

Only those D.C. Characteristics that require special attention by the user are presented in this section. The reader is referred to the 2716 device data sheet for further details. The pertinent D.C. device specifications are tabulated in Table II.

The range of the leakage currents shown in Table II apply for all inputs and outputs, including the outputs (O_0-O_7) when they are serving as data inputs for programming.

I_{PP1} is the current required by the V_{PP} pin (pin 21) when the V_{PP} supply is set to 5V, as it would be for normal read operations. The device specification requires a ±5% tolerance on the V_{CC} supply. In anticipation that users will couple pin 21 to pin 24 by way of a diode, the tolerance on V_{PP} has been relaxed to ±0.6V to allow for the forward drop of the diode.

I_{PP} is only applicable to the current drawn by pin 21 when the PD/PGM pulse is low; when it is high (as in the case of the program pulse) the current drawn by this pin will be 30mA.

I_{CC1} is the power supply current when PD/PGM is high and V_{PP} is at a nominal 5V, and represents 25% of the total maximum I_{CC} current. As was discussed previously, the outputs are automatically placed in the high impedance state when the PD/PGM pin is raised to V_{IH}. I_{CC2} is the maximum power supply current required by a 2716 in read mode, and reaches this maximum of 500mW (30µW/bit) at maximum temperature.

Table II. 2716 D.C. and Operating Characteristics.

$T_A = 0°C$ to $70°C$, $V_{CC}{}^{[1,2]} = +5V \pm 5\%$, $V_{PP}{}^{[2]} = V_{CC} \pm 0.6V{}^{[3]}$

Symbol	Parameter	Limits Min.	Limits Typ.[4]	Limits Max.	Unit	Conditions
I_{LI}	Input Load Current			10	µA	$V_{IN} = 5.25V$
I_{LO}	Output Leakage Current			10	µA	$V_{OUT} = 5.25V$
$I_{PP1}{}^{[2]}$	V_{PP} Current			5	mA	$V_{PP} = 5.85V$
$I_{CC1}{}^{[2]}$	V_{CC} Current (Standby)		10	25	mA	PD/PGM = V_{IH}, $\overline{CS} = V_{IL}$
$I_{CC2}{}^{[2]}$	V_{CC} Current (Active)		57	100	mA	$\overline{CS}$ = PD/PGM = V_{IL}
V_{IL}	Input Low Voltage	-0.1		0.8	V	
V_{IH}	Input High Voltage	2.2		$V_{CC}+1$	V	
V_{OL}	Output Low Voltage			0.45	V	I_{OL} = 2.1 mA
V_{OH}	Output High Voltage	2.4			V	I_{OH} = –400 µA

NOTES:
1. V_{CC} must be applied simultaneously or before V_{PP} and removed simultaneously or after V_{PP}.
2. V_{PP} may be connected directly to V_{CC} except during programming. The supply current would then be the sum of I_{CC} and I_{PP1}.
3. The tolerance of 0.6V allows the use of a driver circuit for switching the V_{PP} supply pin from V_{CC} in read to 25V for programming.
4. Typical values are for $T_A = 25°C$ and nominal supply voltages.
5. This parameter is only sampled and is not 100% tested.
6. t_{ACC2} is referenced to PD/PGM or the addresses, whichever occurs last.

All inputs are TTL compatible, requiring a VIL between –.01 and 0.8V and a V_{IH} of 2.2V minimum. Care should be exercised in selecting address buffers to ensure that the minimum V_{IH} level is met by use of appropriate TTL circuit elements or pull up resistors to V_{CC}.

The outputs are also TTL compatible, producing a V_{OL} of 0.45V maximum at 2.1mA and a V_{OH} of 2.4V with -400mA capability.

A.C. Characteristics

Figure 7, the read mode timing indicates the maximum or minimum timing for the various timing parameters. Particular attention should be paid to tDF, chip deselect to output float time. This parameter indicates that the output buffers of the 2716 are not guaranteed to reach the high impedance state until 100ns after $\overline{CS}$ reaches V_{IH}. If another device takes control of the output node before the first device output is in the high impedance state, excessive I_{CC} current will be drawn. See the Applications Section for further discussion.

Power Down Mode

The 2716 is the first MOS EPROM to have a completely static power down mode. This mode is activated by raising the PD/PGM input to a TTL high level, with $V_{PP} = 5V$.

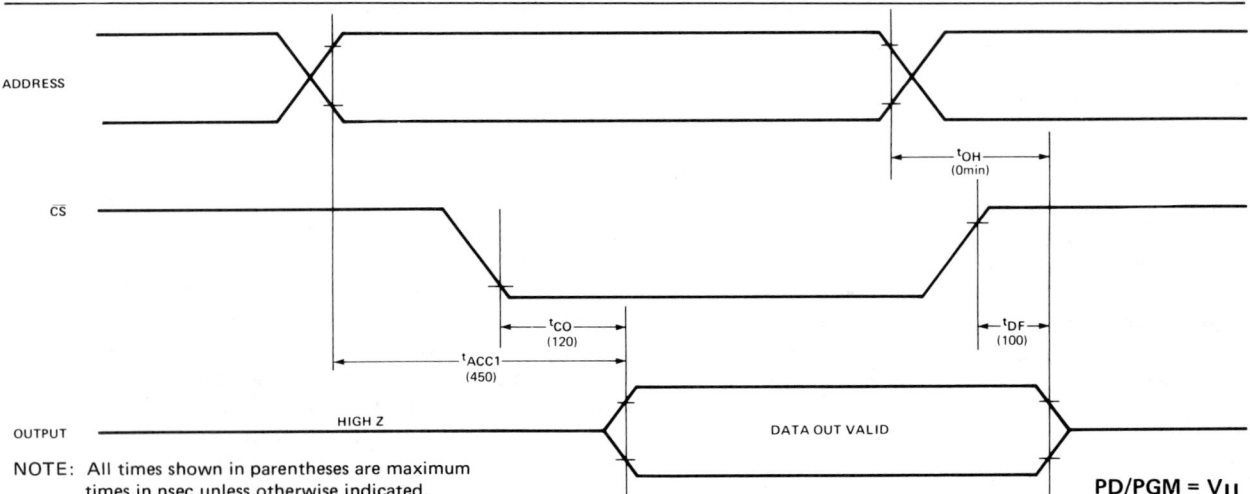

Figure 7. 2716 Read Waveforms.

DESIGNING WITH 2716 EPROM

The power is reduced by 75% (from 500mW to 125mW) during the time PD/PGM is high.

When the PD/PGM pin is lowered to a TTL low level, the access time (t_{ACC2}) of 450ns is met as shown in Figure 8. Of course, t_{ACC2} is referenced to either the addresses becoming stable or to the rising edge of PD/PGM, whichever occurs last. Table III summarizes the A.C. Characteristics for both normal and power down read cycles.

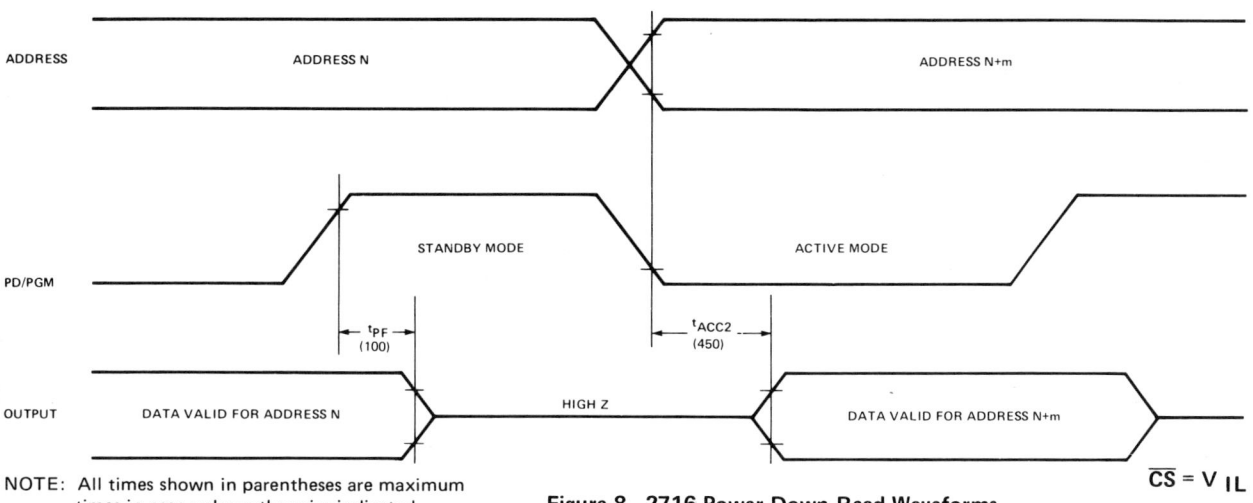

NOTE: All times shown in parentheses are maximum times in nsec unless otherwise indicated.

Figure 8. 2716 Power Down Read Waveforms.

Table III. 2716 A.C. Characteristics.

$T_A = 0°C$ to $70°C$, $V_{CC}[1] = +5V \pm 5\%$, $V_{PP}[2] = V_{CC} \pm 0.6V[3]$

Symbol	Parameter	Limits			Unit	Test Conditions
		Min.	Typ.[4]	Max.		
t_{ACC1}	Address to Output Delay		250	450	ns	PD/PGM = $\overline{CS}$ = V_{IL}
t_{ACC2}	PD/PGM to Output Delay		280	450	ns	$\overline{CS}$ = V_{IL}
t_{CO}	Chip Select to Output Delay			120	ns	PD/PGM = V_{IL}
t_{PF}	PD/PGM to Output Float	0		100	ns	$\overline{CS}$ = V_{IL}
t_{DF}	Chip Deselect to Output Float	0		100	ns	PD/PGM = V_{IL}
t_{OH}	Address to Output Hold	0			ns	PD/PGM = $\overline{CS}$ = V_{IL}

PROGRAM MODE
D.C. Characteristics

The 2716 requires a single TTL level pulse to program each address with the V_{PP} supply set to 25V. Addresses can be programmed in any sequence. The V_{PP} supply can be left at +25V continuously while programming; it can also be left at +25V for program verify cycles, but must be returned to the +5 volt level for normal read cycles to reduce power dissipation. A maximum of 30mA will be drawn from the V_{PP} supply when the PD/PGM pulse is high and $\overline{CS}$ is high; during read operations the I_{PP1} specification of 5mA applies.

The address and data inputs are TTL compatible during programming, with the same requirements of V_{IL} and V_{IH} as for the Read Mode. The D.C. Characteristics for programming are shown in Table IV. To enable the device for programming, the $\overline{CS}$ pin is taken to V_{IH} and the correct address and data inputs provided. After the appropriate set up times, (see Figure 9) a single pulse from V_{IL} to V_{IH} on the PD/PGM input for 50ms programs the desired address.

During program operation, the outputs become the data inputs and should be treated as a three state bus. The same leakage, as well as V_{IL} and V_{IH} specifications apply to the outputs as for the inputs during normal read operations.

The program pulse, which is a TTL pulse of 50ms duration, is applied to the PD/PGM input. During the time that this pulse is high, a maximum of 30mA (I_{CC2}) will be required from the V_{PP} power supply. V_{PP} can be left high (at +25V) to verify the programmed data, however, it must be returned to the 5V level to reduce power dissipation. Also, in order to reduce power dissipation, the PD/PGM pulse must not be left high longer than 55ms when

Table IV. 2716 D.C. Programming Characteristics.

$T_A = 25°C \pm 5°C$, $V_{CC}{}^{[2]} = 5V \pm 5\%$, $V_{PP}{}^{[2,3]} = 25V \pm 1V$

Symbol	Parameter	Min.	Typ.	Max.	Units	Test Conditions
I_{LI}	Input Current (for Any Input)			10	µA	V_{IN} = 5.25V/0.45
I_{PP1}	V_{PP} Supply Current			5	mA	PD/PGM = V_{IL}
I_{PP2}	V_{PP} Supply Current During Programming Pulse			30	mA	PD/PGM = V_{IH}
I_{CC}	V_{CC} Supply Current			100	mA	
V_{IL}	Input Low Level	−0.1		0.8	V	
V_{IH}	Input High Level	2.2		V_{CC}+1	V	

NOTES:
1. Intel's standard product warranty applies only to devices programmed to specifications described herein.
2. V_{CC} must be applied simultaneously or before V_{PP} and removed simultaneously or after V_{PP}. The 2716 must not be inserted into or removed from a board with V_{PP} at 25 ±1V to prevent damage to the device.
3. The maximum allowable voltage which may be applied to the V_{PP} pin during programming is +26V. Care must be taken when switching the V_{PP} supply to prevent overshoot exceeding this 26V maximum specification.

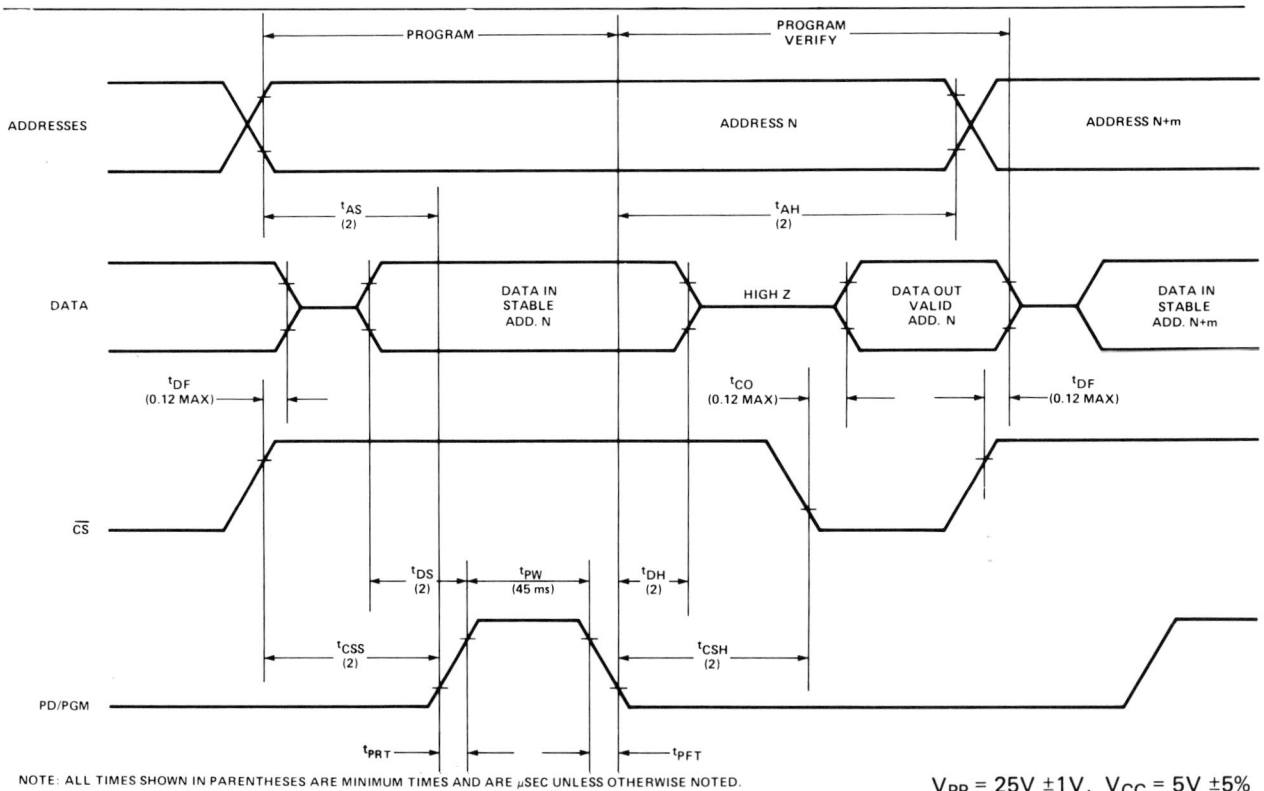

Figure 9. 2716 Programming Waveforms.

the V_{PP} supply is at +25V; it can be left high only with V_{PP} at +5V, which deselects the output and places the device in the low power standby mode.

The tolerance on the V_{PP} supply is 25V ±1V. When switching the V_{PP} supply from +5V to +25V, particular care should be taken to ensure that there is no overshoot above 26V; exceeding this can be destructive to the programming circuits on the device. It is also not permitted to "hot socket" the device in a programmer (with respect to the V_{PP} supply) as the resulting transients could cause the V_{PP} supply to exceed the maximum of 26V.

A.C. Characteristics

Figure 9 indicates the program mode timing, while Table V tabulates the various programming A.C. parameters.

To program a 2716, the address, data and $\overline{CS}$ signals must all be stable 2µs before the PD/PGM pin is pulsed high for 50ms ±5ms. This is shown in

DESIGNING WITH 2716 EPROM

Figure 9 as t_{AS}, t_{DS} and t_{CS}. After the falling edge of the program pulse, these same signals must be held stable for 2µs (t_{AH}, t_{DH} and t_{CSH}); then the next address and data can be presented, sequentially or not according to the ease of system implementation, and the next address programmed. In this manner it is possible to program an entire 2716 in approximately 100 seconds, while a single address requires only 50 ms to program.

PROGRAMMING

A number of programmers are commercially available that will properly program the 2716. (see Table VI) Intel maintains a service whereby commercial programmer manufacturers obtain design approval prior to marketing their device, in order to assure compatibility with Intel specifications. This approval should be verified with the particular manufacturer prior to purchase.

For those users who want to build their own programmer, a design is included at the end of this section.

Figure 10 illustrates a typical 2716 programmer block diagram. The address & data inputs can come from a system bus, or from toggle or thumbwheel switches. If system inputs are used, the Address Input Buffer should be a latch to allow the system bus to be free during the 50ms program time per address. The Data Input/Output Buffer should be of the bi-directional type to allow both programming and data verification.

The start control activates the timing chain to generate the required address and data setup and hold times, as well as the program pulse.

The program timer latches the address and data inputs stable and raises CS to V_{IH}, while the address and data setup timer delays the start of the program pulse for at least 2µs, which is the minimum re-

Table V. 2716 A.C. Programming Characteristics.

$T_A = 25°C \pm 5°C$, $V_{CC}^{[2]} = 5V \pm 5\%$, $V_{PP}^{[2,3]} = 25V \pm 1V$

Symbol	Parameter	Min.	Typ.	Max.	Units	
t_{AS}	Address Setup Time	2			µs	
t_{CSS}	$\overline{CS}$ Setup Time	2			µs	
t_{DS}	Data Setup Time	2			µs	
t_{AH}	Address Hold Time	2			µs	
t_{CSH}	$\overline{CS}$ Hold Time	2			µs	
t_{DH}	Data Hold Time	2			µs	
t_{DF}	Chip Deselect to Output Float Delay	0		120	ns	PD/PGM = V_{IL}
t_{CO}	Chip Select to Output Delay			120	ns	PD/PGM = V_{IL}
t_{PW}	Program Pulse Width	45	50	55	ms	
t_{PRT}	Program Pulse Rise Time	5			ns	
t_{PFT}	Program Pulse Fall Time	5			ns	

NOTES: 1. Intel's standard product warranty applies only to devices programmed to specifications described herein.
2. V_{CC} must be applied simultaneously or before Vpp and removed simultaneously or after Vpp. The 2716 must not be inserted into or removed from a board with Vpp at 25 ±1V to prevent damage to the device.
3. The maximum allowable voltage which may be applied to the Vpp pin during programming is +26V. Care must be taken when switching the Vpp supply to prevent overshoot exceeding this 26V maximum specification.

Table VI. Approved Programmers.

	1602A/1702A Family	2708 Family	2716 Family	3601 Family	3602/3622 Family	3604/3624 Family
Intel MDS-UPP-100 Santa Clara, Calif.	X	X	X	X	X	X
Data I/O Model V Issaquah, Wash.	X	X	X	X	X	X
Prolog Series 90 Monterey, Calif.	X	Note 1	Note 1	X	X	X
Spectrum Dynamics Series 550 Burlington, Mass.	X	Note 1	Note 1	Note 1	Note 1	Note 1

Note 1. This programming card is pending Intel approval.

quired address and data setup time (t_{AS} and t_{DS}). The program pulse timer is activated by the falling edge of the address and data setup timer, and generates the required 50ms program pulse. The falling edge of the program pulse activates the address and data hold timer, (2µs minimum) and the falling edge of the data hold timer resets the program times, releasing the latch on the address and data in buffers, freeing the system for either a verify cycle or a program cycle on another address.

On board programming is also very easily implemented with the 2716, as the PD/PGM pin functions as a program inhibit, i.e., if a given device has $\overline{CS}$ high, V_{PP} = 25V, and PD/PGM low, it will not be programmed. A system showing how on-board programming could be implemented is shown in Figure 11. In the figure, device #4 will have address IFF_H programmed with F4H, while the contents of address IFF in devices #1, #2 and #3 will be unaffected.

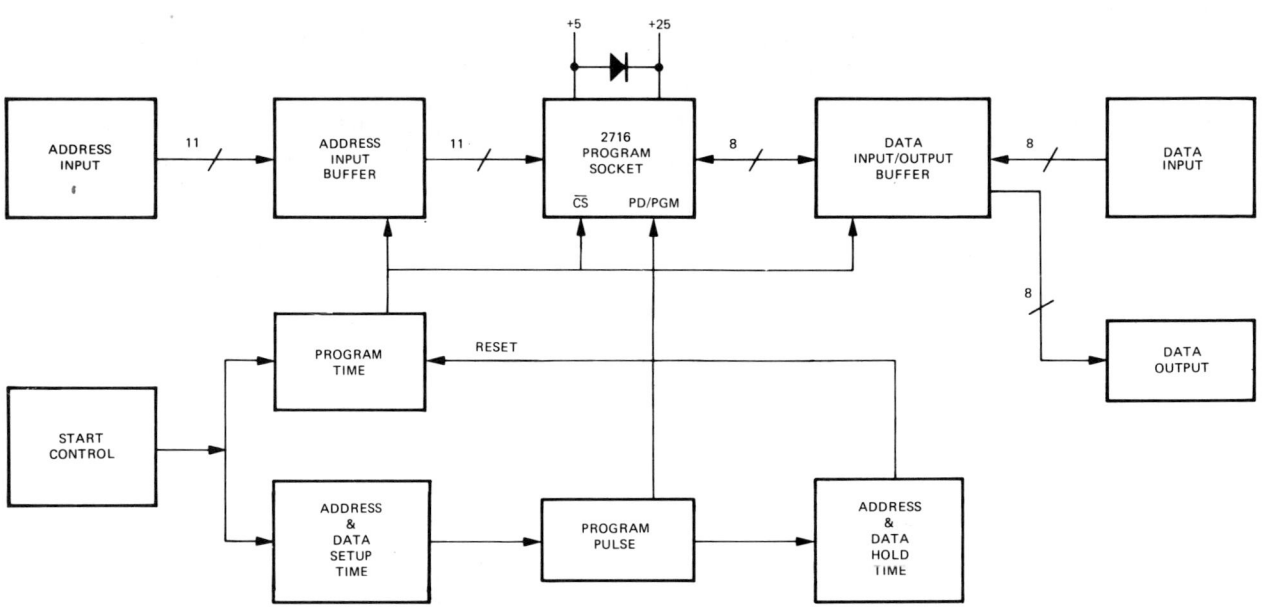

Figure 10. 2716 Programmer Block Diagram.

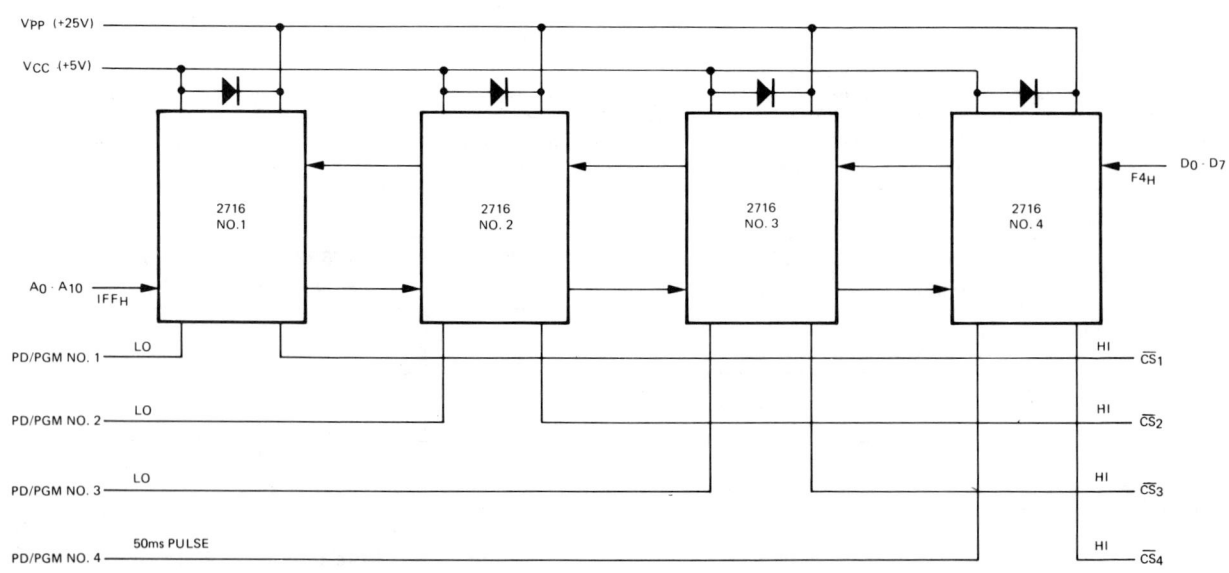

Figure 11. 2716 On Board Programming.

DESIGNING WITH 2716 EPROM

2716 Mini Programmer

Figure 12 presents the schematic for a 2716 programmer which is based on the block diagram shown in the previous section. This programmer has been design approved by Intel, by the same procedure used for commercial programmer manufacturers. The programmer has several features that make it useful for small development labs.

Manual Programming

Selecting any Hex address with the 3 address input thumb wheel switches and entering it by depressing the load button will cause the selected address to be displayed in Hex. The data is then entered by way of the 2 Hex thumb wheel data switches. When programming the data, the PROGRAM button is depressed, the location indicated by the address display is programmed and the address incremented to the next sequential location. For verification a verify mode is included that will automatically slowly step through all addresses, allowing for manual, visual verification of the programmed data. The rate at which it sequences through the addresses is adjustable, and can be started at any location by way of the ADDRESS INPUT and LOAD ADDRESS switches.

Duplicate Mode

By selecting the duplicate mode, a 2716 placed in the READ ONLY socket will be duplicate and automatically compared with a 2716 placed in the PROGRAM socket. After verification a green "PASS" or a red "FAIL" LED will indicate the completion of the program cycle. A blank check is not performed.

The design described here does not include a power supply design—the user must provide appropriate +5 volt and +25 volt power supplies. Current requirements, as measured on the prototype board, are about 1 A at +5V and 60 mA at 25 V.

The design also includes a transistor switch to prevent hot socketing of the 2716. As was mentioned in the programming section, it is not permitted to install a 2716 in a socket with the +25 volts present: it must be switched on after the 2716 is in the socket and +5 volts is applied.

ERASING

Erasure begins to appear when the 2716 is exposed to light with wavelengths shorter than approximately 4000 Angstroms (Å). It should be noted that sunlight and certain types of fluorescent lamps have wavelengths in the 3000-4000Å range. Constant exposure to room level fluorescent lighting could erase the typical 2716 in approximately 3.5 years while it would take approximately 1 month to cause erasure when exposed to direct sunlight. If the 2716 is to be exposed to these types of lighting conditions for extended periods of time, opaque labels are available from Intel which should be placed over the 2716 window to prevent unintentional erasure.

The recommended erasure procedure for the 2716 is exposure to shortwave ultraviolet light which has a wavelength of 2537 Angstroms (Å). The integrated dose (i.e., UV intensity x exposure time) for erasure should be a minimum of 15 W-sec/cm^2. The erasure time with this dosage is approximately 20 minutes using an ultraviolet lamp with a 12000 μW/cm^2 power rating. The 2716 should be placed within one inch from the lamp tubes during exposure. Some lamps have a filter on their tubes and this filter should be removed before erasure.

The 2716 should not be under bias during erasure as current paths exist that will effectively cancel the energy being provided by the UV light.

UV Sources

There are several models of UV lamps that can be used to erase 2716's (see Table VII). The model numbers in the table refer to lamps manufactured by Ultra Violet Products of San Gabriel, Calif. In addition there are several other manufacturers, including Data I/O, PRO-LOG, Prometrics, and Turner Designs. The individual manufacturers should be consulted for detailed product descriptions.

Table VII. 2716 Erase Time.

MODEL	POWER RATING	REQUIRED TIME FOR INDICATED DOSAGE 15 W-sec 2716
R-52	13000μW/cm^2	19.2 min
S-52	12000μW/cm^2	20.7 min
S-68	12000μW/cm^2	20.7 min
UVS-54	5700μW/cm^2	43.8 min
UVS-11	5500μW/cm^2	45.6 min

According to the manufacturers, the output of the UV lamp bulb decreases with age. The output of the lamp should be verified periodically to ensure that adequate intensities are maintained. If this is not done, bits may be partially erased which will interfere with later programming and/or operation at high temperature.

For lamps other than those listed, the erase time can be determined by using a UV intensity meter, such as the Ultra Violet Products model J-225. When a meter is used, the intensity should be measured at the same position (distance from the lamp) as the EPROMs to be erased. This will require careful positioning to insure that the sensor will receive the same amount of UV light that the window of the EPROM will receive.

The sensors used with most UV intensity meters showed reduced output with constant exposure to UV light. Therefore they should not be permanently placed inside the erasure enclosure; they should only be used for periodic measurements.

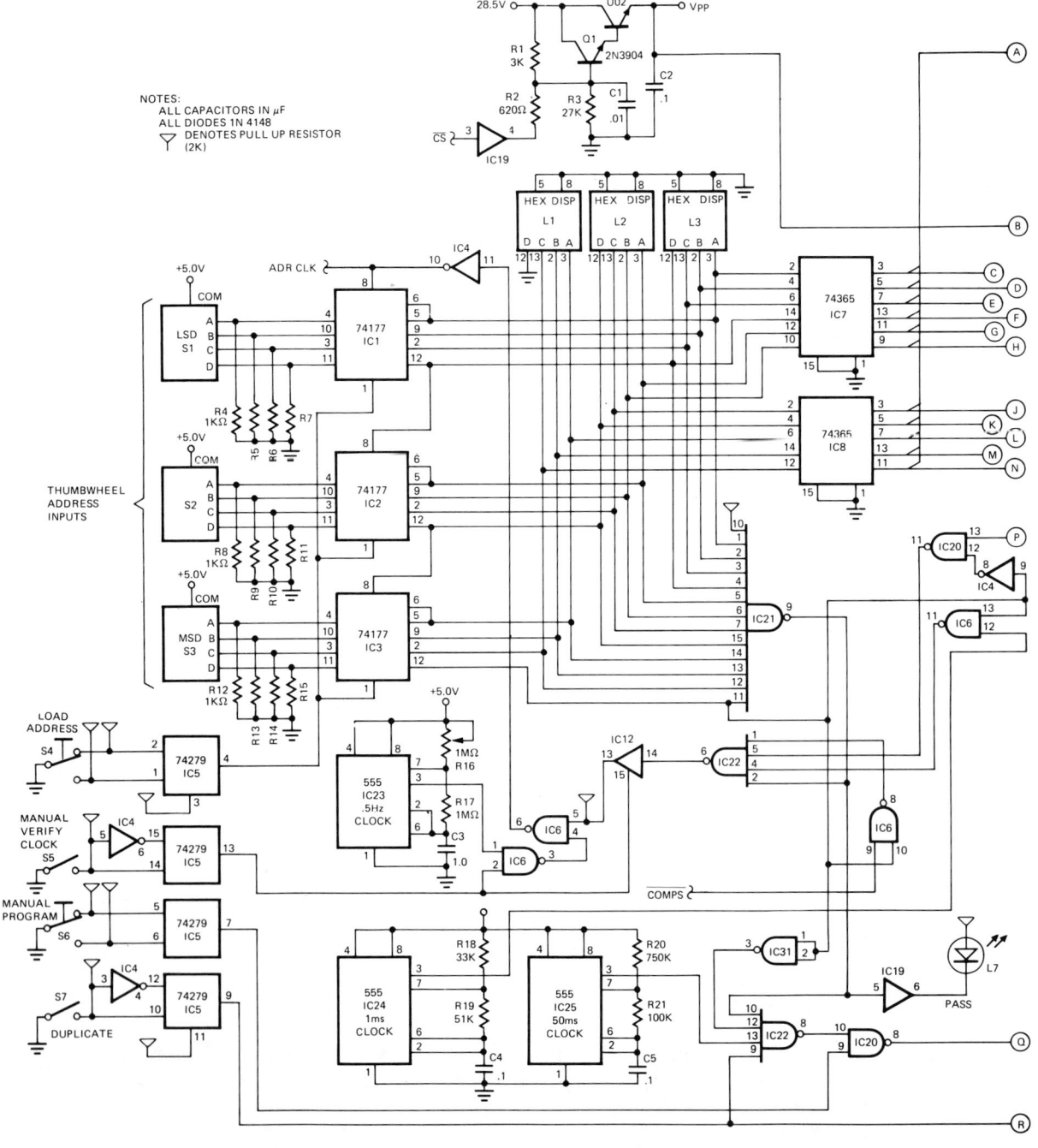

Figure 12. 2716 Mini Programmer Schematic.

DESIGNING WITH 2716 EPROM

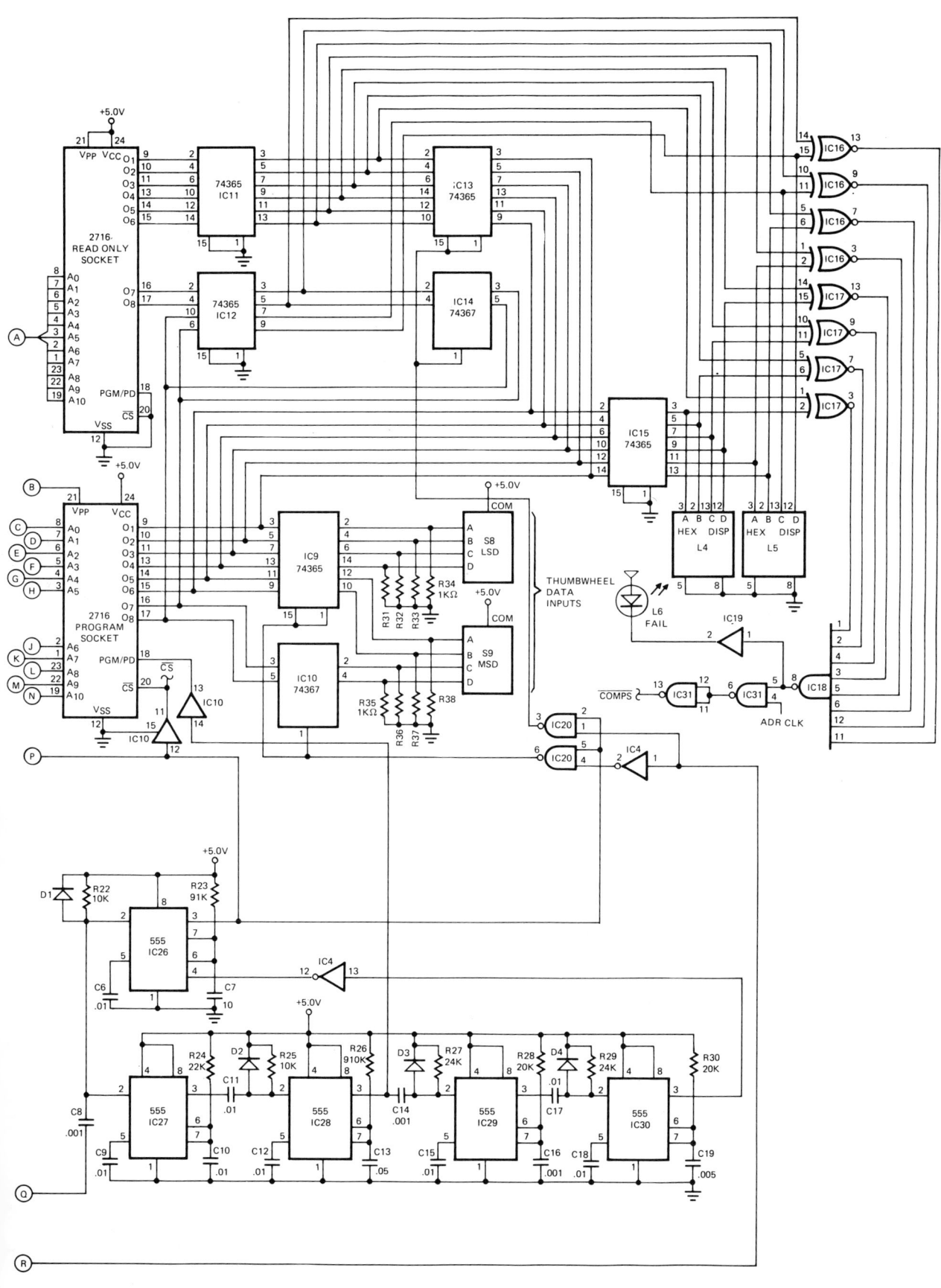

Under Programming And Under Erasing

It is possible to "under program" the 2716 the same as it is with the 2708, such that the cell characteristic crosses the sense threshold. The result is that the cell apparently drops or picks up bits. As can be seen in Figure 13, the threshold characteristic has been shifted such that small changes in voltage or temperature will cause a "1" or a "0" to be sensed. This is always the result of insufficient erasing or programming. For programming to cause this problem, the device has only been partially programmed, and the characteristic curve has been shifted to the sense threshold point and the device will again seem to either pick up or drop bits. For erasure to cause the problem, the device has only been partially erased, such that the characteristic curve has only been shifted (right to left in the figure) to the threshold.

The cure in either case is to: 1) adequately erase by providing the required 15 W-sec/cm^2 of UV light at a frequency of 2537Å or; 2) program in accordance with the specifications.

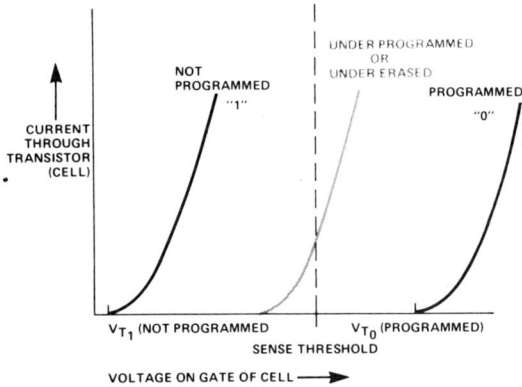

Figure 13. Effect of Under Programming or Under Erasure

2716 Mini Programmer

The Mini Programmer shown on the previous pages has been design approved by Intel and can be built as shown, or portions of the circuit can be modified to fit a specific user circuit application.

Circuit Description

The Mini Programmer has several modes of operation which are described below.

Manual Program — Controlled by pushbutton switch S6, this mode allows the user to program the address displayed by the address input displays (L1-L3) with the data that is entered in the data input thumbwheels (S8 & S9). The desired address to be programmed is entered by way of the LOAD ADDRESS switch, S4. This transfers the contents of the address input thumbwheel switches (S1-S3) to the address input buffers and the address display LEDs, L1-L3.

The desired data is entered in the form of two hexadecimal characters by way of the data input thumbwheel switches, S8 & S9. Prior to programming, the data output display will read FF_H, indicating that the addressed location contains all highs, i.e., is erased.

After the displayed address is programmed, the output display will momentarily display the contents of the programmed address, and then increment the address by 1 count, thus preparing the next sequential address to be programmed. Should other than the next sequential be desired, it is only necessary to dial in the new address and depress the LOAD ADDRESS pushbutton.

Manual Verify — In order to assure the user that the correct data pattern has been entered in an entire program, a manual verify function has been included. In this mode, the address counter will slowly cycle through addresses starting with the address that was loaded by the LOAD ADDRESS switch. The rate at which the counter will cycle is controlled by R16, and should be set for convenient visual recognition of the programmed data.

Duplicate Mode — Duplicate mode allows the contents of another 2716 to be programmed into an erased device that is inserted in the program socket. Each location is programmed and verified, and the next sequential location is programmed. Upon completion, PASS-FAIL indication is provided by way of LEDs L6 and L7.

Transistors Q1 and Q2 provide for switching V_{PP} between 26V and 5V, while assuring that proper sequence and overshoot control is maintained.

DESIGNING WITH 2716 EPROM

Table VIII. 2716 Mini Programmer Parts List.

IC1-3	74177	4-Bit Counter
IC4	7404	Hex Driver
IC5	74279	Quad Set/Reset Latch
IC6, 20, 31	7400	Quad NAND
IC7-15	74367	Hex Tristate Driver
IC16, 17	74135	Quad Exclusive OR/NOR Gates
IC18	7430	8-Input NAND
IC19	7407	Open Collector, High Voltage Driver
IC21	74133	13-Input NAND
IC22	7420	Dual 4-Input NAND
IC23-30	NE555	Timer
Q1	MPS U02	Transistor
Q2	2N3904	Transistor
R1	3KΩ	¼W Resistor
R2	820Ω	¼W Resistor
R3	27KΩ	¼W Resistor
R4-15, 31-38	1KΩ	¼W Resistor
R16	1MΩ	Potentiometer (VERIFY Clock Rate)
R4-15	1KΩ	¼W Resistor
R31-38	1KΩ	¼W Resistor
R16	1MΩ	Potentiometer
R17	1MΩ	¼W Resistor
R18	33KΩ	¼W Resistor
R19	51KΩ	¼W Resistor
R20	750KΩ	¼W Resistor
R21	100KΩ	¼W Resistor
R22	10KΩ	¼W Resistor
R23	91KΩ	¼W Resistor
R24	22KΩ	¼W Resistor
R25	10KΩ	¼W Resistor
R26	910KΩ	¼W Resistor
R27, 29	24KΩ	¼W Resistor
R28, 30	20KΩ	¼W Resistor
C1, 6, 9-12, 15, 17, 18	0.01µF	Capacitor 20 wvdc (min)
C2, 4, 5	0.1µF	Capacitor 20 wvdc (min)
C3	1.0µF	Capacitor 20 wvdc (min)
C7	10µF	Capacitor 20 wvdc (min)
C8, 14, 16	0.001µF	Capacitor 20 wvdc (min)
C13	0.05µF	Capacitor 20 wvdc (min)
C19	0.005µF	Capacitor 20 wvdc (min)
S1-S3		(LSD-MSD): Address Input Switches (Cherry T-10 Thumbwheel)
S4		Address Load (Pushbutton)
S5		1Hz Verify Clock SPST Switch
S6		Program Button (Pushbutton)
S7		Duplicate Mode SPST Switch
S8, S9		(LSD-MSD): Data Input (Cherry T-10 Thumbwheel)
PROM Sockets		Textool 24-Pin ZIP DIP
L1-L5		TIL311 Hexadecimal Display
L6		MV5025 (Red LED)
L7		MV5253 (Green LED)

1702A

2K (256 x 8) UV ERASABLE PROM

1702A-2	0.65 us Max.
1702A	1.0 us Max.
1702A-6	1.5 us Max.

- **Fast Access Time: Max. 650 ns (1702A–2)**
- **Fast Programming: 2 Minutes for all 2048 Bits**
- **All 2048 Bits Guaranteed* Programmable: 100% Factory Tested**
- **Static MOS: No Clocks Required**
- **Inputs and Outputs DTL and TTL Compatible**
- **Three-State Output: OR-tie Capability**

The 1702A is a 256 word by 8-bit electrically programmable ROM ideally suited for uses where fast turn-around and pattern experimentation are important. The 1702A undergoes complete programming and functional testing prior to shipment, thus insuring 100% programmability.

Initially all 2048 bits of the 1702A are in the "0" state (output low). Information is introduced by selectively programming "1"'s (output high) in the proper bit location. The 1702A is packaged in a 24 pin dual in-line package with a transparent lid. The transparent lid allows the user to expose the 1702A to ultraviolet light to erase the bit pattern. A new pattern can then be written into the device.

The circuitry of the 1702A is completely static. No clocks are required. Access times from 650ns to 1.5μs are available. A 1702AL family is available (see 1702AL data sheets for specifications) for those systems requiring lower power dissipation than the 1702A.

A pin-for-pin metal mask programmed ROM, the Intel 1302, is also available for large volume production runs of systems initially using the 1702A.

The 1702A is fabricated with silicon gate technology. This low threshold technology allows the design and production of higher performance MOS circuits and provides a higher functional density on a monolithic chip than conventional MOS technologies.

*Intel's liability shall be limited to replacing any unit which fails to program as desired.

PIN CONFIGURATION

```
         A₂  — 1    24 — V_DD
         A₁  — 2    23 — V_CC
         A₀  — 3    22 — V_CC
*DATA OUT 1 — 4 (LSB) 21 — A₃
*DATA OUT 2 — 5    20 — A₄
*DATA OUT 3 — 6    19 — A₅
*DATA OUT 4 — 7    18 — A₆
*DATA OUT 5 — 8    17 — A₇
*DATA OUT 6 — 9    16 — V_GG
*DATA OUT 7 — 10   15 — V_BB
*DATA OUT 8 — 11 (MSB) 14 — CS̄
         V_CC — 12   13 — PROGRAM
```

*THIS PIN IS THE DATA INPUT LEAD DURING PROGRAMMING

PIN NAMES

$A_0 - A_7$	Address Inputs
$\overline{CS}$	Chip Select Input
$D_{OUT1} - D_{OUT8}$	Data Outputs

BLOCK DIAGRAM

DATA OUT 1 ... DATA OUT 8 → OUTPUT BUFFERS ← $\overline{CS}$

OUTPUT BUFFERS ← 2048 BIT ROM MATRIX (256 x 8) ← PROGRAM

ROM MATRIX ← DECODER ← INPUT DRIVERS ← $A_0, A_1, ..., A_7$

NOTE: In the read mode a logic 1 at the address inputs and data outputs is a high and logic 0 is a low.

U.S. Patent No. 3660819

1702A

PIN CONNECTIONS

The external lead connections to the 1702A differ, depending on whether the device is being programmed or used in read mode (see following table). In the programming mode, the data inputs 1-8 are pins 4-11 respectively. *The programming voltages and timing are shown in the ROM and PROM Programming instructions section, page 3-55.*

MODE \ PIN	12 (V_{CC})	13 (Program)	14 ($\overline{CS}$)	15 (V_{BB})	16 (V_{GG})	22 (V_{CC})	23 (V_{CC})	24 (V_{DD})
Read	V_{CC}	V_{CC}	GND	V_{CC}	V_{GG}	V_{CC}	V_{CC}	V_{DD}
Programming	GND	Program Pulse	GND	V_{BB}	Pulsed V_{GG}	GND	GND	Pulsed V_{DD}

Absolute Maximum Ratings*

Ambient Temperature Under Bias .... $-10°C$ to $+80°C$
Storage Temperature .... $-65°C$ to $+125°C$
Soldering Temperature of Leads (10 sec) .... $+300°C$
Power Dissipation .... 2 Watts
Read Operation: Input Voltages and Supply
 Voltages with respect to V_{CC} .... $+0.5V$ to $-20V$
Program Operation: Input Voltages and Supply
 Voltages with respect to V_{CC} .... $-48V$

*COMMENT

Stresses above those listed under "Absolute Maximum Ratings" may cause permanent damage to the device. This is a stress rating only and functional operation of the device at these or at any other condition above those indicated in the operational sections of this specification is not implied. Exposure to Absolute Maximum Rating conditions for extended periods may affect device reliability.

D.C. and Operating Characteristics
READ OPERATION

$T_A = 0°C$ to $70°C$, $V_{CC} = +5V \pm 5\%$, $V_{DD} = -9V \pm 5\%$, $V_{GG} = -9V \pm 5\%$, unless otherwise noted.

Symbol	Test	1702A, 1702A-6 Limits			1702A-2 Limits			Unit	Conditions
		Min.	Typ.[1]	Max.	Min.	Typ.[1]	Max.		
I_{LI}	Address and Chip Select Input Load Current			1			1	μA	$V_{IN} = 0.0V$
I_{LO}	Output Leakage Current			1			1	μA	$V_{OUT} = 0.0V$, $\overline{CS} = V_{IH2}$
I_{DD1}[1]	Power Supply Current		35	50		40	60	mA	$\overline{CS} = V_{IH2}$, $I_{OL} = 0.0mA$, $T_A = 25°C$, Continuous
I_{DD2}	Power Supply Current		32	46		37	55	mA	$\overline{CS} = 0.0V$, $I_{OL} = 0.0mA$, $T_A = 25°C$, Continuous
I_{DD3}	Power Supply Current		38	60		43	65	mA	$\overline{CS} = V_{IH2}$, $I_{OL} = 0.0mA$, $T_A = 0°C$, Continuous
I_{CF1}	Output Clamp Current		8	14		7	13	mA	$V_{OUT} = -1.0V$, $T_A = 0°C$, Continuous
I_{CF2}	Output Clamp Current		7	13		6	12	mA	$V_{OUT} = -1.0V$, $T_A = 25°C$, Continuous
I_{GG}	Gate Supply Current			1			1	μA	
V_{IL1}	Input Low Voltage for TTL Interface	-1		0.65	-1		0.65	V	
V_{IL2}	Input Low Voltage for MOS Interface	V_{DD}		$V_{CC}-6$	V_{DD}		$V_{CC}-6$	V	
V_{IH1}	Addr. Input High Voltage	$V_{CC}-2$		$V_{CC}+0.3$	$V_{CC}-2$		$V_{CC}+0.3$	V	
V_{IH2}	Chip Sel. Input High Volt.	$V_{CC}-2$		$V_{CC}+0.3$	$V_{CC}-1.5$		$V_{CC}+0.3$	V	
I_{OL}	Output Sink Current	1.6	4		1.6	4		mA	$V_{OUT} = 0.45V$
I_{OH}	Output Source Current	-2.0			-2.0			mA	$V_{OUT} = 0.0V$
V_{OL}	Output Low Voltage		-3	0.45		-3	0.45	V	$I_{OL} = 1.6mA$
V_{OH}	Output High Voltage	3.5	4.5		3.5	4.5		V	$I_{OH} = -200μA$

Note 1: Typical values are at nominal voltages and $T_A = 25°C$.

A.C. Characteristics

$T_A = 0°C$ to $+70°C$, $V_{CC} = +5V \pm 5\%$, $V_{DD} = -9V \pm 5\%$, $V_{GG} = -9V \pm 5\%$ unless otherwise noted

Symbol	Test	1702A Limits Min.	1702A Limits Max.	1702A-2 Limits Min.	1702A-2 Limits Max.	1702A-6 Limits Min.	1702A-6 Limits Max.	Unit
Freq.	Repetition Rate		1		1.6		0.66	MHz
t_{OH}	Previous Read Data Valid	0.1		0.1		0.1		μs
t_{ACC}	Address to Output Delay		1		0.65		1.5	μs
t_{CS}	Chip Select Delay		0.1		0.3		0.6	μs
t_{CO}	Output Delay From $\overline{CS}$		0.9		0.35		0.9	μs
t_{OD}	Output Deselect		0.3		0.3		0.3	μs

Capacitance* $T_A = 25°C$

SYMBOL	TEST	TYPICAL	MAXIMUM	UNIT	CONDITIONS	
C_{IN}	Input Capacitance	8	15	pF	$V_{IN} = V_{CC}$, $\overline{CS} = V_{CC}$	All unused pins are at A.C. ground
C_{OUT}	Output Capacitance	10	15	pF	$V_{OUT} = V_{CC}$, $V_{GG} = V_{CC}$	

*This parameter is periodically sampled and is not 100% tested.

Switching Characteristics

Conditions of Test:
Input pulse amplitudes: 0 to 4V; t_R, $t_F \leq 50$ ns
Output load is 1 TTL gate; measurements made at output of TTL gate ($t_{PD} \leq 15$ ns), $C_L = 15$ pF

A) READ OPERATION

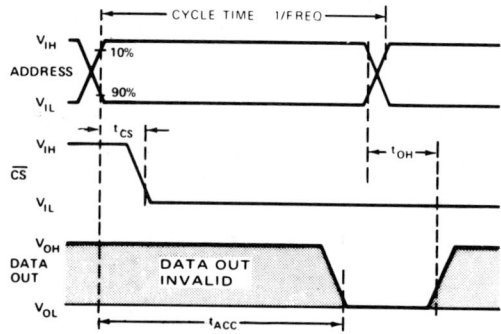

B) DESELECTION OF DATA OUTPUT IN OR-TIE OPERATION

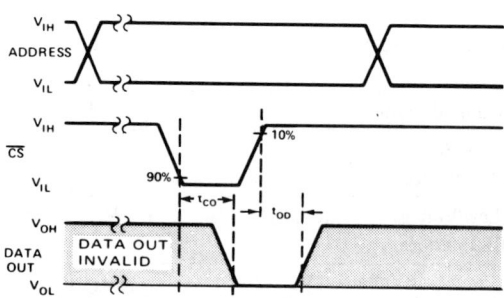

1702A

Typical Characteristics

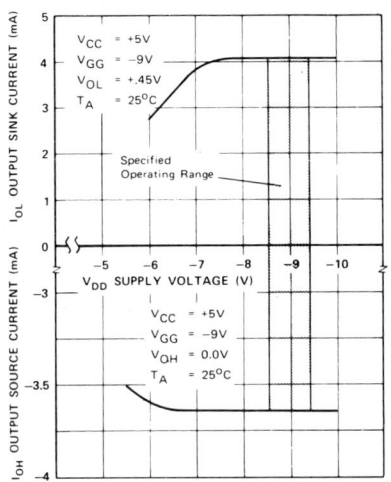
OUTPUT CURRENT VS. V_{DD} SUPPLY VOLTAGE

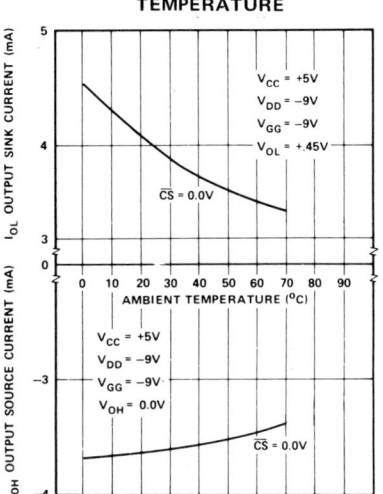
OUTPUT CURRENT VS. TEMPERATURE

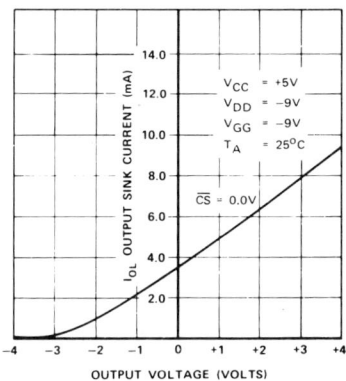
OUTPUT SINK CURRENT VS. OUTPUT VOLTAGE

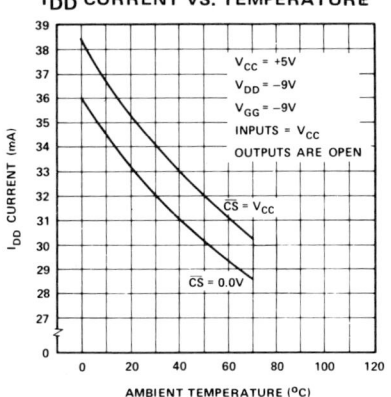
I_{DD} CURRENT VS. TEMPERATURE

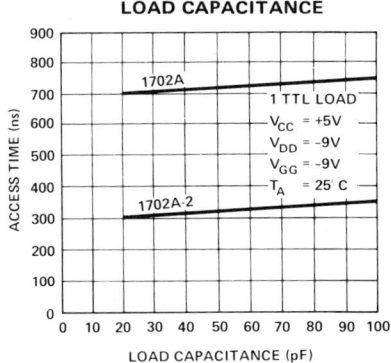
ACCESS TIME VS. LOAD CAPACITANCE

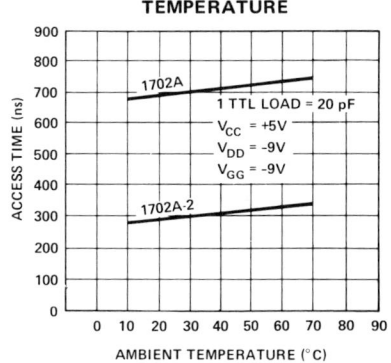
ACCESS TIME VS. TEMPERATURE

1702 AL
2K (256 × 8) LOW POWER UV EPROM

Part No.	MAXIMUM ACCESS (µs)	t_{DVGG} (µs)
1702AL	1.0	0.4
1702AL-2	0.65	0.3

- **Clocked V_{GG} Mode for Low Power Dissipation**
- **Fast Programming: 2 Minutes for all 2048 Bits**
- **All 2048 Bits Guaranteed* Programmable: 100% Factory Tested**
- **Inputs and Outputs DTL and TTL Compatible**
- **Three-State Output: OR-tie Capability**

The 1702AL is a 256 word by 8 bit electrically programmable ROM and is the same chip as the industry standard 1702A. The programming and erasing specifications are identical to the 1702A. The 1702AL operates with the V_{GG} clocked to reduce the power dissipation.

Initially all 2048 bits of the 1702AL are in the "0" state (output low). Information is introduced by selectively programming "1"s (output high) in the proper bit location. The 1702AL is packaged in a 24 pin dual in-line package with a transparent lid. The transparent lid allows the user to expose the 1702AL to ultraviolet light to erase the bit pattern. A new pattern can then be written into the device.

The 1702AL is fabricated with silicon gate technology. This low threshold technology allows the design and production of high performance MOS circuits and provides a higher functional density on a monolithic chip than conventional MOS technologies.

*Intel's liability shall be limited to replacing any unit which fails to program as desired.

PIN CONFIGURATION

```
       A₂  ─┤ 1    24 ├─ V_DD
       A₁  ─┤ 2    23 ├─ V_CC
       A₀  ─┤ 3    22 ├─ V_CC
*DATA OUT 1 ─┤ 4 (LSB) 21 ├─ A₃
*DATA OUT 2 ─┤ 5    20 ├─ A₄
*DATA OUT 3 ─┤ 6    19 ├─ A₅
*DATA OUT 4 ─┤ 7    18 ├─ A₆
*DATA OUT 5 ─┤ 8    17 ├─ A₇
*DATA OUT 6 ─┤ 9    16 ├─ V_GG
*DATA OUT 7 ─┤ 10   15 ├─ V_BB
*DATA OUT 8 ─┤ 11 (MSB) 14 ├─ CS
       V_CC ─┤ 12   13 ├─ PROGRAM
```
*THIS PIN IS THE DATA INPUT LEAD DURING PROGRAMMING

PIN NAMES

A_0–A_7	Address Inputs
$\overline{CS}$	Chip Select Input
D_{OUT1}–D_{OUT8}	Data Outputs

BLOCK DIAGRAM

DATA OUT 1 ... DATA OUT 8
↑
OUTPUT BUFFERS ← $\overline{CS}$
↑
2048 BIT ROM MATRIX (256 × 8) ← PROGRAM
↑
DECODER
↑
INPUT DRIVERS
↑
A_0 A_1 ... A_7

NOTE: In the read mode a logic 1 at the address inputs and data outputs is a high and logic 0 is a low.

U.S. Patent No. 3660819

1702AL

PIN CONNECTIONS

The external lead connections to the 1702AL differ, depending on whether the device is being programmed or used in read mode (see following table). In the programming mode, the data inputs 1-8 are pins 4-11 respectively. *The programming voltages and timing are shown in the ROM and PROM Programming Instructions section, pages 3-55.*

MODE \ PIN	12 (V_{CC})	13 (Program)	14 (CS)	15 (V_{BB})	16 (V_{GG})	22 (V_{CC})	23 (V_{CC})	24 (V_{DD})
Read	V_{CC}	V_{CC}	GND	V_{CC}	Clocked V_{GG}	V_{CC}	V_{CC}	V_{DD}
Programming	GND	Program Pulse	GND	V_{BB}	Pulsed V_{GG}	GND	GND	Pulsed V_{DD}

Absolute Maximum Ratings*

Ambient Temperature Under Bias .... −10°C to +80°C
Storage Temperature .... −65°C to +125°C
Soldering Temperature of Leads (10 sec) .... +300°C
Power Dissipation .... 2 Watts
Read Operation: Input Voltages and Supply
   Voltages with respect to V_{CC} .... +0.5V to −20V
Program Operation: Input Voltages and Supply
   Voltages with respect to V_{CC} .... −48V

*COMMENT

Stresses above those listed under "Absolute Maximum Ratings" may cause permanent damage to the device. This is a stress rating only and functional operation of the device at these or at any other condition above those indicated in the operational sections of this specification is not implied. Exposure to Absolute Maximum Rating conditions for extended periods may affect device reliability.

D.C. and Operating Characteristics
READ OPERATION

T_A = 0°C to 70°C, V_{CC} = +5V ±5%, V_{DD} = −9V ±5%, V_{GG}[1] = −9V ±5%, unless otherwise noted.

Symbol	Test	1702AL Limits			1702AL-2 Limits			Unit	Conditions
		Min.	Typ.[2]	Max.	Min.	Typ.[2]	Max.		
I_{LI}	Address and Chip Select Input Load Current			1			1	μA	V_{IN} = 0.0V
I_{LO}	Output Leakage Current			1			1	μA	V_{OUT} = 0.0V, $\overline{CS}$ = V_{CC}−2
I_{DDO1}[1]	Power Supply Current		7	10		7	10	mA	T_A=25°C $\overline{CS}$=V_{IH}, V_{GG}=V_{CC}
I_{DDO2}	Power Supply Current			15			15	mA	T_A=0°C I_{OL}=0.0mA
I_{DD1}[1]	Power Supply Current		35	50		35	50	mA	$\overline{CS}$ = V_{CC}−2, I_{OL} = 0.0mA, T_A = 25°C, Continuous
I_{DD2}	Power Supply Current		32	46		32	46	mA	$\overline{CS}$ = 0.0V, I_{OL} = 0.0mA, T_A = 25°C, Continuous
I_{DD3}	Power Supply Current		38	60		38	60	mA	$\overline{CS}$ = V_{CC}−2, I_{OL} = 0.0mA, T_A = 0°C, Continuous
I_{CF1}	Output Clamp Current		8	14		5.5	8	mA	V_{OUT} = −1.0V, T_A = 0°C, Continuous
I_{CF2}	Output Clamp Current		7	13		5	7	mA	V_{OUT} = −1.0V, T_A = 25°C, Continuous
I_{GG}	Gate Supply Current			1			1	μA	
V_{IL1}	Input Low Voltage for TTL Interface	−1		0.65	−1		0.65	V	
V_{IL2}	Input Low Voltage for MOS Interface	V_{DD}		V_{CC}−6	V_{DD}		V_{CC}−6	V	
V_{IH}	Address and Chip Select Input High Voltage	V_{CC}−2		V_{CC}+0.3	V_{CC}−2		V_{CC}+0.3	V	
I_{OL}	Output Sink Current	1.6	4		1.6	4		mA	V_{OUT} = 0.45V
I_{OH}	Output Source Current	−2.0			−2.0			mA	V_{OUT} = 0.0V
V_{OL}	Output Low Voltage		−3	0.45		−3	0.45	V	I_{OL} = 1.6mA
V_{OH}	Output High Voltage	3.5	4.5		3.5	4.5		V	I_{OH} = −200μA

NOTES: 1. The 1702AL is operated with the V_{GG} clocked to obtain low power dissipation. The average I_{DD} will vary between I_{DDO} and I_{DD1} (at 25°C) depending on the V_{GG} duty cycle (see curve opposite). 2. Typical values are at nominal voltage and T_A = 25°C.

Typical Characteristics

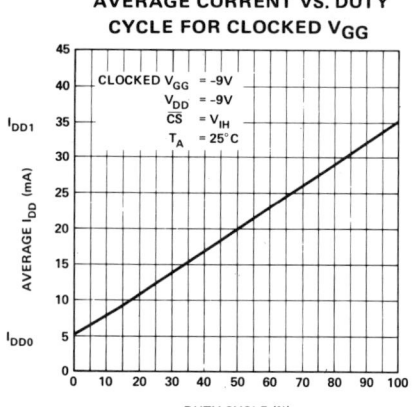
AVERAGE CURRENT VS. DUTY CYCLE FOR CLOCKED V_{GG}

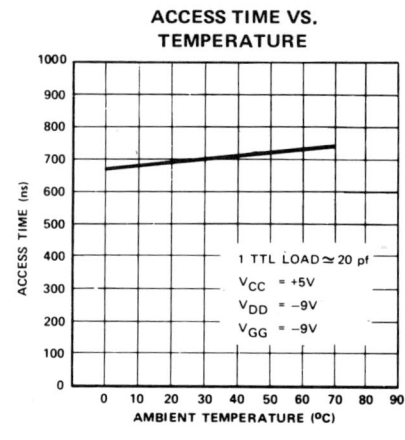
ACCESS TIME VS. TEMPERATURE

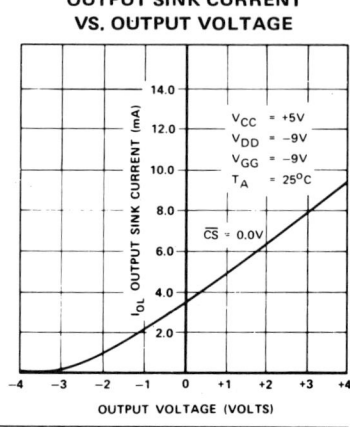
OUTPUT SINK CURRENT VS. OUTPUT VOLTAGE

A.C. Characteristics $T_A = 0°C$ to $+70°C$, $V_{CC} = +5V \pm 5\%$, $V_{DD} = -9V \pm 5\%$ unless otherwise noted

Symbol	Test	1702AL Limits Min.	1702AL Limits Max.	1702AL-2 Limits Min.	1702AL-2 Limits Max.	Unit
Freq.	Repetition Rate		1		1.6	MHz
t_{ACC}	Address to output delay		1		0.65	µs
$t_{DV_{GG}}$	Clocked V_{GG} set up	0.4		0.3		µs
t_{CS}	Chip select delay		0.1		0.3	µs
t_{CO}	Output delay from $\overline{CS}$		0.9		0.35	µs
t_{OD}	Output deselect		0.3		0.3	µs
t_{OHC}	Data out hold in clocked V_{GG} mode	5		5		µs

Capacitance $T_A = 25°C$

SYMBOL	TEST	TYPICAL	MAXIMUM	UNIT	CONDITIONS	
C_{IN}	Input Capacitance	8	15	pF	$V_{IN} = V_{CC}$ $\overline{CS} = V_{CC}$ $V_{OUT} = V_{CC}$ $V_{GG} = V_{CC}$	All unused pins are at A.C. ground
C_{OUT}	Output Capacitance	10	15	pF		
$C_{V_{GG}}$	V_{GG} Capacitance (Note 1)		30	pF		

*This parameter is periodically sampled and is not 100% tested.

Switching Characteristics

Conditions of Test:
Input pulse amplitudes: 0 to 4V; t_R, $t_F \leq 50$ ns
Output load is 1 TTL gate; measurements made at output of TTL gate ($t_{PD} \leq 15$ ns), $C_L = 15$ pF

A. READ OPERATION

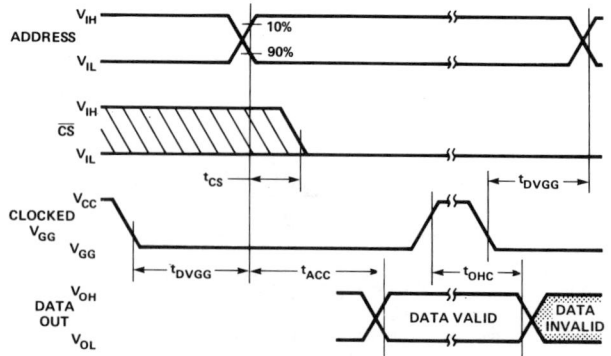

B. DESELECTION OF DATA OUTPUT IN OR-TIE OPERATION

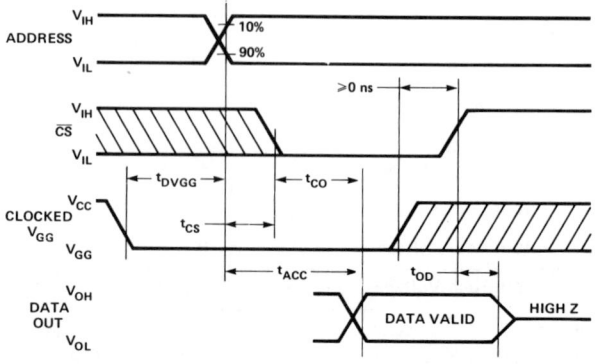

2308
8192 BIT STATIC MOS ROM

- **Fast Access Time: 450 ns**
- **Standard Power Supplies: +12V, ±5V**
- **TTL Compatible: All Inputs and Outputs**
- **Programmable Chip Select Input for Easy Memory Expansion**
- **Three-State Output: OR-Tie Capability**
- **Fully Decoded: On Chip Address Decode**
- **Inputs Protected: All Inputs Have Protection Against Static Charge**
- **Pin Compatible to 2708 PROM**

The Intel 2308 is a 8192 bit static MOS read only memory organized as 1024 words by 8-bits. This ROM is designed for memory applications where high performance, large bit storage, and simple interfacing are important design objectives.

The inputs and outputs are TTL compatible. The chip select input (CS2/$\overline{CS2}$) is programmable. An active high or low level chip select input can be defined by the designer and the desired chip select logic level is fixed at Intel during the masking process. The programmable chip select input, as well as OR-tie compatibility on the outputs, facilitates easy memory expansion. The pin compatible UV erasable 2708 PROM is available for initial system prototyping.

The 2308 read only memory is fabricated with N-channel silicon gate technology. This technology provides the designer with high performance, easy-to-use MOS circuits.

PIN CONFIGURATION

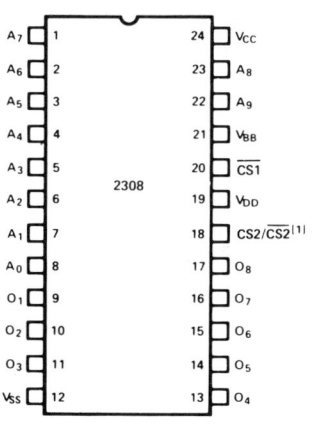

BLOCK DIAGRAM

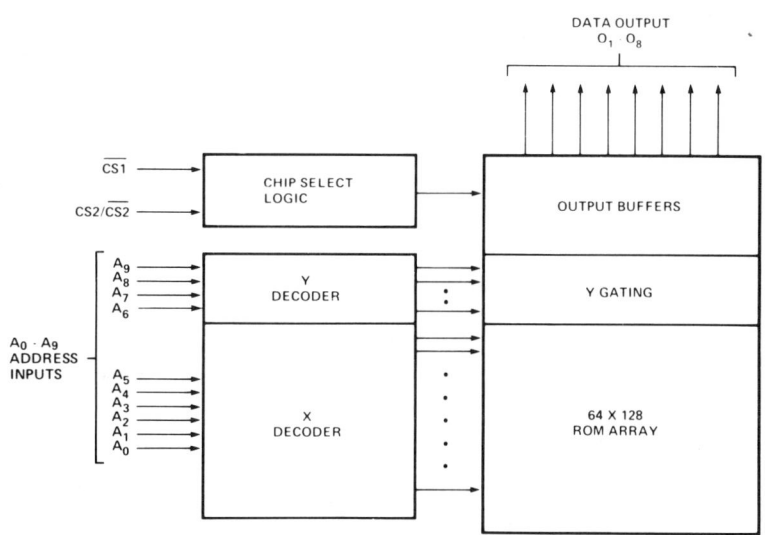

PIN NAMES

A_0-A_9	ADDRESS INPUTS
O_1-O_8	DATA OUTPUTS
$\overline{CS}_1$	CHIP SELECT INPUT
CS2/$\overline{CS2}$ [1]	PROGRAMMABLE CHIP SELECT INPUT

NOTE 1. The CS2/$\overline{CS2}$ LOGIC LEVELS MUST BE SPECIFIED BY THE USER AS EITHER A LOGIC 1 (V_{IH}) OR LOGIC 0 (V_{IL}). A LOGIC 0 SHOULD BE SPECIFIED IN ORDER TO BE COMPATIBLE WITH THE 2708.

Absolute Maximum Ratings*

Ambient Temperature Under Bias . . . . . . -25°C to +85°C
Storage Temperature . . . . -65°C to +150°C
Voltage On Any Pin With Respect
 To V_{BB} . . . . -0.3V to 20V
Power Dissipation . . . . 1.0 Watt

*COMMENT

Stresses above those listed under "Absolute Maximum Ratings" may cause permanent damage to the device. This is a stress rating only and functional operation of the device at these or any other conditions above those indicated in the operational sections of this specification is not implied. Exposure to absolute maximum rating conditions for extended periods may affect device reliability.

D.C. and Operating Characteristics

$T_A = 0°C$ to $+70°C$, $V_{CC} = 5V \pm 5\%$; $V_{DD} = 12V \pm 5\%$, $V_{BB} = -5V \pm 5\%$, $V_{SS} = 0V$ Unless Otherwise Specified.

Symbol	Parameter	Limits			Unit	Test Conditions
		Min.	Typ.[1]	Max.		
I_{LI}	Input Load Current (All Input Pins Except $\overline{CS}_1$)		1	10	µA	V_{IN} = 0 to 5.25V
I_{LCL}	Input Load Current on $\overline{CS}_1$			1.6	mA	V_{IN} = 0.45V
I_{LPC}	Input Peak Load Current on $\overline{CS}_1$			4	mA	$0.8V \leq V_{IN} < 3.3V$
I_{LKC}	Input Leakage Current on $\overline{CS}_1$			10	µA	V_{IN} = 3.3V to 5.25V
I_{LO}	Output Leakage Current			10	µA	Chip Deselected
V_{IL}	Input "Low" Voltage	$V_{SS}-1$		0.8V	V	
V_{IH}	Input "High" Voltage	3.3		$V_{CC}+1.0$	V	
V_{OL}	Output "Low" Voltage			0.45	V	I_{OL} = 2mA
V_{OH1}	Output "High" Voltage	2.4			V	I_{OH} = -4mA
V_{OH2}	Output "High" Voltage	3.7			V	I_{OH} = -1mA
I_{CC}	Power Supply Current V_{CC}		10	15	mA	
I_{DD}	Power Supply Current V_{DD}		32	60	mA	
I_{BB}	Power Supply Current V_{BB}		10µA	1	mA	
P_D	Power Dissipation		460	840	mW	

NOTE 1: Typical values for $T_A = 25°C$ and nominal supply voltage

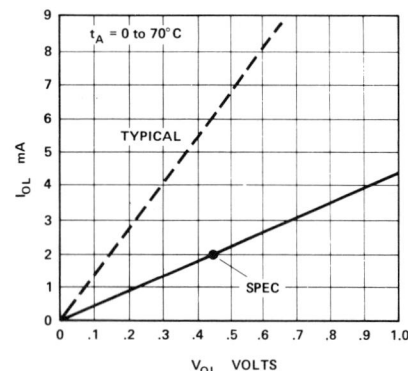

D.C. OUTPUT CHARACTERISTICS

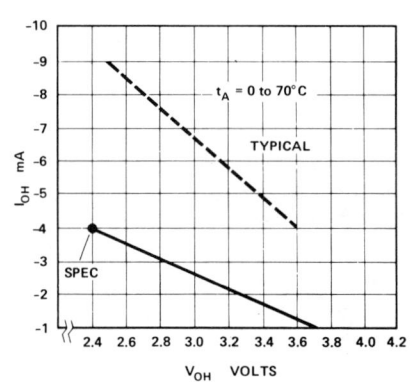

D.C. OUTPUT CHARACTERISTICS

A.C. Characteristics

$T_A = 0°C$ to $+70°C$, $V_{CC} = +5V \pm 5\%$; $V_{DD} = +12V \pm 5\%$, $V_{BB} = -5V \pm 5\%$, $V_{SS} = 0V$, Unless Otherwise Specified.

Symbol	Parameter	Limits[2]		Unit
		Typ.	Max.	
t_{ACC}	Address to Output Delay Time	200	450	ns
t_{CO1}	Chip Select 1 to Output Delay Time	85	160	ns
t_{CO2}	Chip Select 2 to Output Delay Time	125	220	ns
t_{DF}	Chip Deselect to Output Data Float Time	125	220	ns

NOTE 2: Refer to conditions of Test for A.C. Characteristics. Add 50 nanoseconds (worst case) to specified values at $V_{OH} = 3.7V$ @ $I_{OH} = -1mA$, $C_L = 100pF$.

CONDITIONS OF TEST FOR A.C. CHARACTERISTICS

Output Load . . . . 1 TTL Gate, and $C_{LOAD} = 100pF$
Input Pulse Levels . . . . .65V to 3.3V
Input Pulse Rise and Fall Times . . . . 20 nsec
Timing Measurement Reference Level
. . . . 2.4V V_{IH}, V_{OH}; 0.8V V_{IL}, V_{OL}

CAPACITANCE*

$T_A = 25°C$, $f = 1 MHz$, $V_{BB} = -5V$, V_{DD}, V_{CC} and all other pins tied to V_{SS}.

Symbol	Test	Limits	
		Typ.	Max.
C_{IN}	Input Capacitance		6pF
C_{OUT}	Output Capacitance		12pF

*This parameter is periodically sampled and is not 100% tested.

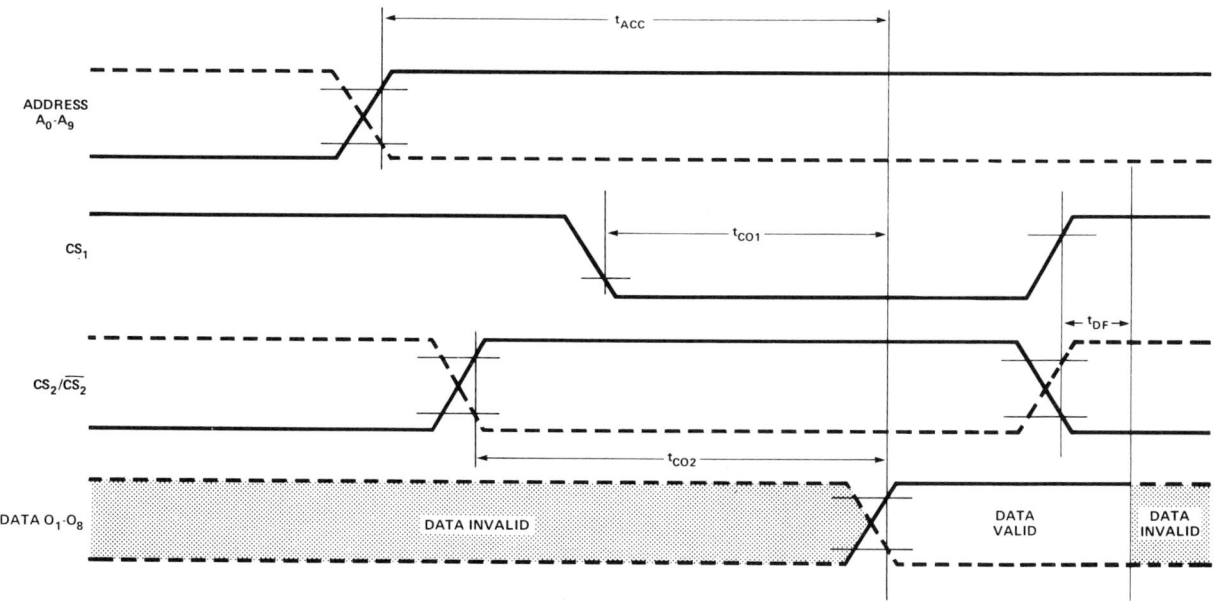

Typical Characteristics (Nominal supply voltages unless otherwise noted.)

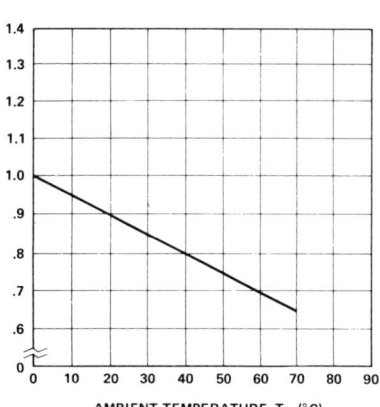

I_{DD} VS. TEMPERATURE
(NORMALIZED)

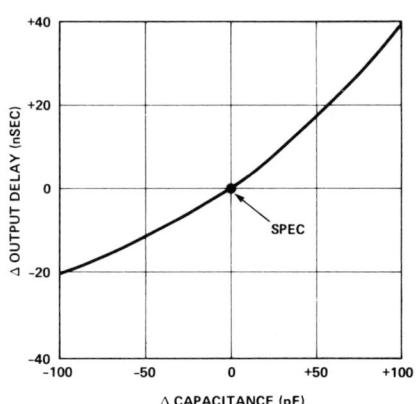

Δ OUTPUT CAPACITANCE
VS. Δ OUTPUT DELAY

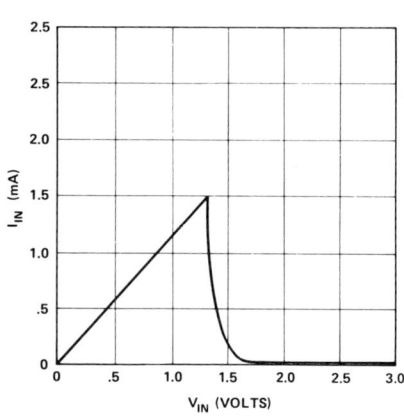

$\overline{CS}_1$ INPUT
CHARACTERISTICS

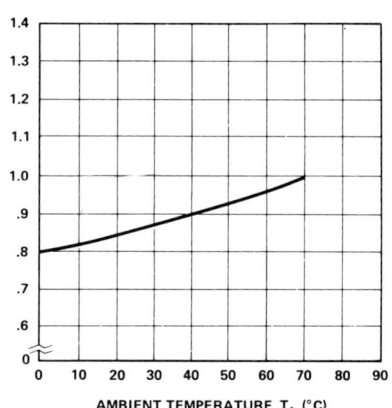

T_{ACC} VS. TEMPERATURE
(NORMALIZED)

2316E
16,384 BIT STATIC ROM

- **Fast Access Time—450 ns Max.**
- **Single +5V ± 10% Power Supply**
- **Intel MCS 80 and 85 Compatible**
- **Three Programmable Chip Selects for Simple Memory Expansion and System Interface**
- **EPROM/ROM Pin Compatible for Cost-Effective System Development**
- **Completly Static Operation**
- **Inputs and Outputs TTL Compatible**
- **Three-State Output for Direct Bus Interface**

The Intel® 2316E is a 16,384-bit static, N-channel MOS read only memory (ROM) organized as 2048 words by 8 bits. Its high bit density is ideal for large, non-volatile data storage applications such as program storage. The three-state outputs and TTL input/output levels allow for direct interface with common system bus structures. The 2316E single +5V power supply and 450 ns access time are both ideal for usage with high performance microcomputers such as the Intel MCS™-80 and MCS™-85 devices.

A cost-effective system development program may be implemented by using the pin compatible Intel 2716 16K UV EPROM for prototyping and the lower cost 2316E ROM for production. The 2716 is fully compatible to the 2316E in all respects. The three 2316E programmable chip selects may be defined by the user and are fixed during the masking process. To simplify the conversion from 2716 prototyping to 2316E production, it is recommended that the 2316E programmable chip select logic levels be defined the same as that shown in the below data sheet pin configuration. This pin configuration and these chip select logic levels are the same as the 2716.

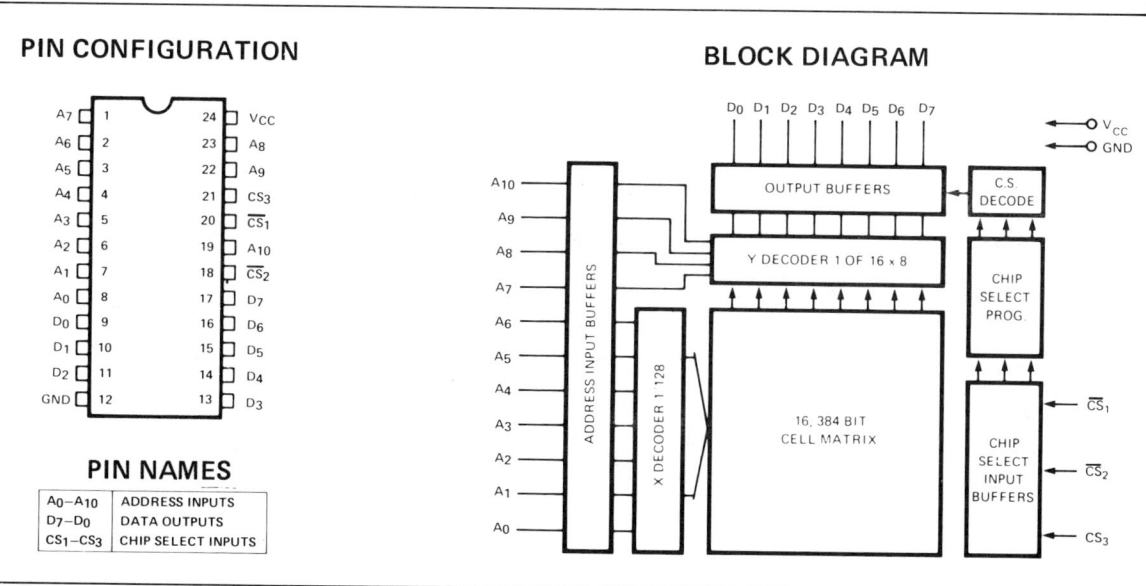

PROMS AND ROMS

ABSOLUTE MAXIMUM RATINGS*

Ambient Temperature Under Bias. . . . .-10°C to 80°C
Storage Temperature . . . .-65°C to +150°C
Voltage On Any Pin With Respect
   to Ground . . . . -0.5V to +7V
Power Dissipation . . . . 1.0 Watt

*COMMENT: Stresses above those listed under "Absolute Maximum Ratings" may cause permanent damage to the device. This is a stress rating only and functional operation of the device at these or at any other conditions above those indicated in the operational sections of this specification is not implied. Exposure to absolute maximum rating conditions for extended periods may affect device reliability.

D.C. AND OPERATING CHARACTERISTICS

T_A = 0°C to +70°C, V_{CC} = 5V ±10%, unless otherwise specified.

SYMBOL	PARAMETER	LIMITS			UNIT	TEST CONDITIONS
		MIN.	TYP.(1)	MAX.		
I_{LI}	Input Load Current (All Input Pins)			10	μA	V_{IN} = 0 to 5.25V
I_{LOH}	Output Leakage Current			10	μA	Chip Deselected, V_{OUT} = 4.0V
I_{LOL}	Output Leakage Current			-20	μA	Chip Deselected, V_{OUT} = 0.4V
I_{CC}	Power Supply Current		70	120	mA	All Inputs 5.25V Data Out Open
V_{IL}	Input "Low" Voltage	-0.5		0.8	V	
V_{IH}	Input "High" Voltage	2.4		V_{CC}+1.0V	V	
V_{OL}	Output "Low" Voltage			0.4	V	I_{OL} = 2.1 mA
V_{OH}	Output "High" Voltage	2.4			V	I_{OH} = -400 μA

NOTE: 1. Typical values for T_A = 25°C and nominal supply voltage.

A.C. CHARACTERISTICS

T_A = 0°C to +70°C, V_{CC} = +5V ±10%, unless otherwise specified.

SYMBOL	PARAMETER	LIMITS		UNIT
		MIN.	MAX.	
t_A	Address to Output Delay Time		450	ns
t_{CO}	Chip Select to Output Enable Delay Time		120	ns
t_{DF}	Chip Deselect to Output Data Float Delay Time	10	100	ns

CONDITIONS OF TEST FOR A.C. CHARACTERISTICS

Output Load . . . . 1 TTL Gate and C_L = 100 pF
Input Pulse Levels . . . . 0.8 to 2.4V
Input Pulse Rise and Fall Times (10% to 90%) . . . . . 20 ns
Timing Measurement Reference Level
   Input . . . . 1V and 2.2V
   Output . . . . 0.8V and 2.0V

CAPACITANCE(2) T_A = 25°C, f = 1 MHz

SYMBOL	TEST	LIMITS	
		TYP.	MAX.
C_{IN}	All Pins Except Pin Under Test Tied to AC Ground	5 pF	10 pF
C_{OUT}	All Pins Except Pin Under Test Tied to AC Ground	10 pF	15 pF

NOTE: 2. This parameter is periodically sampled and is not 100% tested.

A.C. Waveforms

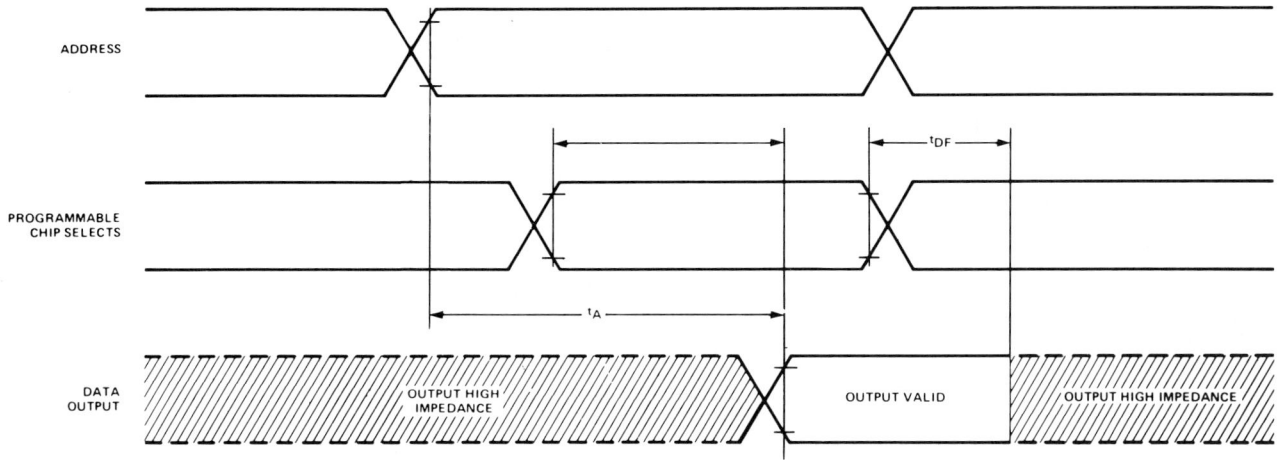

Typical System Application (8K × 8 ROM Memory)

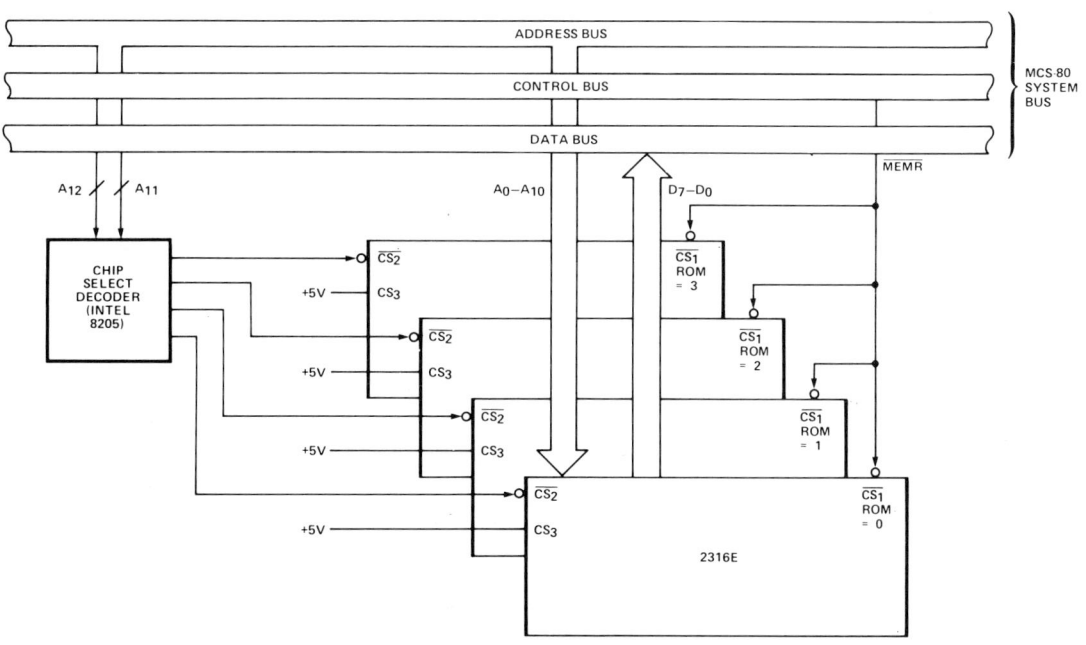

2708
8K AND 4K UV EPROM

- **Fast Access Time--350ns Max. (2708-1)**
- **Pin Compatible to 8K and 16K ROMs For Low Cost Production**
- **Fast Programming-- Typ. 100 sec For All 8K Bits**
- **Static--No Clocks Required**
- **1K×8 (2708), 512×8 (2704) Organizations**
- **Data Inputs and Outputs TTL Compatible During Both Real and Program Modes**
- **Three-State Outputs--or-Tie Capability**

The Intel® 2708 is a 8192-bit ultraviolet light erasable and electrically reprogrammable EPROM ideally suited where fast turnaround and pattern experimentation are important requirements. The electrical characteristics of the 2708 are specified over the 0°C to 70°C operating temperature range and with 5% power supply variation. All data inputs and outputs are TTL compatible during both the read and program mode. Furthermore, the three-state outputs allow for direct interface with common system bus structures. The 2708 is specified at a maximum access time of 450 ns. A higher speed 2708-1 is also available at 350 ns maximum access time. The M2708 is available for −55°C to +125°C operation.

A pin for pin mask programmed ROM, the Intel® 2308, is available for large volume production runs of systems initially using the 2708. For systems requiring higher bit density, the Intel 2716 16K EPROM and its pin compatible 2316E ROM is also available.

The 2704 is a 4096-bit UV EPROM organized as 512 words by 8 bits. It has all the same operating, programming, and erasing specifications of the 2708.

The 2708/2704 is fabricated with the N-channel silicon gate FAMOS technology. They are available in a 24-pin dual in-line package.

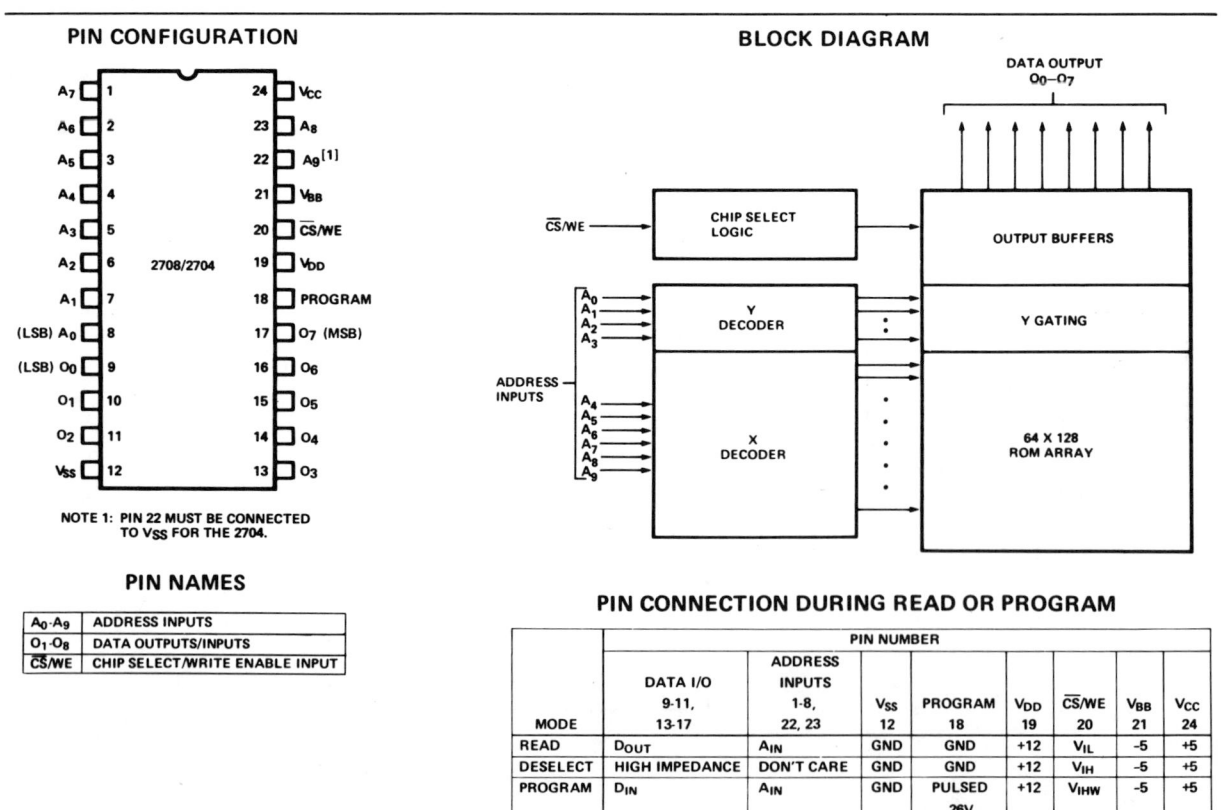

NOTE 1: PIN 22 MUST BE CONNECTED TO V$_{SS}$ FOR THE 2704.

PIN NAMES

A$_0$-A$_9$	ADDRESS INPUTS
O$_1$-O$_8$	DATA OUTPUTS/INPUTS
$\overline{CS}$/WE	CHIP SELECT/WRITE ENABLE INPUT

PIN CONNECTION DURING READ OR PROGRAM

	PIN NUMBER							
MODE	DATA I/O 9-11, 13-17	ADDRESS INPUTS 1-8, 22, 23	V$_{SS}$ 12	PROGRAM 18	V$_{DD}$ 19	$\overline{CS}$/WE 20	V$_{BB}$ 21	V$_{CC}$ 24
READ	D$_{OUT}$	A$_{IN}$	GND	GND	+12	V$_{IL}$	−5	+5
DESELECT	HIGH IMPEDANCE	DON'T CARE	GND	GND	+12	V$_{IH}$	−5	+5
PROGRAM	D$_{IN}$	A$_{IN}$	GND	PULSED 26V	+12	V$_{IHW}$	−5	+5

2708

Absolute Maximum Ratings*

Temperature Under Bias	$-25°C$ to $+85°C$
Storage Temperature	$-65°C$ to $+125°C$
V_{DD} With Respect to V_{BB}	$+20V$ to $-0.3V$
V_{CC} and V_{SS} With Respect to V_{BB}	$+15V$ to $-0.3V$
All Input or Output Voltages With Respect to V_{BB} During Read	$+15V$ to $-0.3V$
$\overline{CS}/WE$ Input With Respect to V_{BB} During Programming	$+20V$ to $-0.3V$
Program Input With Respect to V_{BB}	$+35V$ to $-0.3V$
Power Dissipation	1.5W

*COMMENT

Stresses above those listed under "Absolute Maximum Ratings" may cause permanent damage to the device. This is a stress rating only and functional operation of the device at these or any other conditions above those indicated in the operational sections of this specification is not implied. Exposure to absolute maximum rating conditions for extended periods may affect device reliability.

READ OPERATION
D.C. and Operating Characteristics

$T_A = 0°C$ to $70°C$, $V_{CC} = +5V \pm 5\%$, $V_{DD} = +12V \pm 5\%$, $V_{BB}{}^{[1]} = -5V \pm 5\%$, $V_{SS} = 0V$, unless otherwise noted.

Symbol	Parameter	Min.	Typ.[2]	Max.	Unit	Conditions
I_{LI}	Address and Chip Select Input Sink Current		1	10	μA	$V_{IN} = 5.25V$ or $V_{IN} = V_{IL}$
I_{LO}	Output Leakage Current		1	10	μA	$V_{OUT} = 5.5V$, $\overline{CS}/WE = 5V$
$I_{DD}{}^{[3]}$	V_{DD} Supply Current		50	65	mA	Worst Case Supply Currents:
$I_{CC}{}^{[3]}$	V_{CC} Supply Current		6	10	mA	All Inputs High
$I_{BB}{}^{[3]}$	V_{BB} Supply Current		30	45	mA	$\overline{CS}/WE = 5V$; $T_A = 0°C$
V_{IL}	Input Low Voltage	V_{SS}		0.65	V	
V_{IH}	Input High Voltage	3.0		$V_{CC}+1$	V	
V_{OL}	Output Low Voltage			0.45	V	$I_{OL} = 1.6mA$
V_{OH1}	Output High Voltage	3.7			V	$I_{OH} = -100μA$
V_{OH2}	Output High Voltage	2.4			V	$I_{OH} = -1mA$
P_D	Power Dissipation			800	mW	$T_A = 70°C$

NOTES: 1. V_{BB} must be applied prior to V_{CC} and V_{DD}. V_{BB} must also be the last power supply switched off.
2. Typical values are for $T_A = 25°C$ and nominal supply voltages.
3. The total power dissipation of the 2704/2708 is specified at 800 mW. It is not calculated by summing the various currents (I_{DD}, I_{CC}, and I_{BB}) multiplied by their respective voltages since current paths exist between the various power supplies and V_{SS}. The I_{DD}, I_{CC}, and I_{BB} currents should be used to determine power supply capacity only.

Typical Characteristics

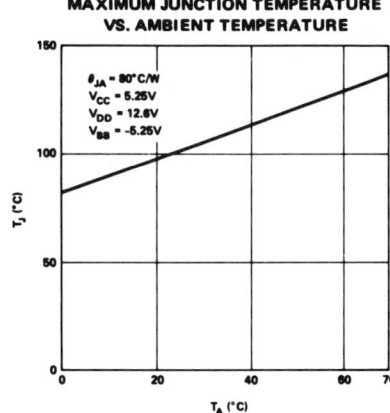

MAXIMUM JUNCTION TEMPERATURE VS. AMBIENT TEMPERATURE

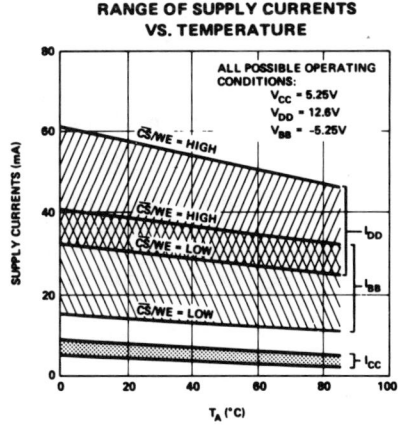

RANGE OF SUPPLY CURRENTS VS. TEMPERATURE

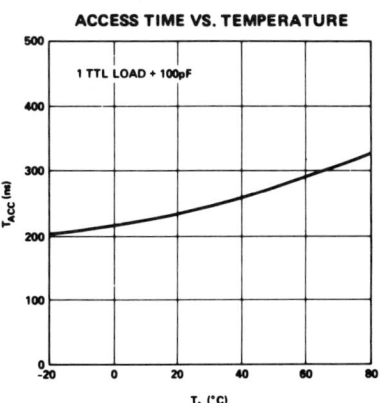

ACCESS TIME VS. TEMPERATURE

A. C. Characteristics

$T_A = 0°C$ to $70°C$, $V_{CC} = +5V \pm 5\%$, $V_{DD} = +12V \pm 5\%$, $V_{BB} = -5V \pm 5\%$, $V_{SS} = 0V$, unless otherwise noted.

Symbol	Parameter	2708-1 Limits			2708 Limits			Units
		Min.	Typ.	Max.	Min.	Typ.	Max.	
t_{ACC}	Address to Output Delay		280	350		280	450	ns
t_{CO}	Chip Select to Output Delay		60	120		60	120	ns
t_{DF}	Chip Deselect to Output Float	0		120	0		120	ns
t_{OH}	Address to Output Hold	0			0			ns

CAPACITANCE[1] $T_A = 25°C$, f = 1 MHz

Symbol	Parameter	Typ.	Max.	Unit.	Conditions
C_{IN}	Input Capacitance	4	6	pF	$V_{IN} = 0V$
C_{OUT}	Output Capacitance	8	12	pF	$V_{OUT} = 0V$

Note: 1. This parameter is periodically sampled and is not 100% tested.

A.C. TEST CONDITIONS:

Output Load: 1 TTL gate and C_L = 100 pF
Input Rise and Fall Times: ≤20 ns
Timing Measurement Reference Levels: 0.8V and 2.8V for inputs; 0.8V and 2.4V for outputs.
Input Pulse Levels: 0.65V to 3.0V

Waveforms

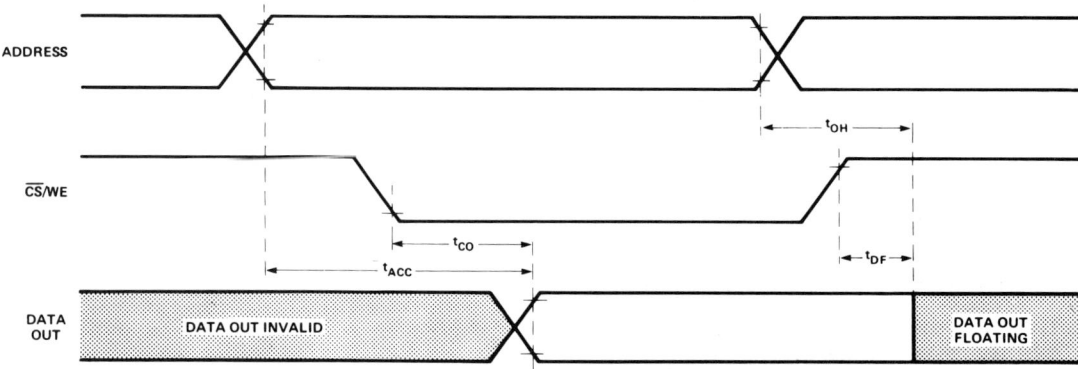

ERASURE CHARACTERISTICS

The erasure characteristics of the 2708 are such that erasure begins to occur when exposed to light with wavelengths shorter than approximately 4000 Angstroms (Å). It should be noted that sunlight and certain types of fluorescent lamps have wavelengths in the 3000–4000Å range. Data show that constant exposure to room level fluorescent lighting could erase the typical 2708 in approximately 3 years, while it would take approximatley 1 week to cause erasure when exposed to direct sunlight. If the 2708 is to be exposed to these types of lighting conditions for extended periods of time, opaque labels are available from Intel which should be placed over the 2708 window to prevent unintentional erasure.

The recommended erasure procedure for the 2708 is exposure to shortwave ultraviolet light which has a wavelength of 2537 Angstroms (Å). The integrated dose (i.e., UV intensity X exposure time) for erasure should be a minimum of 15 W-sec/cm². The erasure time with this dosage is approximately 15 to 20 minutes using an ultraviolet lamp with a 12000 μW/cm² power rating. The 2708 should be placed within 1 inch of the lamp tubes during erasure. Some lamps have a filter on their tubes which should be removed before erasure. The 1977 Intel Data Catalog should be consulted for further erasure information.

Programming

Initially, and after each erasure, all 8192/4096 bits of the 2708/2704 are in the "1" state (output high). Information is introduced by selectively programming "0" into the desired bit locations. A programmed "0" can only be changed to a "1" by UV erasure.

The circuit is set up for programming operation by raising the $\overline{CS}/WE$ input (pin 20) to +12V. The word address is selected in the same manner as in the read mode. Data to be programmed are presented, 8 bits in parallel, to the data output lines (O_1-O_8). Logic levels for address and data lines and the supply voltages are the same as for the read mode. After address and data set up, one program pulse per address is applied to the program input (pin 18). One pass through all addresses is defined as a program loop. The number of loops (N) required is a function of the program pulse width (t_{PW}) according to N x t_{PW} ≥ 100 ms.

The width of the program pulse is from 0.1 to 1 ms. The number of loops (N) is from a minimum of 100 (t_{PW} = 1 ms) to greater than 1000 (t_{PW} = 0.1 ms). There must be N successive loops throuhg all 1024 addresses. *It is not permitted to apply N program pulses to an address and then change to the next address to be programmed.* Caution should be observed regarding the end of a program sequence. The $\overline{CS}/WE$ falling edge transition must occur before the first address transition when changing from a program to a read cycle. The program pin should also be pulled down to V_{ILP} with an active instead of a passive device. This pin will source a small amount of current (I_{ILL}) when $\overline{CS}/WE$ is at V_{IHW} (12V) and the program pulse is at V_{ILP}.

Programming Examples (Using N x t_{PW} ≥ 100 ms)

Example 1: All 8096 bits are to be programmed with a 0.5 ms program pulse width.

The minimum number of program loops is 200. One program loop consists of words 0 to 1023.

Example 2: Words 0 to 100 and 500 to 600 are to be programmed. All other bits are "don't care". The program pulse width is 0.75 ms.

The minimum number of program loops is 133. One program loop consists of words 0 to 1023. The data entered into the "don't care" bits should be all 1's.

Example 3: Same requirements as example 2, but the PROM is now to be *updated* to include data for words 750 to 770.

The minimum number of program loops is 133. One program loop consists of words 0 to 1023. The data entered into the "don't care" bits should be all 1's. Addresses 0 to 100 and 500 to 600 must be re-programmed with their original data pattern.

2704, 2708
PROGRAM CHARACTERISTICS

T_A = 25°C, V_{CC} = 5V ±5%, V_{DD} = +12V ±5%, V_{BB} = -5V ±5%, V_{SS} = 0V, Unless Otherwise Noted.

D.C. Programming Characteristics

Symbol	Parameter	Min.	Typ.	Max.	Units	Test Conditions
I_{LI}	Address and $\overline{CS}/WE$ Input Sink Current			10	µA	V_{IN} = 5.25V
I_{IPL}	Program Pulse Source Current			3	mA	
I_{IPH}	Program Pulse Sink Current			20	mA	
I_{DD}	V_{DD} Supply Current		50	65	mA	Worst Case Supply Currents: All Inputs High $\overline{CS}/WE$ = 5V; T_A = 0°C
I_{CC}	V_{CC} Supply Current		6	10	mA	
I_{BB}	V_{BB} Supply Current		30	45	mA	
V_{IL}	Input Low Level (except Program)	V_{SS}		0.65	V	
V_{IH}	Input High Level for all Addresses and Data	3.0		V_{CC}+1	V	
V_{IHW}	$\overline{CS}/WE$ Input High Level	11.4		12.6	V	Referenced to V_{SS}
V_{IHP}	Program Pulse High Level	25		27	V	Referenced to V_{SS}
V_{ILP}	Program Pulse Low Level	V_{SS}		1	V	$V_{IHP} - V_{ILP}$ = 25V min.

A.C. Programming Characteristics

Symbol	Parameter	Min.	Typ.	Max.	Units
t_{AS}	Address Setup Time	10			µs
t_{CSS}	CS/WE Setup Time	10			µs
t_{DS}	Data Setup Time	10			µs
t_{AH}	Address Hold Time	1			µs
t_{CH}	CS/WE Hold Time	.5			µs
t_{DH}	Data Hold Time	1			µs
t_{DF}	Chip Deselect to Output Float Delay	0		120	ns
t_{DPR}	Program To Read Delay			10	µs
t_{PW}	Program Pulse Width	.1		1.0	ms
t_{PR}	Program Pulse Rise Time	.5		2.0	µs
t_{PF}	Program Pulse Fall Time	.5		2.0	µs

NOTE: Intel's standard product warranty applies only to devices programmed to specifications described herein.

2704, 2708
Programming Waveforms

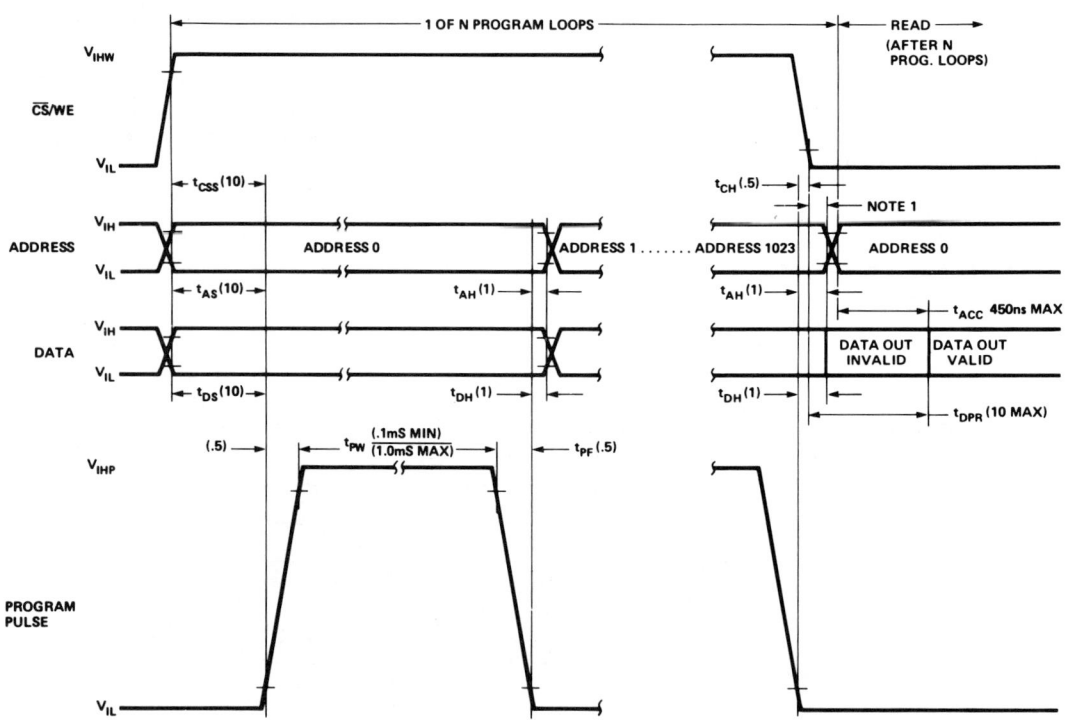

NOTE 1. THE CS/WE TRANSITION MUST OCCUR AFTER THE PROGRAM PULSE TRANSITION AND BEFORE THE ADDRESS TRANSITION.
NOTE 2. NUMBERS IN ( ) INDICATE MINIMUM TIMING IN µS UNLESS OTHERWISE SPECIFIED.

2716
16K (2K × 8) UV EPROM

- **Fast Access Time**
  - 350 ns Max. 2716-1
  - 390 ns Max. 2716-2
  - 450 ns Max. 2716

- **Single +5V Power Supply**

- **Low Power Dissipation**
  - 525 mW Max. Active Power
  - 132 mW Max. Standby Power

- **Pin Compatible to Intel® 5V ROMs (2316E, 2332, and 2364) and 2732 EPROM**

- **Simple Programming Requirements Single Location Programming Programs with One 50 ms Pulse**

- **Inputs and Outputs TTL Compatible during Read and Program**

- **Completely Static**

The Intel® 2716 is a 16,384-bit ultraviolet erasable and electrically programmable read-only memory (EPROM). The 2716 operates from a single 5-volt power supply, has a static standby mode, and features fast single address location programming. It makes designing with EPROMs faster, easier and more economical. For production quantities, the 2716 user can convert rapidly to Intel's pin-for-pin compatible 16K ROM (the 2316E) or the new 32K and 64K ROMs (the 2332 and 2364 respectively).

The 2716, with its single 5-volt supply and with an access time up to 350 ns, is ideal for use with the newer high performance +5V microprocessors such as Intel's 8085 and 8086. The 2716 is also the first EPROM with a static standby mode which reduces the power dissipation without increasing access time. The maximum active power dissipation is 525 mW while the maximum standby power dissipation is only 132 mW, a 75% savings.

The 2716 has the simplest and fastest method yet devised for programming EPROMs — single pulse TTL level programming. No need for high voltage pulsing because all programming controls are handled by TTL signals. Now, it is possible to program on-board, in the system, in the field. Program any location at any time — either individually, sequentially or at random, with the 2716's single address location programming. Total programming time for all 16,384 bits is only 100 seconds.

PIN CONFIGURATION*

```
       2716                          2732†
   ┌────┐                         ┌────┐
A7 │1  24│ Vcc                 A7 │1  24│ Vcc
A6 │2  23│ A8                  A6 │2  23│ A8
A5 │3  22│ A9                  A5 │3  22│ A9
A4 │4  21│ Vpp                 A4 │4  21│ A11
A3 │5  20│ OE                  A3 │5  20│ OE/Vpp
A2 │6 16K 19│ A10              A2 │6 32K 19│ A10
A1 │7  18│ CE                  A1 │7  18│ CE
A0 │8  17│ O7                  A0 │8  17│ O7
O0 │9  16│ O6                  O0 │9  16│ O6
O1 │10 15│ O5                  O1 │10 15│ O5
O2 │11 14│ O4                  O2 │11 14│ O4
GND│12 13│ O3                  GND│12 13│ O3
```

†Refer to 2732 data sheet for specifications

PIN NAMES

A0–A9	ADDRESSES
CE/PGM	CHIP ENABLE/PROGRAM
OE	OUTPUT ENABLE
O0–O7	OUTPUTS

MODE SELECTION

PINS MODE	CE/PGM (18)	OE (20)	Vpp (21)	Vcc (24)	OUTPUTS (9-11, 13-17)
Read	V_{IL}	V_{IL}	+5	+5	D_{OUT}
Standby	V_{IH}	Don't Care	+5	+5	High Z
Program	Pulsed V_{IL} to V_{IH}	V_{IH}	+25	+5	D_{IN}
Program Verify	V_{IL}	V_{IL}	+25	+5	D_{OUT}
Program Inhibit	V_{IL}	V_{IH}	+25	+5	High Z

BLOCK DIAGRAM

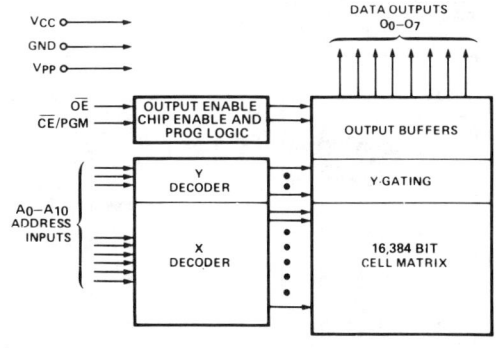

*Pin 18 and pin 20 have been renamed to conform with the entire family of 16K, 32K, and 64K EPROMs and ROMs. The die, fabrication process, and specifications remain the same and are totally uneffected by this change.

Absolute Maximum Ratings*

Temperature Under Bias	–10°C to +80°C
Storage Temperature	–65°C to +125°C
All Input or Output Voltages with Respect to Ground	+6V to –0.3V
V_{PP} Supply Voltage with Respect to Ground	+28V to –0.3V

*COMMENT: Stresses above those listed under "Absolute Maximum Ratings" may cause permanent damage to the device. This is a stress rating only and functional operation of the device at these or any other conditions above those indicated in the operational sections of this specification is not implied. Exposure to absolute maximum rating conditions for extended periods may affect device reliability.

READ OPERATION

D.C. and Operating Characteristics

$T_A = 0°C$ to $70°C$, V_{CC}[1,2] = +5V ±5%, V_{PP}[2] = V_{CC} ±0.6V[3]

Symbol	Parameter	Limits			Unit	Conditions
		Min.	Typ.[4]	Max.		
I_{LI}	Input Load Current			10	µA	V_{IN} = 5.25V
I_{LO}	Output Leakage Current			10	µA	V_{OUT} = 5.25V
I_{PP1} [2]	V_{PP} Current			5	mA	V_{PP} = 5.85V
I_{CC1} [2]	V_{CC} Current (Standby)		10	25	mA	PD/PGM = V_{IH}, $\overline{CS}$ = V_{IL}
I_{CC2} [2]	V_{CC} Current (Active)		57	100	mA	$\overline{CS}$ = PD/PGM = V_{IL}
A_R [5]	Select Reference Input Level	–0.1		0.8	V	I_{IN} = 10 µA
V_{IL}	Input Low Voltage	–0.1		0.8	V	
V_{IH}	Input High Voltage	2.2		V_{CC} + 1	V	
V_{OL}	Output Low Voltage			0.45	V	I_{OL} = 2.1 mA
V_{OH}	Output High Voltage	2.4			V	I_{OH} = –400 µA

NOTES FOR PAGES 2 AND 3:
1. V_{CC} must be applied simultaneously or before Vpp and removed simultaneously or after Vpp.
2. Vpp may be connected directly to V_{CC} except during programming. The supply current would then be the sum of I_{CC} and I_{PP1}.
3. The tolerance of 0.6V allows the use of a driver circuit for switching the Vpp supply pin from V_{CC} in read to 25V for programming.
4. Typical values are for T_A = 25°C and nominal supply voltages.
5. A_R is a reference voltage level which requires an input current of only 10 µA. The 2758 S1865 is also available which has a reference voltage level of V_{IH} instead of V_{IL}.
6. This parameter is only sampled and is not 100% tested.
7. t_{ACC2} is referenced to PD/PGM or the addresses, whichever occurs last.

Typical Characteristics

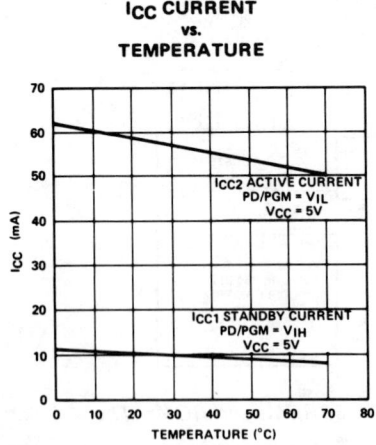

I_{CC} CURRENT vs. TEMPERATURE

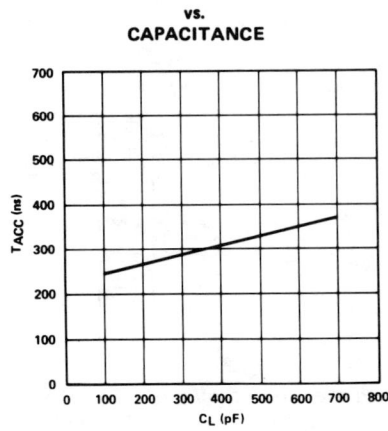

ACCESS TIME vs. CAPACITANCE

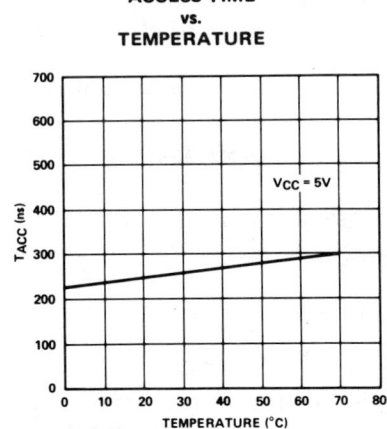

ACCESS TIME vs. TEMPERATURE

A.C. Characteristics

$T_A = 0°C$ to $70°C$, $V_{CC}^{[1]} = +5V \pm 5\%$, $V_{PP}^{[2]} = V_{CC} \pm 0.6V^{[3]}$

Symbol	Parameter	Limits Min.	Limits Typ.[4]	Limits Max.	Unit	Test Conditions
t_{ACC1}	Address to Output Delay		250	450	ns	PD/PGM = $\overline{CS}$ = V_{IL}
t_{ACC2}[7]	PD/PGM to Output Delay		280	450	ns	$\overline{CS}$ = V_{IL}
t_{CO}	Chip Select to Output Delay			120	ns	PD/PGM = V_{IL}
t_{PF}	PD/PGM to Output Float	0		100	ns	$\overline{CS}$ = V_{IL}
t_{DF}	Chip Deselect to Output Float	0		100	ns	PD/PGM = V_{IL}
t_{OH}	Address to Output Hold	0			ns	PD/PGM = $\overline{CS}$ = V_{IL}

Capacitance[6] $T_A = 25°C$, f = 1 MHz

Symbol	Parameter	Typ.	Max.	Unit	Conditions
C_{IN}	Input Capacitance	4	6	pF	V_{IN} = 0V
C_{OUT}	Output Capacitance	8	12	pF	V_{OUT} = 0V

NOTE: Please refer to page 2 for notes.

A.C. Test Conditions:

Output Load: 1 TTL gate and C_L = 100 pF
Input Rise and Fall Times: ≤20 ns
Input Pulse Levels: 0.8V to 2.2V
Timing Measurement Reference Level:
   Inputs   1V and 2V
   Outputs 0.8V and 2V

WAVEFORMS

A. Read Mode
PD/PGM = V_{IL}

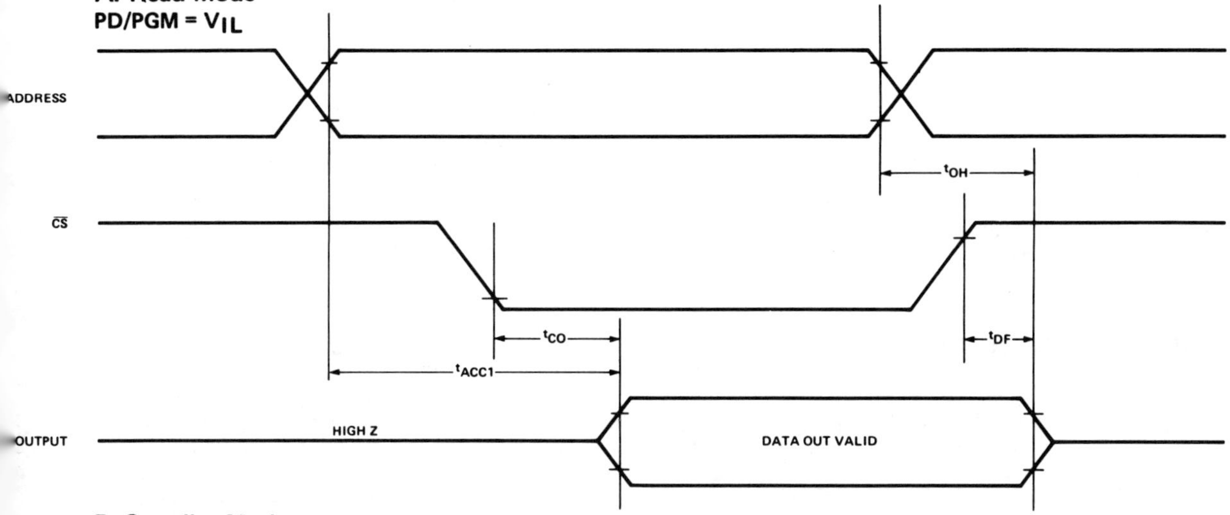

B. Standby Mode
$\overline{CS}$ = V_{IL}

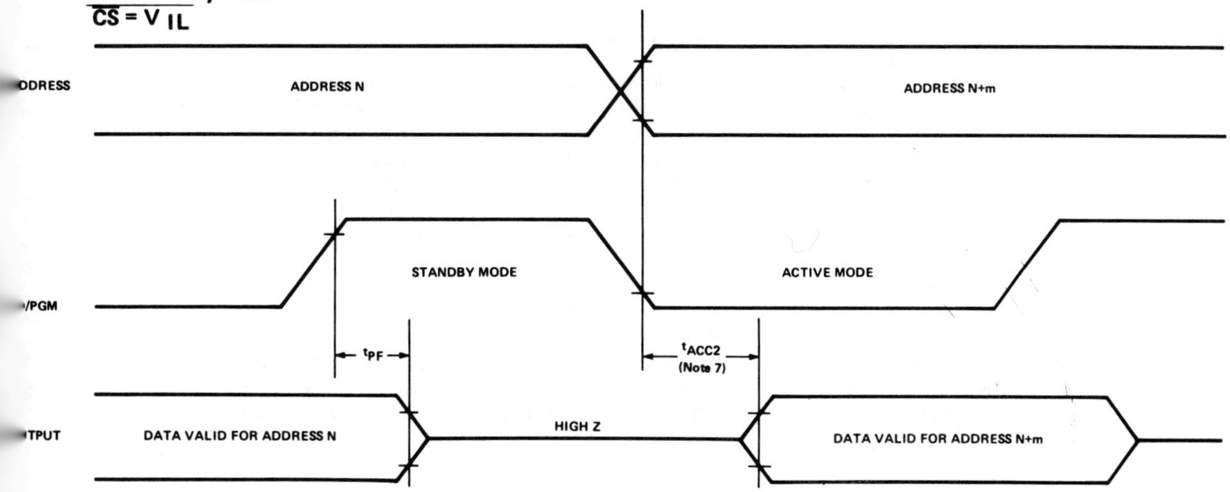

ERASURE CHARACTERISTICS

The erasure characteristics of the 2758 are such that erasure begins to occur when exposed to light with wavelengths shorter than approximately 4000 Angstroms (Å). It should be noted that sunlight and certain types of fluorescent lamps have wavelengths in the 3000–4000Å range. Data show that constant exposure to room level fluorescent lighting could erase the typical 2758 in approximately 3 years, while it would take approximately 1 week to cause erasure when exposed to direct sunlight. If the 2758 is to be exposed to these types of lighting conditions for extended periods of time, opaque labels are available from Intel which should be placed over the 2758 window to prevent unintentional erasure.

The recommended erasure procedure for the 2758 is exposure to shortwave ultraviolet light which has a wavelength of 2537 Angstroms (Å). The integrated dose (i.e., UV intensity × exposure time) for erasure should be a minimum of 15 W-sec/cm^2. The erasure time with this dosage is approximately 15 to 20 minutes using an ultraviolet lamp with a 12,000 μW/cm^2 power rating. The 2758 should be placed within 1 inch of the lamp tubes during erasure. Some lamps have a filter on their tubes which should be removed before erasure.

DEVICE OPERATION

The six modes of operation of the 2758 are listed in Table I. It should be noted that all inputs for the six modes are at TTL levels. The power supplies required are a +5V V_{CC} and a V_{PP}. The V_{PP} power supply must be at 25V during the three programming modes, and must be at 5V in the other three modes. In all operational modes, A_R must be at V_{IL} (except for the 2758 S1865 which has A_R at V_{IH}).

TABLE I. MODE SELECTION

PINS / MODE	PD/PGM (18)	A_R (19)	$\overline{CS}$ (20)	V_{PP} (21)	V_{CC} (24)	OUTPUTS (9-11, 13-17)
Read	V_{IL}	V_{IL}	V_{IL}	+5	+5	D_{OUT}
Deselect	Don't Care	V_{IL}	V_{IH}	+5	+5	High Z
Power Down	V_{IH}	V_{IL}	Don't Care	+5	+5	High Z
Program	Pulsed V_{IL} to V_{IH}	V_{IL}	V_{IH}	+25	+5	D_{IN}
Program Verify	V_{IL}	V_{IL}	V_{IL}	+25	+5	D_{OUT}
Program Inhibit	V_{IL}	V_{IL}	V_{IH}	+25	+5	High Z

READ MODE

Data is available at the outputs in the read mode. Data is available 450 ns (t_{ACC}) from stable addresses with $\overline{CS}$ low or 120 ns (t_{CO}) from $\overline{CS}$ with addresses stable.

DESELECT MODE

The outputs of two or more 2758 may be OR-tied together on the same data bus. Only one 2758 should have its outputs selected ($\overline{CS}$ low) to prevent data bus contention between 2758 in this configuration. The outputs of the other 2758 should be deselected with the $\overline{CS}$ input at a high TTL level.

POWER DOWN MODE

The 2758 has a power down mode which reduces the active power dissipation by 75%, from 525 mW to 132 mW. Power down is achieved by applying a TTL high signal to the PD/PGM input. In power down the outputs are in a high impedance state, independent of the $\overline{CS}$ input.

PROGRAMMING

Initially, and after each erasure, all bits of the 2758 are in the "1" state. Data is introduced by selectively programming "0's" into the desired bit locations. Although only "0's" will be programmed, both "1's" and "0's" can be presented in the data word. The only way to change a "0" to a "1" is by ultraviolet light erasure.

The 2758 is in the programming mode when the V_{PP} power supply is at 25V and $\overline{CS}$ is at V_{IH}. The data to be programmed is applied 8 bits in parallel to the data output pins. The levels required for the address and data inputs are TTL.

When the addresses and data are stable, a 50 msec, active high, TTL program pulse is applied to the PD/PGM input. A program pulse must be applied at each address location to be programmed. You can program any location at any time — either individually, sequentially, or at random. The program pulse has a maximum width of 55 msec. The 2758 must not be programmed with a DC signal applied to the PD/PGM input.

Programming of multiple 2758s in parallel with the same data can be easily accomplished due to the simplicity of the programming requirements. Like inputs of the parallelled 2758s may be connected together when they are programmed with the same data. A high level TTL pulse applied to the PD/PGM input programs the paralleled 2758s.

PROGRAM INHIBIT

Programming of multiple 2758s in parallel with different data is also easily accomplished. Except for PD/PGM, all like inputs (including $\overline{CS}$) of the parallel 2758s may be common. A TTL level program pulse applied to a 2758's PD/PGM input with V_{PP} at 25V will program that 2758. A low level PD/PGM input inhibits the other 2758 from being programmed.

PROGRAM VERIFY

A verify should be performed on the programmed bits to determine that they were correctly programmed. The verify may be performed with V_{PP} at 25V. Except during programming and program verify, V_{PP} must be at 5V.

2758
8K (1K × 8) LOW POWER +5V UV EPROM

- **Single +5V Power Supply**
- **Simple Programming Requirements**
  Single Location Programming
  Programs With One 50ms Pulse
- **Low Power Dissipation**
  525mW Max Active Power
  132mW Max Standby Power
- **Fast Access Time: 450ns Max In Active and Standby Power Mode**
- **Inputs and Outputs TTL Compatible During Read and Program**
- **Three-State Outputs For or-Ties**

The Intel® 2758 is a 8192-bit ultraviolet erasable and electrically programmable read-only-memory (EPROM). The 2758 operates from a single 5V power supply, has a static power-down mode, and features fast, single address location programming. It makes designing with EPROMs faster, easier and more economical. The total programming time for all 8192 bits is 50 seconds.

The 2758 has a static power-down mode which reduces the power dissipation without increasing access time. The maximum active power dissipation is 525 mW, while the maximum standby power dissipation is only 132 mW, a 75% savings. Power-down is achieved by applying a TTL-high signal to the PD/PGM input.

A 2758 system may be designed for total upwards compatibility with Intel's 16K 2716 EPROM (see Applications Note 30). The 2758 maintains the simplest and fastest method yet devised for programming EPROMs — single pulse TTL-level programming. There is no need for high voltage pulsing because all programming controls are handled by TTL signals. Now it is possible to program on-board, in the system, in the field. Program any location at any time — either individually, sequentially, or at random, with the single address location programming.

PIN CONFIGURATION

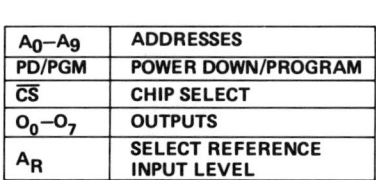

MODE SELECTION

PINS MODE	PD/PGM (18)	A_R (19)	$\overline{CS}$ (20)	V_{PP} (21)	V_{CC} (24)	OUTPUTS (9-11, 13-17)
Read	V_{IL}	V_{IL}	V_{IL}	+5	+5	D_{OUT}
Deselect	Don't Care	V_{IL}	V_{IH}	+5	+5	High Z
Power Down	V_{IH}	V_{IL}	Don't Care	+5	+5	High Z
Program	Pulsed V_{IL} to V_{IH}	V_{IL}	V_{IH}	+25	+5	D_{IN}
Program Verify	V_{IL}	V_{IL}	V_{IL}	+25	+5	D_{OUT}
Program Inhibit	V_{IL}	V_{IL}	V_{IH}	+25	+5	High Z

PIN NAMES

A_0–A_9	ADDRESSES
PD/PGM	POWER DOWN/PROGRAM
$\overline{CS}$	CHIP SELECT
O_0–O_7	OUTPUTS
A_R	SELECT REFERENCE INPUT LEVEL

BLOCK DIAGRAM

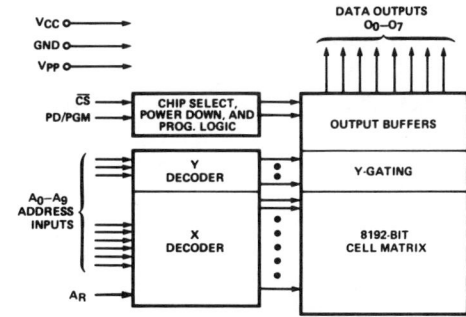

PROGRAMMING

The programming specifications are described in the Data Catalog PROM/ROM Programming Insutrctions on Page 4-83.

Absolute Maximum Ratings*

Temperature Under Bias . . . . −10°C to +80°C
Storage Temperature . . . . −65°C to +125°C
All Input or Output Voltages with
 Respect to Ground . . . . +6V to −0.3V
V_{PP} Supply Voltage with Respect
 to Ground During Program . . . . +26.5V to −0.3V

*COMMENT: Stresses above those listed under "Absolute Maximum Ratings" may cause permanent damage to the device. This is a stress rating only and functional operation of the device at these or any other conditions above those indicated in the operational sections of this specification is not implied. Exposure to absolute maximum rating conditions for extended periods may affect device reliability.

DC and AC Operating Conditions During Read

	2716	2716-1	2716-2
Temperature Range	0°C − 70°C	0°C − 70°C	0°C − 70°C
V_{CC} Power Supply [1,2]	5V ± 5%	5V ± 10%	5V ± 5%
V_{PP} Power Supply [2]	V_{CC} ± 0.6V [3]	V_{CC} ± 0.6V [3]	V_{CC} ± 0.6V [3]

READ OPERATION

D.C. and Operating Characteristics

Symbol	Parameter	Limits Min.	Limits Typ.[4]	Limits Max.	Unit	Conditions
I_{LI}	Input Load Current			10	μA	V_{IN} = 5.25V
I_{LO}	Output Leakage Current			10	μA	V_{OUT} = 5.25V
I_{PP1} [2]	V_{PP} Current			5	mA	V_{PP} = 5.85V
I_{CC1} [2]	V_{CC} Current (Standby)		10	25	mA	$\overline{CE} = V_{IH}$, $\overline{OE} = V_{IL}$
I_{CC2} [2]	V_{CC} Current (Active)		57	100	mA	$\overline{OE} = \overline{CE} = V_{IL}$
V_{IL}	Input Low Voltage	−0.1		0.8	V	
V_{IH}	Input High Voltage	2.0		V_{CC}+1	V	
V_{OL}	Output Low Voltage			0.45	V	I_{OL} = 2.1 mA
V_{OH}	Output High Voltage	2.4			V	I_{OH} = −400 μA

NOTES: 1. V_{CC} must be applied simultaneously or before V_{PP} and removed simultaneously or after V_{PP}.
2. V_{PP} may be connected directly to V_{CC} except during programming. The supply current would then be the sum of I_{CC} and I_{PP1}.
3. The tolerance of 0.6V allows the use of a driver circuit for switching the V_{PP} supply pin from V_{CC} in read to 25V for programming.
4. Typical values are for T_A = 25°C and nominal supply voltages.
5. This parameter is only sampled and is not 100% tested.

Typical Characteristics

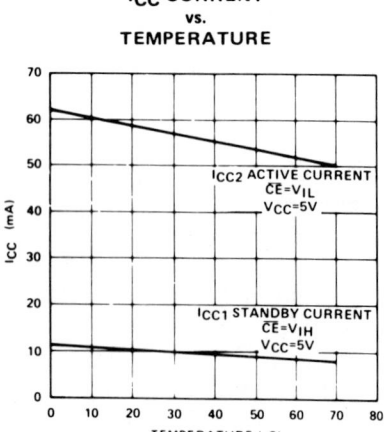

I_{CC} CURRENT vs. TEMPERATURE

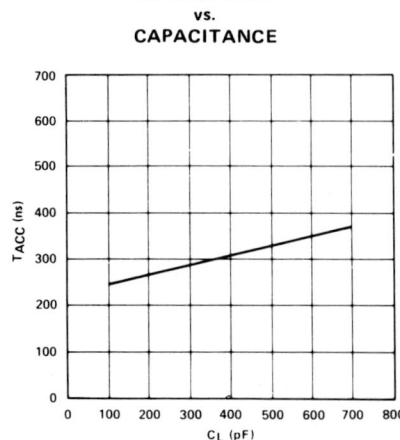

ACCESS TIME vs. CAPACITANCE

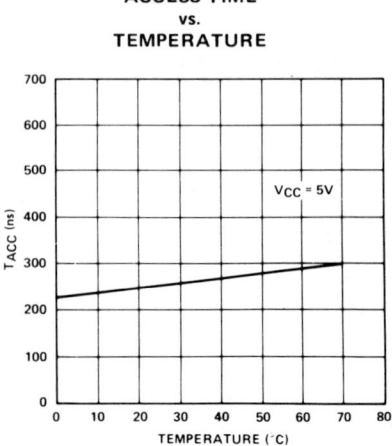

ACCESS TIME vs. TEMPERATURE

2758

A.C. Characteristics

Symbol	Parameter	2716 Limits			2716-1 Limits			2716-2 Limits			Unit	Test Conditions
		Min	Typ[4]	Max	Min	Typ[4]	Max	Min	Typ[4]	Max		
tACC	Address to Output Delay			450			350			390	ns	$\overline{CE} = \overline{OE} = V_{IL}$
tCE	$\overline{CE}$ to Output Delay			450			350			390	ns	$\overline{OE} = V_{IL}$
tOE	Output Enable to Output Delay			120			120			120	ns	$\overline{CE} = V_{IL}$
tDF	Output Enable High to Output Float	0		100	0		100	0		100	ns	$\overline{CE} = V_{IL}$
tOH	Address to Output Hold	0			0			0			ns	$\overline{CE} = \overline{OE} = V_{IL}$

Capacitance[5] $T_A = 25°C$, f = 1 MHz

Symbol	Parameter	Typ.	Max.	Unit	Conditions
C_{IN}	Input Capacitance	4	6	pF	$V_{IN} = 0V$
C_{OUT}	Output Capacitance	8	12	pF	$V_{OUT} = 0V$

A.C. Test Conditions:

Output Load: 1 TTL gate and C_L = 100 pF
Input Rise and Fall Times: ≤20 ns
Input Pulse Levels: 0.8V to 2.2V
Timing Measurement Reference Level:
  Inputs   1V and 2V
  Outputs   0.8V and 2V

A.C. Waveforms (1)

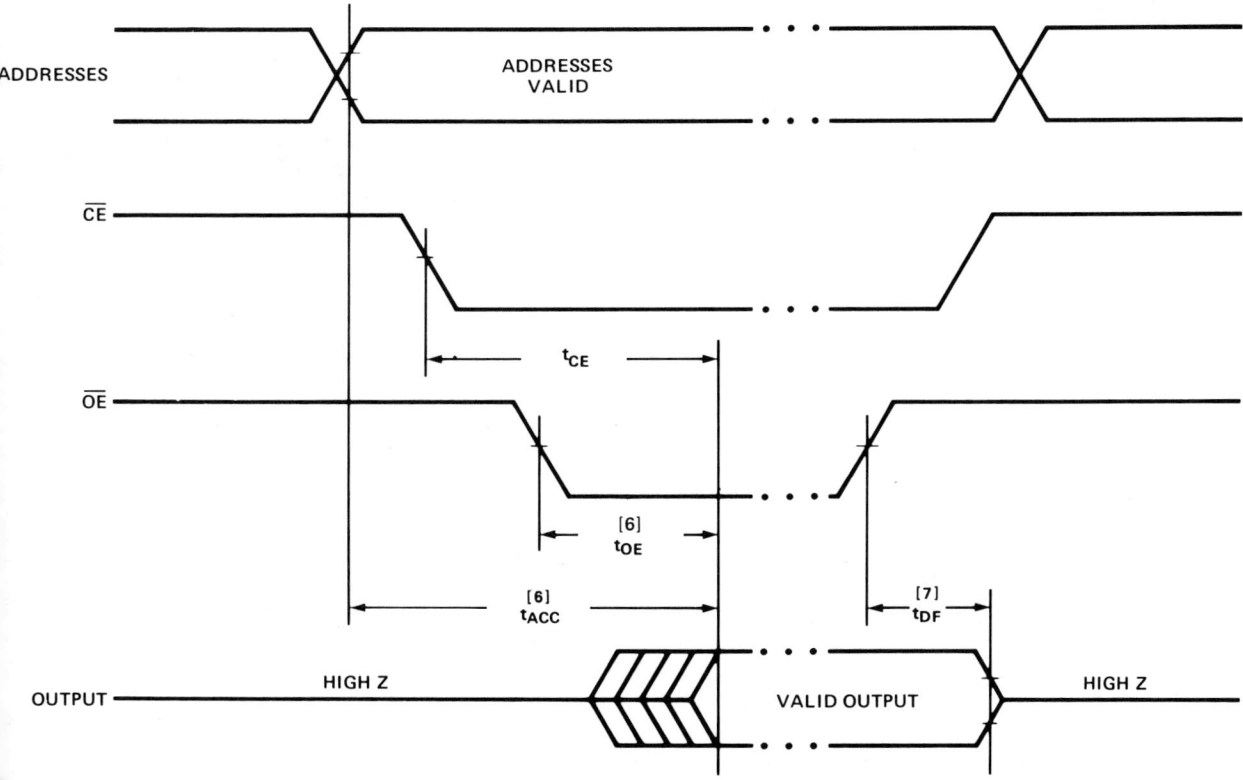

NOTE:
1. V_{CC} must be applied simultaneously or before V_{PP} and removed simultaneously or after V_{PP}.
2. V_{PP} may be connected directly to V_{CC} except during programming. The supply current would then be the sum of I_{CC} and I_{PP1}.
3. The tolerance of 0.6V allows the use of a driver circuit for switching the V_{PP} supply pin from V_{CC} in read to 25V for programming.
4. Typical values are for $T_A = 25°C$ and nominal supply voltages.
5. This parameter is only sampled and is not 100% tested.
6. $\overline{OE}$ may be delayed up to $t_{ACC} - t_{OE}$ after the falling edge of $\overline{CE}$ without impact on t_{ACC}.
7. t_{DF} is specified from $\overline{OE}$ or $\overline{CE}$, whichever occurs first.

ERASURE CHARACTERISTICS

The erasure characteristics of the 2716 are such that erasure begins to occur when exposed to light with wavelengths shorter than approximately 4000 Angstroms (Å). It should be noted that sunlight and certain types of fluorescent lamps have wavelengths in the 3000–4000Å range. Data show that constant exposure to room level fluorescent lighting could erase the typical 2716 in approximately 3 years, while it would take approximatley 1 week to cause erasure when exposed to direct sunlight. If the 2716 is to be exposed to these types of lighting conditions for extended periods of time, opaque labels are available from Intel which should be placed over the 2716 window to prevent unintentional erasure.

The recommended erasure procedure (see Data Catalog page 4-83) for the 2716 is exposure to shortwave ultraviolet light which has a wavelength of 2537 Angstroms (Å). The integrated dose (i.e., UV intensity $\times$ exposure time) for erasure should be a minimum of 15 W-sec/cm^2. The erasure time with this dosage is approximately 15 to 20 minutes using an ultraviolet lamp with a 12000 μW/cm^2 power rating. The 2716 should be placed within 1 inch of the lamp tubes during erasure. Some lamps have a filter on their tubes which should be removed before erasure.

DEVICE OPERATION

The five modes of operation of the 2716 are listed in Table I. It should be noted that all inputs for the five modes are at TTL levels. The power supplies required are a +5V V_{CC} and a V_{PP}. The V_{PP} power supply must be at 25V during the three programming modes, and must be at 5V in the other two modes.

TABLE I. MODE SELECTION

PINS MODE	$\overline{CE}$/PGM (18)	$\overline{OE}$ (20)	V_{PP} (21)	V_{CC} (24)	OUTPUTS (9-11, 13-17)
Read	V_{IL}	V_{IL}	+5	+5	D_{OUT}
Standby	V_{IH}	Don't Care	+5	+5	High Z
Program	Pulsed V_{IL} to V_{IH}	V_{IH}	+25	+5	D_{IN}
Program Verify	V_{IL}	V_{IL}	+25	+5	D_{OUT}
Program Inhibit	V_{IL}	V_{IH}	+25	+5	High Z

READ MODE

The 2716 has two control functions, both of which must be logically satisfied in order to obtain data at the outputs. Chip Enable ($\overline{CE}$) is the power control and should be used for device selection. Output Enable ($\overline{OE}$) is the output control and should be used to gate data to the output pins, independent of device selection. Assuming that addresses are stable, address access time (t_{ACC}) is equal to the delay from $\overline{CE}$ to output (t_{CE}). Data is available at the outputs 120 ns (t_{OE}) after the falling edge of $\overline{OE}$, assuming that $\overline{CE}$ has been low and addresses have been stable for at least $t_{ACC} - t_{OE}$.

STANDBY MODE

The 2716 has a standby mode which reduces the active power dissipation by 75%, from 525 mW to 132 mW. The 2716 is placed in the standby mode by applying a TTL high signal to the $\overline{CE}$ input. When in standby mode, the outputs are in a high impedence state, independent of the $\overline{OE}$ input.

OUTPUT DESELECTION

The outputs of two or more 2716s may be OR-tied together on the same data bus. Only one 2716 should have its output selected ($\overline{OE}$ low) to prevent data bus contention between 2716s in this configuration. The outputs of the other 2716s should be deselected by raising the $\overline{OE}$ input to a TTL high level.

PROGRAMMING

Initially, and after each erasure, all bits of the 2716 are in the "1" state. Data is introduced by selectively programming "0's" into the desired bit locations. Although only "0's" will be programmed, both "1's" and "0's" can be presented in the data word. The only way to change a "0" to a "1" is by ultraviolet light erasure.

The 2716 is in the programming mode when the V_{PP} power supply is at 25V and $\overline{OE}$ is at V_{IH}. The data to be programmed is applied 8 bits in parallel to the data output pins. The levels required for the address and data inputs are TTL.

When the address and data are stable, a 50 msec, active high, TTL program pulse is applied to the $\overline{CE}$/PGM input. A program pulse must be applied at each address location to be programmed. You can program any location at any time — either individually, sequentially, or at random. The program pulse has a maximum width of 55 msec. The 2716 must not be programmed with a DC signal applied to the $\overline{CE}$/PGM input.

Programming of multiple 2716s in parallel with the same data can be easily accomplished due to the simplicity of the programming requirements. Like inputs of the paralleled 2716s may be connected together when they are programmed with the same data. A high level TTL pulse applied to the $\overline{CE}$/PGM input programs the paralleled 2716s.

PROGRAM INHIBIT

Programming of multiple 2716s in parallel with different data is also easily accomplished. Except for $\overline{CE}$/PGM, all like inputs (including $\overline{OE}$) of the parallel 2716s may be common. A TTL level program pulse applied to a 2716's $\overline{CE}$/PGM input with V_{PP} at 25V will program that 2716. A low level $\overline{CE}$/PGM input inhibits the other 2716 from being programmed.

PROGRAM VERIFY

A verify should be performed on the programmed bits to determine that they were correctly programmed. The verify may be performed wth V_{PP} at 25V. Except during programming and program verify, V_{PP} must be at 5V.

PROGRAMMING PROCEDURES

Initially, and after each erasure, all 8,192 bits of the 2758 are in the "1" state. Information is introduced by selectively programming "0" into the desired bit locations. A programmed "0" can only be changed to a "1" by UV erasure.

The 2758 is programmed by applying a 50 ms, TTL programming pulse to the PD/PGM pin with the $\overline{CS}$ input high and the V_{PP} supply at 25V ±1V. Any location may be programmed at any time — either individually, sequentially, or randomly. The programming time for a single bit is only 50 ms and for all 8,192 bits is approximately 50 sec. The detailed programming specifications and timing waveforms are given in the following tables and figures.

CAUTION: The V_{CC} and V_{PP} supplies must be sequenced on and off such that V_{CC} is applied simultaneously or before V_{PP} and removed simultaneously or after V_{PP} to prevent damage to the 2758. The maximum allowable voltage during programming which may be applied to the V_{PP} with respect to ground is +26V. Care must be taken when switching the V_{PP} supply to prevent overshoot exceeding the 26-volt maximum specification. For convenience in programming, the 2758 may be verified with the V_{PP} supply at 25V ± 1V. During normal operation, however, V_{PP} must be at V_{CC}.

2758 PROGRAM CHARACTERISTICS [1]

T_A = 25°C ±5°C, V_{CC}[2] = 5V ±5%, V_{PP}[2,3] = 25V ±1V

D.C. Programming Characteristics

Symbol	Parameter	Min.	Typ.	Max.	Units	Test Conditions
I_{LI}	Input Current (for Any Input)			10	µA	V_{IN} = 5.25V/0.45
I_{PP1}	V_{PP} Supply Current			5	mA	PD/PGM = V_{IL}
I_{PP2}	V_{PP} Supply Current During Programming Pulse			30	mA	PD/PGM = V_{IH}
I_{CC}	V_{CC} Supply Current			100	mA	
V_{IL}	Input Low Level	−0.1		0.8	V	
V_{IH}	Input High Level	2.2		V_{CC}+1	V	
A_R	Select Reference Input Level	−0.1		0.8	V	I_{IN} = 10 µA

A.C. Programming Characteristics

Symbol	Parameter	Min.	Typ.	Max.	Units	
t_{AS}	Address Setup Time	2			µs	
t_{CSS}	$\overline{CS}$ Setup Time	2			µs	
t_{DS}	Data Setup Time	2			µs	
t_{AH}	Address Hold Time	2			µs	
t_{CSH}	$\overline{CS}$ Hold Time	2			µs	
t_{DH}	Data Hold Time	2			µs	
t_{DF}	Chip Deselect to Output Float Delay	0		120	ns	PD/PGM = V_{IL}
t_{CO}	Chip Select to Output Delay			120	ns	PD/PGM = V_{IL}
t_{PW}	Program Pulse Width	45	50	55	ms	
t_{PRT}	Program Pulse Rise Time	5			ns	
t_{PFT}	Program Pulse Fall Time	5			ns	

NOTES:
1. Intel's standard product warranty applies only to devices programmed to specifications described herein.
2. V_{CC} must be applied simultaneously or before V_{PP} and removed simultaneously or after V_{PP}. The 2758 must not be inserted into or removed from a board with V_{PP} at 25 ±1V to prevent damage to the device.
3. The maximum allowable voltage which may be applied to the V_{PP} pin during programming is +26V. Care must be taken when switching the V_{PP} supply to prevent overshoot exceeding this 26V maximum specification.

A.C. Conditions of Test:

V_{CC} . . . . 5V ±5%	Input Pulse Levels . . . . 0.8V to 2.2V
V_{PP} . . . . 25V ±1V	Input Timing Reference Level . . . . 1V and 2V
Input Rise and Fall Times (10% to 90%) . . . . 20 ns	Output Timing Reference Level . . . . 0.8V and 2V

PROGRAMMING WAVEFORMS

V_{PP} = 25V ±1V, V_{CC} = 5V ±5%

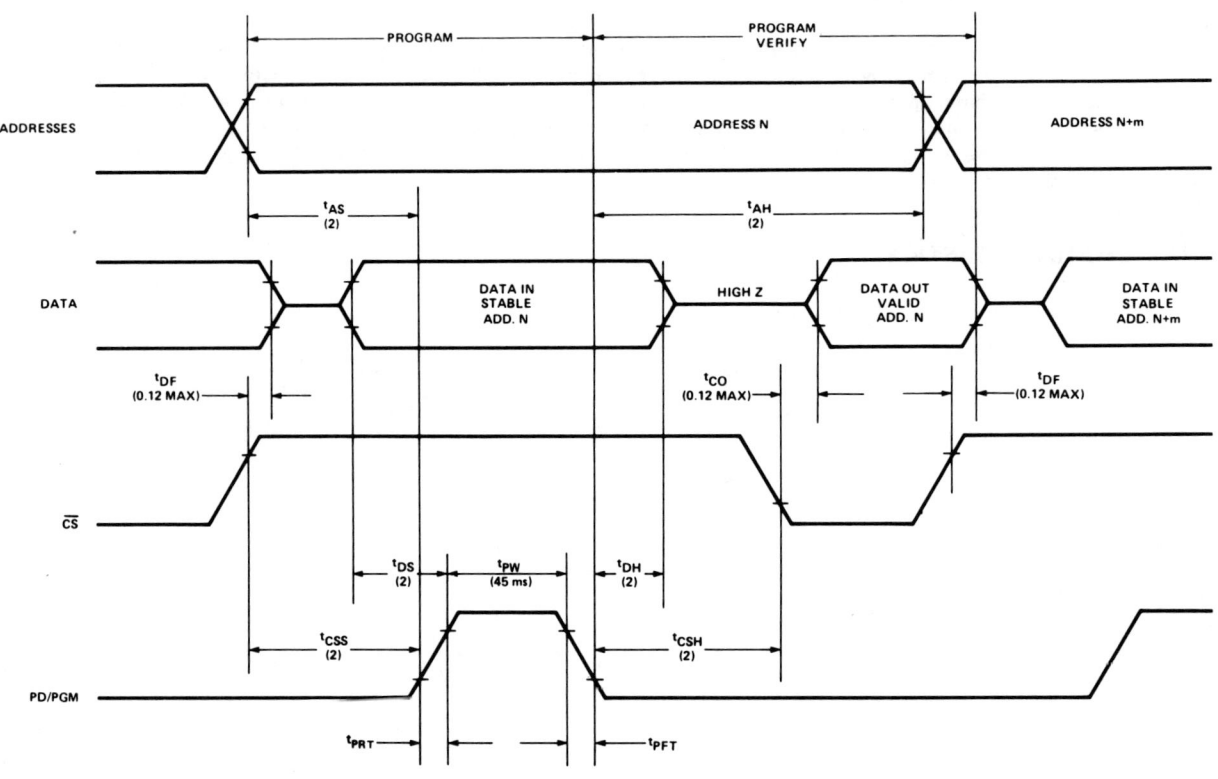

NOTE: ALL TIMES SHOWN IN PARENTHESES ARE MINIMUM TIMES AND ARE μSEC UNLESS OTHERWISE NOTED.

3602A, 3622A
3602, 3622
2048 BIT (512×4) HIGH SPEED PROM

	3602A-2 3622A-2	3602A 3622A	3602 3622
Typ. T_A(ns)	45	55	60
Max. T_A(ns)	60	70	70

- Low Power Dissipation --0.3mW/Bit
- Open Collector (3602A, 3602) or Three State (3622A, 3622) Outputs
- Simple Memory Expansion-- Chip Select Input Lead
- Replaces Two 256×4 PROMs Without Increasing Board Area
- Polycrystalline Silicon Fuse For Higher Reliability
- Hermetic 16-Pin DIP

The Intel® 3602A/3622A and 3602/3622 device families are 2048-bit bipolar PROMs organized as 512 words by 4 bits. The fast second generation 3602A/3622A joins its Intel predecessor, the 3602/3622, featuring 70 ns. A higher speed version, the 3602A-2/3622A-2, is now available at 60 ns. All 3602A/3622A specifications, except programming, are the same as the 3602/3622. Once programmed, the 3602A/3622A are interchangeable with the 3602/3622.

The PROMs are manufactured with all outputs initially logically high. Logic low levels can be electrically programmed in selected bit locations. Both open collector and three-state outputs are available. The power dissipation is typically 0.3 mW/bit.

The pin configuration of the PROMs is the same as the popular 1K bit, 256 × 4 PROMs with the exception that CS_2 (pin 14) is address A_8. The bit density of existing 256 × 4 PROM systems can be easily doubled without an increase in area with the 3602A/3622A or 3602/3622. These PROMs, like the 256 × 4 PROMs, are in 16-pin dual in-line package.

A pin compatible, mask programmable 3302/3322 ROM is available for large volume production systems initially using the 3602/3622. Please contact Intel directly for details on these ROMs.

PIN CONFIGURATION

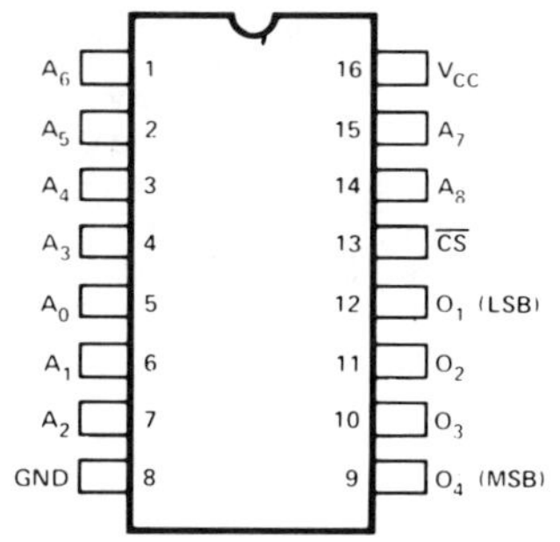

A_6	1	16	V_{CC}
A_5	2	15	A_7
A_4	3	14	A_8
A_3	4	13	$\overline{CS}$
A_0	5	12	O_1 (LSB)
A_1	6	11	O_2
A_2	7	10	O_3
GND	8	9	O_4 (MSB)

LOGIC SYMBOL

Inputs: $\overline{CS}$, A_0, A_1, A_2, A_3, A_4, A_5, A_6, A_7, A_8

Outputs: O_1, O_2, O_3, O_4

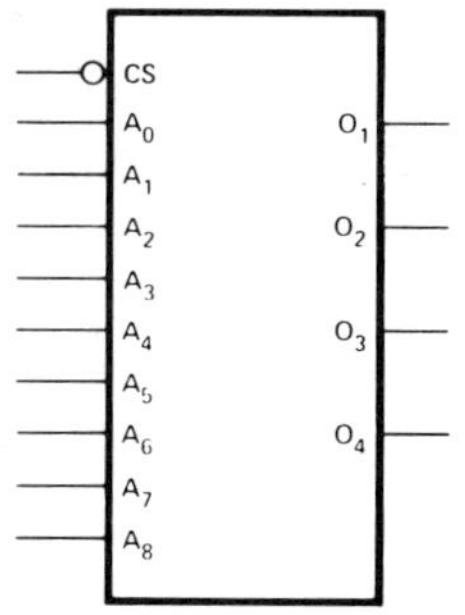

PROMS AND ROMS

PROGRAMMING

The programming specifications are described in the PROM/ROM Programming Instructions on page 3-55.

Absolute Maximum Ratings*

Temperature Under Bias	–65°C to +125°C
Storage Temperature	–65°C to +160°C
Output or Supply Voltages	–0.5V to 7 Volts
All Input Voltages	–1.6V to 5.6V
Output Currents	100mA

*COMMENT

Stresses above those listed under "Absolute Maximum Rating" may cause permanent damage to the device. This is a stress rating only and functional operation of the device at these or at any other condition above those indicated in the operational sections of this specification is not implied.

D. C. Characteristics: All Limits Apply for V_{CC}= +5.0V ±5%, T_A= 0°C to +75°C

Symbol	Parameter	Min.	Typ.[1]	Max.	Unit	Test Conditions
I_{FA}	Address Input Load Current		–0.05	–0.25	mA	V_{CC} = 5.25V, V_A = 0.45V
I_{FS}	Chip Select Input Load Current		–0.05	–0.25	mA	V_{CC} = 5.25V, V_S = 0.45V
I_{RA}	Address Input Leakage Current			40	µA	V_{CC} = 5.25V, V_A = 5.25V
I_{RS}	Chip Select Input Leakage Current			40	µA	V_{CC} = 5.25V, V_S = 5.25V
V_{CA}	Address Input Clamp Voltage		–0.9	–1.5	V	V_{CC} = 4.75V, I_A = –10mA
V_{CS}	Chip Select Input Clamp Voltage		–0.9	–1.5	V	V_{CC} = 4.75V, I_S = –10mA
V_{OL}	Output Low Voltage		0.3	0.45	V	V_{CC} = 4.75V, I_{OL} = 15mA
I_{CEX}	Output Leakage Current			40	µA	V_{CC} = 5.25V, V_{CE} = 5.25V
I_{CC}	Power Supply Current		110	140	mA	V_{CC}=5.25V, $V_{A0}\rightarrow V_{A8}$ = 0V, $\overline{CS}$ = 0V
V_{IL}	Input "Low" Voltage			0.85	V	V_{CC} = 5.0V
V_{IH}	Input "High" Voltage	2.0			V	V_{CC} = 5.0V

3622A, 3622A-2, 3622 ONLY

Symbol	Parameter	Min.	Typ.[1]	Max.	Unit	Test Conditions
$|I_O|$	Output Leakage for High Impedance Stage			40	µA	V_O=5.25V or 0.45V, V_{CC}=5.25V, $\overline{CS}$=2.4V
I_{SC}[2]	Output Short Circuit Current	–15	–25	–60	mA	V_{CC}=5.00V, T_A=25°C, V_O = 0V
V_{OH}	Output High Voltage	2.4			V	I_{OH}=–2.4mA, V_{CC}=4.75V

NOTES: 1. Typical values are at 25°C and at nominal voltage.
2. Unmeasured outputs are open during this test.

A. C. Characteristics $V_{CC} = +5V \pm 5\%$, $T_A = 0°C$ to $+75°C$

SYMBOL	PARAMETER	MAX. LIMIT			UNIT	CONDITIONS
		3602A-2 3622A-2	3602A 3622A	3602 3622		
t_{A++}, t_{A--} t_{A+-}, t_{A-+}	Address to Output Delay	60	70	70	ns	$\overline{CS} = V_{IL}$ to Select the PROM
t_{S++}	Chip Select to Output Delay	30	30	30	ns	
t_{S--}	Chip Select to Output Delay	30	30	30	ns	

Capacitance [1] $T_A = 25°C$, $f = 1$ MHz

SYMBOL	PARAMETER	LIMITS		UNIT	TEST CONDITIONS
		TYP.	MAX.		
C_{INA}	Address Input Capacitance	4	10	pF	$V_{CC} = 5V$ $V_{IN} = 2.5V$
C_{INS}	Chip-Select Input Capacitance	6	10	pF	$V_{CC} = 5V$ $V_{IN} = 2.5V$
C_{OUT}	Output Capacitance	7	12	pF	$V_{CC} = 5V$ $V_{OUT} = 2.5V$

NOTE 1: This parameter is only periodically sampled and is not 100% tested.

Switching Characteristics

Conditions of Test:
Input pulse amplitudes - 2.5V
Input pulse rise and fall times of 5 nanoseconds between 1 volt and 2 volts
Speed measurements are made at 1.5 volt levels
Output loading is 15 mA and 30 pF
Frequency of test - 2.5 MHz

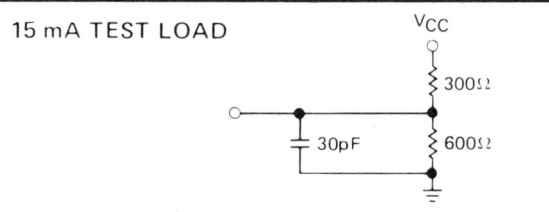

15 mA TEST LOAD

Waveforms

ADDRESS TO OUTPUT DELAY

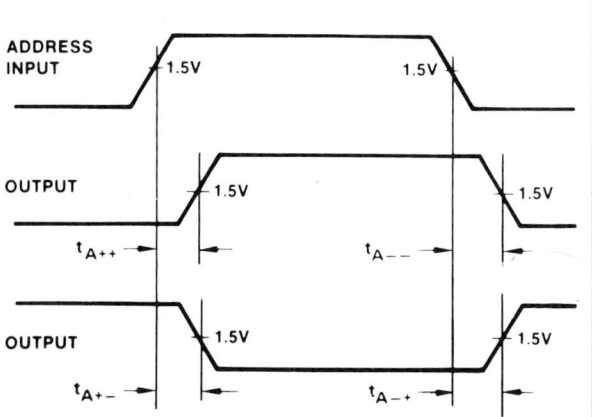

CHIP SELECT TO OUTPUT DELAY

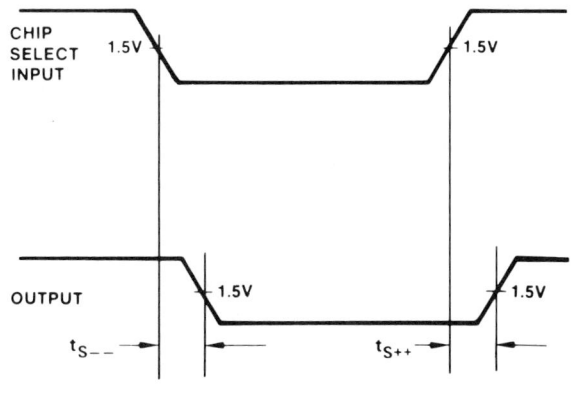

3604A, 3624A AND 3604, 3624
4096 BIT (512×8) HIGH SPEED PROM

	3604A-2 3624A-2	3604A 3624A	3604AL	3604 3624	3604-4 3624-4	3604L-6
Max. T_A(ns)	60	70	90	70	90	90
Max. I_{CC}(mA)	175	175	130/25*	190	190	140/45*

*Standby Current When The Chip is Deselected.

- **Fast Access Time**
  --60ns Max (3604A-2, 3624A-2)
- **Low Standby Power Dissipation**
  (3604AL) --32 μW/Bit Max
- **Open Collector (3604A, 3604) or Three State (3624A, 3624) Outputs**
- **Four Chip Select Inputs For Easy Memory Expansion**
- **Polycrystalline Silicon Fuse For Higher Reliability**
- **Hermetic 24 Pin DIP**

The Intel® 3604A/3624A and 3604/3624 device families are 4096-bit bipolar PROMs organized as 512 words by 8 bits. The fast second generation 3604A/3624A joins its Intel predecessor, the 3604/3624, featuring 70 ns. Higher speed PROMs, the 3604A-2/3624A-2, are now available at 60 ns. The 3604A/3624A families are lower in power dissipation than the 3604/3624 families. All 3604A/3624A specifications, except programming, are the same as or better than the 3604/3624. Once programmed, the 3604A/3624A are interchangeable with the 3604/3624.

The PROMs are manufactured with all outputs initially logically high. Logic low levels can be electrically programmed in selected bit locations. Both open collector and three-state outputs are available. Low standby power dissipation can be achieved with either the 3604AL or 3604L-6. The standby power dissipation is approximately 15% of the active power dissipation.

The 3604A/3624A and 3604/3624 families are available in a hermetic 24-pin dual in-line package. These PROMs are manufactured with the time-proven polycrystalline silicon fuse technology. A pin compatible, mask programmable ROM is available for large volume production of systems initially using the 3604/3624. Please contact Intel directly for details on these ROMs.

Mode/Pin Connection		Pin 22	Pin 24
READ:	3604A, 3604A-2, 3604, 3604-4, 3624A, 3624A-2, 3624, 3624-4	No Connect or 5V	5V
	3604AL, 3604L-6	+5V	Must be Left Open
PROGRAM:	3604A, 3604A-2, 3604, 3604-4, 3624A, 3624A-2, 3624, 3624-4	Pulsed 12.5V	Pulsed 12.5V
	3604AL, 3604L-6	Pulsed 12.5V	Pulsed 12.5V
STANDBY:	3604AL, 3604L-6	Power dissipation is automatically reduced whenever the 3604AL or 3604L-6 is deselected.	

PIN NAMES

$A_0 - A_8$	ADDRESS INPUTS
$\overline{CS_1} - \overline{CS_2}$ $CS_3 - CS_4$	CHIP SELECT INPUTS [1]
$O_1 - O_8$	DATA OUTPUTS

[1] To select the PROM $\overline{CS_1} = \overline{CS_2} = 0$ and $CS_3 = CS_4 = 1$.

PIN CONFIGURATION

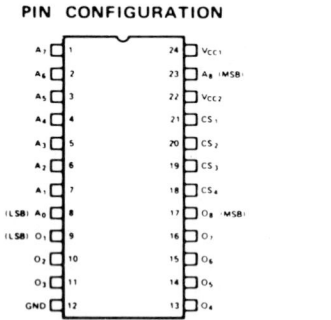

BLOCK DIAGRAM

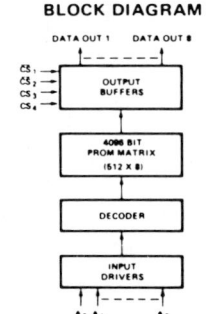

LOGIC SYMBOL

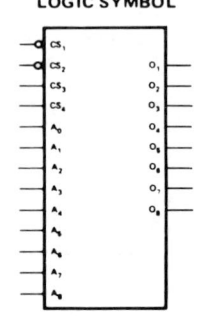

3604A/3604

PROGRAMMING

Absolute Maximum Ratings*

Temperature Under Bias . . . . -65°C to +125°C
Storage Temperature . . . . -65°C to +160°C
Output or Supply Voltages . . . . -0.5V to 7 Volts
All Input Voltages . . . . -1.6 to 5.5V
Output Currents 100mA

*COMMENT

Stresses above those listed under "Absolute Maximum Rating" may cause permanent damage to the device. This is a stress rating only and functional operation of the device at these or at any other condition above those indicated in the operational sections of this specification is not implied.

D. C. Characteristics: All Limits Apply for V_{CC}= +5.0V ±5%, T_A = 0°C to +75°C

Symbol	Parameter	Min.	Typ.[1]	Max.	Unit	Test Conditions
I_{FA}	Address Input Load Current		-0.05	-0.25	mA	V_{CC} = 5.25V, V_A = 0.45V
I_{FS}	Chip Select Input Load Current		-0.05	-0.25	mA	V_{CC} = 5.25V, V_S = 0.45V
I_{RA}	Address Input Leakage Current			40	µA	V_{CC} = 5.25V, V_A = 5.25V
I_{RS}	Chip Select Input Leakage Current			40	µA	V_{CC} = 5.25V, V_S = 5.25V
V_{CA}	Address Input Clamp Voltage		-0.9	-1.5	V	V_{CC} = 4.75V, I_A = -10 mA
V_{CS}	Chip Select Input Clamp Voltage		-0.9	-1.5	V	V_{CC} = 4.75V, I_S = -10 mA
V_{OL}	Output Low Voltage		0.3	0.45	V	V_{CC} = 4.75V, I_{OL} = 15 mA
I_{CEX}	Output Leakage Current			100	µA	V_{CC} = 5.25V, V_{CE} = 5.25V
I_{CC1}	Power Supply Current (3604A, 3604A-2, 3624A, and 3624A-2)		130	175	mA	V_{CC1} = 5.25V, $V_{A0} \to V_{A8}$ = 0V, $\overline{CS}_1 = \overline{CS}_2$ = 0V, $CS_3 = CS_4$ = 5.25V
I_{CC2}	Power Supply Current (3604, 3604-4, 3624, and 3624-4)		160	190	mA	V_{CC1} = 5.25V, $V_{A0} \to V_{A8}$ = 0V, $\overline{CS}_1 = \overline{CS}_2$ = 0V, $CS_3 = CS_4$ = 5.25V
I_{CC3}	Power Supply Current (3604AL) Active		100	130	mA	V_{CC2} = 5.25V, V_{CC1} = Open, $\overline{CS}_1 = \overline{CS}_2$ = 0.45V, $CS_3 = CS_4$ = 2.4V
	Standby		15	25	mA	$\overline{CS}_1 = \overline{CS}_2$ = 2.5V
I_{CC4}	Power Supply Current (3604L-6) Active			140	mA	V_{CC2} = 5.25V, V_{CC1} = Open, $\overline{CS}_1 = \overline{CS}_2$ = 0.45V, $CS_3 = CS_4$ = 2.4V
	Standby			45	mA	$\overline{CS}_1 = \overline{CS}_2$ = 2.5V
V_{IL}	Input "Low" Voltage			0.85	V	V_{CC} = 5.0V
V_{IH}	Input "High" Voltage	2.0			V	V_{CC} = 5.0V

3624A AND 3624 FAMILY ONLY

Symbol	Parameter	Min.	Typ.[1]	Max.	Unit	Test Conditions
$|I_O|$	Output Leakage for High Impedance Stage			100	µA	V_O = 5.25V or 0.45V, V_{CC} = 5.25V, $\overline{CS}_1 = \overline{CS}_2$ = 2.4V
I_{SC}[2]	Output Short Circuit Current	-15	-25	-60	mA	V_{CC} = 5.00V, T_A = 25°C, V_O = 0V
V_{OH}	Output High Voltage	2.4			V	I_{OH} = -2.4mA, V_{CC} = 4.75V

NOTES: 1. Typical values are at 25°C and at nominal voltage.
2. Unmeasured outputs are open during this test.

A. C. Characteristics $V_{CC} = +5V \pm 5\%$, $T_A = 0°C$ to $+75°C$

SYMBOL	PARAMETER	3604A, 3624A FAMILY MAXIMUM LIMITS (ns)			3064, 3624 FAMILY MAXIMUM LIMITS (ns)		
		3604A-2 3624A-2	3604A 3624A	3604AL	3064 3624	3604-4 3624-4	3604L-6
t_{A++}, t_{A--} t_{A+-}, t_{A-+}	Address to Output Delay	60	70	90	70	90	90
t_{S++}	Chip Select to Output Delay	30	30	30	30	30	30
t_{S--}	Chip Select to Output Delay	30	30	120	30	30	120

Capacitance[1] $T_A = 25°C$, $f = 1$ MHz

SYMBOL	PARAMETER	LIMITS		UNIT	TEST CONDITIONS
		TYP.	MAX.		
C_{INA}	Address Input Capacitance	4	10	pF	$V_{CC} = 5V$ $V_{IN} = 2.5V$
C_{INS}	Chip-Select Input Capacitance	6	10	pF	$V_{CC} = 5V$ $V_{IN} = 2.5V$
C_{OUT}	Output Capacitance	7	15	pF	$V_{CC} = 5V$ $V_{OUT} = 2.5V$

NOTE 1: This parameter is only periodically sampled and is not 100% tested.

Switching Characteristics

Conditions of Test:
Input pulse amplitudes - 2.5V
Input pulse rise and fall times of
  5 nanoseconds between 1 volt and 2 volts
Speed measurements are made at 1.5 volt levels
Output loading is 15 mA and 30 pF
Frequency of test - 2.5 MHz

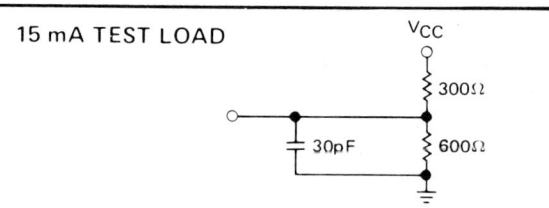
15 mA TEST LOAD

Waveforms

ADDRESS TO OUTPUT DELAY

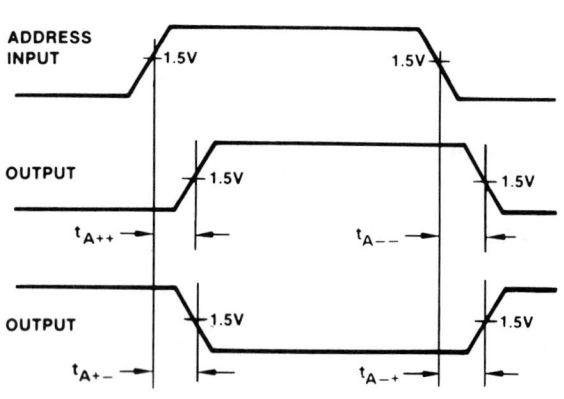

CHIP SELECT TO OUTPUT DELAY

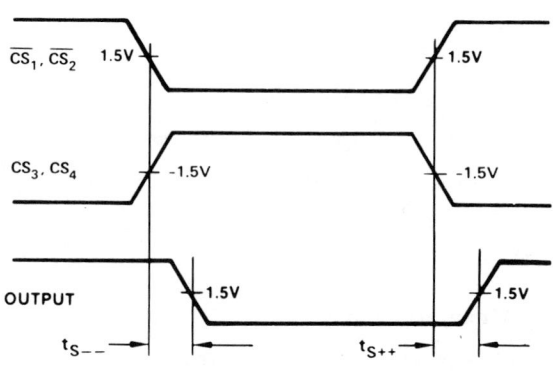

3605, 3625
HIGH SPEED 1K x 4 PROM

3605-2, 3625-2	60 ns Max.
3605, 3625	70 ns Max.

- **Fast Access Time: 45ns Typically**
- **Low Power Dissipation: 0.14mW/Bit Typically**
- **Simple Memory Expansion Two Chip Select Inputs**
- **Open Collector (3605) and Three-State (3625) Outputs**
- **Polycrystalline Silicon Fuse For Higher Reliability**
- **Hermetic 18 Pin DIP**

The Intel® 3605 and 3625 families are high density, 4096-bit bipolar PROMs organized as 1024 words by 4 bits. The 1024 by 4 organization gives ideal word or bit modularity for memory array expansion. The 3605 has open collector outputs and the 3625 has three-state outputs. The 3605 and 3625 are fully specified over the 0°C to 75°C temperature range with ±5% power supply variation. Maximum access times of 60 ns (3605-2/3625-2) and 70 ns (3605/3625) are available. The typical power dissipation is 0.14 mW/bit.

The 3605/3625 are packaged in an 18-pin dual in-line hermetic package with 300 milli-inch centers. Thus, twice the bit density can be achieved with the 3605/3625 in the same memory board area as 512 by 8-bit PROMs in 24-pin packages.

The highly reliable polycrystalline silicon fuse technology is used in the manufacturing of the 3605 and 3625 families. All outputs are initially a logical high and logic low levels can be electrically programmed in selected bit locations.

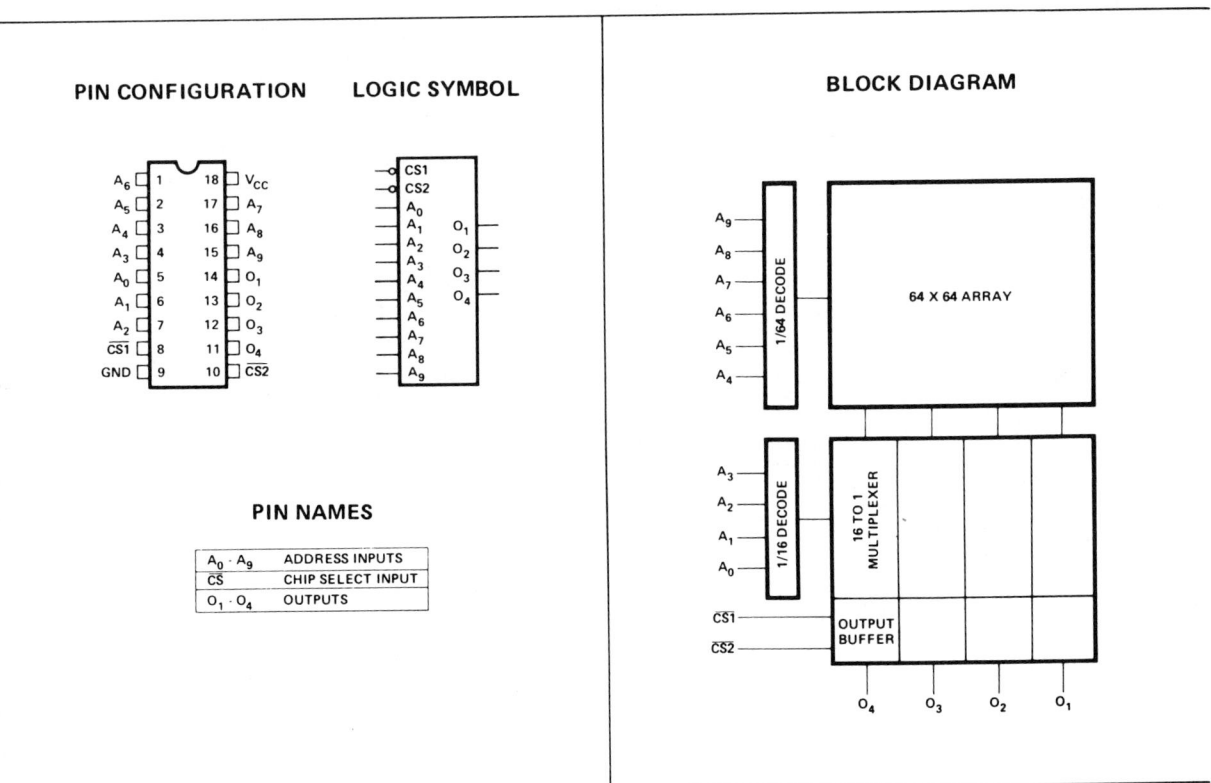

PROMS AND ROMS

PROGRAMMING

Absolute Maximum Ratings*

Temperature Under Bias . . . . −65°C to +125°C
Storage Temperature . . . . −65°C to +160°C
Output or Supply Voltages . . . . −0.5V to 7 Volts
All Input Voltages . . . . −1V to 5.5V
Output Currents 100mA

*COMMENT

Stresses above those listed under "Absolute Maximum Rating" may cause permanent damage to the device. This is a stress rating only and functional operation of the device at these or at any other condition above those indicated in the operational sections of this specification is not implied.

D. C. Characteristics: All Limits Apply for V_{CC} = +5.0V ±5%, T_A = 0°C to +75°C

Symbol	Parameter	Min.	Typ.[1]	Max.	Unit	Test Conditions
I_{FA}	Address Input Load Current		−0.05	−0.25	mA	V_{CC}=5.25V, V_A=0.45V
I_{FS}	Chip Select Input Load Current		−0.05	−0.25	mA	V_{CC}=5.25V, V_S=0.45V
I_{RA}	Address Input Leakage Current			40	μA	V_{CC}=5.25V, V_A=5.25V
I_{RS}	Chip Select Input Leakage Current			40	μA	V_{CC}=5.25V, V_S=5.25V
V_{CA}	Address Input Clamp Voltage		−0.9	−1.5	V	V_{CC}=4.75V, I_A=−10mA
V_{CS}	Chip Select Input Clamp Voltage		−0.9	−1.5	V	V_{CC}=4.75V, I_S=−10mA
V_{OL}	Output Low Voltage		0.3	0.45	V	V_{CC}=4.75V, I_{OL}=15mA
I_{CEX}	3605 Output Leakage Current			40	μA	V_{CC}=5.25V, V_{CE}=5.25V
I_{CC}	Power Supply Current		110	150	mA	V_{CC}=5.25V, V_{A0}→V_{A9}−0V, $\overline{CS}_1$=$\overline{CS}_2$=V_{IH}
V_{IL}	Input "Low" Voltage			0.85	V	V_{CC}=5.0V
V_{IH}	Input "High" Voltage	2.0			V	V_{CC}=5.0V

3625, 3625-2 ONLY

Symbol	Parameter	Min.	Typ.[1]	Max.	Unit	Test Conditions
$|I_O|$	Output Leakage for High Impedance Stage			40	μA	V_O=5.25V or 0.45V, V_{CC}=5.25V, $\overline{CS}_1$=$\overline{CS}_2$=2.4V
I_{SC}[2]	Output Short Circuit Current	−15	−25	−60	mA	V_O = 0V
V_{OH}	Output High Voltage	2.4			V	I_{OH}=−2.4mA, V_{CC}=4.75V

NOTES: 1. Typical values are at 25°C and at nominal voltage.
2. Unmeasured outputs are open during this test.

3605, 3625

A. C. Characteristics
V_{CC} = +5V ±5%, T_A = 0°C to +75°C

Symbol	Parameter	Max. Limits 3605-2 3625-2	Max. Limits 3605 3625	Unit	Conditions
t_{A++}, t_{A--} t_{A+-}, t_{A-+}	Address to Output Delay	60	70	ns	$\overline{CS}_1 = \overline{CS}_2 = V_{IL}$ to select the PROM.
t_{S++}	Chip Select to Output Delay	30	30	ns	
t_{S--}	Chip Select to Output Delay	30	30	ns	

Capacitance [1]
T_A = 25°C, f = 1 MHz

SYMBOL	PARAMETER	LIMITS TYP.	LIMITS MAX.	UNIT	TEST CONDITIONS
C_{INA}	Address Input Capacitance	4	10	pF	V_{CC} = 5V, V_{IN} = 2.5V
C_{INS}	Chip-Select Input Capacitance	6	10	pF	V_{CC} = 5V, V_{IN} = 2.5V
C_{OUT}	Output Capacitance	7	12	pF	V_{CC} = 5V, V_{OUT} = 2.5V

NOTE 1: This parameter is only periodically sampled and is not 100% tested.

Switching Characteristics

Conditions of Test:
Input pulse amplitudes - 2.5V
Input pulse rise and fall times of 5 nanoseconds between 1 volt and 2 volts
Speed measurements are made at 1.5 volt levels
Output loading is 15 mA and 30 pF
Frequency of test - 2.5 MHz

15mA TEST LOAD

Waveforms

ADDRESS TO OUTPUT DELAY

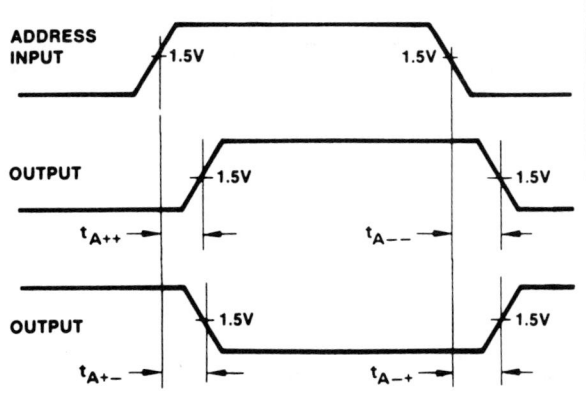

CHIP SELECT TO OUTPUT DELAY

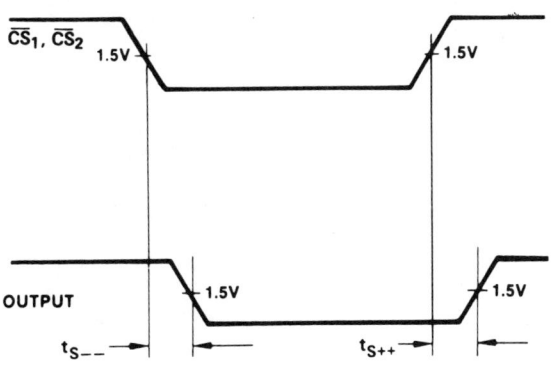

3608, 3628
8K (1K X 8) BIPOLAR PROM

3608, 3628	80 ns Max.
3608-4, 3628-4	100 ns Max.

- **Fast Access Time: 65 ns Typically**
- **Low Power Dissipation: 0.09mW/Bit Typically**
- **Four Chip Select Inputs for Easy Memory Expansion**
- **Open Collector (3608) and Three-State (3628) Outputs**
- **Hermetic 24-Pin DIP**
- **Polycrystalline Silicon Fuses for Higher Fuse Reliability**

The Intel® 3608/3628 are fully decoded 8192-bit PROMs organized as 1024 words by 8 bits. The worst case access time of 80 ns is specified over the 0°C to 75°C temperature range and 5% V_{CC} power supply tolerances. There are four chip selects provided to facilitate expanding 3608/3628s into larger PROM arrays. The PROMs use Schottky clamped TTL technology with polycrystalline silicon fuses. All outputs are initially high and logic low levels can be electrically programmed in selected bit locations.

Prior to the 8192 bit 3608/3628, the highest density bipolar PROM available was 4096 bits. The high density of the 3608/3628 now easily doubles the capacity without an increase in area on existing designs currently using 512 words by 8 bit PROMs. There is also little, if any, penalty in power since the 3608/3628 power/bit is approximately one-half that of 4K PROMs. The 3608/3628 are packaged in a hermetic 24-pin dual in-line package.

PIN CONFIGURATION

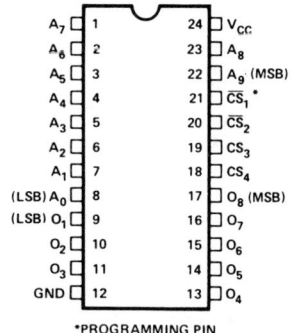

*PROGRAMMING PIN

BLOCK DIAGRAM

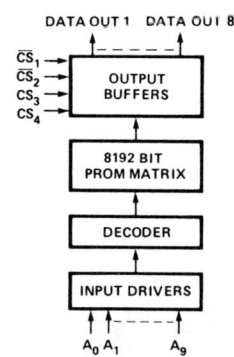

LOGIC SYMBOL

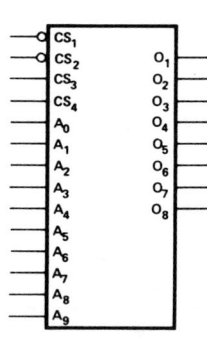

PIN NAMES

$A_0 - A_9$	ADDRESS INPUTS
$\overline{CS}_1 - \overline{CS}_2$ $CS_3 - CS_4$	CHIP SELECT INPUTS [1]
$O_1 - O_8$	DATA OUTPUTS

[1] To select the PROM $\overline{CS}_1 = \overline{CS}_2 = V_{IL}$ and $CS_3 = CS_4 = V_{IH}$

PROGRAMMING

ABSOLUTE MAXIMUM RATINGS*

Temperature Under Bias .... $-65°C$ to $+125°C$
Storage Temperature .... $-65°C$ to $+160°C$
Output or Supply Voltages .... $-0.5V$ to 7 Volts
All Input Voltages .... $-1V$ to $5.5V$
Output Currents 100mA

*COMMENT

Stresses above those listed under "Absolute Maximum Rating" may cause permanent damage to the device. This is a stress rating only and functional operation of the device at these or at any other condition above those indicated in the operational sections of this specification is not implied.

D.C. CHARACTERISTICS: All Limits Apply for V_{CC} = +5.0V ±5%, T_A = 0°C to +75°C

Symbol	Parameter	Min.	Typ.[1]	Max.	Unit	Test Conditions
I_{FA}	Address Input Load Current		-0.05	-0.25	mA	V_{CC}=5.25V, V_A=0.45V
I_{FS}	Chip Select Input Load Current		-0.05	-0.25	mA	V_{CC}=5.25V, V_S=0.45V
I_{RA}	Address Input Leakage Current			40	µA	V_{CC}=5.25V, V_A=5.25V
I_{RS}	Chip Select Input Leakage Current			40	µA	V_{CC}=5.25V, V_S=5.0V
V_{CA}	Address Input Clamp Voltage		-0.9	-1.5	V	V_{CC}=4.75V, I_A=-10mA
V_{CS}	Chip Select Input Clamp Voltage		-0.9	-1.5	V	V_{CC}=4.75V, I_S=-10mA
V_{OL}	Output Low Voltage		0.3	0.45	V	V_{CC}=4.75V, I_{OL}=10mA
I_{CEX}	3608 and 3608-4 Output Leakage Current			100	µA	V_{CC}=5.25V, V_{CE}=5.25V
I_{CC}	Power Supply Current		150	190	mA	V_{CC}=5.25V, V_{A0}→V_{A9}=0V, PROM deselected
V_{IL}	Input "Low" Voltage			0.85	V	V_{CC}=5.0V
V_{IH}	Input "High" Voltage	2.0			V	V_{CC}=5.0V

3628, 3628-4 ONLY

Symbol	Parameter	Min.	Typ.[1]	Max.	Unit	Test Conditions
$|I_O|$	Output Leakage for High Impedance State			100	µA	V_O=5.25V or 0.45V, V_{CC}=5.25V, $\overline{CS}_1$=$\overline{CS}_2$=2.4V
I_{SC}[2]	Output Short Circuit Current	-20	-25	-80	mA	V_O = 0V
V_{OH}	Output High Voltage	2.4	3.4		V	I_{OH}=-2.4mA, V_{CC}=4.75V

NOTES: 1. Typical values are at 25°C and at nominal voltage.
2. Unmeasured outputs are open during this test.

A.C. CHARACTERISTICS $V_{CC} = +5V \pm 5\%$, $T_A = 0°C$ to $+75°C$

SYMBOL	PARAMETER	MAX. LIMITS 3608 3628	MAX. LIMITS 3608-4 3628-4	UNIT	CONDITIONS
t_A	Address to Output Delay	80	100	ns	$\overline{CS}_1 = \overline{CS}_2 = V_{IL}$
t_{EN}	Output Enable Time	40	45	ns	and $CS_3 = CS_4 = V_{IH}$
t_{DIS}	Output Disable Time	40	45	ns	to select the PROM.

CAPACITANCE[1] $T_A = 25°C$, f = 1 MHz

SYMBOL	PARAMETER	TYP.	MAX.	UNIT	TEST CONDITIONS
C_{INA}	Address Input Capacitance	4	10	pF	$V_{CC} = 5V$ $V_{IN} = 2.5V$
C_{INS}	Chip-Select Input Capacitance	6	10	pF	$V_{CC} = 5V$ $V_{IN} = 2.5V$
C_{OUT}	Output Capacitance	7	15	pF	$V_{CC} = 5V$ $V_{OUT} = 2.5V$

NOTE 1: This parameter is only periodically sampled and is not 100% tested.

SWITCHING CHARACTERISTICS

Conditions of Test:
Input pulse amplitudes - 2.5V
Input pulse rise and fall times of
  5 nanoseconds between 1 volt and 2 volts
Speed measurements are made at 1.5 volt levels
Output loading is 15 mA and 30 pF
Frequency of test - 2.5 MHz

15mA TEST LOAD

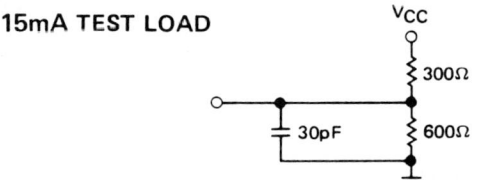

WAVEFORMS

ADDRESS TO OUTPUT DELAY

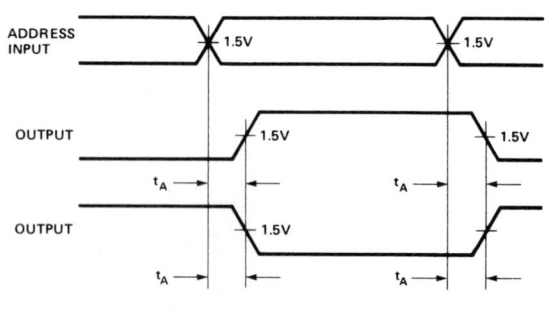

CHIP SELECT TO OUTPUT DELAY

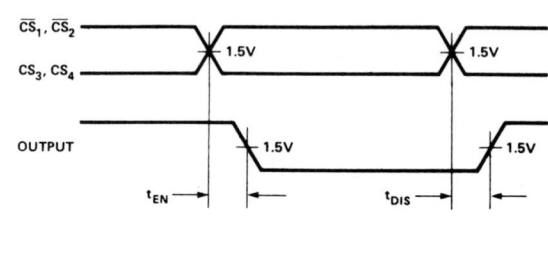

ROM/PROM PROGRAMMING INSTRUCTIONS

MOS EPROMs

A. Erasure Procedure

As stated in the EPROM related data sheets, the recommended erasure procedure to use with EPROMs is to illuminate the window with a UV lamp which has a wavelength of 2537 Angstroms (Å). The data sheets specify a distance of 1 inch and erase times of 10–45 minutes, depending on the type of device and UV lamp. Actually, the amount of time required to erase a device can be concisely stated in terms of the amount of UV energy incident to the window, expressed in Watt-seconds per square centimeter (W-sec/cm^2). Table III lists the required integrated dosgae (UV intensity × exposure time) for the EPROMs currently in production by Intel.

Table III. Required Erase Energy for Device Types

Device Type	2537Å Erase Energy
1702A/4702A	6 W-sec/cm^2
2708/8708	15 W-sec/cm^2
2716	15 W-sec/cm^2
8748	15 W-sec/cm^2
8755	15 W-sec/cm^2

The erase energy expressed in Table III includes a guardband to ensure complete erasure of all bits. *It is not sufficient to monitor "first bit" erasure to determine erasure time, as some other bits in the array may not be erased.*

A1. UV Sources

There are several models of UV lamps that can be used to erase EPROMs (see Table IV). The model numbers in the table refer to lamps manufactured by Ultra Violet Products of San Gabriel, California. In addition, there are several other manufacturers, including Data I/O (Issaquah, Wash.), PROLOG (Monterey, Calif.), Prometrics (Chicago, Ill.), and Turner Designs (Mt. View, Calif.). The individual manufacturers should be consulted for detailed product descriptions. Also shown in the table are typical erase times for various combinations of Intel PROMs and lamp intensities.

Table IV.

Model	Power Rating	Minimum Erase Time for Indicated Dosage Without a Filter Over the Bulb	
		6 W-sec 1702A, 4702A	15 W-sec 2708, 8708, 8755 2716, 8748
R-52	13000 µW/cm^2	7.7 min	19.2 min
S-52	12000 µW/cm^2	8.3 min	20.7 min
S-68	12000 µW/cm^2	8.3 min	20.7 min
UVS-54	5700 µW/cm^2	17.5 min	43.8 min
UVS-11	5500 µW/cm^2	18.2 min	45.6 min

According to the manufacturers, the output of the UV lamp bulbs decrease with age. The output of the lamp should be verified periodically to ensure that adequate intensities are maintained. If this is not done, bits may be partially erased which will interfere with later programming and/or operation at high temperature.

For lamps other than those listed, the erase time can be determined by using a UV intensity meter, such as the Ultra Violet Products Model J-225. When a meter is used, the intensity should be measured at the same position (distance from the lamp) as the EPROMs to be erased. This will require careful positioning to insure that the sensor will receive the same amount of UV light that the window of the EPROM will receive.

The sensors used with most UV intensity meters show reduced output with constant exposure to UV light. Therefore, they should not be permanently placed inside the erasure enclosure, they should only be used for periodic measurements.

B. 1702A/1702AL Family Programming

The 1702A/1702AL is erased by exposure to high intensity short wave ultraviolet light at a wavelength of 2537Å. The recommended integrated dose (i.e., UV intensity × exposure time) is 6 W-sec/cm². An example of an ultraviolet source which can erase the 1702A/1702AL in 10 to 20 minutes is the Model S52 short wave ultraviolet lamp. The lamp should be used without short wave filters and the PROM should be placed within 1 inch away from the lamp tubes.

Initially, all 2048 bits of the PROM are in the "0" state (output low). Information is introduced by selectively programming "1"s (output high) in the proper bit locations.

Word address selection is done by the same decoding circuitry used in the READ mode. All 8 address bits must be in the binary complement state when pulsed V_{CC} and V_{GG} move to their negative levels. The addresses must be held in their binary complement state for a minimum of 25 μsec after V_{DD} and V_{GG} have moved to their negative levels. The addresses must then make the transition to their true state a minimum of 10 μsec before the program pulse is applied. The addresses should be programmed in the sequence 0 through 255 for a minimum of 32 times. The eight output terminals are used as data inputs to determine the information pattern in the 8 bits of each word. A low data input level (–48V) will program a "1" and a high data input level (ground) will leave a "0". All 8 bits of one word are programmed simultaneously by setting the desired bit information patterns on the data input terminals.

During the programming, V_{GG}, V_{DD} and the Program Pulse are pulsed signals. See page 2 of the data sheet for required pin connections during programming.

1702A, 1702AL
D.C. and Operating Characteristics for Programming Operation

$T_A = 25°C$, $V_{CC} = 0V$, $V_{BB} = +12V \pm 10\%$, $\overline{CS} = 0V$ unless otherwise noted

Symbol	Test	Min.	Typ.	Max.	Unit	Conditions
I_{LI1P}	Address and Data Input Load Current			10	mA	$V_{IN} = -48V$
I_{LI2P}	Program and V_{GG} Load Current			10	mA	$V_{IN} = -48V$
I_{BB}[1]	V_{BB} Supply Load Current		10		mA	
I_{DDP}[2]	Peak I_{DD} Supply Load Current		200		mA	$V_{DD} = V_{PROG} = -48V$, $V_{GG} = -35V$
V_{IHP}	Input High Voltage			0.3	V	
V_{IL1P}	Pulsed Data Input Low Voltage	-46		-48	V	
V_{IL2P}	Address Input Low Voltage	-40		-48	V	
V_{IL3P}	Pulsed Input Low V_{DD} and Program Voltage	-46		-48	V	
V_{IL4P}	Pulsed Input Low V_{GG} Voltage	-35		-40	V	

Notes: 1. The V_{BB} supply must be limited to 100mA max. current to prevent damage to the device.
2. I_{DDP} flows only during V_{DD}, V_{GG} on time. I_{DDP} should not be allowed to exceed 300mA for greater than 100μsec. Average power supply current I_{DDP} is typically 40mA at 20% duty cycle.

ROM/PROM PROGRAMMING INSTRUCTIONS

1702A, 1702AL
A.C. Characteristics for Programming Operation

$T_{AMBIENT}$ = 25°C, V_{CC} = 0V, V_{BB} = +12V ±10%, $\overline{CS}$ = 0V unless otherwise noted

Symbol	Test	Min.	Typ.	Max.	Unit	Conditions
	Duty Cycle (V_{DD}, V_{GG})			20	%	
$t_{\phi PW}$	Program Pulse Width		2	3	ms	V_{GG} = –35V, V_{DD} = V_{PROG} = –48V
t_{DW}	Data Set-Up Time	25			µs	
t_{DH}	Data Hold Time	10			µs	
t_{VW}	V_{DD}, V_{GG} Set-Up	100			µs	
t_{VD}	V_{DD}, V_{GG} Hold	10		100	µs	
t_{ACW}	Address Complement Set-Up	25			µs	
t_{ACH}	Address Complement Hold	25			µs	
t_{ATW}	Address True Set-Up	10			µs	
t_{ATH}	Address True Hold	10			µs	

PROGRAM WAVEFORMS

Conditions of Test:

Input pulse rise and fall times ≤ 1µsec
$\overline{CS}$ = 0V

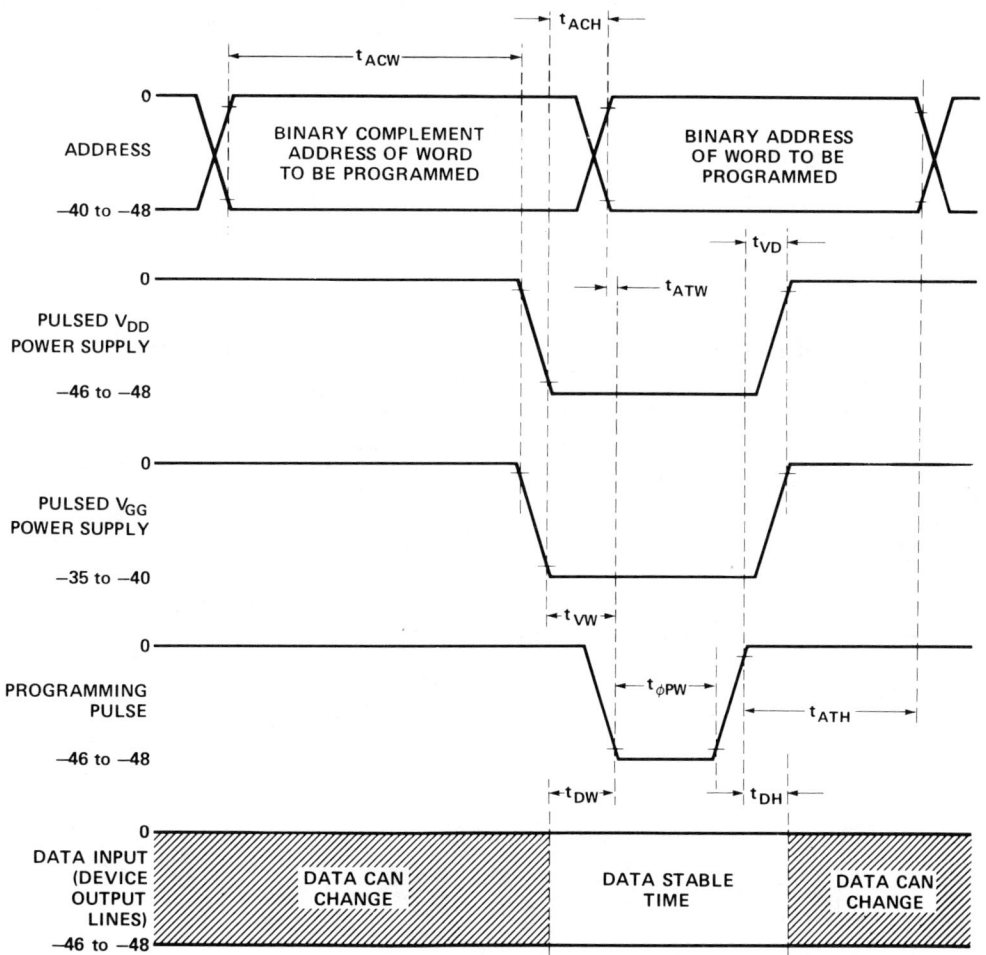

C. 2708/2704 Family Programming

Initially, and after each erasure, all 8192/4096 bits of the 2708/2704 are in the "1" state (output high). Information is introduced by selectively programming "0" into the desired bit locations. A programmed "0" can only be changed to a "1" by UV erasure.

The circuit is set up for programming operation by raising the CS/WE input (pin 20) to +12V. The word address is selected in the same manner as in the read mode. Data to be programmed are presented, 8 bits in parallel, to the data output lines (O_1-O_8). Logic levels for address and data lines and the supply voltages are the same as for the read mode. After address and data set up, one program pulse per address is applied to the program input (pin 18). One pass through all addresses is defined as a program loop. The number of loops (N) required is a function of the program pulse width (t_{PW}) according to N x t_{PW} ≥ 100 ms.

The width of the program pulse is from 0.1 to 1 ms. The number of loops (N) is from a minimum of 100 (t_{PW} = 1 ms) to greater than 1000 (t_{PW} = 0.1 ms). There must be N successive loops throuhg all 1024 addresses. *It is not permitted to apply N program pulses to an address and then change to the next address to be programmed.* Caution should be observed regarding the end of a program sequence. The CS/WE falling edge transition must occur before the first address transition when changing from a program to a read cycle. The program pin should also be pulled down to V_{ILP} with an active instead of a passive device. This pin will source a small amount of current (I_{ILL}) when CS/WE is at V_{IHW} (12V) and the program pulse is at V_{ILP}.

Programming Examples (Using N x t_{PW} ≥ 100 ms)

Example 1:   All 8096 bits are to be programmed with a 0.5 ms program pulse width.

The minimum number of program loops is 200. One program loop consists of words 0 to 1023.

Example 2:   Words 0 to 100 and 500 to 600 are to be programmed. All other bits are "don't care". The program pulse width is 0.75 ms.

The minimum number of program loops is 133. One program loop consists of words 0 to 1023. The data entered into the "don't care" bits should be all 1's.

Example 3:   Same requirements as example 2, but the PROM is now to be *updated* to include data for words 750 to 770.

The minimum number of program loops is 133. One program loop consists of words 0 to 1023. The data entered into the "don't care" bits should be all 1's. Addresses 0 to 100 and 500 to 600 must be re-programmed with their original data pattern.

2704, 2708
PROGRAM CHARACTERISTICS

T_A = 25°C, V_{CC} = 5V ±5%, V_{DD} = +12V ±5%, V_{BB} = -5V ±5%, V_{SS} = 0V, Unless Otherwise Noted.

D.C. Programming Characteristics

Symbol	Parameter	Min.	Typ.	Max.	Units	Test Conditions
I_{LI}	Address and $\overline{CS}$/WE Input Sink Current			10	µA	V_{IN} = 5.25V
I_{IPL}	Program Pulse Source Current			3	mA	
I_{IPH}	Program Pulse Sink Current			20	mA	
I_{DD}	V_{DD} Supply Current		50	65	mA	Worst Case Supply Currents: All Inputs High $\overline{CS}$/WE = 5V; T_A = 0°C
I_{CC}	V_{CC} Supply Current		6	10	mA	
I_{BB}	V_{BB} Supply Current		30	45	mA	
V_{IL}	Input Low Level (except Program)	V_{SS}		0.65	V	
V_{IH}	Input High Level for all Addresses and Data	3.0		V_{CC}+1	V	
V_{IHW}	$\overline{CS}$/WE Input High Level	11.4		12.6	V	Referenced to V_{SS}
V_{IHP}	Program Pulse High Level	25		27	V	Referenced to V_{SS}
V_{ILP}	Program Pulse Low Level	V_{SS}		1	V	V_{IHP} - V_{ILP} = 25V min.

ROM/PROM PROGRAMMING INSTRUCTIONS

A.C. Programming Characteristics

Symbol	Parameter	Min.	Typ.	Max.	Units
t_{AS}	Address Setup Time	10			µs
t_{CSS}	$\overline{CS}$/WE Setup Time	10			µs
t_{DS}	Data Setup Time	10			µs
t_{AH}	Address Hold Time	1			µs
t_{CH}	$\overline{CS}$/WE Hold Time	.5			µs
t_{DH}	Data Hold Time	1			µs
t_{DF}	Chip Deselect to Output Float Delay	0		120	ns
t_{DPR}	Program To Read Delay			10	µs
t_{PW}	Program Pulse Width	.1		1.0	ms
t_{PR}	Program Pulse Rise Time	.5		2.0	µs
t_{PF}	Program Pulse Fall Time	.5		2.0	µs

NOTE: Intel's standard product warranty applies only to devices programmed to specifications described herein.

2704, 2708 Programming Waveforms

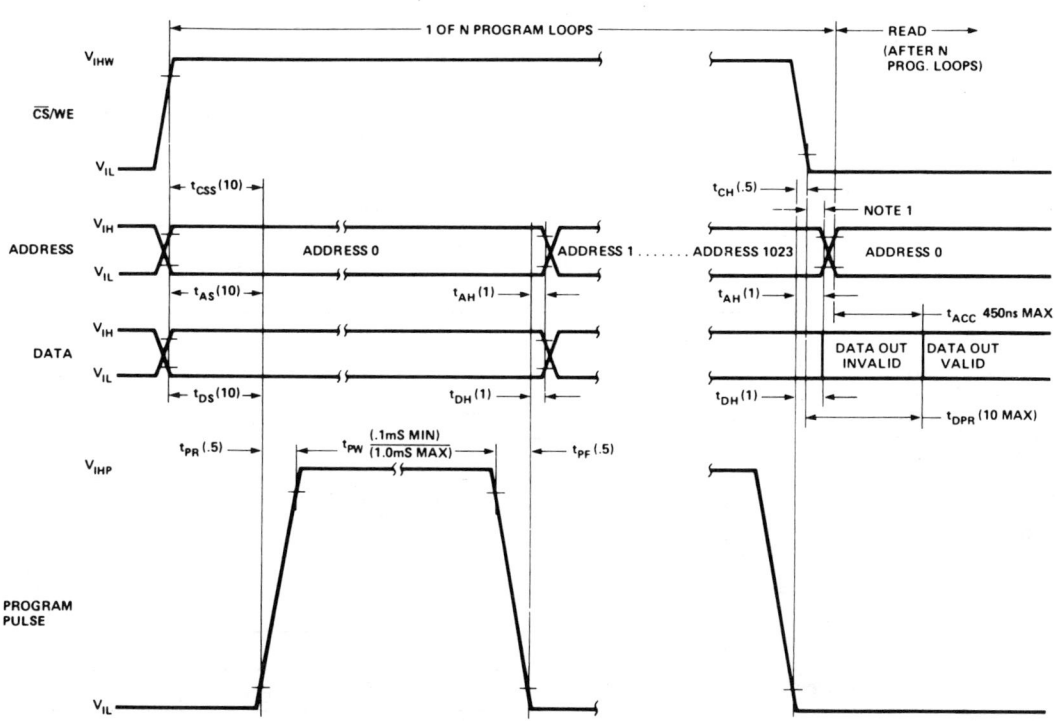

NOTE 1. THE $\overline{CS}$/WE TRANSITION MUST OCCUR AFTER THE PROGRAM PULSE TRANSITION AND BEFORE THE ADDRESS TRANSITION.
NOTE 2. NUMBERS IN ( ) INDICATE MINIMUM TIMING IN µS UNLESS OTHERWISE SPECIFIED.

D. 2716 Programming

Initially, and after each erasure, all 16,384 bits of the 2716 are in the "1" state. Information is introduced by selectively programming "0" into the desired bit locations. A programmed "0" can only be changed to a "1" by UV erasure.

The 2716 is programmed by applying a 50 ms, TTL programming pulse to the PD/PGM pin with the $\overline{CS}$ input high and the V_{PP} supply at 25V ±1V. Any location may be programmed at any time — either indvidually, sequentially, or randomly. The programming time for a single bit is only 50 ms and for all 16,384 bits is approximately 100 sec. The detailed programming specifications and timing waveforms are given in the following tables and figures.

CAUTION: The V_{CC} and V_{PP} supplies must be sequenced on and off such that V_{CC} is applied simultaneously or before V_{PP} and removed simultaneously or after V_{PP} to prevent damage to the 2716. The maximum allowable voltage during programming which may be applied to the V_{PP} with respect to ground is +26V. Care must be taken when switching the V_{PP} supply to prevent overshoot exceeding the 26-volt maximum specification. For convenience in programming, the 2716 may be verified with the V_{PP} supply at 25V ±1V. During normal read operation, however, V_{PP} must be at V_{CC}.

2716 PROGRAM CHARACTERISTICS[1]

T_A = 25°C ±5°C, V_{CC}[2] = 5V ±5%, V_{PP}[2,3] = 25V ±1V

D.C. Programming Characteristics

Symbol	Parameter	Min.	Typ.	Max.	Units	Test Conditions
I_{LI}	Input Current (for Any Input)			10	µA	V_{IN} = 5.25V/0.45
I_{PP1}	V_{PP} Supply Current			5	mA	PD/PGM = V_{IL}
I_{PP2}	V_{PP} Supply Current During Programming Pulse			30	mA	PD/PGM = V_{IH}
I_{CC}	V_{CC} Supply Current			100	mA	
V_{IL}	Input Low Level	−0.1		0.8	V	
V_{IH}	Input High Level	2.2		V_{CC}+1	V	

A.C. Programming Characteristics

Symbol	Parameter	Min.	Typ.	Max.	Units	
t_{AS}	Address Setup Time	2			µs	
t_{CSS}	$\overline{CS}$ Setup Time	2			µs	
t_{DS}	Data Setup Time	2			µs	
t_{AH}	Address Hold Time	2			µs	
t_{CSH}	$\overline{CS}$ Hold Time	2			µs	
t_{DH}	Data Hold Time	2			µs	
t_{DF}	Chip Deselect to Output Float Delay	0		120	ns	PD/PGM = V_{IL}
t_{CO}	Chip Select to Output Delay			120	ns	PD/PGM = V_{IL}
t_{PW}	Program Pulse Width	45	50	55	ms	
t_{PRT}	Program Pulse Rise Time	5			ns	
t_{PFT}	Program Pulse Fall Time	5			ns	

NOTES:
1. Intel's standard product warranty applies only to devices programmed to specifications described herein.
2. V_{CC} must be applied simultaneously or before V_{PP} and removed simultaneously or after V_{PP}. The 2716 must not be inserted into or removed from a board with V_{PP} at 25 ±1V to prevent damage to the device.
3. The maximum allowable voltage which may be applied to the V_{PP} pin during programming is +26V. Care must be taken when switching the V_{PP} supply to prevent overshoot exceeding this 26V maximum specification.

ROM/PROM PROGRAMMING INSTRUCTIONS

A.C. Conditions of Test:

V_{CC} .... 5V ±5%
V_{PP} .... 25V ±1V
Input Rise and Fall Times (10% to 90%) .... 20 ns
Input Pulse Levels .... 0.8V to 2.2V
Input Timing Reference Level .... 1V and 2V
Output Timing Reference Level .... 0.8V and 2V

PROGRAMMING WAVEFORMS

V_{PP} = 25V ±1V, V_{CC} = 5V ±5%

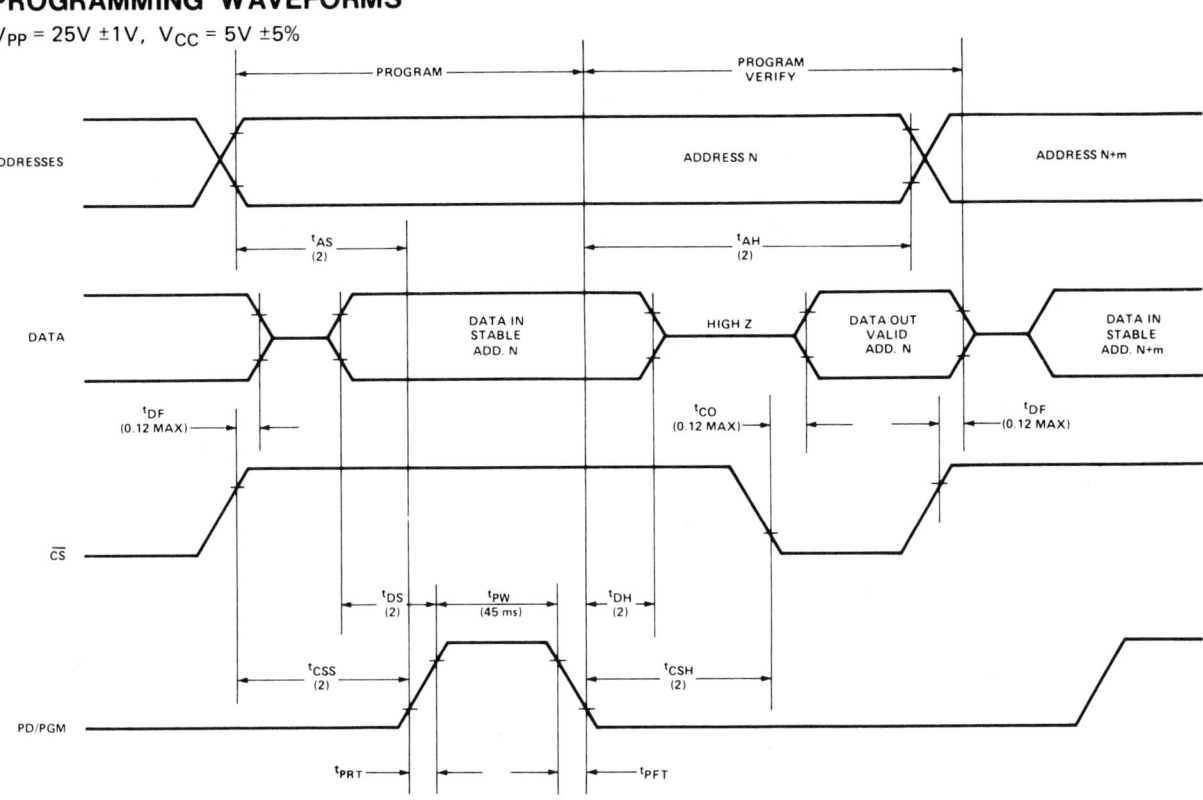

NOTE: ALL TIMES SHOWN IN PARENTHESES ARE MINIMUM TIMES AND ARE μSEC UNLESS OTHERWISE NOTED.

E. 8748/8755 Programming

Initially, and after each erasure, all bits of the EPROM portions of the 8748 and 8755 are in the "1" state. Information is introduced by selectively programming "0" into the desired bit locations. A programmed "0" can only be changed to a "1" by UV erasure.

The EPROM portions of the 8748 and 8755 are programmed on the Intel® Universal PROM Programmer (UPP). The UPP and its related personality cards for the 8748 and 8755 are described beginning on page 13-45 of this catalog.

III. BIPOLAR PROM PROGRAMMING

A. 3621, 2K, 4K, and 8K PROM Programming

All Intel bipolar PROMs except for the 3601/3601-1 are programmed with the algorithm described below. (The 3601/3601-1 programming algorithm is described in the below Section IIIB.) This algorithm was developed specifically to program the 3602A/3622A, 3604A/3624A, 3605/3625, and 3608/3628. *The algorithm described in this section must be used on the aforementioned PROMs to insure properly and reliably programmed fuses.* This algorithm may also be used to program the 3621, 3602/3622, and 3604/3624 PROM families. It is preferred over previously published Intel algorithms for these PROMs for increased programming yields.

Initially, all bits are in a logic 1 (high) state. To program a bit to a logic 0 (low) state, it is necessary to force 5 mA into the output to be programmed. A series of program pulses must also be applied to the V_{CC} power supply and to any one of the logically low true chip select ($\overline{CS}$) inputs. The logic level of the other chip selects, in the case of PROMs with multiple chip selects, should be such that the PROM is selected during verification.

Program pulses are applied to all outputs of a word in a cycle time. The program pulses are multiplexed during a cycle time to each output of the word to be programmed. If desired, a N word by 8-bit PROM may have its words programmed in two separate groups — the four lower order bits (O_1 to O_4) and the four higher order bits (O_5 to O_8). The operation in this manner is the same as for a N word by 4-bit PROM. For fastest programming time, it is preferred that all eight outputs be programmed at the same time.

The programming specifications are given in Table V and the programming waveforms are shown in Figure 1. The programming procedure (described with nominal specifications) is as follows:

1. A 5 mA current must be forced into the output to be programmed by a current source. The current source must be clamped to V_{CC} by a silicon diode. All the other outputs must be floating until it is their turn for programming. The V_{CC} power supply and the chip select ($\overline{CS}$) input is pulsed as shown in Figures 1 and 2. The width of V_{CC} is linearly increased from 0.2 μs to 8 μs according to the ramp time shown in Figure 3. The total ramp time for a group of four outputs is 180 ms and 360 ms for a group of eight outputs.

   The V_{CC} program pulses are multiplexed during a cycle time to the outputs of the word to be programmed. The cycle time (t_{CYC}) between the V_{CC} program pulses to the same output will increase as the V_{CC} program pulse width increases from 0.2 μs to 8 μs. The time (t_D) between V_{CC} pulses of two different outputs is constant at 1.8 μs.

2. All outputs must be continuously monitored for programming verification. This verification must occur after V_{CC} has been at 4.5V for 90% of t_D and prior to V_{CC} rising to 12.5V. The program/verification cycles must still be applied (with the pulse width still linearly increasing to a maximum of 8 μs) even though the output has been sensed as being programmed. An additional 128 verifications (i.e., 128 program/verify cycles) on each output must be obtained to insure a correctly programmed output. This additional 128 verification is a minimum number and must occur *after* all the bits of the word are sensed as being programmed. Please refer to Figure 1 for the timing waveforms.

   More than 128 program/verify cycles may be required to achieve the 128 verifications on each bit. The cycles should still continue even if one bit fails, since the verifications are not required to be in consecutive sequence. After the 128 verifications have occurred for all bits, a final V_{CC} and CS pulse at a width of 2.5 ms is simultaneously applied to all outputs. Programming should cease if the 128 verifications are not achieved in 800 ms.

3. A 4 mA ±50% current must also be forced into CS_3 (pin 19) of the 3608/3628 family and into CS_4 (pin 18) of the 3604A/3624A family during programming. If desired for commonality the 4 mA may also be forced into CS_4 of the 3604/3624 family.

4. The 4 mA current into the chip select input may be easily accomplished by using a 1.2K resistor connected to a +15V power supply. The voltage on the chip select input will be approximately 10V with the 1.2K resistor.

Table V. Programming Characteristics

$T_A = 25°C$

Symbol	Parameter	Limits			Units	Conditions
		Min.	Nom.	Max.		
V_{IH1}	V_{CC} Program Pulse Amplitude	12	12.5	13	V	
V_{IH2}	$\overline{CS}$ Program Pulse Amplitude	3	5	5.5	V	
V_{IL1}	V_{CC} During Verify	4.25	4.5	4.75	V	
V_{IH2}	$\overline{CS}$ During Verify	0	0.2	0.4	V	
t_{PW1}	V_{CC} Pulse Width at Beginning of Pulse Train	160	200	240	ns	Measured at 12V
t_{PW2}	V_{CC} Pulse Width at End of Pulse Train	7.2	8	8.8	μs	Measured at 12V
T_{CSS}	Chip Select Setup Time	0			ns	Measured from 1.5V on rising edge of $\overline{CS}$ to 5.0V on rising edge of V_{CC}
T_{CSH}	Chip Select Hold Time	100			ns	Measured from 5.0V on falling edge of V_{CC} to 1.5V on falling edge of $\overline{CS}$
T_R	V_{CC} Rise Time	300	400	500	ns	Measured from 5V to 12V on V_{CC}
T_F	V_{CC} Fall Time	50	100	200	ns	Measured from 12V to 5V on V_{CC}
T_{CYC}	Time Between Pulses to Same Output	9	10		μs	Measured at 5V on V_{CC}
T_{OP}	DC Program Time After Verification Has Been Obtained	2.2	2.5	2.8	ms	Measured at 12V
T_D	Time Between V_{CC} Pulses to Successive Outputs	1.5	1.8		μs	Measured at 5V on V_{CC}
T_{RAMP}	Time During Which V_{CC} Pulse Width is Increased Linearly from t_{PW1} to t_{PW2} — 4 outputs	160	180	200	ms	
	8 outputs	320	360	400	ms	
I_{CS}	Current to CS_3 of 3608/3628 or to CS_4 of 3604A/3624A	2	4	6	mA	CS_3 or CS_4 should be driven with a 1.2K resistor from a 15V power supply

ROM/PROM PROGRAMMING INSTRUCTIONS

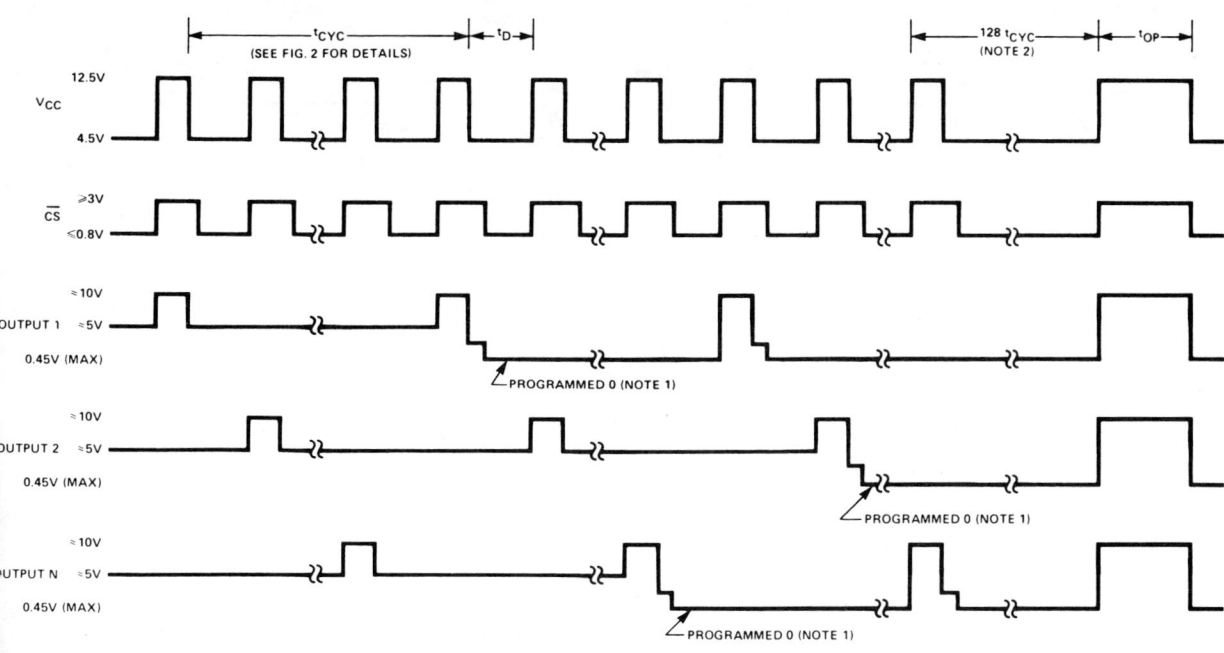

Figure 1. Programming Cycles.

NOTES: 1. PROGRAM VERIFICATION MUST OCCUR AFTER V_{CC} HAS BEEN AT 4.5V FOR 90% OF t_D AND PRIOR TO V_{CC} RISING TO 12.5V. THE PROGRAMMED OUTPUT IS <0.45V WHEN $\overline{CS}$ ≤0.8V AND FLOATING WHEN $\overline{CS}$ ≥ 3V.
2. AFTER THE LAST BIT HAS BEEN PROGRAMMED, 128 ADDITIONAL VERIFICATIONS ARE REQUIRED FOR EACH OUTPUT TO BE CORRECTLY PROGRAMMED.
3. AFTER THE 128 PROGRAM VERIFICATIONS, A FINAL 2.5 ms V_{CC} AND $\overline{CS}$ PULSE SHOULD BE APPLIED WHILE SIMULTANEOUSLY ENABLING THE CURRENT SOURCES TO ALL OUTPUTS WHICH ARE TO BE PROGRAMMED.

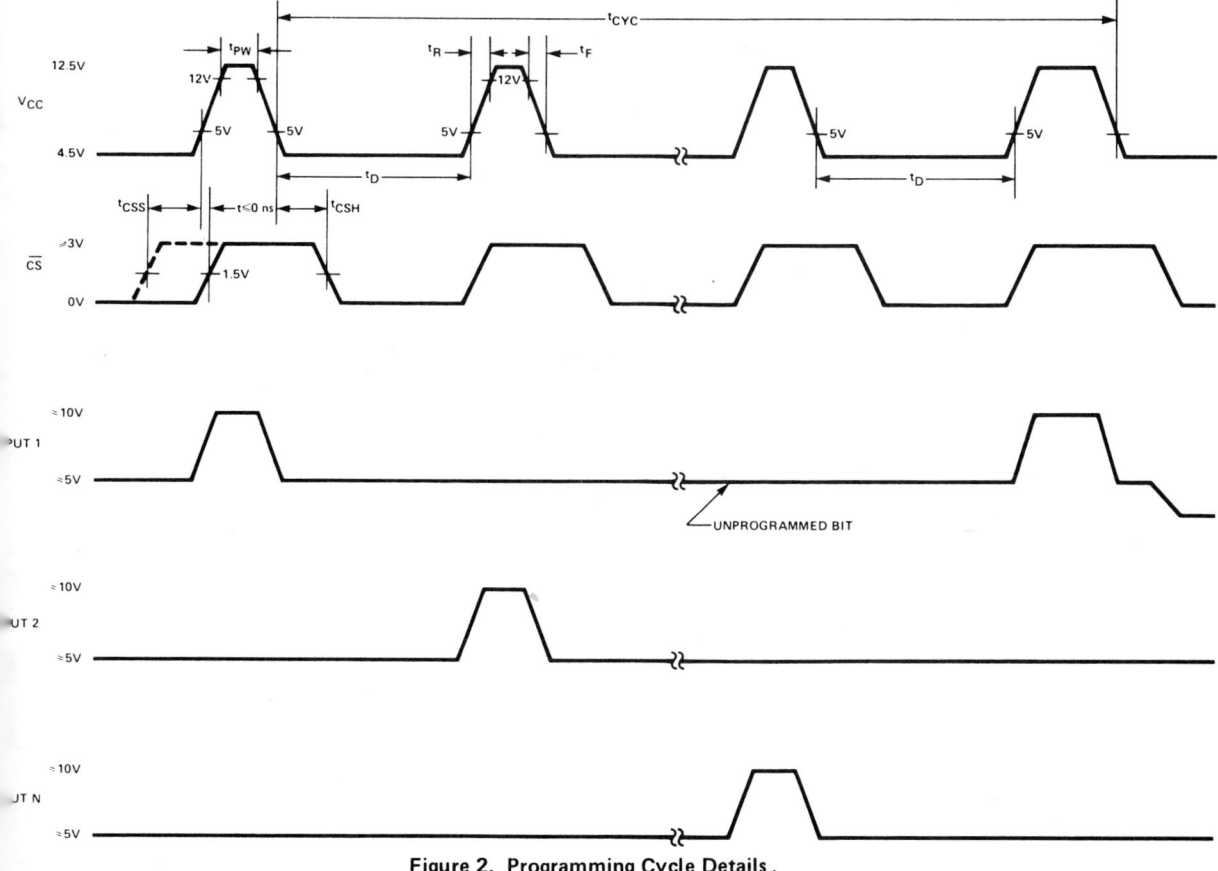

Figure 2. Programming Cycle Details.

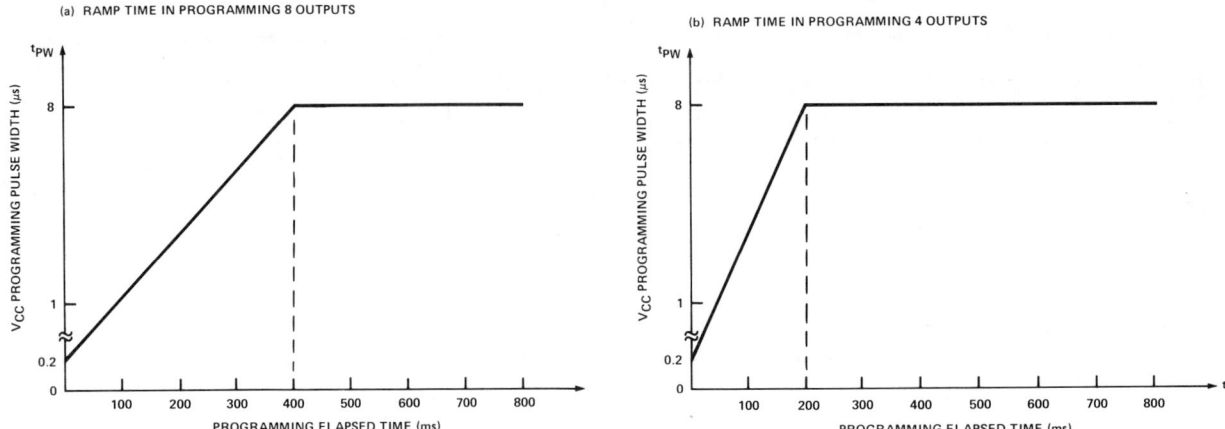

Figure 3. V_{CC} Pulse Width vs. Programming Time.

B. 3601 Programming

The 3601 may be programmed using the basic circuit of Figure 1. Address inputs are at standard TTL levels. Only one output may be programmed at a time. The output to be programmed must be connected to V_{CC} through a 300Ω resistor. This will force the proper programming current (3–6 mA) into the output when the V_{CC} supply is later raised to 10V. All other outputs must be held at a TTL low level (0.4V).

The programming pulse generator produces a series of pulses to the 3601 V_{CC} and CS_2 leads. V_{CC} is pulsed from a low of 4.5 ±0.25V to a high of 10 ±0.25V, while CS_2 is pulsed from a low of ground (TTL logic 0) to a high of 15 ±0.5V. It is important to accurately maintain these voltage levels, otherwise, improper programming may result. The pulses applied must maintain a duty cycle of 50 ±10% and start with an initial width of 1 (±10%) μs, and increase linearly over a period of approximately 100 ms to a maximum width of 8 (±10%) μs. Typical devices have their fuse blown within 1 ms, but occasionally a fuse may take up to 400 ms. During the application of the program pulse, current to CS_2 must be limited to 100 mA. The output of the 3601 is sensed when CS_2 is at a TTL low level output. A programmed bit will have a TTL high output. After a fuse is blown, the V_{CC} and CS_2 pulse trains must be applied for another 500 μs. The characteristics of the pulse train are shown in Figure 2.

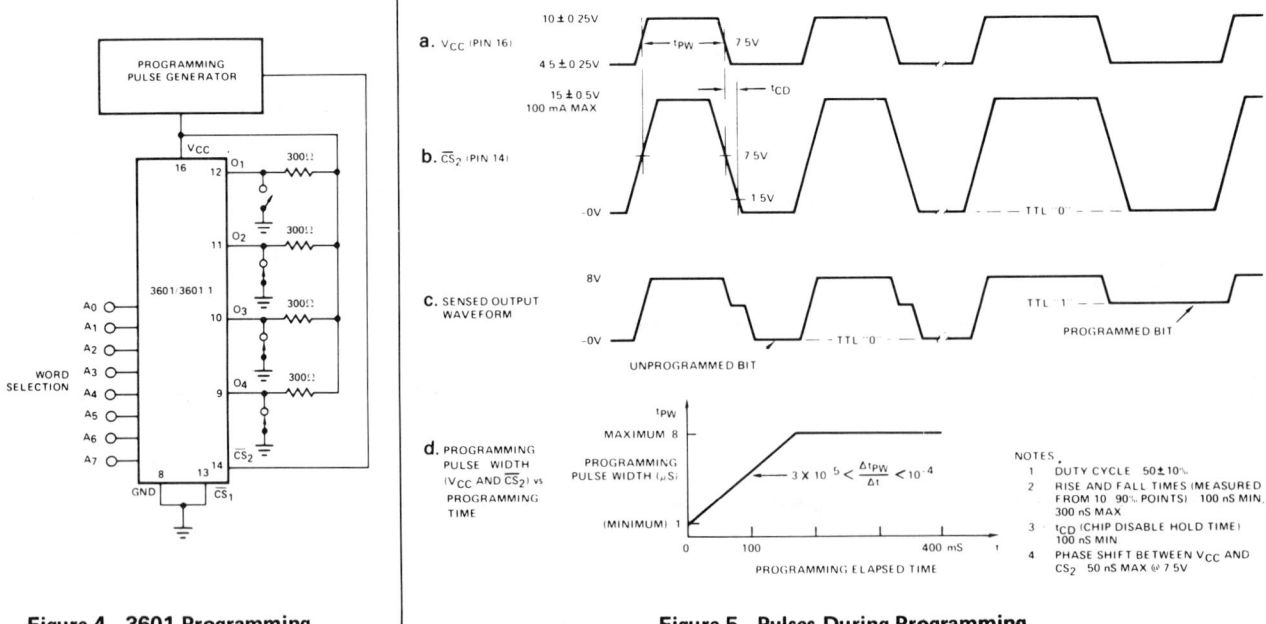

Figure 4. 3601 Programming.

Figure 5. Pulses During Programming.

CHAPTER FOUR
Charged Coupled Devices

Design and Application of 2416 CCD Memory	**374**
2416 16,384 Bit CCD Serial Memory	**403**

DESIGN AND APPLICATION OF 2416 CCD MEMORY

BOB PAPENBERG

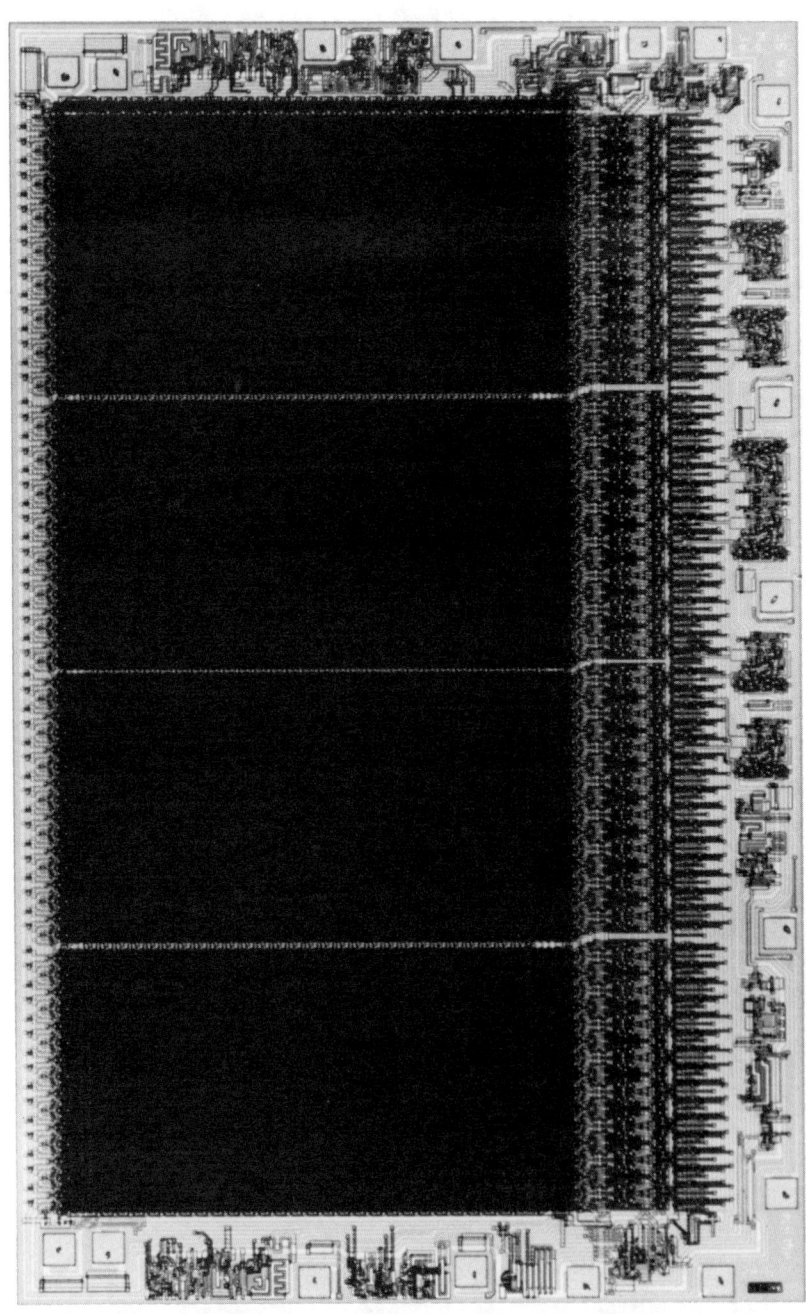

Photomicrograph of 2416 16,384 Word x 1 Bit CCD

APPLICATION OF 2416-16K CCD MEMORY

INTRODUCTION

The Intel® 2416 is a 16,384 word X 1-bit CCD serial memory designed for very low-cost memory applications. The memory is configured as 64 independent recirculating shift registers of 256 bits each. Access to any one of the 64 internal shift registers is done by applying the appropriate code to the 6 address inputs. The 2416 is fabricated using Intel's advanced high voltage n-channel silicon gate MOS process.

The 2416 memory device utilizes the simple surface channel structure and inherent very high density of a charge coupled device. This, in addition to a unique memory organization, provides an extremely versatile, dense and reliable memory unit. The purpose of this application note is to provide the system design engineer with an insight into the organization, structure, technology and operation of the 2416 device.

This application note is divided into three major sections: 1). Internal device organization, operation and specifications; 2). Device operation in a system; and 3). System organization examples. It is particularly important to users unfamiliar with the 2416 to carefully review the first section on organization and operation. A thorough understanding of the device will increase the versatility of the device to the user.

Information is also presented on interfacing clock and control signals to the 2416 in a system environment. Several specific applications are shown to illustrate the versatility of the 2416.

2416 INTERNAL ORGANIZATION AND OPERATION

The 2416 operates with the industry standard power supplies for memory components: V_{DD} = 12.0V and V_{BB} = -5.0V. The output is implemented with an open drain device which allows OR tieing of the outputs. For TTL operation the output pin is usually tied to a resistor which is returned to V_{CC} (+5V). The pin configuration for the 18 and 22 pin versions of the 2416 are shown in Figure 1.

The 2416 internal memory organization combines both serial and random address memory functions. As shown in Figure 2, the 2416 is arranged as 64-256 bit charge coupled device (CCD) shift registers. The data in these registers is simultaneously shifted by exercising the four-phase clock signals ϕ_1 through ϕ_4. After a shift cycle, each of the 64 CCD registers can be selected for an input/output (I/O) function by applying the appropriate 6-bit address code and applying cenable, chip select and write-enable signals in the required manner.

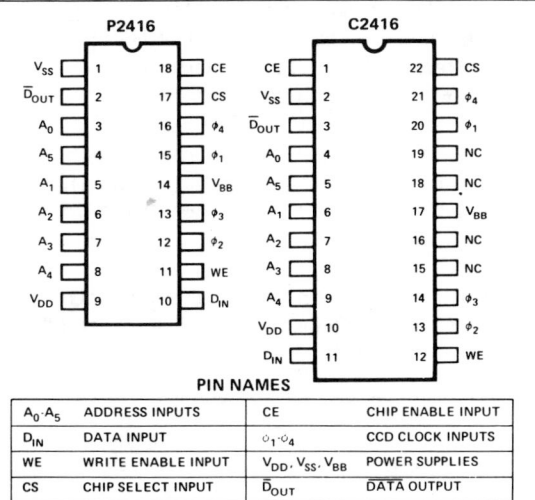

Figure 1. 2416 Pin Configuration.

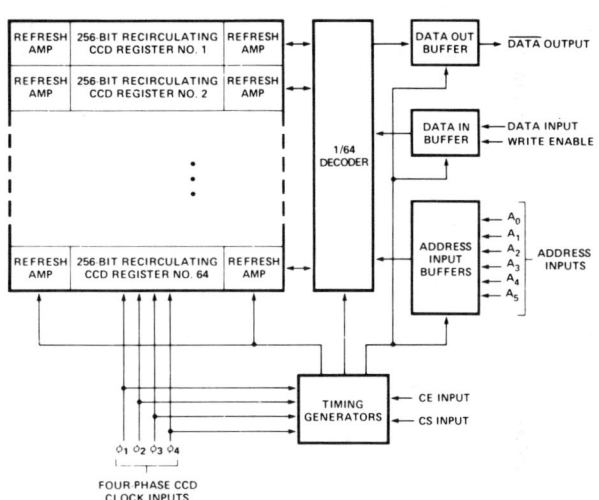

Figure 2. 2416 Block Diagram.

The flexibility of the 2416 internal memory organization in memory systems applications cannot be overemphasized. It is necessary for the designer to have a clear understanding of this organization to be able to take maximum advantage of the capability of the 2416.

The organization of the 2416 is most easily seen by referring to the diagram in Figure 3. In this diagram, the CCD is visualized as a cylinder comprised of 64 "tracks" (representing the 64 CCD recirculating shift registers) with each track divided into 256 "sectors" (representing the 256 CCD data storage cells). The "rate of rotation" of the cylinder is controlled by the four-phase clocks and is in the direction indicated by the "shift direction" arrow shown in Figure 3. (Note that the four-phase clocks *always* shift the cylinder in the same direction. The clocks cannot be manipulated to reverse the shift direction).

Read/Write capability in the CCD is performed by 64 bi-directional data buffers (one data-buffer per track). These buffers are located in position A shown in Figure 3 as the shaded column. The cylinder is considered to rotate *through* the buffers so that each shift of the cylinder (controlled by the four-phase clocks) places the next sequential sector of each track "in" the buffer. The buffers shown in column A also provide a refresh function to each cell in addition to performing read/write functions. (Note that an additional refresh-only buffer is shown in column B of Figure 3. These buffers are located half way around the cylinder as shown.)

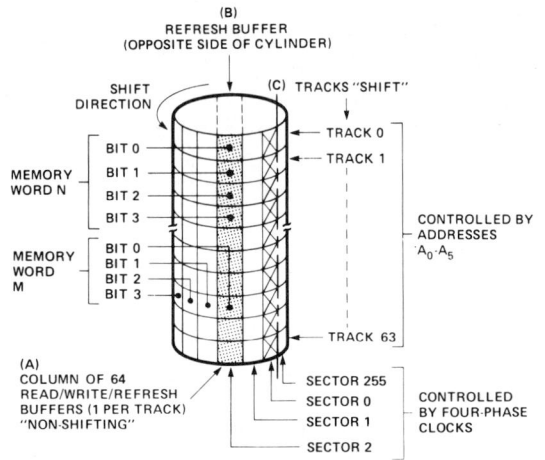

Figure 3. Symbolic 2416 Organization.

Two basic addressing methods may be used to store data words in the 2416:

1. In a given sector.
2. Around a given track.

In the first method, the desired word is accessed by shifting the cylinder (using the four-phase clocks) until the sector (0-255) containing the word is coincident with the read/write buffers (shown as column A). The word is then accessed one bit at a time by addressing the appropriate track with addresses A_0-A_5. An example of this addressing technique is shown as the four bit memory word N shown in Figure 3. The second addressing method places a word sequentially around the cylinder in a given track. Access to a particular word requires both a four-phase clock shift followed by a data access cycle for each bit of the word. (Note that for this case, A_0-A_5 do not change once the desired track is accessed.) An example of this addressing technique is shown as four bit memory word M in Figure 3.

Because of system addressing problems it is not generally desirable to combine the two addressing methods at once (although it is certainly possible). As is shown in the Systems Considerations section, addressing method 1 (sector addressing) is usually the more preferable technique. A major advantage of this data organization is the low four-phase clock driver power required to achieve the maximum serial data transfer rate of 2 megabits/sec from a single 2416. In most serial applications, the four-phase clock signals are only required to operate at less than 55 kHz rate to obtain a 2 MHz I/O data rate. *This is because the four-phase clocks are used solely to shift/refresh data and are not used to perform input/output functions.* For each shift of the clock, 64 "new" data bits are available in the 64 internal data registers for access through the address, chip enable and read/write control signals. These data control signals have a low input capacitance which makes them very easy to drive.

An alternate method of visualizing the organization of the 2416 is shown in Figure 4. This diagram is derived from the cylinder shown in Figure 3 by imagining that the cylinder is cut along the line marked C (between sector 0 and 255) and laying the cylinder out flat as shown in Figure 4.

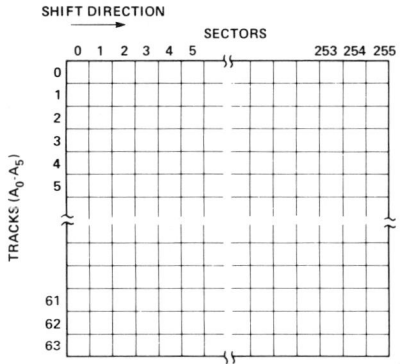

Figure 4. Planar View Symbolic 2416 Organization.

CCD Structure

There are two common CCD types referred to as surface channel and buried channel. The surface channel is characterized by the storing and transferring of charge (data) along the surface of the substrate. The buried channel type, because of additional substrate doping, stores and transfers the charge (data) further into the bulk of the substrate.

The primary differences in characteristics between the surface channel and buried channel is that the surface channel has: (1) higher total charge carrying capability, (2) lower charge transfer efficiency at extremely high charge transfer rates. (However, it is noted that the loss of charge transfer efficiency occurs at a frequency much higher than the maximum shift frequency of the 2416.) (3) simpler fabrication process. Charge transfer efficiency, number

APPLICATION OF 2416-16K CCD MEMORY

2 above, is defined as the percentage of the total charge packet (data) which is actually shifted or transferred per shift (the efficiency is typically greater than 99.9% per shift).

The 2416 internal memory array is comprised of four-phase surface channel charge-coupled structures. The CCD structure is formed by a series of MOS thinfield gate oxide devices placed as shown in Figure 5. Note that these MOS devices do not have the source/drain diffusions usually associated with other MOS structures. Figure 5(a) is the top view of the storage array and illustrates that the clock phases are laid out perpendicular to the shift register channels. Electrical isolation between shift register channels is obtained by channel stop diffusions and thick film oxide methods. Data input/output connections to the registers are obtained from n+ diffusions at the ends of the registers.

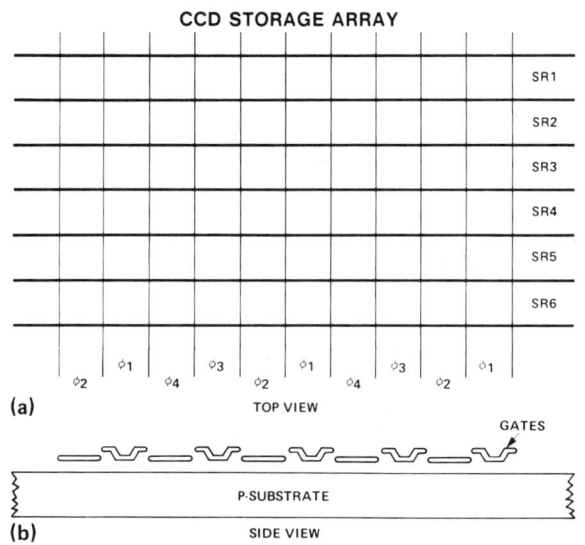

Figure 5. CCD Storage Array Layout.

Data Storage

The CCD stores data in the form of charge, as do all dynamic MOS memory devices. Indeed, in many respects the storage mechanism of the 2416 is very similar to the 4096 bit random access memories implemented with single transistor cells (such as Intel's 2107B). The storage element is most easily understood if it is considered to resemble a "potential well." This potential well is formed when a positive voltage potential is applied on the clock gates. The positive voltage repels the majority substrate carriers (holes) from the vicinity of the gate and forms a charge depletion area under it. This depleted region has the capability of accepting and storing a negative charge packet as long as the gate forming the well remains sufficiently positive with respect to the substrate.

The CCD structure is inherently dynamic and therefore must be refreshed periodically to maintain data. The dynamic nature of a CCD device is the result of thermally generated carriers (traditionally called "dark current effect") which acts to fill an uncharged potential well with charge thereby changing that particular cell's logic state.

Data Transfer

Figure 6 shows the relationship between the 2416 four-phase clock sequence and the CCD data storage and transfer mechanism.

The position of potential wells relative to the four-phase clock levels is shown in Figure 6(a). When the clocks are sequenced in the manner outlined in Figure 6(b), the potential wells generated provide a "low impedance" path for the charge packets to follow.

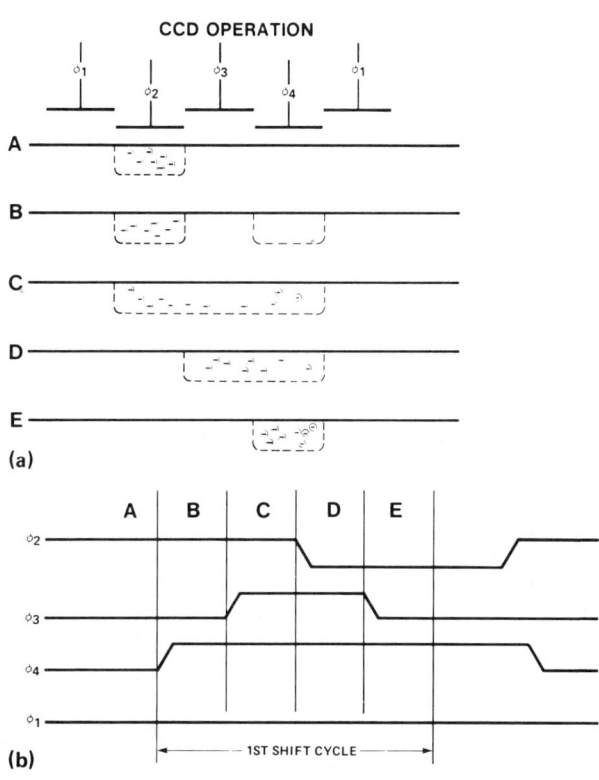

Figure 6. 2416 Charge Transfer Mechanism.

At time A, only the ϕ_2 gates are at a high level forming a storage well under the ϕ_2 gates. The storage well is assumed to contain an externally injected charge packet. The origin of the charge packet will be discussed later. At time B, both ϕ_2 and ϕ_4 gates are high and an additional storage well is formed in the substrate under the ϕ_4 gates. Note that the storage wells under the ϕ_4 gates do not now contain charge packets. At time C, ϕ_2, ϕ_3 and ϕ_4

gates are all high which forms ϕ_3 storage wells overlapping both the ϕ_2 and ϕ_4 storage wells. Thus a continuous storage well is formed from the ϕ_2 gates to the ϕ_4 gates which allows charge packets under ϕ_2 gates to disperse throughout the charge wells of all three gates. At time D, the ϕ_2 gate goes to a low level eliminating the storage well under it. This forces the charge packet into the remaining storage wells under the ϕ_3 and ϕ_4 gates. At time E, the charge transfer is complete when the ϕ_3 gate voltage goes low which forces the charge packet into the remaining storage well under the ϕ_4 gate. The charge packet (data) has now been shifted by one bit position. Note that the shift execution time shown in Figure 6 is the time that data is being shifted as defined by periods B, C, and D.

Applying clocks in the above manner (ϕ_3 shift) results in a parallel shift of all data. Another shift cycle can then begin by utilizing ϕ_1 and ϕ_4 (ϕ_1 shift) thus completing a full cycle on the four-phase clocks. The shifting mechanism using the ϕ_1 and ϕ_4 clocks is identical to that described for the ϕ_3 and ϕ_2 clocks.

CCD Internal Data Interface

Each of the 256-bit CCD shift registers is comprised of two 128-bit registers. Each of the two 128-bit registers is further multiplexed into dual 64-bit registers (making the 256-bit register a quad 64-bit register). This allows data operation on either ϕ_1 or ϕ_3 shift. A simplified diagram of an internal 256-bit register is shown in Figure 7.

Data is written into the internal CCD register by the Write Data Amplifier which either injects or removes charge from the N+ regions as shown in Figure 7. The data will then be multiplexed through register 1 or register 2 (in each 128-bit half) depending on the state of ϕ_1 and ϕ_3. A read of the data is performed in a similar manner except the N+ region is either charged or discharged by the state of the CCD cell adjacent to the buffer. Data is read from either register 3 or register 4 depending on the state of ϕ_1 and ϕ_3. A sense amplifier, connected as shown, senses the state of the data after it passes through the refresh amplifier.

Data Refresh

As shown in Figure 7, each of sixty-four 256-bit shift registers is arranged as four 64-bit shift registers (as far as refresh is concerned) connected by an inverting refresh amplifier at each end to form a continuous data loop. Therefore, it requires 128 shift cycles (clock phases 1 through 4) to completely refresh the memory. The refresh amplifiers serve to restore the integrity of the data charge which is reduced through the dark current effect and shift transfer losses inherent in the CCD struc-

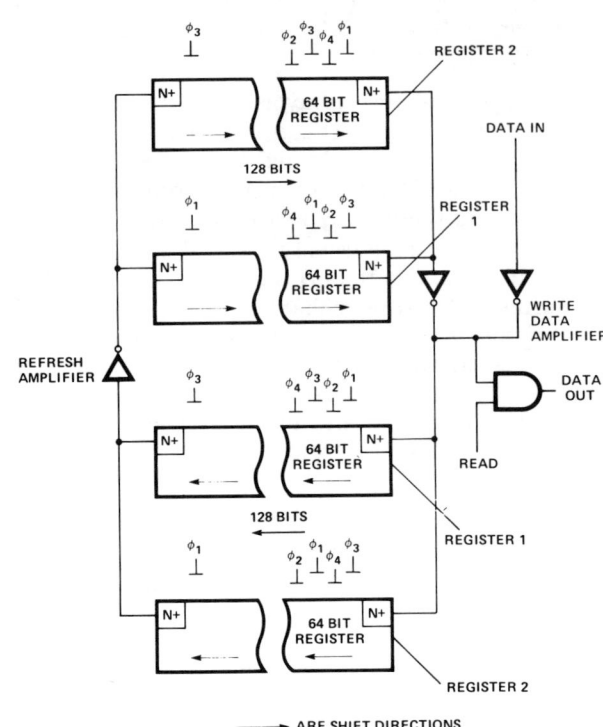

Figure 7. Simplified Diagram 2416 256-Bit Register.

ture. The refresh amplifiers shown on the right side of the array in Figure 7 include an input/output gating function controlled by the address decoders and write enable lines. These refresh amplifiers serve as read/write amplifiers to the associated 256-bit channel.

2416 DEVICE SPECIFICATIONS

D.C. Characteristics

The D.C. and Operating characteristics of the 2416 are shown in Table I. Although the table is self explanatory, several items (marked as (3) in Table I) deserve special attention. First, note that the maximum average V_{DD} supply current (I_{DDAV}) is very low (25mA max.) at minimum cycle timing. This results in very low device power during operation at maximum data rate or shift rate. I_{DDAV} is inversely proportional to the cycle time of shift or data access cycles.

The input levels for the four-phase clocks (V_{ILC}, V_{IHC1}, V_{IHC2}) show the margin available for clock drivers and for the control inputs (addresses, read/write, chip enable, etc). Each of these limits will be discussed in detail in the Systems Considerations section along with driver designs which meet the 2416 input requirements.

APPLICATION OF 2416-16K CCD MEMORY

Table I. 2416 D.C. and Operating Characteristics: $T_A = 0°C$ to $70°C$, $V_{DD} = +12V \pm 5\%$, $V_{BB}[1] = -5V \pm 5\%$, $V_{SS} = 0V$, unless otherwise specified.

Symbol	Parameter	Min.	Max.	Unit	Test Conditions
I_{LI}	Input Leakage		10	μA	$V_{IN} = 0V$
I_{LO}	Output Leakage Current		10	μA	$CE = 0V$, $V_{OUT} = 0V$
I_{DD1}	Standby V_{DD} Supply Current		2	mA	$CE = 0V$, $\phi_2 = V_{DD}$, $\phi_4 = 0V$ (or $\phi_2 = 0V$, $\phi_4 = V_{DD}$), $\phi_1 = \phi_3 = 0V$.
$I_{DDAV}[3]$	Average V_{DD} Supply Current [4]		25	mA	Minimum Cycle Timing
I_{BB}	Average V_{BB} Supply Current		200	μA	
V_{IL1}	Input Low Voltage, all Inputs except D_{IN} and $\phi_1 \ldots \phi_4$	-1.0	0.8	V	
$V_{IH1}[3]$	Input High Voltage, all Inputs except D_{IN} and $\phi_1 \ldots \phi_4$	$V_{DD}-1$	$V_{DD}+1$	V	
$V_{ILC}[3]$	$\phi_1 \ldots \phi_4$ Input Low Voltage	-2.0	0.6	V	Note 2
$V_{IHC1}[3]$	ϕ_1 and ϕ_3 Input High Voltage	$V_{DD}-1.0$	$V_{DD}+2$	V	
$V_{IHC2}[3]$	ϕ_2 and ϕ_4 Input High Voltage	$V_{DD}-.6$	$V_{DD}+2$	V	
V_{ILD}	D_{IN} Input Low Voltage	-1.0	0.8	V	
V_{IHD}	D_{IN} Input High Voltage	3.5	$V_{DD}+1$	V	
I_{OL}	Output Low Current	3		mA	$V_{OL} = .45V$
I_{OH}	Output High Current		10	μA	$V_{OH} = +5V$

NOTES:
1. The only requirement for the sequence of applying voltage to the device is that V_{DD} and V_{SS} should never be 0.3V more negative than V_{BB}.
2. The difference in the low level reference voltages between all four clock phases must not exceed 0.5 volts.
3. See Text.
4. Combined shift and Data I/O. For shift only mode $I_{DD} = 2.0 + 15/t_\phi/2$ ($t_\phi/2$ is in μsec).

Data Cycles

The 2416 has two basic modes of operation: (1) data and (2) shift. In normal operation, the 2416 will use both of these modes. For clarity, however, the data mode will be treated separately from the shift mode. In the following sections, the discussion will describe writes, reads, and read-modify-writes to the 2416, *before or after a shift operation has been performed.* Figure 8 shows a detailed block diagram of the 2416 as it relates to data I/O cycles.

WRITE CYCLE

The write cycle of the 2416 is explained with the aid of the diagram of Figure 9 and term definitions shown in Table II.

As shown in Figure 9, write cycles may only be performed after a delay time (t_{TC}) from the trailing edge of ϕ_1 or ϕ_3 and continue until a time t_{CP} prior to the leading edges of ϕ_4 or ϕ_2. During the intervals between, ϕ_1 and ϕ_4 or ϕ_3 and ϕ_2, the data is not shifted and remains stationary in the 256 discrete locations of each of the 64 shift registers. Any of the 64 register input/output buffers can be accessed during this time through addresses A_0-A_5 and chip enable.

After the address lines are stable and chip select (CS) signal is high, a write cycle can start with the leading edge of the chip enable (CE) pulse. The CE and CS signals trigger an internal timing generator which generates internal enable and precharge signals to the address decoders and data-in buffers. The addresses are then decoded to activate one of the 64 decode lines which in turn enables the write amplifier for the selected channel. The write enable signal (WE) is then set to a high state after the data-in signal is stable (t_{DW}) and the CE to WE set up time (T_{cw}) has lapsed. The write enable signal enables the data-in buffer which in turn gates the input data to the selected write amplifier (WRT) via the data-in bus line. The selected write amp stores the data in the form of a charge "packet" at the input bit location of the selected buffer register (see CCD Internal Data Interface section). By selecting new address combinations and maintaining the chip enable off time requirement (t_{CC}), additional data bits can be stored in the other registers before a new shift execution cycle ($t_\phi/2$) is required.

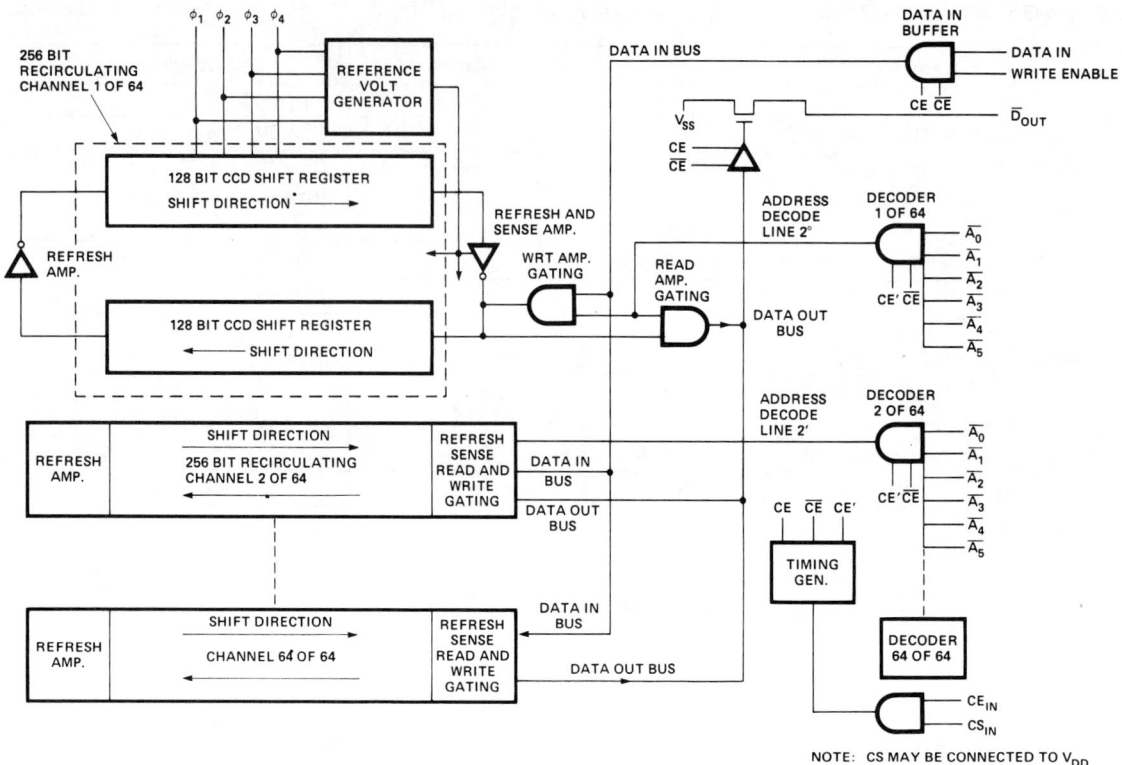

Figure 8. 2416 Internal Detailed Block Diagram.

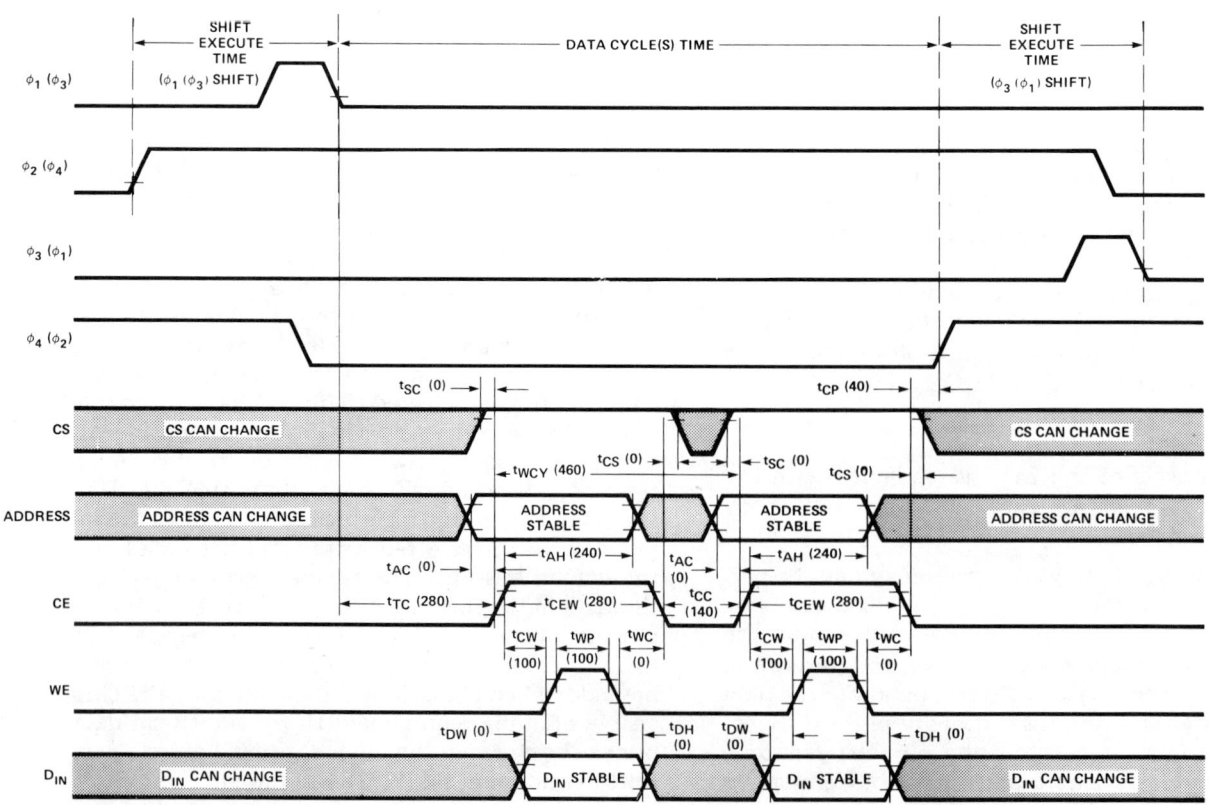

Figure 9. 2416 Write Cycle Timing.

(Numbers in parentheses are for minimum cycle timing in ns.)

APPLICATION OF 2416-16K CCD MEMORY

Table II. Definition of Terms.

Symbol	Parameter
t_{WCY}	WRITE Cycle Time
t_{PT}	ϕ_2 On to ϕ_1 On Time, ϕ_4 On to ϕ_3 On Time
t_{TD}	ϕ_1 to ϕ_4 Overlap, ϕ_3 to ϕ_2 Overlap
t_{DT}	ϕ_4 to ϕ_1 Hold Time, ϕ_2 to ϕ_3 Hold Time
$t_{\phi/2}$	Half Clock Period for $\phi_1 \ldots \phi_4$
t_{T1}	Transition Times for $\phi_1 \ldots \phi_4$
t_{T2}	Transition Times for Inputs Other than $\phi_1 \ldots \phi_4$
t_{TC}	ϕ_1 or ϕ_3 Off to CE On
t_{SC}	CS to CE Setup Time
t_{AC}	Address to CE Setup Time
t_{AH}	Address Hold Time
t_{CS}	CE to CS Hold Time
t_{CC}	CE Off Time
t_{CP}	CE Off to ϕ_2 or ϕ_4 On
t_{CEW}	CE On Time
t_{CW}	CE to WE Setup Time
t_{DW}	D_{IN} to WE Set Up
t_{WP}	WE Pulse Width
t_{WC}	WE Off to CE Off
t_{DH}	D_{IN} Hold Time
t_{RCY}	READ Cycle Time
t_{CER}	CE On Time
t_{CF}	CE Off to Output High Impedance State
t_{CO}	CE to $\bar{D}_{OUT}$ Valid
t_{RWC}	READ–MODIFY–WRITE Cycle Time
t_{WD}	CE On to WE On
t_{WF}	WE to $\bar{D}_{OUT}$ Undefined

READ CYCLE

The read cycle timing (shown in Figure 10 and Table II) is identical to the write cycle for the CE, address and four-phase clock inputs. The only difference in operation between the read and write cycle is that the write enable (WE) signal must remain at a low state. In a read cycle the data-in line is electrically disconnected from the internal circuitry by a low level on the write enable input. Data is presented at the output pin at or before t_{CO} time.

The detailed block diagram shown in Figure 8 shows that a low level write enable signal inhibits the write amplifier gates. This allows valid register data to be present at the read amplifier inputs. The data from the read amplifier selected by the address decoder is gated to the data-out buffer via the data-out bus line. The data-out buffer amplifies the stored data voltage level and provides an open drain output signal from the memory device. The organization and CCD shift structure of the 2416 inherently contribute to a high internal signal-to-noise ratio at the data-out buffer as the result of the following:

1. Relatively small number of shift cycles (128) required between refresh. (This compensates for transfer losses and provides a high input signal level to the sense amplifier.)

2. The CCD shift structure minimizes the interconnection length from the data cell to the read amplifiers.

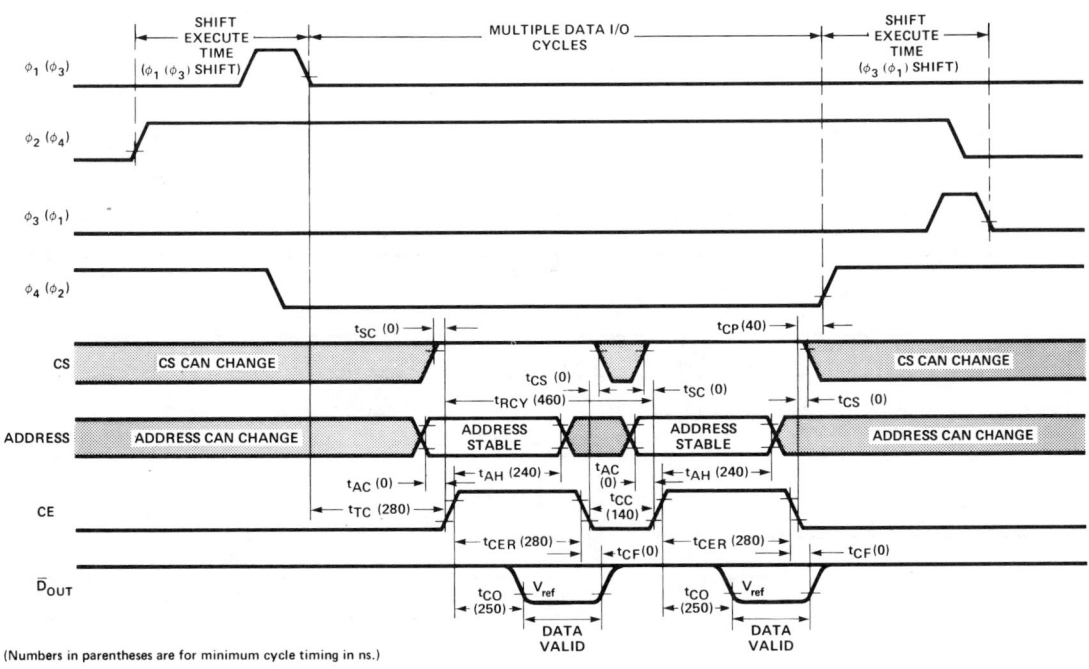

(Numbers in parentheses are for minimum cycle timing in ns.)

Figure 10. 2416 Read Cycle Timing.

Reducing the line lengths reduces the RC time constant effect on the signals to the read amplifier. This allows the signal to switch through the threshold point of the amplifier at a very fast rate, thus providing a very definite and easily sensed data-out signal.

READ-MODIFY-WRITE CYCLE

The read-modify-write cycle (RMW) shown in Figure 11 (see Table II for symbol explanation) combines both a read cycle and a write cycle, but requires less than the sum of the two cycle times to execute. The cycle time reduction is attributed to the condition that one, not two, CE off time intervals (T_{CC}) is required for a RMW cycle. Another advantage of the RMW cycle is that only one address hold time (t_{AH}) is required. Control of the RMW cycle is up to the user in that on an individual cycle a RMW cycle may be initiated by a separate command from the control logic (which extends the CE on time and delays the WE signal from the normal write time) or it can be performed on all data cycles.

SHIFT ONLY CYCLE

The previous section on Data Cycles outlined the timing requirements on the address, data, read-write and chip enable inputs necessary to perform a read or write operation. This section on shift-only cycles outlines the timing conditions on the four clock lines ϕ_1, ϕ_2, ϕ_3 and ϕ_4 required to simultaneously shift the 64-256 bit CCD registers.

The shift only mode performs two basic functions: (1) "Searches" for data or blocks of data in the CCD registers (see Systems Considerations section) and (2) Sequentially shifts data through refresh amplifiers (see 2416 Internal Organization and Operation) to perform data refresh.

The timing diagram for shift only mode operation of the 2416 is shown in Figure 12 with symbol definition shown in Table II. (Note that the timing diagram shown in Figure 12 is an extension of the description on charge transfer mechanism Figure 6.) As shown in Figure 12, *a complete clock* cycle (all four phases sequentially exercised) is given by:

$$t_{cyc} = 2\, t_\phi/2 = t_\phi \quad (1)$$

(See Table II for definition of terms.)

Note that a complete clock cycle actually shifts data two locations.

A half clock cycle ($t_\phi/2$) shifts the CCD register one location. The half clock cycle is composed of two parts:

1. Shift execution time (t_{SX}). (See Figure 12.)
2. Clock "low" (t_{TP}). (See Table II.)

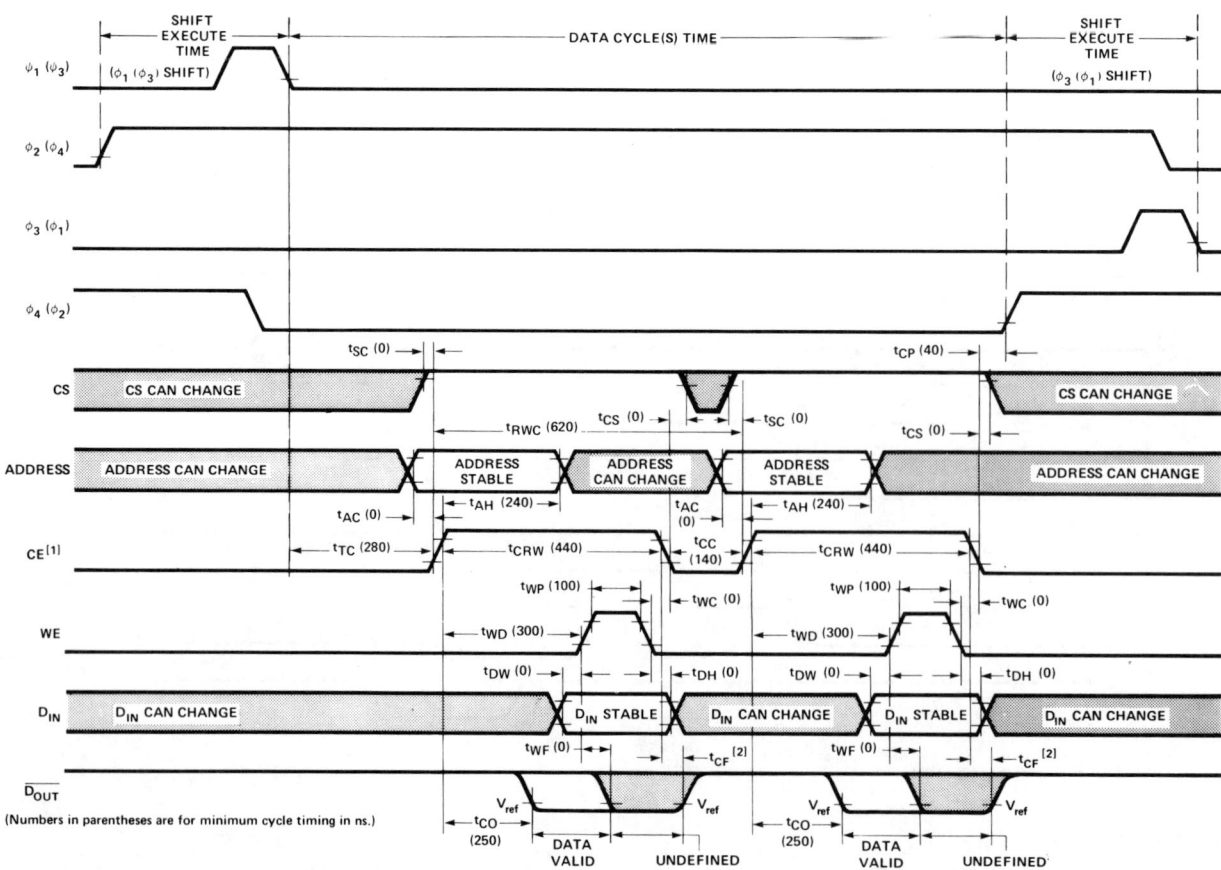

Figure 11. 2416 Read-Modify-Write Cycle Timing.

APPLICATION OF 2416-16K CCD MEMORY

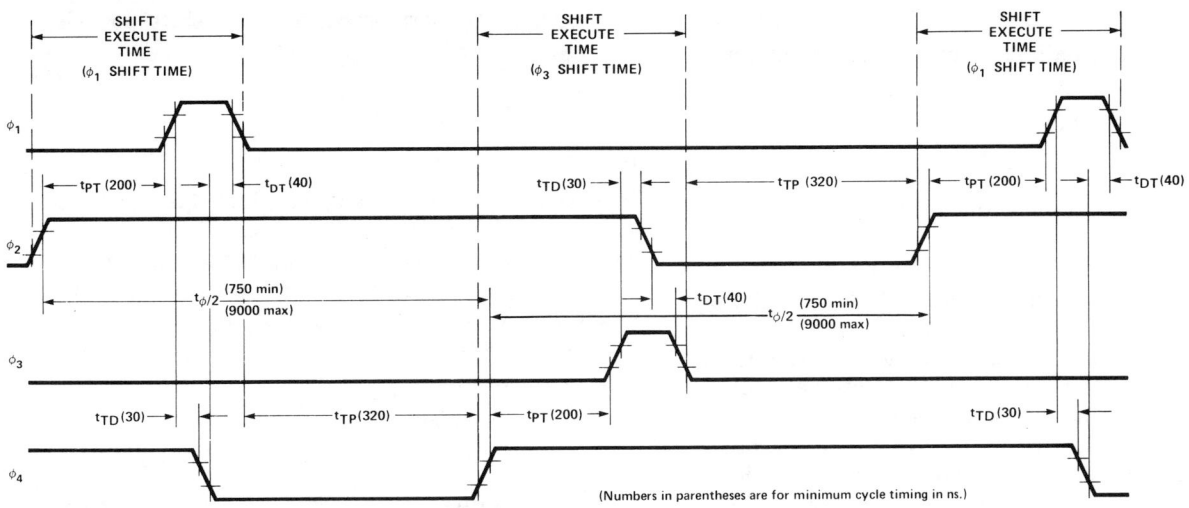

Figure 12. Shift Only Cycle Timing.

Note that the two shift execution times shown in Figure 12 *can be* identical but are relative to different portions of the four-phase clocks. For example, the first shift execution time is timing associated with ϕ_1, ϕ_2, ϕ_4 while the second shift execution time is associated with ϕ_3, ϕ_4 and ϕ_2.

The time required to shift data (shift execution time t_{SX}) is given by:

$$t_{SX} = t_{PT} + t_{TD} + t_{DT} + 4t_T \quad (2)$$

(See Table II and Figure 12 for definition of terms.)

The shift period, t_{SP}, is given by:

$$t_{SP} = t_{SX} + t_{TP} = t_\phi/2 \quad (3)$$

Where:

t_{SX} = shift execution time (equation 2).

t_{TP} = clock off to on time (Table II).

(Note that the term t_{SP} has been substituted for $t_\phi/2$ in equation 3. The reasons for this will be evident in the Systems Considerations section.) The minimum search cycle time is obtained by operating the four-phase clocks at the maximum repetition rate (for this case t_{SP} is 750 nsec). The maximum half cycle time between clocks (for a single shift cycle) is 9000 nsec. The maximum cycle time is most often used for refresh and for obtaining maximum data rates.

SHIFT/MULTIPLE DATA/SHIFT CYCLE

The previous sections discussed the data and shift cycles of the 2416 as separate functions. *This section discusses the combined operation of the shift and data cycles.* Data cycles may be initiated after a minimum of t_{TC} nsec from the completion of a shift execution (end of ϕ_1, see Figure 10). After a shift, the 64 internal data buffers may be accessed in any order by addresses A_0-A_5. The number of data cycles, N, which may be performed between shift execution times is dependent on two criteria:

1. System addressing technique.
2. Refresh rate.

The number of data cycles which can be performed between shift periods is simply the time available between shift cycles divided by the data cycle time. A simple expression for the number of data cycles allowable between shift cycles is determined by inspection from Figures 9, 10, or 11 and is expressed as:

$$N = \frac{t_{SP} - t_{SX} - t_{TC} + (t_{CC} - t_{CP})}{t_{DC}} \quad (4)$$

Where:

N — number of cycles between shifts

t_{SX} — shift execution time (see equation 2 or Figures 9, 10, or 11)

t_{TC}, t_{CC}, t_{CP} — (see Table II or Figures 9, 10, or 11)

t_{DC} — data cycle time (e.g. $t_{DC} = t_{RCY}$ for a read cycle).

For those systems where the time relationship of the last data cycle relative to a shift cycle cannot be predicted, the term $(t_{CC} - t_{CP})$ in equation (4) equals zero. (For this case, the maximum number of cycles is decreased slightly.) A practical maximum (due to system address considerations) of data cycles between shift periods is sixteen for a shift period of 9000 nsec.

DATA RATE

Consider now the data rate for the following conditions:

A. Maximum number of cycles between shifts.
B. One data cycle per shift.

1. The maximum data rate of 2 megabits/sec is obtained when the time between clock cycles is maximized and the maximum number of data cycles possible are inserted between these shift executions. As shown in Figure 10, this rate is actually a maximum *average* data rate because of the shift execution intervals. (Recall that during shift execution, no data cycles are permitted. Therefore, the maximum data rate is the average of the data rate during data cycles and a data rate of zero during shift cycles.)

   Clearly the maximum data rate is proportional to the shift period and approaches $1/t_{DC}$ as the shift period t_{SP} is increased. Also, as the shift period is increased the clock frequency is decreased resulting in lower clock driver power and higher data rates. (The significance of this will be evident in the four-phase driver section.)

2. The data rate is the same as the shift execution rate when there is only one data cycle (N) between shift cycles (see Figure 13). In this special case, clock driver power will increase as the data rate increases. (Remember that minimum driver power and maximum data rate occur when a maximum number of data cycles are performed between shift periods, t_{SP}.)

SYSTEM CONSIDERATIONS

Typical Applications

The combined high density and high speed characteristics of the 2416 make this part ideal for use in many types of systems. Of particular interest to many designers are four general system categories where the 2416 is especially ideal from a cost/performance viewpoint.

These categories are:

1. Drum replacement.
2. Small "rotating" memory applications.
3. Hi-reliability (ruggedized) "rotating" memory.
4. Conventional shift register replacement.

It is useful to briefly review each of the above categories to illustrate the versatility of the 2416.

DRUM REPLACEMENT

The 2416 has several significant system advantages over conventional mechanical drum assemblies. For example, the 2416 is an order of magnitude faster than a high speed drum (average latency time of a 2416 system is 100μsec); the 2416 system is more reliable since there are no mechanical assemblies rotating at high speed; and the 2416 drum type system is cost competitive and more compact than standard drum type systems. In addition, the extremely high data rate of a 2416 system can handle virtually any computer data rate requirement.

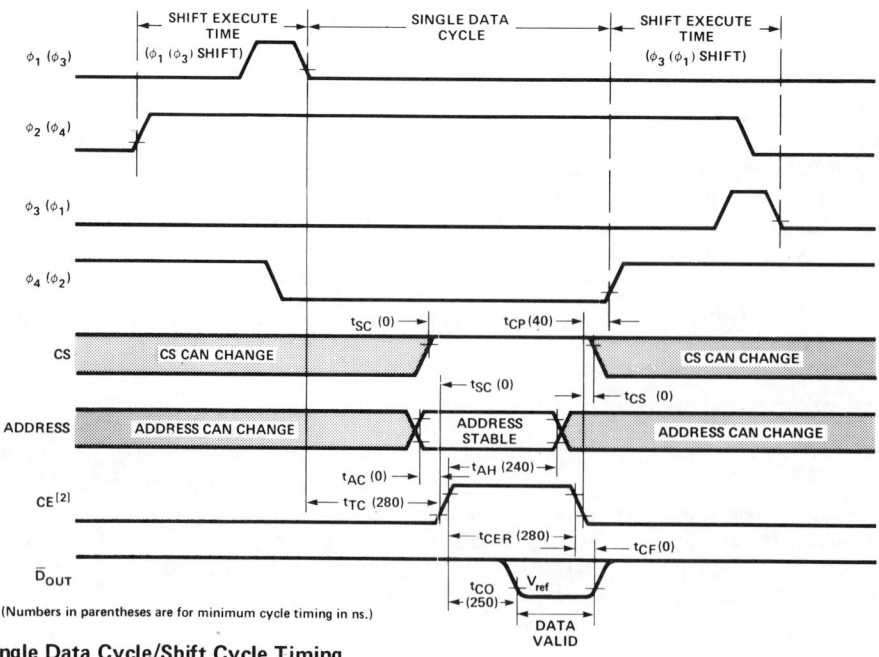

Figure 13. Shift/Single Data Cycle/Shift Cycle Timing.

SMALL "ROTATING" MEMORY APPLICATIONS

The 2416 is very competitive in applications previously favoring various types of rotating memory. The real strength of this CCD device is readily apparent in those types of applications requiring a relatively small amount of rotating memory. In these rotating systems the overhead cost of drive motors, sense heads and other peripherals significantly impact the overall cost per bit at the system level. For these systems the 2416 offers a significant cost advantage over conventional rotating memory devices. As in the case of drum replacement type memories, the 2416 is significantly faster than the small rotating memories it is designed to replace.

HI-RELIABILITY (RUGGEDIZED) "ROTATING" MEMORY

Many applications for mass storage requiring a "ruggedized" rotating memory need significant attention paid to the mechanical mechanisms to assure reliable operation in a hostile mechanical environment. A clear advantage of the 2416 is its lack of any mechanical rotating mechanism which needs to be ruggedized. This CCD device offers high density and speed for most Hi-reliability applications requiring a mechanically rugged support.

SHIFT REGISTER REPLACEMENT

The 2416 can easily be used (as is shown later in this section) as one very long shift register (16,384 stages) or as 64 256-bit shift registers. In either case the density advantage of this CCD device over conventional shift registers is readily apparent. These types of shift register applications include CRT display refresh and communications buffers. In these applications the advantages of speed and density are particularly evident.

The previous sections detailed specific timing requirements and associated data rates of the 2416. In this section, examples of timing, control, and driver interface implementation for a memory system are discussed.

Addressing Considerations and Control

In the previous sections describing the internal organization and operation of the 2416, it was pointed out that this CCD device has both a "sector" type address controlled by the four phase clocks and a "track" type address defined by addresses A_0-A_5. The location of specific "track" addresses is very straight forward with track zero defined by A_0 through A_5 equaling logic zero and track 63 defined by A_0 through A_5 equaling logic one, etc. However, the starting and ending "sector" addresses are not uniquely defined in the same manner as the "track" addresses. Throughout this section on Addressing Considerations it should be remembered that control circuitry for the four-phase clocks must contain logic capable of "recalling" where the previously defined starting location of the sector addresses is positioned and determining how many shifts to perform to reach a desired sector. It is shown later in this section just how simple such interface requirements are. In the following discussion, two basic types of control circuitry will be analyzed:

1. Serial memory applications (shift/single data cycle/shift).
2. "Random" memory applications (shift/multiple data/shift).

(The "random" memory application is actually an extension of the serial mode to include search type operations.)

SHIFT/SINGLE DATA CYCLE/SHIFT CONTROL

A simple shift/single data cycle/shift control circuit which has one data cycle per shift is shown in Figure 14(a) and (b). Basic timing of the control circuit is shown in Figure 14(a). The four-phase clocks (performing the shift) are shown in block form (labeled as shift execution time) with the corresponding data cycle shown below.

Operation is most easily understood with the aid of the diagram shown in Figure 14(c). This figure illustrates the addressing and shifting sequence applied to a 2416 operating in a shift/single data cycle/shift mode as a 16K bit shift register. First a particular CCD register (1 of 64) is accessed by addresses A_0-A_5 and a read or write cycle performed. Then a shift is executed and the next CCD cell accessed (data is moving from IA_0 to IA_1 as shown in Figure 14(c) (refer to Figure 4 for explanation of data sequencing addressing). This sequence is repeated 255 times to access all of the cells in one of the 64 256-bit CCD registers. (Note that during this entire operation addresses A_0-A_5 have not changed.) After the entire 256-bit register has been accessed, the 2416 addresses are incremented (A+1) and the next internal register is sequenced in the same manner. The entire operation is summarized as follows:

1. Access memory.
2. Shift (four-phase clocks) once.
3. Repeat 1 and 2 255 times then:
4. Increment 2416 addresses by 1 (A_0-A_5).
5. Repeat 1 through 4 sixty-four times.

As shown in Figure 14(b), a data cycle is begun with an initial pulse triggering a single-shot S_1. The leading edge of the single-shot output (Q) initiates a read or write to the 2416 at the address defined by A_0-A_5. When the single-shot times out, the trailing edge ($\bar{Q}$) triggers a four-phase clock generator and increments the index counter controlling addresses A_0-A_5 as shown. (Details of the four-phase clock generator are shown in Figure 16.)

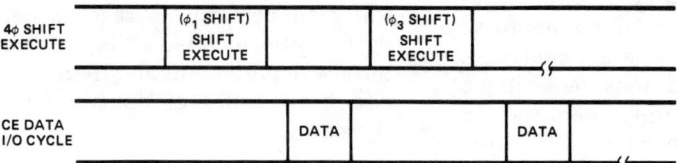

(a) Timing

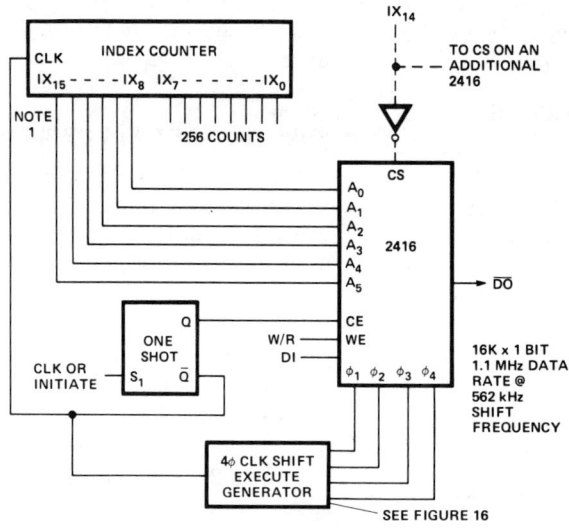

(b) Control Block Diagram

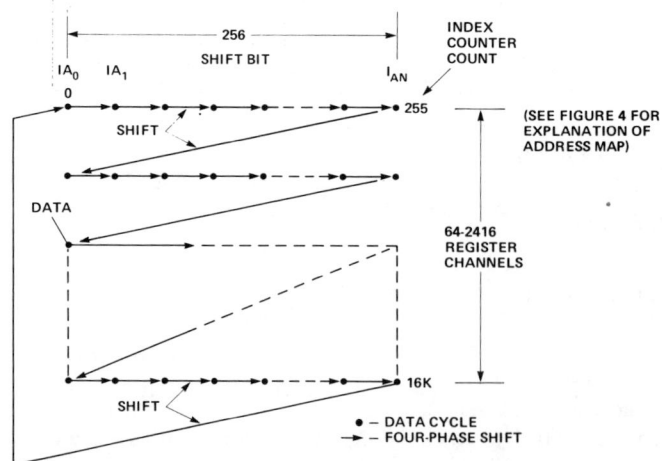

(c) Address Sequence

Figure 14. Control and Address Sequence Shift/Single Data Cycle/Shift.

APPLICATION OF 2416-16K CCD MEMORY

Address Expansion

The control circuit shown in Figure 15(b) is easily expanded in the bit direction (e.g., 16K x 8) by adding more 2416s and paralleling the addresses (A_0-A_5), four-phase clock, and control input (R/W, CE, CS) lines. Further expansion to 32K words is most easily done by using the next high order bit of the index counter and generating a select and $\overline{\text{select}}$ signal which go to respective CS inputs. Figure 15 shows a 64K x 1-bit configuration. (Other lines are paralleled as described.)

SHIFT/MULTIPLE DATA/SHIFT CONTROL

An expansion of the shift/single data cycle/shift mode is the shift/multiple data/shift mode. It is this mode that is most often used in general system applications because of its ability to handle a wide variety of applications. This mode also includes the "search" mode requirement.

A shift/multiple data/shift control interface is given in Figure 17. (The addressing sequence is given in Figure 18.) Note that in this implementation *16* data cycles are performed between shift cycles. The relationship of data cycles to shift cycles is shown in Figure 17(b).

Implementing control for the multiple data mode differs from the implementation used for the single cycle data mode (Figure 14) by simply changing the 2416 address and shift initiate address connections to the index counter as shown. The multiple data mode control requires that, during a shift operation, the data cycle request line must be inhibited. (A method to "hide" the four-phase shift clocks so that the data cycles are not interrupted will be dis-

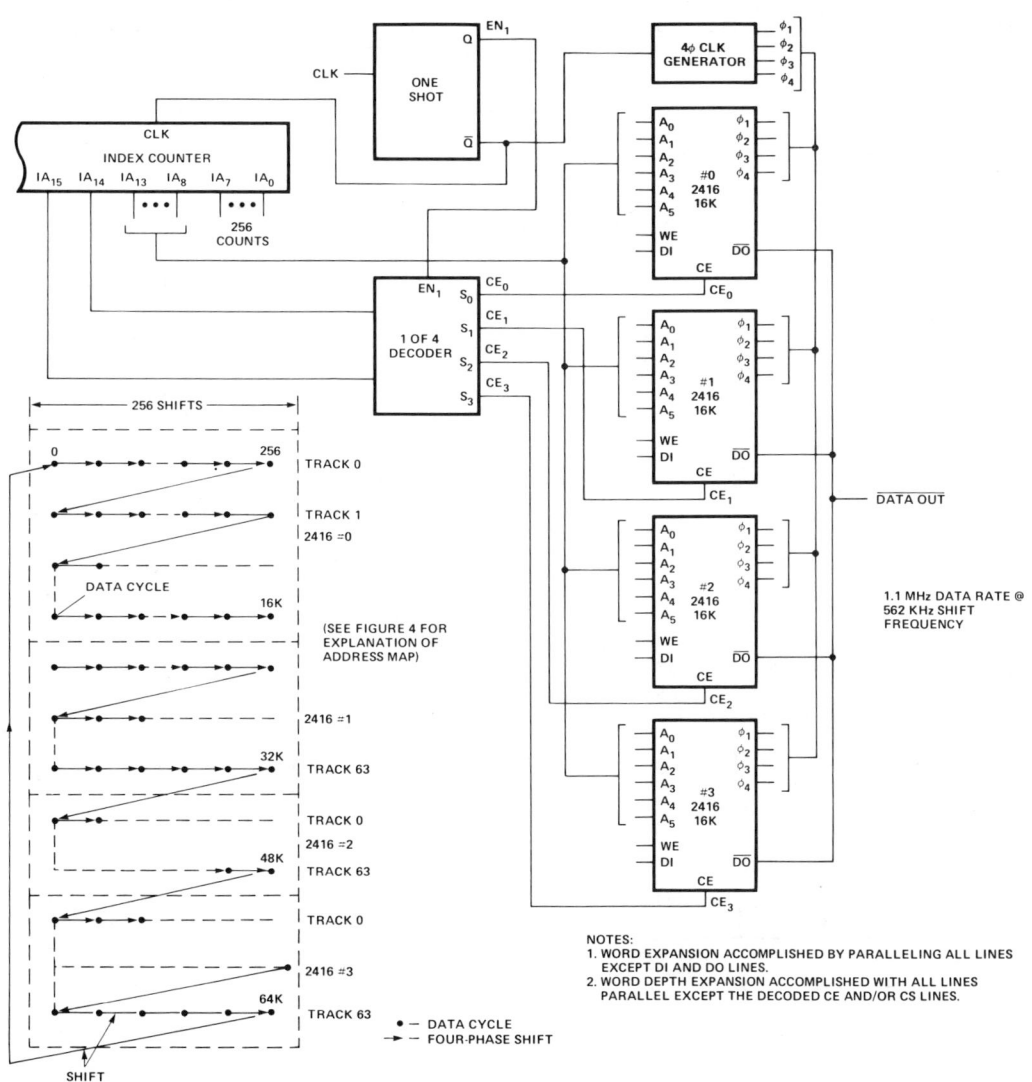

Figure 15. Shift/Single Data Cycle/Shift Address Expansion.

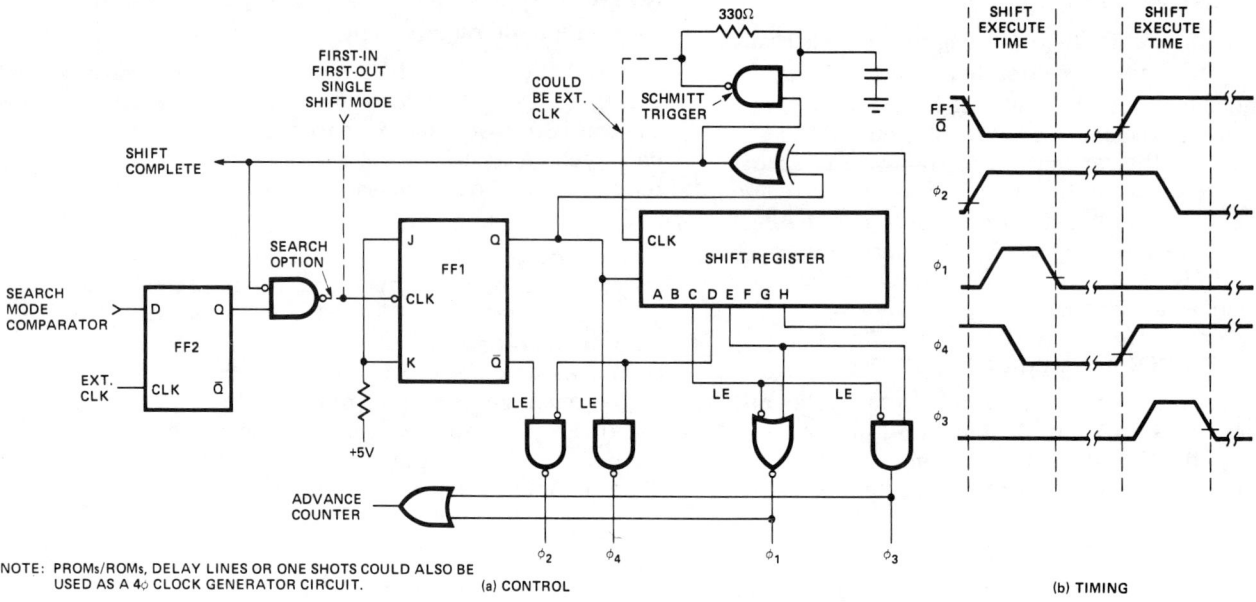

Figure 16. Four-Phase Clock Generator Circuit.

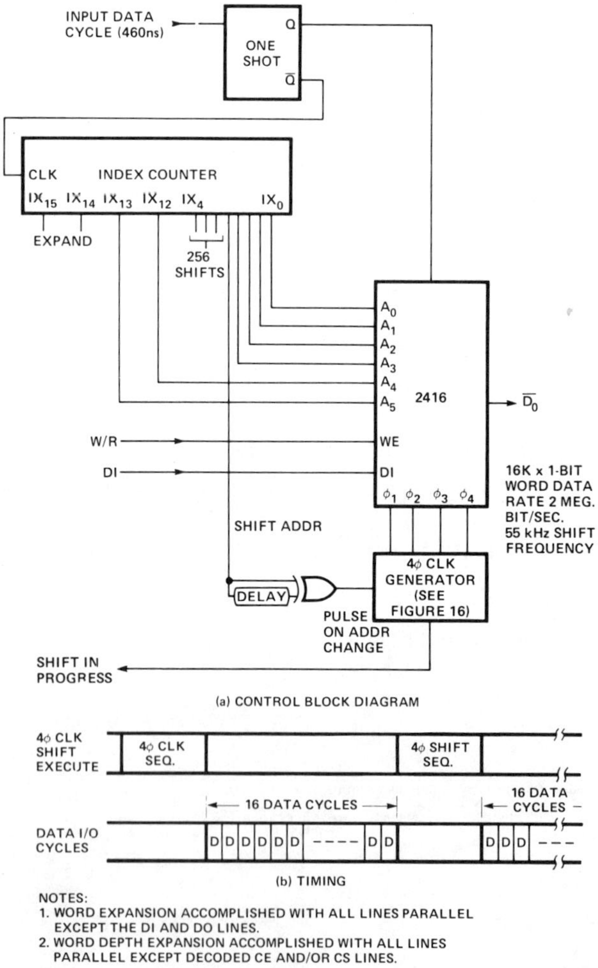

Figure 17. Shift/Multiple Data Cycle/Shift Control.

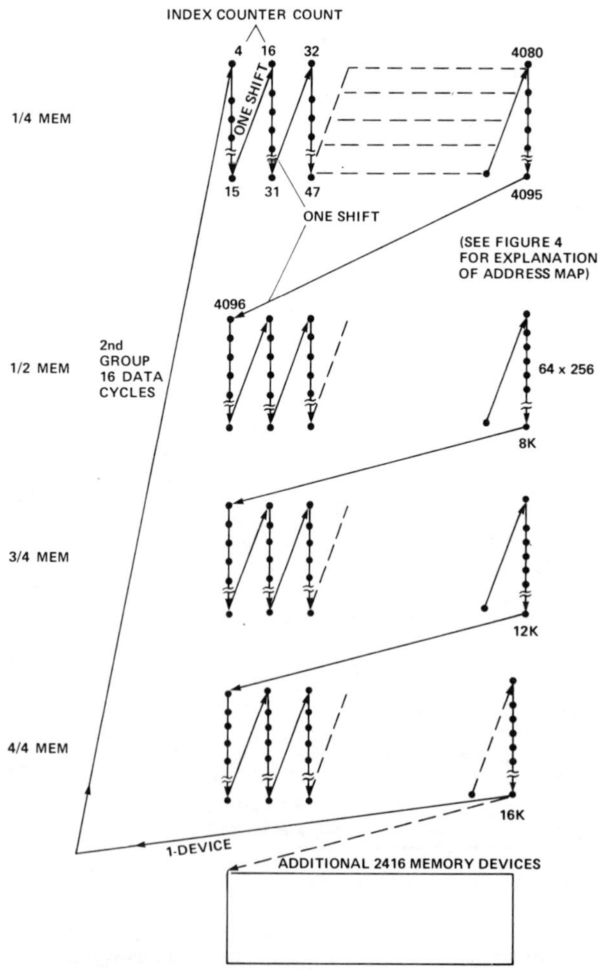

Figure 18. Shift/Multiple Data Cycle/Shift Address Sequence.

APPLICATION OF 2416-16K CCD MEMORY

cussed later.) The data cycle inhibit time gap is shown in Figure 17(b) by the absences of data cycles during the shift cycles. Relating the control schematic (Figure 17b) to the data address sequence chart (Figure 18) shows that 16 (out of 64) of the internal 2416 registers are selected before a shift cycle is initiated by index address 4 (IX_4). The selection of this first group of internal registers is repeated 255 times before a new group of 16 is selected by the change in index counter address 12 (IX_{12}). This sequence is repeated three more times, before the index counter either selects a new 2416 or returns to the original address location.

COMBINED SEARCH AND DATA CYCLE CONTROL

The control circuitry described in Figure 14 and 17 applied primarily to sequential applications which do not require a "search" to find a block of data. A more general control circuit is one that is capable of performing a "search" (or shift at high speeds to locate a block of data) and then accessing data at the maximum data transfer rate and the minimum shift cycle time.

Figure 19 is a block diagram of the control for operation of the 2416 in a random access or search cycle mode. The previously discussed principles of the sequential control modes are applied with the addition of a shift address comparator circuit and a request and acknowledge loop (which provides data synchronization). The search cycle mode occurs when one or more of the 8 shift address lines do not compare to the corresponding 8-bit index counter lines. (This means that the starting address location of a block of data is not in the data out buffer.) A "not compared" condition inhibits a data start cycle signal and enables the four-phase shift generator. The four-phase shift generator shifts the 2416 at the maximum four-phase clock shift rate and increments the 8-bit index counter until a "compare" is obtained. The compare enables a data start cycle which allows data access to the 2416 in the same manner as described in the sequential mode of Figure 17.

"HIDDEN" SHIFT CYCLE CONTROL

Time gaps in input/output data transfers in the previously described control circuit are the result

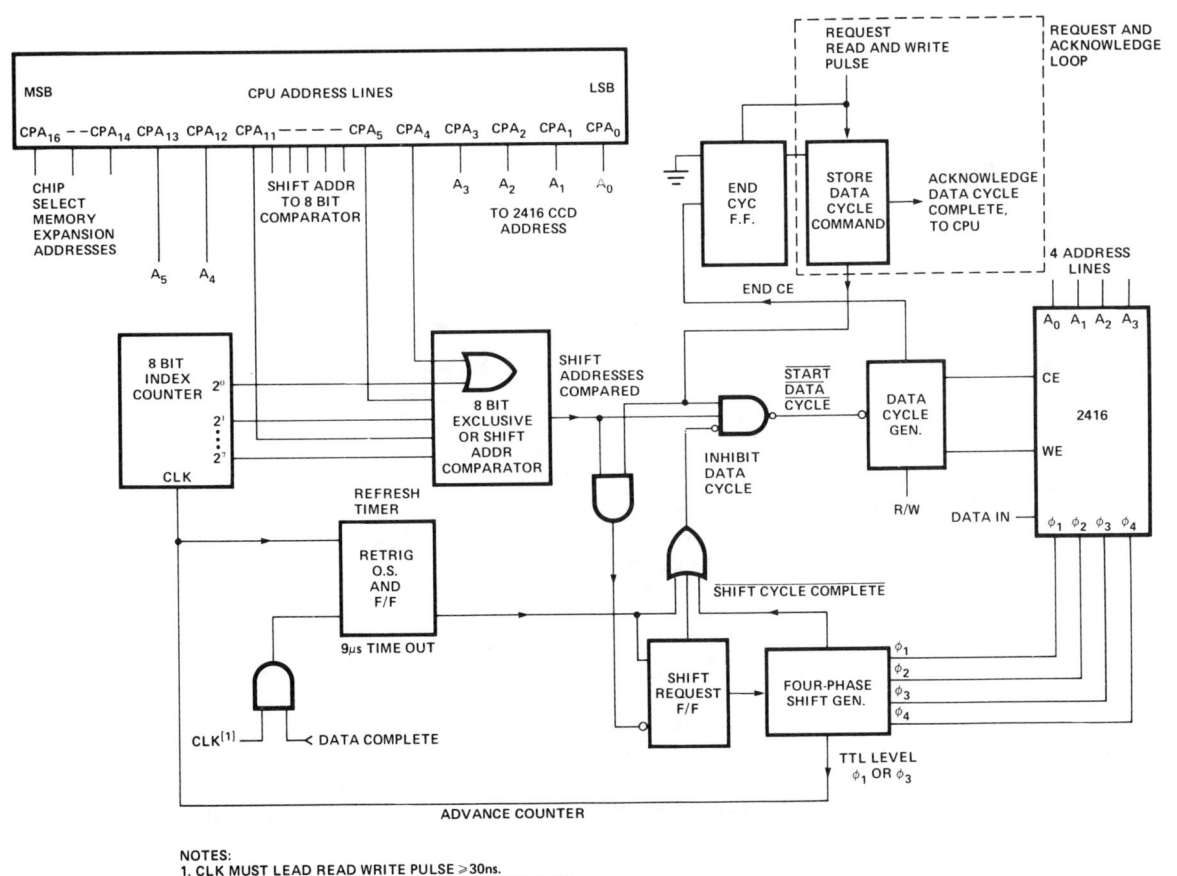

Figure 19. Block Diagram Control Random Access Mode.

of a shift cycle taking place. (Remember that no data I/O operations may be performed during a shift cycle.) In most systems applications, this time gap in the data I/O rate can be ignored or an external one word data buffer added to "hide" the gap. However, for those systems in which neither of the above alternatives is acceptable, the time gap can be hidden by the system controller shown in Figure 20. (The particular example is for a serial access design but can be extended to the search/multiple data mode described previously.)

The circuit in Figure 20 emphasizes the concept of obtaining high data rates with minimum four-phase clock shift rates and expands this concept by interleaving the shift times between two 2416 devices. As shown in the timing diagram included in Figure 20, interleaving the shift times and multiplexing the data out signals from both 2416 devices to a common data out line "hides" the shift time of the device being shifted from the *system* input/output data stream. Note that only one 2416 at a time is being shifted.

Operation of the 2416 in the system shown in Figure 20 is described as follows (see Figure 21). The first input data cycle inhibits the refresh oscillator and generates a chip enable signal from the I/O data cycle generator. The chip enable signal is steered to either device A or B by the state of the 2^3 bit on the index counter. (Read or write is determined by the state of the R/W.) The end of chip enable increments the index counter which establishes a new data location by changing the 2416 address lines. As the index counter is incremented it will select one shift location in one CCD and sequentially access 8 of the 64 CCD internal shift registers. At the end of the 8th cycle, the 2^3 index bit initiates a shift in the device being accessed and enables the data I/O in the other device. In summary, when one device is being shifted the other device is being accessed. Note that the same 8 registers (defined by addresses A_0-A_5) will alternately be selected between devices until 256 shift cycles have occurred, (i.e., the 2^{12} index counter bit changes state). After every 256 shift cycles a new group of 8 internal CCD registers per device (defined by addresses A_0-

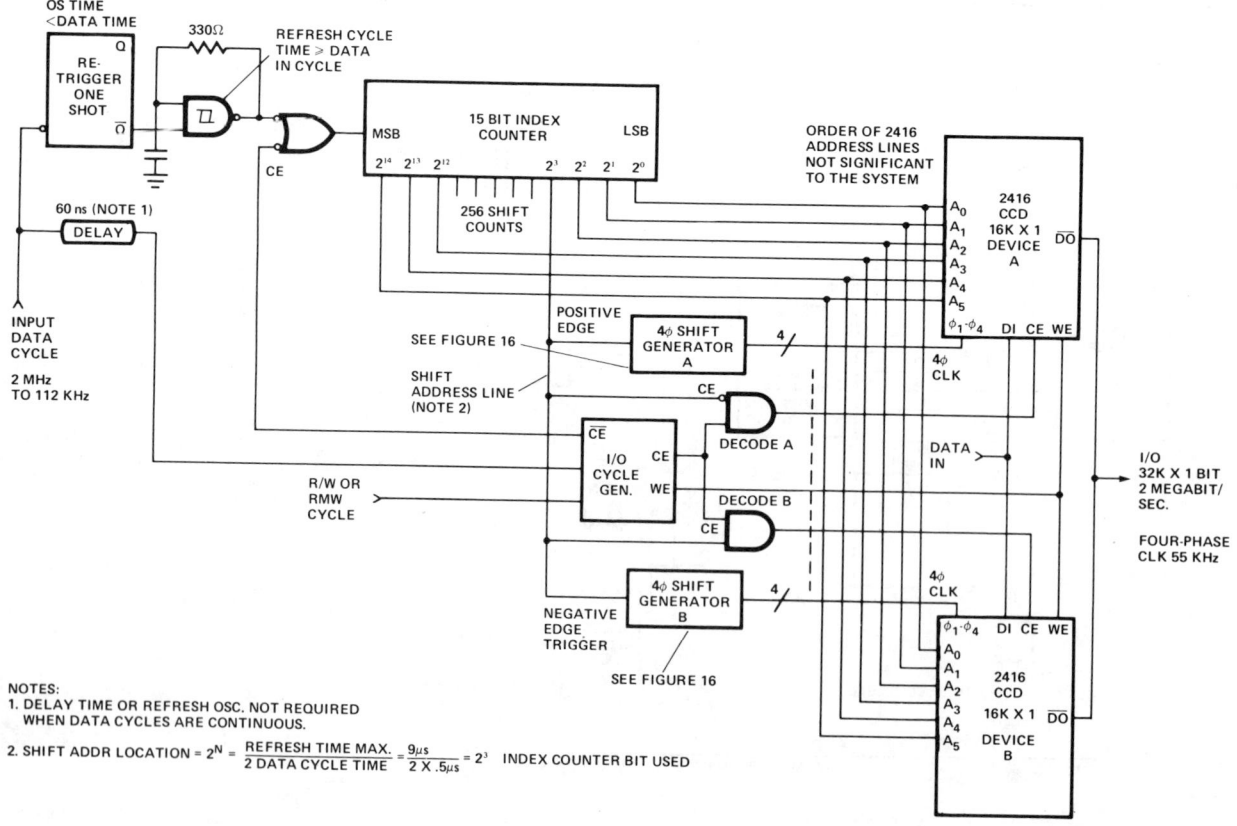

Figure 20. Hidden Shift Cycle Control.

APPLICATION OF 2416-16K CCD MEMORY

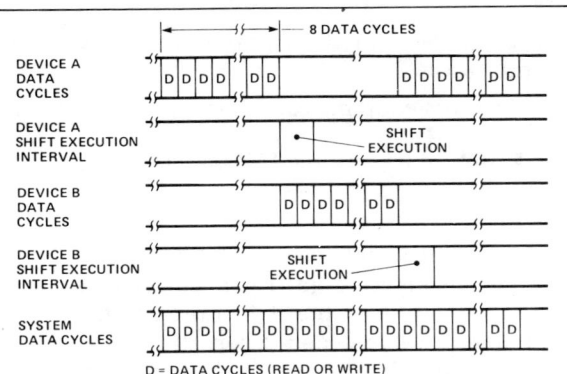

(a) Data

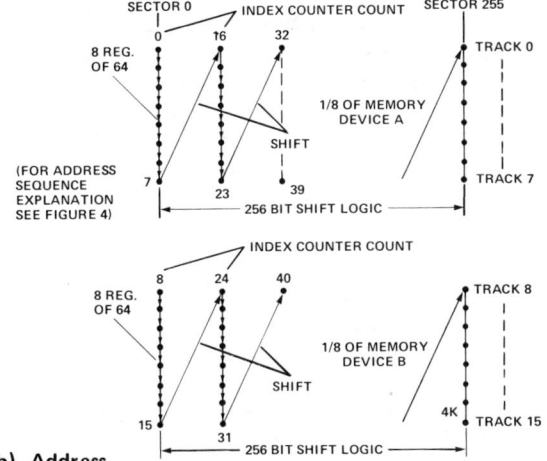

(b) Address

Figure 21. Hidden Shift Cycles Data and Address Sequence.

A_5) will be alternately selected between the two devices for 256 shift locations. This cycle continues until 8 groups of 8 registers (i.e., all 64 CCD internal data registers) are selected. The system described in Figure 19 can be expanded to 128K x 1 as shown in Figure 22.

Driving Considerations

This section discusses the 2416 input signal characteristics, driver requirements and data-out sensing considerations. All 2416 input lines operate from a nominal 12 volt logic swing driving signals. However, there are basic differences in input capacitance and voltage margins between some input signals. We will first consider the driver requirements for the four-phase clock inputs.

FOUR-PHASE CLOCK INPUTS

The four-phase clock inputs are connected internally to thin oxide gates (which overlap other clock lines) and as a result have relatively long lines (Figure 5). These clock lines cover the 64 shift register channels. The long lines and the close proximity of the gates to the substrate result in a high capacitance to substrate on the four-phase clock inputs. In addition, the overlapping on the gate electrode (to adjacent clock lines) produces added coupling capacitance between adjacent four-phase clock inputs.

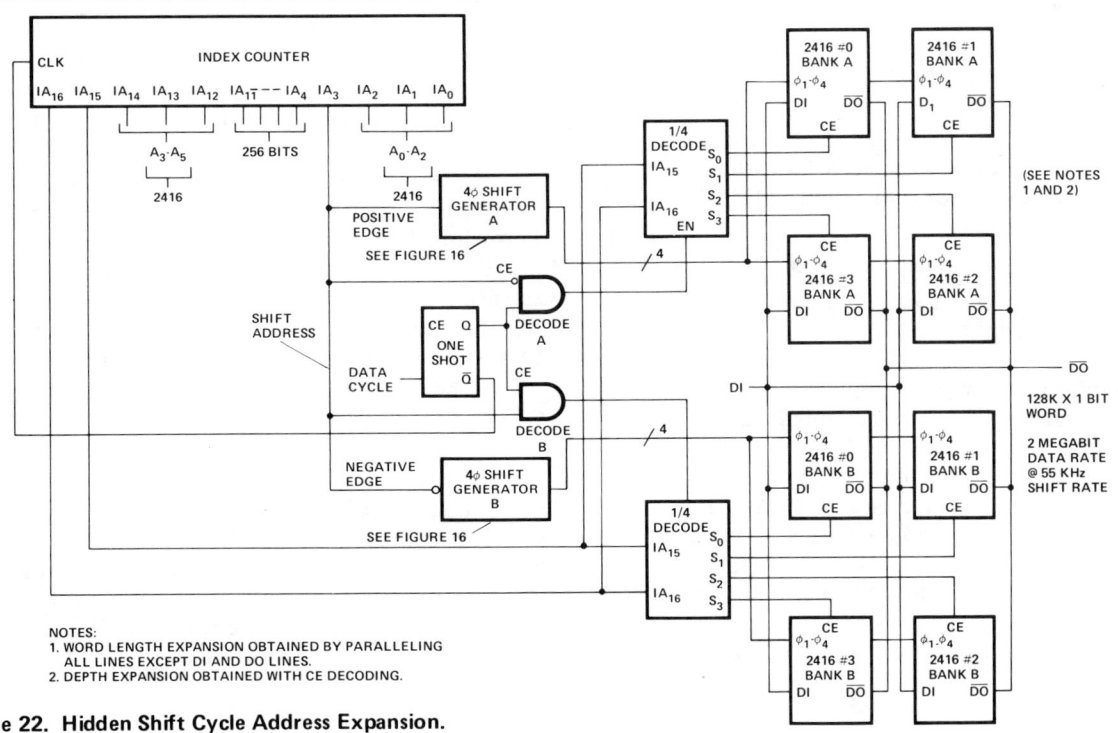

Figure 22. Hidden Shift Cycle Address Expansion.

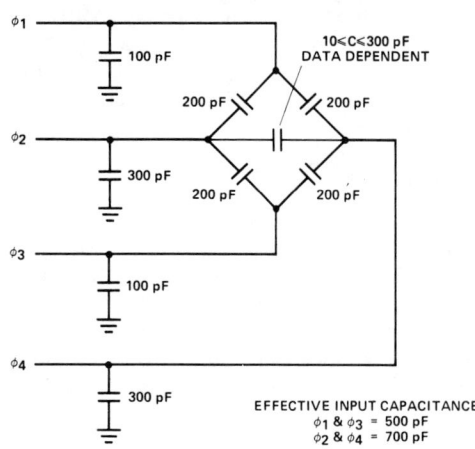

(a) Equivalent Circuit

		Max. pF
$C_{\phi 1}[1], C_{\phi 3}[1]$	ϕ_1, ϕ_3 Input Capacitance	500
$C_{\phi 2}[1], C_{\phi 4}[1]$	ϕ_2, ϕ_4 Input Capacitance	700
$C_{\phi 1 - \phi 2}$	Clock ϕ_1 To Clock ϕ_2 Capacitance	200
$C_{\phi 1 - \phi 4}$	Clock ϕ_1 To Clock ϕ_4 Capacitance	200
$C_{\phi 3 - \phi 2}$	Clock ϕ_3 To Clock ϕ_2 Capacitance	200
$C_{\phi 3 - \phi 4}$	Clock ϕ_3 To Clock ϕ_4 Capacitance	200

Note 1: The $C_{\phi 1}\ldots C_{\phi 4}$ input clock capacitance includes the clock to clock capacitance.

(b) Capacitance Values

Figure 23. Four-Phase Clock Input Equivalent Circuit.

Figure 23 shows the four-phase clock input equivalent circuit with the maximum capacitance values.

The equivalent circuit of Figure 23 suggests two clock driver requirements which must be considered in most clock driver designs for a particular system. These two requirements are:

1. Ability to drive high capacitance loads.
2. Ability to suppress cross-coupling current transients.

Of the two design requirements, number two is the most difficult to control. The cross-coupled current affects the ability of an adjacent clock driver to maintain the required high or low voltage margins while the adjacent phase driver is switching. The cross-coupled current that the quiescent driver must sink is proportional to the coupling capacitance and the slope of the active driver transitions (expressed as nsec per volt). The cross-coupled current is expressed by the equation for a linear charge of a capacitor:

$$I = C \frac{dv}{dt} \qquad (5)$$

Where:

I is the current for the duration of the signal transition.

C is the cross-coupling capacitance.

$\frac{dv}{dt}$ is the slope of the voltage transition across the capacitor.

The coupling capacitor between clock phases two and four shown in Figure 23(a) has a capacitance value that is a function of the data stored in the 2416. Its minimum value occurs when all data results in no charge stored in the potential wells under the phase 2 and phase 4 devices. Its maximum value occurs when the data under phase 2 and phase 4 devices has stored charge in the potential wells. (Remember that the refresh and buffer amplifiers in the 2416 are inverting, see Figure 7, so that all potential wells contain stored charge *only if* the original input data is a low level for 128 shifts and then a high level for 128 shifts. Complete absence of charge is the opposite logic condition.) This capacitance generally has a negligible effect on the overall design of the clock driver and is included only for completeness of the discussion on drivers.

Examining the clock input equivalent circuit and the above equation indicates a contradictory driver output impedance requirement. For the quiescent driver to hold the coupling voltage to a minimum requires that the driver have a very low output impedance. However, when that driver becomes active this low output impedance increases the slope of the transitions which in turn increases coupling currents to the other drivers. The above conditions suggest that a driver have a controlled output transition time and a low output impedance characteristic in the quiescent state (high or low level). Doubling the clock transition time (t_T) results in halving the cross-coupled currents (a very desirable effect). The clock transitions ($4t_T$) in the shift execution time expression, t_{SX}, (equation 2) have a minimal effect on data rate. Therefore, a large change in clock transition time will not appreciably effect the shift cycle, data rates, and latency time of the 2416. The effect of doubling the clock transition time decreases the maximum data I/O rate by less than 1% and increases the latency time by less than 20%.

FOUR-PHASE VOLTAGE MARGINS

The clock voltage margins and optimum "low" levels are also driver considerations. All four-phase clock low levels, V_{ILC}, are specified at V_{SS} +0.6/−2.0V for the 2416, including cross-coupling and over shoot transients. Another clock margin requirement is that the difference in the low level average reference

APPLICATION OF 2416-16K CCD MEMORY

voltage between all four-phase clocks must not exceed 0.5 volts. This means that all four-phase clock drivers should use the same DC power supply voltages. (Although MOS drivers usually take power from the same power supplies, it is emphasized here because of the 0.5 volt restriction.)

The high level margin (V_{IHC1}) for the transfer gates, ϕ_1 and ϕ_3 is $V_{DD} \pm 2.0V$, and the high level margin (V_{IHC2}) for the storage gates ϕ_2 to ϕ_4, is V_{DD} +2.0/-0.6V.

The power dissipated by a clock driver when driving a capacitive load is given by:

$$P = P_{dc} + P_{ac}, \quad (6)$$

where:

P_{dc} — is the average dc power dissipated by the driver when in quiescent state (high or low).

P_{ac} — is the power associated with driving capacitive loads.

and:

$$P_{ac} = CV^2 f \quad (7)$$

where:

C is load capacitance

f is clock frequency.

V is clock voltage swing.

The term P_{dc} can be considered a constant for a given clock driver design (to a first order approximation) and attention focused on P_{ac}. As shown in the equation for P_{ac} for a given capacitive load and drive voltage, the transient power is a function solely of the clock frequency. Therefore, to minimize driver power, the clock frequency must be minimized. As a result, maximum input/output data rates are achieved with minimum clock driver power. (Remember that the maximum data rate is obtained at minimum clock frequency.)

CLOCK DRIVER POWER VS. 2416 OPERATING MODE

A.C. clock driver power is calculated for several 2416 operating modes. In these calculations the d.c. component of the driver power is neglected as a first order approximation.

The four basic operating modes of the 2416 which effect the clock shift frequency and hence the clock driver power are:

1. Continuous search (maximum shift rate).
2. Refresh mode (minimum shift rate).
3. Shift/multiple data/shift mode.
4. Search data block transfer mode (combination of maximum and minimum shift rates).

Continuous Search Driver Power

This mode results in the maximum driver power dissipation with minimum (zero) input/output data rate. The clock driver power for the continuous search mode is expressed by the following equation:

$$P_S = C_T V^2 f_S \quad (8)$$

Where:

P_S — driver power in search mode.

C_T — total driver load capacitance.

f_s — clock frequency in search mode = $\frac{1}{t_{CYC}}$ (equation 1).

V — clock voltage swing.

For maximum loading conditions and search frequencies, equation (8) is solved as follows:

$$P_S = 2400 \,(10^{-12})(12)^2 \frac{1}{(2)(750)(10^{-9})} \quad (9)$$

or

$$P_S = .23 \text{ watts dissipated in clock driver per 2416 device.} \quad (10)$$

The search-after-every-data-cycle mode of operation (similar to the search only mode) results in a driver power dissipation of approximately that derived in equation 10 for maximum shift rates.

Refresh Cycle Driver Power

Derivation of the driver power for system operating in the refresh only mode is similar to the power derived for a continuous search mode. The power is calculated as follows:

$$P_{REF} = 2400 \,(10^{-12})(12)^2 \frac{1}{2(9000)(10^{-9})} \quad (11)$$

or

$$P_{REF} = .019 \text{ watts per 2416 device.} \quad (12)$$

Equations 11 and 12 clearly show that clock driver power is a reciprocal of the clock cycle time. Remember, to minimize clock driver power and simultaneously maximize data input/output rate, the four-phase clock cycle time should be maximized.

Shift/Multiple Data/Shift Driver Power

The continuous shift / multiple data / shift mode driver power depends on how many multiple data cycles occur between shift intervals. In this mode, the driver power can range from approximately 73% of the continuous search power (when only one data cycle between a shift interval is implemented) to as low as the refresh power (when 16 data cycles are implemented between shift intervals).

The calculation of four-phase driver power for a general system operating in a shift/multiple data/shift mode is a combination of the Search cycle and Refresh cycle power previously calculated. Since the actual power dissipated is a function of a particular system, the user is left to make the calculation for his particular system.

Search/Data Block Transfer Driver Power

The clock driver power in the search/data block transfer mode is the time averaged power between the high driver power during a search mode and the low power of the shift/multiple data/shift mode during a data block I/O transfer. The following example will better illustrate the power and time magnitudes involved in this mode of operation.

In this example, the data block length is assumed to be 4K data cycles and the maximum search latency time is assumed to be 200 μsec. The maximum 2416 shift/multiple data/shift rate is 2 megabits/sec. The average driver power including the search latency time and total data I/O time is expressed by the following equation:

$$P_{SDB} = P_{REF} \frac{(D_{bt})}{D_{bt} + L_{at}} + P_S \frac{(L_{at})}{D_{bt} + L_{at}} \quad (13)$$

Where:

P_{SDB} — the driver power in a search/data block transfer mode.

P_S — is the search mode driver power.

P_{REF} — is the previously determined refresh power.

D_{bt} — the time required to transfer a block of data, and is expressed by:

D_{bt} = number of Data Cycles/Data Block
   or   Average Data Rate

$$D_{bt} = \frac{4K}{2 \text{meg. bit}} = 2\text{ms}$$

L_{at} — maximum latency time ($255 \times t\phi/2$ — see equation 1).

$$P_{SDB} = .019w \left(\frac{2\text{ms}}{2.2\text{ms}}\right) + .23w \left(\frac{2\text{ms}}{2.2\text{ms}}\right) = .038w/ \quad (14)$$
2416 device.

The above example indicates that even at the highest search rates, a search/data block retrieval time ratio as low as 1 to 10 results in very low clock driver power dissipation.

Table III gives a summary comparison between the driver power requirements, data rates and mode of operation as calculated in the previous sections.

DRIVING THE 2416

The 4ϕ clock driving requirements of the 2416 determine the type of drivers that must be used. This driver must have the ability to drive a large capacitance as well as be able to maintain voltage levels during other clock transitions. (The four phase clock equivalent circuit is shown in Figure 23.) Two basic types of drivers which can be used to drive the clock inputs are those made with discrete components and integrated drivers. The complexity of discrete drivers virtually eliminate them from consideration. The problem now reduces to the selection of a suitable integrated circuit driver.

There are many integrated circuit drivers capable of driving a high capacitive load. However, the additional requirement of being able to suppress clock coupling transients make these drivers unsatisfactory for use in large 2416 systems. A driver designed especially for CCD devices is the Intel® 5244. The 5244 is a quad CCD clock driver capable of driving high capacitance loads *and* suppressing clock coupling transients.

THE 5244 QUAD CCD CLOCK DRIVER

The 5244 is a CMOS driver capable of driving four 2416's. This driver requires a single +12V supply and has fully TTL compatible inputs. The 5244 is designed specifically to drive CCD devices and as

Table III. Four-Phase Clock Driver Power Summary.

Mode of Operation	Symbol	Data Rates Bits/Sec.	4-ϕ Driver Power (mW)	Comments
Continuous Search	P_S	0 to 5000[1]	230	Maximum Driver Power
Refresh Only Mode	P_{REF}	0	19	—
Shift/Multiple Data/Shift (MIN)	P_{SMS} (MIN)	2 Megabit	19	Maximum Data Rate (16 Data Cycles between Shifts)
Shift/Multiple Data/Shift (MAX)	P_{SMS} (MAX)	970 Kilo Bit[2]	167	Shift after each Data Cycle
Search with Block Transfer	P_{SDB}	2 Megabit[3]	38	Data Block 4K Words

Notes:
1. Worst case decrement pattern on shift locations.
2. Input data rate is actual data rate and not average data rate.
3. Does not include search time.

such has internal circuitry designed to minimize the cross-coupling transients during clock transitions. The pin configuration and block diagram are shown in Figure 24 and 25 respectively. As shown in Figure 25, the output signal is fed back to an output transition control to assure that the clock transition times do not fall below the minimum required by CCD devices.

In most memory systems, and certainly in large CCD memory systems the power dissipation of any drivers is very important. Because the 5244 is implemented by CMOS devices, the quiescent power dissipation is very low. The DC characteristics of the 5244 are shown in Table IV. I_{DD0} and I_{DD1} (standby and operating currents respectively) are defined for zero frequency ($t\phi/2$) and for a frequency of f = 0.67MHZ respectively. The readers attention is directed to the equation for operating current shown in note 1, Table IV. This equation gives the expected operating current as a function of shift time ($t\phi/2$) and can be used accordingly.

Driver Characteristics

The 5244 is specified to drive four 2416's and have the characteristics required by the 2416. These requirements are placed in two categories:

1.) Transition time
2.) Cross coupled voltage suppression

The transition time of the 5244 is specified between a minimum of 30nsec and a maximum of 75nsec for phases 1 and 3 and a minimum of 30nsec and maximum of 90nsec for phases 2 and 4 when driving four 2416's. When using the driver in this configuration, no additional components (such as resistors) are necessary to be added in the output. However, if fewer than 4 2416's are driven by the

PIN CONFIGURATION

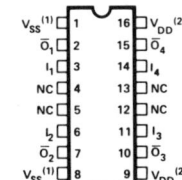

NOTES: 1. BOTH PIN 1 AND 8 MUST BE CONNECTED TO V_{SS}.
2. BOTH PIN 9 AND 16 MUST BE CONNECTED TO V_{DD}.

PIN NAMES

$I_1 - I_4$	TTL INPUT
$\overline{O}_1 - \overline{O}_4$	DRIVER OUTPUT
V_{DD}	+12V POWER SUPPLY
NC	NOT CONNECTED
V_{SS}	GROUND

Figure 24. 5244 Pin Configuration.

BLOCK DIAGRAM

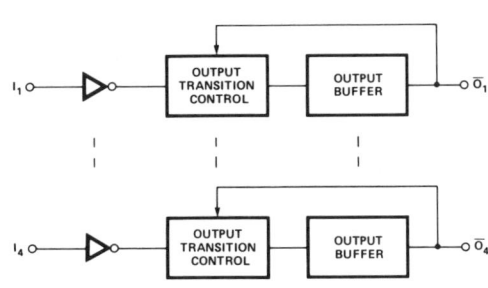

Figure 25. 5244 Block Diagram.

Table IV. 5244 D.C. and Operating Characteristics.

$T_A = 0°C$ to $70°C$, $V_{DD} = +12V \pm 5\%$, $V_{SS} = 0V$

Symbol	Parameter	Limits			Unit	Test Conditions
		Min.	Typ.	Max.		
I_{IL}	Low Level Input Current	-10	±0.1	10	μA	$V_{IN} \leq V_{IL}$
I_{IH}	High Level Input Current	-10	±0.1	10	μA	$V_{IN} \geq V_{IH}$
V_{IL}	Input Low Voltage		+1.2	+0.85	V	
V_{IH}	Input High Voltage	+2.0	+1.5	V_{DD}+1.0	V	
V_{OL}	Output Low Voltage	0	0.03	+0.1	V	I_{OL} = 5mA
V_{OH}	Output High Voltage	V_{DD}-0.1	V_{DD}-.03	V_{DD}	V	I_{OH} = -5mA
I_{DD0}	Standby Current		2.0	4.0	mA	$V_{IN} \geq V_{IH}$, $V_{IN} \leq V_{IL}$, f = 0 MHz
I_{DD1}	Operating Current		75	105[1]	mA	$V_{IN} \geq V_{IH}$ or $V_{IN} \leq V_{IL}$, f = 0.67MHz [2]

[1] $I_{DD1} = 4.0mA + \dfrac{75.4mA}{t\phi/2 \text{ (in } \mu s)}$

[2] Output load = four 2416 clock inputs or equivalents per Figure 23.

5244, an external capacitor must be added to each phase driver as shown in Figure 26. These capacitors must be added to assure that the driver transition time is not less than the minimum specified by the 2416.

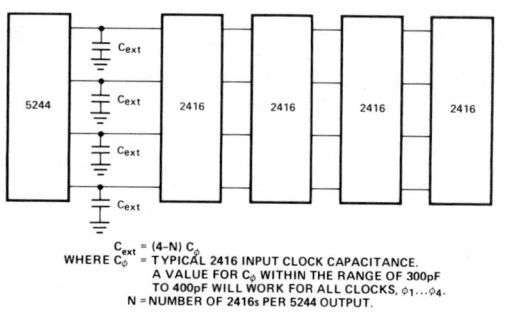

Figure 26. External Loading Requirements When Driving Fewer Than Four 2416's.

A more difficult parameter to specify is the cross-coupled voltage transient resulting from driving four 2416's. Figure 27 shows the cross-coupling to be expected (vertical scale is exaggerated). The cross coupled noise suppression is specified both in level above and below quiescent and in time. The designer is reminded that the coupling transients shown assume a reasonable signal distribution the printed circuit board of the clock inputs. A typical distribution technique acceptable for CCD devices is shown in the Memory Array Layout section.

The relationship between the 5244 driver output specification, and the 2416 input requirements are shown in Figure 28. As shown in these diagrams, the specifications associated with the 5244 allow an adequate noise margin when operating with the 2416's

Typical Waveforms of the 5244

Typical waveforms of the 5244 driving 4 2416's are shown in Figure 29. The driver placement shown in this figure is that described in Figure 36.

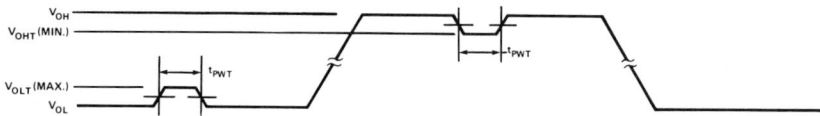

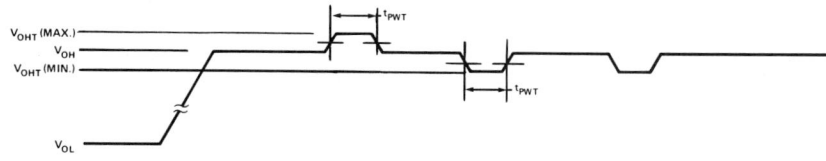

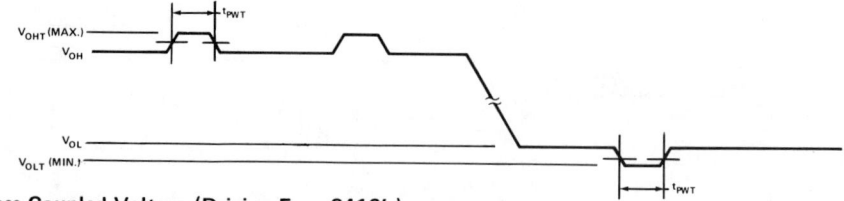

Figure 27. 5244 Output Cross-Coupled Voltage (Driving Four 2416's)

APPLICATION OF 2416-16K CCD MEMORY

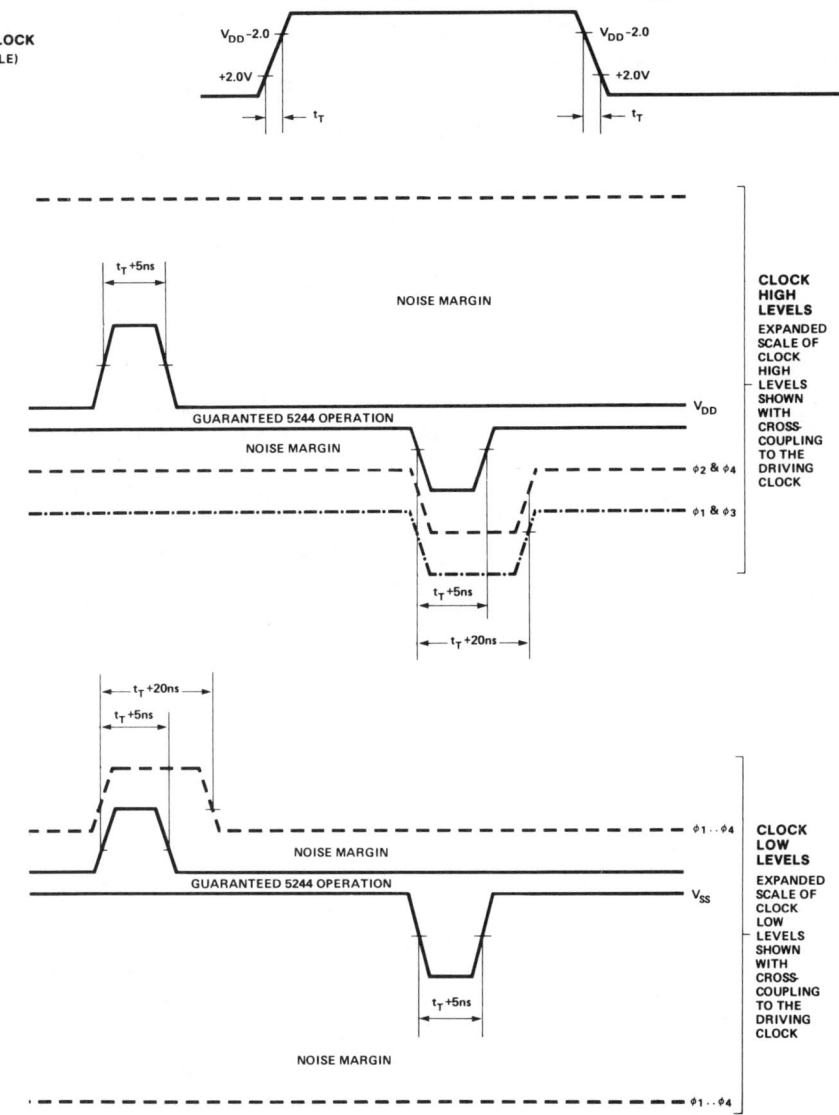

Figure 28. Noise Margins Between 5244 Output Specs and 2416 $\phi_1 \ldots \phi_4$ Input Requirements.

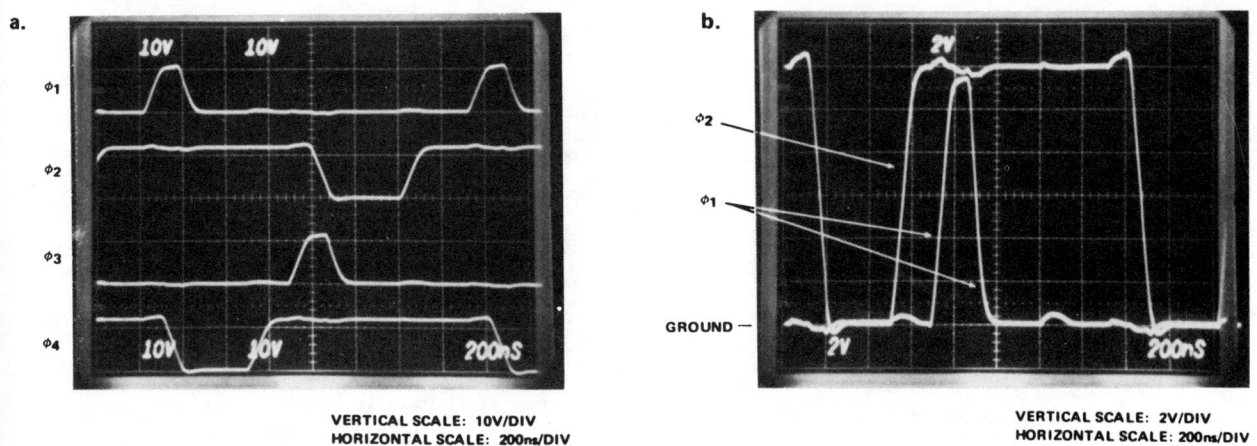

Figure 29. 5244 Typical Waveforms Driving Four 2416's.

DRIVING CS, CE, ADDRESS AND DATA-IN LINES

The remaining 2416 input lines, i.e. chip enable, chip select, address and data-in, exhibit a capacitive input of 4pF each. The low level margin (V_{IL1}) for these inputs is V_{SS} +.8V/–1.0V. The high level margin (V_{IH1}) for these signals is V_{DD} ±1.0 volts, except for data-in which has a V_{IHD} from 3.5 volts to V_{DD}+1 volt. The wide voltage margin on the data-in line allows it to be driven by a CMOS circuit or a TTL with a 470Ω pull-up resistor or the same type of driver used for the CE, CS, and address lines.

Maintaining Voltage Levels

The internal line to line coupling capacitance between the low capacitance inputs is less than 1pF. In addition, coupling can exist between signals at the card level. To suppress this total cross coupling effect, and thus maintain the required voltage margins, a driver with a low output impedance to V_{SS} and/or V_{DD} is required. Generally, drivers utilizing CMOS, complementary collector, and the totem pole type configurations, with an over-driven emitter follower, will suppress or recover from the coupling transients with sufficient margin. A driver employing a passive pull-up resistor or an emitter follower without over-drive voltage produces marginal results.

In addition to coupling, the over-shoot tendencies associated with the fast signal transition times also affect the voltage margins. It is important to locate the driver as close as possible to the memory array, usually split or branched from the center. Inserting a 10Ω (for multilayer board) or a 20Ω (for a two-sided printed circuit board) series damping resistor suppresses these over-shoot tendencies. The number of memory devices connected to the driver increases the coupling between inputs in addition to increasing the driver delay. These effects are shown in Figure 30 for an Intel® 3245 and 5235 quad drivers driving 4, 8, 32 and 64 2416 devices on a two sided printed circuit board array.

Sensing and Data-Out Characteristics

The 2416 data-out line is driven from an open drain MOS circuit which allows "OR" tying of the outputs. The access time of the 2416 is specified with a 5K pull-up resistor on the data-out pin to a 5 volt supply and a capacitive load of 50pF. The 50pF represents eight 2416 data-out lines OR tied. (i.e., 5pF per device and approximately .5pF/device stray capacitance.)

The waveforms in Figure 31 show the results of connecting 4, 8 and 16 2416 data out lines to the input of a series 74 type TTL gate. The recovery time of the data-out line is determined from the RC time constant of the data out line pull up resistor (including the sensing device input resistance) and the total data-out line load capacitance. This time constant should be less than 60% of the data cycle time to allow the bus to recharge from the previous cycle. The minimum value of the pull-up resistor is determined from the 2416 (I_{OL}) data out 3mA low sink current capability while maintaining less than +.45 volts above V_{SS} (GND). Limiting the data out sink current to 3mA results in an effective minimum pullup resistance of 1.7K ohms when connected to 5 volts and 4K ohms when connected to 12 volts.

The output of the 2416 goes low *only* when a logic "0" is read out. The output is held at a high level at all other times by the pull-up resistor. The advantage of this arrangement is that the time constant of the load capacitance and pull-up resistance has a minor effect on the access time.

Care should be taken when using a large value of pull-up resistance to assure that noise coupled into the output during sense time is not excessive.

In summary, the 2416 data out sensing characteristics are suitable to both high and low level CMOS

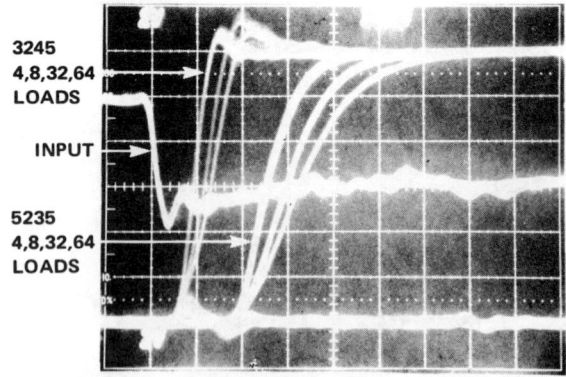

HORIZONTAL: 20NSEC/DIV
VERTICAL: 2V/DIV

Figure 30. Driver Waveforms as a Function of Loading.

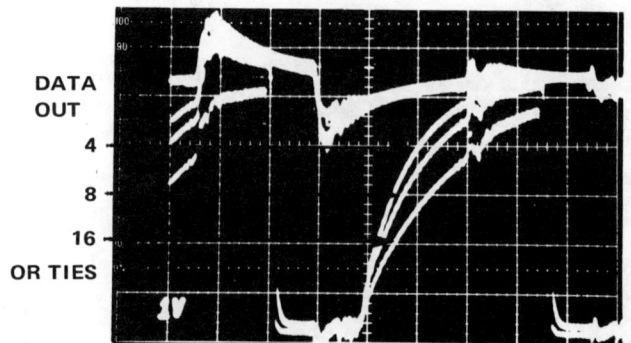

VERTICAL: 1V/DIV
HORIZONTAL: 100NSEC/DIV

Figure 31. 2416 Data Out Waveforms OR Tied.

APPLICATION OF 2416-16K CCD MEMORY

inputs and TTL inputs. The Intel® 3212 high input impedance 8-bit latch with three state output or the 3404 6-bit latch also provides system advantages when used as a 2416 sensing device.

CARD AND SYSTEM ORGANIZATION

The optimum organization of a CCD memory card is determined by the memory application, card expansion capability and memory word size requirements of the system. When a simple parity check or single error correction is used, it is desirable to organize the memory card to minimize multiple bit errors by avoiding common drivers and sensing circuits to more than one bit in a word. This is easy to accomplish where large memory systems are required. If the card is organized in a one or two bit configuration, the number of bits per word is obtained by adding additional cards. For example, such a card might be organized as 512K words by 1 bit. For this case, 8 cards would be required to obtain a word size of 8 bits. This system is capable of a data rate of 2 megabytes/sec.

Additional memory depth expansion is accomplished by additional basic storage units which also become very adaptable to four-phase clock bank switching techniques. Figure 32 shows the basic card organization for the 512K x 1-bit card.

Megabit Storage Card

In many previous systems, a requirement for a large amount of memory usually meant the necessity of having several printed circuit cards to achieve the storage requirements. With the introduction of a 16K CCD device, very high memory densities can be achieved on a single printed circuit card. As an example, the high density storage card shown in Figure 33 stores one million bits of information. This card is self contained in that all clock drivers, sense amplifiers, and data bus drivers are included on the card.

This card is organized as 128K words x 8 bits (and can be modified to 64K x 16) as shown in Figure 35. The data rate of this memory is two megabytes per second (i.e. sixteen megabits per second) as configured. (This high data rate is achieved with the four-phase clocks cycling at minimum frequency of 55 KHz.) The combination of high density and high data rates make this type of card ideal for use in many types of applications. The 128K x 8 CCD storage card operates in parallel on the eight bits of a given word to achieve a data rate of sixteen megabits per second. If the data stream is to enter the memory system at a *serial* data rate of sixteen megabits per second then the memory interface can be

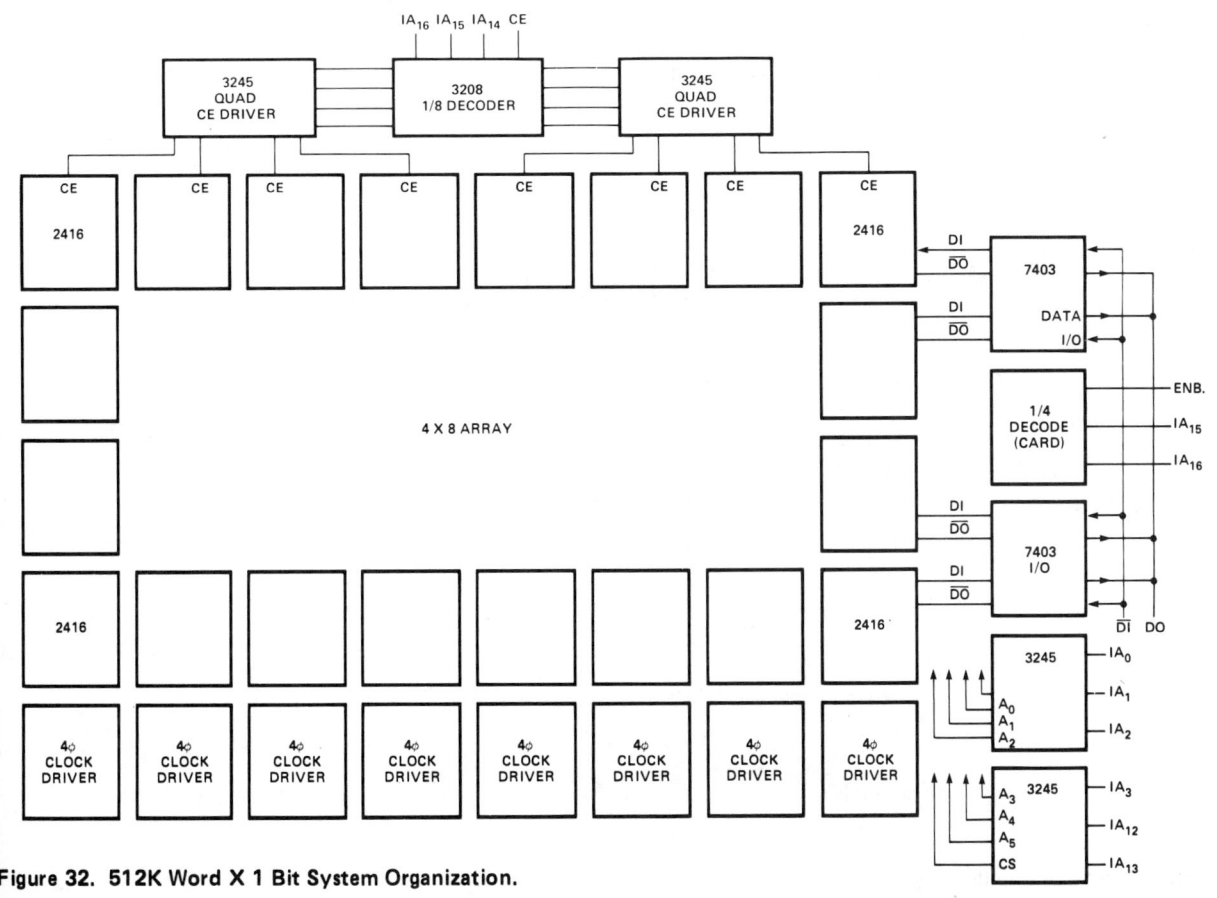

Figure 32. 512K Word X 1 Bit System Organization.

slightly modified to eliminate undesirable interruptions to the input and output data streams. Such a modification is shown in Figure 34 (for an eight megabit data rate).

Memory Array Layout

A well grided power distribution in the memory array is a very important layout consideration. Both

Figure 33. 128K Word X 8 Bit CCD Memory System (Megabit Card).

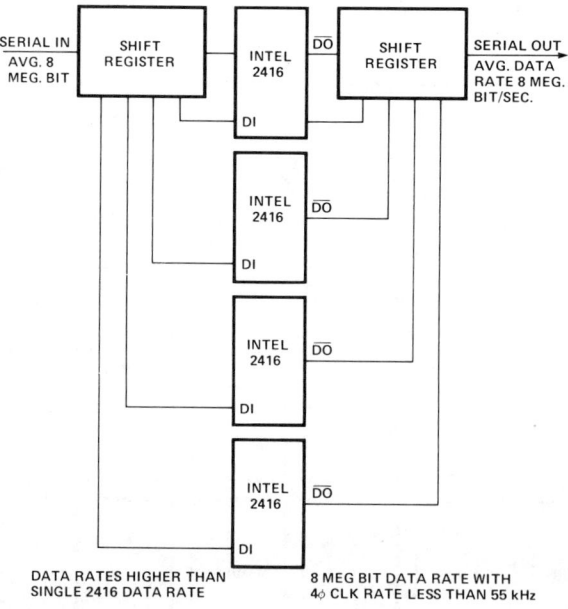

Figure 34. Paralleling 2416 to Handle Very High Data Rates.

APPLICATION OF 2416-16K CCD MEMORY

voltage and ground buses should be bused in the horizontal and vertical directions through every memory component. Generally, the width of the bus traces is not as critical as the separation between the grid construction. *Even in multilayer construction, an internally grided or continuous structure is important to minimize the charge and discharge paths from the array to the drivers.*

All memory array signal traces are usually 15 mils wide, which allows them to fit between the device pins with sufficient margin. However, at least 50 mil clock traces are recommended because of the peak currents involved with 2416 four-phase clock lines when several 2416s are driven with minimum transition times. It is recommended that these clock lines be run next to a GND line back to the driver as shown in Figure 36. These wider traces and the GND separation between them lowers the series inductance and coupling properties of these clock lines. However, the ground trace between the clock lines is not required when an internal ground plane or voltage plane is incorporated into the card.

The memory array layout of the 64K x 16 memory described previously is shown in Figure 37.

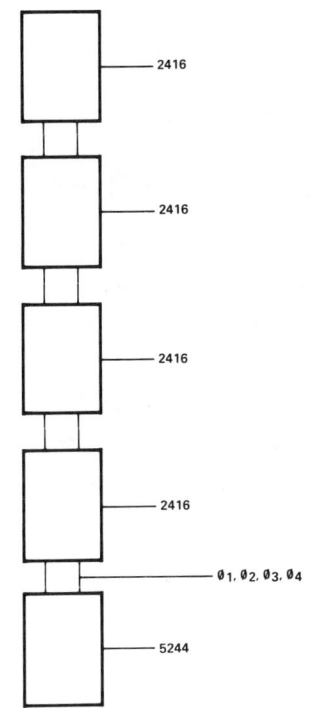

Figure 36. Four-Phase Clock Layout.

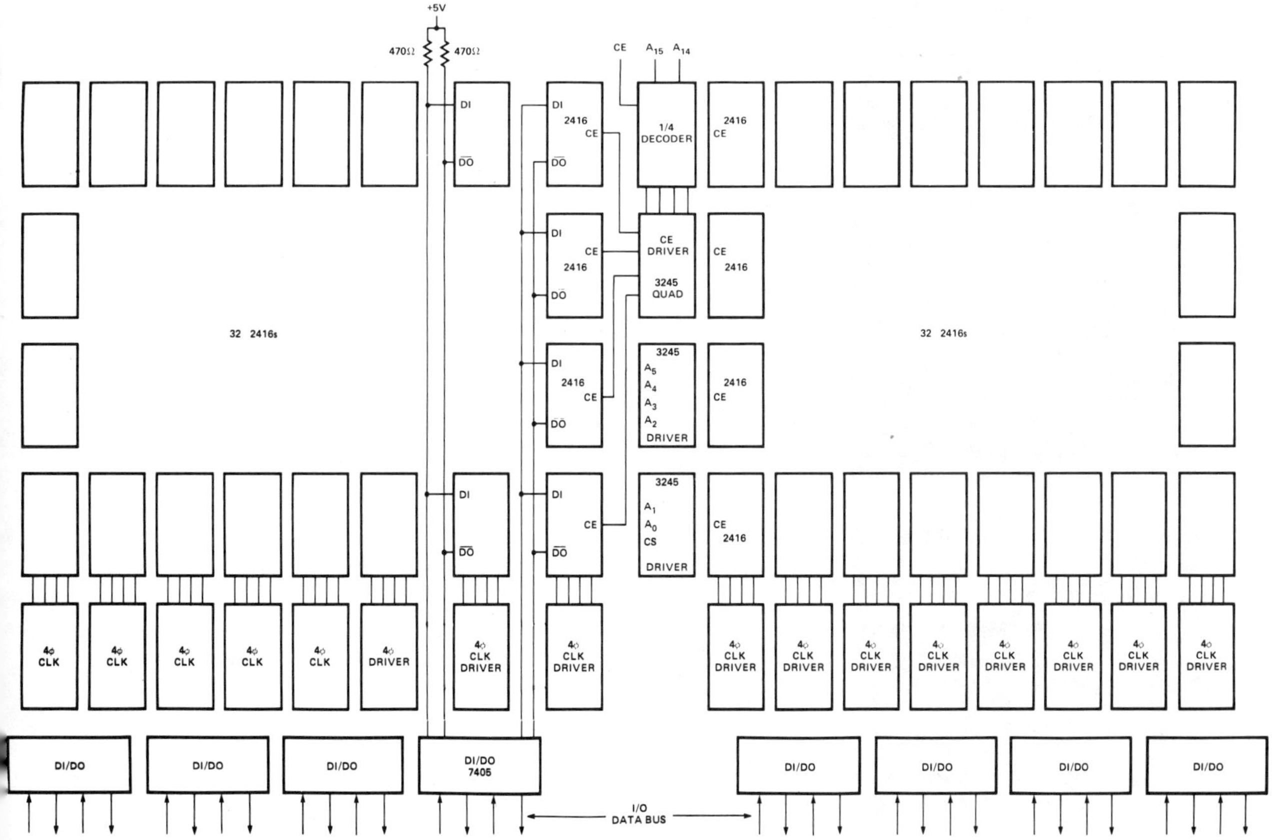

Figure 35. 128K Word x 8 Bit Card Organization.

Memory Array Decoupling

The 2416 decoupling requirements, as shown by the transient current waveforms in Figure 38, are very moderate. Tests show, from a 64 device 1 million bit board, that placing .1µF decoupling capacitors from V_{DD} to V_{SS} at every other device location in the array and a .1µF from V_{BB} to V_{SS} at the other devices will suppress the transient voltage spikes to less than 200mV.

However, on the four-phase clock drivers, a 1µF from V_{DD} to V_{SS} and 1µF from V_{BB} to V_{SS} is recommended for every two four-phase drivers.

Tantalum capacitors (~100µF) should also be added for low frequency decoupling.

SUMMARY

The 2416 has been shown to be a versatile and flexible memory device. This flexibility is maximized when a thorough understanding of the internal storage organization is achieved. Interface drivers and control circuits have been discussed for several of the more typical applications to demonstrate the ease with which the 2416 can be used.

ACKNOWLEDGMENT

Appreciation is extended to Jim Oliphant of the Application Engineering Department for his review and comments on this chapter.

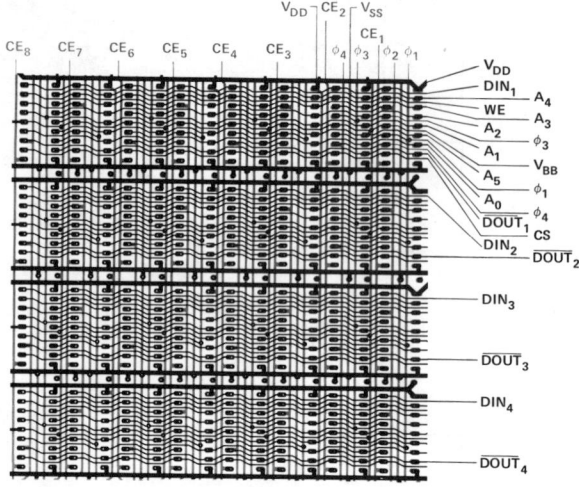

Figure 37. 2416 Memory Array Layout.

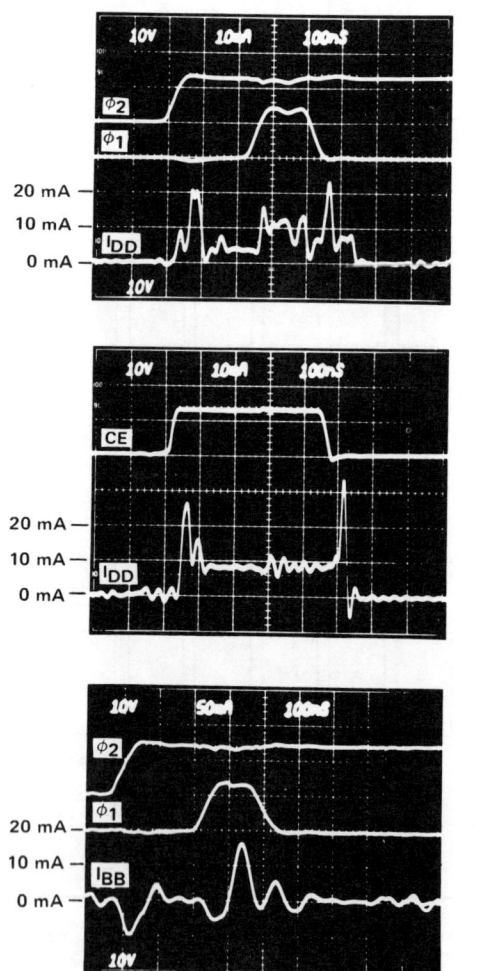

Figure 38. Transient Currents

2416

16,384 BIT CCD SERIAL MEMORY

- **Organization: 64 Recirculating Shift Registers of 256 Bits Each**

- Avg. Latency Time Under 100 μs
- Max. Serial Data Transfer Rate —2 mega bits/sec.
- Address Registers Incorporated on Chip
- Standard Power Supplies— +12V, −5V
- Open Drain Output
- Combined Read/Write Cycles Allowed
- Compatible to Intel® 5244 CCD Driver

The Intel® 2416 is a 16,384 bit CCD serial memory designed for low-cost memory applications requiring average latency times to under 100 μs. To achieve low latency time the memory was organized in the form of 64 independent recirculating shift registers of 256 bits each. Any one of the 64 shift registers can be accessed by applying an appropriate 6-bit address input.

The shift registers recirculate data automatically as long as the four-phase CCD clocks ($\phi_1 \ldots \phi_4$) are continuously applied and no write command is given. A one-bit shift is initiated in all 64 registers following a low-to-high transition of either ϕ_2 or ϕ_4. After the shift operation the contents of the 64 registers at the bit location involved are available for non-destructive reading, and/or for modification. I/O functions are accomplished in a manner similar to that of a 64-bit dynamic RAM. At the next shift cycle, the contents of the 64 accessible bits (whether modified or not) are transferred forward into the respective registers and the contents of the next bit of each register become accessible. No I/O function can be performed during the shift operation itself.

The Intel 2416 generates and uses an internal reference voltage which requires some time to stabilize after the power supplies and four phase clocks have been turned on. No I/O functions should be performed until the four-phase CCD clocks have executed at least 4000 shift cycles with power supplies at operating voltages. After this start-up period, no special action is needed to keep the internal reference voltage stable.

The 2416 is fabricated using Intel's advanced high voltage N-channel Silicon Gate MOS process.

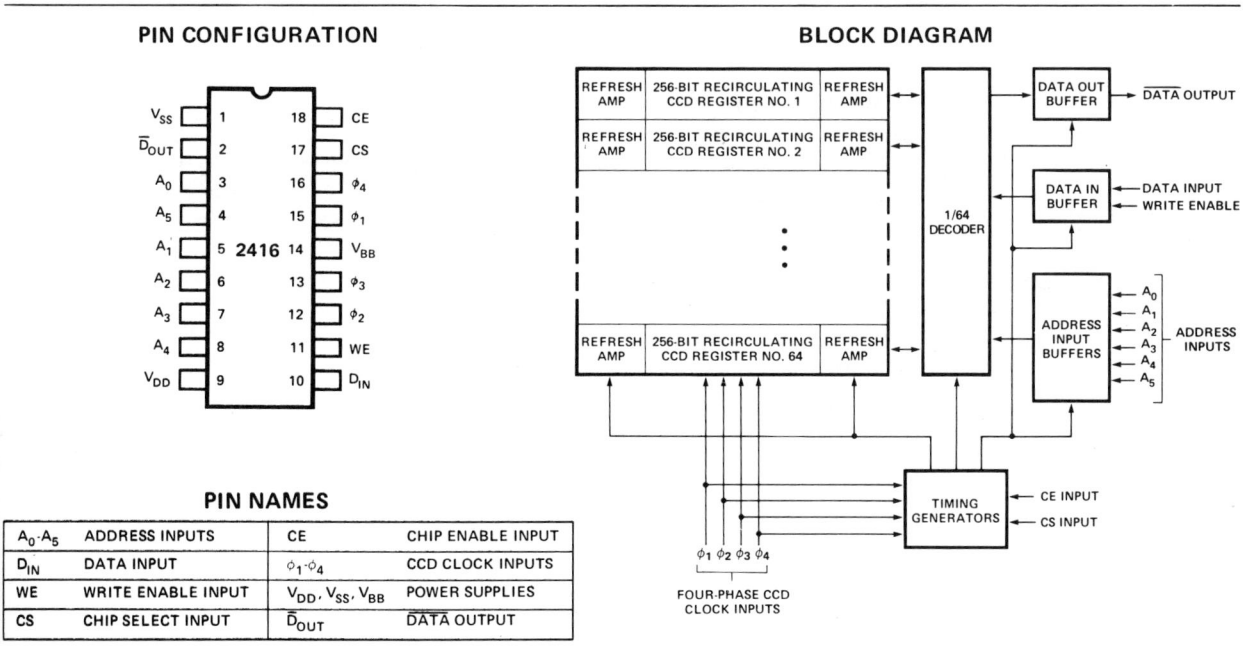

PIN NAMES

A_0-A_5	ADDRESS INPUTS	CE	CHIP ENABLE INPUT
D_{IN}	DATA INPUT	ϕ_1-ϕ_4	CCD CLOCK INPUTS
WE	WRITE ENABLE INPUT	V_{DD}, V_{SS}, V_{BB}	POWER SUPPLIES
CS	CHIP SELECT INPUT	$\overline{D}_{OUT}$	DATA OUTPUT

Absolute Maximum Ratings*

Temperature Under Bias	$-10°C$ to $80°C$
Storage Temperature	$-65°C$ to $+150°C$
All Input or Output Voltages with Respect to the most Negative Supply Voltage, V_{BB}	$+25V$ to $-0.3V$
Supply Voltages V_{DD} and V_{SS} with Respect to V_{BB}	$+20V$ to $-0.3V$
Power Dissipation	1.0W

*COMMENT:

Stresses above those listed under "Absolute Maximum Ratings" may cause permanent damage to the device. This is a stress rating only and functional operation of the device at these or any other conditions above those indicated in the operational sections of this specification is not implied. Exposure to absolute maximum rating conditions for extended periods may affect device reliability.

D.C. and Operating Characteristics

$T_A = 0°C$ to $70°C$, $V_{DD} = +12V \pm 5\%$, V_{BB}[1] $= -5V \pm 5\%$, $V_{SS} = 0V$, unless otherwise specified.

Symbol	Parameter	Min.	Typ.	Max.	Unit	Test Conditions
I_{LI}	Input Leakage Current		1	10	µA	$V_{IN} = 0V$
I_{LO}	Output Leakage Current		1	10	µA	$CE = 0V$, $V_{OUT} = 0V$
I_{OL}	Output Low Current	3			mA	$V_{OL} = .45V$
I_{OH}	Output High Current			10	µA	$V_{OH} = +5V$
I_{DDAV1}	Average V_{DD} Supply Current for Shift Cycles Only			Note 2	mA	
I_{DDAV2}[3]	Average V_{DD} Supply Current		15	25	mA	
I_{BB}	Average V_{BB} Supply Current		100	200	µA	
V_{IL}	Input Low Voltage, All Inputs Except $\phi_1 \ldots \phi_4$	-1.0		0.8	V	
V_{IH1}	Input High Voltage, All Inputs Except D_{IN} and $\phi_1 \ldots \phi_4$	$V_{DD}-1$		$V_{DD}+1$	V	
V_{IHD}	D_{IN} Input High Voltage	3.5		$V_{DD}+1$	V	
V_{ILC}[4]	$\phi_1 \ldots \phi_4$ Input Low Voltage dc	-2.0		0.6	V	
V_{ILCT}	$\phi_1 \ldots \phi_4$ Input Low Voltage w/Coupling	-2.0[5]		1.2[6]	V	
V_{IHC1}	ϕ_1 and ϕ_3 Input High Voltage dc	$V_{DD}-1$		$V_{DD}+2$	V	
V_{IHCT1}	ϕ_1 and ϕ_3 Input High Voltage w/Coupling	$V_{DD}-1.6$[6]		$V_{DD}+2$[5]	V	
V_{IHC2}	ϕ_2 and ϕ_4 Input High Voltage dc	$V_{DD}-0.6$		$V_{DD}+2$	V	
V_{IHCT2}	ϕ_2 and ϕ_4 Input High Voltage w/Coupling	$V_{DD}-1.2$[6]		$V_{DD}+2$[5]	V	
t_{PWT}	Cross Coupling Voltage Pulse Width			Note 7	ns	Pulse width measured at 0.8V and $V_{DD}-1.2V$ (ϕ_1 and ϕ_3) or $V_{DD}-0.8V$ (ϕ_2 and ϕ_4)

Notes:
1. The only requirement for the sequence of applying voltage to the device is that V_{DD} and V_{SS} should never be 0.3V more negative than V_{BB}.
2. For shift only mode $I_{DD} = 2.0mA + \dfrac{15mA}{t_\phi/2 \text{ (in } \mu s)}$
3. I_{DDAV2} is for combined shift and data I/O cycles.
4. The difference in the low level reference voltages between all four clock phases must not exceed 0.5 volts.
5. These voltage levels with coupling are within the specified dc range and are not, therefore, subject to t_{PWT} restrictions.
6. These voltage levels with coupling are outside specified dc ranges and must be restricted to t_{PWT} pulse widths.
7. The maximum clock cross coupled pulse width is the sum of the clock transition time (t_T) plus 20ns.

$\phi_1 \ldots \phi_4$ CROSS-COUPLING

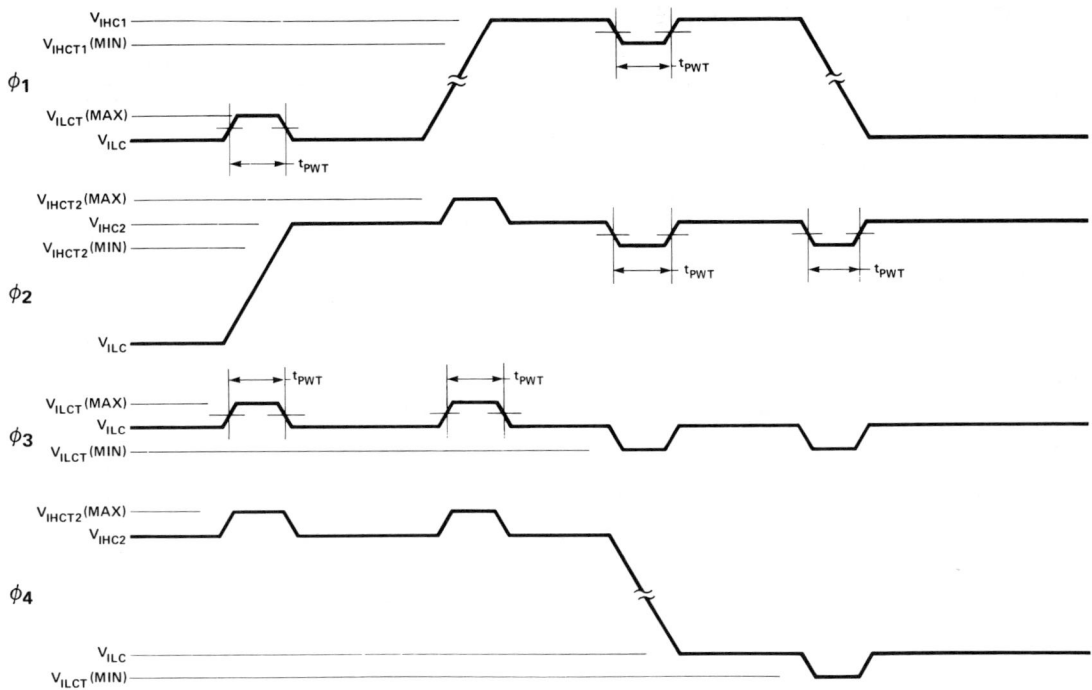

A.C. Characteristics
$T_A = 0°C$ to $70°C$, $V_{DD} = 12V \pm 5\%$, $V_{BB} = -5V \pm 5\%$, $V_{SS} = 0V$, unless otherwise specified.

SHIFT ONLY CYCLES

Symbol	Parameter	Min.	Max.	Unit	Conditions
$t_{\phi/2}$	Half Clock Period for $\phi_1 \ldots \phi_4$	750[1]	10,000	ns	$t_T = 40$nsec
t_{PT}	ϕ_2 On to ϕ_1 On Time, ϕ_4 On to ϕ_3 On Time	200		ns	
t_{TD}	ϕ_1 to ϕ_4 Overlap, ϕ_3 to ϕ_2 Overlap	30		ns	
t_{DT}	ϕ_4 to ϕ_1 Hold Time, ϕ_2 to ϕ_3 Hold Time	40		ns	
t_{TP}	ϕ_1 Off to ϕ_4 On, ϕ_3 Off to ϕ_2 On	320		ns	
t_T	Transition Times for $\phi_1 \ldots \phi_4$	30	200	ns	

Note: 1. The 750ns Half Clock Period will be met for $30ns \leq t_T \leq 40ns$. Values of $t_T > 40ns$ lengthen $t_{\phi/2}$.

WAVEFORMS (Numbers in parentheses are for minimum cycle timing in ns)

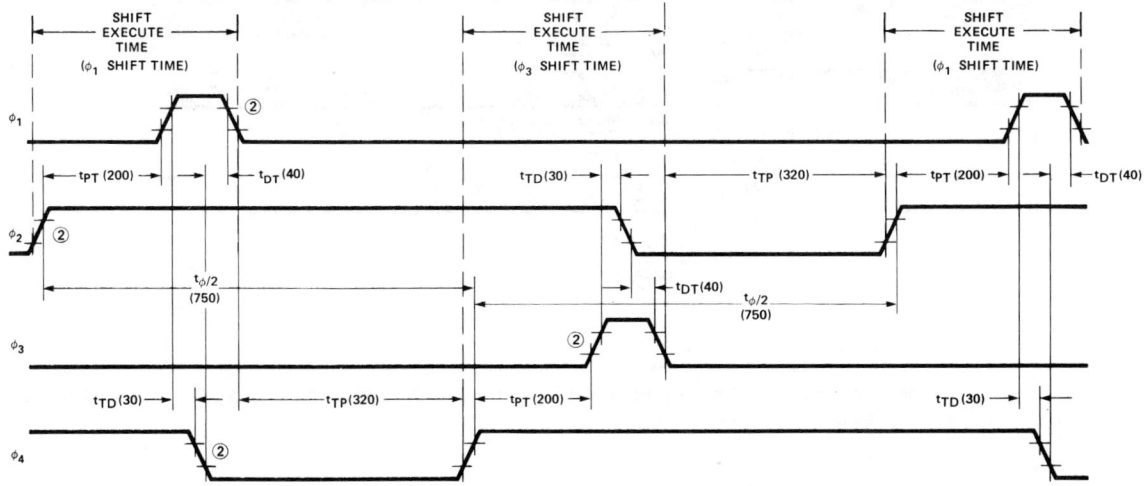

Note: 2. +2.0V and V_{DD}-2.0V are the reference low and high level respectively for measuring the timing of ϕ_1, ϕ_2, ϕ_3 and ϕ_4.

A.C. Characteristics

SHIFT–READ–READ–...–READ–SHIFT CYCLE

Symbol	Parameter	Min.	Max.	Unit	Conditions
t_{RCY}	READ Cycle Time	460		ns	
t_{PT}	ϕ_2 On to ϕ_1 On Time, ϕ_4 On to ϕ_3 On Time	200		ns	t_T = 40ns
t_{TD}	ϕ_1 to ϕ_4 Overlap, ϕ_3 to ϕ_2 Overlap	30		ns	t_{T1} = 20ns
t_{DT}	ϕ_4 to ϕ_1 Hold Time, ϕ_2 to ϕ_3 Hold Time	40		ns	
$t_{\phi/2}$	Half Clock Period for $\phi_1 \ldots \phi_4$		10,000	ns	
t_T	Transition Times for $\phi_1 \ldots \phi_4$	30	200	ns	
t_{T1}	Transition Times for Inputs Other Than $\phi_1 \ldots \phi_4$		100	ns	
t_{TC}	ϕ_1 or ϕ_3 Off to CE On	280		ns	
t_{SC}	CS to CE Set-Up Time	0		ns	
t_{AC}	Address to CD Set-Up Time	0		ns	
t_{AH}	Address Hold Time	240		ns	
t_{CS}	CE to CS Hold Time	0		ns	
t_{CC}	CE Off Time	140		ns	
t_{CP}	CE Off to ϕ_2 or ϕ_4 On	40		ns	
t_{CER}	CE On Time	280		ns	
t_{CF}	CE Off to Output High Impedance State	0		ns	
t_{CO}	CE to $\overline{D}_{OUT}$ Valid		250	ns	

WAVEFORMS[1] (Numbers in parentheses are for minimum cycle timing in ns)

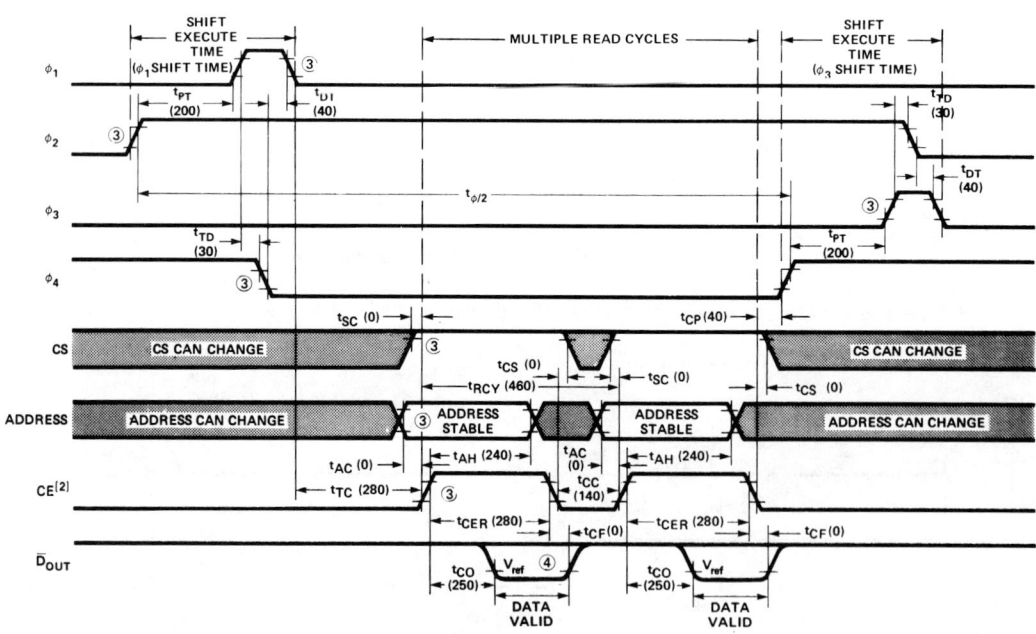

NOTES: 1. WE must be continuously low during the READ cycle.
2. When CE is off, the 2416 output level is determined by the external output termination.
3. +2.0V and V_{DD}-2.0V are the reference low and high level respectively for measuring the timing of $\phi_1 \ldots \phi_4$, CE, CS and addresses.
4. +0.8V is the reference level for measuring the timing of $\overline{D}_{OUT}$.

A.C. Characteristics

SHIFT—WRITE—WRITE—...—WRITE—SHIFT CYCLE

Symbol	Parameter	Min.	Max.	Unit	Conditions
t_{WCY}	WRITE Cycle Time	460		ns	
t_{PT}	ϕ_2 On to ϕ_1 On Time, ϕ_4 On to ϕ_3 On Time	200		ns	t_T = 40ns
t_{TD}	ϕ_1 to ϕ_4 Overlap, ϕ_3 to ϕ_2 Overlap	30		ns	t_{T1} = 20ns
t_{DT}	ϕ_4 to ϕ_1 Hold Time, ϕ_2 to ϕ_3 Hold Time	40		ns	
$t_{\phi/2}$	Half Clock Period for $\phi_1 \ldots \phi_4$		10,000	ns	
t_T	Transition Times for $\phi_1 \ldots \phi_4$	30	200	ns	
t_{T1}	Transition Times for Inputs Other Than $\phi_1 \ldots \phi_4$		100	ns	
t_{TC}	ϕ_1 or ϕ_3 Off to CE On	280		ns	
t_{SC}	CS to CE Set-Up Time	0		ns	
t_{AC}	Address to CE Set-Up Time	0		ns	
t_{AH}	Address Hold Time	240		ns	
t_{CS}	CE to CS Hold Time	0		ns	
t_{CC}	CE Off Time	140		ns	
t_{CP}	CE Off to ϕ_2 or ϕ_4 On	40		ns	
t_{CEW}	CE On Time	280[1]		ns	
t_{CW}	CE to WE Set-Up Time	100[1]		ns	
t_{DW}	D_{IN} to WE Set-Up	0		ns	
t_{WP}	WE Pulse Width	100[1]		ns	
t_{WC}	WE Off to CE Off	0[1]		ns	
t_{DH}	D_{IN} Hold Time	0		ns	

Note: 1. The minimum t_{CW}, t_{WP} and t_{WC} times with appropriate transitions do not necessarily add up to the minimum t_{CEW}. This allows the user flexibility in setting the WE Pulse Width edges without affecting either t_{CEW} or the WRITE Cycle Time, t_{WCY}.

WAVEFORMS (Numbers in parentheses are for minimum cycle timing in ns)

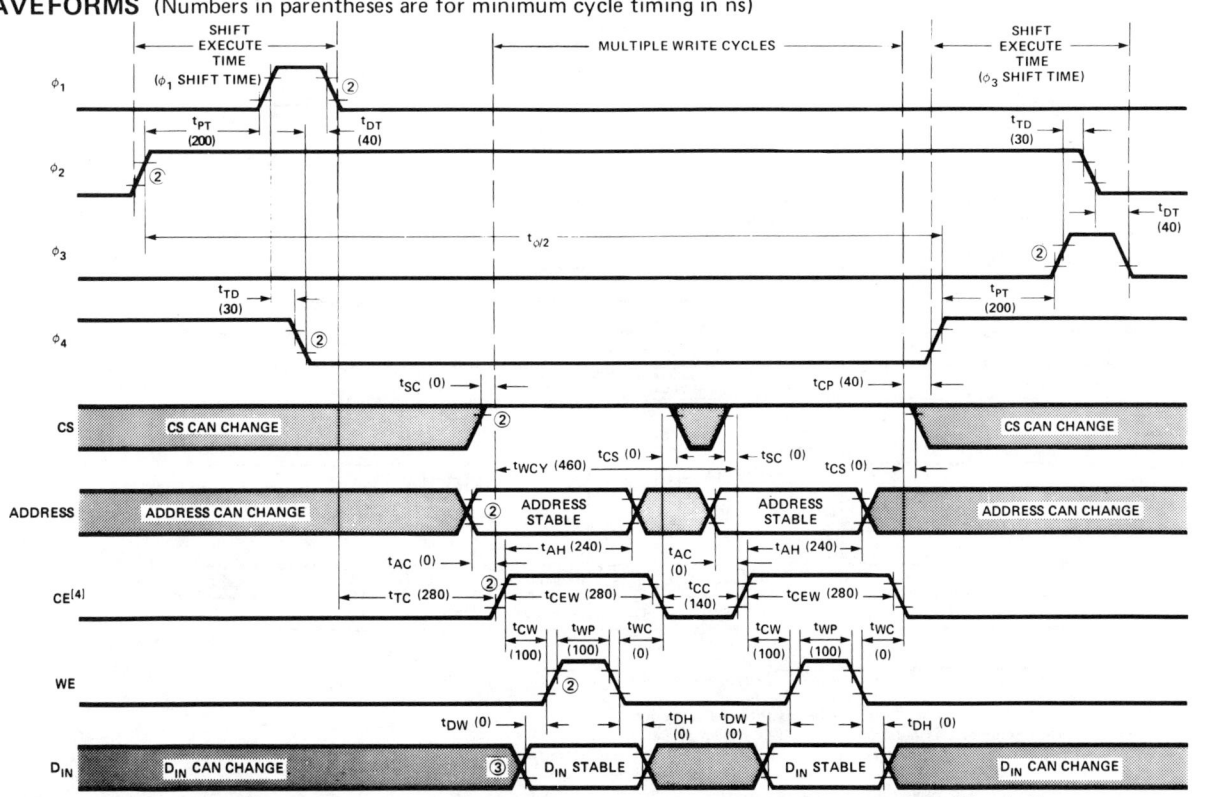

Notes: 2. +2.0V and V_{DD}−2.0V are the reference low and high level respectively for measuring the timing of $\phi_1 \ldots \phi_4$, CE, CS, WE, and addresses.
3. +1.5V and +3.0V are the reference low and high level respectively for measuring the timing of D_{IN}.

A.C. Characteristics SHIFT–RMW–RMW– ... –RMW–SHIFT CYCLE

Symbol	Parameter	Min.	Max.	Unit	Conditions
t_{RWC}	READ-MODIFY-WRITE Cycle Time	620		ns	t_T = 40ns
t_{PT}	ϕ_2 On to ϕ_1 On Time, ϕ_4 On to ϕ_3 On Time	200		ns	t_{T1} = 20ns
t_{TD}	ϕ_1 to ϕ_4 Overlap, ϕ_3 to ϕ_2 Overlap	30		ns	
t_{DT}	ϕ_4 to ϕ_1 Hold Time, ϕ_2 to ϕ_3 Hold Time	40		ns	
$t_{\phi/2}$	Half Clock Period for $\phi_1 \ldots \phi_4$		10,000	ns	
t_T	Transition Times for $\phi_1 \ldots \phi_4$	30	200	ns	
t_{T1}	Transition Times for Inputs Other Than $\phi_1 \ldots \phi_4$		100	ns	
t_{TC}	ϕ_1 or ϕ_3 Off to CE On	280		ns	
t_{SC}	CS to CE Set-Up Time	0		ns	
t_{AC}	Address to CE Set-Up Time	0		ns	
t_{AH}	Address Hold Time	240		ns	
t_{CS}	CE to CS Hold Time	0		ns	
t_{CC}	CE Off Time	140		ns	
t_{CP}	CE Off to ϕ_2 or ϕ_4 On	40		ns	
t_{CRW}	CE On Time	440[1]		ns	
t_{CO}	CE On to $\overline{D}_{OUT}$ Valid		250	ns	
t_{DW}	D_{IN} to WE Set-Up Time	0		ns	
t_{WP}	WE Pulse Width	100[1]		ns	
t_{WC}	WE Off to CE Off	0		ns	
t_{DH}	D_{IN} Hold Time	0		ns	
t_{WD}	CE On to WE On	300[1]		ns	
t_{WF}	WE to $\overline{D}_{OUT}$ Undefined	0		ns	

Note: 1. The minimum t_{WD} and t_{WP} times with appropriate transitions do not necessarily add up to the minimum t_{CRW}. This allows the user flexibility in setting the WE Pulse Width edges without affecting either t_{CRW} or the READ-MODIFY-WRITE Cycle Time, t_{RWC}.

WAVEFORMS (Numbers in parentheses are for minimum cycle timing in ns)

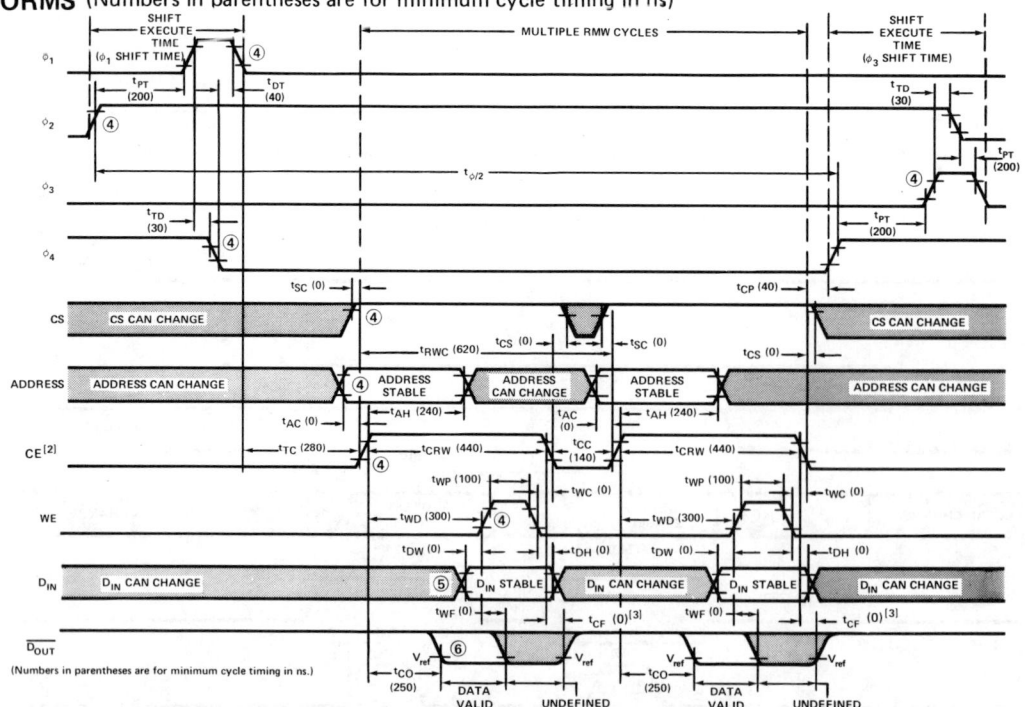

Notes: 2. When CE is off, the 2416 output level is determined by the external output termination.
3. The parameter t_{CF} is the same as in the Shift-Read-Shift Cycle on page 4.
4. +2.0V and V_{DD}–2.0V are the reference low and high level respectively for measuring the timing of $\phi_1 \ldots \phi_4$, CE, CS, WE, and addresses.
5. +1.5V and +3.0V are the reference low and high level respectively for measuring the timing of D_{IN}.
6. +0.8V is the reference level for measuring the timing of $\overline{D}_{OUT}$.

A.C. Characteristics

CAPACITANCE[1] $T_A = 25°C$

Symbol	Parameter	Typ.	Max.	Unit	Conditions
C_{IN}	Address, D_{IN}, CS, CE, WE Capacitance	4	6	pF	$V_{IN} = V_{SS}$
C_{OUT}	$\overline{D}_{OUT}$ Capacitance	3	5	pF	$V_{OUT} = V_{SS}$
$C_{\phi 1}$[1], $C_{\phi 3}$[2]	ϕ_1, ϕ_3 Input Capacitance	350	500	pF	$V_\phi = V_{SS}$
$C_{\phi 2}$[1], $C_{\phi 4}$[2]	ϕ_2, ϕ_4 Input Capacitance	480	700	pF	$V_\phi = V_{SS}$
$C_{\phi 1 - \phi 2}$	Clock ϕ_1 To Clock ϕ_2 Capacitance	120	175	pF	$V_\phi = V_{SS}$
$C_{\phi 1 - \phi 4}$	Clock ϕ_1 To Clock ϕ_4 Capacitance	150	200	pF	$V_\phi = V_{SS}$
$C_{\phi 3 - \phi 2}$	Clock ϕ_3 To Clock ϕ_2 Capacitance	150	200	pF	$V_\phi = V_{SS}$
$C_{\phi 3 - \phi 4}$	Clock ϕ_3 To Clock ϕ_4 Capacitance	120	175	pF	$V_\phi = V_{SS}$

Notes: 1. This parameter is periodically sampled and is not 100% tested.
2. The $C_{\phi 1} \ldots C_{\phi 4}$ input clock capacitance includes the clock to clock capacitance. The equivalent input capacitance is given below.

Four-Phase Clock Inputs

The four-phase clock inputs are internally connected to long electrodes used for several thin-oxide gates, resulting in high capacitance to the substrate on the clock inputs. In addition, considerable cross-coupling between adjacent clock exists due to the overlapping structure of the electrodes. The figure to the right shows the circuit equivalent of the clock inputs, indicating maximum capacitance values.

The equivalent circuit suggests two opposed clock driver requirements:

1. Ability to drive high-capacitance loads quickly.
2. Ability to suppress cross-coupled current transients.

The first requirement could ordinarily be met rather easily, if it weren't for the fact that the cross-coupled current, I, is proportional to the rate of change of the voltage, i.e., $I = C \frac{dv}{dt}$. For the quiescent driver to hold the coupled voltage to a minimum, the driver must have very low output impedance. However, when this driver becomes active the low output impedance increases the slope of the transitions which in turn increases coupling currents to the other drivers. This suggests that a driver have a controlled output transition time and a low output impedance characteristic in the quiescent state (high or low level). The Intel® 5244 meets these requirements.

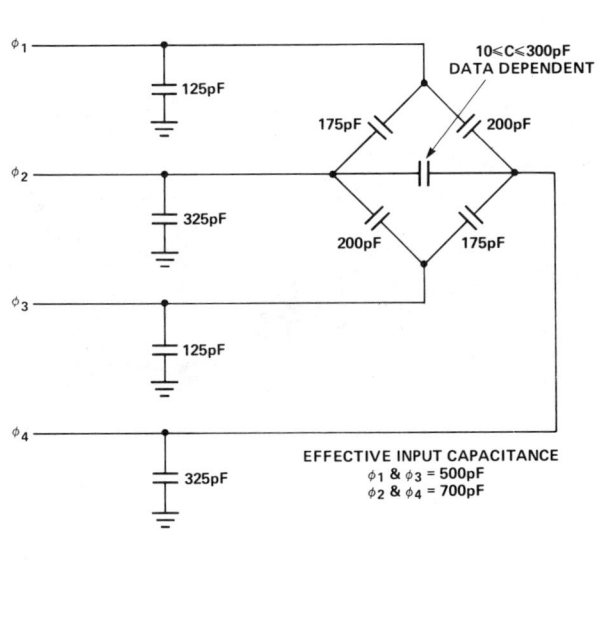

5244 — CCD Clock Driver

The Intel® 5244 is a CMOS implemented fully TTL input compatible high voltage MOS driver, designed especially for the four phase clock inputs of the 2416. The device features very low DC power dissipation from a single +12V supply with output characteristics directly compatible with the 2416 clock input requirements.

The 5244 uses internal circuitry to control the cross-coupled voltage transients between the clock phases generated by the 2416. This internal circuitry limits the transition time to a specified range so that excessively fast transitions (<30ns) do not occur on the clock line. The entire operation is transparent to the user

The 5244 is designed to drive four 2416s, but can drive fewer devices when loaded with additional capacitance to prevent a speedup in the transition times. Additional information on this and other aspects of the 5244 can be found on the 5244 data sheet.

Application Information

The Intel® 2416 is a charge coupled device (CCD) containing 16,384 bits of dynamic shift register storage available in a standard 18 pin plastic package. To minimize latency time (access time to any given bit in the device), the 2416 has been organized as 64 registers containing 256 bits each and, therefore, any bit can be accessed with a maximum of 255 shift operations. Since the minimum shift cycle requires 750 ns, the maximum latency time for the 2416 is less than 200μsec.

Access to the 64 recirculating registers is performed in a random access mode. A six bit address selects one of the 64 registers for read, write, or read/modify/write operations. These random access operations are performed between shift operations, and can be performed in any number or sequence as long as the basic shift frequency is maintained.

Because of substrate leakage currents the charge coupled storage mechanism is dynamic in nature. To satisfy the refresh requirements of the 2416, one shift operation must be performed every **ten microseconds.** A shift operation is completed on the falling edge of clock phase ϕ_1 or ϕ_3 and random access cycles may occur only between (1) the falling edge of ϕ_1 and the rising edge of ϕ_4 or (2) the falling edge of ϕ_3 and the rising edge of ϕ_2. This refresh requirement limits the number of random access cycles between successive shift operations to a maximum of 16.

Random access operations are performed in a manner which is very similar to any random access memory (RAM). All random access cycles are initiated with the rising edge and terminated with the falling edge of CE (Chip Enable). Read operations are performed when WE (Write Enable) remains low throughout a CE cycle. Data is strobed into the memory whenever WE is strobed high during a CE cycle as illustrated in the appropriate timing diagrams. CS (Chip Select) controls only the input and output circuits and is only effective when CE is high.

Typical Current Transients vs. Time

The oscilloscope photos in Figures 1 and 2 show typical I_{DD} current transients during shift and I/O cycles. The typical I_{BB} current during a shift cycle is shown in Figure 3.

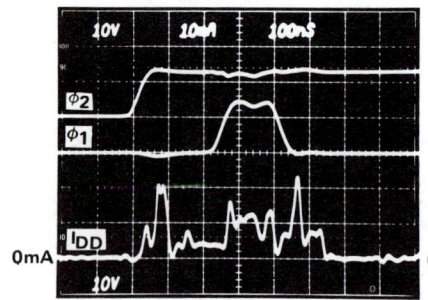

Figure 1. I_{DD} transient current during shift cycles.
I_{DD} scale: 10mA/div.

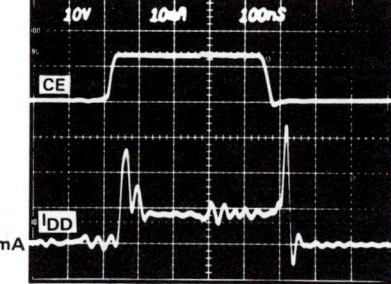

Figure 2. I_{DD} transient current during I/O cycles.
I_{DD} scale: 10mA/div.

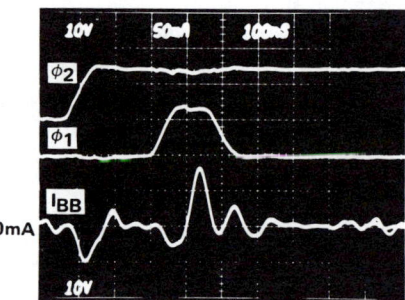

Figure 3. I_{BB} transient current during a shift cycle.
I_{BB} scale: 50mA/div.

CHAPTER FIVE
Memory Support Circuits for Dynamic RAMs

Using Support Circuits for Dynamic RAMs	412
3205 High Speed 1 out of 8 Binary Decoder, 3404 High Speed 6 Bit Latch	435
3207A Quad Bipolar-to-MOS Level Shifter and Driver	439
3207A-1 Quad Bipolar-to-MOS Level Shifter and Driver	443
3208A, 3408A Hex Bipolar Sense Amplifiers for MOS Circuits	445
3222 Refresh Controller for 4K Dynamic RAMs	451
3232 Address Multiplexer and Refresh Counter for 4K Dynamic RAMs	457
3242 Address Multiplexer and Refresh Counter for 16K Dynamic RAMs	461
3245 Quad TTL-to-MOS Driver	465
5235, 5235–1 Quad TTL-to-MOS Driver	469
5244 Quad CCD Clock Driver	473

USING SUPPORT CIRCUITS FOR DYNAMIC RAMS

JIM OLIPHANT

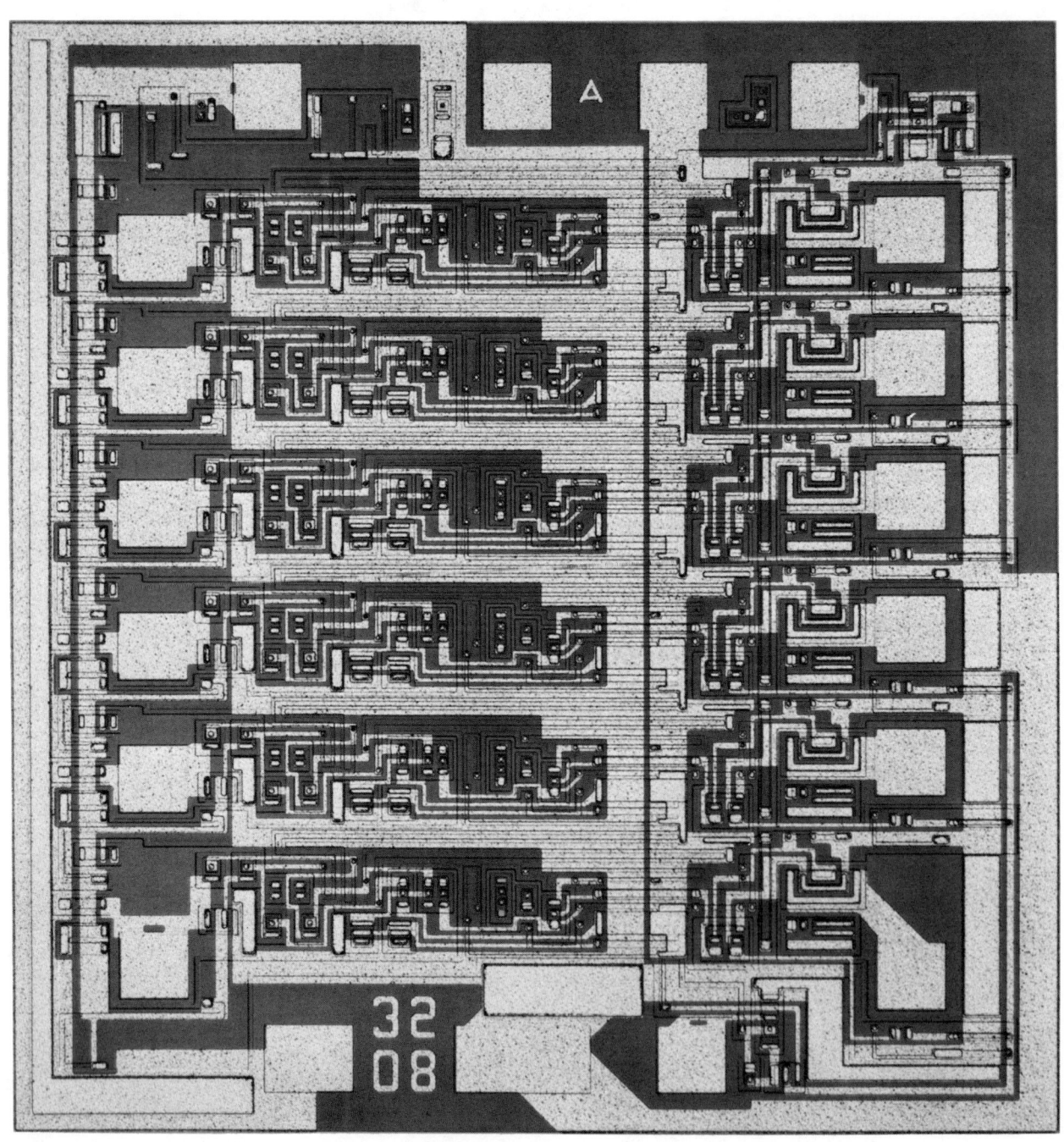

Photomicrograph of the Intel 3208.

USING SUPPORT CIRCUITS FOR DYNAMIC RAMS

INTRODUCTION

The evolution of semiconductor dynamic Random Access Memories has resulted in devices which are very easy to use in system applications. These devices have evolved to the point that, today, some are fully TTL compatible including clocks, while others have all but the clock input TTL compatible. Most random access memory devices are treated as "just another component" by system designers — a very desirable situation.

Although there are many ways to design a solid and reliable memory system, care must be exercised in the implementation of the support circuits which "surround" the memory devices. In many cases, marginal memory system operation can be traced directly to marginal or inadequate peripheral components. This is especially true in those memory systems which exhibit "soft" failures. ("Soft" failures are usually not repeatable and are almost completely random.) These "soft" failures can result from timing glitches causes by refresh interference, inadequate high or low input levels to the memory device, or very tight timing constraints in the system. Using the reasonably conservative design techniques discussed in this Application Brief allows the memory system designer to obtain maximum system speed with minimum peripheral power. This, in turn, allows the system into which the memory goes to treat its memory as just another "black box."

Common characteristics of dynamic RAMs are the requirements for:

1. Refresh
2. Signal drive (TTL or MOS level)

The first requirement allows high density, high speed, and low power RAMs to be designed in the first place. The second requirement is the result of the large number of memory devices (and therefore high capacitance) usually contained on a printed circuit board.

The pupose of this Application Brief is to describe support circuits which are used to perform refresh control and multiplexing functions and drivers used to drive the memory array. The devices described are used primarily with 16 and 22-pin 4K and 16-pin 16K RAMs. For reference, those devices to be described in this Brief are shown in Table I.

This Application Brief is divided into two major sections. The first section describes Refresh Controllers and Address Multiplexers and the second section describes TTL and MOS level drivers (for clocks, address lines, etc.).

REFRESH SUPPORT CIRCUITS

Two relatively new types of memory support circuits have been made available recently, Refresh Controllers and Address Multiplexers. The devices in this category which will be discussed are Refresh Controller/Address Multiplexer — Intel® 3222, and Address Multiplexers — Intel® 3232 and 3242. As shown in Table I, the 3222 is used primarily with 22-pin 4K RAMs, while the 3232 and 3242 are used with the 16-pin 4K and 16K RAMs, respectively.

Refresh Controllers/Address Multiplexers

In any memory system utilizing dynamic RAMs, some method must be provided to periodically refresh the contents of memory. Although the design of a refresh controller using standard TTL logic gates is not difficult, care is required to avoid refresh interference (especially in asynchronous systems).

Refresh interference is usually caused by the inability of system logic to distinguish between a simultaneous memory cycle and refresh cycle request. The cause is most likely the result of using a latch improperly in trying to distinguish between the two types of cycles. An example of the improper use of a latch is shown in Figure 1. In this figure, the D input is asynchronous from the clock input C. If both should occur simultaneously, the set-up time required between the clock and data inputs is violated and the latch state is indeterminate for an undefined period of time. This indeterminate state can cause *both* a refresh and memory cycle to be started almost simultaneously, causing errors to occur.

Table I. Support Circuit Characteristics

FUNCTIONS PERFORMED	DEVICE		
	3222	3232	3242
Refresh Controller	X		
Refresh Counter	X	X	X
12 Two-Way Multiplexers (used with 22-pin 4K RAM)	X		
6 Three-Way Multiplexers (used with 16-pin 4K RAM)		X	
7 Three-Way Multiplexers (used with 16-pin 16K RAM)			X
Driver Capability		X	X
Zero Refresh Address Detect		X	X

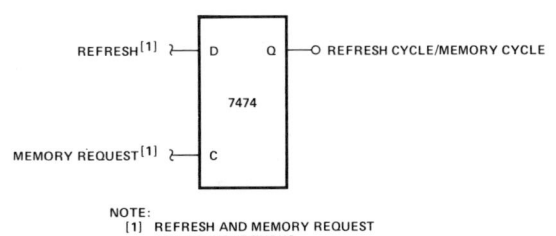

Figure 1. Improper Use of Latch

In addition to logic controller functions, the system refresh controller is required to have a sequential counter for the refresh addresses and an address multiplexer to multiplex between refresh and system addresses. Most controller designs able to handle all of the above requirements require a minimum of 12 IC packages in addition to a moderate amount of design and debug time.

The Intel® 3222 is designed to perform the functions associated with a system refresh controller. The 3222 is designed especially for systems using 22-pin 4K RAMs such as the Intel® 2107B.

The 3222 performs the following functions:

1. Selection between a Refresh and Read/Write cycle (system control)
2. 64 refresh address counter
3. 6-bit refresh and system address multiplexer
4. Refresh timing control generator

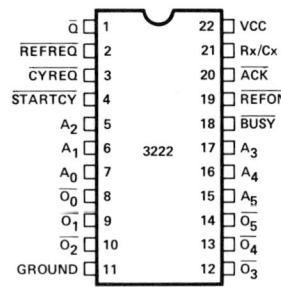

Figure 2. 3222 Pin Configuration

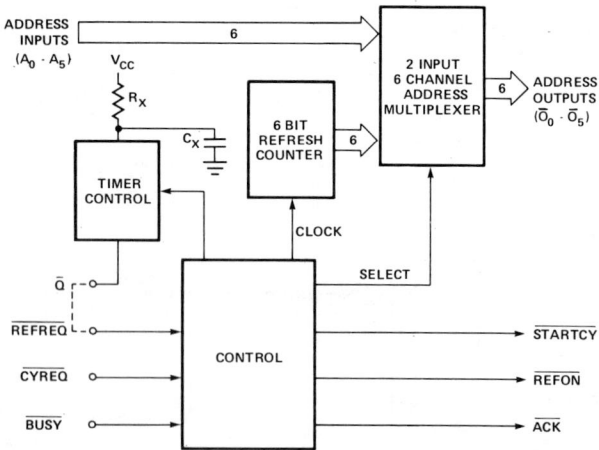

Figure 3. 3222 Block Diagram

The pin configuration for the 3222 is shown in Figure 2. An internal block diagram of this device is shown in Figure 3, outlining the four functions described previously. The use of the 3222 is made easier if the designer understands the internal logic circuits of the device. The internal logic diagram of the 3222 is shown in Figure 4. Each of the four device functions is described using the internal logic diagram.

System Control

The system control logic internal to the 3222 performs two functions:

1. Selects either a Refresh or Read/Write Cycle (depending on input).
2. Provides external control signals back to the system.

The first function — selection between a refresh and read/write cycle — is most important to the designer because it eliminates the chance of refresh interference associated with many new system designs. This function is performed by the priority latch shown in Figure 4. This latch has been designed so that the simultaneous occurrence of a cycle request ($\overline{\text{CYREQ}}$) and refresh request ($\overline{\text{REFREQ}}$) does not cause the latch to enter a long period of indecisiveness. (This problem may occur when such a latch is implemented with standard TTL logic gates.)

The second function — providing external control signals to the system — is implemented by the generation of the three control signals. These signals and their functions are:

1. Start Cycle: ($\overline{\text{STARTCY}}$) — Occurs shortly after a Refresh or Read/Write cycle is initiated. This signal is used by external control logic to start memory system timing.
2. Refresh On: ($\overline{\text{REFON}}$) — This signal is a logic low *only* when refresh cycle is beginning or is in progress.
3. Acknowledge: ($\overline{\text{ACK}}$) — $\overline{\text{ACK}}$ is a logic low *only* when the system is in a read or write cycle.

6-Bit Refresh Address Counter

An internal 6-bit refresh address counter provides for the refresh of the first 64 low-order addresses required for 4K RAMs. The address counter is automatically incremented by one at the end of every Refresh cycle. In addition, the counter is wrapped around so that after the 64th count the counter automatically resets to the first refresh address.

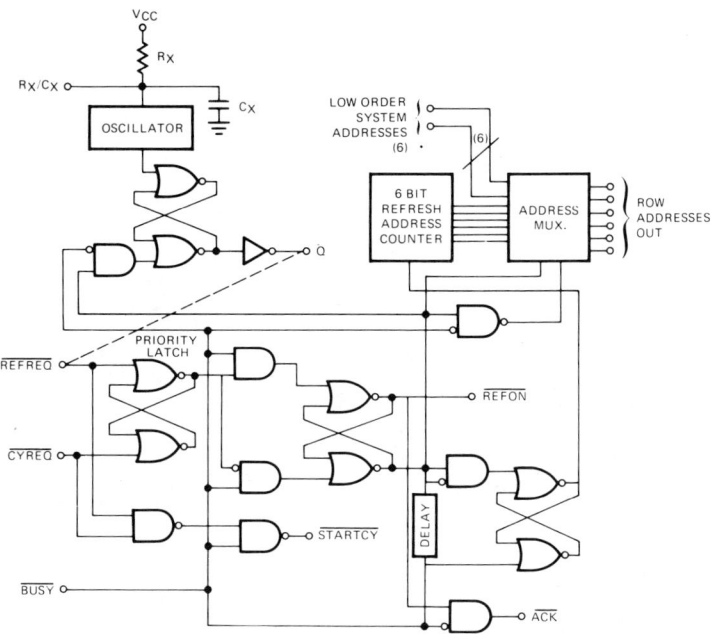

Figure 4. 3222 Internal Logic Diagram Support Circuits

Address Multiplexing

An internal 2-input-, 6-channel address multiplexer selects either the 6-bit refresh address or the 6 low-order system addresses. To allow minimum access time systems to be designed using the 3222, the 6 low-order system addresses are selected (i.e., available at the output pins) at all times except for refresh.

Refresh Timing Control

To round out the capability of the 3222, a refresh timing one-shot is incorporated in the device. This one-shot allows a distributed refresh mode to be used with no external circuits added. (If burst refresh is desired, an external one-shot is added, as will be explained later.)

The timing of the refresh interval is controlled by a simple RC network connected as shown in Figure 5. The relationship between the RC time constant and the time between refresh is given by:

1. $\dfrac{t_{REF}}{r} = 0.63 R_x C_x$

where:

t_{REF} = Total time between refreshes in msec (e.g., 2 msec).

r = number of device addresses to be refreshed

R_x = external timing resistor in kΩ

C_x = external timing capacitor in μF

The range of values associated with R_x and C_x are:

2. $3 \text{ k}\Omega \leq R_x \leq 10 \text{ k}\Omega$

and

3. $0.005 \text{ }\mu\text{F} \leq C_x \leq 0.02 \text{ }\mu\text{F}$

These conditions on R_x and C_x result in a range of refresh intervals (assuming r = 64) of:

4. $0.6 \text{ msec} \leq t_{REF} \leq 8.1 \text{ msec}$

This refresh range includes virtually all system refresh requirements for 64 refresh address RAMs.

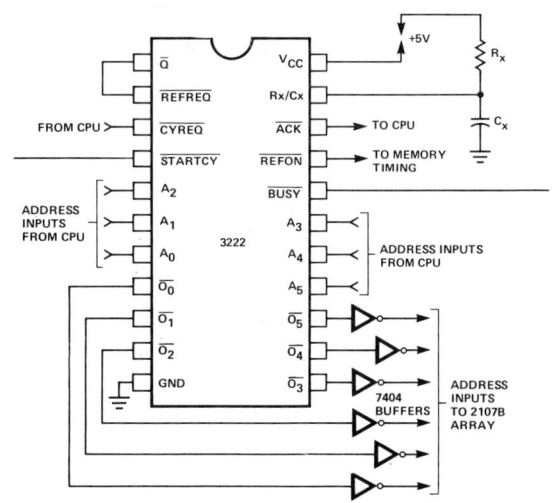

Figure 5. Refresh Timing Control

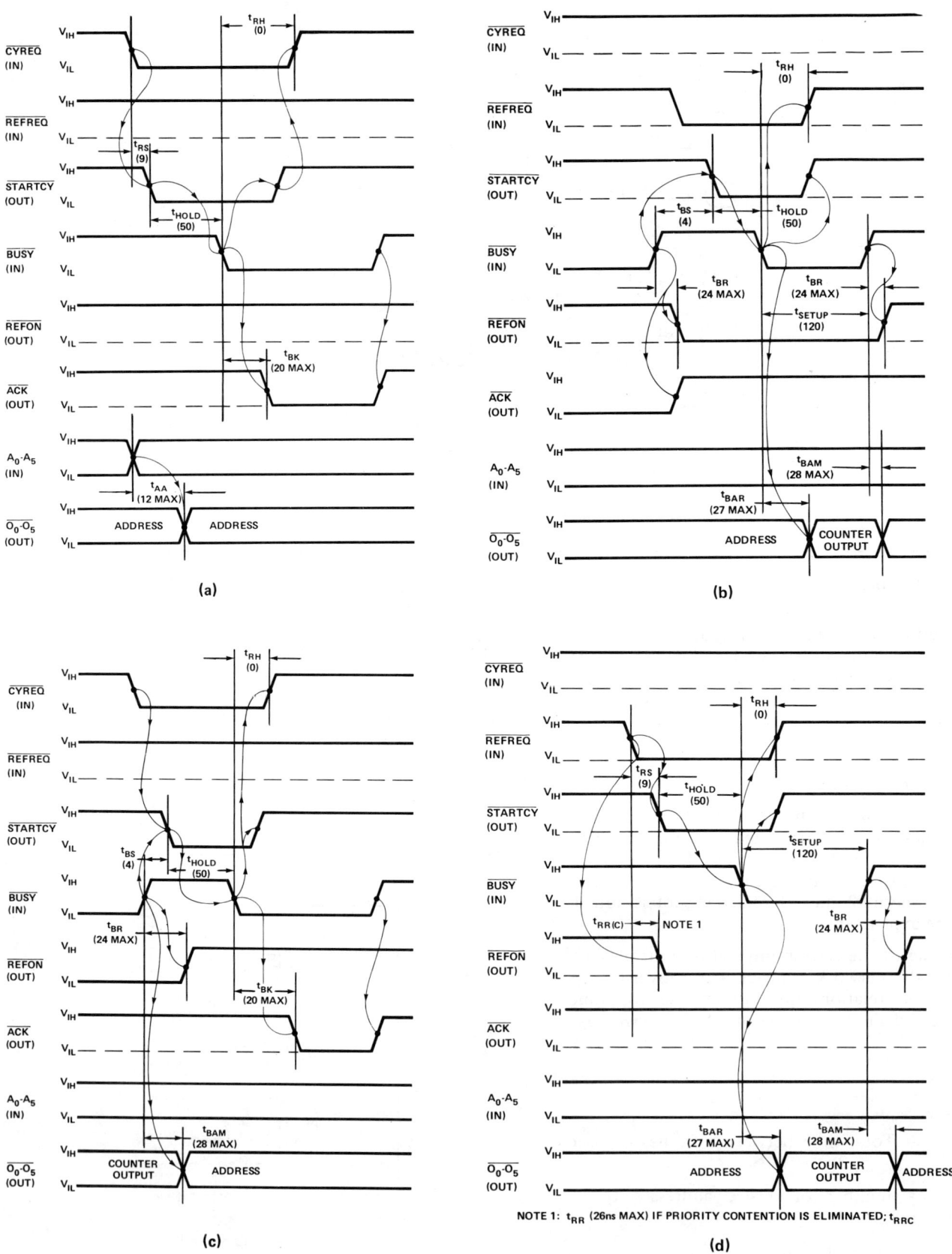

Figure 6. 3222 Timing States

USING SUPPORT CIRCUITS FOR DYNAMIC RAMS

3222 System Operation

The 3222 Refresh Controller/Address Multiplexer is designed for memory systems using 22-pin RAMs such as the 2107B. The following discussion concentrates on the use of the 3222 in a system using just such a RAM. Because of the plethora of RAM timing specifications available, the discussion will be limited to the operation of the device with only those timing parameters critical to the 3222 being mentioned.

The two timing diagrams showing the combinations of system memory cycles/refresh cycles for the two Busy states are shown in Figure 6. A schematic of the logic required to implement timing/control for the 3222 is shown in Figure 7. In order to simplify the explanation of circuit operation, the discussion is limited to the following examples:

1. System Memory Cycle with Memory Not Busy
2. Refresh Cycle with Memory Busy (following System Cycle)
3. System Memory Cycle with Memory Busy (following Refresh Cycle)
4. Refresh Cycle with Memory Not Busy

The above four conditions are shown in Figure 8. using Cycle Request ($\overline{CYREQ}$) and Refresh Request ($\overline{REFREQ}$). In all system memory cycle examples, it is assumed that the system addresses are valid at the 3222 inputs (A_0–A_5) coincident with cycle request.

System Memory Cycle with Memory Not Busy

The first example is for a system memory cycle with memory not busy (refer to Figure 6a). The control function is followed by the arrows labeled 1–4. When $\overline{CYREQ}$ goes low, start cycle ($\overline{STARTCY}$) goes low at or before t_{RS} time (arrow 1). The $\overline{STARTCY}$ output is used to trigger a Busy latch at or after t_{HOLD} time. When the externally generated Busy signal goes low, it automatically sends $\overline{STARTCY}$ high and issues an acknowledge ($\overline{ACK}$) command from the 3222. The $\overline{ACK}$ output of the 3222 is used to signal the system controller that a memory cycle has been initiated and accepted by the processor. It is important to note that the system controller cannot issue a memory cycle request and not monitor the acknowledge output in an asynchronous system, since there is no other way to assure that the command has been accepted by the 3222. In a asynchronous system, however, the acknowledge out need not be monitored if refresh is designed *not* to occur during a data cycle.

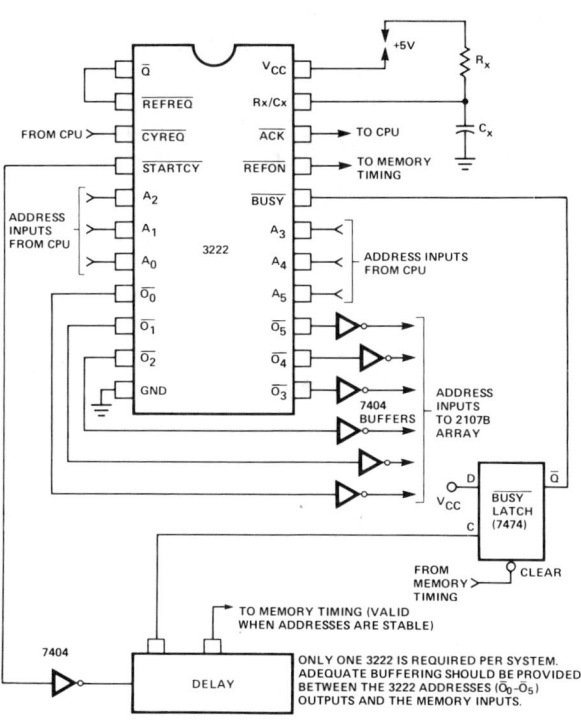

Figure 7. Timing/Control Logic for 3222

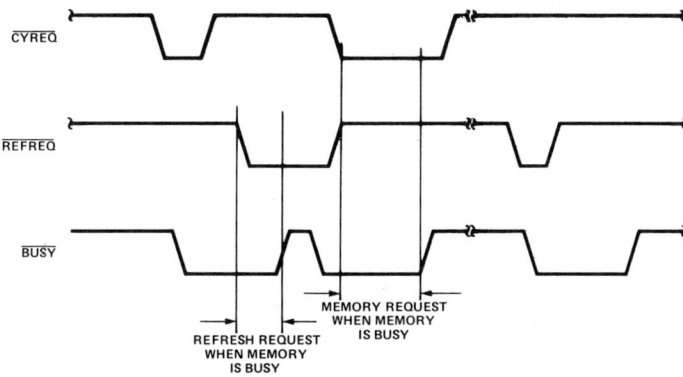

Figure 8. Four 3222 "Wait" States

Refresh Cycle with Memory Busy

The second example (for a refresh cycle with the memory busy) shows the timing expected to/from the 3222 for a refresh cycle requirement during a system access cycle. In this case, refresh has been requested while $\overline{BUSY}$ is low; i.e., the system is busy. After $\overline{BUSY}$ is set high (the $\overline{REFREQ}$ is still low) the $\overline{STARTCY}$ out goes low a maximum of t_{RS} later. After $\overline{BUSY}$ goes high, the refresh on output ($\overline{REFON}$) goes low, indicating the 3222 has accepted the refresh request. The $\overline{STARTCY}$ output going low is again used to set the external $\overline{BUSY}$ input low. Shortly after $\overline{BUSY}$ goes low, $\overline{STARTCY}$ and $\overline{REFREQ}$ are reset (i.e., go high) after the time indicated in Figure 6b, and the refresh addresses are valid at or after t_{BAR} time.

System Memory Cycle with Memory Busy

The third example assumes that the memory system is busy with a refresh cycle when a system access is requested. The major difference between this example and example 1 is that the system addresses at the output of the 3222 are not valid until after t_{BAM} time. It is noted that t_{BAM} is much greater than t_{AA} (see Figure 6c). Care should be exercised to assure that addresses are valid at the memory device at or before the clock (chip enable for the 2107B) goes high. More will be discussed about these requirements in the 3222 systems considerations section.

Refresh Memory Cycle with Memory Not Busy

The last example gives the timing for a lonesome refresh out in the middle of nowhere. The major difference between this example and example 2 is the occurrence of $\overline{REFON}$. When the memory is not busy, refresh is delayed by t_{RRC} relative to the refresh request input. All other timing conditions are as described in example 2.

3222 System Considerations

There are many ways to interface a 3222 to a memory system as there are creative designers. Therefore, it is useless to describe in detail the cleverness of a particular design. However, several hints regarding what to watch out for (or-how-not-to-foul-things-up-before-you-even-get-started) in the interface are useful. These hints are the requirements of the memory component used and not because of the 3222. The following hints should be observed:

1. Do not allow start cycle to begin a memory cycle before the addresses are valid at the memory component (see Figure 9). In an asynchronous system (where a system cycle or refresh cycle can be requested while the memory is busy) this means using the start cycle to address output delay maximum of t_{BAM} and not t_{AA}.

2. If delay lines are used as the timing element in the $\overline{STARTCY}$ path, care should be exercised to assure that the $\overline{STARTCY}$ output is long enough to propagate through the delay line with minimum distortion. The $\overline{STARTCY}$ output can be very short if a fixed pulse width is used for $\overline{CYREQ}$ and $\overline{REFREQ}$ (see Figure 10). This condition may result in the delay line not transmitting the signal properly (see Figure 11). For this reason, externally generated

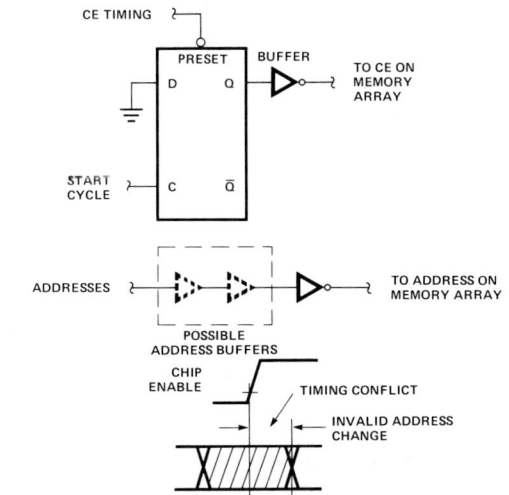

Figure 9. Possible Timing Conflict in Chip Enable Address Path

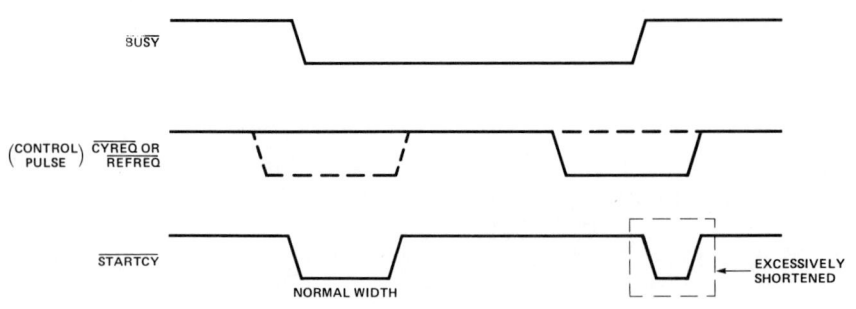

Figure 10. STARTCY Pulse Width

USING SUPPORT CIRCUITS FOR DYNAMIC RAMS

$\overline{\text{CYREQ}}$ and $\overline{\text{REFREQ}}$ inputs should be generated as shown in Figure 12. As shown in Figure 12, a latch is used to form the cycle and refresh requests $\overline{\text{CYREQ}}$ and $\overline{\text{REFREQ}}$, respectively. When a cycle is requested (either for an access or refresh), the request remains valid until it is serviced. An Acknowledge ($\overline{\text{ACK}}$) for a memory request or a refresh on ($\overline{\text{REFON}}$) are required to remove the request from the line.

3. If a delay line is used on $\overline{\text{STARTCY}}$, the line should not be driven directly from the 3222 because of insufficient output drive for this application.

4. Note that the address outputs (O_0–O_5) are buffered before driving a memory array. The 3222 output drive is not normally sufficient to drive the memory array directly.

5. A power-on reset must be provided to the 3222 to assure proper start-up operation. This reset should be a negative-going pulse on the $\overline{\text{BUSY}}$ line and is best provided by momentarily clamping the $\overline{\text{BUSY}}$ input to ground. The time constant selected should assure that the $\overline{\text{BUSY}}$ input is held at or below $V_{IL(MAX)}$ until the V_{CC} (+5V) supply has stabilized.

Burst Refresh Timing Generation

The 3222 is capable of generating both sequential and burst refresh cycles. Sequential refresh has been previously discussed. Consider the requirements for burst refresh generation.

Burst refresh is used primarily in systems which cannot be interrupted for refresh during data operation. These systems must have a time period when no memory accesses are required so that the entire memory can be refreshed. A circuit which allows for burst refresh cycles is shown in Figure 13.

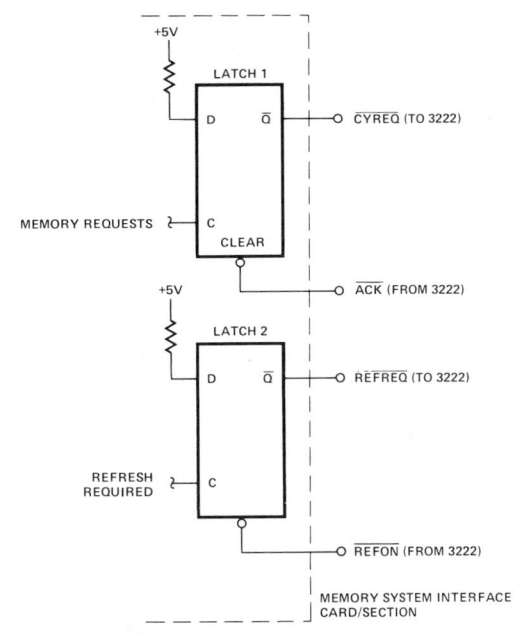

Figure 12. External Generation of $\overline{\text{CYREQ}}$ and $\overline{\text{REFREQ}}$

Operation of the circuit shown in Figure 13 is as follows. The refresh timing interval (usually 2 ms) is generated by timing circuits internal to the 3222 by using appropriate values of R_x and C_x. When the 3222 timer signals for refresh ($\overline{Q}$ output on the 3222 goes low), single-shot S_1 is triggered. The timing interval of S_1 is 64 times the refresh cycle time of the memory devices (for memory devices requiring 64-cycle refresh). S_1 allows astable multivibrator A_1 to cycle through all 64 refresh addresses at the desired cycle time (e.g., 500 ns). Refresh addresses are counted automatically at the completion of each refresh cycle. Note that S_1 is required because the $\overline{Q}$ output of the 3222 goes high after the first refresh cycle.

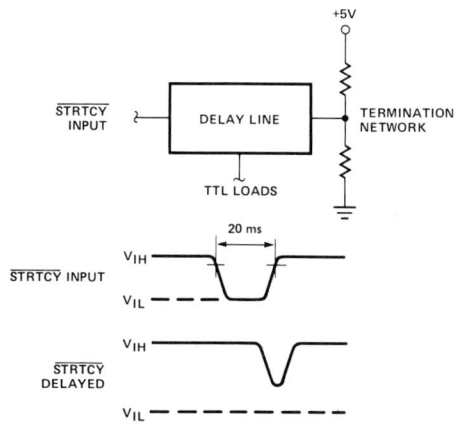

Figure 11. Delay Line Effects

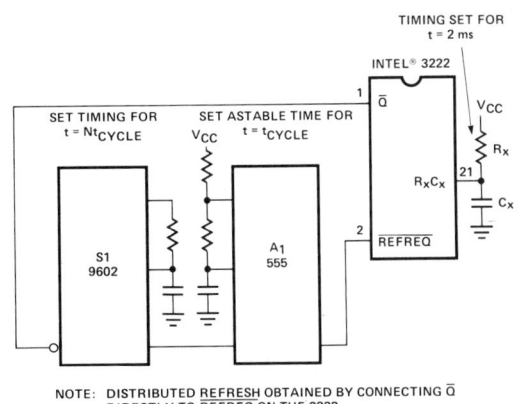

Figure 13. Burst Refresh Connections

16-Pin RAM Support Circuits

The support circuits required for 16-pin dynamic RAMs differ from those required for 22-pin devices. The primary difference for 16-pin RAMs (both 4K and 16K) is the requirement for three-way multiplexing on the address lines. (Recall that a single address line performs the functions of row, column, and refresh address.) The class of devices available for supporting 16-pin RAM memories are called Refresh Counter/Address Multiplexers. The two devices of this type to be discussed in this Application Brief are the Intel® 3232 and 3242. These devices are designed primarily for interface to the memory array when using 16-pin 4K and 16K RAMs, respectively.

The high packaging density realized by using 16-pin 4K RAMs is made possible by multiplexing the 12 (14 for 16K) system addresses on 6 (7 for 16K) pins. Because the addresses are multiplexed, it is necessary to provide two strobe clocks. These clocks are called Row Address Strobe ($\overline{RAS}$) and Column Address Strobe ($\overline{CAS}$).

The relationship of addresses to $\overline{RAS}$ and $\overline{CAS}$ is shown in Figure 14. Operation of these 16-pin RAMs requires a three-way multiplexer to provide for the following functions:

1. low-order system addresses for the Row Address Strobe.
2. high-order system addresses during $\overline{CAS}$.
3. refresh addresses during refresh cycle.

3232/3242 Operation

Operation of the Intel® 3242 is identical to the 3232, with one exception. This difference is that a 7-bit, three-way multiplexer is provided on the 3242 (allowing 14 system addresses to be multiplexed by the device). Otherwise, description of operation of the 3232 applies equally to the 3242.

The pin configuration and logic diagram of the 3232 is shown in Figure 15. For completeness the pin configuration and logic diagram for the 3242 is shown in Figure 16.

The 3232/3242 provides four basic functions:

1. Address multiplexing
2. Refresh address counting
3. Refresh address zero detect output
4. Memory array address drive capability.

Timing considerations for a system memory cycle and refresh cycle are shown in Figures 17 and 18, respectively. The logic operation is evident from Figures 15 and 16.

The zero detect function provides a means of keeping track of refresh addresses during burst mode refresh cycles. When using the 3232/3242 it is important to remember that momentary indications of zero detect are likely when incrementing the refresh address counter.

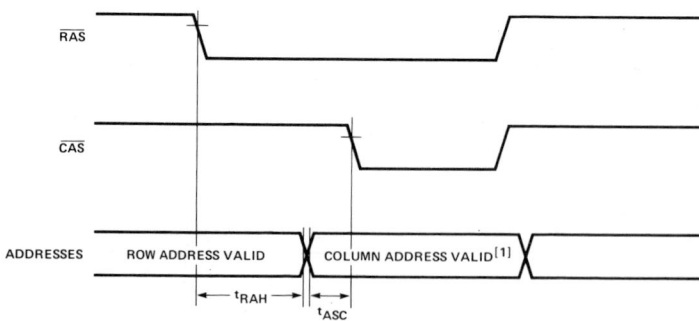

NOTE: [1] HOLDING COLUMN ADDRESSES STABLE THROUGHOUT CAS IS NOT REQUIRED BY DEVICE, BUT IS DESIRABLE FROM SYSTEM STANDPOINT.

Figure 14. Pin Dynamic RAM Clock Timing

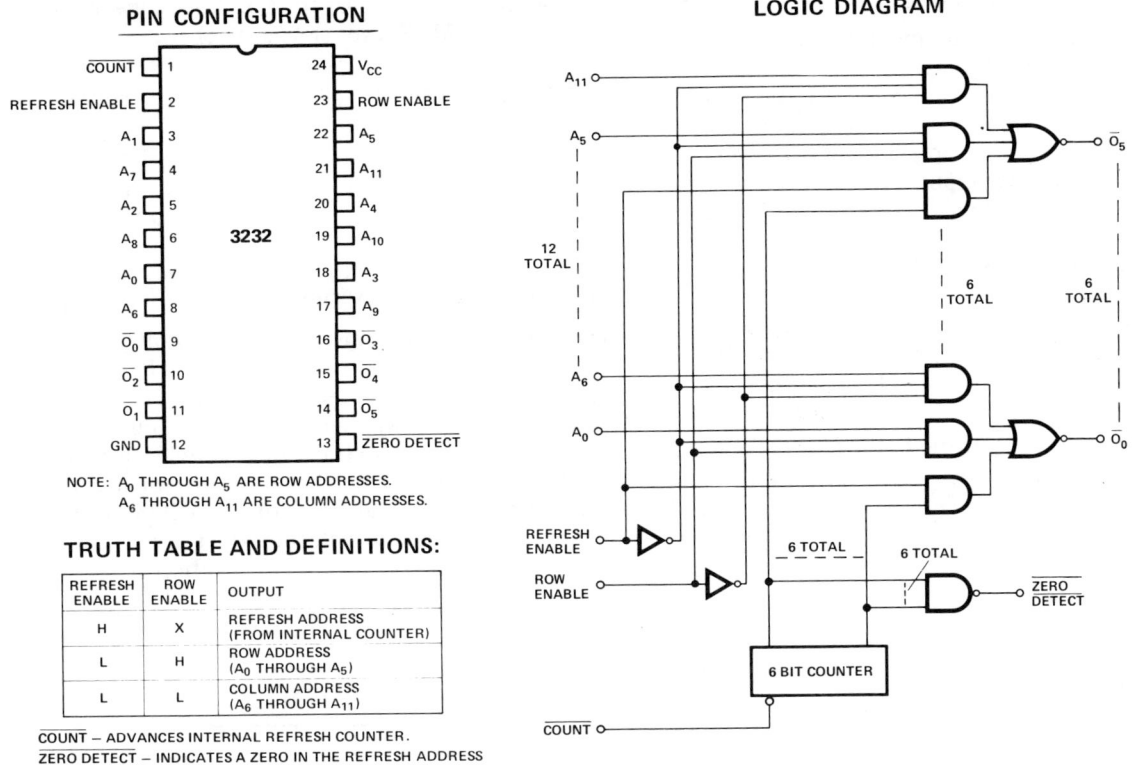

Figure 15. Pin Configuration and Logic Diagram

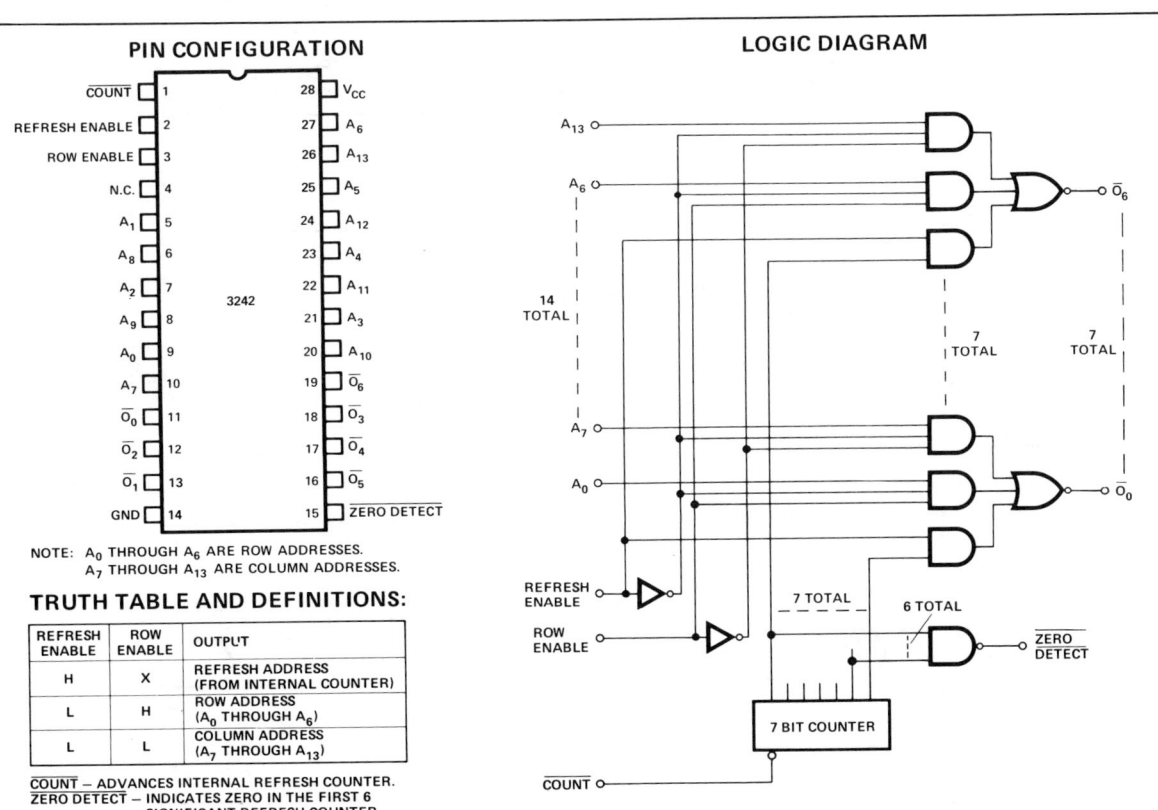

Figure 16. 3242 Pin Configuration and Logic Diagram

3232/3242 System Considerations

Interfacing the 3232/3242 to memory systems is very simple. An example of such an interface is shown in Figure 19. The 3232/3234 is specified for driving a capacitance load of 250 pF. Since the 4K/16K address inputs are implemented with MOS devices, there is only low input leakage current in both the high and low logic states. The address input capacitance is a maximum of 7 pF on the Intel® 2104A (16-pin 4K RAM). This indicates that, if worst case address capacitance is assumed, the 3232/3242 can drive 32 RAM devices, taking into account stray line capacitance. A more detailed description of the 3232/3242 as a driver is found in the Driver Circuits for Memory Arrays Section, along with various other types of drivers.

DRIVER CIRCUITS FOR MEMORY ARRAYS

Drivers for semiconductor memory arrays fall into two general categories:

1. Low level (TTL, ECL) to MOS level converter/drivers (for MOS level clocks, etc.)

2. TTL-TTL buffer drivers (for data lines, addresses, etc.).

These categories are treated separately in this Application Brief.

TTL to MOS Level Converter/Drivers

The continual improvement in TTL to MOS level drivers have kept pace with dynamic RAMs in ease of use. Gone are the days when drivers required external components (such as transistors or capacitors) or an extra power supply (usually 3V above the supply required for the memory). Today, there are several types of TTL-MOS drivers which require no external components nor additional power supplies for proper operation.

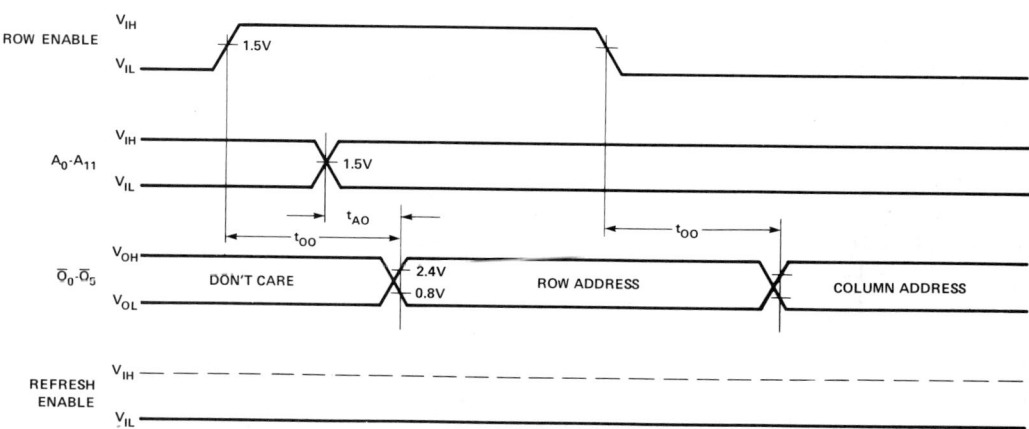

Figure 17. System Cycle Timing Relationships

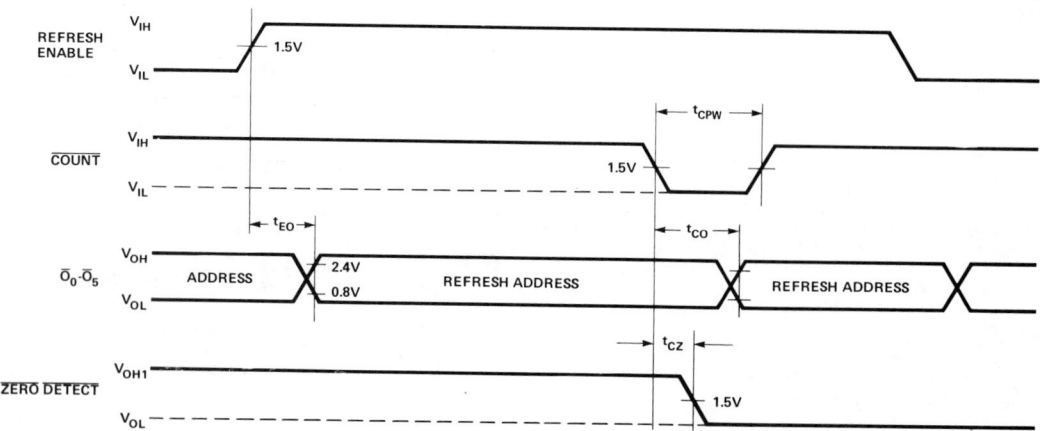

Figure 18. Refresh Cycle Timing Relationships

USING SUPPORT CIRCUITS FOR DYNAMIC RAMS

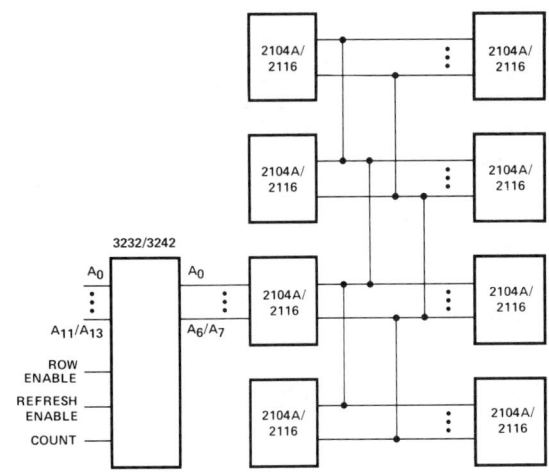

Figure 19. 3232/3242 Memory Array Interface

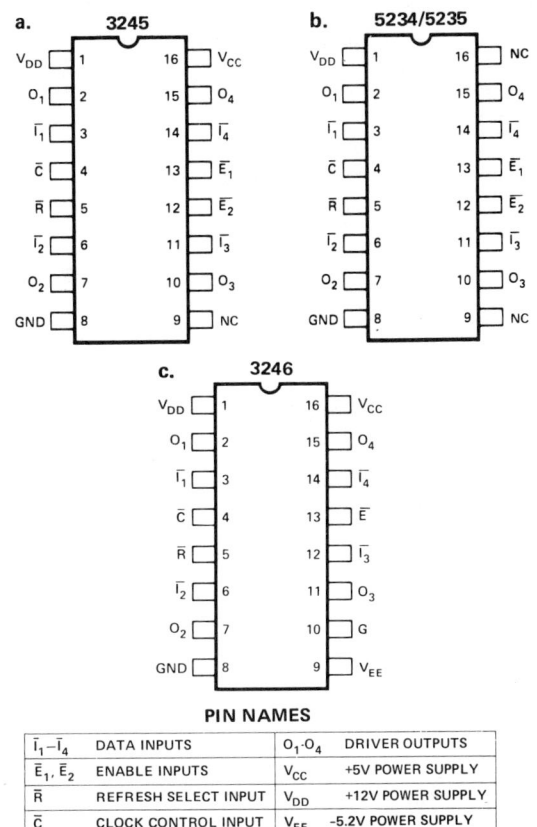

PIN NAMES

$\bar{I}_1 - \bar{I}_4$	DATA INPUTS	$O_1 - O_4$	DRIVER OUTPUTS
$\bar{E}_1, \bar{E}_2$	ENABLE INPUTS	V_{CC}	+5V POWER SUPPLY
$\bar{R}$	REFRESH SELECT INPUT	V_{DD}	+12V POWER SUPPLY
$\bar{C}$	CLOCK CONTROL INPUT	V_{EE}	-5.2V POWER SUPPLY
NC	NO CONNECTION	G	ECL GROUND REFERENCE

Figure 20. Quad Clock Driver Pin Configuration

Intel has produced a complete family of Quad High Voltage Clock Drivers as shown in Figure 20. Figure 20 also shows the pin configuration for each of the members of the driver family.

The device speed, listed in Table II, is the maximum worst case value from the data sheet. For consistency, each driver is specified with minimum, typical, and maximum delays which correspond to load values of 150, 200 and 250pF, respectively, which also correspond to the minimum, typical and maximum capacitance, respectively, of nine 2107B chip enable inputs plus associated stray capacitance.

The two level gating structure, shown in the logic diagram, Figure 21, is common to all but the 3246. The gating structure for the 3246 is shown in Figure 28.

The 3245 is designed specifically to support the Intel® 2107B, although its input and output levels make it compatible with many other N-channel MOS RAMs. A specific application of the 3245 can be seen in the 2107B section of this handbook, where the Row Enable selection is accomplished by using this device.

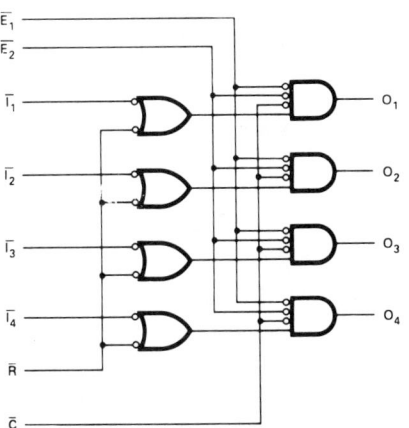

Figure 21. 3245, 5234, 5235 Quad Family Logic Diagram

Table II. Quad Clock Drivers

Device Type	Input Level		Output Level		Speed (1)	Power Dissipation	Power Supplies
	V_{IL}	V_{IH}	V_{OL}	V_{OH}			
3245	0.8	2.0	0.45	$V_{DD}-0.5$	32nS	485mW	+5,+12
3246	-1.500	-1.025	0.45	$V_{DD}-0.5$	30nS	725mW	+5,+12,-5.2
5234	2.0	$V_{DD}-2.0$	0.4	$V_{DD}-0.4$	100nS	1.4µW	+12
5235/-1	0.8	2.0	0.4	$V_{DD}-0.4$	90/120nS	27mW	+12

Note: 1. Maximum Delay and Transition Time Driving 1 TTL Gate and 250pF.

Table III. 3245 D.C. Characteristics

$T_A = 0°C$ to $75°C$, $V_{CC} = 5.0V \pm 5\%$, $V_{DD} = 12V \pm 5\%$.

Symbol	Parameter	Min.	Max.	Unit	Test Conditions
I_{FD}	Input Load Current, $\overline{I}_1, \overline{I}_2, \overline{I}_3, \overline{I}_4$		-0.25	mA	$V_F = 0.45V$
I_{FE}	Input Load Current, $\overline{R}, \overline{C}, \overline{E}_1, \overline{E}_2$		-1.0	mA	$V_F = 0.45V$
I_{RD}	Data Input Leakage Current		10	µA	$V_R = 5.0V$
I_{RE}	Enable Input Leakage Current		40	µA	$V_R = 5.0V$
V_{OL}	Output Low Voltage		0.45	V	$I_{OL} = 5mA, V_{IH} = 2V$
		-1.0		V	$I_{OL} = -5mA$
V_{OH}	Output High Voltage	$V_{DD}-0.50$		V	$I_{OH} = -1mA, V_{IL} = 0.8V$
			$V_{DD}+1.0$	V	$I_{OH} = 5mA$
V_{IL}	Input Low Voltage, All Inputs		0.8	V	
V_{IH}	Input High Voltage, All Inputs	2		V	

Table IV. 3245 Power Supply Current Drain and Power Dissipation

Symbol	Parameter	Typ.	Max.	Unit	Test Conditions — Input states to ensure the following output states:	Additional Test Conditions
I_{CC}	Current from V_{CC}	23	30	mA	High	
I_{DD}	Current from V_{DD}	19	26	mA		
P_{D1}	Power Dissipation	365	485	mW		
	Power Per Channel	91	121	mW		$V_{CC} = 5.25V$
I_{CC}	Current from V_{CC}	29	39	mA	Low	$V_{DD} = 12.6V$
I_{DD}	Current from V_{DD}	12	15	mA		
P_{D2}	Power Dissipation	300	388	mW		
	Power Per Channel	75	97	mW		

Table V. 3245 A.C. Characteristics. $T_A = 0°$ to $75°C$, $V_{CC} = 5.0V \pm 5\%$, $V_{DD} = 12V \pm 5\%$

Symbol	Parameter	Min.[1]	Typ.[2,4]	Max.[3]	Unit	Test Conditions
t_{-+}	Input to Output Delay	5	11		ns	$R_{SERIES} = 0$
t_{DR}	Delay Plus Rise Time		20	32	ns	$R_{SERIES} = 0$
t_{+-}	Input to Output Delay	3	7		ns	$R_{SERIES} = 0$
t_{DF}	Delay Plus Fall Time		18	32	ns	$R_{SERIES} = 0$
t_T	Output Transition Time	10	17	25	ns	$R_{SERIES} = 20\Omega$
t_{DR}	Delay Plus Rise Time		27	38	ns	$R_{SERIES} = 20\Omega$

Notes:
1. $C_L = 150pF$ (minimum C_L for 9 4K RAMs)
2. $C_L = 200pF$ (typical C_L for 9 4K RAMs). Typical values measured at $T_A = 25°C$.
3. $C_L = 250pF$ (maximum C_L for 9 4K RAMs).
4. Refer to Figure 22 for waveforms used.

The 3245, whose pin configuration is shown in Figure 20a has DTL and TTL compatible inputs as shown in Table III. The D.C. characteristics and power dissipation for various outputs states is shown in Table IV.

Table VI. 3245 Input Capacitance

Symbol	Test	Typ.	Max.	Unit
C_{IN}	Input Capacitance, $\bar{I}_1, \bar{I}_2, \bar{I}_3, \bar{I}_4$	5	8	pF
C_{IN}	Input Capacitance, $\bar{R}, \bar{C}, \bar{E}_1, \bar{E}_2$	8	12	pF

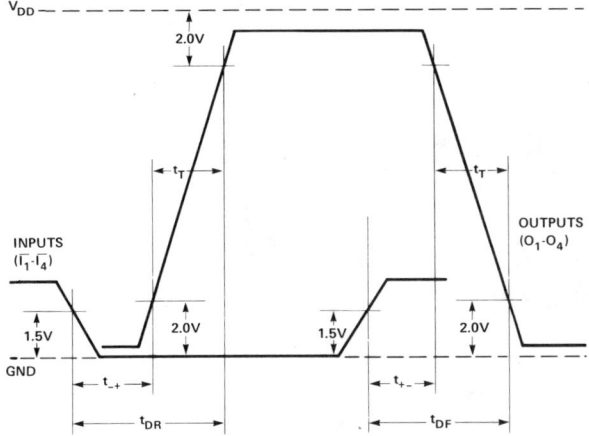

Figure 22. 3245 Waveforms

The A.C. characteristics are shown in Table V and the capacitance of the input pins is shown in Table VI. Figure 22 shows the threshold levels used in determining the delays specified in Table V.

Graphs showing the effect of capacitance loads on delay and rise times are shown in Figures 23a and b.

Waveforms of the 3245 driver in a 2107B system are shown in Figure 24a-d. The driver configuration used is shown in Figure 25.

Figure 24 shows the leading and trailing edge of chip enable at both the beginning and ending of the printed line for an added series resistance R of 10Ω. Note the transition time and overshoot for each of these edges. The overshoot is worst case at the leading edge at the driver end and on the trailing edge at the end of the line. The trailing edge overshoot is 2.2V while the leading edge overshoot is 1.5V. Both values are very marginal for system operation.

The effect of increasing the series resistance to 20Ω for the above driver is shown in Figure 24. Note that the transition time has increased but is still within entirely acceptable limits and the over-

shoots have been cut in half. the driver is now operating in an acceptable mode with minimal overshoot.

The effect of high temperatures (70°C) on the 3245 is shown in Figure 26. A 20Ω series resistor is used with the driver.

The power dissipation values shown in Table II are based on worst case conditions; power supplies at maximum voltage levels and outputs continuously enabled. In reality, the drivers operate on some duty cycle, as determined by the memory system cycle specifications. To realistically determine the power dissipated by the driver at a system level, calculations should be performed as shown below.

Referring to the 2107B chapter of this handbook, the minimum timing for a 2107B is with CE high for 230nS and low for 130nS for a 400nS minimum cycle. The total power dissipated in the driver for each cycle will be the sum of the power as shown by equation (1). For each memory cycle, 1 of the 4 driver elements will execute a cycle, while the other 3 outputs will remain low.

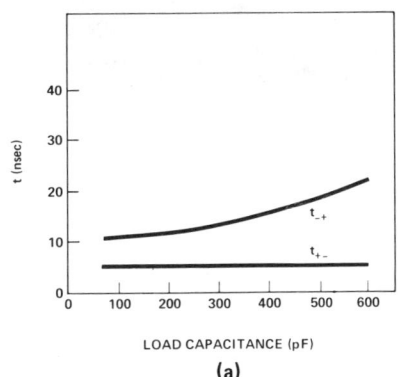

(a)

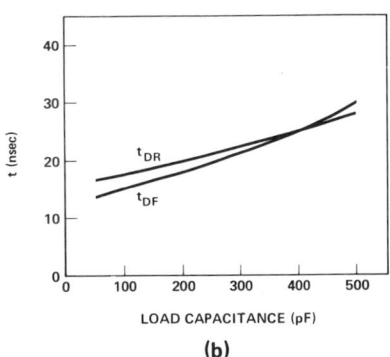

(b)

Figure 23. 3245 Delay and Transition Times as a Function of Load Capacitance

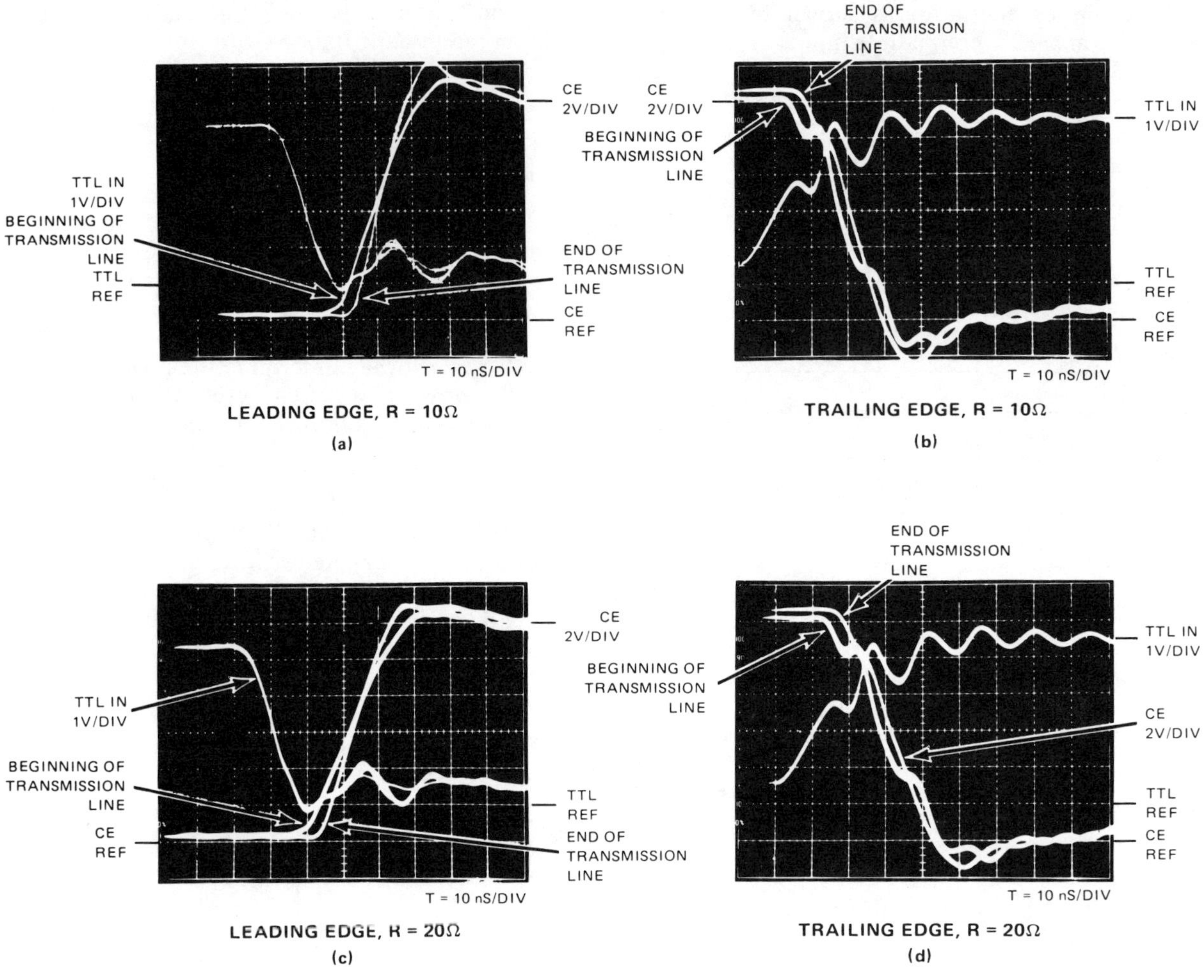

Figure 24. 3245 Typical Driver Waveforms

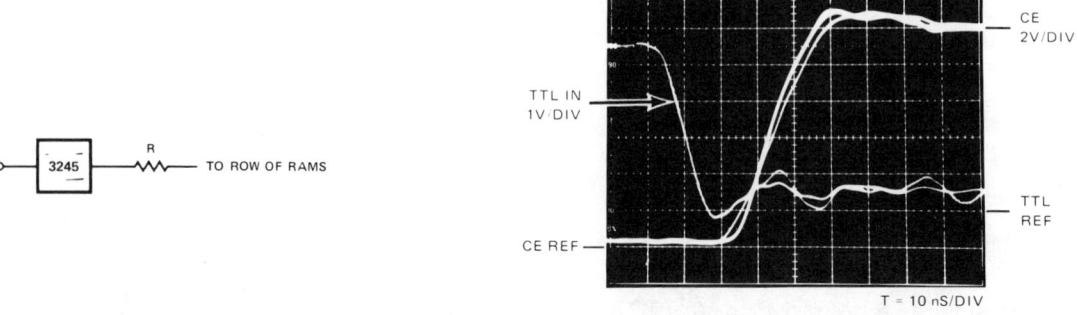

Figure 25. 3245 Driver Configuration

Figure 26. 3245 Driver Waveform with Temperature = 70°C

USING SUPPORT CIRCUITS FOR DYNAMIC RAMS

OTHER MOS LEVEL DRIVERS

3246 Operation

The Intel® 3246 is a QUAD ECL to MOS Driver designed for driving 4K N-channel MOS RAMs. The pin configuration and logic diagrams are shown in Figures 27 and 28. The device requires three power supplies, $V_{DD} = 12V$, $V_{CC} = +5V$ and the ECL reference voltage V_{EE} of $-5.25V$.

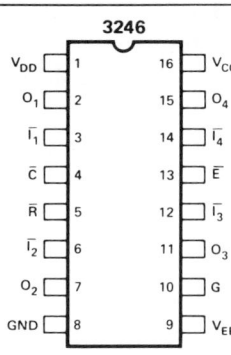

Figure 27. 3246 Pin Configuration

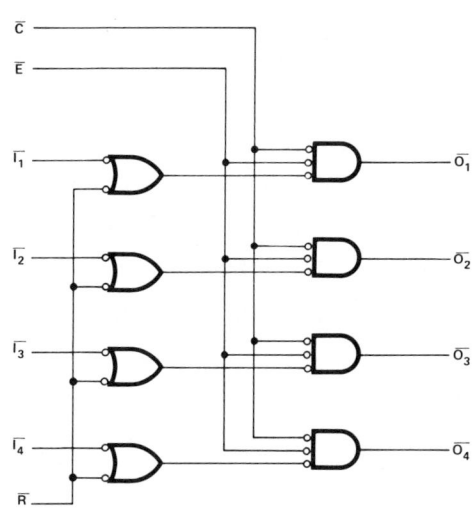

Figure 28. 3246 Logic Diagram

Tables VIII and IX presents the DC characteristics, including the power supply drain and input and output levels. To aid overall system planning, the supply currents are shown for outputs both high and low, so that power dissipation can be calculated as was done in the 2107B chapter.

Table VIII. 3246 D.C. Characteristics

$T_A = 0°C$ to $75°C$, $V_{CC} = 5.0V \pm 5\%$, $V_{DD} = 12V \pm 5\%$, $V_{EE} = -5.2V \pm 5\%$.

Symbol	Parameter	Min.	Max.	Unit	Test Conditions
I_{FD}	Input Load Current, $\overline{I}_1, \overline{I}_2, \overline{I}_3, \overline{I}_4$		0.5	mA	$V_F = -0.8V$
I_{FE}	Input Load Current, $\overline{R}, \overline{C}, \overline{E}_1, \overline{E}_2$		3.0	mA	$V_F = -0.8V$
V_{OL}	Output Low Voltage		0.45	V	$I_{OL} = 10mA$, $V_{IH} = -1.025V$
V_{OH}	Output High Voltage	$V_{DD}-0.5$		V	$I_{OH} = -1mA$, $V_{IL} = -1.500V$
V_{IL}	Input Low Voltage, All Inputs		-1.500	V	
V_{IH}	Input High Voltage, All Inputs	-1.025		V	

Table IX. 3246 Power Supply Current Drain and Power Dissipation

Symbol	Parameter	Typ.	Max.	Unit	Test Conditions — Input states to ensure the following output states:	Additional Test Conditions
I_{CC}	Current from V_{CC}	20	26	mA		
I_{DD}	Current from V_{DD}	22	30	mA		
I_{EE}	Current from V_{EE}	-31	-39	mA	High	
P_{D1}	Power Dissipation	550	725	mW		$V_{CC} = 5.25V$
	Power Per Channel	137	181	mW		$V_{DD} = 12.6V$
I_{CC}	Current from V_{CC}	20	26	mA		$V_{EE} = -5.46V$
I_{DD}	Current from V_{DD}	14	20	mA		
I_{EE}	Current from V_{EE}	-29	-36	mA	Low	
P_{D2}	Power Dissipation	440	585	mW		
	Power Per Channel	110	146	mW		

Table X. 3246 A.C. Characteristics

$T_A = 0°$ to $75°C$, $V_{CC} = 5.0V \pm 5\%$, $V_{DD} = 12V \pm 5\%$, $V_{EE} = -5.2V \pm 5\%$.

Symbol	Parameter	Min.[1]	Typ.[2]	Max.[3]	Unit	Test Conditions
t_{-+}	Input to Output Delay	8	12		ns	$R_{SERIES} = 0$
t_{DR}	Delay Plus Rise Time		18	30	ns	$R_{SERIES} = 0$
t_{+-}	Input to Output Delay	8	14		ns	$R_{SERIES} = 0$
t_{DF}	Delay Plus Fall Time		25	35	ns	$R_{SERIES} = 0$
t_R	Rise Time	10	17	25	ns	$R_{SERIES} = 20\Omega$
t_{DR}	Delay Plus Rise Time			40		$R_{SERIES} = 20\Omega$

Notes: 1. $C_L = 150pF$ (minimum C_L for 9-4K RAMs).
2. $C_L = 200pF$ (typical C_L for 9-4K RAMs). Typical values measured at $T_A = 25°C$.
3. $C_L = 250pF$ (maximum C_L for 9-4K RAMs).
4. Refer to Figure 29 for waveforms.

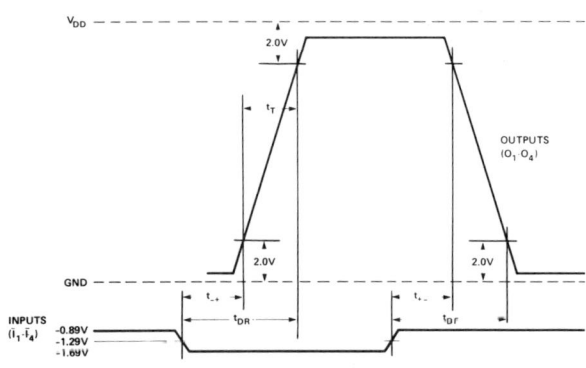

Figure 29. 3246 Waveform

The AC characteristics are shown in Table X, as specified with the waveforms shown in Figure 29.

The capacitive load attached to the output of the driver will affect the effective device speed. For this reason, the AC characteristics are specified with capacitive load values corresponding to typical system configurations. Figures 30a and b show the speed variations as a function of other capacitive loads.

The input capacitance of the various input pins are shown in Table XI.

Table XI. 3246 A.C. Characteristics

$T_A = 25°C$

Symbol	Test	Typ.	Max.	Unit
C_{IN}	Input Capacitance, $\overline{I}_1, \overline{I}_2, \overline{I}_3, \overline{I}_4, \overline{R}$	4	7	pF
C_{IN}	Input Capacitance, $\overline{C}, \overline{E}$	8	12	pF

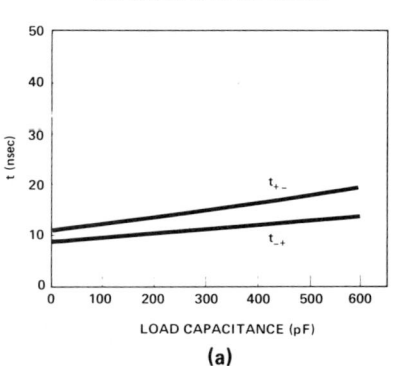

(a)

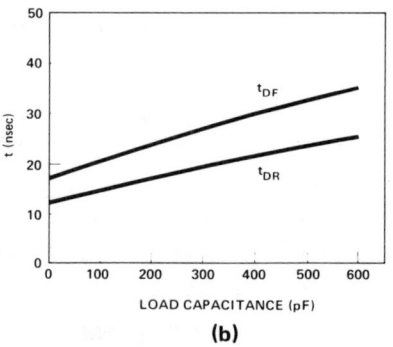

(b)

Figure 30. 3246 Delays vs. Load Capacity

USING SUPPORT CIRCUITS FOR DYNAMIC RAMS

5234 and 5235 Operation

The Intel® 5234 and 5235 are respectively CMOS to MOS and TTL to MOS drivers implemented with Silicon Gate CMOS technology. Because of the very low power drain each device is suitable for systems where battery backup is used. The pin configuration is shown in Figure 20b, and the logic configuration is shown in Figure 21. The DC characteristics are shown in Tables XII and XIII for both devices.

The AC characteristics for the 5234 are shown in Table XIV, with waveforms shown in Figure 31a. In a similar manner Table XV and Figure 31b present the AC characteristics and waveforms for the 5235.

Pertinent input capacitance information is presented in Table XVI.

Table XII. 5234 D.C. Characteristics
$T_A = 0°C$ to $70°C$, $V_{DD} = 12V \pm 5\%$.

Symbol	Parameter	Min.	Typ.[1]	Max.	Unit	Test Conditions
$|I_{LI}|$	Input Load Current, $\bar{I}_1, \bar{I}_2, \bar{I}_3, \bar{I}_4$			0.1	µA	$V_{IN} = 0$ to V_{DD}
V_{OL}	Output Low Voltage		0.15	0.4	V	$I_{OL} = 5mA$
		−1.0	−0.15			$I_{OL} = −5mA$
V_{OH}	Output High Voltage	$V_{DD}-0.4$	$V_{DD}-0.15$		V	$I_{OH} = -5mA$
		$V_{DD}+.15$	$V_{DD}+0.5$			$I_{OH} = 5mA$
V_{IL}	Input Low Voltage, All Inputs			2.0	V	
V_{IH}	Input High Voltage, All Inputs	$V_{DD}-2.0$			V	
I_{DD}	Supply Current		0.1	100	µA	$V_{DD} = 13.2V, f = 0$
I_{DD1}	Supply Current		13	20	mA	$V_{DD} = 13.2V, f = 1MHz$, $C_L = 0$, (See Figure 15a)

Note 1: Typical values are at 25°C and nominal voltage.

Table XIII. 5235 D.C. Characteristics
$T_A = 0°C$ to $70°C$, $V_{DD} = 12V \pm 10\%$.

Symbol	Parameter	Min.	Typ.[1]	Max.	Unit	Test Conditions
$|I_{LI}|$	Input Load Current		0.1	10	µA	$V_{IN} = \leq 0.4V$ or $\geq 2.4V$
V_{OL}	Output Low Voltage		0.15	0.4	V	$I_{OL} = 5mA$
		−1.0	−0.15		V	$I_{OL} = −5mA$
V_{OH}	Output High Voltage	$V_{DD}-0.4$	$V_{DD}-0.15$		V	$I_{OH} = -5mA$
		$V_{DD}+0.15$	$V_{DD}+0.5$			$I_{OH} = 5mA$
V_{IL}	Input Low Voltage, All Inputs			0.8	V	
V_{IH}	Input High Voltage, All Inputs	2.0			V	
I_{DD}	Supply Current		1.0	2.0	mA	$f = 0MHz$, $V_{DD}=13.2V$
I_{DD1}	Supply Current		12	20	mA	$f = 1MHz$ (See Figure 15b), $V_{IN} \leq 0.4V$ or $V_{IN} \geq 2.4V$, $C_L = 0$.

Note 1: Typical values are at 25°C and nominal voltage.

Table XIV. 5234 A.C. Characteristics
$T_A = 0°C$ to $70°C$, $V_{DD} = \pm 5\%$.

Symbol	Parameter	Min.[1]	Typ.[2,4]	Max.[3]	Unit	Test Conditions
t_{-+}	Input to Output Delay	20	45		ns	$R_{SERIES} = 0$
t_{DR}	Delay Plus Rise Time		70	100	ns	$R_{SERIES} = 0$
t_{+-}	Input to Output Delay	20	45		ns	$R_{SERIES} = 0$
t_{DF}	Delay Plus Fall Time		70	100	ns	$R_{SERIES} = 0$
t_T	Output Transition Time	10	25	40	ns	$R_{SERIES} = 0$

NOTES: 1. $C_L = 150pF$ — These values represent a range of
2. $C_L = 200pF$ — total stray plus clock capacitance
3. $C_L = 250pF$ — for nine 4K RAMs.
4. Typical values are measured at 25°C.

Table XV. 5235 A.C. Characteristics

$T_A = 0°C$ to $70°C$, $V_{DD} = 12V \pm 5\%$.

Symbol	Parameter	5235-1			5235			Unit	Test Conditions
		Min.[1]	Typ.[2,4]	Max.[3]	Min.[1]	Typ.[2,4]	Max.[3]		
t_{-+}	Input to Output Delay	20	55		20	70		ns	$R_{SERIES} = 0$
t_{DR}	Delay Plus Rise Time		75	90		95	120	ns	$R_{SERIES} = 0$
t_{+-}	Input to Output Delay	20	55		20	70		ns	$R_{SERIES} = 0$
t_{DF}	Delay Plus Fall Time		75	90		95	120	ns	$R_{SERIES} = 0$
t_T	Transition Time	10	20	40	10	25	40	ns	$R_{SERIES} = 0$

NOTES:
1. $C_L = 150pF$
2. $C_L = 200pF$
3. $C_L = 250pF$

These values represent a range of total stray plus clock capacitance for nine 4K RAMs.

4. Typical values are measured at 25°C, and nominal voltage.

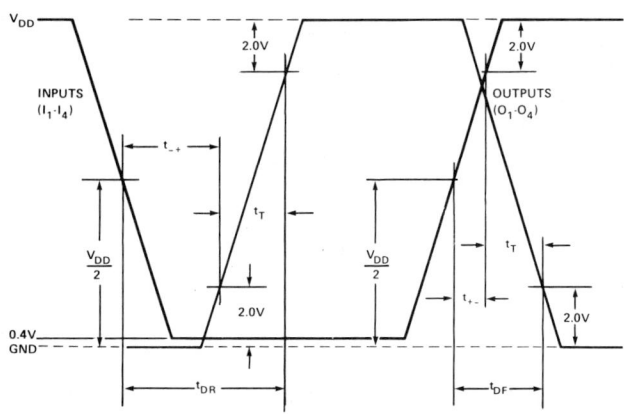

a. 5234 Waveforms

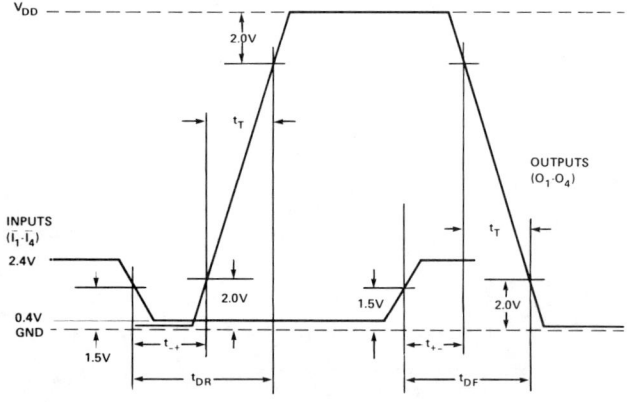

b. 5235 Waveforms

Figure 31. CMOS Driver Waveforms

Table XVI. 5234/35 Capacitance

Symbol	Test	Typ.	Max.	Unit
C_{IN}	Input Capacitance	8	14	pF

As has been pointed out before, the effective device speed will vary as a function of the capacitive load which the device is driving. For determining driver speed with capacitive loads different from those mentioned in Table XIV or XV, Figure 32 or 33 can be used to find typical delays.

Because the 5234 and 5235 are implemented with CMOS technology, the power supply current will vary as a function of frequency. Figure 34 will aid in the calculation of power supply requirements, as it correlates the input frequency to I_{DD}.

Figure 35 shows the variation in V_{IL} and V_{IH} as a function of t_{DR} and t_{DF}.

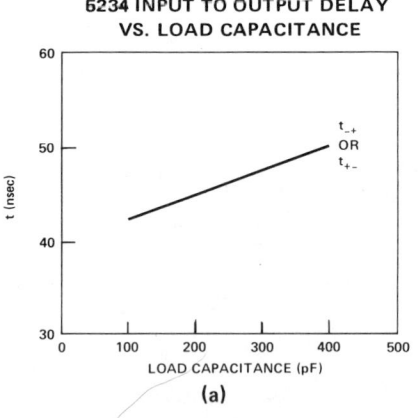

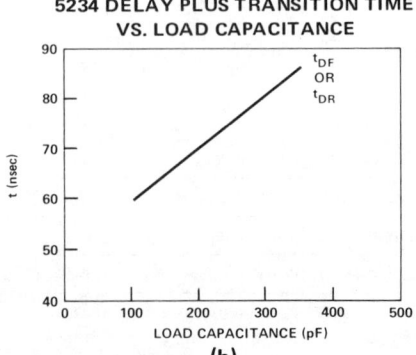

Figure 32. Output Characteristics

USING SUPPORT CIRCUITS FOR DYNAMIC RAMS

a.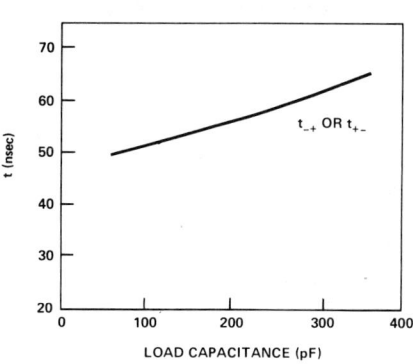

b.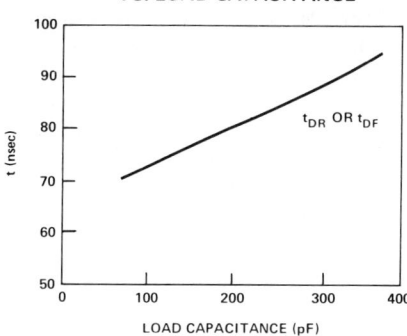

Figure 33. 5235 Delay vs. Capacitance

a.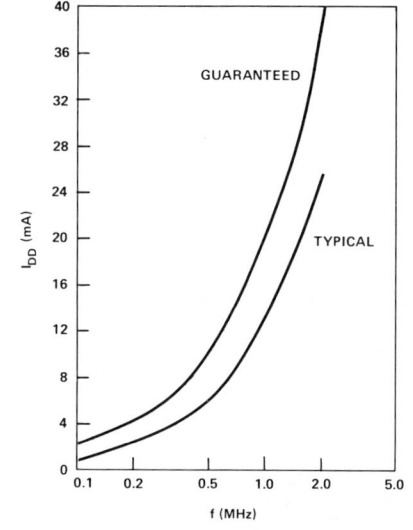

b.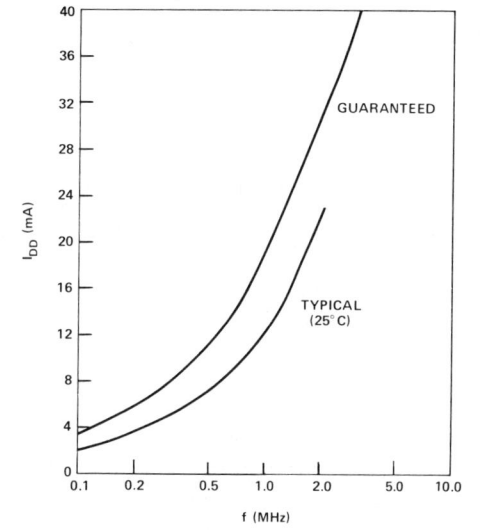

Figure 34. 5234/5235 Switching Frequency vs. I_{DD} Current

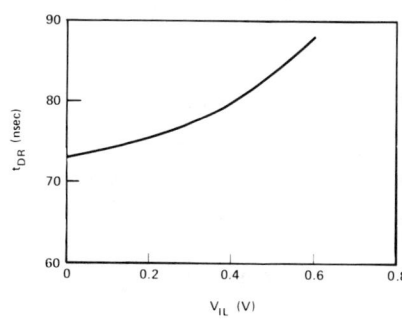

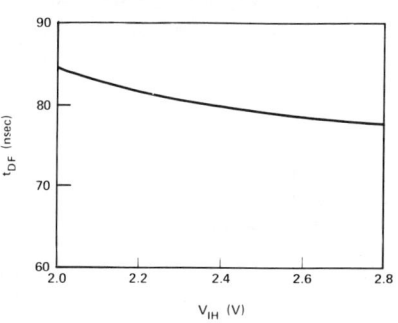

Figure 35. 5235 Delay Plus Transition Time vs. V_{IL} and V_{IH}.

MEMORY SUPPORT CIRCUITS FOR DYNAMIC RAMS

SYSTEMS CONSIDERATIONS

Because MOS level drivers are used primarily to drive the timing clock on 4K RAMs, the clock drivers are subjected to more stringent requirements than TTL level drivers. These requirements are usually on driver transition time and high and low output levels. In addition, drivers for MOS level clocks should have minimum power dissipation when their output is in the low or inactive state to minimize system standby power.

A typical system configuration using the 3245 in a 2107B memory system is shown in Figure 36. Note that each driver output drives 9 devices with the drivers placed in the middle of the array. It is important to place the drivers as close as possible to the clock inputs. This minimizes adverse transmission line effects.

When several memory cards are used in a system, the drivers can be used as logic gates inhibiting those memory devices not in use. The necessary logic to perform this function is shown in Figure 36. Using this method of decoding, up to four memory cards can be accommodated in the system.

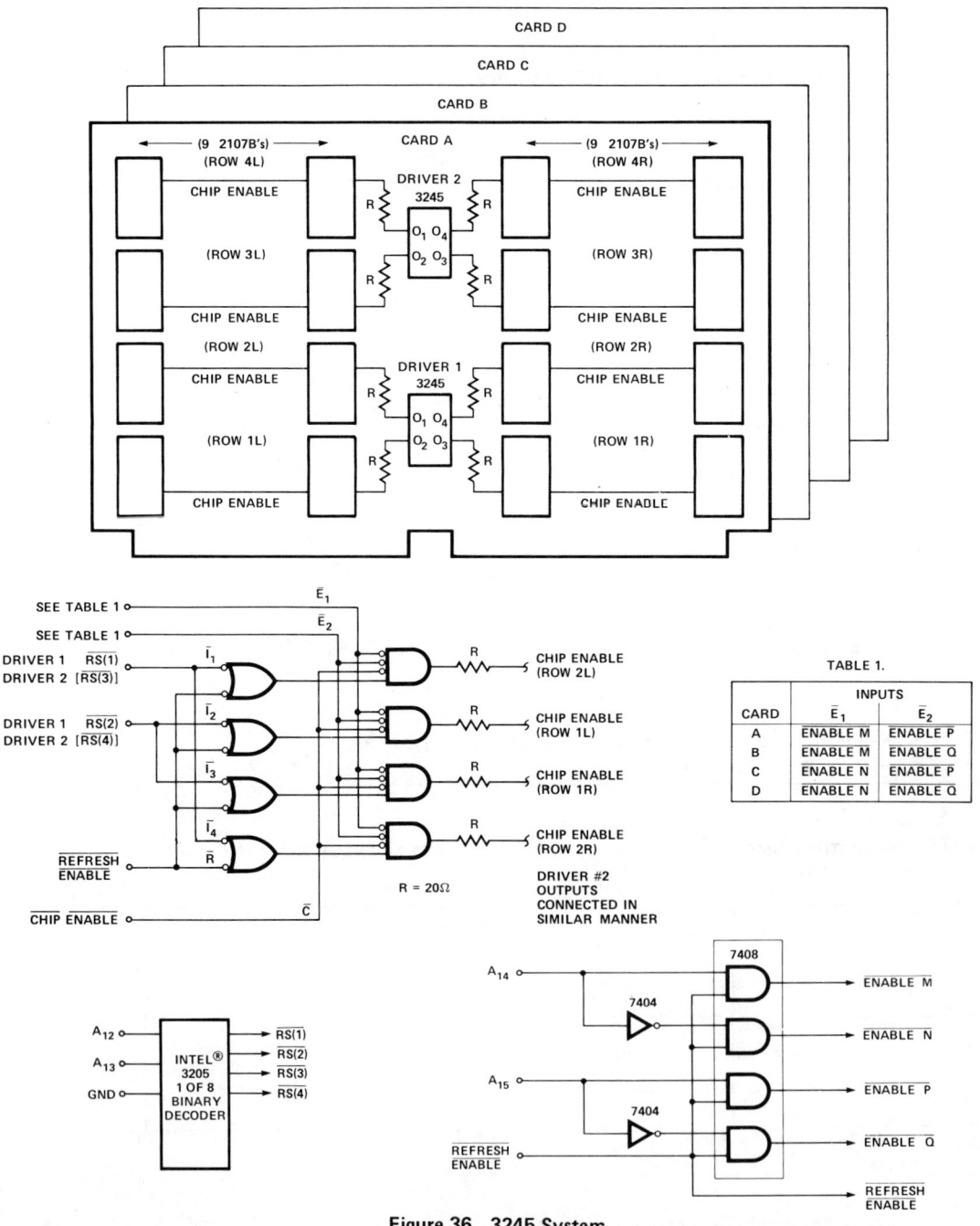

Figure 36. 3245 System.

USING SUPPORT CIRCUITS FOR DYNAMIC RAMS

The characteristics of the MOS level waveform generated by the 3245 depends on several factors: load capacitance, transmission line characteristics, driver placement relative to memory array, etc. In most systems, it is necessary to add a series terminating resistor on the output of the driver (see Figure 36) to more closely match the characteristics of the transmission line. The effect of adding a series terminating resistor is shown in Figure 37. The two extremes of resistance ($R_s = 0\Omega$ and $R_s = 51\Omega$) is clearly evident for a load of 250 pF. With $R_s = 0\Omega$, the overshoot on the clock driver is excessive and will lead to marginal system operation. The over/undershoots shown will not damage MOS devices. If the series resistance is too high, the $R_s = 51\Omega$, the effect is to significantly slow the clock transition times. Since 4K dynamic devices use the chip enable clock to initiate internal timing, the rising transistion of this external clock has a maximum limit. On the other hand, too fast a transition can introduce noise or cause marginal device operation. It is therefore necessary to control the transition time so that it is neither too fast nor too slow.

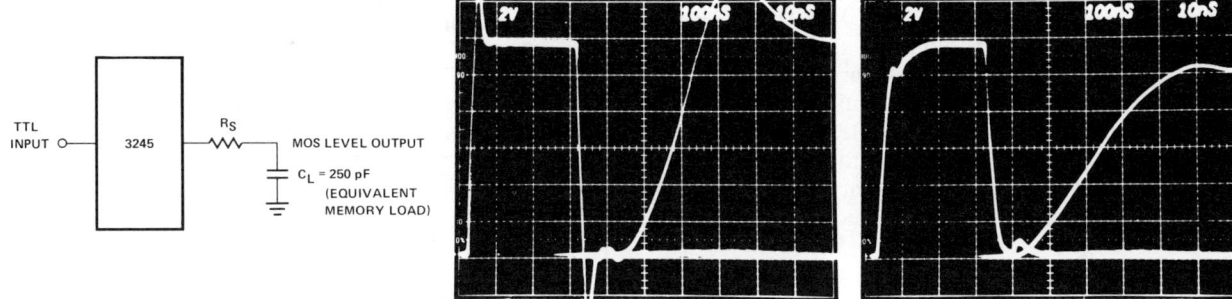

Figure 37. Effect of Series Terminating Resistor on MOS Level Driver

Table XVII. Summary of 3245 Driver Board Delay Measurements

NUMBER 2107B LOADS AND CIRCUIT CONFIGURATION		MEASURED CONDITIONS INPUT TO OUTPUT DELAY				MEASURED DELAY[3] PLUS RISE		MEASURED DELAY[4] PLUS FALL	
		t_{-+}[1]		t_{+-}[2]					
		TYP.	WORST[5] CASE	TYP.	WORST[5] CASE	TYP.	WORST[5] CASE	TYP.	WORST[5] CASE
3245 18 LOADS[6]	R = 20Ω	12	12	10	10	34	37	33	35
3245 9 LOADS	R = 20Ω	11		10		30	33[7]	25	27[7]

NOTES:
1. TTL 1.5 to V_{SS} +1 volt
2. TTL 1.5 to V_{DD} −1 volt
3. TTL 1.5 to V_{DD} −1 volt
4. TTL 1.5 to V_{SS} +1 volt
5. Worst case driver on board at 70°C and 5% power supply variation.
6. 18 loads 20Ω split resistor (see Figure 25).
7. Projected from 18 load delay.

(1)

$$P_{Total} = \frac{CV^2}{t_r} + \frac{t_{ce}}{t_{cyc}} \times P_{OH} + \frac{t_{\overline{ce}}}{t_{cyc}} \times P_{OL} + 3 \times P_{OL}$$

$$= \frac{(153)(8)^2}{20} + \frac{250}{400} \times 121 + \frac{150}{400} \times 97 +$$

$3 \times 97 = 892.6$ MW/Cycle for 36 2107B's; 1 row of 9 Enabled

Where:

P_{TOTAL} = Total power per cycle dissipated per 3245 driver.

C = input capacitance for 9 2107B CE inputs = 17pF x 9 = 153pF.

$V = (V_{DD} - 2) - (V_{SS} + 2) = 8V$.

t_r = CE transition time = 20nS.

t_{ce} = CE high time = 230nS + transition time.

$t_{\overline{ce}}$ = CE low time = 130ns + transition time.

t_{cyc} = memory cycle time = 400nS.

P_{OH} = 3245 power per driver for output high = 121mW.

P_{OL} = 3245 power per driver for output low = 97mW.

All values used are based on the 2107B system described in Section II. Nominal supply voltages and typical values were used where applicable.

The Motorola MC3460 and the National DS3644/3674 are alternate sources to the 3245.

The results of tests made on the board shown in Figure 36, of a typical 3245 driver driving 18 loads and 9 loads is shown in Table XVII. Note that the delay does not change appreciably with temperature but the transition time increased approximately 2-3 nsec from 25°C to 70°C. Calculation of driver power is shown in Equation 1.

Transmission Line Effect

The physical placement of the clock driver in close proximity to the memory array is an important design requirement. An equivalent circuit of the transmission line from the driver through the memory array is shown in Figure 38. The first section of transmission line, L, consists solely of the printed trace etch. At the memory array, the transmission line characteristics are significantly modified by the load of the chip enable inputs. While this load effects both the delay and impedance characteristics of the line, the primary effect observed by the system designer is the effect of the line impedance variation. The mismatch between the two sections of transmission lines is approximated by:

$$Z_1 = \sqrt{\frac{L}{C_1}}$$

and

$$Z_2 = \sqrt{\frac{L}{C_2}}$$

where:

- Z_1, Z_2 = impedance of sections L_1 and L_2, respectively
- L = per unit length inductance of line (assume constant for entire length of line)
- C_1 = per unit length capacitance of L_1
- C_2 = per unit length capacitance of L_2

Since the per unit length inductance is assumed constant for the transmissions line, the impedance mismatch is approximated by:

$$\frac{Z_1}{Z_2} = \sqrt{\frac{C_2}{C_1}}$$

Assuming the memory devices are on 0.5-in. centers in the array with each chip enable input typically 17 pF, C_2 is equivalent to:

$$C_2 = C_1 + 2(17 \text{ pF}) = C_1 + 34 \text{ pF/in.}$$

Therefore, the impedance ratio is:

$$\frac{Z_1}{Z_2} = \sqrt{34} = 5.8 \quad \text{Since } C_1 \ll 34 \text{ pF/in.}$$

There is, therefore, almost a 6:1 impedance difference between the two lines.

It is evident that if the clock driver is far removed from the memory array (L_1 is large) the effect of the mismatch is greatly magnified. Figure 39 shows a typical clock driver waveform. The "step" shown in the middle of the rising transition is the result of line mismatch. The severity of the "step" is a direct function of the length of segment L_1.

The placement of the MOS level clock driver close to the memory array is important to minimize ground (V_{SS}) and power supply (V_{DD}) noise. It is important to have ground traces between the memory array and MOS level drivers as short as possible. An example of an acceptable and marginal power distribution is shown in Figure 40.

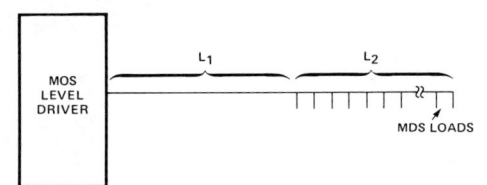

Figure 38. Typical Transmission Line on MOS Level Driver

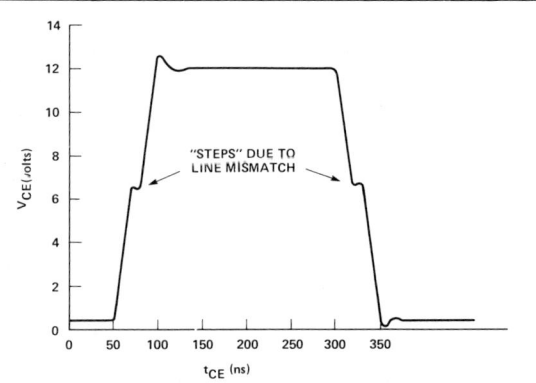

Figure 39. Effect of Transmission Line Mismatch

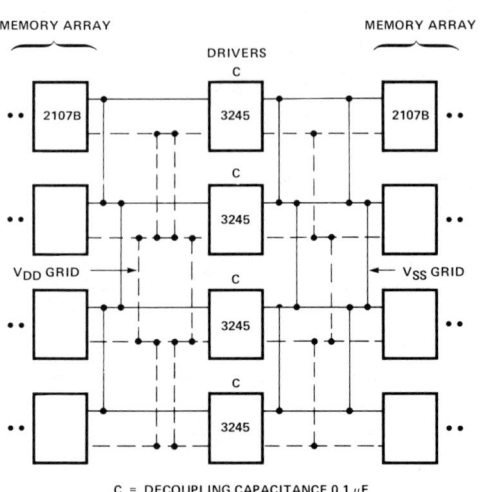

Figure 40. Driver Power Distribution

3205, 3404

3205 HIGH SPEED 1 OUT OF 8 BINARY DECODER
3404 HIGH SPEED 6-BIT LATCH

- 18ns Max. Delay Over 0°C to 75°C Temperature: 3205
- 12ns Max. Data to Output Delay Over 0°C to 75°C Temperature: 3404
- Directly Compatible With DTL and TTL Logic Circuits
- Totem-Pole Output
- Low Input Load Current: .25mA Max., 1/6 Standard TTL Input Load
- Minimum Line Reflection: Low Voltage Diode Input Clamp
- Outputs Sink 10mA Min.
- 16-Pin Dual In-Line Package
- Simple Expansion: Enable Inputs

3205
The 3205 decoder can be used for expansion of systems which utilize memory components with active low chip select input. When the 3205 is enabled, one of its eight outputs goes "low", thus a single row of a memory system is selected. The 3 chip enable inputs on the 3205 allow easy memory expansion. For very large memory systems, 3205 decoders can be cascaded such that each decoder can drive 8 other decoders for arbitrary memory expansions.

3404
The Intel 3404 contains six high speed latches organized as independent 4-bit and 2-bit latches. They are designed for use as memory data registers, address registers, or other storage elements. The latches act as high speed inverters when the "Write" input is "low".

The Intel 3404 is packaged in a standard 16-pin dual-in-line package; and its performance is specified over the temperature range of 0°C to +75°C, ambient. The use of Schottky barrier diode clamped transistors to obtain fast switching speeds results in higher performance than equivalent devices made with a gold diffusion process.

PIN CONFIGURATION

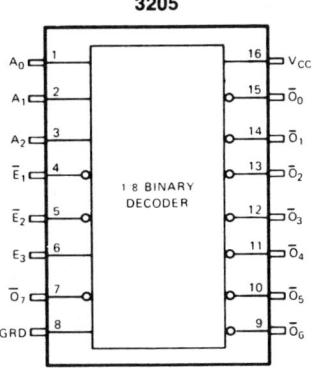
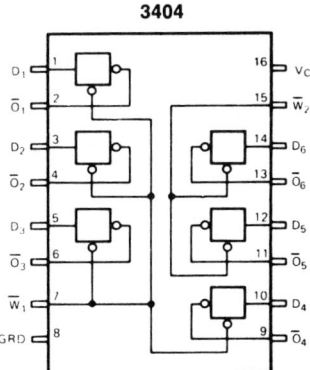

Absolute Maximum Ratings*

Temperature Under Bias:	Ceramic	−65°C to +125°C
	Plastic	−65°C to +75°C
Storage Temperature		−65°C to +160°C
All Output or Supply Voltages		−0.5 to +7 Volts
All Input Voltages		−1.0 to +5.5 Volts
Output Currents		125 mA

*COMMENT
Stresses above those listed under "Absolute Maximum Rating" may cause permanent damage to the device. This is a stress rating only and functional operation of the device at these or at any other condition above those indicated in the operational sections of this specification is not implied. Exposure to absolute maximum rating conditions for extended periods may affect device reliability.

D.C. Characteristics $T_A = 0°C$ to $+75°C$, $V_{CC} = 5.0V \pm 5\%$

3205, 3404

SYMBOL	PARAMETER	LIMIT MIN.	LIMIT MAX.	UNIT	TEST CONDITIONS
I_F	INPUT LOAD CURRENT		−0.25	mA	$V_{CC} = 5.25V$, $V_F = 0.45V$
I_R	INPUT LEAKAGE CURRENT		10	μA	$V_{CC} = 5.25V$, $V_R = 5.25V$
V_C	INPUT FORWARD CLAMP VOLTAGE		−1.0	V	$V_{CC} = 4.75V$, $I_C = −5.0$ mA
V_{OL}	OUTPUT "LOW" VOLTAGE		0.45	V	$V_{CC} = 4.75V$, $I_{OL} = 10.0$ mA
V_{OH}	OUTPUT HIGH VOLTAGE	2.4		V	$V_{CC} = 4.75V$, $I_{OH} = −1.5$ mA
V_{IL}	INPUT "LOW" VOLTAGE		0.85	V	$V_{CC} = 5.0V$
V_{IH}	INPUT "HIGH" VOLTAGE	2.0		V	$V_{CC} = 5.0V$
I_{SC}	OUTPUT HIGH SHORT CIRCUIT CURRENT	−40	−120	mA	$V_{CC} = 5.0V$, $V_{OUT} = 0V$
V_{OX}	OUTPUT "LOW" VOLTAGE @ HIGH CURRENT		0.8	V	$V_{CC} = 5.0V$, $I_{OX} = 40$ mA

3205 ONLY

SYMBOL	PARAMETER	LIMIT MIN.	LIMIT MAX.	UNIT	TEST CONDITIONS
I_{CC}	POWER SUPPLY CURRENT		70	mA	$V_{CC} = 5.25V$, Outputs Open

3404 ONLY

SYMBOL	PARAMETER	LIMIT MIN.	LIMIT MAX.	UNIT	TEST CONDITIONS
I_{CC}	POWER SUPPLY CURRENT		75	mA	$V_{CC} = 5.25V$, Outputs Open
I_{FW1}	WRITE ENABLE LOAD CURRENT PIN 7		−1.00	mA	$V_{CC} = 5.25V$, $V_W = 0.45V$
I_{FW2}	WRITE ENABLE LOAD CURRENT PIN 15		−0.50	mA	$V_{CC} = 5.25V$, $V_W = 0.45V$
I_{RW}	WRITE ENABLE LEAKAGE CURRENT		10	μA	$V_R = 5.25V$

Typical Characteristics

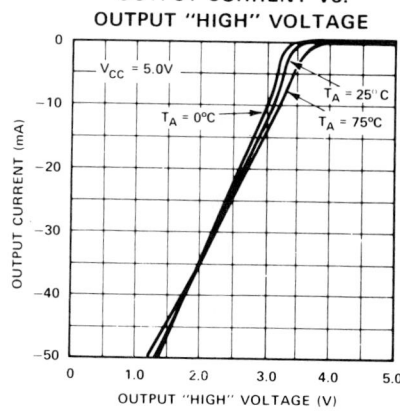

OUTPUT CURRENT VS. OUTPUT "LOW" VOLTAGE

OUTPUT CURRENT VS. OUTPUT "HIGH" VOLTAGE

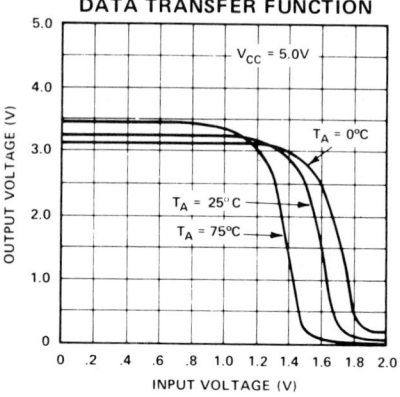

DATA TRANSFER FUNCTION

3205 - HIGH SPEED 1 OUT OF 8 BINARY DECODER
Switching Characteristics

CONDITIONS OF TEST:

Input pulse amplitudes: 2.5V

Input rise and fall times: 5 nsec between 1V and 2V

Measurements are made at 1.5V

TEST LOAD:

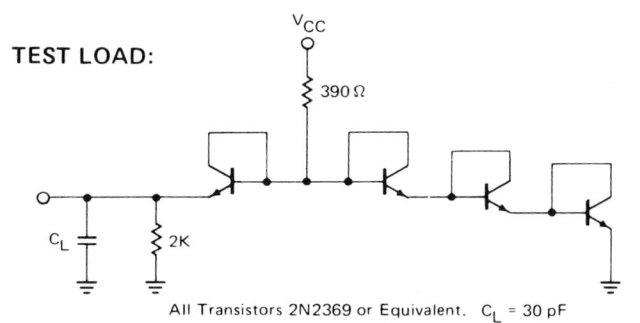

All Transistors 2N2369 or Equivalent. C_L = 30 pF

TEST WAVEFORMS

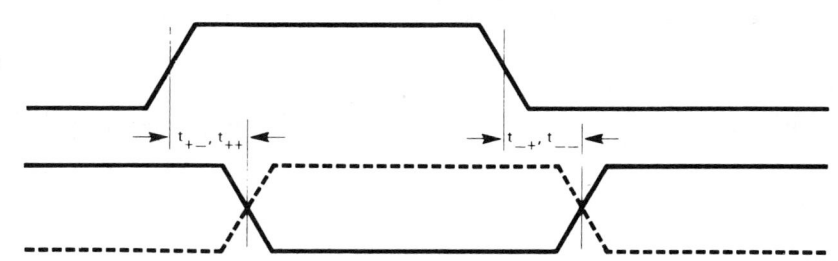

A.C. Characteristics T_A = 0°C to +75°C, V_{CC} = 5.0V ±5% unless otherwise specified.

SYMBOL	PARAMETER	MAX. LIMIT	UNIT	TEST CONDITIONS
t_{++}	ADDRESS OR ENABLE TO OUTPUT DELAY	18	ns	
t_{-+}		18	ns	
t_{+-}		18	ns	
t_{--}		18	ns	
C_{IN} [1]	INPUT CAPACITANCE P3205	4 (typ.)	pF	f = 1 MHz, V_{CC} = 0V
	C3205	5 (typ.)	pF	V_{BIAS} = 2.0V, T_A = 25°C

1. This parameter is periodically sampled and is not 100% tested.

Typical Characteristics

ADDRESS OR ENABLE TO OUTPUT DELAY VS. LOAD CAPACITANCE

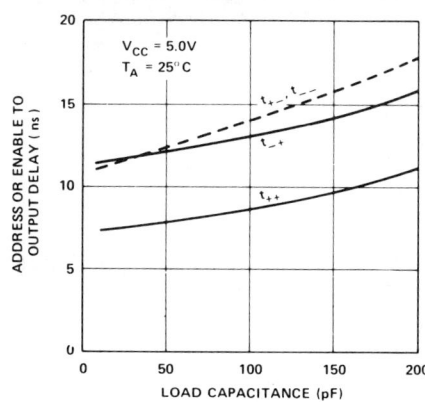

ADDRESS OR ENABLE TO OUTPUT DELAY VS. AMBIENT TEMPERATURE

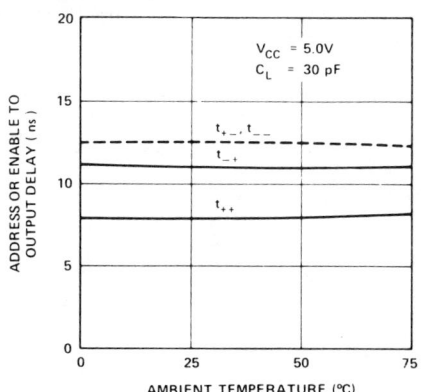

3404 - 6-BIT LATCH
Switching Characteristics

CONDITIONS OF TEST:

Input pulse amplitudes: 2.5V

Input rise and fall times: 5 nsec between 1V and 2V

Measurements are made at 1.5V

TEST LOAD:

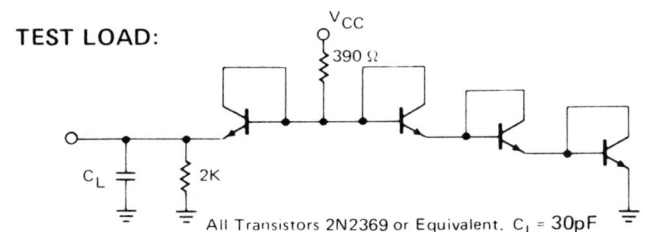

All Transistors 2N2369 or Equivalent. C_L = 30pF

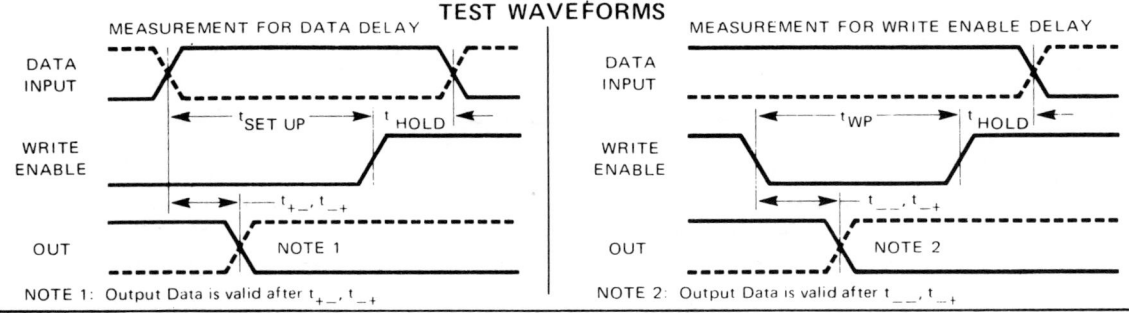

TEST WAVEFORMS

NOTE 1: Output Data is valid after t_{+-}, t_{-+}

NOTE 2: Output Data is valid after t_{--}, t_{-+}

A.C. Characteristics T_A = 0°C to +75°C, V_{CC} = 5.0V ±5%; unless otherwise specified.

SYMBOL	PARAMETER		LIMITS		UNIT	TEST CONDITIONS
		MIN.	TYP.	MAX.		
t_{+-}, t_{-+}	DATA TO OUTPUT DELAY			12	ns	
t_{--}, t_{-+}	WRITE ENABLE TO OUTPUT DELAY			17	ns	
$t_{SET\ UP}$	TIME DATA MUST BE PRESENT BEFORE RISING EDGE OF WRITE ENABLE	12			ns	
t_{HOLD}	TIME DATA MUST REMAIN AFTER RISING EDGE OF WRITE ENABLE	8			ns	
t_{WP}	WRITE ENABLE PULSE WIDTH	15			ns	
C_{IND}(3)	DATA INPUT CAPACITANCE P3404		4		pF	f = 1 MHz, V_{CC} = 0V
	C3404		5		pF	V_{BIAS} = 2.0V, T_A = 25°C
C_{INW}(3)	WRITE ENABLE CAPACITANCE P3404		7		pF	f = 1 MHz, V_{CC} = 0V
	C3404		8		pF	V_{BIAS} = 2.0V, T_A = 25°C

NOTE 3: This parameter is periodically sampled and is not 100% tested.

Typical Characteristics

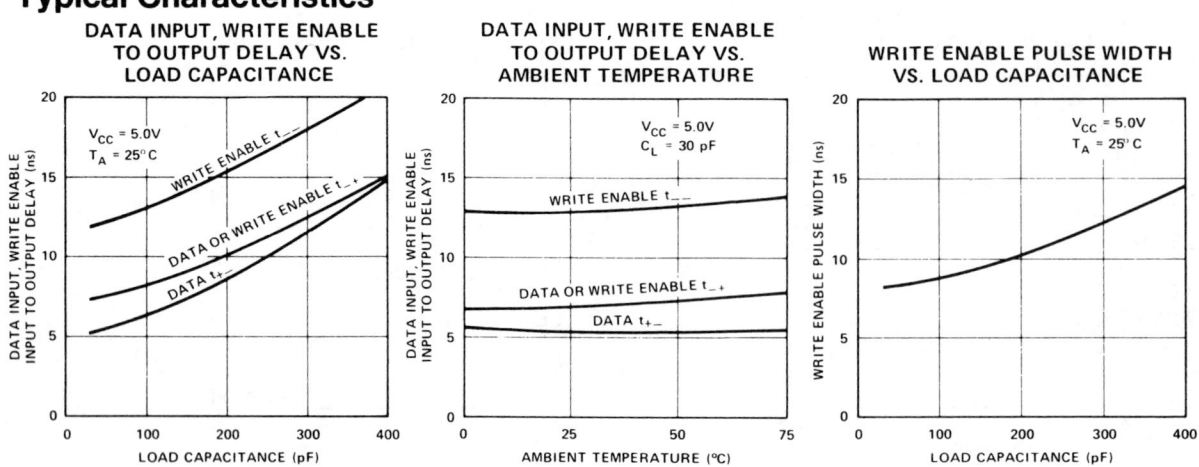

3207A
QUAD BIPOLAR-TO-MOS LEVEL SHIFTER AND DRIVER

- **High Speed, 45 nsec Max.-- Delay + Transition Time Over Temperature with 200 pF Load**
- **TTL & DTL Compatible Inputs**
- **1103 and 1103A Memory Compatible at Output**
- **Simplifies Design -- Replaces Discrete Components**
- **Easy to Use -- Operates from Standard Bipolar and MOS Supplies**
- **Minimum Line Reflection -- Input and Output Clamp Diodes**
- **High Input Breakdown Voltage-- 19 Volts**
- **CerDIP Package -- 16 Pin DIP**

The 3207A is a Quad Bipolar-to-MOS level shifter and driver which accepts TTL and DTL input signals, and provides high output current and voltage suitable for driving MOS circuits. It is particularly suitable for driving the 1103 and 1103A memory chips. The circuit operates from a 5 volt TTL power supply, and V_{SS} and V_{BB} power supplies from the 1103 and 1103A.

The device features two common enable inputs per pair of devices which permits some logic to be done at their inputs, such as cenable and precharge decoding for the 1103 and 1103A.

For the TTL inputs a logic "1" is V_{IH} and a logic "0" is V_{IL}. The 3207A outputs correspond to a logic "1" as V_{OL} and a logic "0" as V_{OH} for driving MOS inputs.

The 3207A is packaged in a hermetically sealed 16 pin ceramic dual-in-line package. The device performance is specified over the same temperature range as the 1103 and 1103A, i.e. from 0°C to +70°C.

PIN CONFIGURATION

```
           V_SS  [ 1    16 ] V_CC
      OUTPUT O_1 [ 2    15 ] O_4  OUTPUT
  DATA INPUT D_1 [ 3    14 ] D_4  DATA INPUT
ENABLE INPUT E_1 [ 4    13 ] E_4  ENABLE INPUT
ENABLE INPUT E_2 [ 5    12 ] E_3  ENABLE INPUT
  DATA INPUT D_2 [ 6    11 ] D_3  DATA INPUT
      OUTPUT O_2 [ 7    10 ] O_3  OUTPUT
           GND  [ 8     9 ] V_BB
```

LOGIC SYMBOL

MEMORY SUPPORT CIRCUITS FOR DYNAMIC RAMS

Absolute Maximum Ratings*

Temperature Under Bias .... 0°C to +70°C
Storage Temperature........... −65°C to +160°C
All Input Voltages and V_{SS} .... −1.0 to +21V
Supply Voltage V_{CC} .... −1.0 to +7V
All Outputs and Supply Voltage
V_{BB} with respect to GND .... −1.0 to +25V
Power Dissipation at 25°C .... 2 Watts [1]

*COMMENT

Stresses above those listed under "Absolute Maximum Ratings" may cause permanent damage to the device. This is a stress rating only and functional operation of the device at these or at any other condition above those indicated in the operational sections of this specification is not implied. Exposure to absolute maximum rating conditions for extended periods may affect device reliability.

[1] Refer to the graph of Junction Temperature versus Total Power Dissipation on page 5-10 for other temperatures.

D. C. Characteristics

T_A = 0°C to 70°C, V_{CC} = 5V ± 5%, V_{SS} = 16V ± 5%, $V_{BB} - V_{SS}$ = 3.0V to 4.0V

SYMBOL	TEST	LIMIT MIN.	LIMIT MAX.	UNIT	CONDITIONS
I_{FD}	DATA INPUT LOAD CURRENT		−0.25	mA	V_D = .45V, V_{CC} = 5.25V, All Other Inputs at 5.25V, V_{SS} = 16V, V_{BB} = 19V
I_{FE}	ENABLE INPUT LOAD CURRENT		−0.50	mA	V_E = .45V, V_{CC} = 5.25V, All Other Inputs at 5.25V, V_{SS} = 16V, V_{BB} = 19V
I_{RD}	DATA INPUT LEAKAGE CURRENT		20	µA	V_D = 19V, V_{CC} = 5.0V, All Other Inputs Grounded, V_{SS} = 16V, V_{BB} = 19V
I_{RE}	ENABLE INPUT LEAKAGE CURRENT		20	µA	V_E = 19V, V_{CC} = 5.0V, All Other Inputs Grounded, V_{SS} = 16V, V_{BB} = 19V
V_{OL}	OUTPUT "LOW" VOLTAGE		.8 .7 .6	V(0°C) V(25°C) V(70°C)	I_{OL} = 500µA, V_{CC} = 4.75V, V_{SS} = 16V, V_{BB} = 19V, All Inputs at 2.0V
V_{OH} (MIN.)	OUTPUT "HIGH" VOLTAGE	V_{SS} − .7 V_{SS} − .6 V_{SS} − .5		V(0°C) V(25°C) V(70°C)	I_{OH} = −500µA, V_{CC} = 5.0V, V_{SS} = 16V, V_{BB} = 19V, All Inputs at 0.85V
V_{OH} (MAX.)			V_{SS} + 1.0	V	I_{OH} = 5mA, V_{CC} = 5.0V, V_{SS} = 16V, V_{BB} = 19V
I_{OL}	OUTPUT SINK CURRENT	100		mA	V_O = 4V, V_{CC} = 5.0V, V_{SS} = 16V, V_{BB} = 19V, V_E = V_D = 2.0V
I_{OH}	OUTPUT SOURCE CURRENT	−100		mA	V_O = V_{SS} − 4V, V_{CC} = 5.0V, V_{SS} = 16V, V_{BB} = 19V, V_E = V_D = 0.85V
V_{IL}	INPUT "LOW" VOLTAGE		1.0	V	V_{CC} = 5.0V, V_{SS} = 16V, V_{BB} = 19V
V_{IH}	INPUT "HIGH" VOLTAGE	2.0		V	V_{CC} = 5.0V, V_{SS} = 16V, V_{BB} = 19V
C_{IN}	INPUT CAPACITANCE		8 (Typical)	pF	V_{BIAS} = 2.0V, V_{CC} = 0V

POWER SUPPLY CURRENT DRAIN:

All Outputs "Low"

Symbol	Parameter	Min.	Max.	Unit	Conditions
I_{CC}	Current from V_{CC}		83	mA	V_{CC} = 5.25V, V_{SS} = 16.8V, V_{BB} = 20.8V, All Inputs Open
I_{SS}	Current from V_{SS}		250	µA	
I_{BB}	Current from V_{BB}		21	mA	
P_{TOTAL}	Total Power Dissipation		900	mW	

All Outputs "High"

Symbol	Parameter	Min.	Max.	Unit	Conditions
I_{CC}	Current from V_{CC}		33	mA	V_{CC} = 5.25V, V_{SS} = 16.8V, V_{BB} = 20.8V, All Inputs Grounded
I_{SS}	Current from V_{SS}		250	µA	
I_{BB}	Current from V_{BB}		3	mA	
P_{TOTAL}	Total Power Dissipation		250	mW	

Standby Condition with V_{CC} = 0V, V_{SS} = V_{BB}

Symbol	Parameter	Min.	Max.	Unit	Conditions
I_{CC}	Current from V_{CC}		0	mA	V_{CC} = 0V, V_{SS} = 16.8V, V_{BB} = 16.8V
I_{SS}	Current from V_{SS}		250	µA	
I_{BB}	Current from V_{BB}		250	µA	
P_{TOTAL}	Total Power Dissipation		10	mW	

3207A QUAD DRIVER

Switching Characteristics
A.C. Characteristics

T_A = 0°C to 70°C, V_{CC} = 5V ±5%, V_{SS} = 16V ±5%, V_{BB} = V_{SS} +3 to 4V, f = 2 MHz, 50% Duty Cycle

SYMBOL	TEST	LIMITS (ns)				DELAY DIFFERENTIAL[1] C_L = 200 pF MAX.
		C_L = 100 pF		C_L = 200 pF		
		MIN.	MAX.	MIN.	MAX.	
t_{+-}	INPUT TO OUTPUT DELAY	5	15	5	15	5
t_{-+}	INPUT TO OUTPUT DELAY	5	25	5	25	10
t_r	OUTPUT RISE TIME	5	20	5	30	10
t_f	OUTPUT FALL TIME	5	20	10	30	10
t_D	DELAY + RISE OR FALL TIME	10	35	20	45	10

(1) This is defined as the maximum skew between any output in the same package, eg., all the input to output delays for the t_{-+} parameter are within a maximum of 10 nsec of each other in the same package.

Waveforms

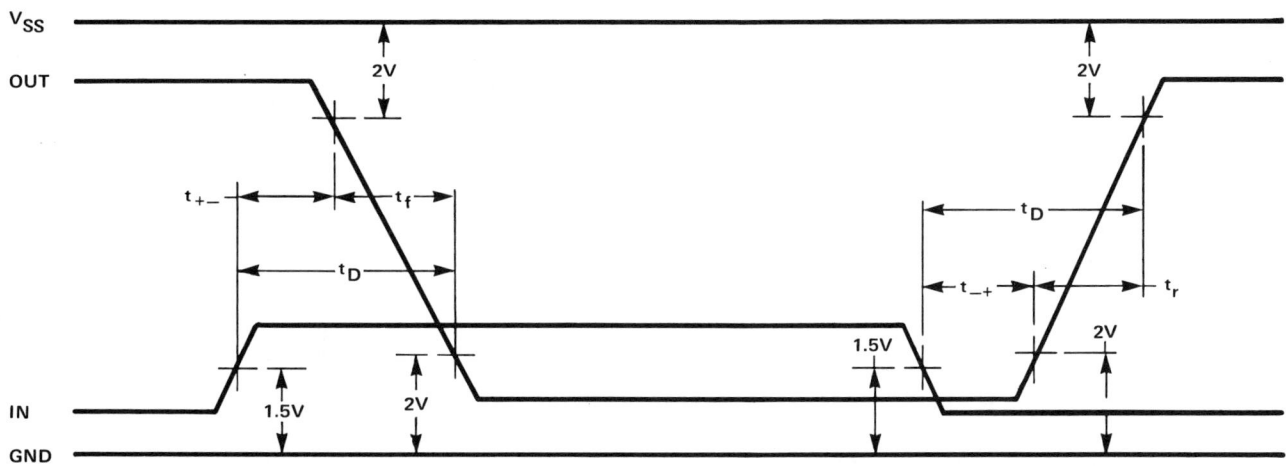

Typical Characteristics

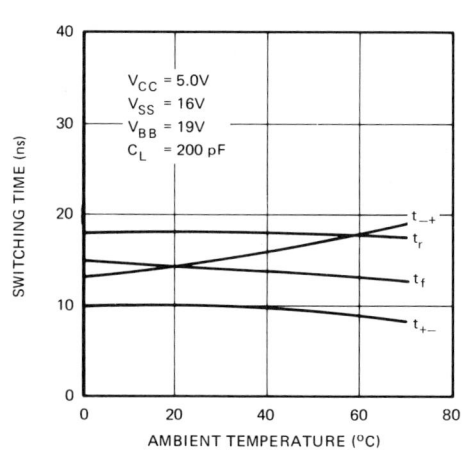

SWITCHING TIME VS. AMBIENT TEMPERATURE

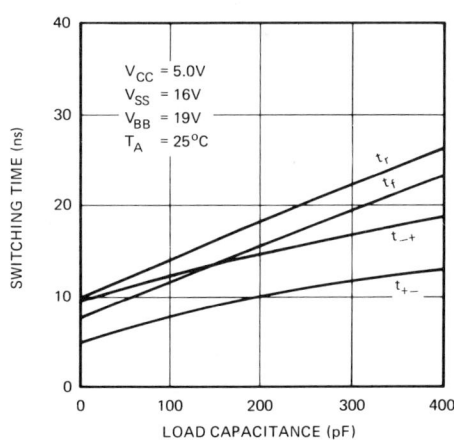

SWITCHING TIME VS. LOAD CAPACITANCE

Power and Switching Characteristics

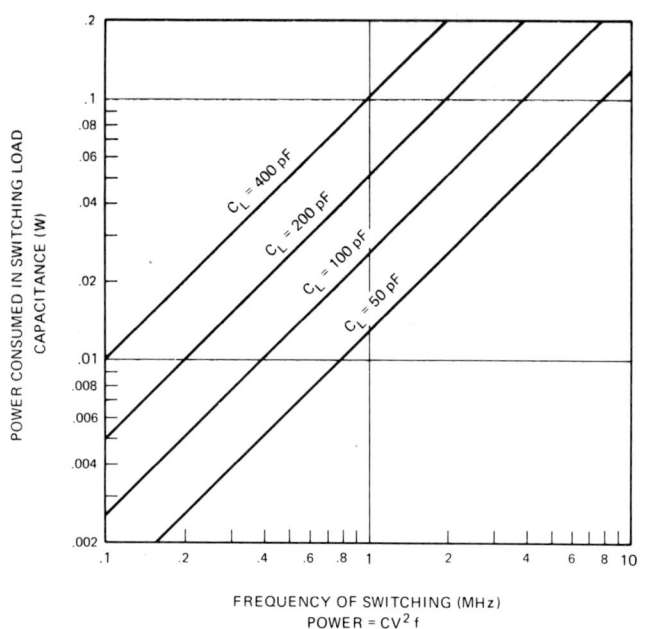

POWER CONSUMED IN CHARGING AND DISCHARGING LOAD CAPACITANCE OVER 0V TO 16V INTERVAL

POWER = $CV^2 f$

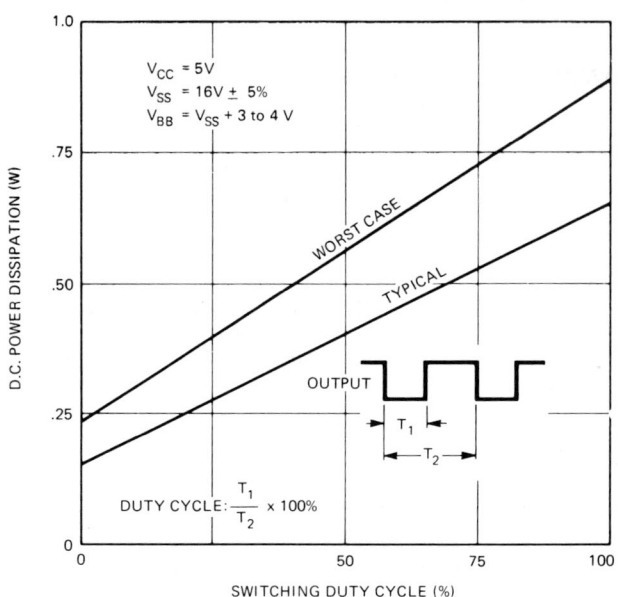

NO LOAD D.C. POWER DISSIPATION VS. OPERATING DUTY CYCLE

$V_{CC} = 5V$
$V_{SS} = 16V \pm 5\%$
$V_{BB} = V_{SS} + 3$ to 4 V

DUTY CYCLE: $\frac{T_1}{T_2} \times 100\%$

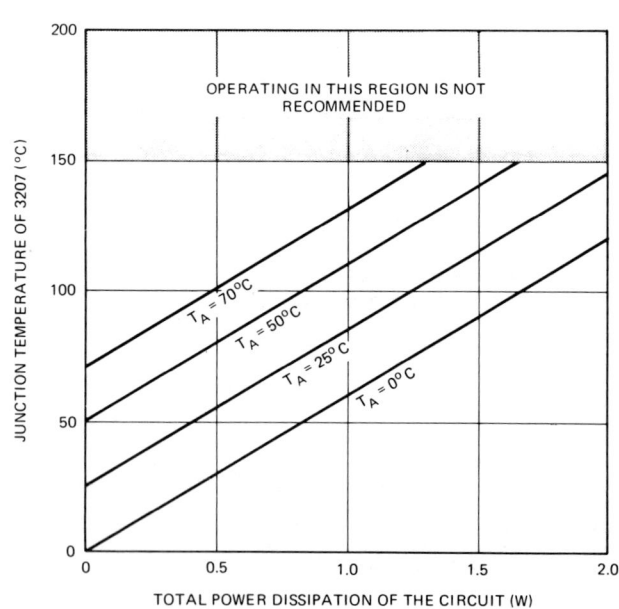

JUNCTION TEMPERATURE VS. TOTAL POWER DISSIPATION OF THE CIRCUIT

TOTAL POWER = D.C. POWER + POWER CONSUMED IN CHARGING AND DISCHARGING LOAD CAPACITANCE.

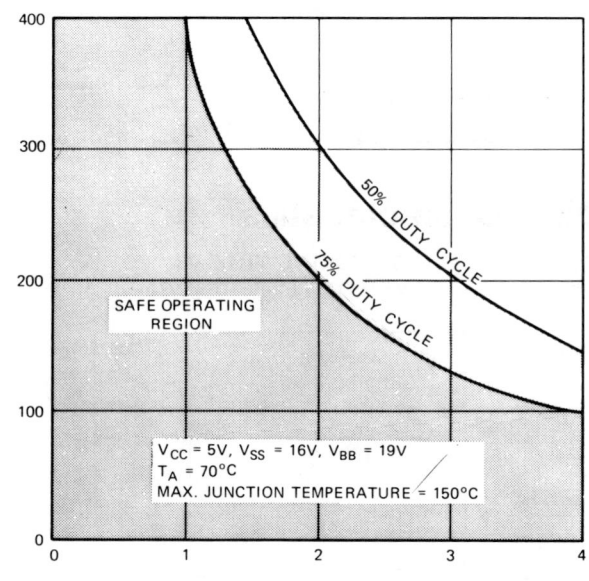

WORST CASE LOAD CAPACITANCE ON EACH OUTPUT VS. FREQUENCY OF SWITCHING

$V_{CC} = 5V$, $V_{SS} = 16V$, $V_{BB} = 19V$
$T_A = 70°C$
MAX. JUNCTION TEMPERATURE = 150°C

3207A-1
QUAD BIPOLAR-TO-MOS LEVEL SHIFTER AND DRIVER

- **Power Supply Voltage Compatible with the High Voltage 1103-1**
- **1103-1 Memory Compatible at Output**

The Intel 3207A-1 is the high voltage version of the standard 3207A, and is compatible with the 1103-1. The 3207A-1 has all the same features as the standard 3207A. The absolute maximum ratings and pin configuration are repeated below for convenience, while the DC and AC characteristics appear below and on the next page.

PIN CONFIGURATION

V_{SS}	1	16	V_{CC}
OUTPUT O_1	2	15	O_4 OUTPUT
DATA INPUT D_1	3	14	D_4 DATA INPUT
ENABLE INPUT E_1	4	13	E_4 ENABLE INPUT
ENABLE INPUT E_2	5	12	E_3 ENABLE INPUT
DATA INPUT D_2	6	11	D_3 DATA INPUT
OUTPUT O_2	7	10	O_3 OUTPUT
GND	8	9	V_{BB}

LOGIC SYMBOL

ABSOLUTE MAXIMUM RATINGS*

Temperature Under Bias .... 0°C to +55°C
Storage Temperature .... –65°C to +160°C
All Input Voltages .... –1.0 to +21 Volts
Supply Voltage V_{CC} .... –1.0 to +7.0 Volts
All Outputs and Supply Voltages V_{BB} and V_{SS}
with respect to GND .... –1.0 to +25 Volts
Power Dissipation at 25°C .... 2 Watts

COMMENT:
Stresses above those listed under "Absolute Maximum Ratings" may cause permanent damage to the device. This is a stress rating only and functional operation of the device at these or at any other condition above those indicated in the operational sections of this specification is not implied. Exposure to absolute maximum rating conditions for extended periods may affect device reliability.

D. C. Characteristics
$T_A = 0°C$ to $55°C$, $V_{CC} = 5V \pm 5\%$, $V_{SS} = 19V \pm 5\%$, $V_{BB} - V_{SS} = 3.0V$ to $4.0V$

SYMBOL	TEST	LIMIT MIN.	LIMIT MAX.	UNIT	CONDITIONS
I_{FD}	DATA INPUT LOAD CURRENT		–0.25	mA	$V_D = .45V$, $V_{CC} = 5.25V$, All Other Inputs at 5.25V, $V_{SS} = 19V$, $V_{BB} = 23V$
I_{FE}	ENABLE INPUT LOAD CURRENT		–0.50	mA	$V_E = .45V$, $V_{CC} = 5.25V$, All Other Inputs at 5.25V, $V_{SS} = 19V$, $V_{BB} = 23V$
I_{RD}	DATA INPUT LEAKAGE CURRENT		20	µA	$V_D = 19V$, $V_{CC} = 5.0V$, All Other Inputs Grounded, $V_{SS} = 19V$, $V_{BB} = 23V$
I_{RE}	ENABLE INPUT LEAKAGE CURRENT		20	µA	$V_E = 19V$, $V_{CC} = 5.0V$, All Other Inputs Grounded, $V_{SS} = 19V$, $V_{BB} = 23V$
V_{OL}	OUTPUT "LOW" VOLTAGE		0.8 0.7 0.6	V(0°C) V(25°C) V(55°C)	$I_{OL} = 500µA$, $V_{CC} = 4.75V$, $V_{SS} = 19V$, $V_{BB} = 23V$, All Inputs at 2.0V
V_{OH} (MIN.)	OUTPUT "HIGH" VOLTAGE	$V_{SS} - 0.7$ $V_{SS} - 0.6$ $V_{SS} - 0.5$		V(0°C) V(25°C) V(55°C)	$I_{OH} = -500µA$, $V_{CC} = 5.0V$, $V_{SS} = 19V$, $V_{BB} = 23V$, All Inputs at 0.85V
V_{OH} (MAX.)			$V_{SS} + 1.0$	V	$I_{OH} = 5mA$, $V_{CC} = 5.0V$, $V_{SS} = 19V$, $V_{BB} = 23V$
I_{OL}	OUTPUT SINK CURRENT	100		mA	$V_O = 4V$, $V_{CC} = 5.0V$, $V_{SS} = 19V$, $V_{BB} = 23V$, $V_E = V_D = 2.0V$
I_{OH}	OUTPUT SOURCE CURRENT	–100		mA	$V_O = V_{SS} - 4V$, $V_{CC} = 5.0V$, $V_{SS} = 19V$, $V_{BB} = 23V$, $V_E = V_D = 0.85V$
V_{IL}	INPUT "LOW" VOLTAGE		1.0	V	$V_{CC} = 5.0V$, $V_{SS} = 19V$, $V_{BB} = 23V$
V_{IH}	INPUT "HIGH" VOLTAGE	2.0		V	$V_{CC} = 5.0V$, $V_{SS} = 19V$, $V_{BB} = 23V$
C_{IN}	INPUT CAPACITANCE		8(Typical)	pF	$V_{BIAS} = 2.0V$, $V_{CC} = 0V$

D.C. Characteristics (Continued) $T_A = 0°C$ to $+55°C$, $V_{CC} = 5V \pm 5\%$, $V_{SS} = 19V \pm 5\%$, $V_{BB} - V_{SS} = 3.0V$ to $4.0V$

POWER SUPPLY CURRENT DRAIN:
All Outputs "Low"

Symbol	Parameter	Min.	Max.	Unit	Conditions
I_{CC}	Current from V_{CC}		83	mA	$V_{CC} = 5.25V$, $V_{SS} = 20V$, $V_{BB} = 24V$
I_{SS}	Current from V_{SS}		250	µA	All Inputs Open
I_{BB}	Current from V_{BB}		25	mA	
P_{TOTAL}	Total Power Dissipation		1040	mW	

All Outputs "High"

Symbol	Parameter	Min.	Max.	Unit	Conditions
I_{CC}	Current from V_{CC}		33	mA	$V_{CC} = 5.25V$, $V_{SS} = 20V$, $V_{BB} = 24V$
I_{SS}	Current from V_{SS}		250	µA	All Inputs Grounded
I_{BB}	Current from V_{BB}		5	mA	
P_{TOTAL}	Total Power Dissipation		297	mW	

Standby Condition with $V_{CC} = 0V$, $V_{SS} = V_{BB}$

Symbol	Parameter	Min.	Max.	Unit	Conditions
I_{CC}	Current from V_{CC}		0	mA	$V_{CC} = 0V$, $V_{SS} = 20V$, $V_{BB} = 20V$
I_{SS}	Current from V_{SS}		500	µA	
I_{BB}	Current from V_{BB}		500	µA	
P_{TOTAL}	Total Power Dissipation		15	mW	

A.C. Characteristics
$T_A = 0°C$ to $55°C$, $V_{CC} = 5V \pm 5\%$, $V_{SS} = 19V \pm 5\%$, $V_{BB} = V_{SS} + 3$ to $4V$, $f = 2$ MHz, 50% Duty Cycle

SYMBOL	TEST	LIMITS (ns)				DELAY DIFFERENTIAL[1]
		$C_L = 100$ pF		$C_L = 200$ pF		$C_L = 200$ pF
		MIN.	MAX.	MIN.	MAX.	MAX.
t_{+-}	INPUT TO OUTPUT DELAY	5	15	5	15	5
t_{-+}	INPUT TO OUTPUT DELAY	5	25	5	25	10
t_r	OUTPUT RISE TIME	5	20	5	30	10
t_f	OUTPUT FALL TIME	5	25	10	35	10
t_D	DELAY + RISE OR FALL TIME	10	35	20	45	10

(1) This is defined as the maximum skew between any output in the same package, eg., all the input to output delays for the t_{-+} parameter are within a maximum of 10 nsec of each other in the same package.

Waveforms

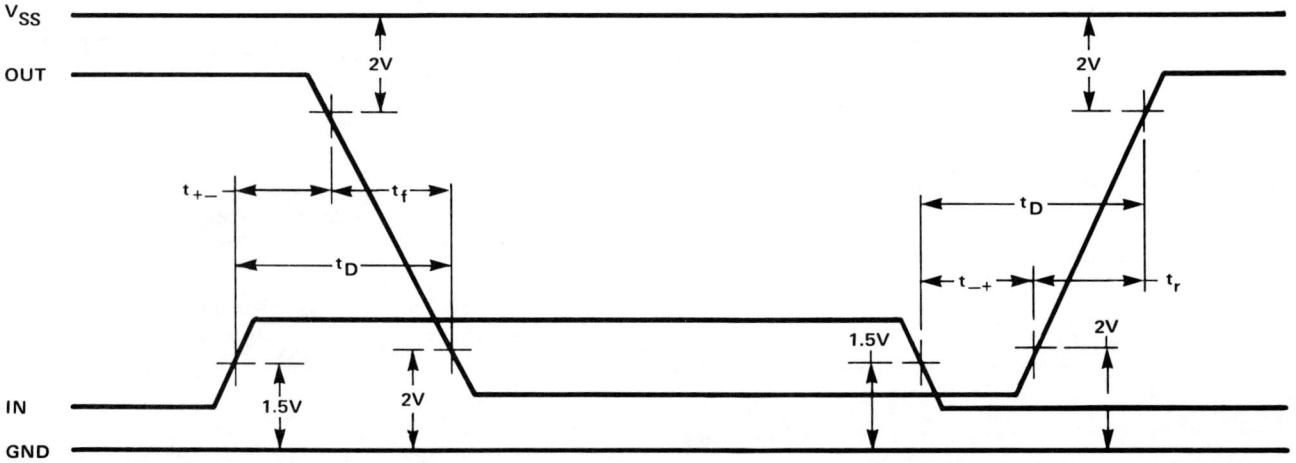

3208A, 3408A
HEX BIPOLAR SENSE AMPLIFIERS FOR MOS CIRCUITS
3208A HEX SENSE AMPLIFIER
3408A HEX SENSE AMPLIFIER WITH LATCHES

- High Speed—20 nsec. max.
- Wire-OR Capability—
  Open Collector Output .. 3208A
  Three-State Output 3408A
- Single 5 V Power Supply
- Input Level Compatible with 1103 Output
- Two Enable Inputs
- Minimum Line Reflection .... Low Voltage Diode Input Clamp
- Plastic 18 Pin Dual In-Line Package
- Schottky TTL

The Intel 3208A is a high speed hex sense amplifier designed to sense the output signals of the 1103 memory. The device features two separate enable inputs each controlling the output state of three sense amplifiers, and a common voltage reference input. OR-tie capability is available with the 3208A open collector TTL compatible output.

The 3408A is a hex sense amplifier with a latch circuit connected to each amplifier. The sensed data may be stored in the latches through application of a write pulse. The 3408A has three-state TTL outputs, hence in the non-enabled state the outputs float allowing wire-OR memory expansion. The latches may be bypassed by grounding the write input pin. Under this condition, the 3408A functions as a hex sense amplifier.

The 3208A and 3408A operate from a single +5 volt power supply. Device performance is specified over the complete ambient temperature range of 0°C to 70°C and over a V_{CC} supply voltage range of 5 volts ±5%. The 3208A and 3408A are packaged in an 18 pin plastic dual in-line package.

PIN CONFIGURATIONS

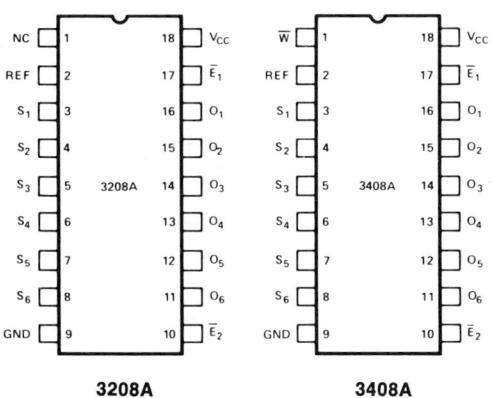

BLOCK DIAGRAMS

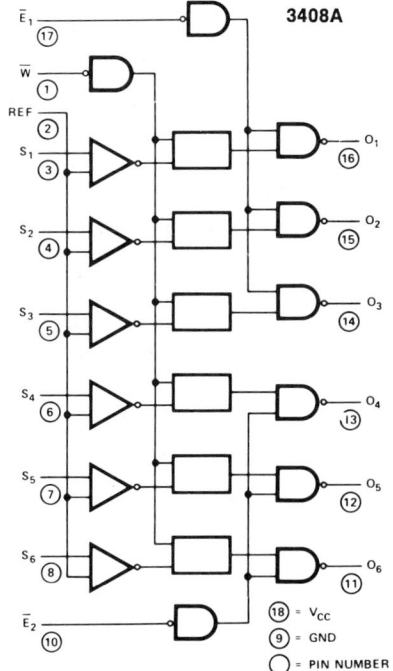

PIN NAMES

$S_1, S_2, S_3, S_4, S_5, S_6$	SENSE AMP INPUTS
$\overline{E}_1, \overline{E}_2$	ENABLE INPUTS
REF	REFERENCE INPUT
$O_1, O_2, O_3, O_4, O_5, O_6$	OUTPUTS (Non-inverting)
$\overline{W}$	WRITE INPUT (3408A only)

445

Absolute Maximum Ratings*

Temperature Under Bias	−55°C to +125°C
Storage Temperature	−65°C to +160°C
All Outputs or Supply Voltage	−0.5 to +7 Volts
All TTL Input Voltages	−1 to +5.5 Volts
All Sense Input Voltages	−1 to +1 Volt
Output Currents Total	300 mA
Input Current	125 mA

*COMMENT:

Stresses above those listed under "Absolute Maximum Rating" may cause permanent damage to the device. This is a stress rating only and functional operation of the device at this or at any other condition above those indicated in the operational sections of this specification is not implied.

D. C. Characteristics for 3208A T_A = 0°C to 70°C, V_{CC} = 5V ±5%

SYMBOL	PARAMETER	LIMITS MIN.	LIMITS TYP.	LIMITS MAX.	UNIT	TEST CONDITIONS
I_{FE}	INPUT LOAD CURRENT ON ENABLE INPUT			−0.25	mA	V_{CC} = 5.25V, V_F = 0.45V
I_{RE}	INPUT LEAKAGE CURRENT ON ENABLE INPUT			20	μA	V_{CC} = 4.75V, V_R = 5.25V
V_{IH}	INPUT "HIGH" VOLTAGE ON ENABLE INPUT	2.0			V	V_{CC} = 5.0V
V_{IL}	INPUT "LOW" VOLTAGE ON ENABLE INPUT			0.85	V	V_{CC} = 5.0V
V_{OL}	OUTPUT "LOW" VOLTAGE			0.45	V	V_{CC} = 4.75V, I_{OL} = 10mA
I_{CEX}	OUTPUT LEAKAGE CURRENT			100	μA	V_{CC} = 5.25V, V_{CEX} = 5.25V
I_{REF}	INPUT CURRENT ON REFERENCE INPUT			−150	μA	V_{CC} = 5.25V, V_{REF} = 100mV
I_S	INPUT CURRENT ON SENSE AMP INPUT			−25	μA	V_{CC} = 5.25V, V_S = 100mV
V_{SH}	INPUT "HIGH" VOLTAGE FOR SENSE AMP INPUT	V_{REF}			mV	V_{CC} = 4.75 to 5.25V, V_{REF} = 100 to 200mV
V_{SL}	INPUT "LOW" VOLTAGE FOR SENSE AMP INPUT			V_{REF} −50	mV	V_{CC} = 4.75 to 5.25V, V_{REF} = 100 to 200mV
V_{REF}	OPERATING RANGE OF REFERENCE VOLTAGE	100		200	mV	V_{CC} = 4.75 to 5.25V
I_{CC}	POWER SUPPLY CURRENT			120	mA	V_{CC} = 5.25V
V_C	INPUT CLAMP VOLTAGE ON ALL INPUTS			−1.0	V	V_{CC} = 4.75V, I_C = −5.0mA
V_{SD}	SENSE INPUT CLAMP DIODE VOLTAGE			1.0	V	V_{CC} = 5.0V, I_D = 5.0mA

3208A TRUTH TABLE

INPUTS Sense Amp	Enable	OUTPUT
<V_{REF} −50mV	L	L
>V_{REF}	L	H
X	H	H

X = Don't care

3208A, 3408A HEX SENSE AMPLIFIERS

D. C. Characteristics for 3408A $T_A = 0°C$ to $+70°C$, $V_{CC} = 5V \pm 5\%$

SYMBOL	PARAMETER	LIMITS MIN.	LIMITS TYP.	LIMITS MAX.	UNIT	TEST CONDITIONS
I_{FE}	INPUT LOAD CURRENT ON ENABLE INPUT			−0.25	mA	$V_{CC} = 5.25V$, $V_F = 0.45V$
I_{RE}	INPUT LEAKAGE CURRENT ON ENABLE INPUT			20	μA	$V_{CC} = 4.75V$, $V_R = 5.25V$
I_{FW}	INPUT LOAD CURRENT ON WRITE INPUT			−0.25	mA	$V_{CC} = 5.25V$, $V_F = 0.45V$
I_{RW}	INPUT LEAKAGE CURRENT ON WRITE INPUT			20	μA	$V_{CC} = 4.75V$, $V_R = 5.25V$
V_{IH}	INPUT "HIGH" VOLTAGE ON ENABLE AND WRITE INPUT	2.0			V	$V_{CC} = 5.0V$
V_{IL}	INPUT "LOW" VOLTAGE ON ENABLE AND WRITE INPUT			0.85	V	$V_{CC} = 5.0V$
V_{OL}	OUTPUT "LOW" VOLTAGE			0.45	V	$V_{CC} = 4.75V$, $I_{OL} = 10mA$
V_{OH}	OUTPUT "HIGH" VOLTAGE	2.4			V	$V_{CC} = 4.75V$, $I_{OH} = -1.5mA$
$|I_O|$	OUTPUT LEAKAGE CURRENT FOR HIGH IMPEDANCE STATE			100	μA	$V_{CC} = 5.25V$, $V_O = 0.45V/5.25V$
I_{SC}	OUTPUT SHORT CIRCUIT CURRENT	−40		−100	mA	$V_{CC} = 5.0V$, $V_O = 0V$
I_{REF}	INPUT CURRENT ON REFERENCE INPUT			−150	μA	$V_{CC} = 5.25V$, $V_{REF} = 100mV$
I_S	INPUT CURRENT ON SENSE INPUT			−25	μA	$V_{CC} = 5.25V$, $V_S = 100mV$
V_{SH}	INPUT "HIGH" VOLTAGE FOR SENSE AMP INPUT	V_{REF}			mV	$V_{CC} = 4.75$ to $5.25V$, $V_{REF} = 100$ to $200mV$
V_{SL}	INPUT "LOW" VOLTAGE FOR SENSE AMP INPUT			$V_{REF} -60$	mV	$V_{CC} = 4.75$ to $5.25V$, $V_{REF} = 100$ to $200mV$
V_{REF}	OPERATING RANGE OF REFERENCE VOLTAGE	100		200	mV	$V_{CC} = 4.75$ to $5.25V$
I_{CC}	POWER SUPPLY CURRENT			125	mA	$V_{CC} = 5.25V$
V_C	INPUT CLAMP VOLTAGE ON ALL INPUTS			−1.0	V	$V_{CC} = 4.75V$, $I_C = -5.0V$
V_{SD}	SENSE INPUT CLAMP DIODE VOLTAGE			1.0	V	$V_{CC} = 5.0V$, $I_D = 5.0mA$

3408A TRUTH TABLE

INPUTS Sense Amp	Enable	Write	OUTPUT
< V_{REF} −60mV	L	L	L
> V_{REF}	L	L	H
X	L	H	Previous Data Stored
X	H	X	High Z*

X = Don't care
*The output of the 3408A is three-state, hence when not enabled the output is a high impedance.

Typical D. C. Characteristics for 3208A/3408A

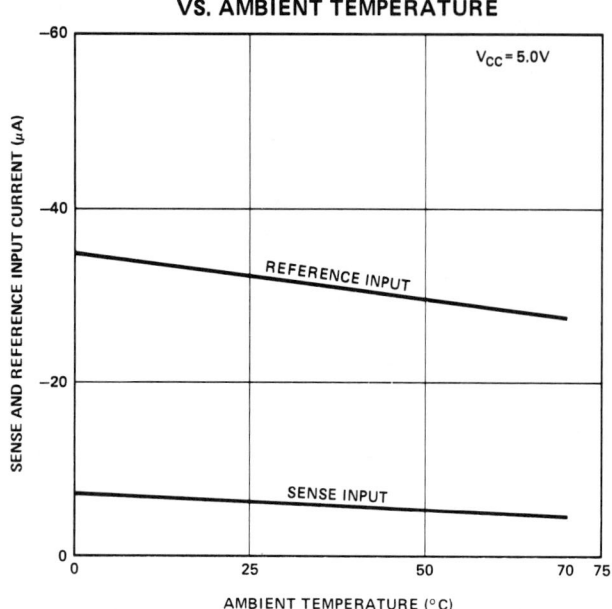

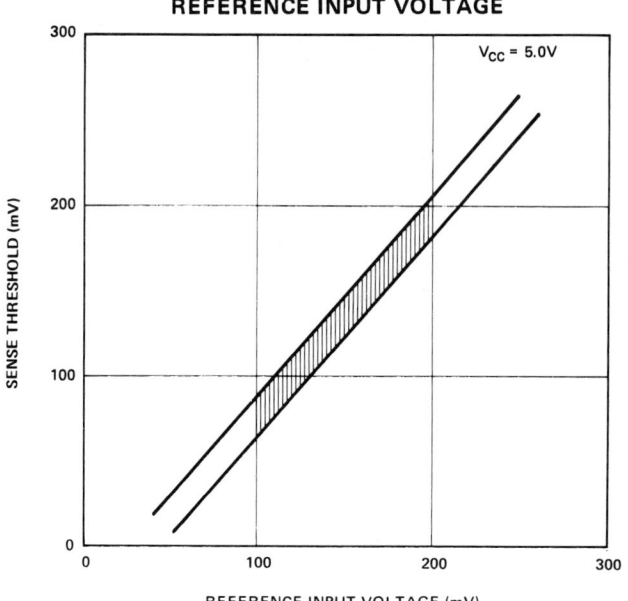

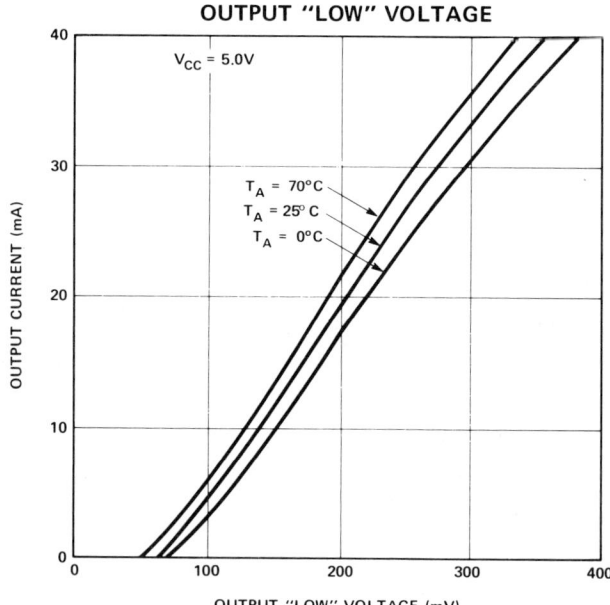

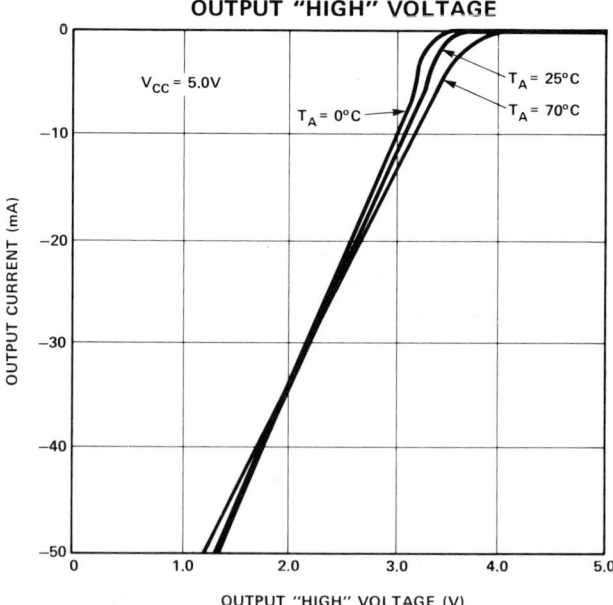

3208A, 3408A, HEX SENSE AMPLIFIERS

A.C. Characteristics $T_A = 0°C$ to $70°C$, $V_{CC} = 5V \pm 5\%$

3208A

SYMBOL	PARAMETER	LIMITS MIN.	LIMITS TYP.	LIMITS MAX.	UNIT	TEST CONDITIONS
t_{S-}	SENSE AMP INPUT TO OUTPUT DELAY			20	ns	D.C. LOAD = 10mA, C_L = 30pF
t_{E-}	ENABLE INPUT TO OUTPUT DELAY			20	ns	D.C. LOAD = 10mA, C_L = 30pF
t_{E+}				25		

3408A

SYMBOL	PARAMETER	LIMITS MIN.	LIMITS TYP.	LIMITS MAX.	UNIT	TEST CONDITIONS
t_{WP}	WRITE PULSE WIDTH	30			ns	D.C. LOAD = 10mA, C_L = 30pF
t_{S-}	SENSE AMP INPUT TO OUTPUT DELAY			25	ns	D.C. LOAD = 10mA, C_L = 30pF
t_{E-}	ENABLE INPUT TO OUTPUT DELAY, LATCH STORES "LOW"			20	ns	D.C. LOAD = 10mA, C_L = 30pF
t_{E+}	ENABLE INPUT TO OUTPUT DELAY, LATCH STORES "HIGH"			25	ns	D.C. LOAD = 10mA, C_L = 30 pF

Capacitance[1] $T_A = 25°C$, $f = 1\,MHz$

SYMBOL	TEST	LIMITS TYP.	LIMITS MAX.
C_O	$V_{CC} = 0V$, $V_{BIAS} = 2.0V$	8	12
C_{INE}	ENABLE INPUT $V_{CC} = 0V$, $V_{BIAS} = 2.0V$	6	10
C_{INS}	SENSE INPUT $V_{CC} = 0V$, $V_{BIAS} = 0V$	6	10

(1) This parameter is periodically sampled and is not 100% tested.

Switching Characteristics
CONDITIONS OF TEST
- Input Pulse amplitude: 2.5V for all TTL compatible inputs and 2.5V through a resistor network as shown below for sense input.
- Input Pulse rise and fall times: 5ns.
- Speed measurements are made at 1.5V for all TTL compatible inputs and outputs, and for sense input, see network and waveforms below. V_{REF} is set at 150 mV.

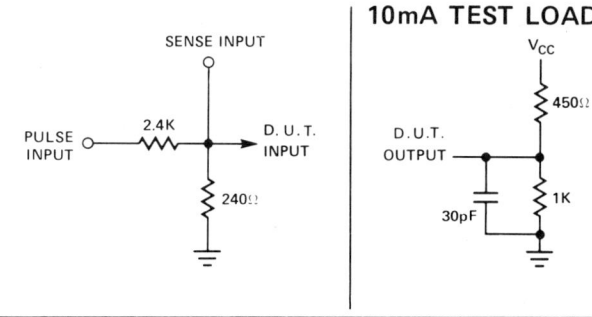

Waveforms
3208A/3408A

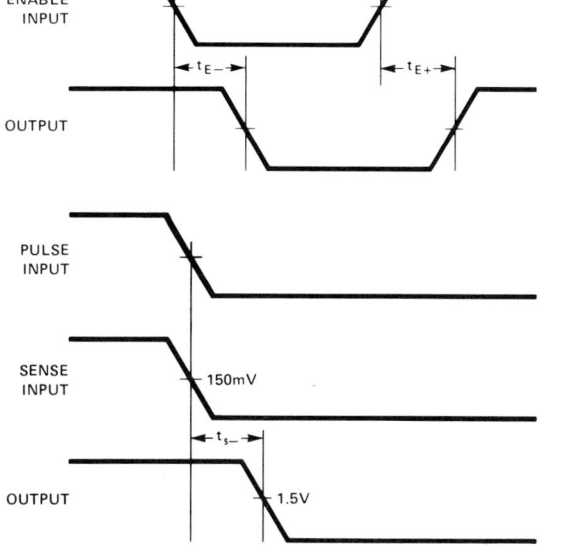

3408A ONLY

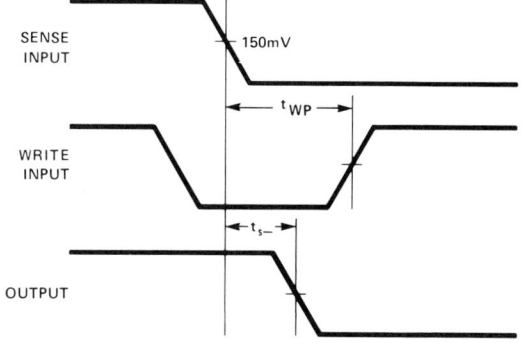

Typical A. C. Characteristics

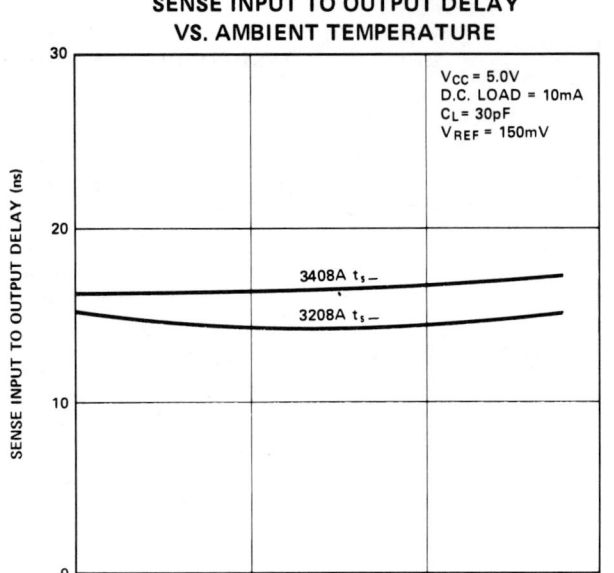

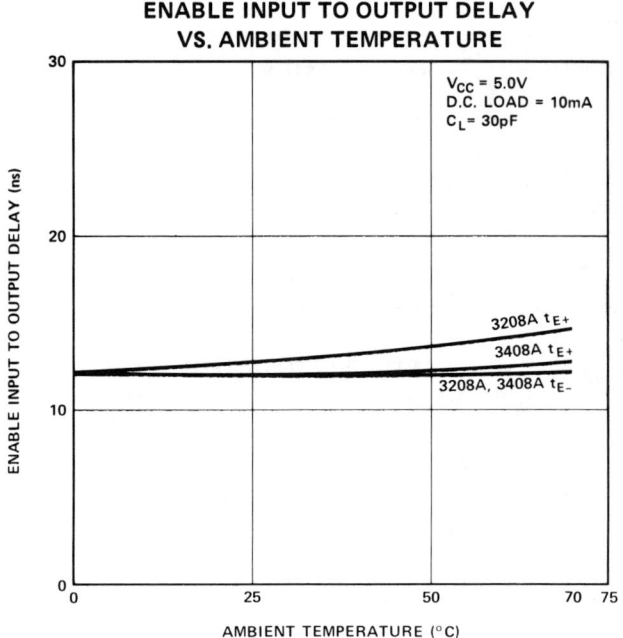

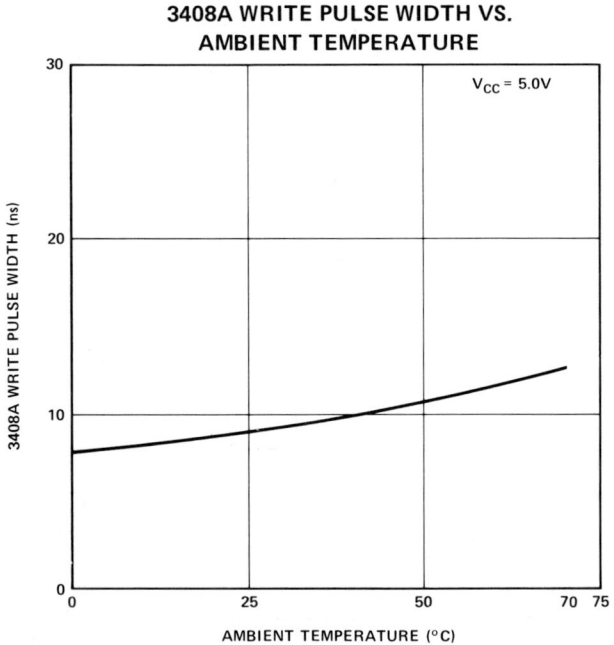

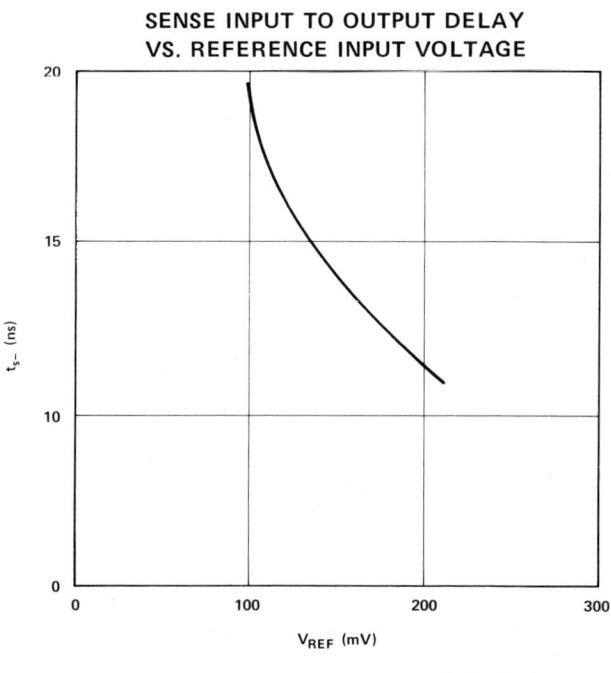

3222

REFRESH CONTROLLER FOR 4K DYNAMIC RANDOM ACCESS MEMORIES

- **Ideal for use in 2107A, 2107B Systems**
- **Simplifies System Design**
- **Reduces Package Count**
- **Standard 22-Pin DIP**
- **Adjustable Refresh Timing Oscillator**
- **6-Bit Address Multiplexer**
- **6-Bit Refresh Address Counter**
- **Refresh Cycle Controller**

The Intel® 3222 is a refresh controller for dynamic RAMs requiring refresh of up to 6 input addresses (or 4K bits for 64 x 64 organization). The device contains an accurate refresh timer (whose frequency can be set by an external resistor and capacitor), plus all necessary control and I/O circuitry to provide for the refresh requirements of dynamic RAMs. The chip's high performance makes it especially suitable for use with high speed N-channel RAMs like the Intel® 2107B. The 3222 is well suited for asynchronous dynamic memory systems.

The 3222 operates from a single +5 volt power supply and is specified for operation over a 0°C to 75°C ambient temperature range. It is fabricated by means of Intel's highly reliable Schottky bipolar process.

PIN CONFIGURATION

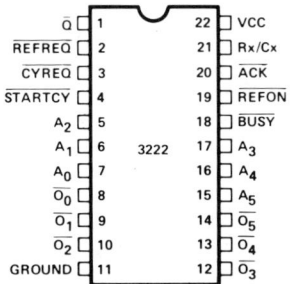

BLOCK DIAGRAM

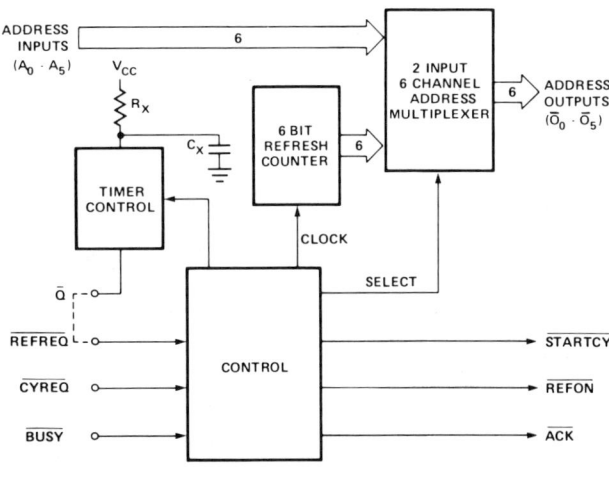

PIN NAMES

$A_0 \cdot A_5$	ADDRESS INPUTS	$\overline{O}_0 \cdot \overline{O}_5$	ADDRESS OUTPUTS
$\overline{ACK}$	ACKNOWLEDGE OUTPUT	$\overline{Q}$	INTERNAL REFRESH REQUEST LATCH OUTPUT
$\overline{BUSY}$	BUSY INPUT	$\overline{REFON}$	REFRESH ON OUTPUT
$\overline{CYREQ}$	CYCLE REQUEST INPUT	$\overline{REFREQ}$	REFRESH REQUEST INPUT
		RxCx	RC TIE POINT
		$\overline{STARTCY}$	START CYCLE OUTPUT
		V_{CC}	+5V SUPPLY

MEMORY SUPPORT CIRCUITS FOR DYNAMIC RAMS

Absolute Maximum Ratings*

Temperature Under Bias . . . . −65° to +125°C
Storage Temperature . . . . −65° to +160°C
All Input, Output or Supply Voltages . . . . −0.5V to +7V
All Input Voltages . . . . −1.0V to +5.5V
Output Currents . . . . 100 mA
Power Dissipation . . . . 1 W

*COMMENT

Stresses above those listed under "Absolute Maximum Rating" may cause permanent damage to the device. This is a stress rating only and functional operation of the device at these or at any other condition above those indicated in the operational sections of this specification is not implied.

D.C. Characteristics All Limits Apply for V_{CC} = +5.0V ±5%, T_A = 0°C to +75°C.

Symbol	Parameter	Min.	Typ.[1]	Max.	Unit	Test Conditions
I_{FB}	Input Load Current $\overline{BUSY}$		0.40	1	mA	V_{IN} = 0.45V
I_{FO}	Input Load Current All Other Inputs		0.05	0.25	mA	V_{IN} = 0.45V
I_{RB}	Input Leakage Current $\overline{BUSY}$		<1	50	µA	V_{IN} = V_{CC}
I_{RO}	Input Leakage Current All Other Inputs		<1	20	µA	V_{IN} = 5.25V
V_{CLAMP}	Input Clamp Voltage		−0.76	−1	V	I_C = −5.0mA
V_{IL}	Input "Low" Voltage			0.8	V	
V_{IH}	Input "High" Voltage	2.0			V	
I_{CC}	Power Supply Current		91	120	mA	V_{CC} = 5.25V
I_{SC}	Output High Short Circuit Current		−48	−70	mA	V_{OUT} = 0V, V_{CC} = 5.25V
V_{OL}	Output Low Voltage		0.32	0.45	V	I_{OL} = 5mA
V_{OH}	Output High Voltage ($\overline{O}_0$-$\overline{O}_5$)	2.6	3.1		V	I_{OH} = −1mA, V_{CC} = 4.75V
V_{OH1}	Output High Voltage (All Other Outputs)	2.4	3.0		V	I_{OH} = −1mA, V_{CC} = 4.75V

Note 1: Typical values are for T_A = 25°C and nominal power supply voltages.

Capacitance[2], T_A = 25°C

Symbol	Test	Typ.	Max.	Conditions
C_{IN} (Address)	Input Capacitance	5	10	V_{bias} = 2.0V
C_{IN} ($\overline{CYREQ}$)	Input Capacitance	6	10	V_{CC} = 0V
C_{IN} ($\overline{BUSY}$)	Input Capacitance	20	30	f = 1 MHz

Note 2: This parameter is periodically sampled and is not 100% tested.

3222 REFRESH CONTROLLER

A.C. Characteristics
All Limits Apply for $V_{CC} = +5.0V \pm 5\%$, $T_A = 0°C$ to $+75°C$. Load = 1 TTL, $C_L = 15pF$.
Conditions of Test: Input pulse amplitude: 3V, Input rise and fall times: 5ns between 1V and 2V. Measurements are made at 1.5V.

Symbol	Parameter	Min.	Typ.[1]	Max.	Unit	Conditions
t_{AA}	Address In to Address Out		7	12	ns	$\overline{BUSY} = V_{IH}$
t_{BAM}	$\overline{BUSY}$ In to Address Out		21	28	ns	
t_{BAR}	$\overline{BUSY}$ In to Counter Out		18	27	ns	
t_{BK}	$\overline{BUSY}$ In to $\overline{ACK}$ Out		14	20	ns	$\overline{REFREQ} = V_{IH}$, $\overline{CYREQ} = V_{IL}$
t_{BR}	$\overline{BUSY}$ In to $\overline{REFON}$ Out		15	24	ns	
t_{BS}	$\overline{BUSY}$ In to $\overline{STARTCY}$ Out	4	7	14	ns	$\overline{CYREQ} = V_{IL}$
t_{HOLD}	$\overline{BUSY}$ Hold Time	50			ns	External Delay between $\overline{STARTCY}$ and $\overline{BUSY}$
t_{RH}	$\overline{CYREQ}$ or $\overline{REFREQ}$ Hold Time	0			ns	External Delay after $\overline{BUSY}$
t_{RR}	$\overline{REFREQ}$ to $\overline{REFON}$		18	26	ns	$\overline{CYREQ}$ and $\overline{BUSY} = V_{IH}$, No priority contention between $\overline{REFREQ}$ and $\overline{CYREQ}$
t_{RRC}	$\overline{REFREQ}$ to $\overline{REFON}$		33	45	ns	$\overline{BUSY} = V_{IH}$
t_{RS}	$\overline{CYREQ}$ or $\overline{REFREQ}$ In to $\overline{STARTCY}$ Out	9	14	21	ns	$\overline{BUSY} = V_{IH}$
t_{Setup}	$\overline{BUSY}$ Setup Time	120			ns	$\overline{BUSY} = V_{IL}$ During Refresh

Note 1: Typical values are for $T_A = 25°C$ and nominal power supply voltages.

A. SYSTEM MEMORY CYCLE WITH MEMORY NOT BUSY

B. SYSTEM MEMORY CYCLE WITH MEMORY BUSY (FOLLOWING REFRESH CYCLE)

(Numbers in parentheses are minimum values in ns unless otherwise specified.)

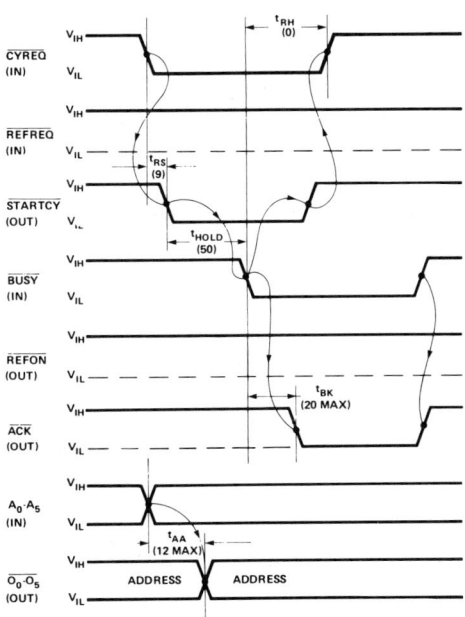
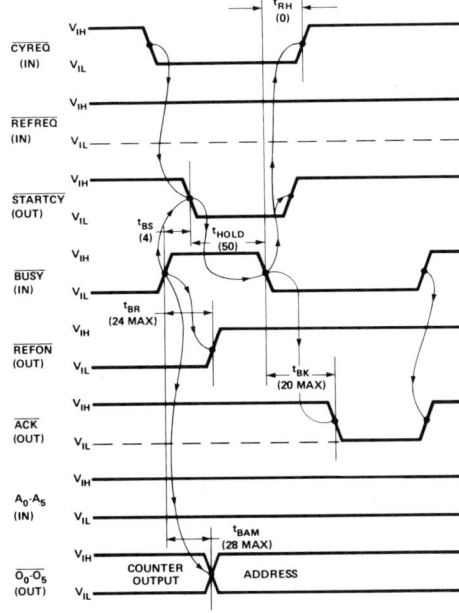

MEMORY SUPPORT CIRCUITS FOR DYNAMIC RAMS

C. REFRESH MEMORY CYCLE WITH MEMORY NOT BUSY

D. REFRESH MEMORY CYCLE WITH MEMORY BUSY (FOLLOWING SYSTEM CYCLE)

(Numbers in parentheses are minimum values in ns unless otherwise specified.)

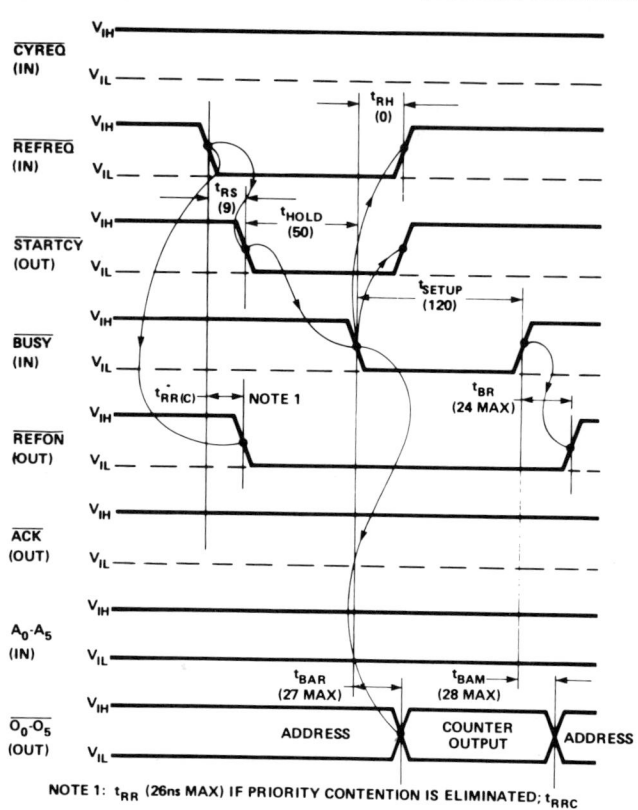

NOTE 1: t_{RR} (26ns MAX) IF PRIORITY CONTENTION IS ELIMINATED; t_{RRC}

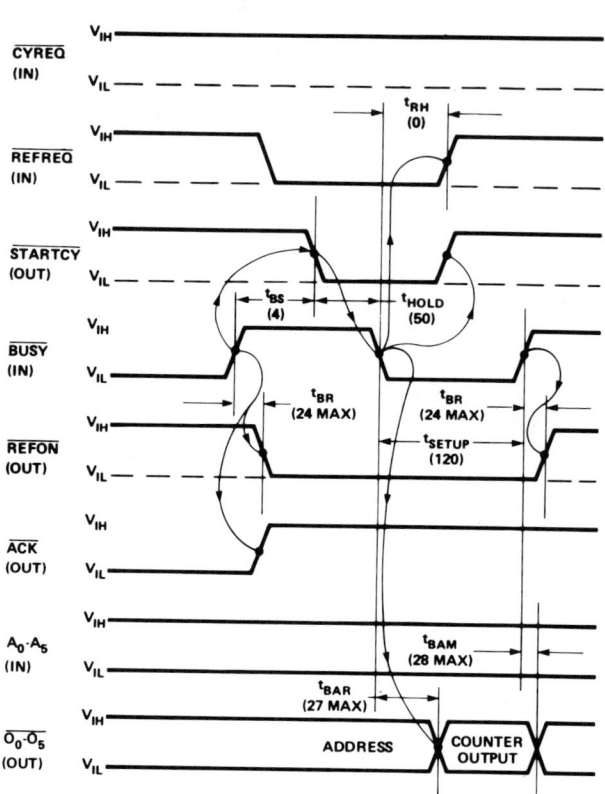

E. TYPICAL APPLICATION OF 3222 REFRESH CONTROLLER IN A 2107B SYSTEM

F. USE OF 3222 FOR REFRESH TIMING AND CONTROL IN A 2104A SYSTEM

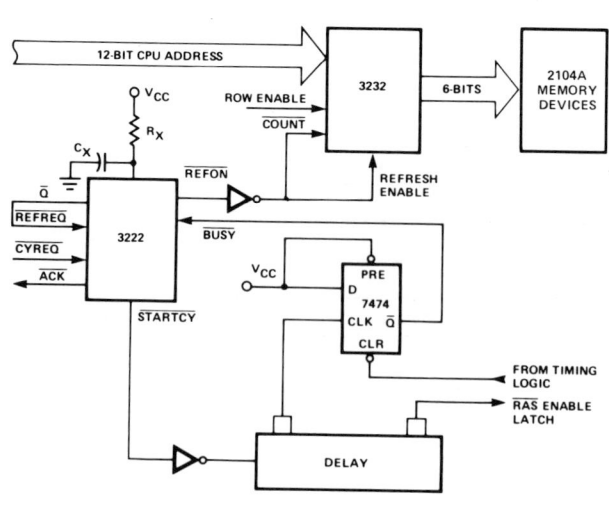

3222 REFRESH CONTROLLER

PIN NAMES AND FUNCTIONS

Pin No.	Pin Name	Function
1	$\overline{Q}$	Output of the internal Refresh Request latch. This pin may be connected to the Refresh Request input ($\overline{REFREQ}$) directly for asynchronous sequential mode refresh or indirectly through control logic for burst mode or synchronous mode refresh (see text).
2	$\overline{REFREQ}$	Refresh Request input (active when low). The request is honored only if the memory is not presently executing a cycle ($\overline{BUSY}$ high) and if a system cycle request did not occur first.
3	$\overline{CYREQ}$	System Memory Cycle Request input (active when low). The request is honored only if the memory is not presently executing a cycle ($\overline{BUSY}$ high) and if a refresh request did not occur first.
4	$\overline{STARTCY}$	Output signal indicating to external circuitry that a memory cycle (system or refresh) is to begin. See text for timing considerations for a refresh cycle.
5-7 15-17	A_0-A_5	Low order system address inputs. These addresses are multiplexed to the address output pins ($\overline{O}_0$-$\overline{O}_5$) during a system cycle.
8-10	$\overline{O}_0$-$\overline{O}_5$	Low order memory address outputs. During a system cycle these outputs give the low order (A_0-A_5) address of a memory access. During a refresh cycle these outputs give the refresh address (generated internal to the 3222).
11	GROUND	Ground.
18	$\overline{BUSY}$	An externally generated signal which the 3222 monitors to determine memory system status. If $\overline{BUSY}$ is high the memory is not busy and a system or refresh cycle may begin. If $\overline{BUSY}$ is low the memory is being accessed for a data I/O or refresh cycle and no other cycle may begin.
19	$\overline{REFON}$	The 3222 output which when low indicates the memory system is either ready to begin or is in a refresh cycle (Refresh On).
20	$\overline{ACK}$	The 3222 output which when low indicates the memory system is either ready to begin or is in a system cycle (system cycle request accepted and acknowledged).
21	RX/CX	Connection point for the RC network which determines the refresh period for sequential refresh mode. (See Refresh Control section).
22	V_{CC}	+5 volt supply.

FUNCTIONAL DESCRIPTION

The Intel® 3222 performs the four basic functions of a refresh controller by:

1. Providing a refresh timing oscillator.
2. Generating six bit refresh addresses.
3. Multiplexing refresh and system addresses to the six low order address inputs ($\overline{O}_0$-$\overline{O}_5$).
4. Providing control signals for both refresh and memory cycle accesses.

As shown in the pin configuration figure, the 3222 has as inputs the six low order (A_0-A_5) system addresses. These addresses are internally multiplexed with six internally generated refresh addresses. The output of these multiplexers provide the six low order addresses to the memory array.

The block diagram shows the four main circuit categories of the 3222. An explanation of the workings of each of these categories is given in the Device Operation section from a users point of view.

DEVICE OPERATION

Operation of the Intel® 3222 Refresh Controller is most easily explained by considering five conditions presented by the three input control lines Cycle Request ($\overline{CYREQ}$), Refresh Request ($\overline{REFREQ}$), and System Busy ($\overline{BUSY}$). These conditions are:

1. System memory cycle request — memory not busy ($\overline{BUSY}$ = High)
2. System memory cycle request — memory busy ($\overline{BUSY}$ = Low)
3. Refresh cycle request — memory not busy ($\overline{BUSY}$ = High)
4. Refresh cycle request — memory busy ($\overline{BUSY}$ = Low)
5. Simultaneous system memory cycle and refresh cycle requests.

Condition 5 is actually a subset of the four previous conditions and is included for completeness.

As is implied in the five conditions, the response of the 3222 to both refresh and memory cycles is dependent on the state of the $\overline{BUSY}$ input. The $\overline{BUSY}$ signal is generated externally to the 3222 and, when low, defines the time when the memory is performing a cycle (refresh or memory access). It is important to assure that $\overline{BUSY}$ is low for the *entire* memory cycle time. Interference may occur in asynchronous memory systems if the $\overline{BUSY}$ input goes high prematurely. (An asynchronous memory system is one in which the refresh and memory cycle requests occur independent of each other.)

System Memory Cycle Request — Memory Not Busy

This section details operation of the 3222 when the memory is not busy and a request for a system memory cycle is made (See Figure A for timing sequences). The request for a memory cycle is made by the $\overline{CYREQ}$ input going low. The Start Cycle output $\overline{STARTCY}$ goes low at t_{RS} after $\overline{CYREQ}$. STARTCY is used for two purposes:

1. To set the external $\overline{BUSY}$ latch. (See Figure E.)
2. To initiate memory system timing (after appropriate delay).

The required delay time depends on system configuration and associated delay paths for both Chip Enable (2107B input signal) and system addresses.

The low going $\overline{BUSY}$ input causes the internally generated Start Cycle output to go high and the Acknowledge output $\overline{ACK}$ to go low (after t_{BK} time). The Acknowledge output confirms that the requested system memory cycle has been accepted by the 3222. Note that the cycle request input may be returned to the high state when the $\overline{BUSY}$ input goes low. However, at the designer's discretion, the cycle request line may remain low until "just prior" to $\overline{BUSY}$ returning high. (If $\overline{BUSY}$ goes high before $\overline{CYREQ}$ goes high, another memory access may inadvertently be started.)

When the memory is not busy and a cycle request has been made, the low order system address delay through the 3222 is t_{AA} nsec. When the 3222 is not busy, the low order system addresses (A_0-A_5) are gated through to the output ($\overline{O}_0$-$\overline{O}_5$) independent of any other input.

System Memory Cycle Request — Memory Busy

The major differences between a system memory cycle request when the system is busy and when it is not busy (as previously described) are:

1. The Start Cycle output $\overline{STARTCY}$ does not go low until t_{BS} after the rising edge of the $\overline{BUSY}$ input. (Even though the $\overline{CYREQ}$ input is low.)
2. Output addresses $\overline{O}_0$-$\overline{O}_5$ change at or before t_{AA} time if the previous cycle was a system cycle request and change at or before t_{BAM} if the previous cycle was a Refresh Cycle request. (Note that the longer delay is after a refresh cycle.) See Figure B for definition of terms.

Note that for a system memory cycle following a refresh cycle, the refresh on output $\overline{REFON}$ goes high at or before t_{BR} relative to $\overline{BUSY}$ going high. Since the Acknowledge output $\overline{ACK}$ can not go low until after t_{HOLD} there is no ambiguity between $\overline{REFON}$ and $\overline{ACK}$. The memory is always defined as being in a refresh cycle, system cycle or no cycle.

Refresh Cycle — Memory Not Busy

Operation of the 3222 for a refresh request with the memory not busy (see Figure C) is similar to a system cycle request under the same condition. A refresh cycle is initiated by the Refresh Request input ($\overline{REFREQ}$) going low. This low going input causes both the Start Cycle output, $\overline{STARTCY}$, and Refresh On output, $\overline{REFON}$, to go low at t and t_{RRC} (or t_{RR}) time respectively. The low going edge of $\overline{STARTCY}$ is used to set the external $\overline{BUSY}$ latch low. As in the previous two cases, the $\overline{BUSY}$ input must remain low for the entire cycle required by the memory. As in the previous two cases, the low going $\overline{BUSY}$ drives the $\overline{STARTCY}$ output high.

Refresh Cycle — Memory Busy

For this condition, it is assumed that the previous cycle was a system access cycle. Timing conditions for this operation are shown in Figure D. Here, the $\overline{STARTCY}$ input goes low t_{BS} after $\overline{BUSY}$ returns high from the previous cycle. As before, $\overline{REFON}$ goes low t_{BR} after $\overline{BUSY}$ goes high. After t_{HOLD}, relative to $\overline{STARTCY}$, $\overline{BUSY}$ again goes low and places the low order refresh addresses on the address outputs ($\overline{O}_0$-$\overline{O}_5$) after t_{BAR} time. Internal refresh timing is performed in a manner identical to that described in Refresh Cycle-Memory Not Busy section.

Simultaneous Refresh and Memory System Cycle Request

The simultaneous request for a refresh and memory system access is almost a certainty in asynchronous systems. It is, therefore, necessary to have circuitry in any refresh controller capable of resolving the attendent ambiguity with minimum additional delay. The Intel® 3222 Refresh Controller has just such a circuit. (All timing parameters specified for asynchronous operation assume that a refresh and memory system request can occur at the same time.) A latch internal to the 3222 decides which signal ($\overline{CYREQ}$ or $\overline{REFREQ}$) it will accept if both occur simultaneously, and conditions the other control circuits appropriately. If a refresh cycle was accepted, $\overline{REFON}$ will go low at the appropriate time. If a memory system access was accepted then $\overline{ACK}$ will go low at the appropriate time.

Refresh Control

The 3222 controls both burst and distributed refresh modes. The burst refresh mode requires that $\overline{REFREQ}$ be generated externally to the 3222 since refresh is completed in 64 consecutive cycles every 2ms. A system requiring distributed refresh timing, however can be controlled either by the 3222 or by external circuitry. If refresh timing is to be controlled by the 3222 the output $\overline{Q}$ is tied to the $\overline{REFREQ}$ input. Timing is controlled by an oscillator internal to the 3222. The desired refresh timing interval is determined by:

1. $t_{REF} = .63\, R_x C_x$
   $\phantom{t_{REF}=}r$

   Where:

   t_{REF} = the total time between refreshes (e.g. 2msec) in msec.

   r = the number of rows to be refreshed on the memory device (for the 2107B r = 64).

   R_x = external timing resistance in $K\Omega$ (3K to 10K)

   C_x = external timing capacitance in μf. (0.005μf to 0.02μf)

The 3222's oscillator stability is guaranteed to be ±2% for a given part and ±6% from part to part, both over the ranges $0°C \leq T_A \leq 75°C$ and $V_{CC} = 5.0V$ ±5%.

Figure F shows how the 3222 may be used to control refresh in a 2104A system.

3232

ADDRESS MULTIPLEXER AND REFRESH COUNTER FOR 4K DYNAMIC RAMs

- **Ideal For 2104A**
- **Simplifies System Design**
- **Reduces Package Count**
- **Standard 24-Pin DIP**
- **Address Input to Output Delay: 9ns Maximum Driving 15pF, 25ns Maximum Driving 250pF**
- **Suitable For Either Distributed Or Burst Refresh**
- **Single Power Supply: +5 Volts ±10%**

The Intel® 3232 contains an address multiplexer and refresh counter for multiplexed address dynamic RAMs requiring refresh of up to 6 input addresses (or 4K bits for 64 x 64 organization). It multiplexes twelve bits of system supplied address to six output address pins. The device also contains a 6 bit refresh counter which is externally controlled so that either distributed or burst refresh may be used. The high performance of the 3232 makes it especially suitable for use with high speed N-channel RAMs like the 2104A.

The 3232 operates from a single +5 volt power supply and is specified for operation over a 0 to +75°C ambient temperature range.

PIN CONFIGURATION

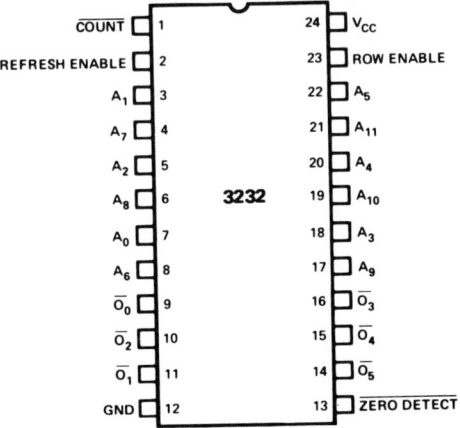

NOTE: A_0 THROUGH A_5 ARE ROW ADDRESSES.
A_6 THROUGH A_{11} ARE COLUMN ADDRESSES.

LOGIC DIAGRAM

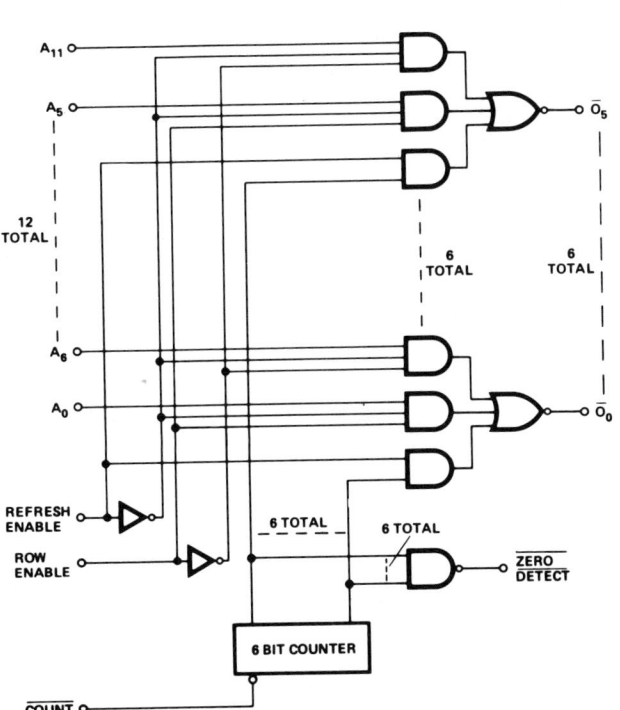

TRUTH TABLE AND DEFINITIONS:

REFRESH ENABLE	ROW ENABLE	OUTPUT
H	X	REFRESH ADDRESS (FROM INTERNAL COUNTER)
L	H	ROW ADDRESS (A_0 THROUGH A_5)
L	L	COLUMN ADDRESS (A_6 THROUGH A_{11})

$\overline{COUNT}$ — ADVANCES INTERNAL REFRESH COUNTER.
$\overline{ZERO\ DETECT}$ — INDICATES A ZERO IN THE REFRESH ADDRESS (USED IN BURST REFRESH MODE).

Absolute Maximum Ratings*

Temperature Under Bias	−65° to +125°C
Storage Temperature	−65° to +160°C
All Input, Output, or Supply Voltages	−0.5V to +7 Volts
Output Currents	100mA
Power Dissipation	1W

*COMMENT:
Stresses above those listed under "Absolute Maximum Ratings" may cause permanent damage to the device. This is a stress rating only and functional operation of the device at these or any other conditions above those indicated in the operational sections of this specification is not implied. Exposure to absolute maximum rating conditions for extended periods may affect device reliability.

D.C. and Operating Characteristics

All Limits Apply for $V_{CC} = 5.0V \pm 10\%$, $T_A = 0°C$ to $+75°C$

SYMBOL	PARAMETER	MIN.	TYP.(1)	MAX.	UNIT	TEST CONDITIONS
I_F	Input Load Current		−0.04	−0.25	mA	$V_{IN} = 0.45V$
I_R	Input Leakage Current		0	10	μA	$V_{IN} = 5.5V$
V_{IH}	Input High Voltage	2.0			V	
V_{IL}	Input Low Voltage			0.8	V	
V_{OL}	Output Low Voltage		0.25	0.40	V	$I_{OL} = 5mA$
V_{OH}	Output High Voltage ($\overline{O}_0$-$\overline{O}_5$)	2.8	4.0		V	$I_{OH} = -1mA$
V_{OH1}	Output High Voltage ($\overline{\text{Zero Detect}}$)	2.4	3.3		V	$I_{OH} = -1mA$
I_{CC}	Power Supply Current		100	150	mA	$V_{CC} = 5.5V$

Note 1. Typical values are for $T_A = 25°C$ and $V_{CC} = 5.0V$.

3232 Address Multiplexer for 4K RAMs

A.C. Characteristics

All Limits Apply for V_{CC} = +5.0V ±10%, T_A = 0°C to 75°C, Load = 1 TTL, C_L = 250pF, Unless Otherwise Specified.

SYMBOL	PARAMETER	MIN.	TYP.[1]	MAX.	UNIT	CONDITIONS
t_{AO}	Address Input to Output Delay		6	9	ns	Refresh Enable = Low [1] [2]
t_{AOI}	Address Input to Output Delay		16	25	ns	Refresh Enable = Low
t_{OO}	Row Enable to Output Delay	7	12	27	ns	Refresh Enable = Low [1] [2]
t_{OOI}	Row Enable to Output Delay	12	28	41	ns	Refresh Enable = Low
t_{EO}	Refresh Enable to Output Delay	7	14	27	ns	Note 1, 2
t_{EOI}	Refresh Enable to Output Delay	12	30	45	ns	
t_{CO}	Count to Output	15	40	60	ns	Refresh Enable = High [1] [2]
t_{COI}	Count to Output	20	55	80	ns	Refresh Enable = High
f_C	Counting Frequency	5			MHz	
t_{CPW}	Count Pulse Width	35			ns	
t_{CZ}	Count to Zero Detect	15		70	ns	Note 2

Note 1: V_{CC} = 5.0V, T_A = 25°C
2: C_L = 15pF

A.C. TIMING WAVEFORMS (Typically used with 2104A)

NORMAL CYCLE

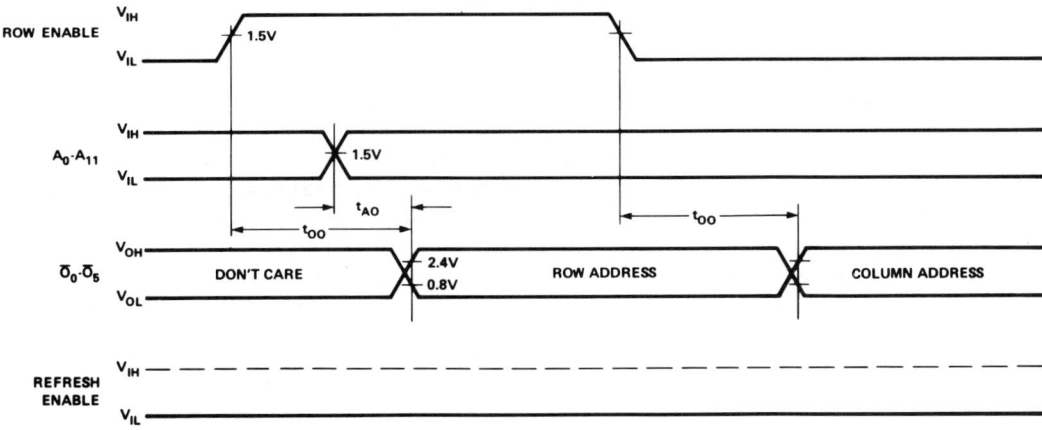

REFRESH CYCLE

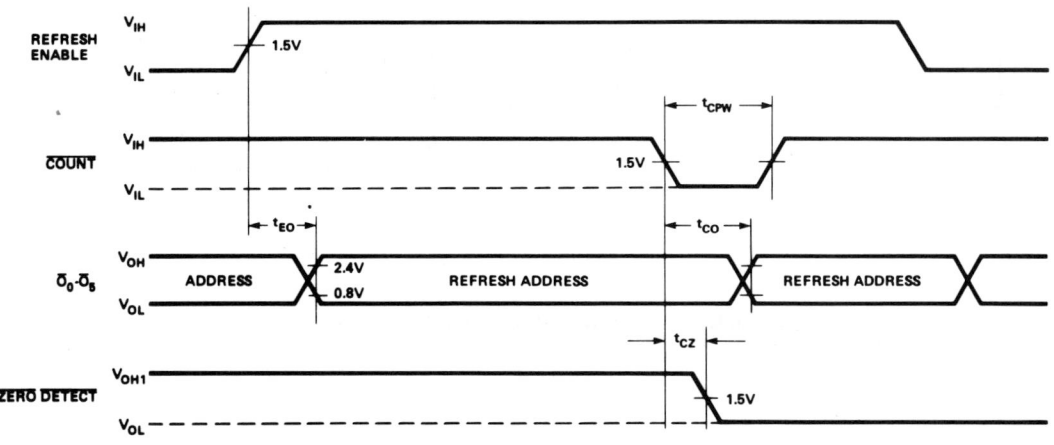

PIN NAMES AND FUNCTIONS

Pin. No.	Pin Name	Function
1	Count Input	Active low input increments internal six bit counter by one for each count pulse in.
2	Refresh Enable Input	Active high input which determines whether the 3232 is in refresh mode (H) or address enable (L).
7,3,5,18, 20,22	A_0-A_5 Inputs	Row Address inputs.
8,4,6,17, 19,21	A_6-A_{11} Inputs	Column address inputs.
9,11,10, 16,15,14	$\overline{O}_0$-$\overline{O}_5$ Outputs	Address outputs to memories. Inverted with respect to address inputs.
12	GND	Power supply ground.
13	Zero Detect Output	Active low output which senses that all six bits of refresh address in the counter are zero. Can be used in the burst mode to sense refresh completion.
23	Row Enable Input	High input selects row, low input selects column addresses of the driven memories.
24	V_{CC}	+5V power supply input.

DEVICE OPERATION

The Intel® 3232 Address Multiplexer/Refresh Counter performs the following functions:

1. Row, Column and Refresh Address multiplexing
2. Address counting for burst or distributed refresh.

These functions are controlled by two signals: Refresh Enable and Row Enable, both of which are active high TTL inputs. The truth table on page 1 shows the levels required to multiplex to the output:

1. Refresh addresses (from internal counter)
2. Row addresses (A_0 through A_5)
3. Column addresses (A_6 through A_{11})

Burst Refresh Mode

When refresh is requested, the refresh enable input is high. This input is ANDed with the 6 outputs of the internal 6 bit counter. At each Count pulse the counter increments by one, sequencing the outputs ($\overline{O}_0$-$\overline{O}_5$) through all 64 row addresses. When the counter sequences to all zeros, the Zero Detect output goes low signaling the end of the refresh sequence. Due to counter decoding spikes, the Zero Detect output is valid only after t_{CZ} following the low going edge of Count.

Distributed Refresh Mode

In the distributed refresh mode, one row is selected for refresh each ($t_{REFRESH}/n$) time where n = number of rows in the device and $t_{REFRESH}$ is the specified refresh rate for the device. For the 2104A $t_{REFRESH}$ = 2msec and n = 64, therefore one row is refreshed each 31 μsec. Following the refresh cycle at row n_x, the Count input is pulsed, advancing the refresh address by one row so that the next refresh cycle will be performed on row n_{x+1}. The Count input may be pulsed following each refresh cycle or within the refresh cycle after the specified memory device address hold time.

Row and Column Address

All twelve system address lines are applied to the inputs of the 3232. When Refresh Enable is low and Row Enable is high, input addresses A_0-A_5 are gated to the outputs and applied to the driven memories. Conversely, when Row Enable is low (with Refresh Enable still low), input addresses A_6-A_{11} are gated to the outputs and applied to the driven memories.

Figure 1 shows a typical connection between the 3232 and the 2104A 4K dynamic RAM. When the memory devices are driven directly by the 3232, the address applied to the memory devices is the inverse of the address at the 3232 inputs due to the inverted outputs of the 3232. This should be remembered when checking out the memory system.

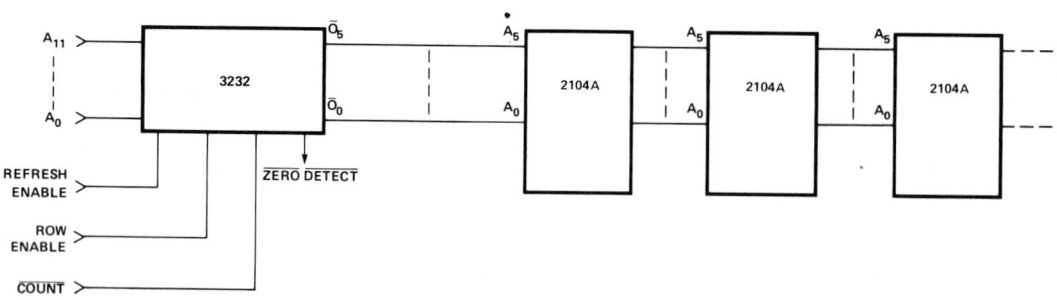

Figure 1. Typical Connection of 3232 and 2104 Memories.

3242
ADDRESS MULTIPLEXER AND REFRESH COUNTER FOR 16K DYNAMIC RAMs

- **Ideal For 2116**
- **Simplifies System Design**
- **Reduces Package Count**
- **Standard 28-Pin DIP**
- **Suitable For Either Distributed Or Burst Refresh**
- **Single Power Supply: +5 Volts ±10%**
- **Address Input to Output Delay: 9ns Driving 15 pF, 25ns Driving 250pF**

The Intel® 3242 is an address multiplexer and refresh counter for multiplexed address dynamic RAMs requiring refresh of 64 or 128 cycles. It multiplexes 14 bits of system supplied address to 7 output address pins. The device also contains a 7 bit refresh counter which is externally controlled so that either distributed or burst refresh may be used. The high performance of the 3242 makes it especially suitable for use with high speed N-channel RAMs like the 2116.

The 3242 operates from a single +5 volt power supply and is specified for operation over a 0 to +75°C ambient temperature range. It is fabricated by means of Intel's highly reliable Schottky bipolar process and is packaged in a hermetically sealed 28 pin Type D package.

PIN CONFIGURATION

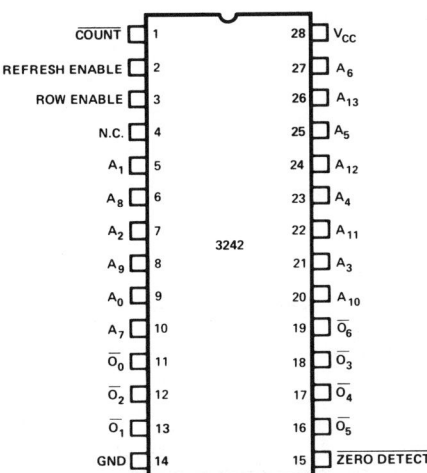

NOTE: A_0 THROUGH A_6 ARE ROW ADDRESSES.
A_7 THROUGH A_{13} ARE COLUMN ADDRESSES.

TRUTH TABLE AND DEFINITIONS:

REFRESH ENABLE	ROW ENABLE	OUTPUT
H	X	REFRESH ADDRESS (FROM INTERNAL COUNTER)
L	H	ROW ADDRESS (A_0 THROUGH A_6)
L	L	COLUMN ADDRESS (A_7 THROUGH A_{13})

$\overline{COUNT}$ – ADVANCES INTERNAL REFRESH COUNTER.
ZERO DETECT – INDICATES ZERO IN THE FIRST 6 SIGNIFICANT REFRESH COUNTER BITS (USED IN BURST REFRESH MODE)

LOGIC DIAGRAM

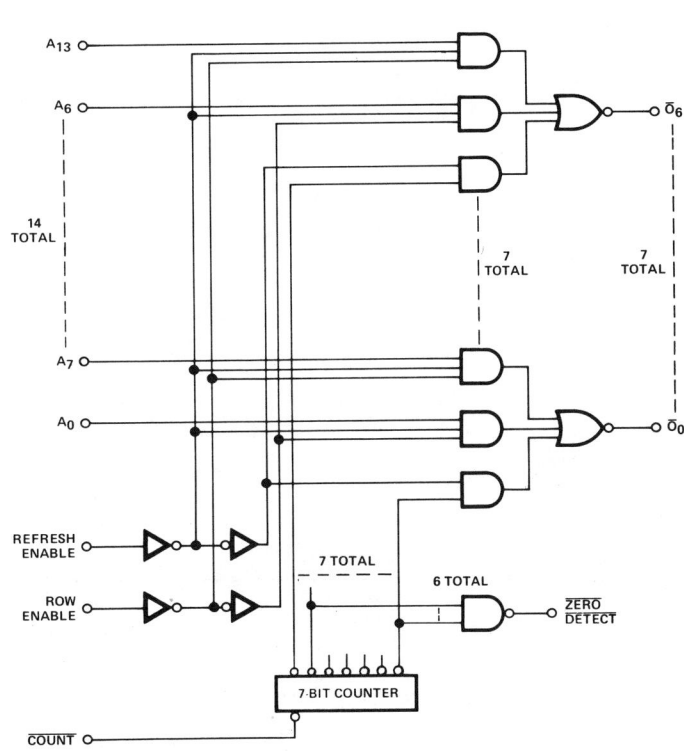

461

MEMORY SUPPORT CIRCUITS FOR DYNAMIC RAMS

A.C. Characteristics

All Limits Apply for V_{CC} = +5.0V ±10%, T_A = 0°C to 75°C, Load = 1 TTL, C_L = 250pF, Unless Otherwise Specified.

SYMBOL	PARAMETER	MIN.	TYP.(1)	MAX.	UNIT	CONDITIONS
t_{AO}	Address Input to Output Delay		6	9	ns	Refresh Enable = Low (2)(3)
t_{AO1}	Address Input to Output Delay		16	25	ns	Refresh Enable = Low
t_{OO}	Row Enable to Output Delay	7	12	27	ns	Refresh Enable = Low (2)(3)
t_{OO1}	Row Enable to Output Delay	12	28	41	ns	Refresh Enable = Low
t_{EO}	Refresh Enable to Output Delay	7	14	27	ns	Notes 2, 3
t_{EO1}	Refresh Enable to Output Delay	12	30	45	ns	
t_{CO}	Count to Output	15	40	60	ns	Refresh Enable = High (2)(3)
t_{CO1}	Count to Output	20	55	80	ns	Refresh Enable = High
f_C	Counting Frequency			5	MHz	
t_{CPW}	Count Pulse Width	35			ns	
t_{CZ}	Count to Zero Detect	15		70	ns	Note 3

Notes: 1. Typical values are for T_A = 25°C and V_{CC} = 5.0V.
2. T_A = 25°C, V_{CC} = 5.0V.
3. C_L = 15 pF.

A.C. TIMING WAVEFORMS (Typically used with 2116)

NORMAL CYCLE

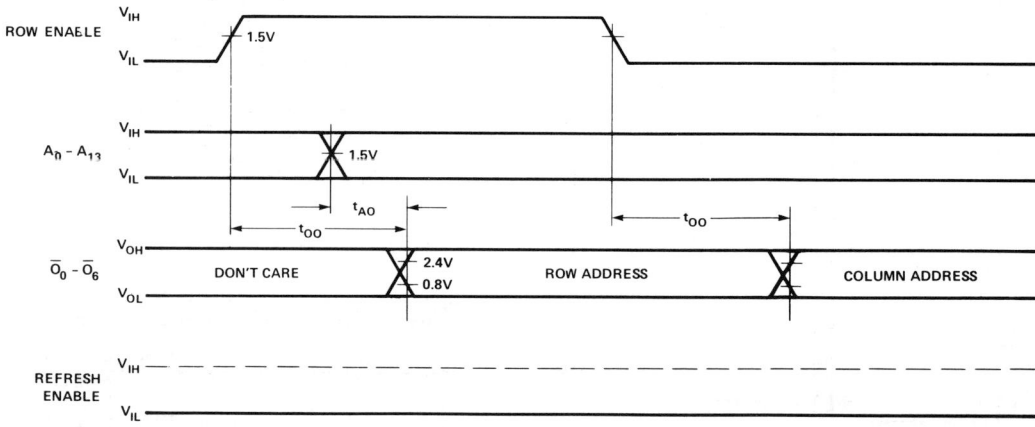

REFRESH CYCLE

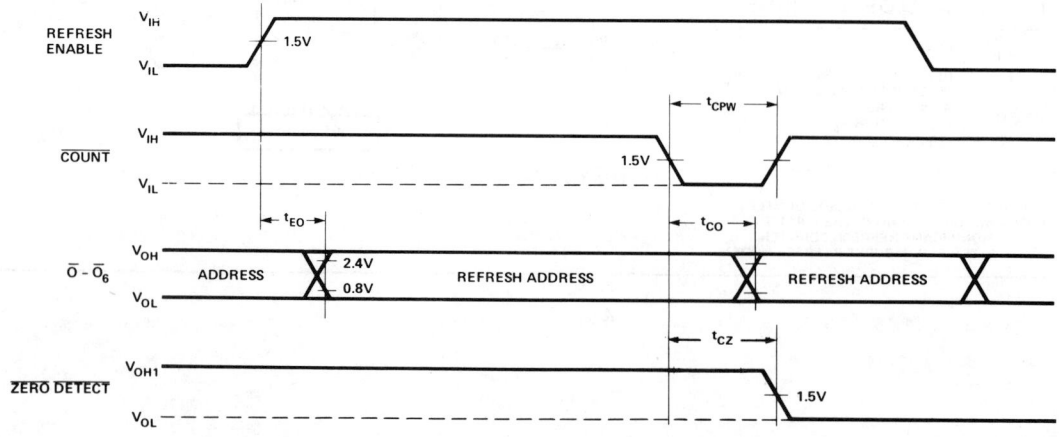

3242 ADDRESS MULTIPLEXER FOR 16K RAMS

Absolute Maximum Ratings*

Temperature Under Bias .... -10° to +85°C
Storage Temperature .... -65° to +150°C
All Input, Output, or
  Supply Voltages .... -0.5V to +7 Volts
Output Currents 100mA
Power Dissipation 1W

*COMMENT:

Stresses above those listed under "Absolute Maximum Ratings" may cause permanent damage to the device. This is a stress rating only and functional operation of the device at these or any other conditions above those indicated in the operational sections of this specification is not implied. Exposure to absolute maximum rating conditions for extended periods may affect device reliability.

D.C. and Operating Characteristics

All Limits Apply for V_{CC} = 5.0V ±10%, T_A = 0°C to +75°C

SYMBOL	PARAMETER	MIN.	TYP.(1)	MAX.	UNIT	TEST CONDITIONS
I_F	Input Load Current		-0.04	-0.25	mA	V_{IN} = 0.45V, Note 2
I_R	Input Leakage Current		0.01	10	µA	V_{IN} = 5.5V
V_{IH}	Input High Voltage	2.0			V	
V_{IL}	Input Low Voltage			0.8	V	
V_{OL}	Output Low Voltage		0.25	0.40	V	I_{OL} = 8mA
V_{OH}	Output High Voltage ($\overline{O}_0$-$\overline{O}_6$)	3.0	4.0		V	I_{OH} = -1mA
V_{OH1}	Output High Voltage (Zero Detect)	2.4	3.3		V	I_{OH} = -1mA
I_{CC}	Power Supply Current		105	165	mA	V_{CC} = 5.5V

Notes: 1. Typical values are for T_A = 25°C and V_{CC} = 5.0V.
2. Inputs are high impedance, TTL compatible, and suitable for bus operation.

Packaging Information

28 LEAD HERMETIC DUAL IN-LINE PACKAGE
TYPE D

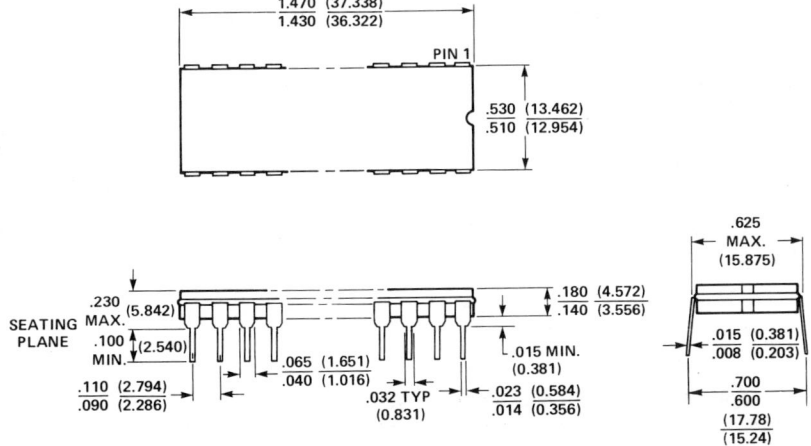

PIN NAMES AND FUNCTIONS

Pin No.	Pin Name	Function
1	$\overline{\text{Count}}$ Input*	Active low input increments internal 7-bit counter by one for each count pulse in.
2	Refresh Enable Input*	Active high input which determines whether the 3242 is in refresh mode (H) or address enable (L).
9,5,7,21, 23,25,27	A_0–A_6 Inputs*	Row address inputs.
10,6,8,20, 22,24,26	A_7–A_{13} Inputs*	Column address inputs.
11,13,12, 18,17,16, 19	$\overline{O}_0$–$\overline{O}_6$ Outputs	Address outputs to memories. Inverted with respect to address inputs.
14	GND	Power supply ground.
15	$\overline{\text{Zero Detect}}$ Output	Active low output which senses that the six low order bits of refresh address in the counter are zero. Can be used in the burst mode to sense refresh completion.
3	Row Enable Input*	High input selects row, low input selects column addresses of the driven memories.
28	V_{CC}	+5V power supply input.

*The inputs are high impedance, TTL compatible, and suitable for bus operation.

DEVICE OPERATION

The Intel® 3242 Address Multiplexer/Refresh Counter performs the following functions:

1. Row, Column and Refresh Address multiplexing.
2. Address Counting for burst or distributed refresh.

These functions are controlled by two signals: Refresh Enable and Row Enable, both of which are active high TTL inputs. The truth table on page 1 shows the levels required to multiplex to the output:

1. Refresh addresses (from internal counter).
2. Row addresses (A_0 through A_6).
3. Column addresses (A_7 through A_{13}).

Burst Refresh Mode

When refresh is requested, the refresh enable input is high. This input is ANDed with the seven outputs of the internal 7-bit counter. At each $\overline{\text{Count}}$ pulse the counter increments by one, sequencing the outputs ($\overline{O}_0$–$\overline{O}_6$) through 128 row addresses. When the first six significant bits of the counter sequence to all zeros, the $\overline{\text{Zero Detect}}$ output goes low, signaling the end of the refresh sequence. Due to counter decoding spikes, the $\overline{\text{Zero Detect}}$ output is valid only after t_{CZ} following the low-going edge of $\overline{\text{Count}}$. The $\overline{\text{Zero Detect}}$ output used in this manner signals the completion of 64 refresh cycles. To use the 128-cycle burst refresh mode, an external flip-flop must be driven by the $\overline{\text{Zero Detect}}$.

Distributed Refresh Mode

In the distributed refresh mode, one row is selected for refresh each ($t_{REFRESH}/n$) time where n = number of refresh cycles required for the device and $t_{REFRESH}$ is the specified refresh rate for the device. For the 2116 $t_{REFRESH}$ = 2 msec and n = 128 or 64, therefore, one row is refreshed each 15.5 or 31 μsec, respectively. Following the refresh cycle at row n_x, the $\overline{\text{Count}}$ input is pulsed, advancing the refresh address by one row so that the next refresh cycle will be performed on row n_{x+1}. The $\overline{\text{Count}}$ input may be pulsed following each refresh cycle or within the refresh cycle after the specified memory device address hold time.

Row and Column Address

All 14 system address lines are applied to the inputs of the 3242. When Refresh Enable is low and Row Enable is high, input addresses A_0–A_6 are gated to the outputs and applied to the driven memories. Conversely, when Row Enable is low (with Refresh Enable still low), input addresses A_7–A_{13} are gated to the outputs and applied to the driven memories. Figure 1 shows a typical connection between the 3242 and the 2116 16K dynamic RAM. When the memory devices are driven directly by the 3242, the address applied to the memory devices is the inverse of the address at the 3242 inputs due to the inverted outputs of the 3242. This should be remembered when checking out the memory system.

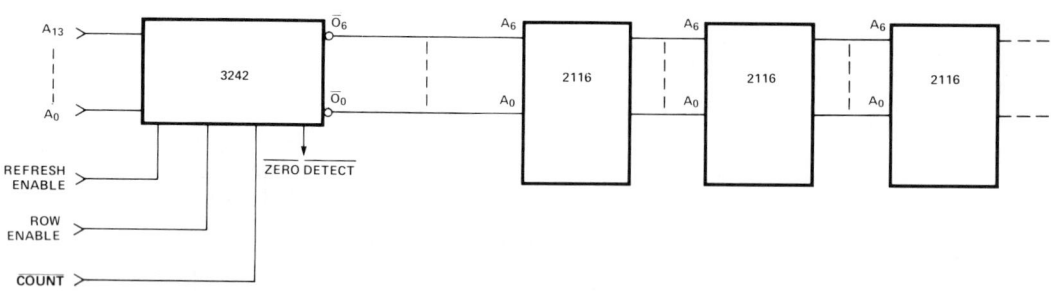

Figure 1. Typical Connection of 3242 and 2116 Memories.

3245

QUAD TTL-TO-MOS DRIVER

For 4K N-Channel MOS RAMs

- **Fully Compatible With 4K RAMs Without Requiring Extra Supply Or External Devices**
- **High Speed, 32 nsec Max. — Delay + Transition Time**
- **Low Power — 75mW Typical Per Channel**
- **High Density — Four Drivers in One Package**
- **TTL & DTL Compatible Inputs**
- **CerDIP Package — 16 Pin DIP**
- **Only +5 and +12 Volt Supplies Required**

The Intel® 3245 is a Quad Bipolar-to-MOS driver which accepts TTL and DTL input signals. It provides high output current and voltage suitable for driving the clock inputs of N-channel MOS memories such as the 2107B. The circuit operates from two power supplies which are 5 and 12 volts. Input and output clamp diodes minimize line reflections.

The device features two common enable inputs, a refresh select input, and a clock control input for simplified system designs. The internal gating structure of the 3245 eliminates gating delays and minimizes package count.

The 3245 is fabricated by means of Intel's highly reliable Schottky bipolar process and is specified for operation over a 0 to +75°C ambient temperature range.

PIN CONFIGURATION

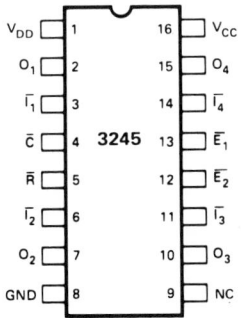

PIN NAMES

$\overline{I_1}$-$\overline{I_4}$	SELECT INPUTS	O_1-O_4	DRIVER OUTPUTS
$\overline{E_1}$, $\overline{E_2}$	ENABLE INPUTS	V_{CC}	+5V POWER SUPPLY
$\overline{R}$	REFRESH SELECT INPUT	V_{DD}	+12V POWER SUPPLY
$\overline{C}$	CLOCK CONTROL INPUT	NC	NOT CONNECTED

LOGIC DIAGRAM

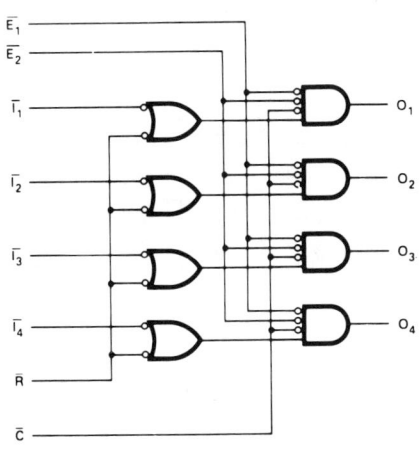

Absolute Maximum Ratings*

Temperature Under Bias . . . . −10°C to 85°C
Storage Temperature. . . . . −65°C to +150°C
Supply Voltage, V_{CC} . . . . −0.5 to +7V
Supply Voltage, V_{DD} . . . . −0.5 to +14V
All Input Voltages . . . . −1.0 to V_{DD}
Outputs for Clock Driver . . . . −1.0 to V_{DD} +1V
Power Dissipation at 25°C 2W

*COMMENT: Stresses above those listed under "Absolute Maximum Ratings" may cause permanent damage to the device. This is a stress rating only and functional operation of the device at these or any other conditions above those indicated in the operational sections of this specification is not implied. Exposure to absolute maximum rating conditions for extended periods may affect device reliability.

D.C. Characteristics

$T_A = 0°C$ to $75°C$, $V_{CC} = 5.0V \pm 5\%$, $V_{DD} = 12V \pm 5\%$

Symbol	Parameter	Min.	Max.	Unit	Test Conditions
I_{FD}	Input Load Current, $\bar{I}_1, \bar{I}_2, \bar{I}_3, \bar{I}_4$		−0.25	mA	$V_F = 0.45V$
I_{FE}	Input Load Current, $\bar{R}, \bar{C}, \bar{E}_1, \bar{E}_2$		−1.0	mA	$V_F = 0.45V$
I_{RD}	Data Input Leakage Current		10	μA	$V_R = 5.0V$
I_{RE}	Enable Input Leakage Current		40	μA	$V_R = 5.0V$
V_{OL}	Output Low Voltage		0.45	V	$I_{OL} = 5mA, V_{IH} = 2V$
		−1.0		V	$I_{OL} = -5mA$
V_{OH}	Output High Voltage	V_{DD}−0.50		V	$I_{OH} = -1mA, V_{IL} = 0.8V$
			V_{DD}+1.0	V	$I_{OH} = 5mA$
V_{IL}	Input Low Voltage, All Inputs		0.8	V	
V_{IH}	Input High Voltage, All Inputs	2		V	

POWER SUPPLY CURRENT DRAIN AND POWER DISSIPATION

Symbol	Parameter	Typ.	Max.	Unit	Test Conditions — Input states to ensure the following output states:	Additional Test Conditions
I_{CC}	Current from V_{CC}	23	30	mA	High	
I_{DD}	Current from V_{DD}	19	26	mA		
P_{D1}	Power Dissipation	365	485	mW		
	Power Per Channel	91	121	mW		$V_{CC} = 5.25V$
I_{CC}	Current from V_{CC}	29	39	mA	Low	$V_{DD} = 12.6V$
I_{DD}	Current from V_{DD}	12	15	mA		
P_{D2}	Power Dissipation	300	388	mW		
	Power Per Channel	75	97	mW		

3245 TTL-MOS DRIVER

A.C. Characteristics $T_A = 0°$ to $75°C$, $V_{CC} = 5.0V \pm 5\%$, $V_{DD} = 12V \pm 5\%$

Symbol	Parameter	Min.[1]	Typ.[2,4]	Max.[3]	Unit	Test Conditions
t_{-+}	Input to Output Delay	5	11		ns	$R_{SERIES} = 0$
t_{DR}	Delay Plus Rise Time		20	32	ns	$R_{SERIES} = 0$
t_{+-}	Input to Output Delay	3	7		ns	$R_{SERIES} = 0$
t_{DF}	Delay Plus Fall Time		18	32	ns	$R_{SERIES} = 0$
t_T	Output Transition Time	10	17	25	ns	$R_{SERIES} = 20\Omega$
t_{DR}	Delay Plus Rise Time		27	38	ns	$R_{SERIES} = 20\Omega$
t_{DF}	Delay Plus Fall Time		25	38	ns	$R_{SERIES} = 20\Omega$

NOTES:
1. $C_L = 150pF$
2. $C_L = 200pF$
3. $C_L = 250pF$ These values represent a range of total stray plus clock capacitance for nine 4K RAMs.
4. Typical values are measured at 25°C.

Capacitance* $T_A = 25°C$

Symbol	Test	Typ.	Max.	Unit
C_{IN}	Input Capacitance, $\overline{I}_1, \overline{I}_2, \overline{I}_3, \overline{I}_4$	5	8	pF
C_{IN}	Input Capacitance, $\overline{R}, \overline{C}, \overline{E}_1, \overline{E}_2$	8	12	pF

*This parameter is periodically sampled and is not 100% tested. Condition of measurement is $f = 1$ MHz, $V_{bias} = 2V$, $V_{CC} = 0V$, and $T_A = 25°C$.

A.C. CONDITIONS OF TEST

Input Pulse Amplitudes: 3.0V
Input Pulse Rise and Fall Times: 5 ns between 1 volt and 2 volts
Measurement Points: See Waveforms

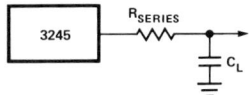

Waveforms

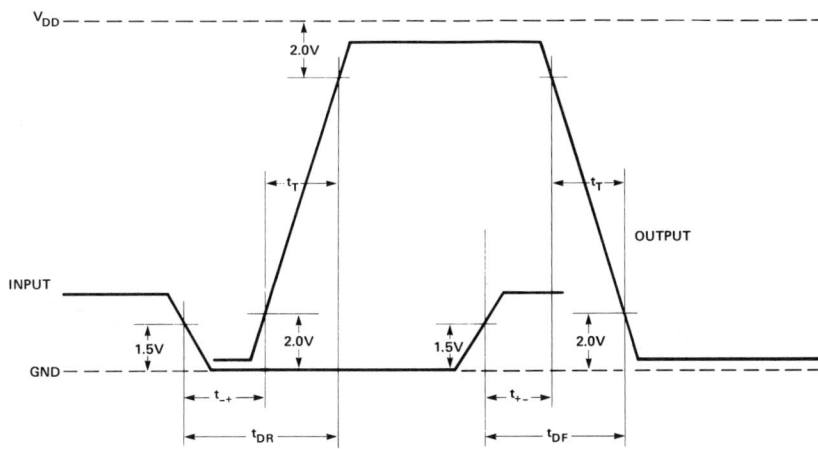

Typical Characteristics

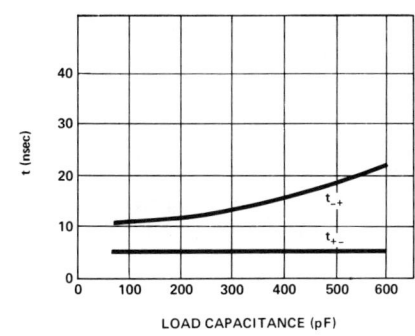

INPUT TO OUTPUT DELAY VS. LOAD CAPACITANCE

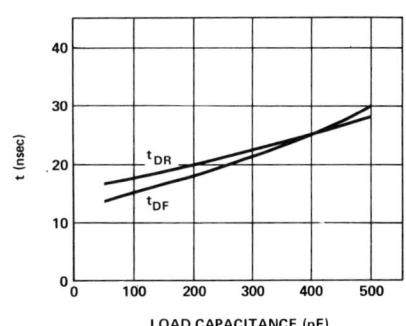

DELAY PLUS TRANSITION TIME VS. LOAD CAPACITANCE

MEMORY SUPPORT CIRCUITS FOR DYNAMIC RAMS

Typical System

Below is an example of a 64K x 18 bit memory system (each card is 16K x 18) employing the 3245 quad high voltage driver for the chip enable inputs. A single 3245 package drives 16K x 9 bits. A_0 through A_{11} are 2107B addresses.

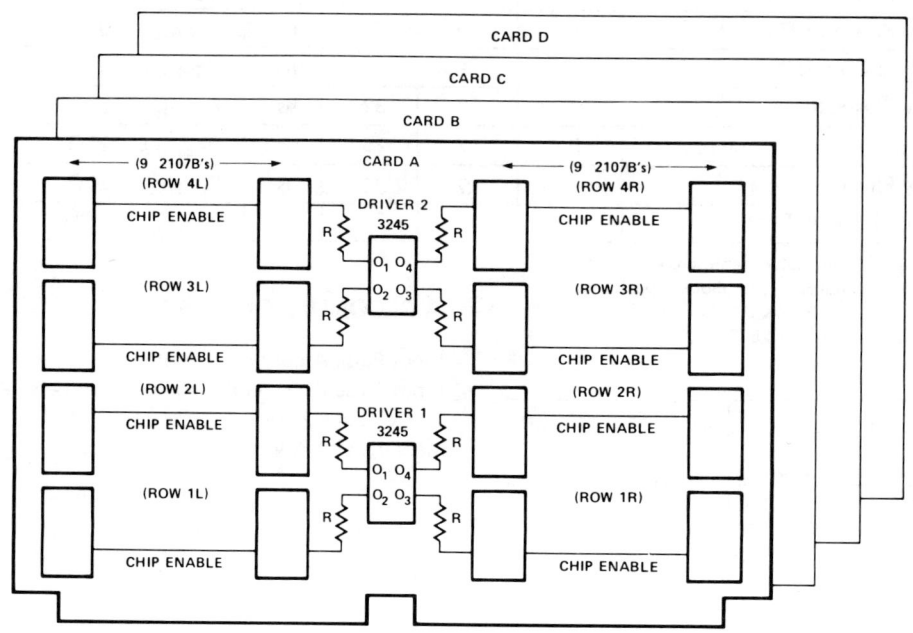

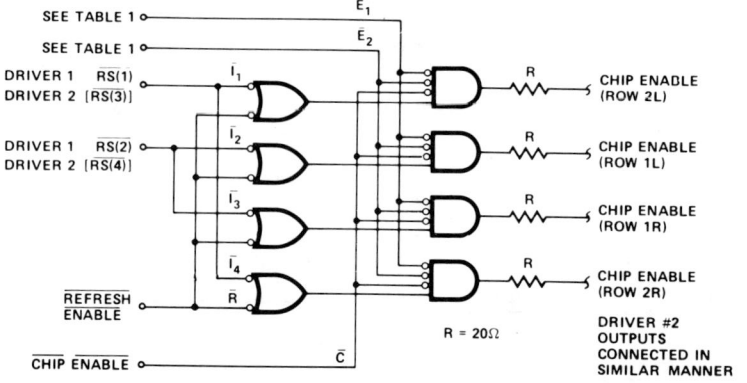

TABLE 1.		
CARD	INPUTS	
	$\bar{E}_1$	$\bar{E}_2$
A	ENABLE M	ENABLE P
B	ENABLE M	ENABLE Q
C	ENABLE N	ENABLE P
D	ENABLE N	ENABLE Q

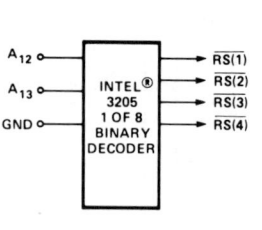

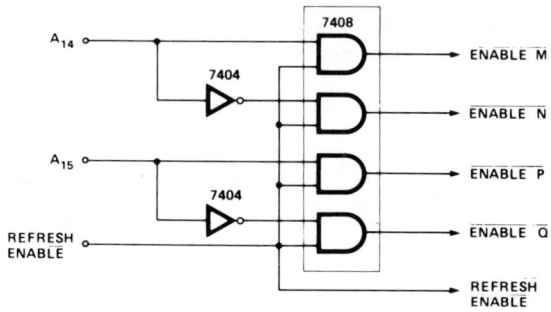

5235, 5235-1
QUAD TTL-TO-MOS DRIVER
For 4K N-Channel MOS RAMs

- **CMOS Technology for Very Low Power: Suitable for Battery Backup**
- **High Density: Four Drivers in One Package**
- **Internal Gating Structure Minimizes Package Count**
- **TTL & DTL Compatible Inputs**
- **CerDIP Package: 16 Pin DIP**
- **Only One Power Supply Required, +12V (±10%)**

The Intel® 5235 and 5235-1 are Low Power Quad TTL-to-MOS drivers which accept TTL and DTL input levels. They provide high output current and voltage suitable for driving the clock inputs of N-channel MOS memories such as the 2107A or 2107B. The circuit operates from a single 12 volt power supply.

The device features two common enable inputs, a refresh select input, and a clock control input for simplified system design.

The 5235-1 is a selection of the 5235 and is guaranteed for 95ns maximum delay plus transition time while driving a 250pF load.

The Intel ion-implanted, silicon gate Complementary MOS (CMOS) process allows the design and production of very low power drivers.

PIN CONFIGURATION

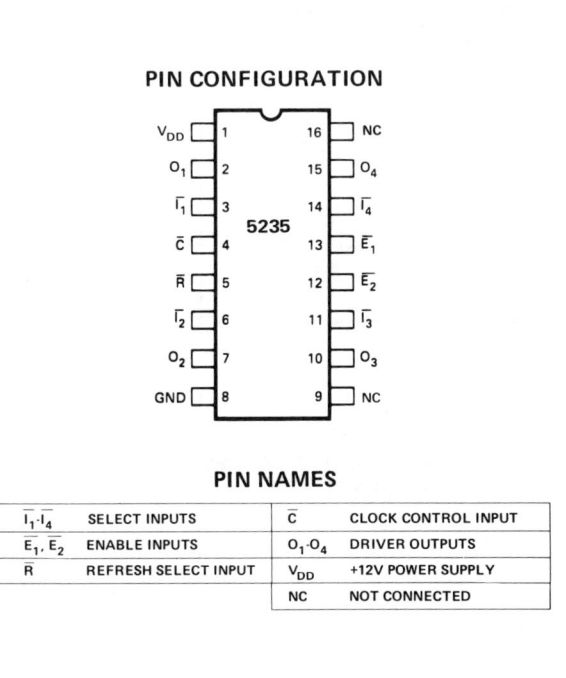

PIN NAMES

$\overline{I_1}$-$\overline{I_4}$	SELECT INPUTS	$\overline{C}$	CLOCK CONTROL INPUT
$\overline{E_1}$, $\overline{E_2}$	ENABLE INPUTS	O_1-O_4	DRIVER OUTPUTS
$\overline{R}$	REFRESH SELECT INPUT	V_{DD}	+12V POWER SUPPLY
		NC	NOT CONNECTED

LOGIC DIAGRAM

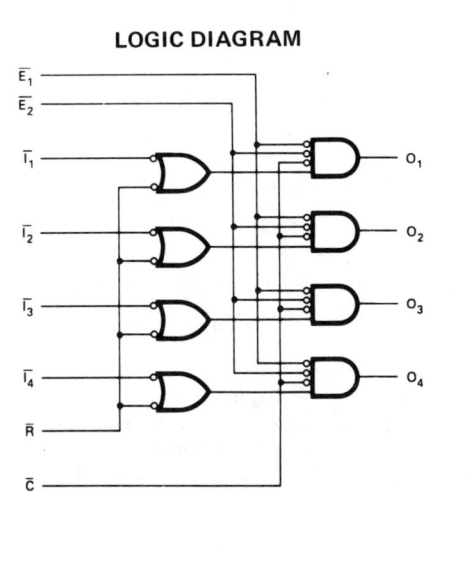

469

MEMORY SUPPORT CIRCUITS FOR DYNAMIC RAMS

Absolute Maximum Ratings*

Temperature Under Bias . . . . –10°C to 80°C
Storage Temperature . . . . –65°C to +150°C
Supply Voltage, V_{DD} . . . . –0.5 to +14V
All Input Voltages . . . . –0.5 to (V_{DD}+0.5V)
Outputs for Clock Driver . . . . . . –0.5 to (V_{DD}+0.5V)
Power Dissipation at 25°C 1W

*COMMENT: Stresses above those listed under "Absolute Maximum Ratings" may cause permanent damage to the device. This is a stress rating only and functional operation of the device at these or any other conditions above those indicated in the operational sections of this specification is not implied. Exposure to absolute maximum rating conditions for extended periods may affect device reliability.

D.C. Characteristics T_A = 0°C to 70°C, V_{DD} = 12V ±10%.

Symbol	Parameter	Min.	Typ.[1]	Max.	Unit	Test Conditions	
$|I_{LI}|$	Input Load Current		0.1	10	µA	V_{IN} = ≤0.4V or ≥2.4V	
V_{OL}	Output Low Voltage		0.15	0.4	V	I_{OL} = 5mA	
		–1.0	–0.15		V	I_{OL} = –5mA	
V_{OH}	Output High Voltage	V_{DD}–0.4	V_{DD}–0.15		V	I_{OH} = –5mA	
			V_{DD}+0.15	V_{DD}+0.5	V	I_{OH} = 5mA	
V_{IL}	Input Low Voltage, All Inputs			0.8	V		
V_{IH}	Input High Voltage, All Inputs	2.0			V		
I_{DD0}	Supply Current		1.0	2.0	mA	f = 0MHz	V_{DD}=13.2V V_{IN}≤0.4V or V_{IN}≥2.4V, C_L = 0pf.
I_{DD1}	Supply Current		12	20	mA	f = 1MHz (See Figure 1)	

Note 1: Typical values are at 25°C and nominal voltage.

Typical Characteristics

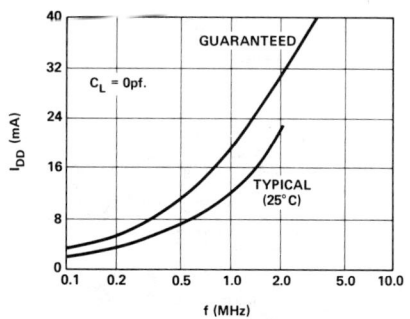

Figure 1.
POWER SUPPLY CURRENT VS. FREQUENCY
(ALL 4 CHANNELS SWITCHING)

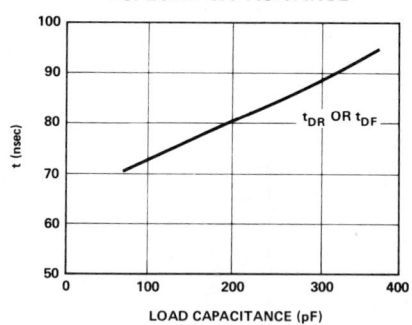

Figure 2.
DELAY PLUS TRANSITION TIME
VS. LOAD CAPACITANCE

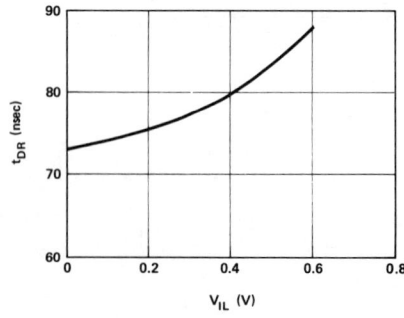

Figure 3.
DELAY PLUS TRANSITION TIME
VS. INPUT VOLTAGE

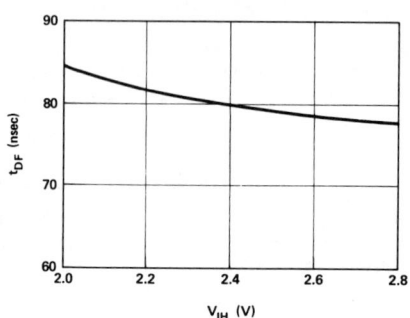

Figure 4.
DELAY PLUS TRANSITION TIME
VS. INPUT VOLTAGE

5235, 5235-1 TTL-MOS DRIVER

A.C. Characteristics $T_A = 0°$ to $70°C$, $V_{DD} = 12V \pm 10\%$.

Symbol	Parameter	5235-1 Min.[1]	5235-1 Typ.[2,4]	5235-1 Max.[3]	5235 Min.[1]	5235 Typ.[2,4]	5235 Max.[3]	Unit
t_{-+}	Input to Output Delay	20	55		20	70		ns
t_{DR}	Delay Plus Rise Time		75	95		95	125	ns
t_{+-}	Input to Output Delay	20	55		20	70		ns
t_{DF}	Delay Plus Fall Time		75	95		95	125	ns
t_T	Transition Time	10	20	40	10	25	40	ns

NOTES:
1. $C_L = 150pF$ — These values represent a range of
2. $C_L = 200pF$ — total stray plus clock capacitance
3. $C_L = 250pF$ — for nine 4K RAMs.
4. Typical values are measured at 25°C, and nominal voltage.

Capacitance* $T_A = 25°C$

Symbol	Test	Typ.	Max.	Unit
C_{IN}	Input Capacitance	8	14	pF

*This parameter is periodically sampled and is not 100% tested. Condition of measurement is $f = 1$ MHz, $V_{bias} = 2V$, $V_{CC} = 0V$, and $T_A = 25°C$.

A.C. CONDITIONS OF TEST

Input Pulse Amplitudes: 2.0V
Input Pulse Rise and Fall Times: 5 ns between 0.9 volt and 1.9 volts
Measurement Points: See Waveforms

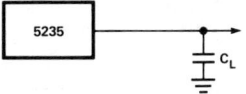

Waveforms

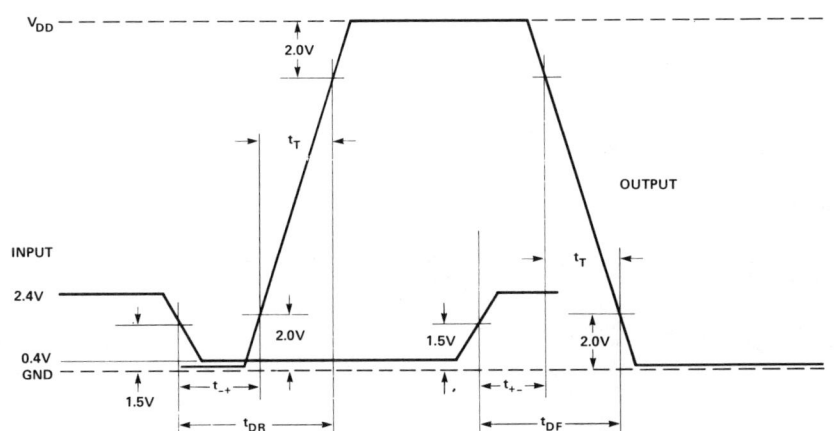

MEMORY SUPPORT CIRCUITS FOR DYNAMIC RAMS

Typical System

Below is an example of a 64K x 18 bit memory system (each card is 16K x 18) employing the 5235 quad high voltage driver for the chip enable inputs. A single 5235 package drives 16K x 9 bits. A_0 through A_{11} are 2107B addresses.

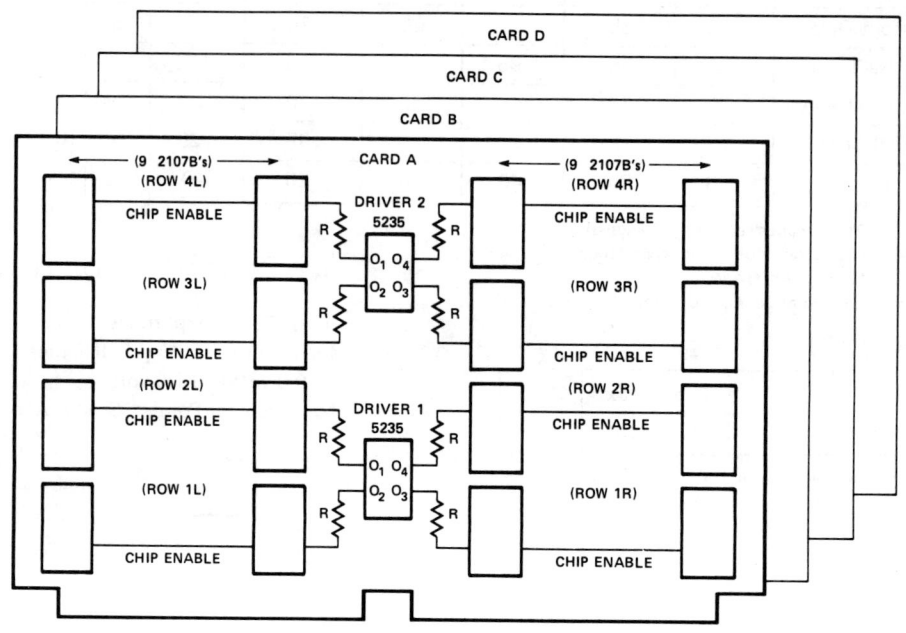

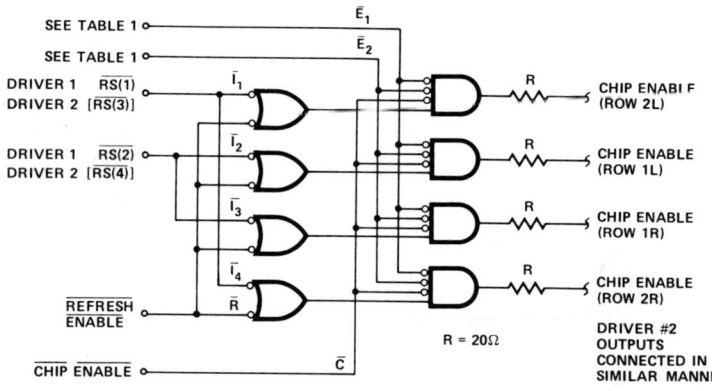

TABLE 1.		
CARD	INPUTS	
	$\overline{E}_1$	$\overline{E}_2$
A	ENABLE M	ENABLE P
B	ENABLE M	ENABLE Q
C	ENABLE N	ENABLE P
D	ENABLE N	ENABLE Q

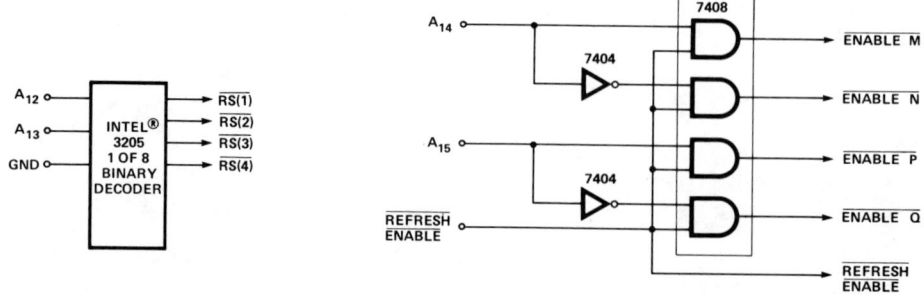

5244

QUAD CCD CLOCK DRIVER

- **Internal Circuitry Minimizes CCD Clock Cross-Coupling Voltage Transients**
- **Drives Four 2416s**
- **Low Standby Power Dissipation: 24mW Typically**
- **TTL Inputs**
- **Single +12V Supply**
- **Standard 16 Pin Dual In-Line Package**

The 5244 is a quad clock driver which provides high capacitive drive suitable for driving charge coupled memories. The 5244 features very low D.C. power dissipation from a single 12V supply with output characteristics directly compatible with the 2416 clock input requirements. Internal circuitry controls the cross-coupled voltage transients between the clock phases generated by the 2416 and limits the transition time so that excessively fast transitions do not occur on the clock line.

The 5244 is fabricated using an advanced ion-implanted, silicon gate, CMOS process.

PIN CONFIGURATION

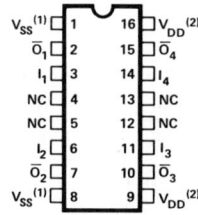

NOTES: 1. BOTH PIN 1 AND 8 MUST BE CONNECTED TO V_{SS}.
2. BOTH PIN 9 AND 16 MUST BE CONNECTED TO V_{DD}.

BLOCK DIAGRAM

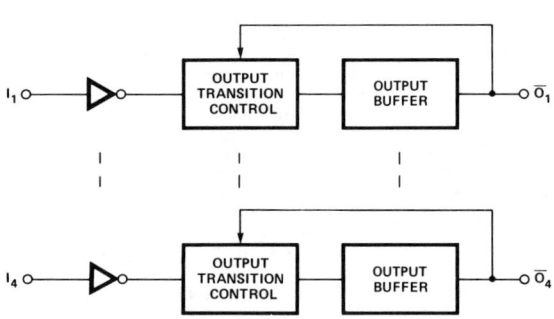

PIN NAMES

I_1 - I_4	TTL INPUT
$\overline{O}_1$ - $\overline{O}_4$	DRIVER OUTPUT
V_{DD}	+12V POWER SUPPLY
NC	NOT CONNECTED
V_{SS}	GROUND

Absolute Maximum Ratings*

Temperature Under Bias .... −10°C to 80°C
Storage Temperature .... −65°C to +150°C
Supply Voltage with Respect to V_{SS} .... −0.5 to +14V
All Input Voltages .... −0.5 to (V_{DD}+1V)
Outputs .... −1V to (V_{DD}+1)
Power Dissipation 1.35W

*COMMENT:
Stresses above those listed under "Absolute Maximum Ratings" may cause permanent damage to the device. This is a stress rating only and functional operation of the device at these or any other conditions above those indicated in the operational sections of this specification is not implied. Exposure to absolute maximum rating conditions for extended periods may affect device reliability.

D.C. and Operating Characteristics

$T_A = 0°C$ to $70°C$, $V_{DD} = +12V \pm 5\%$, $V_{SS} = 0V$

Symbol	Parameter	Limits Min.	Limits Typ.	Limits Max.	Unit	Test Conditions
I_{IL}	Low Level Input Current	−10	±0.1	10	µA	$V_{IN} \leq V_{IL}$
I_{IH}	High Level Input Current	−10	±0.1	10	µA	$V_{IN} \geq V_{IH}$
V_{IL}	Input Low Voltage		+1.2	+0.85	V	
V_{IH}	Input High Voltage	+2.0	+1.5	V_{DD}+1.0	V	
V_{OL}	Output Low Voltage	0	0.03	+0.1	V	I_{OL} = 5mA
V_{OH}	Output High Voltage	V_{DD}−0.1	V_{DD}−.03	V_{DD}	V	I_{OH} = −5mA
I_{DD0}	Standby Current		2.0	4.0	mA	$V_{IN} \geq V_{IH}$, $V_{IN} \leq V_{IL}$, f = 0 MHz
I_{DD1}	Operating Current		75	105[3]	mA	$V_{IN} \geq V_{IH}$ or $V_{IN} \leq V_{IL}$, f = 0.67 MHz [2]

A.C. Characteristics

$T_A = 0°C$ to $70°C$, $V_{DD} = +12V \pm 5\%$, $V_{SS} = 0V$, Note 2

Symbol	Parameter	Limits Driving 4 2416's Min.	Limits Driving 4 2416's Typ.	Limits Driving 4 2416's Max.	Units
V_{OLT}	Transient Cross-Coupled Output Low Voltage	−0.8	±0.5	+0.8	V
V_{OHT}	Transient Cross-Coupled Output High Voltage	V_{DD}−0.8	V_{DD}±0.5	V_{DD}+0.8	V
t_{PWT}	Transient Cross-Coupled Output Pulse Width			Note 1	ns
Δt_D	Differential Delay of t_{DLH} and t_{DHL} for Drivers in the Same Package			15	ns
t_{DLH1}	Input Low to Output High Delay Time, ϕ_1 or ϕ_3	30	50		ns
t_{DHL1}	Input High to Output Low Delay Time, ϕ_1 or ϕ_3	30	50		ns
t_{TLH1}	Output Rise Time, ϕ_1 or ϕ_3	30	50	75	ns
t_{THL1}	Output Fall Time, ϕ_1 or ϕ_3	30	50	75	ns
t_{PLH1}	Input to Output Delay Plus Rise Time, ϕ_1 or ϕ_3		100	160	ns
t_{PHL1}	Input to Output Delay Plus Fall Time, ϕ_1 or ϕ_3		100	150	ns
t_{DLH2}	Input Low to Output High Delay Time, ϕ_2 or ϕ_4	30	55		ns
t_{DHL2}	Input High to Output Low Delay Time, ϕ_2 or ϕ_4	30	55		ns
t_{TLH2}	Output Rise Time, ϕ_2 or ϕ_4	30	55	85	ns
t_{THL2}	Output Fall Time, ϕ_2 or ϕ_4	30	55	90	ns
t_{PLH2}	Input to Output Delay Plus Rise Time, ϕ_2 or ϕ_4		110	175	ns
t_{PHL2}	Input to Output Delay Plus Fall Time, ϕ_2 or ϕ_4		110	170	ns

Notes: 1. The maximum t_{PWT} is the sum of the output transition time (rise or fall) plus 5ns.
2. Output Load = four 2416 clock inputs or equivalent per Figure 2.
3. $I_{DD1} = 4.0 \text{ mA} + \dfrac{75.4 \text{ mA}}{t_{\phi/2} \text{ (in } \mu s)}$

5244 CCD CLOCK DRIVER

CAPACITANCE* $T_A = 25°C$

Symbol	Test	Typ.	Max.	Unit	Conditions
C_{IN}	Input Capacitance	8	14	pf	$f = 1$ MHz, $V_{bias} = 2V$, $V_{DD} = 0V$

*This parameter is periodically sampled and is not 100% tested.

A.C. Test Conditions

1. TTL Input Levels = 0.4V to 2.4V.
2. Input Rise and Fall Times = 5 ns between 0.9V and 1.9V.
3. Output Load = Four 2416 clock inputs or equivalent per Figure 2.
4. Cross Coupled Voltage Pulse Width measured at ±0.4V and V_{DD} ±0.4V.

Waveforms

A. INPUT TO OUTPUT DELAY

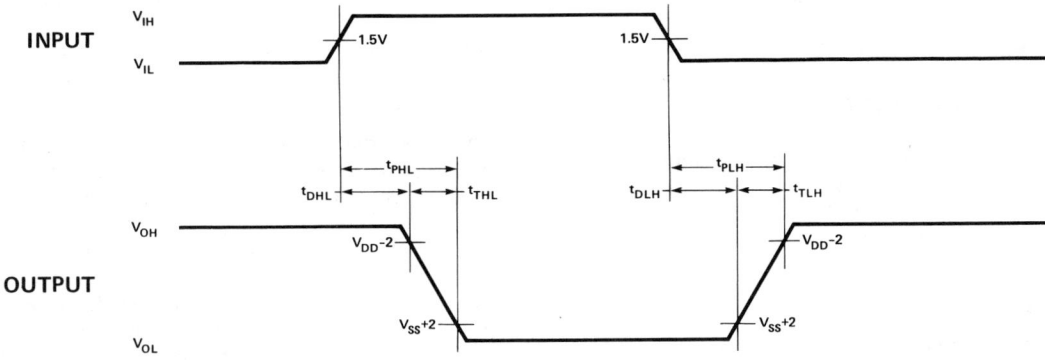

B. 5244 OUTPUT CROSS-COUPLED VOLTAGE (DRIVING FOUR 2416'S)

5244 OUTPUT DRIVING 2416 ϕ_1

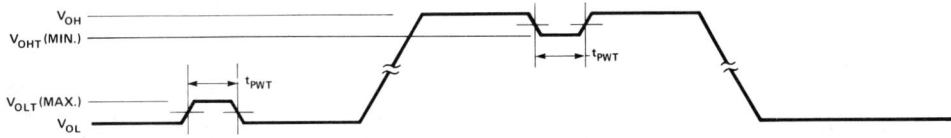

5244 OUTPUT DRIVING 2416 ϕ_2

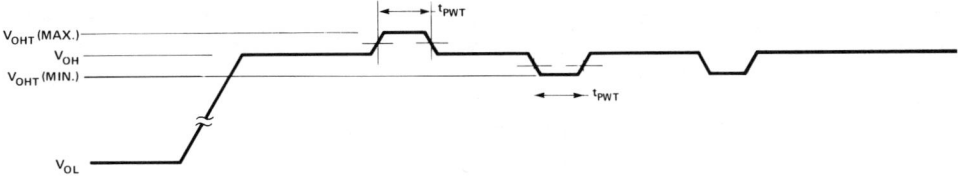

5244 OUTPUT DRIVING 2416 ϕ_3

5244 OUTPUT DRIVING 2416 ϕ_4

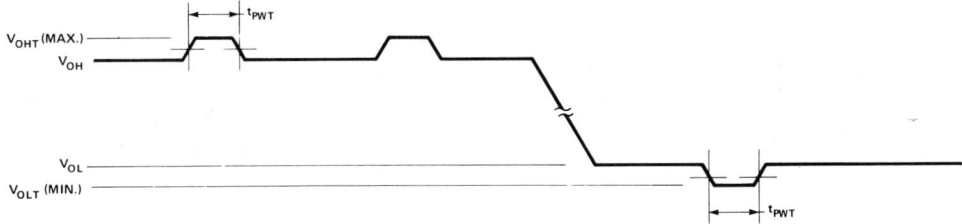

MEMORY SUPPORT CIRCUITS FOR DYNAMIC RAMS

Typical Characteristics

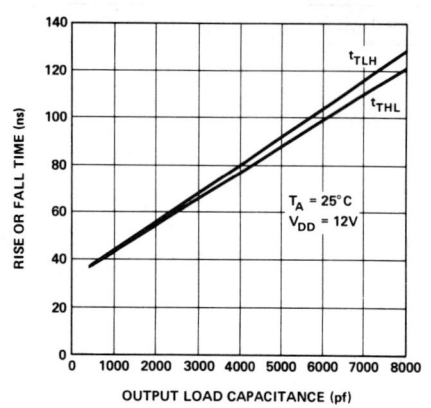

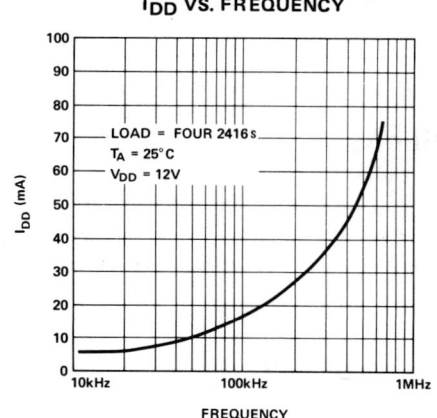

Application Information

The 5244 is a TTL to MOS level converter designed to drive very high capacitive loads with no required additional external components. Its primary application is to drive the clock phase inputs of the Intel® 2416, a 16,384 word x 1 bit charge coupled device.

DRIVING THE 2416

The 5244 is designed to drive the clock phase inputs of four 2416s and meet or exceed the electrical specifications of these inputs. The 2416 clock specifications of special interest to the system designs are:

1. Clock transition time.
2. Clock to clock voltage coupling.

Clock Transition Control

The 5244 will meet the min/max clock transition time requirement of the 2416 when driving four 2416s. However, when driving less than four 2416s an external capacitor (C_{ext}) must be added to assure that the minimum clock transition time (30ns) is adhered to. The maximum clock transition time for the 5244 will not be exceeded if C_{ext} is chosen according to the recommendations in Figure 1.

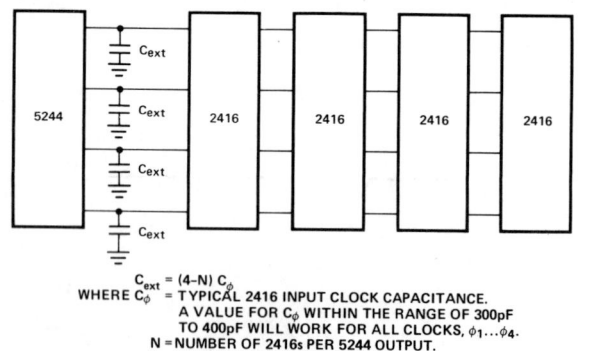

Figure 1. External Loading Requirements When Driving Fewer Than Four 2416s.

Clock Skews

The differential delay of t_{DLH} or t_{DHL} for driver elements in the same package is specified to be Δt_D (15 ns max.). This provides assurance to the system designer that the maximum skew introduced by a 5244 driver package will be limited to Δt_D. As an example, if the fastest t_{DLH} (or t_{DHL}) occurs for I_1 to $\overline{O}_1$ and this is measured to be 45 ns, the output delays for I_2 to $\overline{O}_2$, I_3 to $\overline{O}_3$ or I_4 to $\overline{O}_4$ will be no greater than 60 ns. This should be taken into consideration when designing the TTL source of the four phases required for 2416 operation. To minimize system skew, the four phases associated with any given group of 2416s should be provided from the same 5244 package.

Clock to Clock Voltage Coupling

The equivalent circuit of the 2416 clock phase inputs is shown in Figure 2. The magnitude and duration of the cross-coupling are graphically presented in Waveform B and specified in the A.C. Characteristics. Figure 3, on the next page, shows the noise margin between these specifications and the 2416 input requirements.

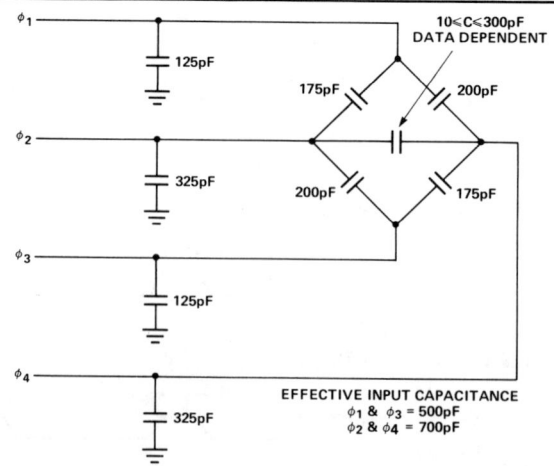

Figure 2. 2416 Equivalent Capacitance Circuit. (Maximum values shown.)

5244 CCD CLOCK DRIVER

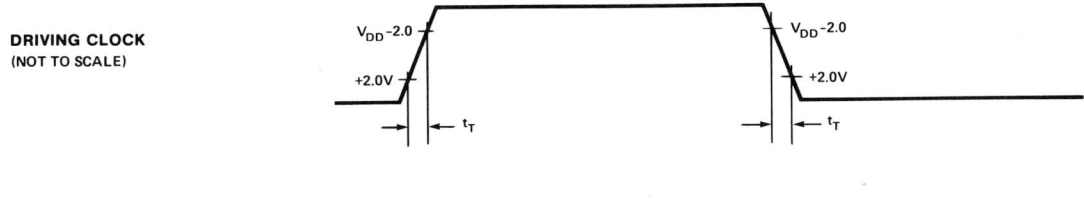
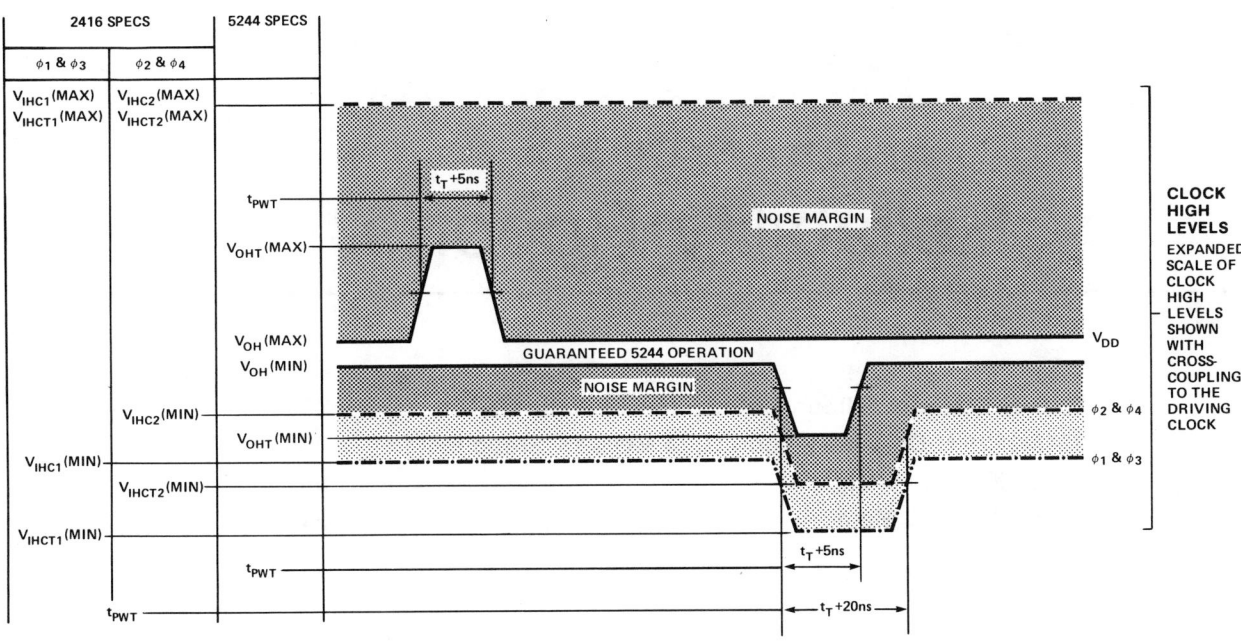
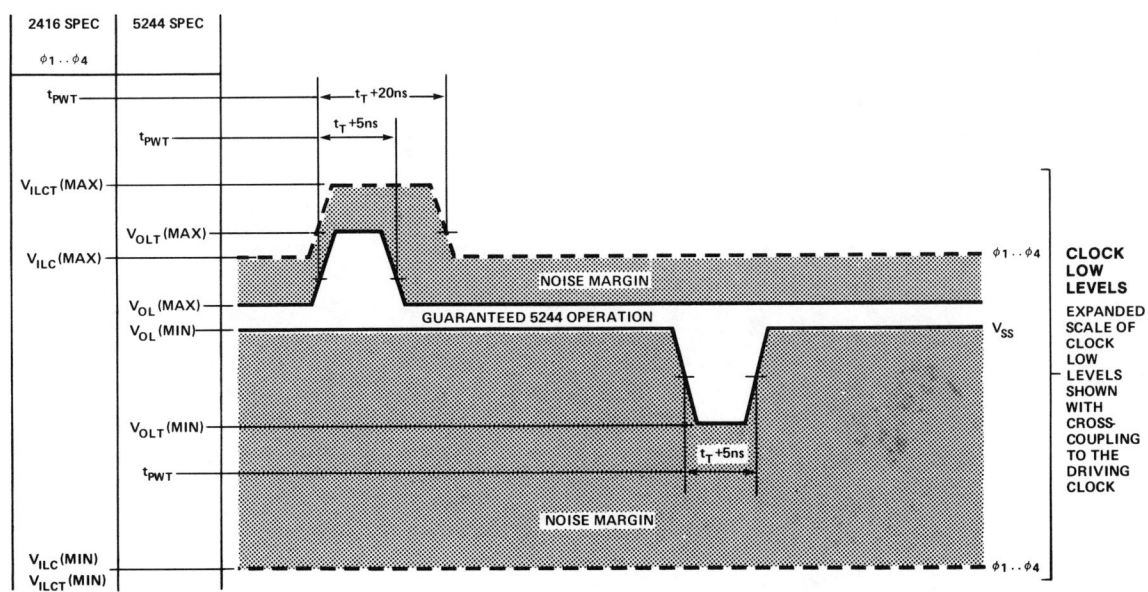

Figure 3. Noise Margins Between 5244 Output Specs and 2416 $\phi_1 \ldots \phi_4$ Input Requirements.

CHAPTER SIX
Reliability Reports

2107 N-Channel Silicon Gate MOS 4K Static RAMs	479
2115/2125 N-Channel Silicon Gate MOS 1K Static RAMs	498
2708 8K UV EPROM	507
2416 16K CCD Memory	514

2107 N-CHANNEL SILICON GATE MOS 4K STATIC RAMS

A Sample of the Equipment in Intel's Reliability Lab.

Vibration Machine

Life Test Chambers

Burn-In Circuit Board

Burn-In Oven

Bit Map Display

INTRODUCTION

Much has been written concerning the reliability of silicon gate MOS devices. While the devices are reliable, readily available reliability information is generally not sufficient to provide the system designer with specific information to determine system reliability. Recognizing the need of system designers for specific component reliability information, Intel has established a system of controls to determine the reliability of its products and to assure that products shipped to customers meet the criteria established.

This report discusses the operation of Silicon Gate N-Channel 4K MOS RAMs, specifically the 2107A and 2107B, (pages 1-5). Typical failure mechanisms are presented with methods used to control or eliminate them (pages 5-7). Intel's reliability and quality assurance programs (pages 7-9) are discussed and data presented to establish the reliability figures of the 2107A and 2107B (pages 9-15).

DESCRIPTION OF THE 2107A AND 2107B

In this section, a brief description of the operation of the 2107A and 2107B is given. Refer to Application Notes AP-4, "Designing Memory Systems with the Intel® 2107A 4K RAM" and AP-10, "Memory System Design with the Intel® 2107B 4K RAM" for more detailed information on the operation and applications of the 2107A and 2107B respectively.

The 2107A and 2107B are 4096 word by 1 bit dynamic N-Channel RAMs. They are packaged using the industry standard 22 pin DIP.

Three power supplies and ground are required: V_{DD} (12V), V_{CC} (5V), V_{SS} (Gnd) and V_{BB} (-5V). The input levels required for Data In, $\overline{\text{Write Enable}}$, $\overline{\text{Chip Select}}$ and all addresses are low voltage compatible and the output $\overline{D}_{OUT}$ is TTL three-state. Logic symbol and pin configuration for the 2107A and 2107B are shown in Figure 1.

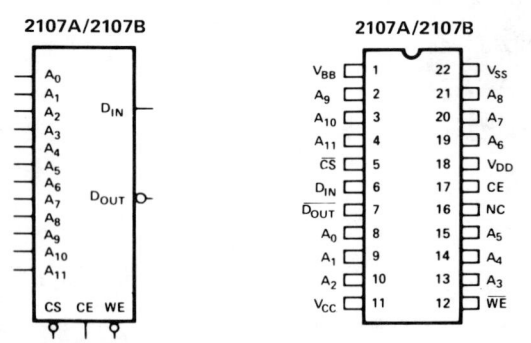

Figure 1. 2107A/2107B Logic Symbol and Pin Configuration

2107A Device Operation

A block diagram of the 2107A is given in Figure 2. As is shown in the diagram, the 4096 storage cells are arranged in a 64 row by 64 column matrix. The six low order addresses (A_0–A_5) are decoded on-chip to select a particular row; the six high order addresses (A_6–A_{11}) select a particular column. A cell is selected by a coincidence of a row and column select. The Chip Select input ($\overline{CS}$) controls only the data in/out buffers and does not effect row or column selection as outlined above. (This allows refresh to be performed with or without $\overline{CS}$.)

The 3 transistor cell is the storage medium used on the 2107A. Optimal cell design has resulted in a very small cell size to allow high storage density to be achieved. A simplified schematic of the storage cell and associated circuitry is shown in Figure 3a.

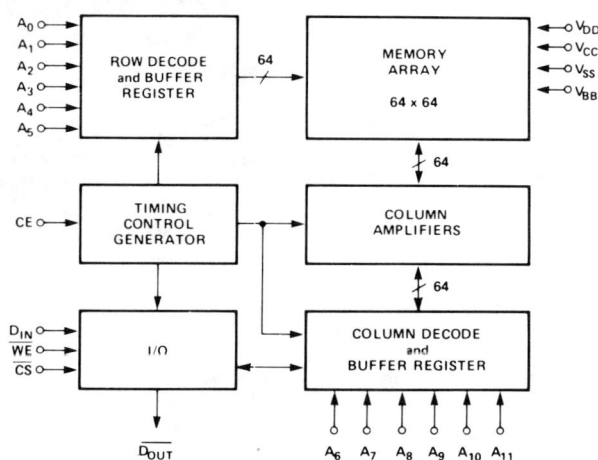

Figure 2. 2107A Block Diagram

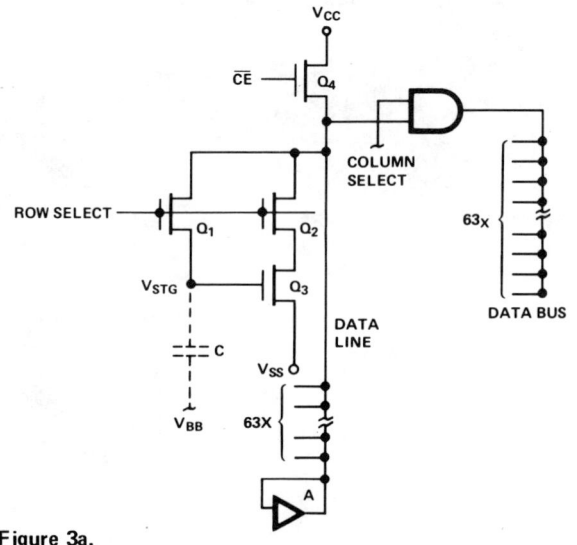

Figure 3a.

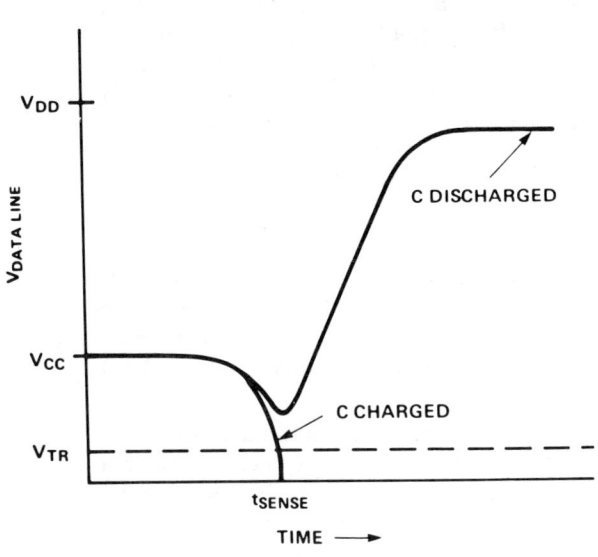

Figure 3b.

The 2107A is activated by Chip Enable going positive. This leading edge of Chip Enable triggers on-chip timing circuits which derive clocks for address latching and decoding, internal control logic and internal data gating.

Storage Cell Operation

The storage mechanism of the 3 transistor cell is the effective capacitance on the node of gate Q_3 (shown as C in Figure 3a). Device Q_1 gates whatever charge is on the Data Line to the storage capacitor and Q_2 gates the inverse of the data stored on C to the Data Line.

Operation of the 3 transistor cell for a Read Access is as follows:

Prior to Chip Enable going positive, the Data Line is precharged to V_{CC} by Q_4. After a decode period (t), the selected Row Select line goes positive (to $\sim V_{CC}$) turning on Q_1 and Q_2. If capacitor C was initially charged, then Q_3 is on and the Data Line begins to discharge through Q_2 and Q_3 as shown in Figure 3b. At a trigger voltage, V_{TR}, sense latch A is triggered forcing the Data Line down to V_{SS}. After the sense latch A has locked the Data Line to V_{SS}, the Row Select line is made more positive ($\sim V_{DD}$) turning Q_1 on harder which completely discharges C. This action results in inverting the data stored in the cell after every row select. (A Data Control cell on each Row Select line keeps track of inversions along the Row Select line and will be explained later). A Column Select line is activated which connects the proper Data Line to a single I/O Data Bus. The Data Bus is gated through control circuitry and buffers to the $\overline{D}_{OUT}$ pin.

A discharged condition on C is sensed in a manner entirely similar to the above. During the rewrite the Data Line is forced high ($\sim V_{DD}$) through a low impedance path to charge C to $\sim V_{DD}$. This low impedance path overrides the effect of turning Q_1, Q_2 and Q_3 on during the rewrite.

A Write operation is performed in a manner similar to a Read but at a later time in the cycle. Modified input data is presented on the Data Bus and gated through Column Select to the proper Data Line where it is latched by the column amplifier associated with that Data Line. The actual write operation is performed by Row Select going to a high state ($\sim V_{DD}$) and transferring the charge on the data line to the gate of Q_3.

Data Control

As explained previously, each access to a given row inverts the data in that row at the end of Chip Enable. It is therefore necessary to know whether to complement the input before writing into a cell or presenting data to the $\overline{D}_{OUT}$ pin. The circuitry which performs this task internally is the Data Control Cell (DCC).

A Data Control Cell is added to the end of each row select line and operates in a manner identical to the storage cell. (The DCC's form a 65th column.) The sole function of the Data Control Cell is to change state after every row select associated with that Data Control Cell. The output of the DCC is sensed and gated to the I/O control circuits. These I/O control circuits invert, if necessary, the data written to the storage cell or data presented to the $\overline{D}_{OUT}$ pin. The entire operation is transparent to the user.

The state of the Data Control Cell is not known when power is applied, hence the actual polarity of the data written into the storage cells is unknown.

2107B Device Operation

The combination of process and device design has resulted in a very small device (see Figure 4) using conservative layout rules (same as 2102A). The small size offers advantages in both large volume production and increased reliability. The 2107B operates with three power supplies relative to ground; V_{DD} (+12V), V_{BB} (-5V), and V_{CC} (+5V). The V_{CC} (+5V) supply is connected only to the output buffer of the 2107B and may be turned off during power down operations.

The 2107B has one MOS level clock (Chip Enable) with all other inputs being low level TTL compatible (+2.4V V_{IH} minimum). The output is capable of driving 1 TTL load.

Figure 4. 2107B Die Photomicrograph.

Internal operation of the 2107B is most easily understood with the aid of the block diagram shown in Figure 5. As is shown in this figure, the memory array is arranged in a 64 row x 64 column matrix of storage cells. The storage cells are implemented with a single transistor and a "storage" capacitor and are called single transistor cells. The operation of the storage cell will be discussed later. The memory cell is accessed by the coincidence of a row select (defined by addresses A_0-A_5) and a column select (defined by addresses A_6-A_{11}) signal at the desired address. An on chip timing and control generator provides for the internal timing signals for decoding, read/write strobing, data gating and output gating. All of the timing circuits in the 2107B are activated by the positive-going edge of Chip Enable.

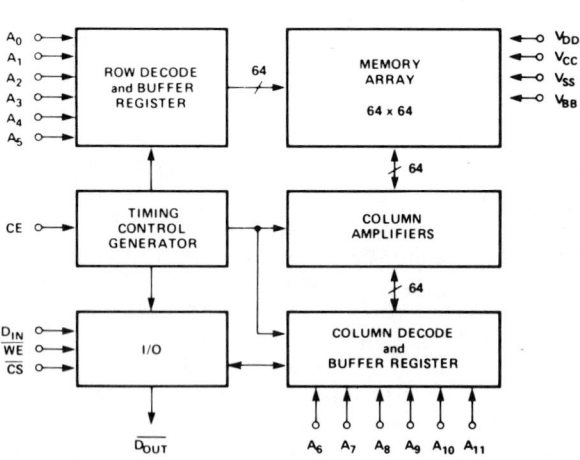

Figure 5. 2107B Block Diagram.

$\overline{\text{Chip Select}}$ controls the data I/O gating circuits internal to the 2107B. When $\overline{\text{Chip Select}}$ is high the output data buffer is in a high impedance state and the data-in buffer is electrically isolated from the Data-in input pin. Since $\overline{\text{Chip Select}}$ controls only the internal data buffers and not the timing generators or address buffers internal to the 2107B, it is possible to refresh the 2107B with $\overline{\text{Chip Select}}$ high by initiating a read/refresh or write cycle.

The address buffer registers consist of latches activated at the leading edge of Chip Enable. Since the addresses are latched shortly after Chip Enable goes high, it is permissible to change the address long before the memory cycle is completed to set up for the next cycle.

The $\overline{\text{Write Enable}}$ input activates the data-in buffer gating data to the selected memory cell. Input data must be valid as the time $\overline{\text{Write Enable}}$ goes low to assure that the proper data is written into memory.

Circuit implementation and operation of each of the major input/output and storage portions are discussed below.

Storage Cell Operation

The storage cell used in the 2107B is implemented with a single transistor and storage capacitor as shown in Figure 6. From this figure it is shown that a charge on a storage cell is gated to the bit sense line by the MOS device connected to the column select line. (Note that for a given column select, 64 storage devices are gated to the respective 64 bit sense lines.)

Consider first a read operation and the case where the storage capacitor C_{STG} is discharged; i.e., node (1) is at V_{SS} (GND). Prior to Chip Enable going high, the bit sense lines have been precharged to V' by device Q_1. [V' is a voltage between V_{DD} (+12V) and V_{SS}.] After the address decoders have stabilized, the proper column select line is brought high, turning on device Q_2. The storage capacitor is then electrically connected to the bit sense line. At this time the charge on $C_{I/O}$ (proportional to the precharge voltage V') is redistributed between $C_{I/O}$ (parasitic capacitance of bit sense line) and C_{STG}. Since C_{STG} was initially discharged (node 1 at V_{SS}) the voltage will distribute between $C_{I/O}$ and C_{STG} according to the following relationship:

$$V_{\text{BIT SENSE}}(t_1) = V_{\text{BIT SENSE}}(t_0) \frac{C_{I/O}}{C_{I/O} + C_{STG}}$$

Since $C_{I/O}$ is very much larger than C_{STG} the change in the voltage on the bit sense line will be very small. The sense amplifier (S/A) is designed to detect very small changes in bit sense line voltage and to latch in a state near V_{SS} (GND) or V_{DD} (+12V), depending on the state of the storage cell.

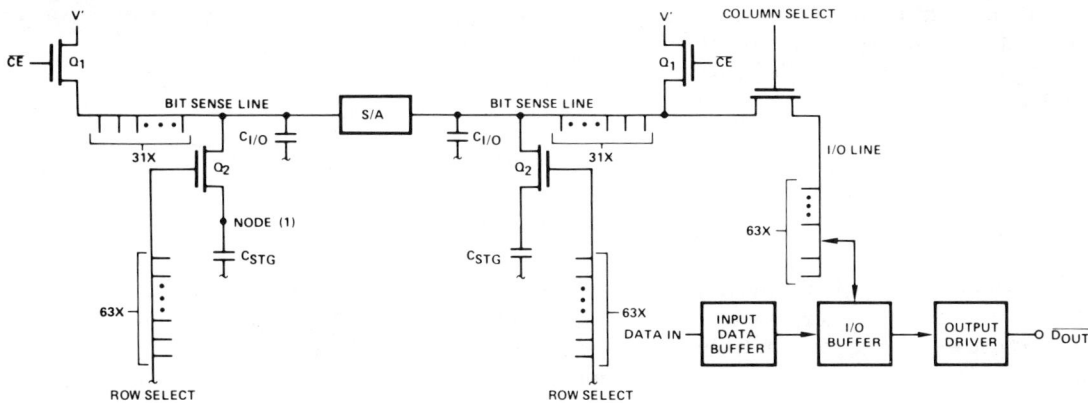

Figure 6. 2107B Memory Cell and Associated I/O Circuitry.

Sensing an initial charge on C_{STG} (proportional to V_X where $V_X = V_{DD} - V_{TH}$, V_{TH} is the effective MOS threshold) is identical to the sequence described above. The only difference is that now the bit sense line is driven above the initial V' precharge voltage. Again the sense amplifier detects the small change in bit sense line voltage and latches in the appropriate state.

Note that during a read operation of the storage cell, the original charge (data) on the storage cell is changed (i.e., the read operation is effectively a destructive read). Data is rewritten back on the storage capacitor C_{STG} by the sense amplifier after it has latched in the proper state. For example, if C_{STG} was initially charged to V_X (~10V), the sense amplifier will latch the bit sense line to V_X and, since the column select line is on (high), the original data is automatically rewritten into C_{STG}. The entire operation is transparent to the user.

A plot of the voltage on the bit sense line for the two cases described above is shown in Figure 7.

A write operation is identical to the rewrite portion of a read cycle. In this case, however, the incoming data "overrides" the state of the sense amplifier (if different from the desired state) and writes into the selected cell. For reference, a low level on the data-in input results in a high level (V_X) being written into the selected storage cell on one side of the Sense-Amp. and a low level on the other side. It is important to remember that the data-output at the output pin is the logical inverse of the data written into memory.

Data Sense/Latch

As discussed previously, a sense amplifier on the bit sense line is necessary to detect the low level data signals generated on the bit sense line during a read cycle. A simplified circuit schematic used for the sense amplifier is shown in Figure 8.

Before Chip Enable is brought high, both sides of the bit sense lines are precharged to V' (as discussed previously). At the proper time (after all data transients have subsided) devices Q_1 and Q_2 are turned on by ϕ_R going positive. At this time,

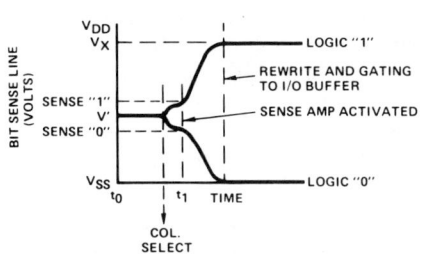

Figure 7. Bit Sense Line Voltage.

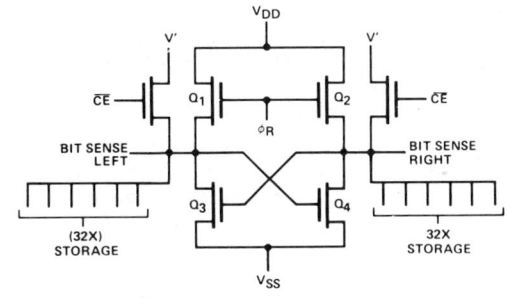

Figure 8. Data Sense/Latch.

the state of bit sense left is compared with bit sense right causing the latch to lock in the appropriate state. For example, if the right bit sense line is at a higher potential than the left bit sense line, device Q_3 will begin to conduct. The cross coupled latch will then fully switch with bit sense left going to V_{SS} and bit sense right to V_X.

2107A and 2107B Bit Maps

Figures 9 and 10 give the location of each cell in the memory matrix for each address. As shown in these figures, the addresses run sequentially starting from the lower left corner (device oriented as shown). The 2107A is shown in Figure 9 and the 2107B is shown in Figure 10.

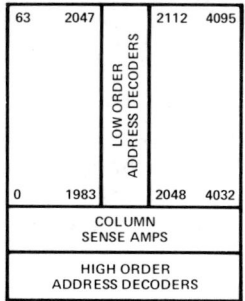

Figure 9. 2107A Bit Map.

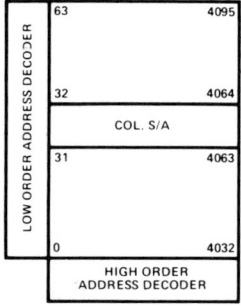

Figure 10. 2107B Bit Map.

FAILURE MECHANISMS

The fundamental principles of reliability engineering predict that the failure rate of any group of devices as a function of time will follow a curve similar to Figure 11.[8] The curve is divided into three regions: Infant Mortality, Random Failures and Wearout Failures. These regions describe the principal classes of failure mechanism encountered in that portion of the life of a device.

Infant Mortality, as the name implies, represents the early life failures of a device. These failures are usually associated with one or more manufacturing defects.

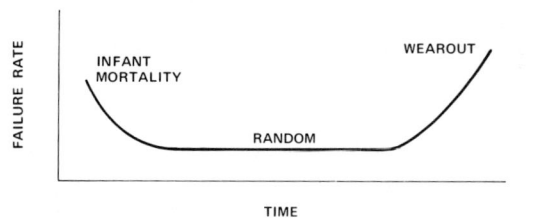

Figure 11. Reliability Life Curve.

After some period (usually in the high tens to low hundreds of hours) the failure rate approaches some constant low value where it remains for a period of hundreds of thousands to millions of hours (for integrated circuits) depending upon temperature, applied voltage, circuit complexity and other factors. This is the random failure portion of the curve and represents the useful portion of device life.

"Wearout" failures occur at the end of the device's useful life and are characterized by a rapidly rising failure rate with time as the devices "wearout" both physically and electrically. This does not occur before the high tens to hundreds of years if at all for integrated circuits.

Associated with each area of the curve are specific failure mechanisms. These failure mechanisms have been well reported in the literature[1,2,3]. Table 1 lists some of the more common mechanisms along with the portion of the reliability curve it effects, the associated thermal activation energy (defined in Section on Lifetesting), the test method used for detection and the preventive measure. A short description of each of these mechanisms follows.

a. Slow trapping represents charge accumulation at the $Si-SiO_2$ (see Figure 12) interface. In P-channel devices, the crystal orientation of the substrate material (silicon) is usually <111> (Figure 13b) and positive charges accumulate at the interface during the application of negative bias on the polysilicon gate. This can cause a threshold voltage (V_{TH}) shift at the interface of approximately 1.0 volt after exposure to 250°C for a period of 30 minutes. (The original threshold voltage for this material is in the 3 to 4 volt range.) The crystal orientation of N-channel devices, such as the 2107A and B is <100> (Figure 13a) which is much less prone to this drift. Under the same conditions, V_{TH} will drift only about 0.01 volt. This failure mechanism is defined as a wearout failure and with an activation energy of 1eV is readily detected by High Temperature Bias Testing.

b. Contamination has been one of the more serious problems in MOS manufacturing. Generally, contamination results in mobile positive charges trapped within the gate oxide which causes V_{TH} drift. The direction on the p-n junction electrostatic fields tends to make N-channel more susceptible to

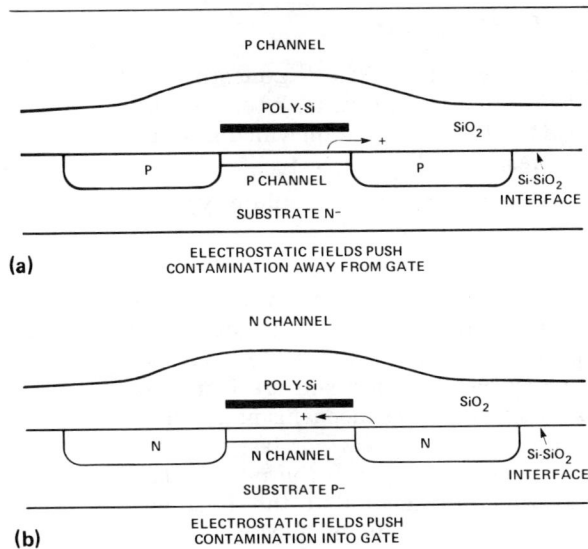

Figure 12. Effect of Contamination.

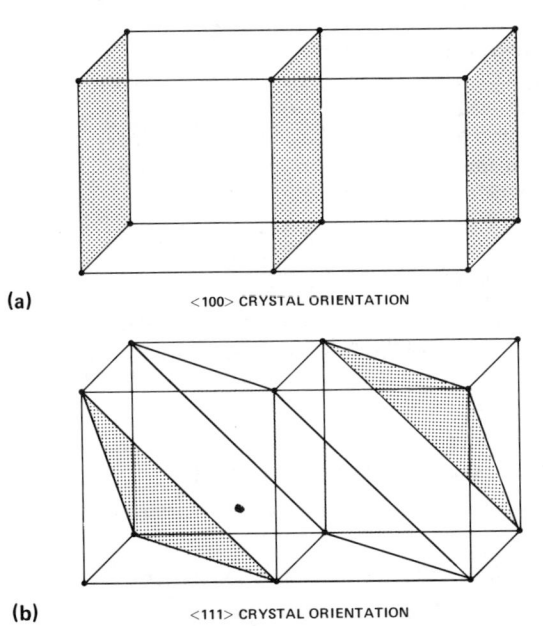

Figure 13. Crystal Orientation.

these mobile ions because they are pushed into the gate region. Figure 12 illustrates the effect of contamination on electrostatic fields. Again, the phenomenon results in wearout of affected devices and can be detected by High Temperature Bias Testing. The silicon gate process, however, allows the device to be subjected to high temperature gettering which "cleans" the oxide and prevents further contamination during subsequent processing.

c. The effect of surface charge phenomena can be observed with suitable test patterns. Charge can spread across the surface of a device and create parasitic transistors or leakage paths. The problem is aggravated by mobile ions on the die surface. Generally, ultra-clean processing combined with raising the field threshold (by process techniques) will prevent this problem. The phenomenon can be detected by bias high temperature testing although the temperature dependence can vary.

d. Polarization causes V_{TH} drift if the gate oxide contains polarizable molecules. This problem is usually present if phosphorous glass is left in the gate region. It is a wearout phenomenon with a strong temperature dependence. Intel's Si gate process, however, uses no phosphorous within the gate dielectric.

e. Electromigration is a well known failure mode[3] prevalent in conductors operated at high current densities. It results in the failure of metallization lines through void migration and eventual burn out. Conservative design rules limiting current to 1×10^5 A/cm^2 are employed in Intel's 4K RAMs. The failure mode results in wearout and high temperature operating life provides a suitable test.

f. Microcracks or open metal over steps have caused reliability problems in the past[1]. Figure 14 illustrates the cross section of such a microcrack. In order to prevent problems the basic process must be such that all steps are contoured and appropriate metal deposition techniques are employed. This random failure mechanism is readily detected by thermal cycling or operating life.

Table 1. Failure Mechanisms in MOS.

Failure Mode	Type	Activation Energy (E_{act})	Detection	Preventive Measure
Slow Trapping	Wearout	1.0 eV	High Temp Bias	Ultra-Clean Processing
Contamination	Wearout/Infant	1.4 eV	High Temp Bias	Ultra-Clean Processing
Surface Charge	Wearout	.5-1.0 eV	High Temp Bias	Ultra-Clean Processing
Polarization	Wearout	1.0 eV	High Temp Bias	Eliminate Phosphorus in Gate Oxide
Electromigration	Wearout	1.0 eV	High Temp Operating Life	$J < 10^5$ A/cm^2
Microcracks	Random	--	Temp Cycling	Contoured Oxide Steps
Contacts	Wearout/Infant	--	High Temp Operating Life	Ultra Clean Processing
Oxide Defects	Infant/Random	.3 eV	High Voltage Operating Life and Cell Stress	Ultra Clean Processing

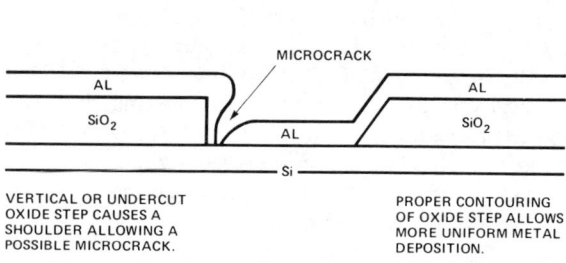

Figure 14. Microcrack Formation.

g. Contact failures can be of several types. Poor cleaning prior to metallization can cause marginal adhesion to silicon and result in opens during life test. Over alloying or excessive heat can drive metal through the metallurgical junction. The latter is more serious on N-channel and is illustrated in Figure 15a. Suitable processing precautions can prevent this failure mechanism. High temperature operating life is the most applicable life test although high temperature bake is also effective.

h. Oxide defects can cause dielectric breakdown in the MOS structures which generally results in an electrical short. High yield clean processing tends to minimize such defects but in practice they represent the reliability limit presently encountered in MOS. This failure mode prevails during the infant mortality and random portions of the reliability curve. An operating life test or high voltage stress is needed to screen the devices with defective oxides. This failure mechanism does not exhibit a strong temperature dependence.

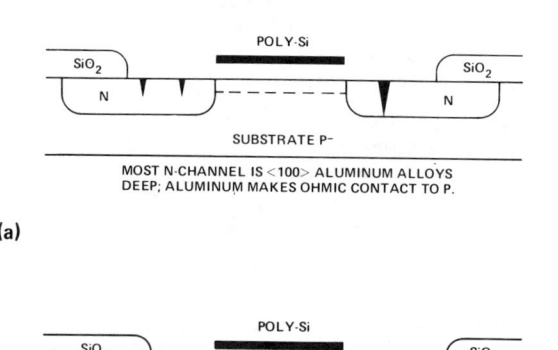

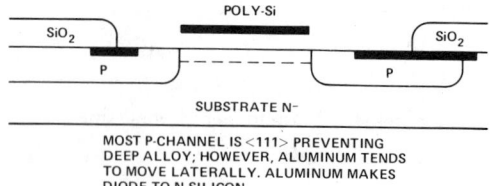

Figure 15. Effect of Over Alloying.

RELIABILITY PROGRAM

Recognizing the system designer's concern about the reliability of memory products, Intel has established a stringent reliability program which has three major areas of responsibility:

1. Qualification of products and failure types.
2. Reliability monitoring.
3. Failure analysis.

In the qualification process, it is essential to determine which failure mechanisms are associated with each portion of the reliability life curve (see Figure 11). Once these mechanisms are identified, methods for reducing these mechanisms are instituted in this design, manufacture or test phases of production. Intel's experience in the manufacture of semiconductors has facilitated the reliability studies on the 4K RAM products. Subsequent testing has proved that the qualifications are effective in reducing the potentially defective devices which are shipped to customers.

Once the failure mechanisms have been identified and the proper controls or tests instituted, Intel's Reliability Group regularly monitors the products. This is accomplished on a sample basis by the use of scanning electron microscopy, burn-in, rotating life tests, high temperature tests (reverse bias, dynamic and storage) and examining wafer processing test data. In addition, the Reliability Group has the responsibility for MIL-STD-883 group C testing.

To aid the qualification and monitoring functions, a well equipped failure analysis laboratory is maintained at Intel. The types of failure analysis which may be performed are:

1. Standard production tests utilizing the same equipment as is used for product testing.
2. Special bench test equipment for non-standard electrical tests.
3. Micromanipulator stations to probe specific die areas.
4. Bit map displays monitor cell condition.
5. Scanning electron microscopes with voltage contrast capability.
6. Electron microprobe.
7. Auger electron spectroscope for material and foreign matter analysis, and a metalurgical lab.

The degree of success of Intel's Reliability Program can be seen in the graph, Figure 16, showing the improvement trends in reliability as a function of time.

Failure Analysis Instrumentation

The Scanning Electron Microscope (SEM)

The SEM is a very powerful tool in failure analysis. Differing from optical microscopes in that it uses

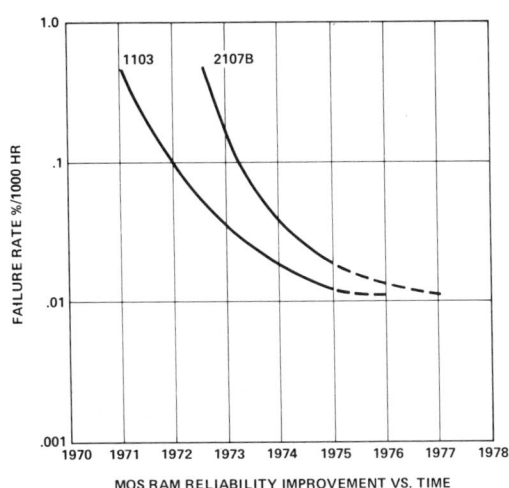

Figure 16. MOS RAM Reliability Improvement Vs. Time.

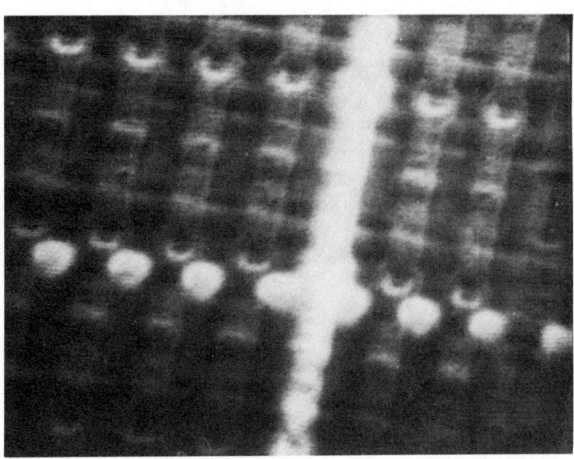

Figure 17. SEM Photomicrograph.

an electron beam instead of light, it exhibits high resolution (≈ 200Å) and a large magnification range (x5 to x50,000). Normally, the SEM is used to examine the surface topography of a device by detection of low energy back-scattered electrons. Another useful technique using the SEM is voltage contrast measurement, in which the presence of changing electric fields are detected in an operating device. Figure 17 shows this technique used to detect an addressed cell failure. These are just two of the many analysis techniques in which the SEM is used[4,5].

Nondispersive Electron Microprobe

This instrument is similar to the SEM, but the sample is bombarded by high energy electrons. The wave length and intensity of the emitted x-rays are then measured. A computer reduces these data and displays them on a C.R.T. Figure 18 shows the display of elemental content. The electron microprobe is used to determine the type and relative amount of impurities or contaminants in a sample. The limitation of the microprobe lies in its inability to see surface contiminants which are in a very thin film (≈ 20Å)[6].

Bit Map Video Display

Used in conjunction with a memory tester, this instrument is used to display the failed cell locations in a memory. The display is topographically the same as the die being examined and thus failure patterns may be easily seen. Figure 19a shows an isolated bit failure and Figure 19b shows a failed row.

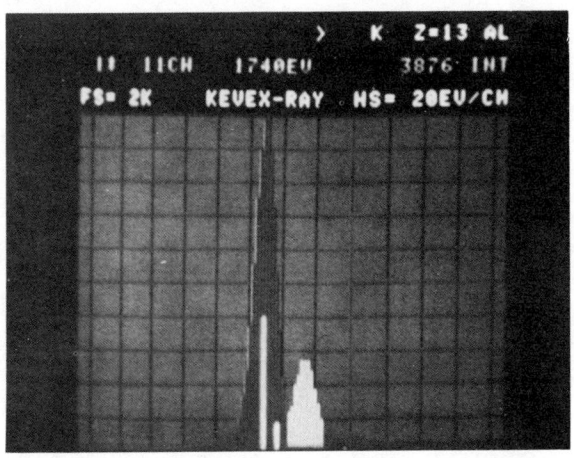

Figure 18. Electron Microprobe Spectrum.

Figure 19a. Bit Map Display Single Bit Failure.

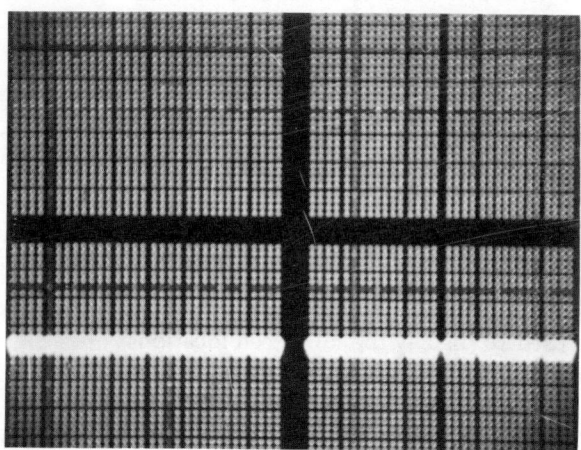

Figure 19b. Bit Map Display Failed Row.

Auger Electron Spectroscope

The Auger Spectroscope is used to detect those surface contaminants which the electron microprobe misses. Low energy (3-5KEV) electrons bombard the test sample and the energy spectrum of the emitted secondary electrons is measured. In this way, surface defects or contamination are examined to a depth of 10 to 50Å. The limitation of the Auger spectroscope is that its accuracy decreases for elements lighter than Boron. For the lighter elements, a similar instrument, the x-ray microprobe is used[7].

LIFE TESTING — THE ARRHENIUS PLOT

One of the most effective ways of gathering failure rate data on a semiconductor device is life testing. Intel employs a variety of these tests to gather initial data and to monitor reliability on a continuing basis. This section discusses some of the aspects of life testing as applied to 2107A and 2107B RAMs.

During the initial phases of a new product, the failure mechanisms must be determined in each portion of the reliability life curve. Since infant mortality failures occur very early in a device's life, no acceleration is necessary. However, when information on wearout failure mechanisms is required, the device must have its life "accelerated". This is accomplished by either subjecting the device to a high temperature, a high voltage or both. If high temperature testing is used and the thermal activation energy of the failure mechanism is well defined (so that the failure can be "accelerated"), a plot as shown in Figure 20 can be made. This is an Arrhenius plot in which time is plotted against the

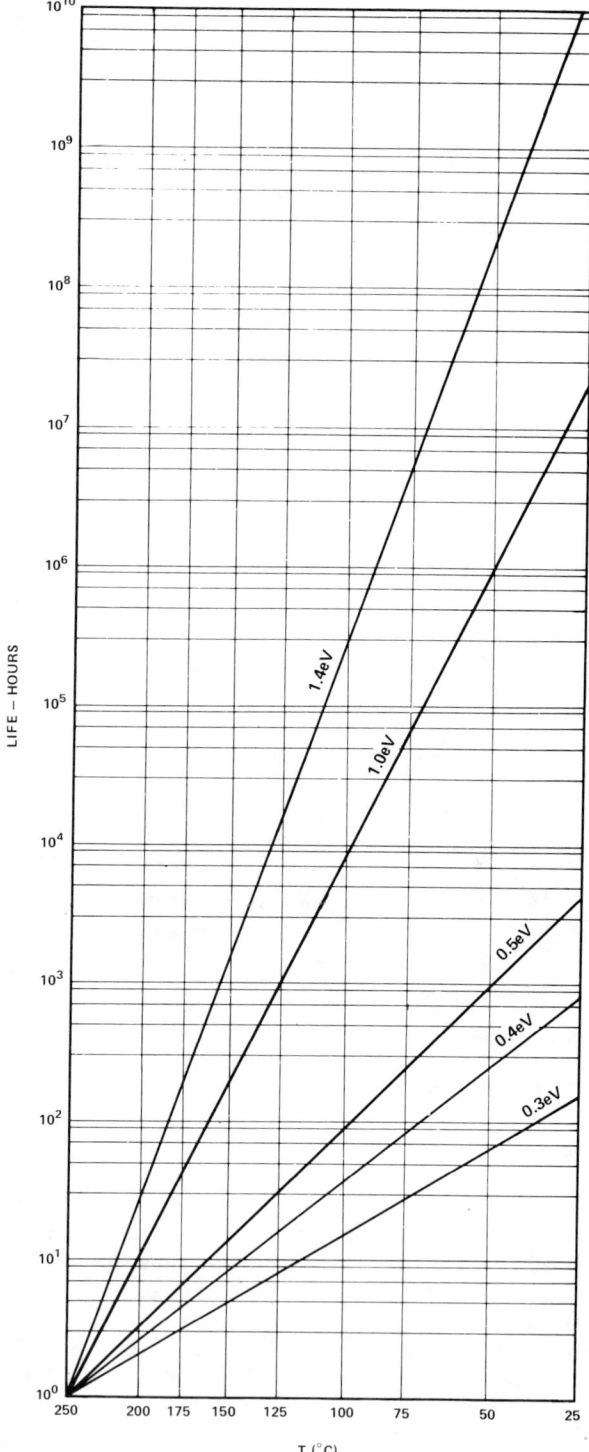

Figure 20. Arrhenius Plot.

reciprocal of temperature and the slope of the curve is the thermal activation energy in electron volts. Using this plot, failure rate data taken at one temperature may be translated to another temperature with reasonable accuracy. In this form the plot follows the relation:

$$\tau = \tau_0 \exp \frac{E}{KT} \quad (1)$$

where τ = MTBF (Mean Time Between Failures) at the desired temperature

τ_0 = MTBF at the test temperature

E = The thermal activation energy (eV)

K = Boltzmanns constant (8.63×10^{-5} eV/°K)

T = Test temperature in °K.

The acceleration factor may be calculated by:

$$F = \exp E/K (1/T_1 - 1/T_2) \quad (2)$$

where T_1 is the test temperature (°K) and T_2 is the desired temperature (°K).

Figure 20 shows an Arrhenius plot with several thermal activation energies for common failure mechanisms (see Table 1).

Thermal activation energy describes the temperature dependence of a failure mechanism and is the slope of the plot of failure rate or its reciprocal, life, versus the reciprocal of temperature (see equations 1 and 2). Equation 1 can be solved for E giving:

$$E = KT \exp^{-1} \frac{\tau}{\tau_0}$$

Forms of Life Testing

The most common forms of life test applied to 2107A and 2107B are:

1. Dynamic Burn-In.
2. High Voltage Cell Stress.
3. High Temperature Bias Testing.
4. System Life Test.

Dynamic Burn-In and High Voltage Cell Stress

During the initial reliability evaluation of the 2107A and 2107B, an Infant Mortality failure rate of approximately 1% was determined. The primary failure mode was oxide breakdown. The failure rate versus time and temperature is shown in Figure 21. Since the failure rate of 1% infant mortality failures was too high, all devices were subjected to a 48 hours 160°C Dynamic Burn-In. This test involves exercising the memory at the elevated temperature. While it can be seen from Figure 21 that this burn-in did accelerate the failure rate to some extent,

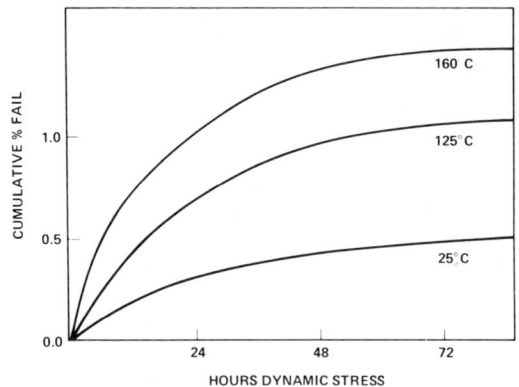

Figure 21. Temperature Dependance of Oxide Breakdown Failures.

its effect in accelerating oxide breakdown failures was not as high as desired. This is because oxide breakdown failures have a thermal activation energy of 0.3eV so that the temperature dependence is relatively small. As a result, a high voltage cell stress test was implemented along with process improvements which succeeded in screening most of the potentially defective devices. Figure 22 and Table 2 show the results of the High Voltage Cell Stress test versus Dynamic Burn-In. Note that when high voltage cell stressed devices are returned to normal operating voltage, the random failure rate decreases. This data shows that cell stress eliminates more oxide defects than Dynamic Burn-In. The cell stress test is accomplished by raising the power supply and input voltages to levels much higher than normal and exercising the device with a standard data pattern for 1.8 sec. and 880ms for the 2107A and 2107B respectively.

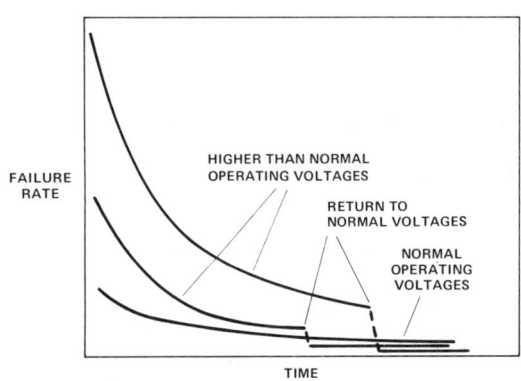

Figure 22. Effect of High Voltage Cell Stress on Infant Mortality Failures.

Table 2. Comparison of Observed Failure Modes After Burn-In vs. the High Voltage Cell Stress Test.

% of Total Cell Stress Failures	Ratio of Cell Stress to Dynamic Burn-In Failures	Failure Mode
11.8%	4.5:1	Total Array
35.3%	3:1	Single Bit
11.8%	1.8:1	Row
17.6%	1:1	Column
23.5%	1:1	Miscellaneous

High Temperature Bias Testing

One of the major areas of concern in MOS devices is V_{TH} stability. As discussed in the section on failure mechanisms, N-channel devices exhibit V_{TH} drift typically two orders of magnitude (.01 vs. 1.0 volt) lower than P-channel devices.

To determine the stability of the 2107A and 2107B threshold voltages, a total of 44 test devices were selected at random from several production lots.

Test devices are discrete transistors which are fabricated on the wafer along with the memories. These devices were subjected to high temperature bias stress under the following conditions:

V_G = 15V
V_{SS} = 0V
V_{DD} = 0V
V_{BB} = -5V

The temperature was 250°C and test time was 40 hours. Of the 44 devices, 22 had a narrow gate of 6μ and 22 had a wide gate of 9μ, representing the ends of the gate width distribution.

Threshold measurements were made initially and at 2, 10, 20 and 40 hours. The average threshold of the two groups fell into two distributions with a mean final drift of less than 10mV as shown in Figures 23 and 24. Also, the first ten hours of stress accounted for the greatest amount of drift. Using the Arrhenius plot, Figure 20, 10 hours at 250°C is equivalent to 10,000 hours at 125°C using a thermal activation energy for threshold drift of 1.0eV. Individual plots of the V_{TH} drift at 40 hours for the 44 test devices are given in Figures 25 and 26 for narrow and wide gate widths, respectively.

A threshold drift of ten millivolts has minimal effect on circuit parameters. To show this, detailed device parameter measurements were made before and after high temperature bias stress. The data shown here are typical of the results obtained. Voltage margins, a sensitive indicator of device degradation are shown in Figure 27a (V_{BB} vs. V_{DD}). These tests were performed on production 2107A and 2107B devices.

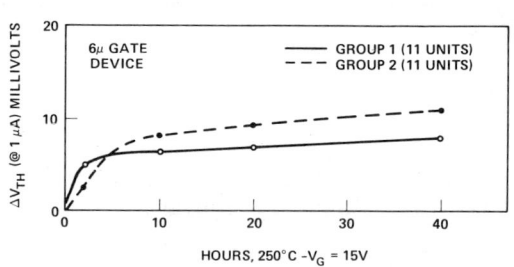

Figure 23. V_{TH} Drift 6μ Gate Devices.

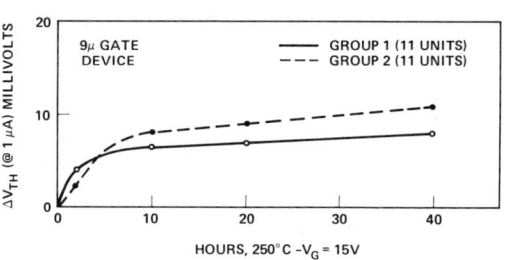

Figure 24. V_{TH} Drift 9μ Gate Devices.

For comparison, margin drifts for 70 hour 160°C dynamic burn-in and 100 hour 125°C life test are shown in Figure 27b and Figure 27c respectively. Substrate bias margin is shown in Figure 28, before and after a 1400 hour 125°C dynamic life test. High temperature bias test connections are shown in Figure 29 and the connections for the dynamic test are shown in Figure 30.

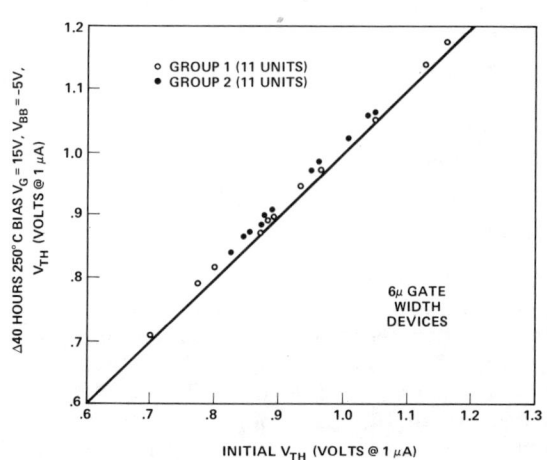

Figure 25. Scatter Plot of V_{TH} Drift 6μ Gate Devices.

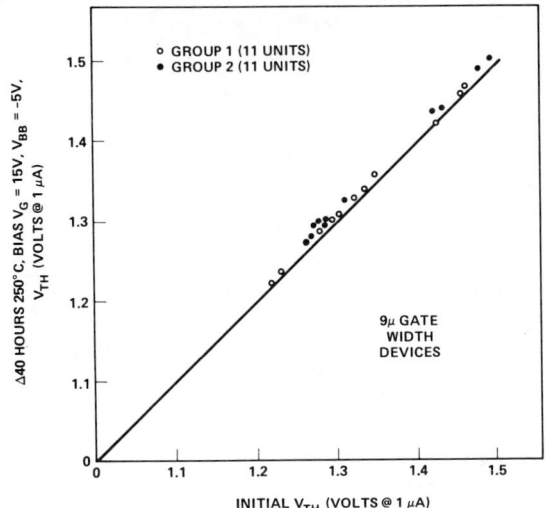

Figure 26. Scatter Plot of V_{TH} Drift. 9μ Gate Devices.

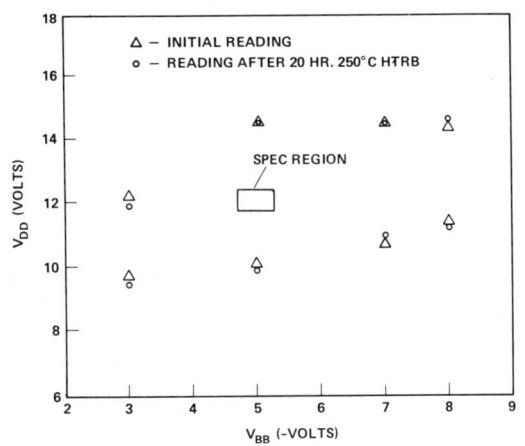

Figure 27a. V_{BB} Vs. V_{DD} HTRB.

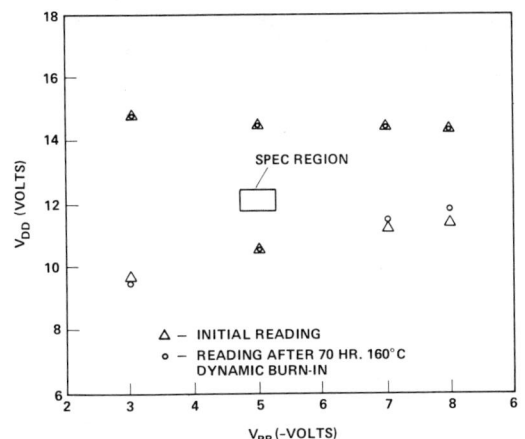

Figure 27b. V_{BB} Vs. V_{DD} Dynamic Burn-In.

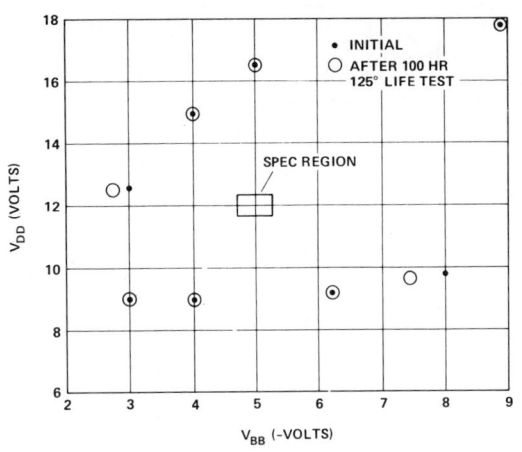

Figure 27c. V_{BB} Vs. V_{DD} 125°C Life Test.

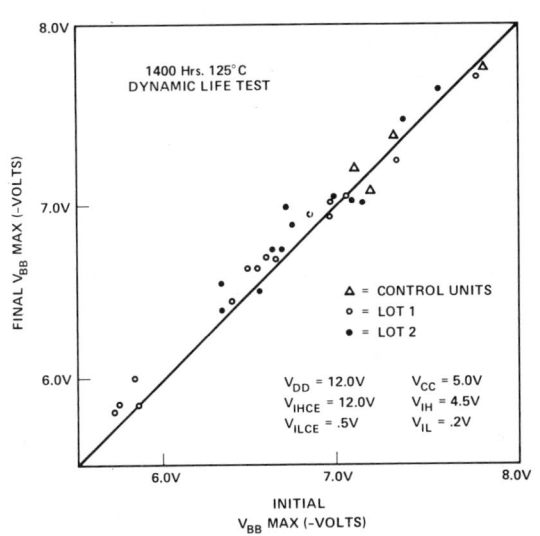

Figure 28. V_{BB} Drift 1400 Hours Dynamic Life Test.

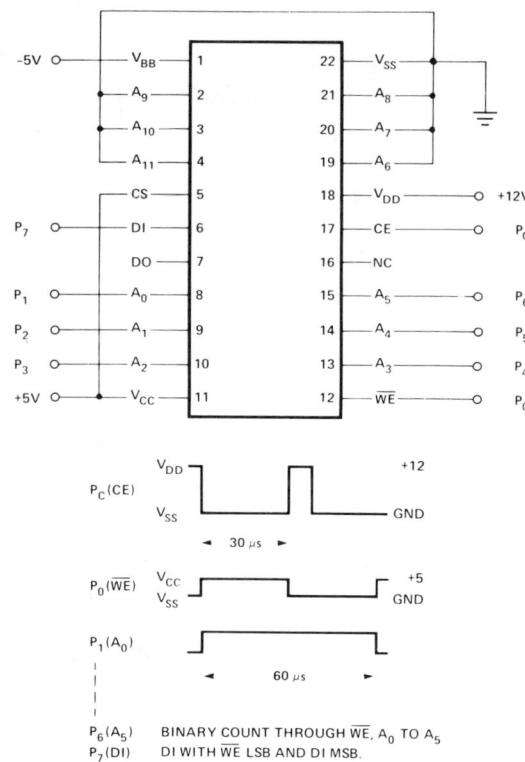

Figure 29. Connections for Dynamic Burn-In.

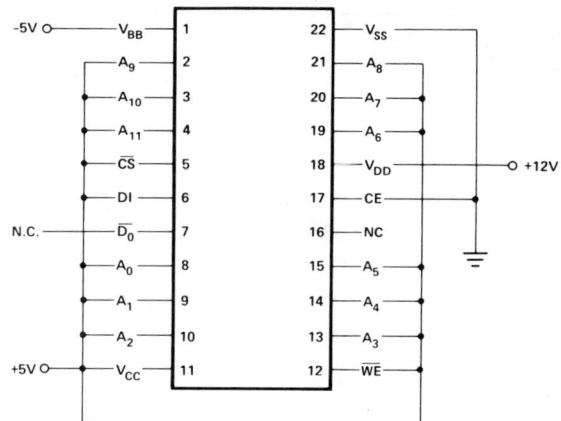

Figure 30. Connections for HTRB.

System Life Test

In addition to the previous tests, an evaluation of 2107A and 2107B failure rates under operating conditions has been made. This is the most useful data to a system designer, since it allows him to determine the reliability of a device under conditions similar to those in which he will use the device. For this purpose, system cards of 32 units each were fabricated and connected to control boards which exercise the devices with selected data patterns. Three test types were performed:

1. Dynamic life test at 125°C, see Figure 28.
2. Continuous life test at 70°C, see Appendix A.
3. Rotating life test — same as continuous except that the devices are rotated out of the system once 1,000 hours of data have been accumulated.

TEST RESULTS

The results of these tests are shown in Tables 3 and 4. Table 5 gives the failure rates at 60% and 90% confidence levels. The failure rates were calculated by the straight line technique, that is, no weight is given to early failures. This conservative technique gives 2107A failure rates of 0.1%/1000 hrs and 0.2%/1000 hrs for 60% and 90% confidence levels, respectively. For the 2107B, .04%/1000 hrs and .05%/1000 hrs for 60% and 90% confidence levels, respectively. This data is based on 70°C ambient temperatures, with 125°C data scaled to 70°C by an activation energy of 0.3eV (oxide breakdown). Although oxide breakdown is no longer the primary failure mode, the use of any higher activation energy associated with the other failure modes would yield a lower failure rate. Confidence levels described above were calculated from the MIL R 38100 nomograph.

Table 3. Results of System Tests for 2107A.

No. Units	Test Type	Test Type	Failures At:			
			168 Hr	1 k Hr	2 k Hr	5 k Hr
255	System C	70°C	0	0[1]	—	—
384	System R	70°C	0	0	—	—
512	System C	70°C	1*	0	—	—
96	Dynamic	125°C	0	0	—	—
95	Dynamic	125°C	0	0	0	—
165	Dynamic	125°C	2**	—	—	—
64	Dynamic	125°C	0	0	0[2]	—
96	Dynamic	125°C	0	0	—	—
160	Dynamic	125°C	1*	—	—	—

Notes:
1. Last reading @ 1700 hr.
2. Last reading @ 3500 hr.
 *Oxide Defect.
**Oxide/Substrate Resistance High.

Table 4. Results of Life Tests 2107B.

No. Units	Test Type	Temp.	Failures At: 168 Hr	1 k Hr	2 k Hr	5 k Hr
256	Dynamic	125 C	0[3]	—	—	—
288	Dynamic	125 C	0	1††	—	—
148	Dynamic	125 C	0	0	—	—
96	Dynamic	125 C	1†	0	0	0
288	System	70 C	0	0	0	0
96	System	70 C	0	0	0[4]	—
148	System	70 C	0	0	0[4]	—
187	Dynamic	125 C	0	1*	0	—
152	Dynamic	125 C	1*	0	—	—
192	Dynamic	125 C	0	0	—	—
96	Dynamic	125 C	0	0	—	—
92	Dynamic	125 C	0	0	—	—
116	Dynamic	125 C	0	0	—	—
96	System C	70 C	0	0	0	—
128	System C	70 C	0	1††	—	0[5]
128	System C	70 C	0	0	0	0[6]
192	System R	70 C	0	0	—	—

Table 5. Failure Rate Predictions.

No. Table	Dev. Hrs.	Activation Energy = 0.3 eV Equiv. Dev. Hrs. 70°C	Total Hours 70°C	Number Failures	Failure Rate 60% C.L.	Failure Rate 90% C.L.
3	660.6K 1329.5K	2242.4K 1329.5K	3971.9K	4	0.1%/K hr.	0.2%/K hr.
4	2072K 3976K	8263K 3976K	12,234K	5	0.04%/K hr.	0.05%/K hr.

Based on thermal activation energy of 0.3eV (oxide breakdown).

Notes: 3. Last reading @ 500 hr.   *Oxide Defect.
       4. Last reading @ 3000 hr.   †Function Failure.
       5. Last reading @ 6000 hr.   ††Fab Defect.
       6. Last reading @ 7000 hr.

ACKNOWLEDGMENT

The author expresses gratitude to the personnel of the Design Engineering, Test, and Reliability groups whose help has been invaluable in preparing this report.

REFERENCES

1. G.L. Schnable and R.S. Keen, Jr., "Failure Mechanisms in Large Scale Integrated Circuits", IEEE Transactions on Electron Devices, Vol. ED-16, pp 322-332, Apr. 1969

2. S.R. Hofstein, "Stabilization of MOS Devices", Solid State Electronics, 10, 657, (1967).

3. J.R. Black, "Mass Transport of Aluminum by Momentum Exchange with Conducting Electrons", 6th Annual Reliability Physics Symposium Proceedings, pp 148-153, Nov. 1967.

4. R.F. Haythornthwaite, A.R. Molozzi, and D.V. Sulway, "Reliability Assurance of Individual Semiconductor Components", Proceedings of the IEEE, Vol. 62, No. 2, Feb. 1974.

5. A.J. Gonzales, "Failure Analysis Applications of the Scanning Electron Microscope", Proceedings of IEEE 11th Annual Reliability Physics Symposium, pp 179-184, 1973.

6. J.W. Colby, "Failure Analysis using the Electron Microprobe", Proceedings of IEEE 11th Annual Reliability Physics Symposium, pp 189-193, 1973.

7. A.J. Gonzales, "Failure Analysis Applications of the Auger Electron Spectroscope", Proceedings of IEEE 11th Annual Reliability Symposium, pp 185-188, 1973.

8. C.G. Prattie, et.al., "Elements of Semiconductor-device Reliability," Proceedings of the IEEE, Vol. 62, no. 2, Feb. 1974.

2107 N-CHANNEL 4K RAMS

APPENDIX A — 2107B SYSTEMS TEST

The life test system for the 2107B consists of a group of storage cards each holding 32 devices. In use, the system exercises four devices in parallel and cycles through the entire storage card in four microseconds.

The basic timing is:

CE on = 1 μsec
CE off = 16.5 μsec
Access strobe = 220ns

and is shown in Figure A1.

Dynamic life tests are run in a controlled environment at 125°C. Rotating life tests are performed at 70°C, also in a controlled environment. Samples are taken from production lots and placed on the rotating life test system every 200 hours. As each card accumulates 1000 hours it is replaced with a new set of devices. See Table A1. The letters indicate the test board positions in the fixture while numbers indicate the sequence of test board insertion. For example, when board 1 accumulates 1000 hours, it is replaced by board 6. Thus each board accumulates 32,000 device hours before being replaced.

In continuous life test, the same conditions prevail except that the units are not rotated at 1000 hours.

As of June 1975, the 2107A and 2107B units undergoing extended life test number 255 and 352 respectively and the cumulative number of devices on rotating life test are 384 and 192 respectively.

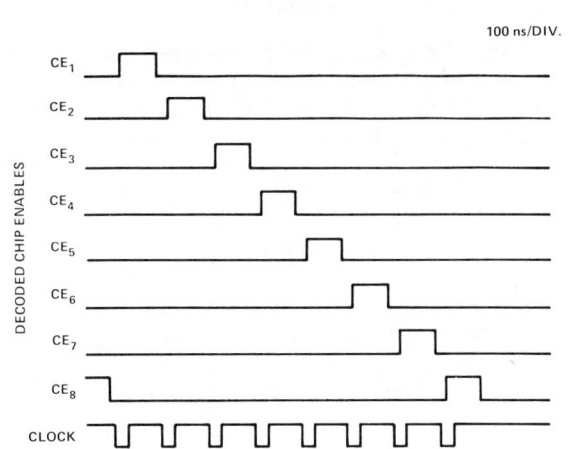

Figure A1(a). System Life Test Timing.

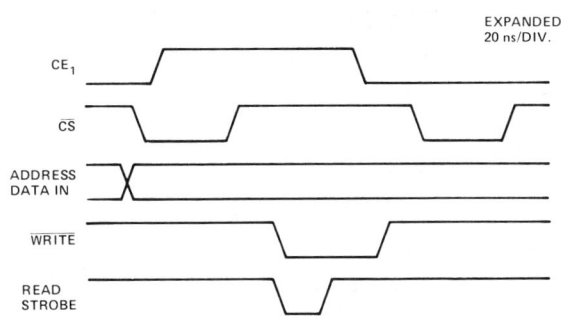

Figure A1(b). Expanded System Life Test Timing.

Table A1. Rotating Life Test.

| Hours | A | Test Position | | | |
		B	C	D	E
0	1	X	X	X	X
200	1	2	X	X	X
400	1	2	3	X	X
600	1	2	3	4	X
800	1	2	3	4	5
1000	6	2	3	4	5
1200	6	7	3	4	5
↓	↓	↓	↓	↓	↓
↓	↓	↓	↓	↓	↓
1800	6	7	8	9	10

X = Blank
Number = Test Board in Position Indicated.

APPENDIX B — QUALITY ASSURANCE

Once a device has been qualified as reliable and transferred to manufacturing, it is imperative that sufficient process controls are in effect to assure the continuation of that qualify. Intel's Q.A. flow centers around a series of acceptance gates between process entities as illustrated in Figure B1 and detailed inspection within the processing areas at critical points. For example, in wafer processing, furnaces are routinely monitored for contamination through the use of capacitance voltage measurements on test chips. Also electrical tests such as breakdown strength measurements are performed on test patterns on each wafer. Routine high magnification scanning electron microscope examinations at critical process steps also provide important process control feedback. A final Q.A. acceptance is performed on all lots prior to shipment to assembly locations. Table B1 details the Q.A. acceptance gates within the assembly flow and identifies the appropriate method employed per MIL-STD 883. During the test and finishing operation Q.A. maintains standards, test tape and calibration control of all production equipment. Table B2 shows the standard test performed on each 2107A and 2107B in production. Final Q.A. inspection is performed after the mark and pack operation.

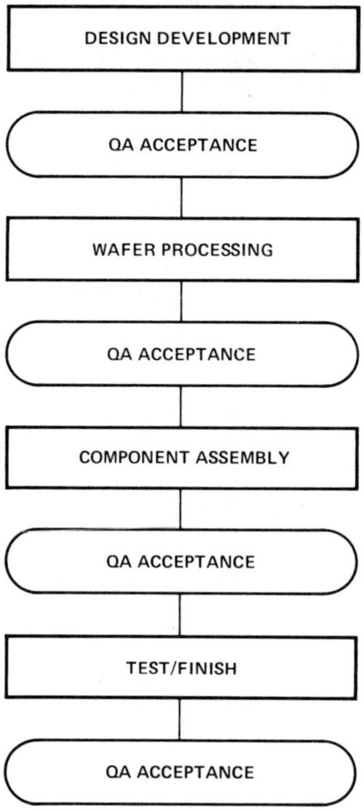

Figure B1. QA Flow Diagram.

Table B1. Assembly Flow.

Operation	Mil Std 883 Method:
Piece Part Inspections	
Scribe and Break	
Q.A. Acceptance	2010.1 B
2nd Optical Insp.	2010.1 B
Lead Bond	
Q.A. Acceptance	2010.1B
3rd Optical Insp.	2010.1B
Seal	
Temperature Cycle	1010 (C)
Q.A. Acceptance	1014 (B)
Fine Leak	1014 (A) 5×10^{-7}
Q.A. Acceptance	1014 (C)
Gross Leak	1014 (C)
Q.A. Acceptance	2009
Final Visual	2009

Table B2. Standard Product Classification Flow

Test Type	No. Tests	Remarks
D. C.	14	Shorts, Opens, Leakage, Currents, etc.
Dynamic	13	5 Basic Patterns
H. V. Cell Stress	4	2 Cell Stress, 2 Function Tests

2107 N-CHANNEL 4K RAMS

APPENDIX C — OTHER N-CHANNEL DATA

The 2107B N-channel technology is based on the successful introduction of several other products. Specifically the 2102, a static 1K RAM; the 2401, a dual 1K shift register; and the 2105, a dynamic 1K RAM have received extensive lifetesting. The data are presented in Table C1, with calculated failure rates shown in Table C2 (60% confidence level).

Table C1. Lifetest Summary.

Product	Lifetest	Temp.	Qty.	Hours	Failures
2102/A	HTRB	125°C	71	9000	0
	HTRB	125°C	150	1000	0
	Dynamic	125°C	105	1000	0
	Dynamic	125°C	88	8500	0
	Dynamic	125°C	71	1000	0
	System	70°C	32	21000	0
	System	70°C	1248	1000	1*
	Dynamic	125°C	105	5000	0
	HTRB	125°C	210	1000	0
	Dynamic	125°C	1090	1000	2**
	System	70°C	448	1000	0
	System	70°C	105	3000	0
	System	70°C	105	6000	0
2401	HTRB	125°C	105	1000	0
	System	70°C	96	22000	0
2105	HTRB	125°C	210	1000	0
	System	70°C	512	1000	1***
8080	Dynamic	125°C	30	2000	0
	Dynamic	125°C	20	3000	0
2101	Dynamic	125°C	150	1000	0
2111	Dynamic	125°C	150	1000	0
2112	Dynamic	125°C	150	1000	0

*The failure was a function failure (Row).
**Catastrophic failures (Row, Column)
***Open

Figure C2. Calculated Failure Rates.

Product	Dev. Hrs. @ °C		Equiv. Dev. Hr. @ 70°C	No. Fail	Failure Rate
2102/A	3,538,000	125	14,152,000	2	
	3,313,000	70	3,313,000	1	
Total			17,465,000	3	0.017%/K hr.
2105	210,000	125	840,000	0	
	512,000	70	512,000	1	.08%/K hr.
Total			1,352,000	1	

2115/2125 N-CHANNEL SILICON GATE MOS 1K STATIC RAMS

BRUCE EUGENT

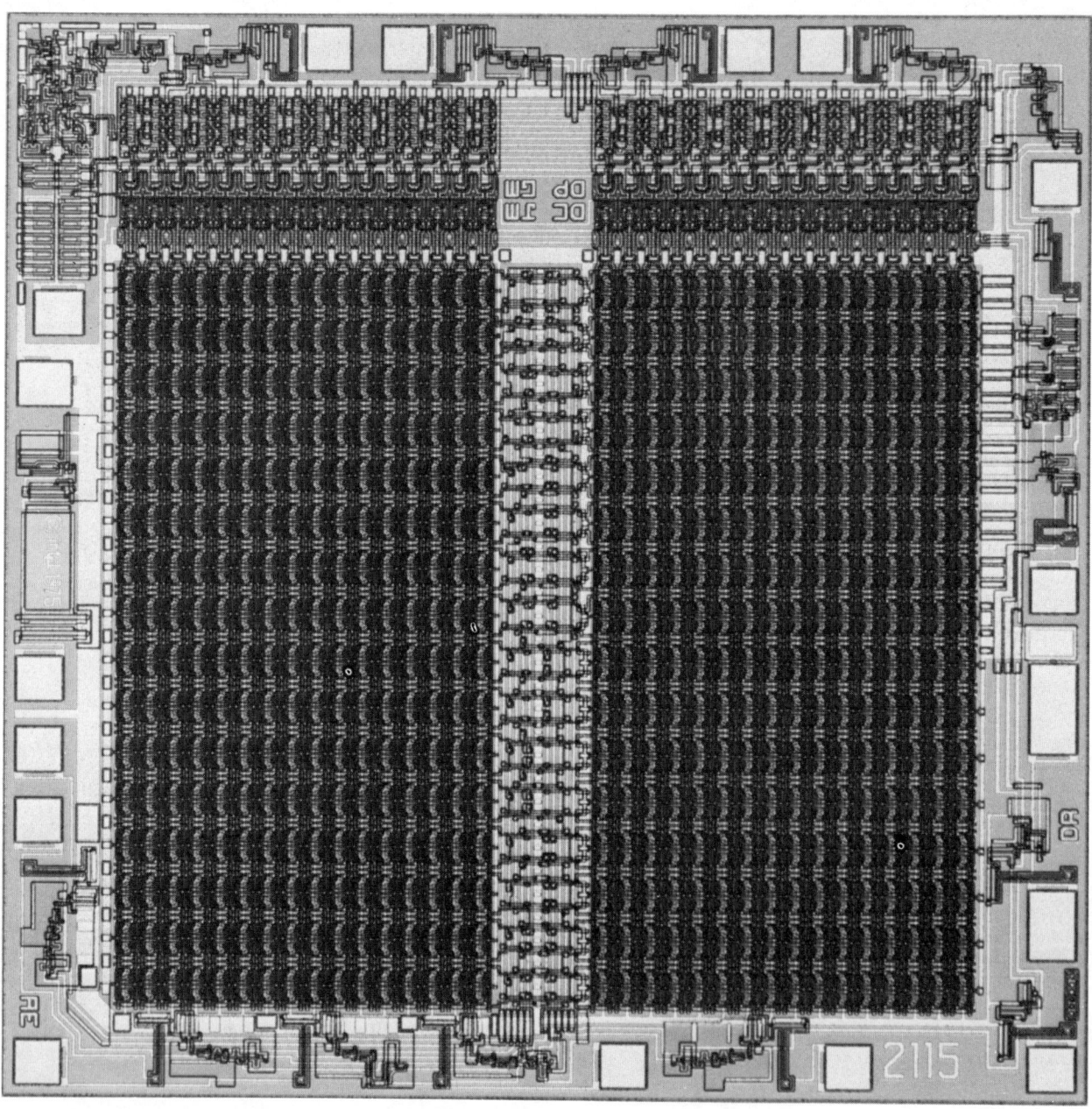

Photomicrograph of the Intel 2115.

2115/2125 STATIC 1K RAMS

INTRODUCTION

This report presents the reliability of the Intel® 2115 and Intel® 2125 high speed, 1Kx1, MOS static RAMs. This report is divided into five sections:

1) Description & device operation
2) Failure mechanisms
3) Reliability program
4) Reliability testing
5) Test results & summary

Based upon data presented in this report, a failure rate prediction of 0.03% per 1000 hours at 75°C with a 90% confidence level can be made.

DESCRIPTION AND DEVICE OPERATION

The Intel® 2115/25 family is a high speed, fully static read/write random access memory, organized 1024 words by 1 bit. They are fully TTL compatible in all respects: inputs, outputs and a single +5 Volt supply. They are packaged in the industry standard 16 pin dual-in-line package. (see Fig. 1).

The Intel® 2115/25 family offers a variety of configurations (see Table I) for output, speed and power considerations.

The Intel® 2115/25 family is fabricated in the N-channel MOS silicon gate technology. This technology allows the design and production of high speed memories which are compatible with the performance of bipolar RAMs but offering the lower power dissipation of MOS.

TABLE I
Intel® 2115/25 Family

Device	TACC	Power (V_{CC} = 5.0V ± 5%)	Output
2115L	95nS	65mA	Open Drain
2115	95nS	100mA	Open Drain
2115-2	70nS	125mA	Open Drain
2125L	95nS	65mA	Three State
2125	95nS	100mA	Three State
2125-2	70nS	125mA	Three State

Device Operation

As shown in the block diagram of Figure 2 the 1024 storage cells are arranged in a 32 row-by 32 column matrix. The five low order addresses (A_0-A_4) are decoded on chip to select a particular column; the five high order addressess (A_5-A_9) select a particular row. A cell is selected by a coincidence of a row and a column select. The sense amps and write drivers are controlled by $\overline{CS}$, $\overline{WE}$, and D_{IN}.

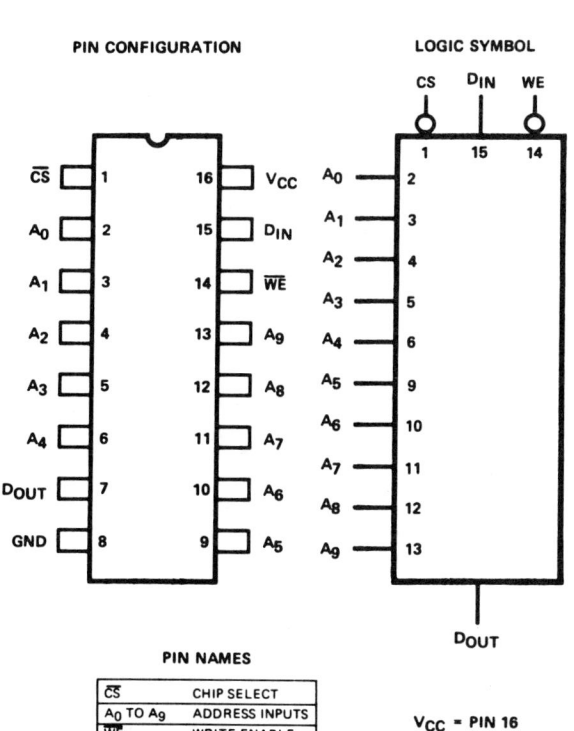

Figure 1. Pin Configuration & Logic Symbol for Intel® 2115/25 Family

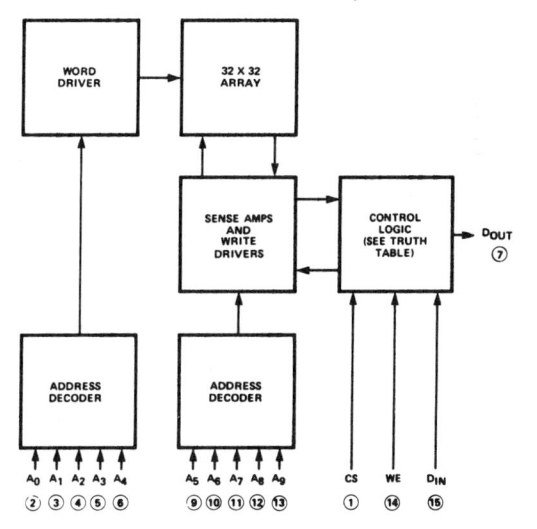

Figure 2. Block Diagram for Intel® 2115/25 Family

A 6-transistor static memory cell is used as the storage medium. This memory cell, shown in Figure 3, uses depletion load pullups to increase speed. This is the same basic memory cell used in the industry standard Intel® 2102A.

The Intel® 2115/25 has a substrate bias generator that provides the correct substrate bias to maintain proper threshold control. The substrate bias generator consists of a free running oscillator and a transfer capacitor to the substrate. The transfer capacitor is clamped below ground to generate the negative substrate bias.

Intel® 2115/25 Bit Map

Figure 4 shows the location of each cell in the memory matrix. Using 0's for address low inputs and 1's for address high inputs, the cell location can be determined by taking the desired address conditions and cross referencing it to the proper column and row designated by that address. The lower order addresses (A_0-A_4) are decoded along the edge of the array. The higher order addresses (A_5-A_9) are decoded up the center of the array. The intersection of the decoded row and column addresses respectively determine the selected cell location.

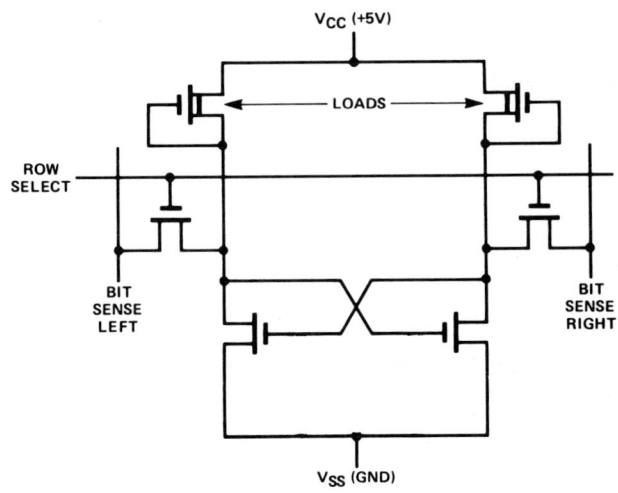

Figure 3. Cell Schematic for Intel® 2115/25 Family

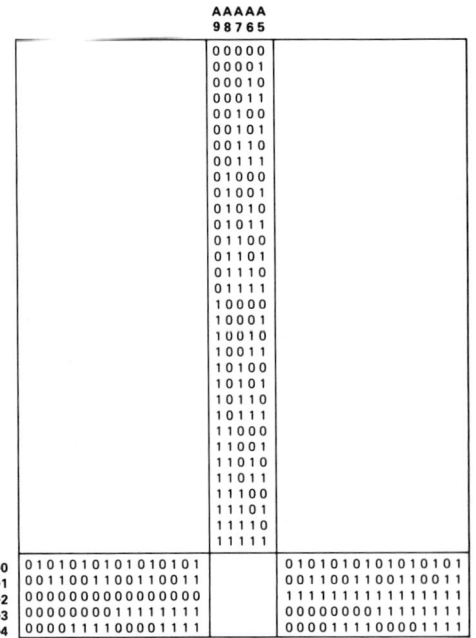

Figure 4. Intel® 2115/25 Family Cell Location Bit Map

FAILURE MECHANISMS

The failure mechanisms associated with N-channel RAM's are applicable to the Intel® 2115/25 family. The fundamental principles of Reliability Engineering predict the failure rate of a group of devices will follow the so called bathtub curve in Figure 5. The curve is divided into three regions: Infant Mortality, Random Failures, and Wearout Failures. All classes of failure mechanisms can be assigned to these regions.

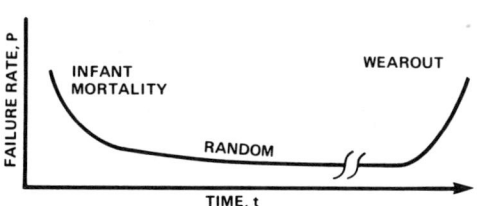

Figure 5. Reliability Life (Bathtub) Curve

Infant Mortality, as the name implies, represents the early life failures of a device. These failures are usually associated with one or more manufacturing defects.

After some period of time (usually in the high tens to low hundreds of hours) the failure rate reaches a low value. This is the random failure portion of the curve that represents the useful portion of device life. This period lasts from hundreds of thousands to millions of hours (for integrated circuits) depending upon temperature, applied voltage, circuit complexity, and other factors. During the random failure portion of the curve there is a decline in the failure rate due to the depletion of potential random failures from the general population.

"Wearout" failures occur at the end of the device's useful life and are characterized by a rapidly rising failure rate with time as the devices "wearout" both physically and electrically. This does not occur before the high tens to hundreds of years if at all for integrated circuits.

Associated with each area of the curve are specific failure mechanisms. Table II lists some of the more common mechanisms along with the portion of the reliability curve it affects, the associated thermal activation energy and the test method used for detection. With the exception of electron injection, each of the failure modes in Table II is discussed in Intel® R.R.7.[1] A brief description of election injection follows.

Electron injection into the gate oxide near the drain produces a shift in device current voltage characteristics. Electron injection is caused by the electron current in the high field region near the drain generating electron – hole pairs from impact ionization. Holes flow to the substrate and electrons are attracted toward the gate and drain. Electrons with enough energy can be injected into the gate oxide, with a small fraction trapped in the oxide causing a negative charge to build up. This can result in an increase in the device threshold and severe changes in the device's I-V characteristic especially when the source and drain are interchanged.[2] A high voltage operating life stress is needed to insure device parameters are such that electrons do not receive enough energy to be injected into the gate oxide in sufficient quantities to affect device characteristics. This failure mechanism is accelerated by low temperature and high voltage.

TABLE II

Failure Mechanisms in MOS

Failure Mode	Type	Activation Energy (E_{act})	Detection
Slow Trapping	Wearout	1.0 eV	High Temp Bias
Contamination	Wearout/Infant	1.4 eV	High Temp Bias
Surface Charge	Wearout	.5-1.0 eV	High Temp Bias
Polarization	Wearout	1.0 eV	High Temp Bias
Electromigration	Wearout	1.0 eV	High Temp Operating Life
Microcracks	Random	—	Temp Cycling
Contacts	Wearout/Infant	—	High Temp Operating Life
Oxide Defects	Infant/Random	.3 eV	High Voltage Operating Life and Cell Stress
$\bar{e}$ injection	Wearout	—	Low Temp, High Voltage Operating Life

RELIABILITY PROGRAM

Recoginizing the system designer's concern about the reliability of Memory Products,[3] Intel has established a stringent reliability program which has three major areas of responsibility:

1. Qualification of products
2. Reliability monitoring
3. Failure analysis

In the qualification process, it is essential to determine which failure mechanisms are associated with each portion of the reliability life curve (see Figure 5). Once these mechanisms are identified, methods for reducing these mechanisms are instituted in the design, manufacture or test phases of production. Intel's experience in the manufacture of semiconductors has facilitated the reliability studies on static RAM products. Subsequent testing has proved that the qualifications are effective in reducing the number of potentially defective devices which are shipped to customers.

Once the failure mechanisms have been identified and the proper controls or tests instituted, Intel's Reliability Group regularly monitors the products. This is accomplished on a sample basis by the use of scanning electron microscopy, burn-in, rotating life tests, high temperature tests (reverse bias, dynamic and storage) and examining wafer processing test data. In addition, the Reliability Group has the responsibility for MIL-STD-883 group C testing.

RELIABILITY TESTING

Six catagories of lifetesting are used to assure the electrical reliability of the Intel® 2115/25 family.

1. High Temperature Lifetest
2. High Temperature Bias
3. High Temperature Storage
4. Low Temperature Lifetest
5. Temperature Cycling
6. Low Temperature Bias

High Temperature Lifetest

This test is used to accelerate failure mechanisms by testing at an elevated temperature. For static RAM's the test temperature is 125°C. The data obtained are translated to a lower temperature using the Arrhenius Plot in Figure 6, giving a large number of equivalent hours of test. The test boards are organized to test 90 devices per board. The memory is sequentially addressed and filled with alternating patterns of ones and zeroes. During this test the outputs are exercised, but not monitored. Figure 7 shows timing and connection diagrams for this test on Intel® 2115/25 family.

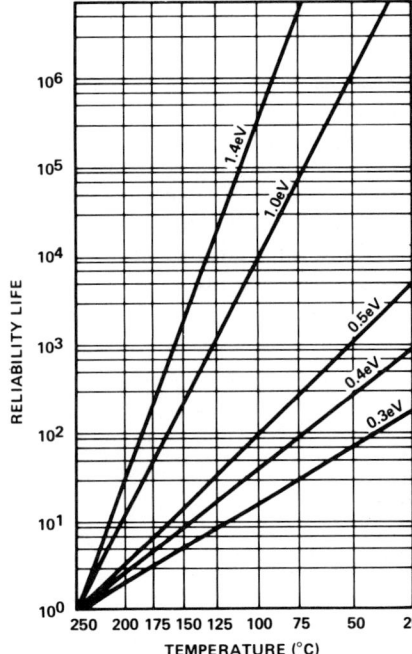

Figure 6. Arrhenius Plot, which assumes a failure rate proportional to exp (−E/kT) where E is the activation energy for the particular failure mechanisms.

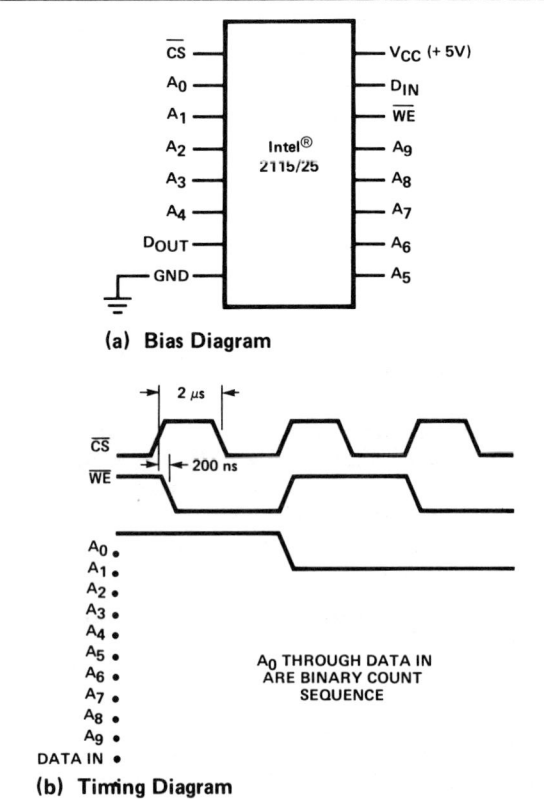

Figure 7. Intel® 2115/25 Family Lifetest

2115/2125 STATIC 1K RAMS

High Temperature Bias (Device Stability)

Since the Intel® 2115/25 family operates in high speed memory systems with critical timing, the stability of device parameters is essential. To insure there is no drift in critical parameters, lifetests are conducted on single transistors and on 1K RAMs. A good measure of a single MOS transistor's performance is its threshold voltage. These transistors are available as test structures on every wafer to monitor process parameters. The transistors are assembled and lifetested under static bias shown in Figure 8. These lifetests are conducted at elevated temperatures to insure there is no shifting of threshold voltage due to slow trapping or contamination. This test is conducted on both thin field and thick field transistors.

The access time of a memory is also a good measure of its performance. The stability of the Intel® 2115/25 family was measured using access time as a measure of device performance before and after standard lifetesting. The access time was measured on a Computest Venture II with a checker-board pattern and data sheet timing and voltages.

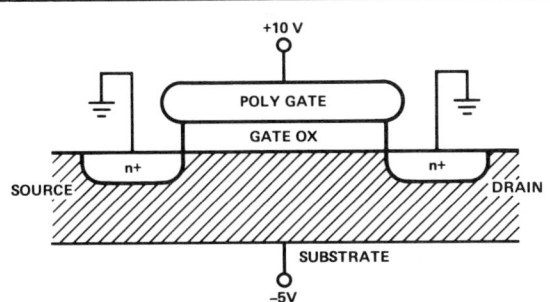

Figure 8. High Temperature Bias of Single Transistors

High Temperature Storage

Another common test is high temperature storage in which devices are subjected to elevated temperatures (160°C to 250°C) with no applied bias. This test is used to detect mechanical reliability problems (e.g., bond integrity) and process stability.

Low Temperature Lifetest

This test is performed at maximum operating frequency to detect the effects of electron injection into the gate oxide. The conditions for electron injection occur during transitions when the transistors are in saturation. This test is performed at −20°C in a bath of Flourinert FC-78 to obtain maximum cooling. Higher than normal power supply voltage can be used to accelerate this effect. Access time measurements are used to insure that there is no degradation.

Temperature Cycling

This test consists of cycling the temperature of the chamber housing the devices from −55°C to 150°C. This test is used to detect mechanical reliability problems and microcracks.

Low Temperature Bias (Device Stability)

This test is run to measure the effect of electron injection on a single MOS transistor. Electron injection, as previously stated (see page 3) will cause an increase in device threshold. Threshold shifts will affect the I-V characteristics of the device.

There are two sections to this test. The first section involves biasing the device (see Fig. 9a) such that the "drain" is grounded, the gate is above threshold and the source to drain voltage places the device well into saturation, i.e. the device is pinched-off. This is done at low temperatures which accelerates the injection mechanism. Next the device "source" and "drain" terminals are interchanged (see Fig. 9b) for a threshold measurement. The injected electrons will accumulate near the "source" terminal. Performing the threshold measurement with the "source" grounded maximizes the threshold shift due to electron injection.

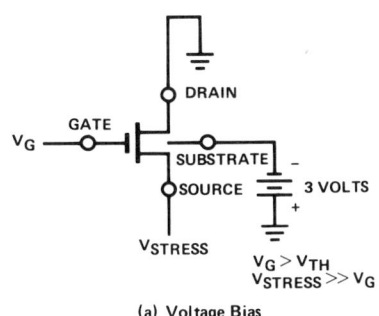

(a) Voltage Bias

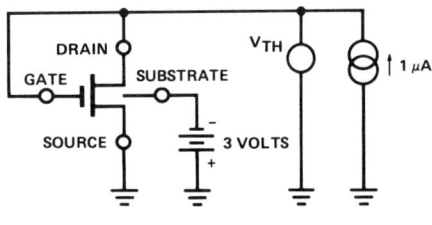

(b) Threshold (V_{TH}) Measurement

Figure 9. Low Temperature Bias for Electron Injection

TEST RESULTS AND SUMMARY

High Temperature Lifetest Results

A summary of the lifetest data for the Intel® 2115/25 family is given in Table III. Only one failure occurred during lifetesting so a dominant failure mechanism could not be determined. Failure rate data are presented in Table IV. The failure rate data are based upon two activation energy levels – 0.3 eV and 1.0 eV. Using an activation energy level of 0.3 eV is a very conservative estimate. Most MOS RAM failure mechanisms (other than oxide breakdown) have activation energies close to 1.0 eV. If the higher activation energy is used (as suggested by MIL Standard 217B) then the predicted failure rate decreases greatly. For example: The failure rate for 75°C at a 90% confidence level would be 0.03%/1000 hours at 0.3 eV activation energy but 0.0016%/1000 hours for a 1.0 eV activation energy. The failure rate difference at 55°C becomes even greater. For a confidence level of 90% the failure rate is 0.02%/1000 hours for 0.3 eV activation energy but the failure rate is 0.0002%/1000 hours for activation energy of 1.0 eV.

TABLE III
Dynamic Lifetest Results[1] – Intel® 2115/25

DEVICE[2]	LIFETEST 125°C			
	168 HRS	500 HRS	1000 HRS	2000 HRS
C2125	0/140	0/140	0/140	
C2125	0/131	0/131	0/131	
C2125	0/137	0/137	0/137	
C2125	0/100	0/100	0/100	
C2125	0/100	0/100	0/100	
C2125	0/90	0/90	0/90	
C2125	0/78	0/78	0/78	
C2115	0/100	0/100	0/100	
C2115	0/100	0/100	0/100	
C2125	0/100	0/100	0/100	
C2125	0/100	0/100	0/100	
P2115	0/20	0/20	0/20	
C2125	0/94	0/94	0/94	
C2115	0/144	0/144	0/144	
C2125	0/100	0/100	0/100	0/100
C2125	0/100	0/100	0/100	0/100
C2125	0/100	0/100	0/100	
C2125	0/100	0/100	0/100	
D2115	0/100	0/100	0/100	
D2115	0/100	0/100	0/100	
C2125	0/342	0/342	0/342	
C2125	0/173	0/173	0/173	0/173
C2115	0/100	0/100	0/100	
C2125	0/100	0/100	0/100	
P2125	0/200	0/200	1/200[3]	
C2125	0/100	0/100	0/100	
D2125	0/100	0/100	0/100	
P2125	0/100	0/100	0/100	
Total	0/3349	0/3349	1/3349	0/373

1. All devices received a production test at the QA conditions.
2. Package Types: C = Ceramic, D = Cerdip, P = Plastic Epoxy
3. Multiple bits in column fail at high V_{CC}, defect in scratch protection above worst bit, too small to be seen and rejected at second optical inspection.

TABLE IV
Calculated Failure Rates

Device Hours at 125°C	Activation Energy Level	Equivalent Hours		Failures	Failure Rate At a 90% Confidence Level (per 1000 hrs)	
		75°C	55°C		75°C	55°C
3,722,000	0.3 eV	1.3×10^7	2.4×10^7	1	0.03%	0.02%
3,722,000	1.0 eV	2.45×10^8	1.87×10^9	1	0.0016%	0.0002%

High Temperature Bias Results

Threshold voltage shifts of single MOS transistors from Intel® 2115 wafers after stressing at 250°C are presented in Figure 10. The results are an average of the increase or decrease in threshold of 12 devices. Figure 11 presents the same data for intermediate field devices.

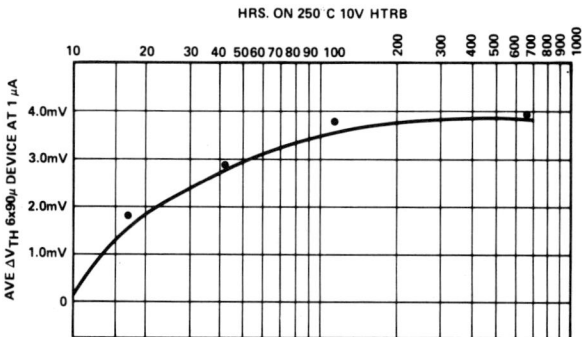

Figure 10. Threshold Shift after High Temperature Reverse Bias (HTRB)

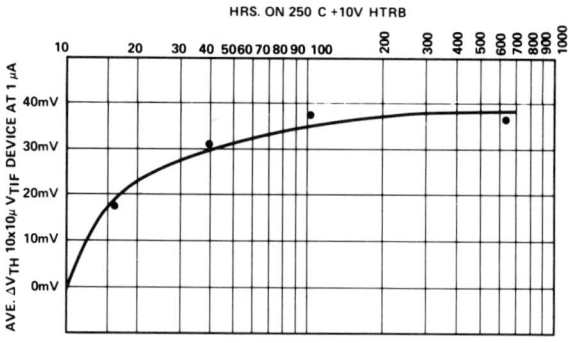

Figure 11. Field Device Threshold Shift after High Temperature Reverse Bias (HTRB)

Since these data show that the shifts observed in single transistors are minimal, it only needs to be shown that they do not cause device degradation. The results of t_{ACC} measurements after 125°C lifetest are reported in Table V and indicate the Intel® 2115/25 family can be expected to remain stable within their operating life. The access time shifts observed are within the production raw class-QA guardbands.

TABLE V
Results of t_{ACC} Measurements

Device	Temp	No. of Devices	Hrs. on Test	Shift in t_{ACC}[1] Max Increase	Avg. Shift
C2125	125°C	10	1000	+2.0ns	+1.0ns
C2125	125°C	10	1500	+3.0ns	−1.0ns
P2125	125°C	10	1000	+1.0ns	−2.0ns
C2125	125°C	25	200	+1.0ns	0ns
C2115	125°C	14	1000	+2.0ns	0ns

[1] Shifts in t_{ACC} were measured relative to control units to account for tester variation. A positive shift indicates the access time increased.

High Temperature Storage Results

The results of 250°C bakes for 168 hours on ceramic and cerdip Intel® 2115/25's are 0/415 failures. Epoxy Intel® 2115/25's were baked at 160°C for 1000 hours with no failures in 180 units baked.

Low Temperature Bias Results

Sixty-two test patterns were subjected to a high voltage stress to check for hot electron injection. Each test pattern contained three different device sizes to check for channel length dependence. No electron injection was seen at V_{DS} = 10 V, which is twice the maximum voltage that can be seen by a device. Table VI shows results of this test.

Temperature Cycle Results

After cycling 435 devices for 200 cycles each, there were no failures in 87,000 device cycles.

TABLE VI
Low Temperature Bias

Stress V_{DS}	No. Devices	Temp.	Device[1] Geometry	AVERAGE V_{TH} After Stress		
				48 Hours	168 Hours	500 Hours
10 V	17	25°C	4 X 90μ	<1mV	<1mV	
10 V	17	25°C	5 X 90μ	<1mV	<1mV	
10 V	17	25°C	7 X 90μ	<1mV	<1mV	
12 V	17	25°C	4 X 90μ	6mV	24mV	
12 V	17	25°C	5 X 90μ	4mV	1mV	
12 V	17	25°C	7 X 90μ	<1mV	<1mV	
9 V	12	25°C	3 X 100μ	<1mV	<1mV	<1mV
9 V	12	25°C	4 X 100μ	<1mV	<1mV	<1mV
9 V	12	25°C	5 X 100μ	<1mV	<1mV	<1mV
7 V	21	25°C	3 X 100μ	<1mV	<1mV	<1mV
7 V	21	25°C	4 X 100μ	<1mV	<1mV	<1mV
7 V	21	25°C	5 X 100μ	<1mV	<1mV	<1mV
7 V	12	−20°C	3 X 100μ	<1mV	<1mV	<1mV
7 V	12	−20°C	4 X 100μ	<1mV	<1mV	<1mV
7 V	12	−20°C	5 X 100μ	<1mV	<1mV	<1mV

[1] Smallest channel length on an Intel® 2115/25 is 4μ.

TABLE VII
−10°C Lifetest Results

Device	No. Hrs.	No. Devices	V_{CC}	Failures	No. Devices Measured	Max. Increase in t_{ACC}
C2125	168	90	+5V	0	10	+1ns
C2115	168	90	+5V	0	10	0ns
C2125	500	90	+7V	0	10	0ns
C2115	500	90	+7V	0	10	0ns
C2125	300	100	+7V	0	20	0ns

Low Temperature Lifetest Results

The results of low temperature lifetesting are presented in Table VII. Included in this table are the results of T_{ACC} measurements. Lifetests were conducted at V_{CC} = +5.0V and +7.0V to accelerate the effects of electron injection and oxide breakdown.

Summary

This report has presented reliability data on the Intel® 2115/25 family that indicate they meet or exceed the reliability standards for semiconductor memories. The experimental data are derived from the early portion of the random failure curve. These data suggest for 90% confidence level, at 75°C, the predicted failure rate for 0.3 eV activation energy is 0.03%/1000 hours and for 1.0 eV activation energy the predicted failure rate is 0.0016%/1000 hours. It is expected that, with extended time, the observed failure rate will continue to decrease.

REFERENCES

1. "Intel® 2107A/2107B N-Channel Silicon Gate MOS 4K RAMS" Reliability Report RR7, Intel Corporation September 1975.

2. S. A. Abbas and R. C. Dockerty, "N-channel IGFET Design Limitations due to Hot Electron Trapping", IEEE IEDM Technical Digest, pp. 35-38, 1975.

3. Sharan, R., "Matrix Computations Forecast Computer Mainframe Reliability, *Computer Design*, pp. 95-99, Aug. 1976.

For further information, call your local field sales office:

2708 8K UV EPROM

GARY GEAR

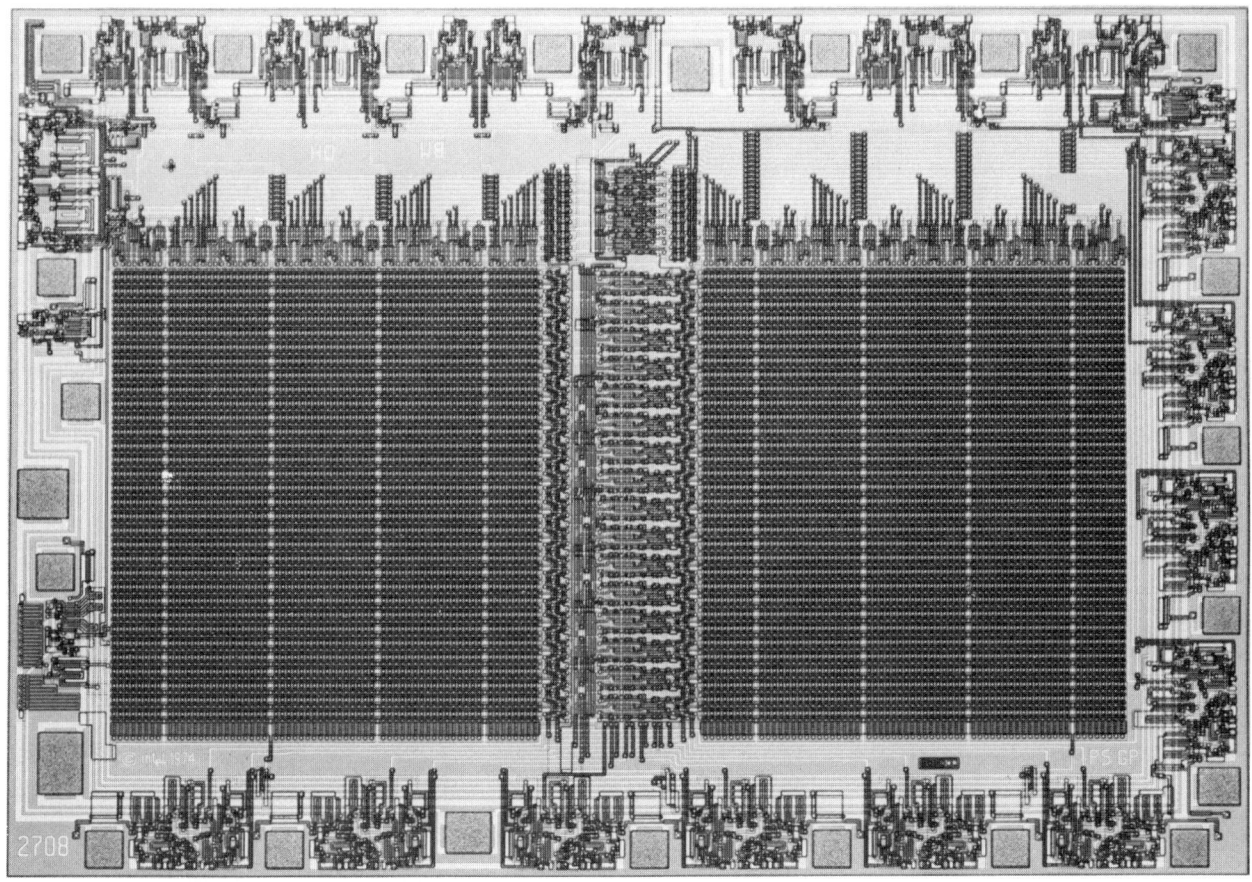

Photomicrograph of the 2708 8K EPROM.

INTRODUCTION

The Intel® 2708 is a 1024-word × 8-bit reprogrammable read-only-memory. This report describes the reliability and factors that effect reliability of the Intel® 2708 UV Erasable PROM (EPROM). Since the Intel® 2708 utilizes the FAMOS[1] technology EPROM cell, failure mechanisms associated with data retention that do not exist in standard N-channel MOS technologies become important. These failure mechanisms are explored in depth to illustrate the completeness of the testing philosophy used on the Intel® 2708. Following the introduction, the report is divided into seven additional sections. The second section describes the FAMOS technology EPROM cell and how it works. Section three discusses integrated circuit failure mechanisms. The reliability testing program is detailed in section four. The life test results and failure rate predictions are discussed in sections five and six. A discussion of production screening techniques follows in section seven, and the eighth section summarizes the data presented.

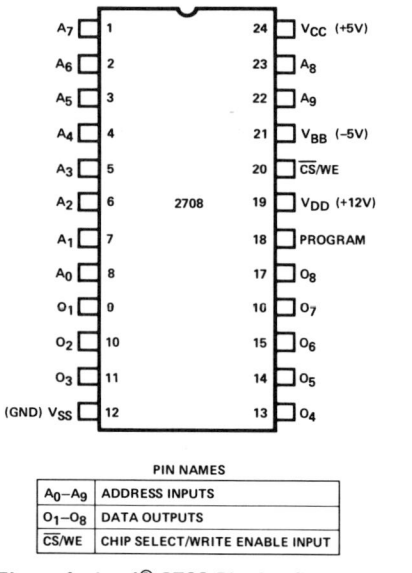

Figure 1. Intel® 2708 Pin Configuration

DEVICE DESCRIPTION

The device is packaged in an industry standard 24-pin package as shown in Figure 1. The program pin (18) receives 26V pulses during programming; during read operations it must be connected to V_{SS} (GND) or held at or below V_{IL}. Pin 20, the $\overline{CS}/WE$ connection, serves three functions:

1. When at V_{IL} (0V) the device is selected for normal read operation.

2. When at V_{IH} (3.0V, min.) the device is deselected and the outputs are placed in the high impedance state.

3. When at V_{IHW} (11.4V, min.) the device is Write Enabled and ready to receive program pulses.

A block diagram of the Intel® 2708 is shown in Figure 2. The low-order address bits (A_0–A_3) perform column (or Y) selection, while the high-order address bits (A_4–A_9) perform the row (or X) selection. A more detailed description of the Intel® 2708 operation is found in Intel's Application Note AP-17.[2]

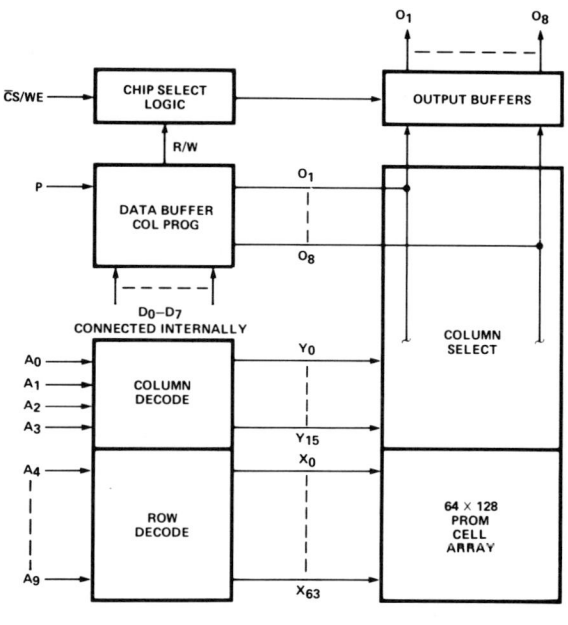

Figure 2. Detailed Block Diagram

Cell Description

The storage medium of the Intel® 2708 is a single transistor stacked-gate cell, implemented with two layers of polycrystalline silicon. The cell consists of a bottom floating gate and a top select gate, as shown in Figure 3. The top gate is connected to the row decoder, while the floating gate is used for charge storage. The cell is programmed by injection of high energy electrons through the oxide, onto the floating gate. Once there the charge is trapped, as there are no electrical connections to this floating gate. The presence of charge on the floating gate causes a shift of the cell threshold, as shown in Figure 4. In the initial state (no charge on the floating gate) the cell has a low threshold and selection of the cell, by way of the top select gate, turns on the transistor Q_1 shown in Figure 3(b). Programming (storing charge on the floating gate) shifts the threshold of device Q_1 to a higher level so that Q_1 will not turn on when selected. The storage cell is interrogated when the select gate is driven high to the sense threshold. If a "1" (no charge on floating gate) is programmed into the cell, cell selection allows a higher current to flow between the source and drain of Q_1 than if a "0" (floating gate charged) is programmed into the cell.

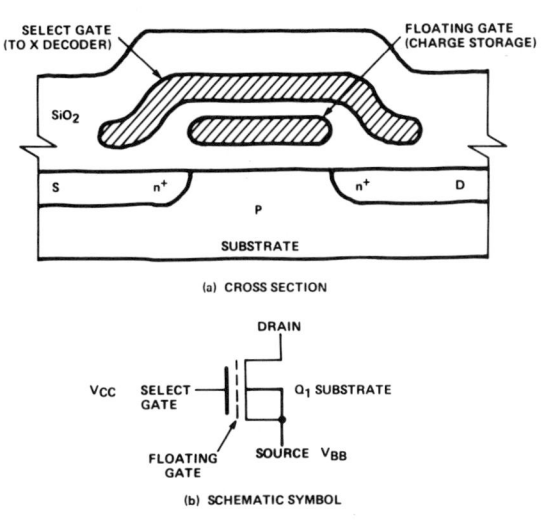

Figure 3. Intel® 2708 Storage Cell

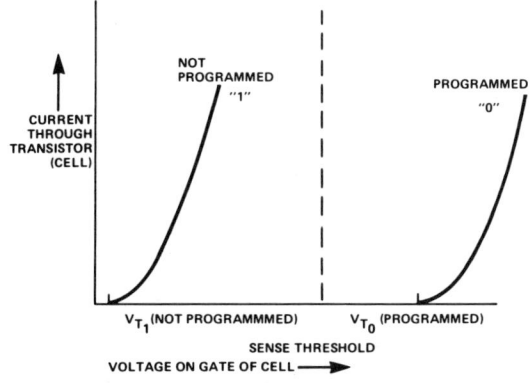

Figure 4. Storage Cell Threshold Shift

There are no electrical connections to the floating gate, therefore stored charge must be removed by non-electrical means. Illumination of the cell with ultraviolet light of the correct wave length (2537Å) and energy (10 watt seconds/cm^2) will impart sufficient photon energy to the trapped electrons to allow the floating gate to be fully discharged.

FAILURE MECHANISMS

The fundamental principles of reliability engineering predict that the failure rate of any group of devices as a function of time will follow a curve similar to Figure 5. The curve is divided into three regions: Infant Mortality, Random Failures, and Wearout Failures. Infant Mortality, as the name implies, represents the early life failures of a device. These failures are usually associated with one or more manufacturing defects.

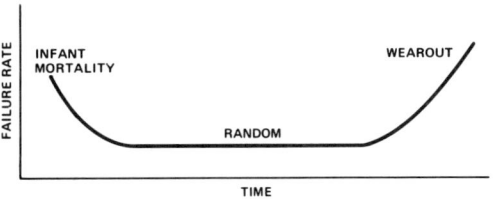

Figure 5. Reliability Life Curve

After some period (usually in the high tens to low hundreds of hours) the failure rate approaches some constant low value where it remains for a period of hundreds of thousands to millions of hours (for integrated circuits), depending upon temperature, applied voltage, circuit complexity, and other factors. This is the random failure portion of the curve and represents the useful portion of device life.

"Wearout" failures occur at the end of the device's useful life and are characterized by a rapidly rising failure rate with time as the devices "wearout" both physically and electrically. This does not occur before the high tens to hundreds of years, if at all, for integrated circuits.

RELIABILITY TESTING

A series of reliability tests have been conducted on the Intel® 2708. This section describes the tests conducted, as well as failure mechanisms involved. The tests focus on charge loss characteristics and long-term degradation; in addition, operating life tests have been conducted.

Activation Energy Experiments

Cell charge loss characteristics were studied both on single FAMOS transistors on a test die and on the EPROM arrays. The single transistor used is the N-channel, two-layer polysilicon cell. By measuring the two-terminal threshold voltage before and after programming, the offset voltage ($V_{TP}-V_T$) on the floating gate can be determined. Figure 6 shows the two-terminal V_T bias configuration and the resulting current-voltage characteristics. These programmed parts were then baked at 200, 250 and 300°C to accelerate the charge loss. The degradation rates are plotted assuming an Arrhenius relationship, in Figure 7, yielding an activation energy of 0.8 eV.[3]

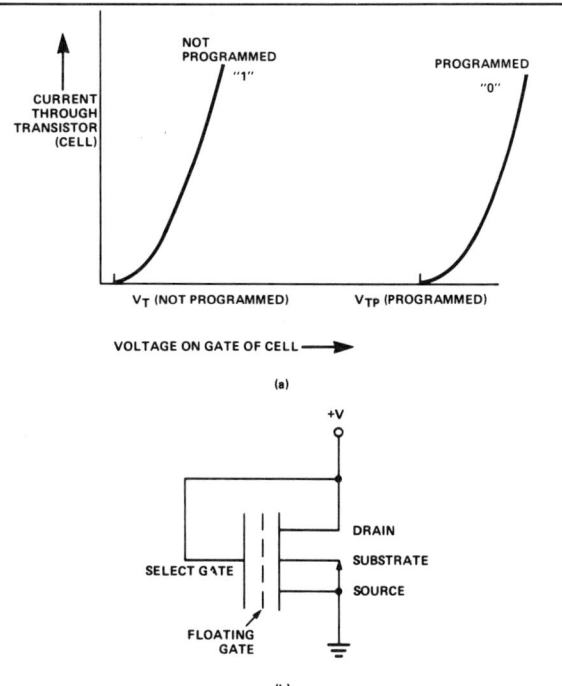

Figure 6. Two-Terminal Threshold Voltage Test

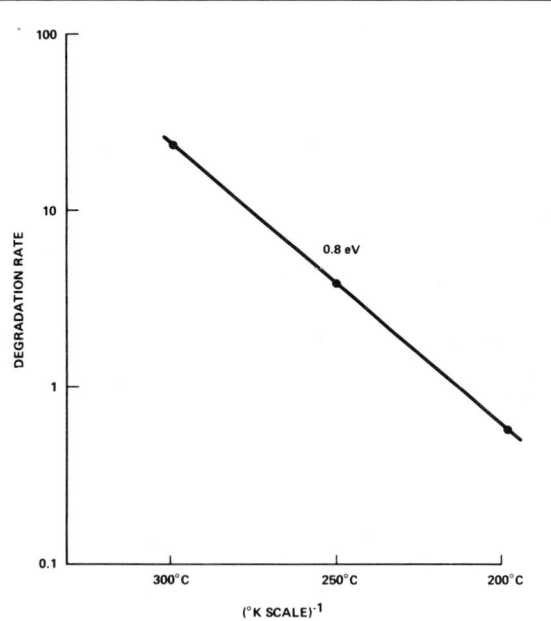

Figure 7. Arrhenius Relationship

Data Retention

Data retention in the Intel® 2708 was evaluated using high temperature bakes. A bit pattern was selected that programmed about 90% of the EPROM cells but contained some unprogrammed bits in each row and column so a complete functional test could be performed. These parts were then baked at high temperatures and the cumulative percent failures plotted with time. Figure 8 shows the 250°C retention characteristics of the Intel® 2708 arrays. Using the 0.8 eV activation energy derived from single transistor measurements, the time to 5% failure at 70°C is estimated at 100 years.

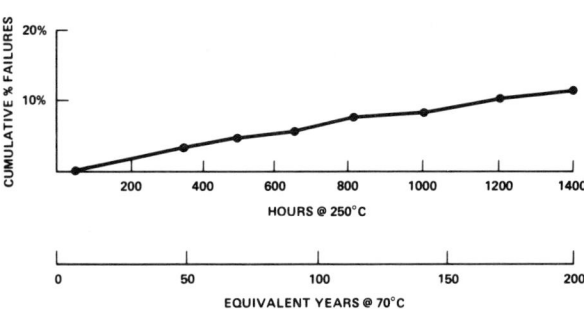

Figure 8. Intel® 250°C Bake Failure Rate

The useful life of most semiconductor products is between 10 and 20 years, and as a result, the portions of Figure 8 past 100 hours at 250°C become unimportant to most real applications of FAMOS technologies.

Voltage Shmoo Tests

Referring to Figure 3, it can be seen that by changing the V_{BB} voltage, V_{GS} will be changed and result in a corresponding change in drain current in the storage cell. This technique was used to determine the amount of charge on the floating gate in the following experiments.

Programming Margin Tests

A programming margin experiment was designed to show how much charge is lost from a "programmed" floating gate during normal use. This experiment consists of monitoring the voltage shmoos on the Intel® 2708. Since the discharge mechanism is known to follow an Arrhenius relationship with a 0.8 eV activation energy, a 250°C bake for 168 hours can be used to simulate greater than 10 years at 70°C. One hundred and twenty Intel® 2708's were programmed on an Intel® MDS Programmer and voltage shmoo margins were measured on each device. The parts were then placed on 250°C bake to accelerate the discharge of electrons from the floating gate. The devices were taken out of the 250°C bake oven and shmoo margins remeasured. Figure 9 shows the average degradation in shmoo margin with time on 250°C bake.

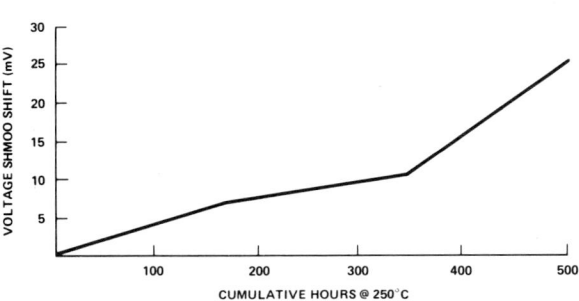

Figure 9. Intel® 2708 Voltage Shmoo Shift vs. Time

"Sunburn" Effect

An experiment to determine the effects of prolonged exposure to UV light, "Sunburn", was run by exposing eight Intel® 2708's to UV light at 6000 $\mu W/cm^2$ for prolonged periods of time (10 watt-seconds in a normal erasure). Periodic readings were taken by programming the parts on an Intel® MDS Programmer and measuring voltage shmoo margins. Figure 10 shows the average shmoo value as a function of UV exposure time.

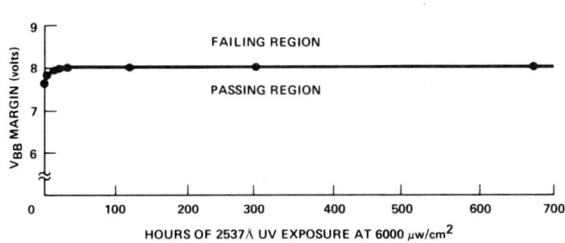

Figure 10. Voltage Shmoo Margin Change as a Function of Time

As can be seen, the shmoo margin actually increases, leaving a net improvement in programming characteristics. Although no specific studies have been undertaken to determine exactly what device parameter is changing with UV exposures, it is expected that excessive degrees of UV light causes radiation damage that results in an increase in threshold voltage.

Multiple Program/Erase

To determine if any degradation in the Intel® 2708 occurs after a number of program/erase cycles, the following test sequence is used:

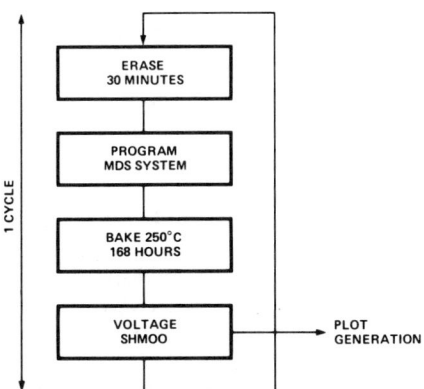

Figure 11. Program Erase Cycles

The results of ten cycles are shown in Figure 12. It is significant to note that the programming margin loss during 250°C, 168-hour bake (5–10 mV) is completely recoverable.

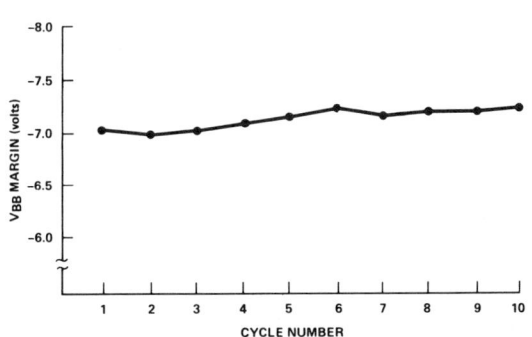

Figure 12. Multiple Program Erase Experiment

LIFE TESTS

Test Set-Up

The electrical reliability of MOS LSI circuits is most realistically evaluated using MIL STD 883, method 1005, condition F, dynamic burn-in and life test. The life test set-up is as follows: The address inputs of the individual devices are parallel-wired on a printed circuit board with output load resistors at each device test socket. The devices are selected by an appropriate bias on the chip select inputs which allows the output drivers to be exercised. The printed circuit boards are then placed in an oven stabilized at an elevated temperature for the appropriate stress period. At each read point, the entire oven is cooled down with all electrical signals still applied to maintain worse case stress conditions. The devices are then tested for complete functionality on a commercially available automatic tester.

Production samples of Intel® 2708 EPROMs were taken at various times throughout 1975 and placed on 160°C operating life tests using the test technique described. These production samples received no special screening or preconditioning prior to the start of life tests. The units were programmed approximately 90% with the same data pattern as that used for retention bakes. Life test results are shown tabulated in Table I. There were 10 functional failures in the Intel® 2708 test group at 160°C. Each of these failures was caused by charge loss in a storage cell that resulted in output data error. (The errors observed were generally single bit errors, although 2 and 3-bit errors have also been observed.) The failed bit locations are random. The devices which failed were given a microscopic examination at the failed bit location, revealing that there were no physical defects present that might have influenced the charge loss rate. In all cases the failed bit locations were reprogrammable, indicating the failure mechanism was pure charge loss.

The life test results on the Intel® 2708's (Table I) show that there are no failures due to shorted dielectrics. This is expected since the device is subjected to the high programming voltages both at wafer probe and final electrical test. This initial programming acts as a good screen for weak gate dielectric devices.

FAILURE RATE PREDICTIONS

Failure rate predictions are made by assuming the total device failure rate is equal to the sum of the failure rate contributions due to each failure mode. In addition, each failure mode is assumed to have temperature dependence. The life test data set for the Intel® 2708 contains a number of failures due to charge loss but no failures due to the other known MOS failure modes. The failure rates were calculated using an activation energy of 0.8 eV for charge loss and 0.4 eV for all other causes. Table II summarizes the device failure rates computed at 55°C and 70°C.

Table II.

TEMPERATURE	FAILURE RATE 60% CONFIDENCE (%/1000 hours)	FAILURE RATE 90% CONFIDENCE (%/1000 hours)
70°C	0.013	0.027
55°C	0.006	0.013

PRODUCTION SCREENS

The family of erasable PROMs has a distinct advantage over bipolar fuse link and avalanche injection type PROMs, in that each memory cell can be tested then erased to guarantee programmability as a part of normal production test flow. This feature allows a manufacturer to implement screens not possible with PROMs manufactured using other semiconductor technologies. This section describes the production screening techniques unique to EPROMs.

Each Intel® 2708 receives two comprehensive functional tests. The first occurs at wafer sort where a data pattern is programmed into the Intel® 2708 die to detect bad devices. A limited number of bits are programmed at this point. The wafers are then UV erased and the die packaged. The first packaged test is a "zero check" where each of the bits programmed at wafer sort is verified as erasable. Each Intel® 2708 is then programmed with the complement of the data pattern

Table I. Operating Life Test Results

TEMPERATURE	SAMPLE SIZE	HOURS	EQUIVALENT DEVICE HOURS @ 70°C	FAILURES	FAILURE MODE
160°C	64	2243	39.9 × 10⁶	1	Charge Loss
160°C	49	2028	27.6 × 10⁶	0	
160°C	51	2028	28.7 × 10⁶	1	Charge Loss
160°C	40	2830	31.4 × 10⁶	2	Charge Loss
160°C	80	1176	26.1 × 10⁶	1	Charge Loss
160°C	77	1176	25.1 × 10⁶	4	Charge Loss
160°C	79	984	21.6 × 10⁶	1	Charge Loss

used at wafer sort. Following this step, each cell in the memory array has been programmed once, and the memory array subjected to high programming voltages twice.

After the second programming step the devices are subjected to a high temperature bake. This bake accelerates the charge loss rate on marginal devices, allowing the subsequent electrical test to screen them out. The final electrical test uses the normal guard band voltage with the addition of a voltage margin test. Following the final electrical test, each part is UV erased and sample zero checked to verify complete erasure. This test sequence guarantees that each cell is programmable and erasable, and has normal charge retention characteristics.

SUMMARY

As a result of the Intel® 2708 reliability tests and Intel's reliability program, the Intel® 2708 has demonstrated a failure rate of 0.013%/1000 hours at 70°C. The Intel® 2708 was first sampled in early 1975. Since that time the Intel® 2708 has established itself as a viable memory storage element both in temporary data storage applications and as "permanent" memory in applications where mask programmable ROM would normally be used.

Quoted device failure rates assume proper erasure and programming. Failures due to improper erase or programming are recoverable and not considered device failures. For a further discussion see Intel's Application Note AP-17.[2]

REFERENCES

1. D. Frohman-Bentchkowsky, "A Fully Decoded 2048 Bit Electrically Programmable MOS Read Only Memory," 1971 ISCC, Philadelphia, Feb. 17–19, 1971.

2. B. Greene, "Application of the Intel® 2708 8K Erasable PROM," Applications Note AP-17, Intel Corporation, 1976.

3. B. Pascoe, "2107A/2107B N-Channel Silicon Gate MOS 4K RAMs," Reliability Report RR-7, Intel Corporation, 1975.

For further information, call your local field sales office.

2416 16K CCD MEMORY

GARY GEAR AND BILL PASCOE

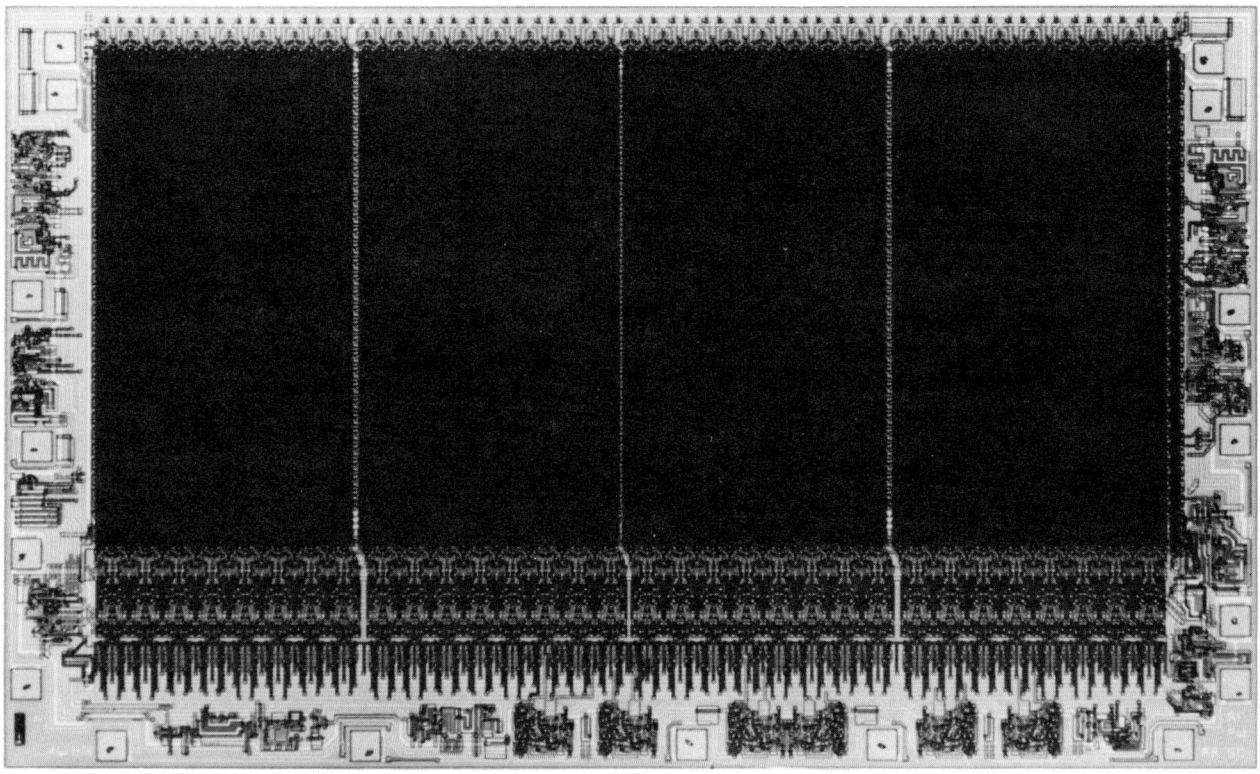

Photomicrograph of the 2416 16K CCD Memory.

2416 CCD MEMORY

INTRODUCTION

This report examines the reliability and related topics of the 2416 CCD serial memory. Included are sections on device description, failure modes, reliability considerations, reliability tests and test results. The report concludes with a brief summary of reliability numbers and references on related topics.

DEVICE DESCRIPTION

The 2416 is a 16,384 word x 1 bit serial CCD memory. It is manufactured using Intel's silicon gate N-channel MOS process. Internally the device is organized as 64 independent recirculating shift registers each with a word length of 256 bits. Any one of these 64 registers may be addressed via a 6 bit address. A block diagram showing the organization of the 2416 is given in Figure 1. A "cell" of the 2416 consists of 2 thin field gate oxide devices, one for data storage and one for data transfer. Each gate is connected to one of the 4 clock phases as shown in Figure 2. Each of the sixty-four 256 bit registers is organized for refresh purposes as two 128 bit registers connected by a refresh amplifier at each end to form a continucus loop. 128 shift cycles within 10 μsec are required to completely refresh the memory. The only diffusion used in the 2416 is for the N+ data input/output connections at the ends of the registers and the peripheral circuits in the memory. There are no diffusions in the array itself.

There are two common CCD types referred to as surface channel and buried channel. The surface channel is characterized by the storing and transferring of charge (data) along the surface of the substrate. The buried channel type, because of

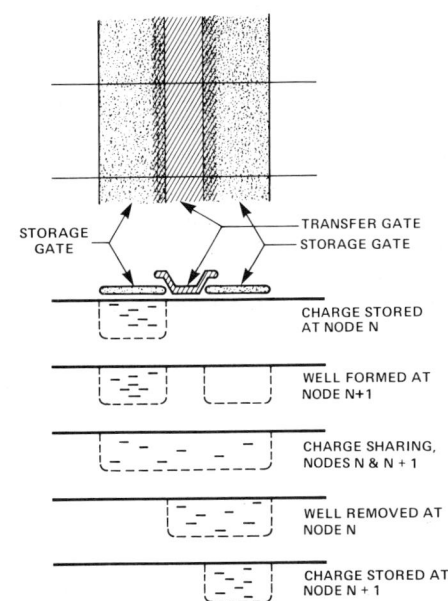

Figure 2. CCD Storage Array Layout.

additional substrate doping, stores and transfers the charge (data) further into the bulk of the substrate. The surface channel technique has the advantage of higher charge storage capacity than the alternate buried channel technique. The charge storage mechanism is very similar to that of a dynamic RAM using a single cell storage design. Indeed, the entire memory is comprised of less than 300 gates of random logic and 32,768 MOS capacitors. As will be seen later, this makes possible a highly reliable and very dense memory. Further information on the operation and use of the 2416 is found in Reference 1.

FAILURE MECHANISMS

Failure mechanisms in the 2416 are similar to those encountered in any N-channel MOS device and have been discussed at length in References 2 and 3. These failure mechanisms are summarized in Table 1. Oxide breakdown is the most prevalent failure mode in this type of device because it consists for the most part of MOS capacitors. Special testing techniques are used to eliminate these failures from shipped parts and will be described in the next section.

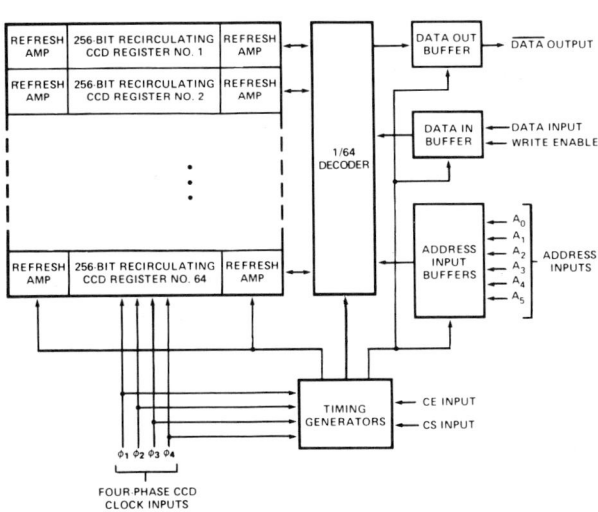

Figure 1. 2416 Block Diagram.

RELIABILITY REPORTS

Table 1. Failure Mechanisms in MOS.

Failure Mode	Type	Activation Energy (E_{act})	Detection
Slow Trapping	Wearout	1.0 eV	High Temperature Bias
Contamination	Wearout/Infant	1.4 eV	High Temperature Bias
Surface Charge	Wearout	.5-1.0 eV	High Temperature Bias
Polarization	Wearout	1.0 eV	High Temperature Bias
Electromigration	Wearout	1.0 eV	High Temperature Operating Life
Microcracks	Random	—	Temperature Cycling
Contacts	Wearout/Infant	—	High Temperature Operating Life
Oxide Defects	Infant/Random	.3 eV	High Voltage Operating Life and Cell Stress

RELIABILITY CONSIDERATIONS

Reliability Program

One of the interesting phenomenons which has occurred in the semiconductor industry is the concurrent increase in device complexity with a decrease in device failure rate. At Intel, a rigid system of reliability controls has consistently improved reliability of Intel products with time. Figure 3 shows this trend with three products, the P-channel 1103 1K RAM, the 2107B 4K RAM and the 2416 CCD memory. Although the complexity of each product increases by a factor of four, the reliability has improved to about the same level. The 2416 follows the same trend.

This reduction in failure rate is accomplished by careful qualification testing of new products, which includes testing and identification of prevalent failure modes. Tests are initiated to detect and eliminate these failure mechanisms from production lots. Monitor tests to determine the on-going quality of production parts are also performed along with reliability analysis of returned parts.

PRODUCTION TESTING

In-process testing also performs a valuable contribution to device reliability. During the manufacturing process various tests are performed which either stress the part or verify that a part will indeed operate under all data sheet conditions. In addition, various optical and electrical inspections are performed. Table 2 lists the various electrical tests performed on the 2416 memory and Table 3 lists some of the inspections performed during the assembly process.

Of the tests listed in Table 2, there are two which require comment. Since the CCD memory is primarily a group of MOS capacitors, a high voltage cell stress test is an excellent test for the manufacturer to determine the presence of cells which have a weak dielectric (SiO_2) between the gates (Polysilicon) and the substrate (P-silicon). See Figure 2. In this test, the voltages to all pins are raised to approximately twice normal and the memory is exercised. Thus every cell is exposed to the high applied voltages. This test accelerates the oxide breakdown failure mode to a point well into the random failure portion of the traditional reliability curve (see Figure 4). This eliminates

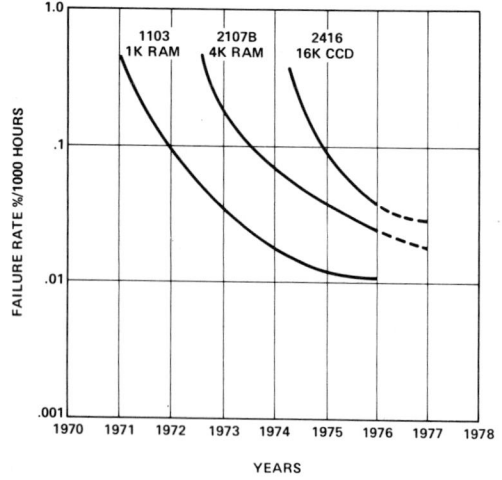

Figure 3. Device Reliability Improvement Vs. Time.

Table 2. Electrical Tests CCD Memory.

Test	Conditions
DC Function	70°C
AC Function	70°C
Refresh Time	70°C
V_{BB} Margin	V_{BB} = -3.5V
H.V. Cell Stress	≈ 2 x Normal Voltages

2416 CCD MEMORY

Table 3. Component Assembly QA Acceptance Steps.

Operation	Mil Std 883 Method:
Piece Part Inspections	
Scribe and Break	
Q.A. Acceptance	
2nd Optical Insp.	2010.1B
Lead Bond/Die Attach	
Q.A. Acceptance	
3rd Optical Insp.	2010.1B
Precap Visual	2010.1B
Seal	
Temperature Cycle	1010 (C)
Q.A. Acceptance	
Fine Leak	1014 (A)
Q.A. Acceptance	
Gross Leak	1014 (C)
Q.A. Acceptance	
Final Visual	2009

Note: All tests in this table are 100% tests.

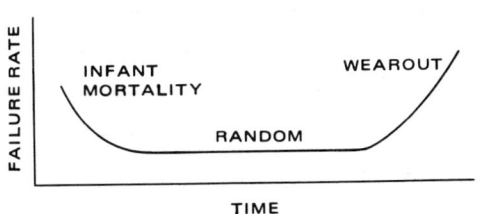

Figure 4. Reliability Life Curve.

virtually all infant mortality failures of this type and also reduces the incidence of this failure mode during normal device life. Once the test is accomplished no additional cell stress tests are necessary since each device has "proven" its capability to withstand voltages far in excess of those normally applied.

The V_{BB} margin test was initiated by Intel's Reliability Department and eliminates most potential failures due to parameter degradation. In this test, V_{BB} is reduced to -3.5V and the device is exercised. Any part failing to meet the prescribed parameters is rejected.

RELIABILITY TESTING

There are two tests normally performed on devices on a sampling basis by Intel's Reliability Department.

Dynamic Burn-In
Extended Life Test

These two tests are the source of reliability data presented in this report.

The dynamic burn-in test is taken at 125°C and is shown schematically with timing in Figure 5B. This test has been run for periods up to 1000 hours.

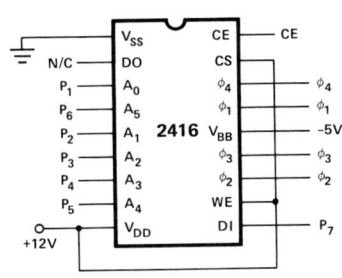

Figure 5A. Dynamic Burn-In Connections.

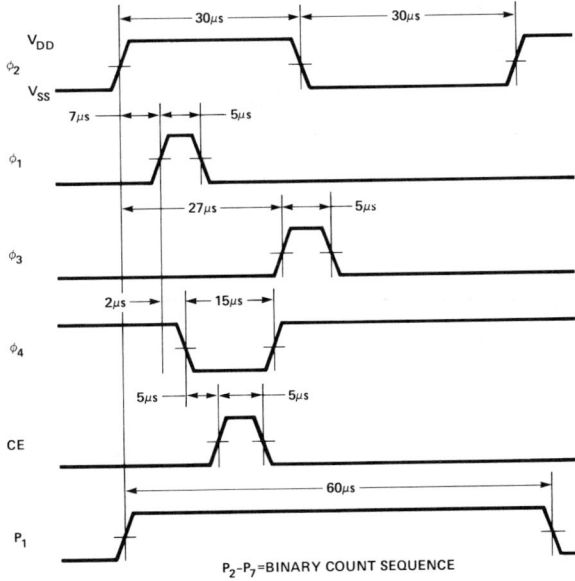

Figure 5B. Dynamic Burn-In Timing.

The extended life test is run at 160°C and is shown schematically with timing in Figure 6B. This test uses more complex timing than the dynamic burn-in circuit and except for temperature, closely approximates a continuous read-modify-write cycle in actual use. In this way the entire memory is continuously exercised.

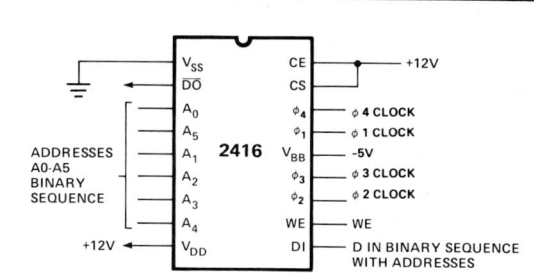

Figure 6A. Connections.

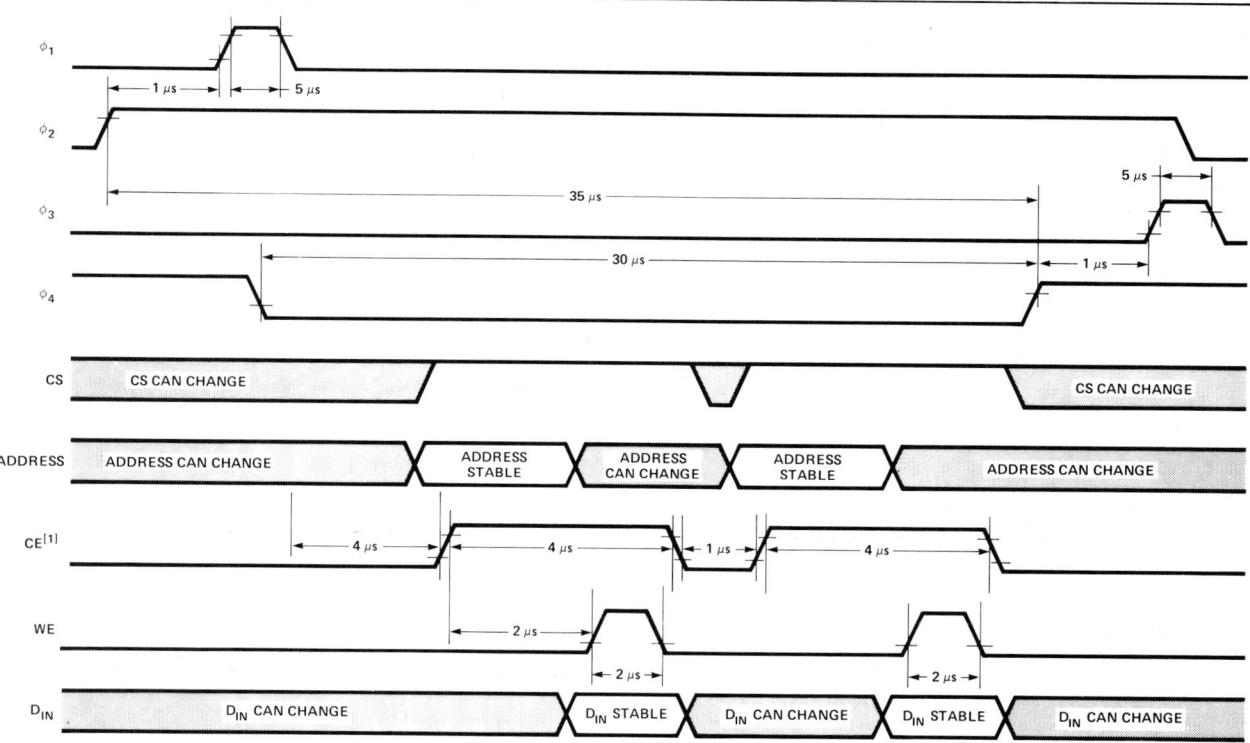

Figure 6B. Extended Life Test Connections and Timing.

TEST RESULTS

The results of the tests described above are given in Tables 4 and 5. Table 4 shows the results of tests made on 264 devices for 1000 hours each at 125°C. Extended life tests on 69 devices for 2508 hours has been made at a temperature of 160°C with data given in Table 5. Using these figures, Table 6 gives the reliability calculations at 55°C and 70°C for an activation energy of 0.5eV at the 60% and 90% upper confidence level. The 90% upper confidence level failure rates are .05%/1000 hours and 0.15%/1000 hours for 55° and 70°C, respectively. Since the equivalent time at 70°C is only 7.7 million hours, the failure rates, calculated by the Chi-squared function,[3] seem somewhat high. The 2416 is a new device, on which a large number of hours have

Table 4. 2416 Dynamic Burn-in Test Results (125°C)

No. Devices	Failures At 500 Hours	No. Devices	Failures At 1000 Hours
51	2	49	1
95	2	93	0
118	1	100*	1
—	—	—	—
264	5	242	2

Note: 500 hour failures (5) were catastrophic function failures.
1000 hour failures (2) were refresh time failures.
*17 devices were removed from test for non-device failure related reasons.

not yet been accumulated. Once the equivalent hours exceed 10-12 million hours, especially on long term tests (>1000 hours) this failure rate will drop to the .03 to .06 percent per 1000 hours rate.

Table 5. 2416 Extended Life Test Results (160°C).

No. Devices	Failures At (Hours):			
	500	1000	1500	2508
69	0	0	0	0

Table 6. Reliability Calculations.

Device Hours At °C		Equivalent Device Hours At		Failure Rate			
				90% U.C.L.		60% U.C.L.	
		55°C	70°C	55°C	70°C	55°C	70°C
253,000	125	8,349,000	2,530,000				
172,500	160	17,250,000	5,175,000				
425,000		25,599,000	7,705,000	.05%/K Hr.	.15%/K Hr.	.03%/K Hr.	.11%/K Hr.

SUMMARY

The structure and failure mechanisms of the 2416 CCD serial memory have been discussed along with the reliability considerations and testing associated with this device. The test results show a failure rate of .05%/1000 hours at 55°C, 0.5eV activation energy and 90% U.C.L.

REFERENCES

1. Papenberg, Bob, "Design and Applications of Intel's 2416 16K Charge Coupled Device", Intel Corp., Memory Design Handbook, 1975.
2. Pascoe, Bill, "2107A/2107B N-Channel Silicon Gate MOS 4K RAMs", Intel Corp., RR-7, Sept. 1975.
3. Pascoe, Bill, "8080/8080A Microcomputer" Intel Corp., RR-10, March 1976.

ACKNOWLEDGEMENT

The authors express their gratitude to Intel's Reliability Department for the test data used in this report and to all the personnel who reviewed the report and offered many valuable suggestions.

Index

1103:	1024 x 1 bit dynamic RAM	*182*
1103-1:	1024 x 1 bit dynamic RAM	*187*
1103A:	1024 x 1 bit dynamic RAM	*190*
1103A-1:	1024 x 1 bit dynamic RAM	*195*
1103A-2:	1024 x 1 bit dynamic RAM	*200*
1702A:	2K (256 x 8) UV EPROM	*322*
1702AL:	2K (256 x 8) low power UV EPROM	*326*
2101A:	256 x 4 bit static RAM	*44*
2102A:	1K x 1 bit static RAM	*48*
2104A:	4096 x 1 bit dynamic RAM	*204*
2107A:	4096 x 1 bit dynamic RAM	*212*
2107B:	4096 x 1 bit dynamic RAM	*218*
2107C:	4096 x 1 bit dynamic RAM	*224*
2108:	8192 x 1 bit dynamic RAM	*229*
2111A:	256 x 4 bit static RAM	*52*
2112A:	256 x 4 bit static RAM	*56*
2114:	1024 x 4 bit static RAM	*61*
2115A/2125A:	high speed 1K x 1 bit static RAM	*65*
2116:	16,384 x 1 bit dynamic RAM	*237*
2117:	16,384 x 1 bit dynamic RAM	*245*
2141:	4096 x 1 bit static RAM	*70*
2142:	1024 x 4 bit static RAM	*74*
2147:	4096 x 1 bit static RAM	*78*
2308:	8K (1K x 8) ROM	*329*
2316E:	16K (2K x 8) ROM	*333*
2416:	16,384 x 1 bit CCD serial memory	*403*
2708:	8K and 4K UV EPROM	*336*
2716:	16K (2K x 8) UV EPROM	*341*
2758:	8K (1K x 8) UV low power EPROM	*345*
3205/3404:	high speed decoder and 6-bit latch	*435*
3207A:	quad bipolar-to-MOS level shifter, driver	*439*
3207A-1:	quad bipolar-to-MOS level shifter, driver	*443*

3208A/3408A:	hex bipolar amplifiers for MOS circuits	*445*
3222:	refresh controller for 4K dynamic RAMs	*451*
3232:	address multiplexer, refresh counter for 4K dynamic RAMS	*457*
3602/3622:	2K (512 x 4) high speed PROM	*351*
3604/3624:	4K (512 x 8) high speed PROM	*354*
3605/3625:	4K (1K x 4) PROM	*357*
3608/3628:	8K (1K x 8) bipolar PROM	*360*
5101:	256 x 4 bit static CMOS RAM	*84*

EPROMs:
256 x 8:	1702A	*322*
256 x 8:	1702AL	*326*
1K x 8:	2708	*336*
1K x 8:	2758	*345*
2K x 8:	2716	*341*

PROMs:
512 x 4:	3602/3622	*351*
512 x 8:	3604/3624	*353*
1K x 4:	3605/3625	*357*
1K x 8:	3608/3628	*360*

RAMs:
256 x 4:	2101A	*44*
256 x 4:	2111A	*52*
256 x 4:	2112A	*56*
256 x 4:	5101	*84*
1024 x 1:	1103	*182*
1024 x 1:	1103-1	*187*
1024 x 1:	1103A	*190*
1024 x 1:	1103A-1	*195*
1024 x 1:	1103A-2	*200*
1K x 1:	2102A:	*48*
1K x 1:	2115A/2125A	*65, 498*
1024 x 4:	2114	*61*
1024 x 4:	2142	*74*
4096 x 1:	2104A	*204*
4096 x 1:	2107A	*212*
4096 x 1:	2107B	*218*
4096 x 1:	2107C	*224*
4096 x 1:	2141	*70*
4096 x 1:	2147	*78*
8192 x 1:	2108	*229*
16,384 x 1:	2116	*237*
16,384 x 1:	2117	*245*

ROMs:
1K x 8:	2308	*329*
2K x 8:	2316E	*333*

Index

Battery powered memories (nonvolatile):	*20*
Bipolar/MOS Static RAM compatibility:	*38*
Bipolar PROM 3608, 3628:	*360*
BIPOLAR SUPPORT CIRCUITS FOR DYNAMIC RAMs:	
3207A:	*439*
3207A-1:	*443*
3208, 3408:	*445*
CCD memory:	*374*
CHARGE COUPLED DEVICES:	
2416:	*403, 514*
Compatible memories:	*39*
Design, applications for 2416 CCD memory:	*374*
Designing with 16K dynamic RAMs:	*130*
Designing with 2708 EPROM:	*294*
Designing with 2716 EPROM:	*308*
Designing non-volatile semiconductor memory systems:	*20*
Designing with 16 pin, 4096 dynamic RAMs:	*89*
Designing with 22 pin, 4096 dynamic RAMs:	*103*
Designing with PROMs, ROMs:	*258*
Designing with static MOS RAMs:	*2*
DYNAMIC RAMs:	
1103:	*182*
1103-1:	*187*
1103A:	*190*
1103A-1:	*195*
1103A-2:	*200*
2104A:	*204*
2107A:	*212*
2107B:	*218*
2107C:	*224*
2108:	*229*
2116:	*237*
2117:	*245*
Dynamic RAMs used with microprocessors:	*139*
Electrical erasable PROMs:	*294, 308*
EPROM erasure characteristics:	*294, 308*
EPROMs:	
1702A:	*322*
1702AL:	*326*
2708:	*336, 507*
2716:	*341*
2758:	*345*
HIGH SPEED RAMs:	
2115A/2125A:	*65*
2147:	*78*
Memory support circuits for dynamic RAMs:	*411*
Microprocessor memories:	*139*

Nonvolatile memories: *20*

PROM programming: *363*
PROMs:
  3602/3622: *351*
  3604/3624: *354*
  3605/3625: *357*
  3608/3628: *360*

Reliability reports: *478*
ROMs:
  2308: *329*
  2316E: *333*

STATIC RAMs:
  2101A: *44*
  2102A: *48*
  2111A: *52*
  2112A: *56*
  2114: *61*
  2115A/2125A: *65*
  2141: *70*
  2142: *74*
  2147: *78*
  5101: *84*
Support circuits: *411*

Using support circuits: *412*
UV EPROMs:
  1702A: *322*
  1702AL: *326*
  2708: *336*
  2716: *341*
  2758: *345*